COLLECTED WORKS OF

WILLIAM P. THURSTON

WITH COMMENTARY

COLLECTED WORKS OF

WILLIAM P. THURSTON

WITH COMMENTARY

III

Dynamics, Computer Science and General Interest

Benson Farb
David Gabai
Steven P. Kerckhoff
Editors

American Mathematical Society

Providence, Rhode Island

2020 *Mathematics Subject Classification.* Primary 20F10, 37F15, 68P05.

Library of Congress Cataloging-in-Publication Data

Names: Farb, Benson, editor. | Gabai, David, editor. | Kerckhoff, Steve, editor.

Title: Collected works of William P. Thurston with commentary / Benson Farb, David Gabai, Steven P. Kerckhoff, editors.

Description: Providence, Rhode Island : American Mathematical Society, [2022] | Includes bibliographical references. | Contents: Volume I. Foliations, surfaces and differential geometry – Volume II. 3-manifolds, complexity and geometric group theory – Volume III. Dynamics, computer science and general interest.

Identifiers: LCCN 2021037684 | ISBN 9781470451646 (hardcover-set) | ISBN 9781470463885 (hardcover–vol. I) | ISBN 9781470463892 (hardcover–vol. II) | ISBN 9781470463908 (hardcover–vol. III) | ISBN 9781470468330 (ebook–vol. I) | ISBN 9781470468347 (ebook–vol. II) | ISBN 9781470468354 (ebook–vol. III)

Subjects: LCSH: Thurston, William P., 1946–2012. | Differential topology. | Geometry, Differential. | Dynamics. | AMS: Manifolds and cell complexes – Low-dimensional topology in specific dimensions – 2-dimensional topology (including mapping class groups of surfaces, Teichmüller theory, curve complexes, etc.). | Manifolds and cell complexes – Low-dimensional topology in specific dimensions – Foliations in differential topology; geometric theory. | Manifolds and cell complexes – Low-dimensional topology in specific dimensions – Hyperbolic 3-manifolds. | Differential geometry – Global differential geometry – General geometric structures on manifolds (almost complex, almost product structures, etc.). | Group theory and generalizations – Special aspects of infinite or finite groups – Geometric group theory. | Group theory and generalizations – Special aspects of infinite or finite groups – Word problems, other decision problems, connections with logic and automata (group-theoretic aspects). | Dynamical systems and ergodic theory – Dynamical systems over complex numbers – Dynamics of complex polynomials, rational maps, entire and meromorphic functions; Fatou and Julia sets. | Computer science – Theory of data – Data structures.

Classification: LCC QA611 .C645 2021 | DDC 514/.22–dc23

LC record available at https://lccn.loc.gov/2021037684

AMS Softcover ISBN (Volume III): 978-1-4704-7470-6

11 10 9 8 7 6 5 4 3 2 28 27 26 25 24 23

Contents

Volume I

PREFACE

William Paul Thurston was born on October 30, 1946 and died on August 21, 2012 at the age of 65. During his lifetime Thurston changed the landscape of mathematics in at least two ways. First, his original ideas changed and connected whole subjects in mathematics, from low-dimensional topology to the theory of rational maps to hyperbolic geometry and far beyond. But, just as importantly, through both his written and non-written work Thurston changed the way we think about and encounter mathematics.

One hope in bringing (almost) all of Thurston's written work together in one place is that it might shed light on the long intellectual journey of a unique thinker: how Thurston developed his viewpoint; what it brought to the subjects he wrote about; and how he applied insights gained in one topic to understand others. Just as important, perhaps, are the countless gems contained in these papers, many well-known but perhaps some still undiscovered by the general mathematical community.

A central theme running through all of Thurston's work is his emphasis on understanding and imagination. We invite and challenge the reader to find others.

Contents. Thurston's holistic approach to mathematics makes it difficult to organize his papers in a way that does not seem to erect artificial dividing lines between different topics. Of course one must pick some ordering, and hence some groupings. We have done our best.

We have organized Thurston's collected work into three volumes, with a fourth consisting of his famous and highly influential 1977-8 Princeton Course notes. Volume I contains Thurston's papers on foliations, on surfaces and mapping class groups, and on differential geometry. Volume II contains Thurston's papers on the geometry and topology of 3-manifolds; on complexity, constructions and computers; and on geometric group theory. Volume III contains Thurstons papers on dynamics and on computer science; it also contains his papers written for general audiences, as well as a few miscellaneous papers, including his 1967 New College undergraduate thesis, a fascinating document that foreshadows Thurston's broad view of mathematics.

At the start of each grouping of Thurston's papers we give an introduction, both as a warmup discussion and as a means of placing the papers in a broader context. We have tried to abide by the philosophy that "less is more", as Thurston's papers truly stand on their own.

Acknowledgements. We would like to thank Joan Birman and Bill Veech for initiating this project. We thank Eriko Hironaka and the American Mathematical Society for their support and help. Finally, we are extremely grateful to Julian Thurston for allowing both preprints and published papers to be used for these volumes. Without her this project would not have been possible.

Permissions & Acknowledgments

The American Mathematical Society gratefully acknowledges the kindness of the following individuals and institutions in granting permission to reprint material in this volume:

Ian Agol and Joel Hass

(with Ian Agol and Joel Hass) "The computational complexity of knot genus and spanning area," *Trans. Amer. Math. Soc.* **358** (2006), no. 9, 3821–3850.

American Mathematical Society

"Noncobordant foliations of S^3," *Bull. Amer. Math. Soc.* **78**, no. 4, (1972), 511–514; ©1972, American Mathematical Society.

"Foliations and groups of diffeomorphisms," *Bull. Amer. Math. Soc.* **80**, no. 2, (1974), 304–307; ©1974, American Mathematical Society.

"The theory of foliations of codimension greater than one," *Differential geometry* (Proc. Sympos. Pure Math., Vol. XXVII, Stanford Univ., Stanford, Calif., 1973), Part 1, p. 321. Amer. Math. Soc., Providence, R.I., 1975; ©1975, American Mathematical Society.

"A local construction of foliations for three-manifolds," *Differential geometry* (Proc. Sympos. Pure Math., Vol. XXVII, Stanford Univ., Stanford, Calif., 1973), Part 1, pp. 315–319. Amer. Math. Soc., Providence, R.I., 1975; ©1975, American Mathematical Society.

(with H. E. Winkelnkemper) "On the existence of contact forms," *Proc. Amer. Math. Soc.* **52** (1975), 345–347; ©1975, American Mathematical Society.

"Some simple examples of symplectic manifolds," *Proc. Amer. Math. Soc.* **55** (1976), no. 2, 467–468; ©1976, American Mathematical Society.

"Three dimensional manifolds, Kleinian groups and hyperbolic geometry," *Bull. Amer. Math. Soc. (N.S.)* **6** (1982), no. 3, 357–381; ©1982, American Mathematical Society.

"A norm for the homology of 3-manifolds," Mem. Amer. Math. Soc. **59** (1986), no. 339, i–vi and 99–130; ©1986, American Mathematical Society.

"Zippers and univalent functions. The Bieberbach conjecture," (West Lafayette, Ind., 1985), 185–197, *Math. Surveys Monogr.*, **21**, Amer. Math. Soc., Providence, RI, 1986; ©1986, American Mathematical Society.

(with Daniel D. Sleator and Robert E. Tarjan) "Rotation distance, triangulations, and hyperbolic geometry," *J. Amer. Math. Soc.* **1** (1988), no. 3, 647–681; ©1988, American Mathematical Society.

"On the geometry and dynamics of diffeomorphisms of surfaces," *Bull. Amer. Math. Soc.* (N.S.) **19** (1988), no. 2, 417–431.

"On proof and progress in mathematics," *Bull. Amer. Math. Soc. (N.S.)* **30** (1994), no. 2, 161–177; ©1994, American Mathematical Society.

(with Yakov M. Eliashberg) *Confoliations,* University Lecture Series **13**, American Mathematical Society, Providence, RI, 1998; ©1998, American Mathematical Society.

(with Ian Agol and Peter A. Storm) "Lower bounds on volumes of hyperbolic Haken 3-manifolds. With an appendix by Nathan Dunfield," *J. Amer. Math. Soc.* **20** (2007), no. 4, 1053–1077; ©2007, Ian Agol, Peter A. Storm, and William Thurston.

Annals of Mathematics

(with Morris W. Hirsch) "Foliated bundles, invariant measures and flat manifolds," *Ann. Math.* (2) **101** (1975), 369–390.

"Existence of codimension-one foliations," *Ann. of Math.* (2) **104** (1976), no. 2, 249–268.

(with Frederick J. Almgren, Jr.) "Examples of unknotted curves which bound only surfaces of high genus within their convex hulls," Ann. of Math. (2) **105** (1977), no. 3, 527–538.

"Hyperbolic structures on 3-manifolds I: Deformation of acylindrical manifolds," *Ann. of Math.* (2) **124** (1986), no. 2, 203–246.

Association for Computing Machinery

Republished with permission of the Association for Computing Machinery, "Separators for sphere-packings and nearest neighbor graphs," with Gary L. Miller, Shang-Hua Teng, and Stephen A. Vavasis, *J. ACM* **44** (1997), no. 1, 1–29; permission conveyed through Copyright Clearance Center, Inc.

Republished with permission of the Association for Computing Machinery, "3-manifold knot genus is NP-complete," with Ian Agol and Joel Hass, *Proceedings of the Thirty-Fourth Annual ACM Symposium on Theory of Computing,* 761–766,ACM, New York, 2002; permission conveyed through Copyright Clearance Center, Inc.

Cambridge University Press

"Hyperbolic geometry and 3-manifolds," *Low-dimensional topology* (Bangor, 1979), pp. 9–25, London Math. Soc. Lecture Note Ser., **48**, Cambridge Univ. Press, Cambridge-New York, 1982. ©1979 Cambridge University Press and reproduced with permission.

Canadian Mathematical Society

"On the construction and classification of foliations," *Proceedings of the International Congress of Mathematicians* (Vancouver, B.C., 1974), Vol. 1, pp. 547–549. Canad. Math. Congress, Montreal, Que., 1975.

James W. Cannon

(with James W. Cannon) "Group invariant Peano curves," *Geom. Topol.* **11** (2007), 1315–1355.

Nathan M. Dunfield

(with Nathan M. Dunfield) "The virtual Haken conjecture: experiments and examples," *Geom. Topol.* **7** (2003), 399–441.

Electronic Library of Mathematics

(with John H. Conway, Olaf Delgado Friedrichs and Daniel H. Huson) "On three-dimensional space groups," *Beiträge Algebra Geom.* **42** (2001), no. 2, 475–507.

Elsevier

Reprinted from "Anosov flows and the fundamental group," with J. F. Plante, *Topology* **11** (1972), 147–150; ©1972 with permission from Elsevier.

Reprinted from "Some remarks on foliations," with H. Rosenberg, *Dynamical systems* (Proc. Sympos., Univ. Bahia, Salvador, 1971), pp. 463–478. Academic Press, New York, ©1973 with permission from Elsevier.

Reprinted from "A generalization of the Reeb stability theorem," *Topology* **13** (1974), 347–352; ©1974 with permission from Elsevier.

Reprinted from "Every connected space has the homology of a $K(\pi, 1)$," with D. M. Kan, *Topology* **15** (1976), no. 3, 253–258; ©1976 with permission from Elsevier.

Reprinted from "A presentation for the mapping class group of a closed orientable surface," with A. Hatcher, *Topology* **19** (1980), no. 3, 221–237; ©1980 with permission from Elsevier.

Reprinted from "New proofs of some results of Nielsen," with Michael Handel, *Adv. in Math.* **56** (1985), no. 2, 173–191; ©1985 with permission from Elsevier.

Reprinted from "Soliton-like behavior in automata," with James K. Park and Kenneth Steiglitz, *Phys. D* **19** (1986), no. 3, 423–432; ©1986 with permission from Elsevier.

Reprinted from "On K_1-theory of the Euclidean space," with L. Vaserstein, *Topology Appl.* **23** (1986), no. 2, 145–148; ©1986 with permission from Elsevier.

Reprinted from "Triangulating 3-manifolds using 5 vertex link types," with D. Cooper, *Topology* **27** (1988), no. 1, 23–25; ©1988 with permission from Elsevier.

Fondation L'Enseignement Mathématique

(with J. Milnor) "Characteristic numbers of 3-manifolds," *Enseign. Math.* (2) **23** (1977), no. 3-4, 249–254.

(with Dennis Sullivan) "Manifolds with canonical coordinate charts: some examples," *Enseign. Math.* (2) **29** (1983), no. 1–2, 15–25.

Joel Hass and Abigail Thompson

(with Joel Hass and Abigail Thompson) "Stabilization of Heegaard splittings," *Geom. Topol.* **13** (2009), no. 4, 2029–2050

Institut des Hautes Études Scientifiques

(with M. Gromov and H. B. Lawson, Jr.) "Hyperbolic 4-manifolds and conformally flat 3-manifolds," *Inst. Hautes Études Sci. Publ. Math.* **68** (1988), 27–45 (1989).

International Press of Boston, Inc.

(with Joel Hass and Jeffrey C. Lagarias) "Area inequalities for embedded disks spanning unknotted curves," *J. Differential Geom.* **68** (2004), no. 1, 1–29. Courtesy of International Press of Boston, Inc.

IOP Publishing, Ltd.

Republished with permission of IOP Publishing Ltd., from "How to see 3-manifolds"; Topology of the Universe Conference (Cleveland, OH, 1997), *Classical Quantum Gravity* **15** (1998), no. 9, 2545–2571, ©1998; permission conveyed through the Copyright Clearance Center, Inc.

John Wiley and Sons

(with D. B. A. Epstein) "Transformation groups and natural bundles," *Proc. London Math. Soc.* (3) **38** (1979), no. 2, 219–236. ©1979 John Wiley and Sons, all rights reserved.

Mathematical Sciences Research Institute

"The Eightfold Way: a mathematical sculpture by Helaman Ferguson," *The Eightfold Way,* 1–7, Math. Sci. Res. Inst. Publ. **35**, Cambridge Univ. Press, Cambridge, 1998.

Princeton University Press

"Entropy in dimension one," *Frontiers in complex dynamics: In Celebration of John Milnor's 80th Birthday,* Araceli Bonifant, Misha Lyubich, and Scott Sutherland, 339–384, Princeton Math. Ser., **51**, Princeton Univ. Press, Princeton, NJ, 2014. Reprinted with permission of Princeton University Press; permission conveyed through Copyright Clearance Center, Inc.

"Degree-d-invariant laminations," *What's Next?: The Mathematical Legacy of William P. Thurston,* with Hyungryul Baik, Gao Yan, John H. Hubbard, Tan Lei, Kathryn A. Lindsey and Dylan P. Thurston, Dylan Thurston (ed.), *Annals of Mathematics Studies* **205**, 259–325, Princeton University Press, Princeton, NJ, 2020; ©2020. Republished with permission of Princeton University Press; permission conveyed through the Copyright Clearance Center, Inc.

Tim Riley

(with T. R. Riley) "The absence of efficient dual pairs of spanning trees in planar graphs," *Electronic Journal of Combinatorics,* **13** 2006, #N13.

Society for Industrial and Applied Mathematics

(with Daniel D. Sleator and Robert E. Tarjan) "Short encodings of evolving structures," *SIAM J. Discrete Math.* **5** (1992), no. 3, 428–450.

(with Gary L. Miller, Shang-Hua Teng, and Stephen A. Vavasis) "Geometric separators for finite-element meshes," *SIAM J. Sci. Comput.* **19** (1998), no. 2, 364–386.

Société Mathématique de francede Gazette des Mathematiciens

(with Jean-Pierre Bourguignon) "Interview de William Thurston," *Gaz. Math. No.* **65** (1995), 11–18.

Springer Nature

Reprinted by permission from Springer Nature, "Anosov flows on new three manifolds," with Michael Handel, *Invent. Math.* **59** (1980), no. 2, 95–103; ©1980.

Reprinted by permission from Springer Nature, "Incompressible surfaces in 2-bridge knot complements," with A. Hatcher, *Invent. Math.* **79** (1985), no. 2, 225–246; ©1985.

Reprinted by permission from Springer Nature, "Extending holomorphic motions," with Dennis P. Sullivan, *Acta Math.* **157** (1986), no. 3-4, 243–257; ©1986.

Reprinted by permission from Springer Nature, "Pinching constants for hyperbolic manifolds," with M. Gromov, *Invent. Math.* **89** (1987), no. 1, 1–12.

Reprinted by permission from Springer Nature, "Solvgroups are not almost convex." with J. W. Cannon, W. J. Floyd and M. A. Grayson, *Geom. Dedicata* **31** (1989), no. 3, 291–300; ©1989.

Reprinted by permission from Springer Nature, (with John Milnor) "On iterated maps of the interval," with John Milnor, *Dynamical systems,* (College Park, MD, 198687), 465–563, Lecture Notes in Math. **1342**, Springer, Berlin, 1988; ©1988.

Reprinted by permission from Springer Nature, "Non-continuity of the action of the modular group at Bers' boundary of Teichmuller space," with Steven P. Kerckhoff, *Invent. Math.* **100** (1990), no. 1, 25–47; ©1990.

Reprinted by permission from Springer Nature, "Automatic mesh partitioning. Graph theory and sparse matrix computation," with Gary L. Miller, Shang-Hua Teng, and Stephen A. Vavasis, 57–84, *IMA Vol. Math. Appl.*, **56**, Springer, New York, 1993; ©1993.

Reprinted by permission from Springer Nature, "The symbolic dynamics of tiling the integers," with Ethan M. Coven, William Geller and Sylvia Silberger, *Israel J. Math.* **130** (2002), 21–27; ©2002.

Reprinted by permission from Springer Nature, "The size of spanning disks for polygonal curves," with Joel Hass and Jack Snoeyink, *Discrete Comput. Geom.* **29** (2003) no. 1, 1–17; ©2003.

Reprinted by permission from Springer Nature, (with Nathan M. Dunfield) "Finite covers of random 3-manifolds," with Nathan M. Dunfield, *Invent. Math.* **166** (2006), no. 3, 457–521©2006.

Swiss Mathematical Society

"The theory of foliations of codimension greater than one," *Comment. Math. Helv.* **49** (1974), 214–231.

(with J. F. Plante) "Polynomial growth in holonomy groups of foliations," *Comment. Math. Helv.* **51** (1976), no. 4, 567–584.

Taylor & Francis

"Conway's tiling groups," *Amer. Math. Monthly* **97** (1990), no. 8, 757–773.

Julian Thurston

"Three-manifolds, foliations and circles, I," December 1997 preprint.

"Hyperbolic structures on 3-manifolds, II: Surface groups and 3-manifolds which fiber over the circle," 1986 preprint, 1998 eprint.

"Hyperbolic structures on 3-manifolds, III: Deformations of 3-manifolds with incompressible boundary," 1986 preprint, 1998 eprint.

"Hyperbolic structures on 3-manifolds: Overall logic," 1980 preprint.

"Three-manifolds with symmetry," 1982 preprint.

"Minimal stretch maps between hyperbolic surfaces," 1986 preprint, 1998 eprint.

"Groups, Tilings and Finite State Automata," Summer 1989 AMS Colloquium Lectures, Research Report GCG 1, 1989 preprint.

"Three-manifolds, foliations and circles, II: the Transverse Asymptotc Geometry of Foliations," January 1998 preprint.

"A Constructive Foundation for Topology," Senior Thesis, New College, June 14, 1967.

"Foliations of three-manifolds which are circle bundles," Ph.D. Thesis, University of California, Berkeley, 1972.

"On the dynamics of iterated rational maps," February 1984 preprint.

"Earthquakes in 2-dimensional hyperbolic geometry," *Low-dimensional topology and Kleinian groups*, (Coventry/Durham, 1984), 269–289, London Math. Soc. Lecture Note Ser., 112, Cambridge Univ. Press, Cambridge, 1986.

"Finite state algorithms for the braid groups," February 1988 preprint.

"Shapes of polyhedra and triangulations of the sphere," The Epstein birthday schrift, 511–549, *Geom. Topol. Monogr.*, **1** Geom. Topol. Publ., Coventry, 1998.

Nathaniel Thurston

Figure 1. Lines of the form $nx + my = 1/2$ where n and m are integers. Any convex polygon in this network which is symmetric in the origin is the unit sphere in $H_2(M)$, for some 3-manifold M. This computer drawn picture was prepared by Nathaniel Thurston. Appeared in : Thurston, William P., A norm for the homology of 3-manifolds, *Mem. Amer. Math. Soc.* **59** (1986), no. 339, i–vi and 99–130.

Turkish Journal of Mathematics

(with Yakov M. Eliashberg) "Contact structures and foliations on 3-manifolds," *Turkish J. Math.* **20** (1996), no. 1, 19–35.

Università degli Studi di Cagliari

(with D. B. A. Epstein) "Combable groups," Conference on Differential Geometry and Topology (Sardinia, 1988). *Rend. Sem. Fac. Sci. Univ. Cagliari* **58** (1988), suppl., 423–439.

Part 1.

Dynamics and Complex Analysis

DYNAMICS

Dynamical systems are a common thread running through almost all of Thurston's work. They are crucial to his study of the mapping class group, Teichmuller space, and measured foliations. They play a central role in his work on geometrizing Haken 3-manifolds. His manuscripts on maps of the interval and on rational maps of S^2 that circulated widely for years before appearing in print have had an outsized influence on those topics.

Thurston's joint work with Milnor ([MT]) began circulating in the late 1970's before being published in 1988. It is concerned with maps of an interval to itself that are *piecewise monotonic*, i.e are strictly increasing or decreasing on a finite collection of subintervals. A heavily studied case is when the function is *unimodular*, when it has a unique maximum or minimum. Of particular interest is the family of quadratic maps $f(x) = \frac{x^2-a}{2}$, where $-1 \leq a \leq 8$. The family exhibits a fascinating behavior of period doubling as a increases towards a value (5.6046...) related to the Feigenbaum constant. Beyond that value the maps become more chaotic.

The Milnor-Thurston paper introduced new invariants, called *kneading invariants*, that provided a new viewpoint on these special examples and applied in general to the class of piecewise monotone maps. It connected the invariants to the entropy of the maps and to zeta functions that record the number of periodic points of a given period. The number of maximal subintervals ("laps") on which the function is monotonic is denoted by $l(f)$. The limit $\lim_{n\to\infty} l(f^n)^{\frac{1}{n}} = s$ is called the growth rate of f. The topological entropy h of f equals $\log s$.

The monotonic subintervals are denoted by $I_j, j = 1, \cdots, l$. The local max and min points, called *turning points*, are denoted by $c_i, i = 1, \cdots, l-1$. The itinerary of a point x is a sequence of symbols $\{I_j\}, \{c_i\}$ describing the position of $f^n(x)$. The *kneading matrix* $[N_{ij}]$ of f is an $(l-1) \times l$ matrix which is determined by the itineraries of the turning points c_i; the entries are power series with integer coefficients. The *kneading determinant* $D(t)$ is a single power series obtained from determinants of $(l-1) \times (l-1)$ submatrices. It determines a holomorphic function in the unit disk whose smallest zero equals $\frac{1}{s}$, where s is the growth rate of f. In particular, the entropy $h = \log \mathrm{s}$ can be computed from the kneading determinant.

The Artin-Mazur zeta function is a formal power series that equals the exponential of $\Sigma_{n=1}^{\infty}|\mathrm{fix}(f^n)|\frac{z^n}{n}$, where $|\mathrm{fix}(f^n)|$ is the cardinality of the set of fixed points of f^n (assumed to be finite for all n). Milnor-Thurston show that a variant of this zeta function equals the reciprocal of $D(t)$. This is a remarkable result that shows that the behavior of the set of periodic points of $f(x)$ can be read off from the kneading determinant. The authors prove a number of corollaries in the unimodular case where the kneading determinant can often readily be computed. In particular, they describe the existence of periodic points of various periods for the quadratic maps $f(x) = \frac{x^2-a}{2}$ as the parameter a and the entropy of f vary.

In his final paper ([Th2]), written more than thirty years later, Thurston returned to the dynamics of iterated maps of an interval. He studies the class of *postcritically finite* maps, those with the property that the orbit of the critical points under all iterates of f is finite. In particular, the map itself has only a finite number of critical points so it is piecewise monotonic. Thurston analyzes the possible values of the entropy of such a map. He proves that h is the entropy of a postcritically finite map of the interval if and only if its exponential $s = e^h$ is an algebraic integer whose absolute value is at least as big as that of all of its Galois conjugates.

The first step comes from a result in the Milnor-Thurston paper where they show that any piecewise monotonic map f is semi-conjugate to a piecewise linear map g whose slopes are all equal to $\pm s$, where $h = \log s$ is the entropy of f (and of g). Thurston shows that g is postcritically finite if f is. This reduces the problem to the case when the original map is PL with constant size slope, which Thurston calls a *uniformly expanding* map. The orbits of the critical points divide the interval into a finite number of subintervals $\{J_i\}$. Under the map f each J_j is mapped to some union of the J_i. The associated transition matrix A is a non-negative integer matrix. If some power of A is strictly positive (A is *mixing*), then the Perron-Frobenius theorem states there is a unique (up to scale) eigenvector with positive entries and its eigenvalue is an algebraic integer whose size is strictly bigger than that of its Galois conjugates. Such a number is called a *Perron* number. If A has the weaker property that for every entry there is a power where that entry is positive (A is *ergodic*) , there is a positive eigenvector (not necessarily unique) with eigenvalue whose size is at least as big as that of its Galois conjugates, a *weak Perron* number. Lind ([L]) proved that for any weak Perron number λ, there is some ergodic matrix with λ as its eigenvalue.

It's easy to show that a uniformly expanding, postcritically finite map has a weak Perron eigenvalue. The difficulty in the proof of the converse comes from the fact that not all ergodic matrices can arise as the transition matrix for a uniformly expanding, postcritically finite map. Constructing such a map with a given eigenvalue is quite intricate.

In characteristic fashion Thurston does not simply prove his theorem but also produces many examples and computer simulations describing the statistics of the numbers that arise under certain conditions on the maps. He studies in detail the case when the eigenvalue is a Pisot number.

Thurston generalizes his techniques to include maps of certain graphs to themselves. He applies this to analyze the expansion values for outer automorphisms of a free group. Bestvina and Handel ([BH]) showed that any automorphism that is irreducible in a suitable sense preserves a *train track* τ, which is a finite graph with some extra structure at its vertices. The edges of τ are mapped to unions of edges, and the transition matrix is mixing. They further define relative train tracks to deal with reducible cases where the transition matrix is only ergodic. Thurston proves that that every weak Perron number is the expansion factor for an ergodic train track map of some outer automorphism of a free group.

Because of its analogies to Kleinian groups, its dynamical complexity, and its beautiful computer graphics, the area of rational maps of the 2-sphere had a natural appeal to Thurston. In the introduction to [Th1] he gives a wonderful description of the attraction of the subject. This paper circulated in preprint form during the early 1980's and was immensely influential. It begins with a brief description of the main properties of rational maps and of their Fatou and Julia sets. These play the roles of the domain of discontinuity and the limit set for a Kleinian group. (See [S] for Sullivan's "dictionary" describing analogies between rational maps and Kleinian groups.)

A special case is that of a polynomial map (degree at least two). The point at infinity is a super attracting point and lies in the Fatou set. The basin of the point at infinity, the points that converge to infinity under iteration, is topologically a disk. The Julia set is its boundary in S^2. With a natural normalization, there is a unique holomorphic conjugation between the map z^d on the exterior of the unit disk and the action on the basin of infinity. When the Julia set is connected and locally connected the conjugating map extends to an equivariant map from the boundary of the disk to the Julia set. The map on the boundary circle is not usually one-to-one and a major portion of the paper is spent analyzing possible identification patterns.

Thurston describes the identifications in terms of invariant geodesic laminations. A *geodesic lamination* is a closed set that is a disjoint union of geodesics in the hyperbolic plane (they may intersect on the circle at infinity). The lamination needs to be "invariant" under the map z^d where the notion of invariance is subtle because the map is not a homeomorphism. Many necessary properties for a degree d invariant lamination are given but it is only in the $d = 2$ case that Thurston provides a detailed description of their structure. An edited version of this portion of Thurston's paper was published in [Th4], along with an appendix written by Schleicher. It provides a current perspective of Thurston's preprint.

Much of the dynamics of a rational map can be predicted by the orbits of its critical points. Postcritically finite rational maps, where this set of orbit points is finite, are a particularly interesting class. Thurston's major result here is a necessary and sufficient topological condition for a map of S^2 to be topologically equivalent to a postcritically finite rational map (of degree $d \geq 2$). Such a rational map determines a degree d branched map f from S^2 to itself. The orbit of the branch points is finite and is denoted by P_f. (The branch points themselves are not included.) Two postcritically finite branched maps f, g are considered equivalent if there is a homeomorphism h of S^2 to itself taking P_f to P_g so that f is isotopic to $h^{-1} \circ g \circ h$ rel P_f.

The map f determines an orbifold Q_f whose underlying space is S^2 and whose exceptional points correspond to P_f. A rational map endows the orbifold with a conformal structure induced from S^2 and determines a holomorphic map of the orbifold to itself. Given a postcritically finite branched map f Thurston sets up a map from the Teichmuller space of conformal structures on Q_f to itself. He shows that the map has a fixed point if and only if there is a postcritically finite rational map equivalent to f. While this was a radical idea in the field, it was a favorite technique of Thurston's to set up fixed point problems in surprising contexts. A similar example is in his proof of the geometrization theorem for Haken 3-manifolds.

A complete statement and proof of Thurston's topological characterization does not appear in this paper. Douady and Hubbard provided such a proof in [DH], based on lectures given by Thurston in 1983.

During the description of degree two invariant laminations in the first part of this paper, Thurston wrote "Much of what we shall do seems to have natural generalizations to the cases of higher degree, and I hope that someone will undertake a careful analysis of these cases." As it turned out, someone did undertake such an analysis: Thurston, himself, some twenty-five years later. During the final year of his life, Thurston ran a seminar at Cornell that focused on this topic. He wrote up some, but not all, of his results. Fortunately, his seminar was well-attended by a group of talented and enthusiastic mathematicians who published his unfinished manuscript and provided additional material that they had learned from Thurston in the seminar and by email ([Th3]). It is fitting that this final paper documents not only Thurston's own insights but also his ability to inspire others.

References

[BH] M. Bestvina and M. Handel, Train tracks and automorphisms of free groups. *Ann. of Math. (2)* 135 (1992), no. 1, 1–51.

[DH] A. Douady & J. Hubbard, A proof of Thurston's topological characterization of rational functions. *Acta Math.* 171 (1993), no. 2, 263–297.

[L] D. Lind, The entropies of topological Markov shifts and a related class of algebraic integers, *Ergodic Theory Dynam. Systems* 4 (1984), no. 2, 283–300.

[MT] J. Milnor & W. Thurston, On iterated maps of the interval *in* "Dynamical systems (College Park, MD, 1986-87)", 465–563, Lecture Notes in Math., 1342, Springer, 1988.

[S] D. Sullivan, Quasiconformal homeomorphisms and dynamics. I. Solution of the Fatou-Julia problem on wandering domains. *Ann. of Math. (2)* 122 (1985), no. 3, 401–418.

[Th1] W. Thurston, On the dynamics of iterated rational maps, preprint.

[Th2] W. Thurston, Entropy in dimension one *in* "Frontiers in complex dynamics", 339–384, Princeton Math. Series, 51, Princeton Univ. Press, Princeton, NJ, 2014.

[Th3] W. Thurston, H. Baik, G. Yan, J. Hubbard, K. Lindsey, T. Lei, and D. Thurston, Degree-d-invariant laminations, *in* "What's Next?: The Mathematical Legacy of William P. Thurston", 259–325, Annals of Math Studies (205), edited by Dylan Thurston, Princeton Univ. Press, 2020.

[Th4] W. Thurston, On the geometry and dynamics of iterated rational maps, edited by Dierk Schleicher and Nikita Selinger, with an appendix by Schleicher, *in* "Complex dynamics", 3–137, A K Peters, 2009.

ON ITERATED MAPS OF THE INTERVAL

by

John Milnor
Institute for Advanced Study
Princeton, NJ 08540 USA

and

William Thurston
Princeton University
Princeton, NJ 08544 USA

Dedicated to the Memory of Rufus Bowen and Peter Štefan.

Introduction.

Mappings from an interval to itself provide the simplest possible examples of smooth dynamical systems. Such mappings have been widely studied in recent years since they occur in quite varied applications, and since their surprisingly complex behavior may provide a useful introduction to the study of higher dimensional situations.

The present paper sets up an effective calculus for describing the qualitative behavior of the successive iterates of a piecewise monotone mapping. Let I be a closed interval of real numbers. By definition, a continuous mapping f from I to itself is *piecewise monotone* if I can be subdivided into finitely many subintervals $I_1,\dots,I_\ell$ on which f is alternately strictly increasing or strictly decreasing. Each such maximal interval on which f is monotone is called a *lap* of f, and $\ell = \ell(f)$ is the *lap number*. The separating points $c_1,\dots,c_{\ell-1}$ at which f has a local minimum or maximum are called the *turning points* of f.

Section 1 studies the sequence $\ell(f),\ell(f^2),\ell(f^3),\dots$ of positive integers, where $f^k = f\circ\dots\circ f$ stands for the k-fold composition of the function f with itself. The limit

$$s = \lim_{k\to\infty} \ell(f^k)^{1/k} = \inf_k \ell(f^k)^{1/k}$$

is a real number in the interval $[1,\ell(f)]$ called the *growth number* of f. (By a theorem of J. Rothschild and of Misiurewicz and Szlenk, its logarithm is precisely the "topological entropy" of f.) The special case of a quadratic map f is discussed in some detail.

Sections 2, 3 introduce an invariantly defined "formal coordinate function" $\theta(x)$ for $x \in I$. This is a formal power series of the form $\sum \theta_k(x)t^k$, which depends monotonely on x, under a suitable ordering. If $f^k(x)$ belongs to the interior of the lap I_j, then the coefficient $\theta_k(x)$ is a corresponding formal symbol I_j multiplied by a sign which is either ± 1 or 0 according as the function f^k is increasing, decreasing, or has a turning point at x.

Section 4 introduces a basic invariant called the <u>kneading</u> <u>matrix</u> of f. This is an $(\ell+1)\times\ell$ matrix with entries in the ring $\underset{\sim}{Z}[[t]]$ of integer formal power series. If $k \geq 1$, the coefficient of t^k in its (i,j)-th entry is ± 2 if f^k maps a neighborhood of c_i into the lap I_j, and is zero otherwise. Closely related is the <u>kneading</u> <u>determinant</u>, a formal power series $D(t) = 1 + D_1 t + D_2 t^2 + \ldots$ with odd integer coefficients. In the simplest case, if f has just one turning point c_1, which is say a minimum point, each coefficient D_k is either +1 or −1 according as f^{k+1} takes on a local minimum or a local maximum at c_1.

Section 5 gives an explicit method for computing the sequence of lap numbers $\ell(f^k)$ in terms of the kneading matrix. As an example, if there is only one turning point, and if f maps both endpoints of the interval I to a single endpoint, then the formal power series $L(t) = \sum_1^\infty \ell(f^k)t^{k-1}$ is equal to $(1-t)^{-1} + D(t)^{-1}(1-t)^{-2}$.

Here is an explicit example. For the mapping $f(x) = (x^2-5)/2$ of Figure 1.4, it is easy to check that the iterates $f^k(0)$ belong alternately to the lap I_1 or I_2. According to §4.5, this information determines the kneading determinant:

$$D(t) = 1 - t - t^2 + t^3 + t^4 - - + + \ldots = (1-t)/(1+t^2).$$

We can now compute the series $L(t) = 2(1-t+t^2)/(1-t)^3$, and it follows easily that $\ell(f^k) = k^2 - k + 2$ for $k \geq 1$, and that $s = 1$.

Section 6 studies convergence properties of these formal power series, showing for example that $D(t)$ is holomorphic in the open unit disk $|t| < 1$, and has smallest zero at $t = 1/s$ if $s > 1$, while $L(t)$ is meromorphic in the open unit disk with a pole at $1/s$.

Section 7, under the hypothesis $s > 1$, shows that f is "topologically semi-conjugate" to a piecewise linear map having slope $\pm s$ everywhere (see Figure 7.11). As a typical application, in the case $\ell = 2$ it is shown that f admits a periodic point of odd period $p \geq 3$ if and only if $s > \sqrt{2}$ (compare Štefan).

Section 8 surveys a number of known methods and theorems concerning periodic opints, due to Julia and Fatou, Sarkovskii, Artin and Mazur, Bowen and Franks, Allwright, and D. Singer. In particular, the Artin-Mazur power series $\zeta(t)$ is a convenient way of enumerating the periodic opints of f. Sections 9 and 10 state several versions of our main theorem which computes this Artin-Mazur zeta function in terms of the kneading determinant. As an application of this theorem, using Bowen and Franks, it is shown that $s > 1$ if an only if there exists a periodic point whose period is not a power of two. Furthermore, there are only finitely many distinct periods (all necessarily powers of two) if and only if the sequence of lap numbers $\ell(f^k)$ is bounded by a polynomial function of k.

Section 11 completes the proof of this main theorem. A basic tool is the statement that, under C^1-smooth deformation of f, as the difference $f^p(c_i) - c_i$ changes sign the kneading determinant of f is changed by a factor of $(1-t^p)/(1+t^p)$.

Section 12 characterizes those power series $D(t)$ which can actually occur as kneading determinant for some map f with lap number $\ell(f) = 2$. It is shown that every such admissible kneading determinant already occurs for some quadratic mapping, for example of the form $f(x) = bx(1-x)$. This section also discussed continuity properties of the growth number $s = s(f)$ under smooth deformation of f. (Compare Figure 13.3.) Section 13 proves a monotonicity theorem for quadratic maps. The last section constructs a number of illustrative examples.

An earlier version of this paper was widely circulated in 1977. The present version incorporates both Part I and Part II of the earlier one, as well as additional material. Most of it was written in 1981, but §13 was added in 1983 and a few minor corrections are more recent.

Contents

1. Laps and the growth number

We are concerned with properties of a mapping which are invariant under topological conjugacy. By definition, two maps $f : I \to I$ and $g : J \to J$ are topologically conjugate if there exists a homeomorphism h from I onto J so that $g = h \circ f \circ h^{-1}$. (If h preserves orientation, then one speaks of an orientation preserving topological conjugacy.)

A mapping f from the closed interval I to itself is called piecewise-monotone if I can be subdivided into finitely many subintervals

$$I_1 = [c_0, c_1],\ I_2 = [c_1, c_2], \ldots, I_\ell = [c_{\ell-1}, c_\ell]$$

in such a way that the restriction of f to each interval I_j is strictly monotone, either increasing or decreasing. Here it is to be understood that $c_0 < c_1 < \ldots < c_\ell$ where I is the interval $[c_0, c_\ell]$. We will always assume that each I_j is a maximal interval on which f is strictly monotone. Such a maximal interval is called a lap of the function f, and the number $\ell = \ell(f)$ of distinct laps is called the lap number of f.

It follows that each of the points $c_0, c_1, \ldots, c_\ell$ is either a local minimum of local maximum point of f. The interior local minimum and maximum points $c_1, \ldots, c_{\ell-1}$ will be called the turning points of f, and will play an important role in what follows.

Together with f, we will also study the sequence

$$f^* = (f^0, f^1, f^2, \ldots)$$

of successive iterates, where

$$f^0 = \text{identity map},\ f^1 = f,\ f^2 = f \circ f,$$

and so on. Here we use the symbol $*$ to indicate a non-negative integer which indexes some sequence.

Caution. The reader must take care since our notation for iterates could be confusing: _Superscripts on a function will always stand for iterates, but superscripts on a variable have the customary meaning as exponents._

A crude but powerful invariant of the topological conjugacy class of f is provided by the sequence

$$\ell(f^*) = (1, \ell(f), \ell(f^2), \ldots)$$

of lap numbers associated with the successive iterates of f. Note that the number of laps of a composition of two functions satisfies the inequality

$$\ell(f \circ g) \le \ell(f)\ell(g), \tag{1.1}$$

since $f \circ g$ can have at most $\ell(f)$ laps within each lap of g. Hence $\ell(f^n) \le \ell(f)^n$. Using these inequalities, a standard argument proves the following.

Lemma 1.2. The n-th root $\ell(f^n)^{1/n}$ tends to a well-defined limit s as $n \to \infty$. Furthermore, this limit is precisely equal to the infimum of the numbers $\ell(f^n)^{1/n}$.

It follows that $\ell(f^n) \ge s^n$ for every n. In particular, s always belongs to the interval $[1, \ell(f)]$.

Definition. This limit s will be called the _growth number_ of f.

It can be shown that the logarithm of s is equal to the invariant known as the _topological entropy_ of f. Compare Rothschild, as well as Misiurewicz and Szlenk.

Proof of 1.2. Fixing some positive integer k, any n can be written as $n = kp + q$ with $0 \le q < k$. Then

$$\ell(f^n) \le \ell(f^k)^p \ell(f)^q$$

by repeated application of 1.1; hence

$$\ell(f^n)^{1/n} \le \ell(f^k)^{p/n} \ell(f)^{q/n}.$$

As $n \to \infty$, the right hand side tends to the limit $\ell(f^k)^{1/k}$; hence

$$\limsup \{\ell(f^n)^{1/n}\} \le \ell(f^k)^{1/k}$$

for every k. Therefore

$$\limsup \{\ell(f^n)^{1/n}\} \le \inf \{\ell(f^k)^{1/k}\} \le \liminf \{\ell(f^k)^{1/k}\},$$

which evidently completes the proof. ∎

In order to illustrate these concepts, let us look at a family of explicit mappings.

Example 1.3. Given any constant $1 \le b \le 4$, consider the quadratic mapping

$$g(y) = by(1-y)$$

from the unit interval $[0,1]$ to itself, with a single turning point at $c_1 = \frac{1}{2}$. In spite of its innocuous appearance, this quadratic example can have a surprising rich structure, and has been studied by many authors, including biologists, physicists, meteorologists and mathematicians. See for example May, Collet and Eckmann, Lorenz, and Smale and Williams.

It is often convenient to make a canonical linear change of variable by setting x equal to the derivative $g'(y) = b(1-2y)$. In this way, we see that g is topologically conjugate to a quadratic mapping of the form

$$f(x) = (x^2-a)/2$$

from the interval $-b \le x \le b$ to itself. Note that the derivative $f'(x)$ is identically equal to x. Here the constant

$$a = b^2 - 2b$$

is uniquely determined by b and ranges over the interval $-1 \le a \le 8$. Conversely, the constant $b = 1 + \sqrt{1+a}$ is uniquely and monotonely determined by a if we require that $1 \le b \le 4$. The mapping f is

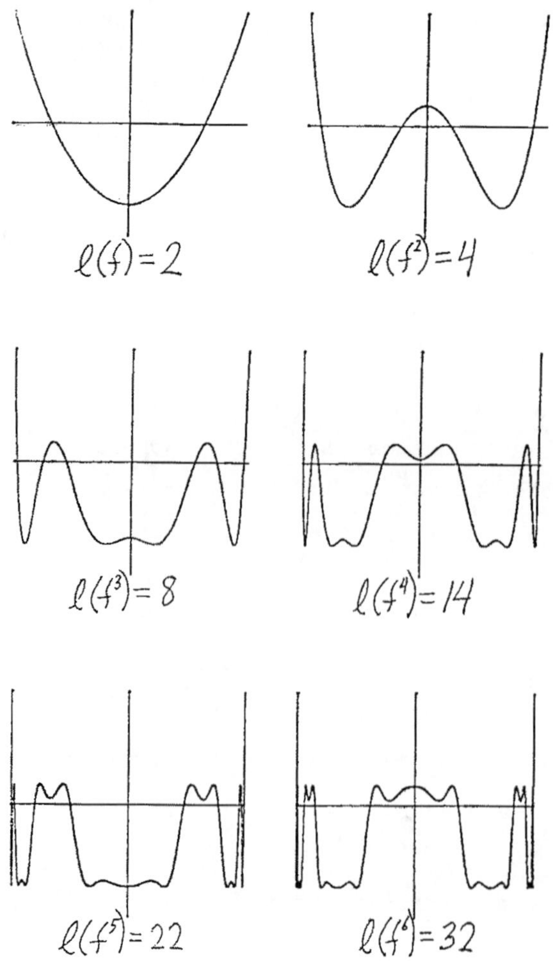

Figure 1.4.

Graphs of iterates of the function $f(y) = (y^2-5)/2$.

monotone decreasing on its first lap $[-b,0]$ and monotone increasing on its second lap $[0,b]$. The special case $a = 5$ (or $b = 1+\sqrt{6}$) is illustrated in Figure 1.4.

We will see in §13 that the behavior of this quadratic function f becomes more and more complicated as the parameter a increases (or as $b = 1+\sqrt{1+a}$ increases). THus, for small values of a, the the lap number $\ell(f^n)$ has the order of magnitude of a polynomial function of n; and it follows that the growth number s is equal to 1. The degree of this polynomial tends to infinity as a increases to the value 5.6046... . Here is a table giving more precise information. For later convenience, we have also listed the "kneading determinant" of f (compare §§4, 5, 12, 13).

Table 1.5. Invariants of $f(x) = (x^2-a)/2$ for $-1 \le a \le 5.6046...$

parameter a	lap number $\ell(f^n)$ for $n > 0$	kneading determinant $D(t)$
$-1 \le a \le 0$	2	$1/(1-t) = 1+t+t^2+...$
$0 < a \le 4$	$2n$	$1/(1+t) = 1-t+t^2-+...$
$4 < a \le 5.2428..$	$n^2 - n + 2$	$(1-t)/(1+t^2) = 1-t-t^2++--...$
$5.2428.. < a \le 5.5261..$	$(2n^3-3n^2+22n+(0 \text{ or } 3))/12$	$(1-t)(1-t^2)/(1+t^4)$
...	polynomial growth of degree $d+1$	$(1-t)...(1-t^{2^{d-1}})/(1+t^{2^d})$
$a = 5.6046..$	faster than polynomial but less than exponential growth	$(1-t)(1-t^2)(1-t^4)(1-t^8)...$

Closely related is the periodicity behavior which is studied beginning in §8. For $-1 \le a \le 3$ the fixed point $x = 1-\sqrt{1+a}$ is stable, and f has no periodic points with period greater than 1. But as a increases past 3 this stable fixed point "bifurcates" into a stable orbit $\{-1 \pm \sqrt{a-3}\}$ of period 2. At $a = 5$ there is a further bifurcation to a stable orbit of period 4, at $a = 5.4723\ldots$ to a stable orbit of period 8, and so on. The stable periodic orbit bifurcates infinitely often as a increases towards the limit $a = 5.6046\ldots$ which was mentioned above (compare Feigenbaum, Lanford).

For $a > 5.6046\ldots$ it seems empirically that $s > 1$, so that the sequence $\ell(f^n)$ grows exponentially with n (compare §13). As a increases, the sequence $\ell(f^*)$ changes uncountably often, and the entire behavior of the iterates of f becomes much more complicated. For the largest parameter value $a = 8$, the map f carries each of its two laps onto the entire interval I, and it follows easily that $\ell(f^n) = 2^n$, hence $s = 2$.

More generally, consider any function f which maps each of its laps $I_1,\ldots,I_\ell$ homeomorphically onto the entire interval I. Then an easy induction on n shows that $\ell(f^n) = \ell^n$, hence $s = \ell = \ell(f)$. The case $\ell = 3$ is illustrated in Figure 1.6.

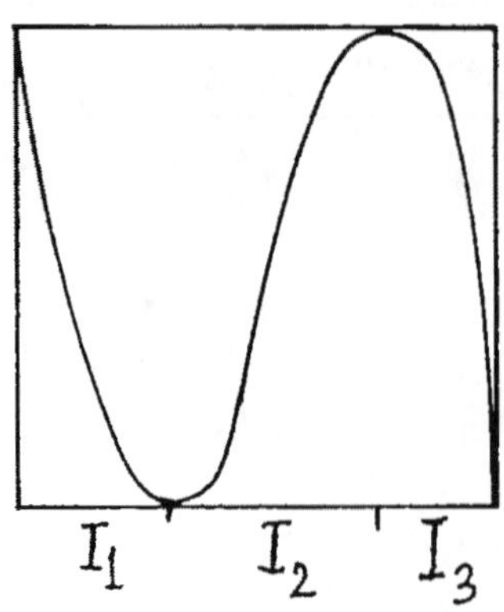

Figure 1.6. Graph of a function f with $\ell(f^n) = 3^n$.

2. The itinerary of a point

Let $I_1,\ldots,I_\ell$ be the laps associated with the mapping f, numbered in their natural order, and let $c_1 < c_2 < \ldots < c_{\ell-1}$ be the turning points of f.

<u>Definition</u>. By the <u>address</u> $A(x)$ of a point $x \in I$ will be meant the formal symbol I_j if x belongs to the lap I_j and is not a turning point, or the formal symbol C_j if x is precisely equal to the turning point c_j. By the <u>itinerary</u> $A(f^*(x))$ will be meant the sequence of addresses

$$(A(x), A(f(x)), A(f^2(x)), \ldots)$$

of the successive images of x. (As usual, the symbol $*$ stands for a variable integer $n \geq 0$.)

Here is a preliminary result which shows that the itinerary $A(f^*(x))$ contains significant qualitative information about the sequence of successive images $f^*(x)$. We will say that the itinerary $A(f^*(x))$ is <u>eventually periodic</u> if there exists an integer $p \geq 1$ so that the address $A(f^n(x))$ is equal to $A(f^{n+p}(x))$ for all large n. The smallest such p will be called the <u>eventual period.</u>

<u>Lemma 2.1</u>. <u>The itinerary $A(f^*(x))$ is eventually periodic if and only if the sequence of points $f^*(x)$ converges towards a periodic orbit of f.</u>

(See §8 for definitions.) If $A(f^*(x))$ has eventual period p, then the proof will show that the limit orbit has period either p or $2p$. Both cases can occur.

Here is an explicit numerical example to illustrate this lemma (compare 1.3).

<u>Example 2.2</u>. Suppose that $f(x) = (x^2-7)/2$. Then $c_1 = 0$ is the only turning point, and a somewhat delicate computation shows that the itinerary

$$A(f^*(0)) = (C_1, I_1, I_2, I_1, I_1, I_2, I_1, \ldots)$$

is periodic of period 3, except for its initial entry. The successive images $f^n(0)$ converge towards the limits

$$-.1099\ldots \quad \text{if} \quad n \equiv 0 \pmod 3$$

$$-3.4939\ldots \quad \text{if} \quad n \equiv 1 \pmod 3$$

$$2.6038\ldots \quad \text{if} \quad n \equiv 2 \pmod 3$$

as $n \longrightarrow \infty$. These three limit points, the roots of the equation $x^3 + x^2 - 9x - 1 = 0$, constitute a (one-sided stable) periodic orbit of period 3.

On the other hand, if $f(x) = (x^2-7.08)/2$, then the itinerary $A(f^*(0))$ again has period 3, except for its initial entry; but in this case the sequence $f^*(0)$ tends towards a periodic orbit of period 6.

<u>Proof of 2.1</u>. Clearly it suffices to consider the special case where the itinerary $A(f^*(x))$ is actually periodic of period p. Define intervals $J_0,\ldots,J_{p-1}$ as follows. Let J_r be the smallest closed interval containing all of the points $f^n(x)$ with $n \equiv r \pmod p$. Then J_r is contained in the single lap $A(f^r(x))$, so it follows that f maps J_r <u>homeomorphically</u> into J_{r+1}. It follows also that f^p maps the interval J_0 homeomorphically into itself. Excluding the trival case where x itself is a periodic point, we may assume that J_0 is a nondegenerate interval. We now distinguish two cases.

<u>Case 1</u>. Suppose that the map f^p from J_0 into itself perserves orientation, i.e., is monotone increasing. If $x \le f^p(x)$, then it follows inductively that

$$x \le f^p(x) \le f^{2p}(x) \le \ldots .$$

Therefore, this subsequence converges to a limit in J_0. Similarly, if $x > f^p(x)$, then the sequence

$$x > f^p(x) > f^{2p}(x) > \ldots$$

converges to a limit in J_0. In both cases, this limit point evidently belongs to a periodic orbit whose period q divides p.

Case 2. Suppose that f^p reverses orientation. Then f^{2p} preserves orientation, and the argument proceeds as before. In this case, the limit orbit has period q dividing $2p$.

Conversely, if the sequence $f^*(x)$ converges to a periodic orbit of period q, then it is quite easy to check that the itinerary $A(f^*(x))$ is eventually periodic with eventual period p dividing q. The argument requires mild care only in the special case where the limit orbit contains a turning point of f.

Since p divides q and q divides $2p$, it follows that q is equal to either p or $2p$. ∎

3. The invariant coordinate $\theta(x)$

In this section we will add just a little bit of information to the itinerary $A(f^*(x))$ in order to obtain a more useful sequence

$$\theta_* = (\theta_0, \theta_1, \theta_2, \ldots)$$

which will be thought of as invariantly defined coordinate for the point x.

First some notations. Let V be the ℓ-dimensional vector space over the rational numbers $\mathbb{Q}$ which has as basis the formal symbols $I_1, \ldots, I_\ell$. (Thus we are introducing a hopefully benign double meaning for I_j.) Our coordinate $\theta_* = \theta_*(x)$ will be a sequence of vectors in this vector space V.

For each symbol I_j, define the sign $\varepsilon(I_j)$ to be either $+1$ or -1 according as f is monotone increasing or monotone decreasing on the lap I_j. These signs alternate, so that we have either $(-1)^j\varepsilon(I_j) = +1$ for all j, or else $(-1)^j\varepsilon(I_j) = -1$ for all j. To each turning point c_j, lying between the laps I_j and I_{j+1}, we assign the vector $C_j = (I_j+I_{j+1})/2$, and the sign $\varepsilon(C_j) = 0$.

Now to each itinerary $A(f^*(x)) = (A_0, A_1, A_2, \ldots)$ we assign the sequence $(\theta_0, \theta_1, \theta_2, \ldots)$ defined by

$$\theta_n = \varepsilon_0 \varepsilon_1 \ldots \varepsilon_{n-1} A_n ,$$

where ε_k is an abbreviation for $\varepsilon(A_k) = \varepsilon(A(f^k(x)))$. Thus

$$\theta_0 = A_0,\ \theta_1 = \varepsilon_0 A_1,\ \theta_2 = \varepsilon_0 \varepsilon_1 A_2, \ldots\ .$$

Remark. The sign $\varepsilon_0 \varepsilon_1 \ldots \varepsilon_{n-1}$ of θ_n can be interpreted geometrically as follows. Note that $\varepsilon(A(x))$ is equal to the local degree of the mapping f at x, being ± 1 or 0 according as f is increasing, decreasing, or has a turning point at x. It follows inductively that the product $\varepsilon_0 \ldots \varepsilon_{n-1} = \varepsilon(A(x)) \ldots \varepsilon(A(f^{n-1}(x)))$ is equal to the local degree of the mapping f^n at x.

Now choose some translation invariant linear ordering of the vector space V (for example, by embedding V $\mathbb{Q}$-linearly into the real numbers) so that the basis vectors satisfy

$$I_1 < \ldots < I_\ell .$$

Order the sequence θ_* lexicographically. Thus $\theta_* < \theta'_*$ if and only if

$$\theta_0 = \theta'_0, \ldots, \theta_{n-1} = \theta'_{n-1},\ \theta_n < \theta'_n$$

for some integer $n \geq 0$. These definitions have been concocted so that the following will be true.

Lemma 3.1. *If* $x < y$, *then* $\theta_*(x) \leq \theta_*(y)$.

In other words, the function $x \longmapsto \theta_*(x)$, from the interval I to certain sequences of vectors, is monotone increasing.

Proof. Suppose, to fix our ideas, that $\theta_k(x) = \theta_k(y)$ for $k < n$. Then we must show that $\theta_n(x) \leq \theta_n(y)$.

Note first that the two points $f^k(x)$ and $f^k(y)$ have the same address, say A_k, for all $k < n$. Let $\varepsilon_k = \varepsilon(A_k)$ be the associated sign. Let J be the closed interval $[x,y]$. Since $J \subset A_0$, it follows that f maps J homeomorphically with degree $\varepsilon_0 \neq 0$ onto an interval $f(J) \subset A_1$. Similarly, f maps $f(J)$ homeomorphically with degree $\varepsilon_1 \neq 0$ onto an interval $f^2(J) \subset A_2$, and so on. <u>It follows inductively that</u> f^n <u>maps</u> J <u>homeomorphically with degree</u> $\varepsilon_0\varepsilon_1\ldots\varepsilon_{n-1} \neq 0$ <u>onto an interval</u> $f^n(J)$. Evidently the two endpoints of this interval satisfy either $f^n(x) < f^n(y)$ or $f^n(x) > f^n(y)$ according as the sign $\varepsilon_0\ldots\varepsilon_{n-1}$ is $+1$ or -1. In either case it follows that

$$\theta_n(x) = \varepsilon_0\ldots\varepsilon_{n-1}A(f^n(x)) \le \varepsilon_0\ldots\varepsilon_{n-1}A(f^n(y)) = \theta_n(y),$$

as required. ■

It will often be convenient to introduce an indeterminate t, and to think of the vectors θ_n as the coefficients of a formal power series

$$\theta = \theta_0 + \theta_1 t + \theta_2 t^2 + \ldots .$$

This series $\theta = \theta(x)$ will be called the <u>invariant coordinate</u> associated with the point x.

Of course, the series θ determines the sequence θ_* of coefficients, and conversely, so that it is just a matter of notational convenience whether we work with one or the other. For ordering purposes we should think of the indeterminate t as being positive, $t > 0$, but infinitesimally small so that a power series $v_0 + v_1 t + v_2 t^2 + \ldots$ will be positive whenever its first nonzero coefficient is positive. With this convention, it follows from 3.1 that the function $x \longmapsto \theta(x)$ is also monotone increasing.

Let $\underset{\sim}{Q}[[t]]$ be the ring of formal power series with rational coefficients, and let $V[[t]]$ be the module consisting of all formal power series with coefficients in the vector space V. Thus each series $\theta = \theta(x)$ is an element of $V[[t]]$, which is a free module with basis $I_1, \ldots, I_\ell$ over the ring $\mathbb{Q}[[t]]$. In other words, we can express θ uniquely as a sum

$$\theta = \Theta_1 I_1 + \cdots + \Theta_\ell I_\ell$$

with coefficients Θ_j which are formal power series with rational coefficients (in fact, usually with integer coefficients).

These coefficients Θ_j satisfy one relation which is a formal consequence of the way in which the series θ was constructed.

Lemma 3.2. _The sum_

$$(1-\varepsilon(I_1)t)\Theta_1 + \ldots + (1-\varepsilon(I_\ell)t)\Theta_\ell$$

in the formal power series ring $\mathbb{Q}[[t]]$ _is identically equal to_ 1.

The coefficient of Θ_j in this relation is either $1-t$ or $1+t$ according as the function f is increasing or decreasing on the lap I_j.

Before giving the proof, let us illustrate these constructions by considering the function $f(x) = (x^2-5)/2$ of Figure 1.4, which maps the interval $|x| \le 1+\sqrt{6}$ to itself. First consider the two fixed points $1 \pm \sqrt{6}$ of f. The interior fixed point $1-\sqrt{6} = -1.44\ldots$ has itinerary $(I_1, I_1, I_1, \ldots)$. (It is often convenient to describe such an itinerary briefly by just listing the subscripts, in this case 1 1 1 1 1 ... .) The corresponding invariant coordinate is

$$\theta(1-\sqrt{6}) = I_1(1-t+t^2-+\ldots) = I_1/(1+t).$$

Similarly, the rightmost point $1+\sqrt{6} = 3.44\ldots$ has itinerary corresponding to the sequence 2 2 2 2 2 ..., with

$$\theta(1+\sqrt{6}) = I_2(1+t+t^2+\ldots) = I_2/(1-t).$$

These fixed points are both unstable: nearby points always have itineraries which are different from these.

The map f has just two other periodic points, constituting an orbit

$$\{-1-\sqrt{2},-1+\sqrt{2}\} = \{-2.41\ldots,0.41\ldots\}$$

of period 2. The itinerary of $-1-\sqrt{2}$ corresponds to the sequence 1 2 1 2 1 2 ..., with

$$\theta(-1-\sqrt{2}) = I_1 - tI_2 - t^2I_1 ++-- \ldots = (I_1-tI_2)/(1+t^2),$$

while the itinerary of $-1+\sqrt{2}$ corresponds to 2 1 2 1 2 1 ..., with

$$\theta(-1+\sqrt{2}) = (tI_1+I_2)/(1+t^2).$$

This orbit is stable, so that any point x which is sufficiently close to $-1-\sqrt{2}$ (or to $-1+\sqrt{2}$) will have the same itinerary and hence the same coordinate θ as $-1-\sqrt{2}$ (or as $-1+\sqrt{2}$).

Listing the four periodic points in their natural order, it is easy to check that

$$\theta(-1-\sqrt{2}) < \theta(1-\sqrt{6}) < \theta(-1+\sqrt{2}) < \theta(1+\sqrt{6}),$$

thus confirming 3.1 for these particular points. Also, it is easy to check that the relation $(1+t)\theta_1 + (1-t)\theta_2 = 1$ of 3.2 is satisfied for each of these points.

<u>Proof of 3.2</u>. Define a $\mathbb{Q}[[t]]$-linear mapping h from the free module $V[[t]]$ to $\mathbb{Q}[[t]]$ by setting

$$h(I_j) = 1-\varepsilon(I_j)t.$$

Then we must prove that $h(\theta) = 1$.

Note first that the vector $C_j = (I_j+I_{j+1})/2$ satisfies the corresponding identity

$$h(C_j) = 1-\varepsilon(C_j)t = 1.$$

Hence for any possible address A_n, with sign $\varepsilon(A_n) = \varepsilon_n$, we have

$$h(A_n) = 1 - \varepsilon_n t.$$

Now setting $\theta = \sum \theta_n t^n = \sum \varepsilon_0 \ldots \varepsilon_{n-1} A_n t^n$, to be summed over all $n \geq 0$, it follows that

$$h(\theta) = \sum \varepsilon_0 \ldots \varepsilon_{n-1}(1-\varepsilon_n t)t^n = (1-\varepsilon_0 t) + \varepsilon_0(1-\varepsilon_1 t)t + \ldots .$$

Evidiently this adds up to $+1$. ∎

Remark 3.3. For the reader who is only interested in the special case $\ell = 2$, much of the above discussion is unnecessary. It would be much easier just to work with the function

$$\bar{\theta}(x) = \theta_2(x) - \theta_1(x),$$

which is a formal power series of the form $\pm 1 \pm t \pm \ldots$, with coefficients ± 1 or 0. Evidiently $\bar{\theta}(x)$ depends monotonely on x, and determines the series $\theta_1(x)$ and $\theta_2(x)$ uniquely and linearly. However, we will continue to develop the somewhat more complicated tools which are needed to deal with the case $\ell > 2$.

4. The kneading matrix $[N_{ij}]$

One basic invariant which describes the qualitative behavior of the map f and its iterates is given by the inineraries $A(f^*(c_1))$, $\ldots, A(f^*(c_{\ell-1}))$ associated with the $\ell-1$ turning points of f. Out of these itineraries, we will construct $\ell-1$ formal power series

$$\nu_1, \ldots, \nu_{\ell-1} \in V[[t]]$$

which will be called the kneading increments of f. These will play a fundamental role in all of our discussions.

Recall that $V[[t]]$ is the free module with basis $I_1, \ldots, I_\ell$ over the ring $\mathbb{Q}[[t]]$. We give $V[[t]]$ the formal power series topology, in which the submodules $t^n V[[t]]$ form a basis for the neighborhoods of zero. As in §3, consider the invariant coordinate function $x \longmapsto \theta(x)$ from I to $V[[t]]$. Since this function is monotone increasing, and since each coefficient $\theta_n(x)$ can take only finitely many distinct values, it follows easily that the <u>right hand limit</u>

$$\theta(x+) = \lim_{y \to x, y>x} \theta(y)$$

exists and is well defined for any x in the interior of the interval I. (To be more explicit, for each $n \geq 0$ there exists $\delta_n > 0$ so that $\theta_n(y)$ takes a constant value, denoted by $\theta_n(x+)$, for all y with $x < y < x+\delta_n$. Then $\theta(x+)$ is equal to the power series $\sum \theta_n(x+)t^n$.) Similarly, the <u>left hand limit</u> $\theta(x-)$ is well defined. The difference $\theta(x+) - \theta(x-)$ measures the discontinuity in the function $x \longmapsto \theta(x)$ at x.

<u>Definition</u>. This measure of discontinuity $\theta(c_i+) - \theta(c_i-)$, evaluated at the turning point c_i, will be called the i-th <u>kneading increment</u> ν_i of f.

Closely related is the <u>kneading matrix</u> $[N_{ij}]$. This is the $(\ell-1)\times\ell$ matrix, with entries in the ring $\mathbb{Z}[[t]]$, which is obtained by setting

$$\nu_i = N_{i1}I_1 + \ldots + N_{i\ell}I_\ell.$$

<u>Remark 4.1</u>. We can of course write this matrix $[N_{ij}]$ as a power series $\sum [N_{ij}^k]t^k$ where the coefficients $[N_{ij}^0], [N_{ij}^1], \ldots$ are integer matrices. Each coefficient matrix $[N_{ij}^k]$ (after being rotated 90° counterclockwise to allow for the difference in orientation conventions for matrices and for graphs) can be thought of as a kind of approximate graph for the function f^k, evaluated only at the turning points $c_1, \ldots, c_{\ell-1}$. Here is a precise statement. For $k \geq 1$, the entry N_{ij}^k is nonzero if and only if the address $A(f^k(c_i+))$ equals I_j, each nonzero entry being either $+2$ or -2 according as the

function f^k has a local minimum or a local maximum at c_i. On the other hand, for $k = 0$, the matrix

$$[N^0_{ij}] = \begin{bmatrix} -1 & 1 & 0 & 0 & & 0 & 0 & 0 \\ 0 & -1 & 1 & 0 & \cdots & 0 & 0 & 0 \\ \cdot & \cdot & \cdot & \cdot & & \cdot & \cdot & \cdot \\ 0 & 0 & 0 & 0 & & 0 & -1 & 1 \end{bmatrix}$$

is completely independent of the mapping f. The proof is not difficult. ∎

Using 3.2 we see easily that the linear relation

$$\Sigma_j N_{ij}(1-\varepsilon(I_j)t) = 0 \tag{4.2}$$

must be satisfied for each i. In other words, the columns $\Gamma_1,\ldots,\Gamma_\ell$ of the kneading matrix $[N_{ij}]$ are linearly dependent, satisfying the relation

$$\Gamma_1(1-\varepsilon t)+\Gamma_2(1+\varepsilon t)+\ldots+\Gamma_\ell(1+(-1)^\ell\varepsilon t) = 0,$$

where $\varepsilon = \varepsilon(I_1)$.

Let $D_i = \det(\Gamma_1,\ldots,\hat{\Gamma}_i,\ldots,\Gamma_\ell)$ denote the determinant of the $(\ell-1)\times(\ell-1)$ matrix which is obtained from N_{ij} by deleting its i-th column. Using the relation 4.2 and elementary linear algebra we obtain the following.

Lemma 4.3. *The ratio* $D = (-1)^{i+1}D_i/(1-\varepsilon(I_i)t)$ *is a fixed element of* $\mathbb{Z}[[t]]$, *independent of the choice of* i. *This power series* D *has leading coefficient* $+1$, *and hence is a unit of the ring* $\mathbb{Z}[[t]]$.

Definition. We will call $D = D(t)$ the *kneading determinant* of f.

In the case $\ell = 2$ these constructions take a particularly simple form. In fact the equations

$$N_{12} = D_1 = (1-\varepsilon t)D$$

$$N_{11} = D_2 = (-1-\varepsilon t)D,$$

where $\varepsilon = \varepsilon(I_1)$, show that the determinant D completely determines the kneading matrix $[N_{ij}]$. This power series D is sometimes called the <u>kneading invariant</u> of f when $\ell = 2$. Note that $N_{12} - N_{11} = 2D$. If we use the notation of §3.3, mapping $\theta(x) \in V[[t]]$ to a power series $\bar{\theta}(x)$ with integer coefficients by mapping the two basic elements I_1 and I_2 to -1 and $+1$ respectively, then $\bar{\theta}(c_1) = 0$, and it follows easily that

$$\bar{\theta}(c_1+) = D$$
$$\bar{\theta}(c_1-) = -D.$$

Evidently D is always an infinite series of the form $1 \pm t \pm t^2 \pm \ldots$.

Here is a more explicit description of how to compute the i-th kneading increment ν_i from the itinerary $A(f^*(c_i+))$. Let us use the notation A_n for the address $A(f^n(c_i+)) = A(f^n(c_i-))$ for $n \geq 1$. Then evidently the right hand limit sequence $\theta_*(c_i+)$ is given by

$$\theta_*(c_i+) = (I_{i+1}, \varepsilon_0 A_1, \varepsilon_0 \varepsilon_1 A_2, \ldots)$$

where the signs

$$\varepsilon_0 = \varepsilon(I_{i+1}), \ \varepsilon_n = \varepsilon(A_n) \quad \text{for} \quad n \geq 1,$$

are all nonzero. Similarly

$$\theta_*(c_i-) = (I_i, -\varepsilon_0 A_1, -\varepsilon_0 \varepsilon_1 A_2, \ldots).$$

Subtracting these two expressions, we find that ν_i is an infinite power series of the form

$$\nu_i = (I_{i+1} - I_i) + 2\varepsilon_0 A_1 t + 2\varepsilon_0 \varepsilon_1 A_2 t^2 + \ldots .$$

These formulas show that the right and left hand limits can be written as

$$(4.4) \qquad \begin{cases} \theta(c_i+) = \theta(c_i) + \nu_i/2 \\ \theta(c_i-) = \theta(c_i) - \nu_i/2. \end{cases}$$

<u>Thus the average of the right and left hand limits is always precisely</u>

<u>equal to the value</u> $\theta(c_1) = (I_i+I_{i+1})/2$ (compare §5.2). Any one of the three series $\theta(c_i+)$, $\theta(c_i-)$ and ν_i determines the other two uniquely.

If we are given the itinerary $A(f^*(c_i))$ together with the sign $\varepsilon(I_1)$, then the addres $A_n = A(f^n(c_i+))$ can be computed as follows by induction on n. If the address $A(f^n(c_i))$ is an interval I_j, then $A(f^n(c_i+))$ equals I_j also, and there is no problem. However, if $f^n(c_i)$ is a critical point c_j, then $A(f^n(c_i+))$ can be either i_{j+1} or I_j. In fact, it will be I_{j+1} whenever the sign $\varepsilon_0\varepsilon_1\ldots\varepsilon_{n-1}$ is +1 and I_j whenever $\varepsilon_0\varepsilon_1\ldots\varepsilon_{n-1} = -1$. For this sign determines whether the function f^n is increasing or decreasing throughout some sufficiently small interval $[c_i, c_i+\delta)$ to the right of c_i.

In the special case $\ell = 2$, these arguments give rise to a quite simple recipe.

<u>Lemma 4.5</u>. <u>If</u> $\ell = 2$, <u>then the kneading determinant can be computed by the formula</u>

$$D = 1 + \varepsilon_1 t + \varepsilon_1\varepsilon_2 t^2 + \varepsilon_1\varepsilon_2\varepsilon_3 t^3 + \ldots .$$

<u>Here the sign</u> $\varepsilon_n = \varepsilon(A(f^n(c_1+)))$ <u>is equal to</u> $\varepsilon(A(f^n(c_1)))$ <u>when</u> $f^n(c_1) \neq c_1$, <u>and satisfies</u> $\varepsilon_1\varepsilon_2\ldots\varepsilon_n = 1$ <u>when</u> $f^n(c_1) = c_1$.

The proof is easily supplied. ∎

<u>Thus, if</u> $f^p(c_1) = c_1$ <u>then the sequence of coefficients</u> $1, \varepsilon_1, \varepsilon_1\varepsilon_2, \ldots$ <u>will be periodic of period</u> p. Briefly we will say that D has period p.

<u>Example 4.6</u>. Let us look at the quadratic map $f(x) = (x^2-a)/2$ for a few representative values 4, 5, 6, 7, 8 of the parameter a. For

$$f(x) = (x^2-4)/2,$$

since $f^2(0) = 0$, the kneading determinant must be periodic of period 2. In fact, computation shows that

$$A(f^*(0)) = (C_1, I_1, C_1, I_1, \ldots);$$

hence $\varepsilon_1 = \varepsilon_2 = \ldots = -1$ and the determinant

$$D = 1 - t + t^2 - t^3 +- \ldots = 1/(1+t)$$

does indeed have period 2. (The kneading matrix $[N_{11}, N_{12}]$ is equal to

$$[-1+t, 1+t]D = [-1+2t-2t^2+-\ldots, 1].)$$

Next consider the map $f(x) = (x^2-5)/2$ of Figure 1.4. In this case

$$A(f^*(0)) = (C_1, I_1, I_2, I_1, I_2, \ldots);$$

and the determinant

$$D = 1 - t - t^2 + t^3 + t^4 --++ \ldots = (1-t)/(1+t^2)$$

has period 4. For the map $f(x) = (x^2-6)/2$, it is not known whether or not the kneading determinant is periodic. Certainly there is no small period. The map $f(x) = (x^2-7)/2$ of §2.2 has determinant

$$D = 1 - t - t^2 + t^3 - t^4 - t^5 +-- \ldots = (1-t-t^2)/(1-t^3)$$

of period 3. Finally, the map $f(x) = (x^2-8)/2$ has a kneading determinant

$$D = 1 - t - t^2 - t^3 - \ldots = (1-2t)/(1-t)$$

which is eventually periodic but not periodic.

In the case $\ell > 2$, the computations are of course more complicated. To give one example the function f graphed in Figure 1.5 has kneading matrix

$$[N_{ij}] = \begin{bmatrix} -1+2t+2t^3+\ldots & 1 & -2t^2-2t^4-\ldots \\ 2t^2+2t^4+\ldots & -1 & 1-2t-2t^3 \end{bmatrix}$$

with kneading determinant

$$D = 1 - 3t + t^2 - 3t^3 + t^4 - 3t^5 +- \ldots = (1-3t)/(1-t^2).$$

5. Counting pre turning points and laps

Let us continue to study the continuity properties of the function $x \longmapsto \theta(x)$ from I to $V[[t]]$, using the formal power series topology in $V[[t]]$. Let x_0 be some interior point of the interval I.

Lemma 5.1. *If none of the successive images $x_0, f(x_0), f^2(x_0), \ldots$ is a turning point of f, then the function θ is continuous at x_0, that is, $\theta(x_0-) = \theta(x_0) = \theta(x_0+)$. In this case $\theta(x_0)$ is an infinite power series in t with all coefficients $\theta_n(x_0)$ different from zero.*

The proof is easily supplied. ∎

Now suppose that some $f^n(x_0)$ is a turning point of f. We will say that x_0 is a *pre turning point* of f. Let $n \geq 0$ be the smallest integer such that $f^n(x_0)$ is a turning point of f, and let $f^n(x_0) = c_i$.

Lemma 5.2. *If x_0 is a pre turning point, where $f^n(x_0) = c_i$ with n minimal, then*

$$\theta(x_0+) = \theta(x_0) + t^n\nu_i/2,$$

$$\theta(x_0-) = \theta(x_0) - t^n\nu_i/2.$$

Thus $\theta(x_0)$ is always equal to the average of the left and right hand limits.

Again, the proof is easily supplied. ∎

Note that x_0 is a turning point of the iterated function f^k if and only if $k > n$. *The integers $n \geq 0$ and i can be uniquely characterized by the fact that the difference $f^n(x) - c_i$ changes sign at x_0.* The equation $f^m(x_0) = c_j$ may have other solutions, with $m > n$, but if this happens, then x_0 will be a turning point of f^m, so that the function $f^m(x)-c_j$ will not change sign at x_0.

Using this lemma, we will give an algorithm for counting the numbers of pre turning points of various kinds in any interval. Suppose that we are given two points $a < b$ of the interval I. *Suppose first, to simplify the discussion, that the function* θ *is continuous at both* a *and* b. Let J denote the interval $[a,b]$.

Lemma 5.3. *The difference* $\theta(b)-\theta(b)$ *can be computed by summing the discontinuities* $\theta(x+)-\theta(x-)$ *over all points* x *in the interior of the interval* J.

In particular, it is asserted that this infinite sum converges in the formal power series topology. Of course, only countably many of the summands can be nonzero.

Proof. First consider the function $\theta(x)$ modulo t^n. Evidently, $\theta(x)$ mod t^n is a step function, with only finitely many discontinuities in the interval J, so the formula

$$\theta(b)-\theta(a) \equiv \sum(\theta(x+)-\theta(x-)) \bmod t^n V[[t]]$$

follows immediately. Taking the liimt as $n \to \infty$, we are done. ∎

With $J = [a,b]$ as above, we will next define formal power series

$$\gamma_1(J), \ldots, \gamma_{\ell-1}(J) \in \underset{\sim}{Z}[[t]],$$

as follows. *For each* $n \geq 0$, *let* $\gamma_{in}(J)$ *be the number of times that the function* $f^n(x) - c_i$ *changes sign in the interior of the interval* J, *and let* $\gamma_i(J) = \sum \gamma_{in}(J)t^n$. Then we will prove the following.

Theorem 5.4. *The difference* $\theta(b)-\theta(a)$ *is equal to* $\gamma_1(J)\nu_1 + \ldots + \gamma_{\ell-1}(J)\nu_{\ell-1}$.

Proof. If $f^n(x) - c_i$ changes sign at x_0, then evidently the point x_0 makes a contribution of t^n to the power series $\gamma_i(J)$, and hence a contribution of $t^n\nu_i$ to the sum $\gamma_1(J)\nu_1 + \ldots + \gamma_{\ell-1}(J)\nu_{\ell-1}$. But, by 5.2 and 5.3, the contribution $\theta(x_0+)-\theta(x_0-)$ of x_0 to the difference $\theta(b)-\theta(a)$ is also precisely $t^n\nu_i$. Summing over all x_0 in the interior of J, the conclusion follows. ∎

<u>Remark 5.5</u>. If we drop the hypothesis that θ is continuous at a and b, then the formula becomes

$$\theta(b-)-\theta(a+) = \gamma_1(J)\nu_1 + \ldots + \gamma_{\ell-1}(J)\nu_{\ell-1},$$

as can easily be established.

<u>Corollary 5.6</u>. *Given the kneading matrix $[N_{ij}]$ and given the difference $\theta(b-)-\theta(a+)$, we can compute the number $\gamma_{in}(J)$ of times that the function $f^n(x)-c_i$ changes sign in the interior of J, and also the number $\ell(f^n|J)$ of laps of the function f^n restricted to J.*

<u>Proof</u>. If we delete say the last column of the matrix $[N_{ij}]$, recall from §4.2 that the determinant D_ℓ of the resulting square matrix is a unit of the ring $\mathbb{Z}[[t]]$. Hence there is a unique two-sided inverse matrix, say $[M_{jk}]$, with entries in $\mathbb{Z}[[t]]$. Writing 5.5 as

$$\theta_j(b-)-\theta_j(a+) = \textstyle\sum_i \gamma_i(J)N_{ij},$$

we can multiply on the right by M_{jk} and sum over j to obtain

$$\sum_{j=1}^{\ell-1}(\theta_j(b-)-\theta_j(a+))M_{jk} = \gamma_k(J). \tag{5.7}$$

This computes the power series $\gamma_k(J)$, and hence its coefficients $\gamma_{kn}(J)$.

But x_0 is a turning point of f^n if and only if $f^m(x)-c_i$ changes sign at x_0 for some $m<n$ and some c_i. The number of such points in the interior of J is $\gamma_{im}(J)$. *Hence the number of turning points of f^n in the interior of J is equal to the sum of $\gamma_{im}(J)$ over all turning points c_i and over all $m<n$.* Adding +1 to this sum, we obtain the required number $(f^n|J)$ of laps of f^n

on J. ∎

For later use, we also record a formal power series version of this last computation.

<u>Corollary 5.8</u>. <u>The formal power series</u> $\sum_1^\infty \ell(f^n|J)t^{n-1}$ <u>is equal to</u>

$$(1+\gamma_1(J)+\ldots+\gamma_{\ell-1}(J))/(1-t).$$

The proof is straightforward. ∎

In particular, these results apply to the case where J is the entire interval I. We use the notations c_0 and c_ℓ for the two endpoints of I and the notation $\dot{I} = \{c_0, c_\ell\}$ for the boundary of I. (Caution: c_0 and c_ℓ are <u>not</u> turning points.) Suppose that the following is satisfied.

<u>Hypothesis</u>. The mapping f carries the boundary $\dot{I}$ into itself.

Then evidently both $\theta(c_0)$ and $\theta(c_\ell)$ can be expressed as linear combinations of the form $\Theta_1 I_1 + \Theta_\ell I_\ell$. Hence 5.7 takes the somewhat simpler form

$$(\Theta_1(c_\ell) - \Theta_1(c_0))M_{1k} = \gamma_k(I).$$

In order to compute $\Theta_1(c_\ell) - \Theta_1(c_0)$, we need only know the precise way in which f maps $\dot{I}$ to itself. There are just four possibilities, which can be listed as follows.

<u>Case</u> Λ. Both end points map to the initial point c_0.

<u>Case</u> N. Both end points are fixed by f.

<u>Case</u> V. Both end points map to c_ℓ.

<u>Case</u> h. The endpoints are interchanged by f.

(The symbols Λ, N, V, and h have been chosen to resemble schematic graphs for the respective functions f.)

In each of these four cases, the computation can easily be carried out. For example, in Case Λ we have

$$\theta(c_0) = I_1/(1-t), \ \theta(c_\ell) = -tI_1/(1-t)+I_\ell;$$

hence

$$\Theta_1(c_\ell)-\Theta_1(c_0) = -(1+t)/(1-t).$$

Similarly, in Case V this difference is -1 and in Cases N and h it is $-1/(1-t)$. (The apparent asymmetry between Cases Λ and V is an artifact of our method of computation, which is based on cancelling the last column of the kneading matrix.)

If $\ell = 2$, then M_{11} is just the reciprocal of $N_{11} = -(1+\varepsilon(I_1)t)D$, so these computations take a particularly simple form.

Corollary 5.9. *If $\ell = 2$, and if $f(\dot{I}) \subset \dot{I}$, then the power series $\gamma_1(I)$ is equal to $(1-t)^{-1}D^{-1}$ where D is the kneading determinant. Hence $L(t) = \sum \ell(f^n)t^{n-1}$ is equal to $(1-t)^{-1}+(1-t)^{-2}D(t)^{-1}$.*

The proof is easily supplied. ∎

Consider for example the function $f(x) = (x^2-4)/2$ mapping the interval $|x| \le b = 1+\sqrt{5}$ to itself, with kneading determinant $D = (1+t)^{-1}$ (compare 1.3, 4.6). Since $f(\pm b) = b$, we are in Case V. By 5.9 we have

$$\gamma_1(I) = (1-t)^{-1}D^{-1} = (1+t)/(1-t) = 1+2t+2t^2+\ldots$$

and

$$L(t) = 2/(1-t)^2 = 2+4t+6t^2+\ldots .$$

In other words $\ell(f^n) = 2n$ for $n \ge 1$. The growth number $s = \lim(2n)^{1/n}$ equals 1.

On the other hand, for the function $f(x) = (x^2-7)/2$ with determinant $D = (1-t-t^2)/(1-t^3)$, we obtain

$$1+\gamma_1(I) = 2/(1-t-t^2) = 2+2t+4t^2+6t^3+10t^4+16t^5+\ldots .$$

The coefficients in this sequence are just the doubles of the Fibonacci numbers

$$1,1,2,3,5,8,\ldots .$$

The associated lap numbers $\ell(f^n)$ can now easily be computed. For example, $\ell(f^6) = 2(1+1+2+3+5+8) = 40$. Using the formula

$$(((1+\sqrt{5})/2)^{n+1} - ((1-\sqrt{5})/2)^{n+1})/\sqrt{5}$$

for the n-th Fibonacci number, or using a radius of convergence argument as in §6, one finds that $s = (1+\sqrt{5})/2 = 1.61\ldots$.

6. Convergent power series

In this section we will re-examine the various formal power series which have been introduced, showing that they can be profitably studied as convergent series. Thus we think of the symbol t no longer as a formal indeterminant, but now rather as a real or complex variable.

As a first example, consider the series $L(t) = \sum \ell(f^n)t^{n-1}$ studied in §§5.8, 5.9. Using the definition $s = \lim \ell(f^n)^{1/n}$, we see that <u>the radius of convergence of this series is precisely</u> $1/s$. In other words, the function $L(t)$ is defined and complex analytic for $|t| < 1/s$, but cannot be extended as an analytic function over any larger disk.

As a second example, consider an entry $N_{ij} = \sum_k N^k_{ij} t^k$ of the kneading matrix. Since the coefficients N^k_{ij} are bounded integers, this series converges for $|t| < 1$. <u>Thus we will now think of the kneading matrix as a matrix</u> $[N_{ij}(t)]$ <u>of functions, all defined and analytic throughout the unit disk</u> $|t| < 1$. It follows that the determinant $D(t)$ is also defined and analytic for $|t| < 1$.

As another example, fixing some point x_0 of I, the coordinate $\theta(x_0) = \sum \theta_j(x_0)I_j$ will now be thought of as a function

$$\theta(x_0,t) = \sum \theta_j(x_0,t)I_j$$

which is defined and analytic for $|t| < 1$, and takes values in the real or complex vector space with basis $I_1,\ldots,I_\ell$.

As a final example, the various power series $\gamma_i(J) = \sum\gamma_{in}(J)t^n$ considered in §5.4 will now be thought of as functions $\gamma_i(J,t)$. Using the easily verified inequality

$$0 \le \gamma_{in}(J) \le \ell(f^n|J) \le \ell(f^n),$$

we see that these functions are all defined and analytic for $|t| < 1/s$. However, we can do better than this.

Lemma 6.1. *For each* i *and* J *this complex function* $\gamma_i(J,t)$, *which a priori is only defined for* $|t| < 1/s$, *actually extends to a function which is defined and meromorphic throughout the disk* $|t| < 1$. *This meromorphic function has poles at most at the zeros of the analytic function* $D(t)$.

Proof. This follows easily from the identity

$$\gamma_k(J,t) = \sum(\theta_j(b-,t) - \theta_j(a+,t))M_{jk}(t)$$

of 5.7., where the θ_j are analytic for $|t| < 1$. Multiplying both sides by $D_\ell(t)$, note that each $D_\ell(t)M_{jk}(t)$ is equal to an appropriate minor determinant of the matrix $[N_{ij}(t)]$, and hence is analytic for $|t| < 1$. Therefore, the product $D(t)\gamma_k(J,t)$ is analytic for $|t| < 1$. ∎

Using 5.8., it follows that the function $L(t) = \sum\ell(f^n)t^{n-1}$ also extends to a function which is meromorphic throughout the disk $|t| < 1$, with $D(t)L(t)$ analytic for $|t| < 1$.

Corollary 6.2. *If* $s > 1$, *then the analytic function* $D(t)$ *has a zero at* $t = 1/s$.

Proof. Since the power series $\sum\ell(f^n)t^{n-1}$ has radius of convergence precisely $1/s$, the meromorphic function $L(t)$ must have a pole somewhere on the circle $|t| = 1/s$. But the coefficients $\ell(f^n)$ are nonnegative; hence

$$|L(t)| \le L(|t|)$$

for $|t| \le 1/s$, and it follows that there must be a pole precisely at $t = 1/s$. ∎

In fact, this is the smallest zero of $D(t)$.

Theorem 6.3. *The function* $D(t)$ *is nonzero for* $t < 1/s$. *Therefore, the first zero of* $D(t)$ *as* t *varies along the interval* $[0,1)$ *occurs precisely at* $t = 1/s$ *if* $s > 1$, *while* $D(t)$ *has no zeros at all in the open unit disk if* $s = 1$.

As examples, consider the five kneading determinants computed in §4.6:

$$D(t) = 1/(1+t), \quad (1-t)/(1+t^2), \quad (1-t-t^2)/(1-t^3),$$

$$(1-2t)/(1-t), \text{ and } (1-3t)/(1-t^2).$$

Looking for the first zero of $D(t)$ along the interval $[0,1)$, we find that the growth numbers s for the corresponding functions f are respectively 1, 1, $(1+\sqrt{5})/2$, 2, and 3.

The proof of 6.3 begins as follows. For the moment we will need the supplementary hypothesis that f maps the boundary $\dot{I}$ into itself. This extra hypothesis will be removed in §7.7.

In the special case $\ell = 2$, the conclusion follows immediately from 5.9 and the definition of s. However, the general case requires more work. We first construct a rather trivial nonsingular matrix.

Lemma 6.4. *Suppose that* $f(\dot{I}) \subset \dot{I}$. *Then the square matrix*

$$[A_{jk}(t)] = [\Theta_j(c_k,t) - \Theta_j(c_0,t)]$$

of analytic functions, where j *and* k *range from* 1 *to* $\ell - 1$, *is nonsingular for every* t *in the disk* $|t| < 1$.

Proof. Introducing the abbreviation $a = -2\Theta_1(c_0,t)$, inspection shows that twice this matrix is equal to

$$\begin{bmatrix} 1+a & a & a & & a & a & a \\ 1 & 1 & 0 & \dots & 0 & 0 & 0 \\ 0 & 1 & 1 & & 0 & 0 & 0 \\ & \dots & & & & \dots & \\ 0 & 0 & 0 & & 1 & 1 & 0 \\ 0 & 0 & 0 & & 0 & 1 & 1 \end{bmatrix}$$

Using elementary row operations to remove an even number of a's, it follows that the determinant of this doubled matrix is equal to $+1$ when ℓ is odd, and is equal to

$$1 + a = 1 - 2\theta_j(c_0, t)$$

when ℓ is even. Computation shows that this last expression equals -1 in Case V, and equals $-(1+t)/(1-t)$ in Case Λ (compare §5). Thus in all cases, the determinant is nonzero for $|t| < 1$. ∎

Proof of 6.3, assuming that $f(\dot{I}) \subset \dot{I}$. In this proof, i,j,k will vary from 1 to $\ell - 1$. Consider the square matrix with entries

$$\Gamma_{ik}(t) = \gamma_i([c_0, c_k], t) + \delta_{ik}/2,$$

where $[\delta_{ik}]$ is the identity matrix. Using 5.5 and 5.2 we easily obtain the formula

$$\theta(c_k, t) - \theta(c_0, t) = \sum \nu_i(t)\Gamma_{ik}(t),$$

or in other words,

$$\theta_j(c_k, t) - \theta_j(c_0, t) = \sum_i N_{ij}(t)\Gamma_{ik}(t), \tag{6.5}$$

where all of these functions are defined and analytic for $|t| < 1/s$. For every such t, the left hand matrix is nonsingular; hence the matrix $[N_{ij}(t)]$ with determinant $D_\ell(t)$ must also be nonsingular. This proves that $D_\ell(t) \neq 0$; hence $D(t) \neq 0$, for $|t| < 1/s$. ∎

7. Piecewise-linear models

Throughout this section we assume that the growth number s is strictly greater than 1. We set $r = 1/s$. Thus $0 < r < 1$.

For any closed interval $J \subset I$, consider the power series

$$L(J,t) = \sum \ell(f^n|J)t^{n-1}$$

of 5.8. Recall from 6.1 that this series converges for $|t| < r$, and extends to a meromorphic function which is defined throughout the disk $|t| < 1$. Taking $J = I$, the meromorphic function $L(I,t)$ definitely does have a pole at $t = r$.

Evidently $0 \le L(J,t) \le L(I,t)$ for $0 \le t < r$. Hence the ratio $L(J,t)/L(I,t)$ remains bounded as t tends to r from the left. This implies that the apparent singularity of this meromorphic function at $t = r$ is removable. The limit

$$\Lambda(J) = \lim_{t\to r} L(J,t)/L(I,t)$$

exists, and satisfies $0 \le \Lambda(J) \le 1$. (Perhaps one should think of this limit intuitively as the probability that a randomly chosen lap of the function f^n, with n very large, will be contained in the interval J.)

Here are some basic properties of this construction.

Lemma 7.1. *If the intervals J_1 and J_2 intersect only at a common end point, then $\Lambda(J_1 \cup J_2) = \Lambda(J_1) + \Lambda(J_2)$.*

Lemma 7.2. *The number $\Lambda(J)$ depends continuously on the end points of the interval J.*

Lemma 7.3. *If J is contained in a single lap of f, or in other words if $f|J$ is a homeomorphism, then $\Lambda(f(J)) = s\Lambda(J)$.*

The first two lemmas imply that the set function Λ gives rise to a measure on the interval I, each point having measure zero and the whole interval having measure $\Lambda(I) = 1$. The third lemma says that f multiplies the measure of any subset of a lap by the constant factor $s > 1$. Thus the growth number s appears again in a somewhat different guise.

Proof of 7.1. Clearly the sum $\ell(f^n|J_1) + \ell(f^n|J_2)$ differs from $\ell(f^n|J_1 \cup J_2)$ by at most 1. Hence the difference

$$L(J_1,t) + L(J_2,t) - L(J_1 \cup J_2,t)$$

remains bounded as $t \to r$. Dividing by $L(I,t)$ and passing to the limit as $t \to r$, we obtain the required equality $\Lambda(J_1) + \Lambda(J_2) - \Lambda(J_1 \cup J_2) = 0$. ∎

Proof of 7.3. If $f|J$ is a homeomorphism, then $\ell(f^{n+1}|J) = \ell(f^n|f(J))$, which implies that

$$L(J,t) = 1 + tL(f(J),t).$$

The required equation follows easily. ∎

Proof of 7.2. If the interval J is small enough to be contained in just one lap of the function f^n, then it follows inductively from 7.3 that

$$\Lambda(J) = s^{-n}\Lambda(f^n(J)) \le s^{-n}.$$

This upper bound tends to zero as $n \to \infty$. Making use of 7.1, the conclusion follows easily. ∎

Now define a map $\lambda : I \to [0,1]$ by setting $\lambda(x) = \Lambda([c_0,x])$. Evidently λ is continuous, monotone, and maps I onto the unit interval.

Theorem 7.4. *There exists one and only one map F from the unit interval $[0,1]$ into itself so that*

$$F(\lambda(x)) = \lambda(f(x))$$

for every x in I. This map F is piecewise-linear, with slope equal to $\pm s$ everywhere. The growth number of F is precisely equal

<u>to</u> s.

(Compare Parry.) Thus the following diagram is commutative:

$$\begin{array}{ccc} I & \xrightarrow{\ f\ } & I \\ {\scriptstyle\lambda}\downarrow & & \downarrow{\scriptstyle\lambda} \\ [0,1] & \xrightarrow{\ f\ } & [0,1], \end{array}$$

where the function λ is continuous, monotone, and onto, but is not necessarily a homeomorphism (see Figure 7.11). Briefly, we will say that f is <u>topological semiconjugate</u> to a piecewise-linear function F having the same growth number s, and having slope $\pm s$.

<u>Remark</u>. The map λ may collapse some entire lap I_j to a point. If this happens, then F will have fewer laps than f. An example, in which λ maps the entire right hand half of the interval I to $+1$, is illustrated in Figure 7.5. In this case the growth number s equals 3, so the slope of F is ± 3 everywhere. The growth number of f restricted to the right hand half of I would be strictly less than 3, so this half gets lost in the piecewise-linear model.

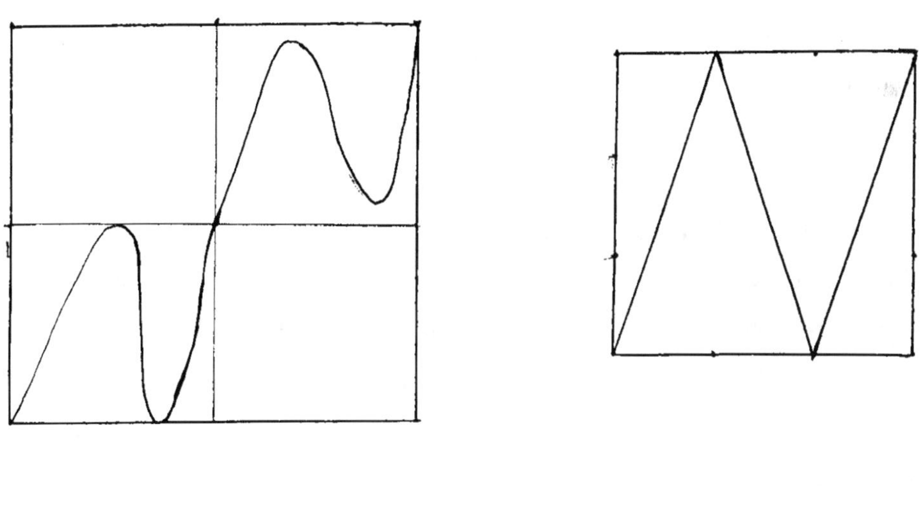

Graph of f Graph of F

Figure 7.5.

<u>Proof of 7.4</u>. We must first show that F exists. That is, if $\lambda(x) = \lambda(y)$, we must show that $\lambda(f(x)) = \lambda(f(y))$ so that $F(\lambda(x)) = \lambda(f(x))$ will be well defined. But if $\lambda(x) = \lambda(y)$ with say $x < y$, then evidently $\Lambda([x,y]) = 0$. Using 7.3, it follows that $\Lambda(f([x,y])) = 0$; hence $\lambda(f(x)) = \lambda(f(y))$, as required.

Next, let us compute $F(\lambda(x))$ as x varies through some lap $I_k = [c_{k-1}, c_k]$ of f. Suppose, to fix our ideas, that $\varepsilon(I_j) = +1$, so that $f|I_k$ preserves orientation. If J denotes the subinterval $[c_{k-1}, x]$, then

$$F(\lambda(x)) = \lambda(f(x)) = \lambda(f(c_{k-1})) + \Lambda(f(J)).$$

But

$$\Lambda(f(J)) = s\Lambda(J) = s\lambda(x) - s\lambda(c_{k-1})$$

by 7.3. Setting A_k equal to the constant $\lambda(f(c_{k-1})) - s\lambda(c_{k-1})$, this proves that

$$F(\lambda(x)) = A_k + s\lambda(x)$$

for all x in I_k. In other words, the function F is linear with slope s throughout the interval $\lambda(I_k)$. If $\varepsilon(I_k) = -1$, then a completely analogous argument shows that F is linear with slope $-s$ throughout $\lambda(I_k)$.

To compute the growth number $s(F)$, we note that F^n has slope $\pm s^n$ everywhere. Hence every lap of F^n has length $\le s^{-n}$, and the number of laps must satisfy $\ell(F^n) \ge s^n$. Therefore, $s(F) \ge s = s(f)$.

On the other hand, λ maps each lap of f^n into a lap of F^n, so it follows that $\ell(F^n) \le \ell(f^n)$, and therefore, $s(F) \le s(f)$. ∎

<u>Remark</u>. More generally, it is true that <u>any</u> piecewise-linear map F with slope $\pm s$ everywhere has growth number s, providing that $s \ge 1$. Compare Misiurewicz and Szlenk.

As an easy corollary of 7.4, we can complete the proof of 6.3.

Consider a map $f_0 : J \longrightarrow J$, with growth number $s_0 \geq 1$, which does <u>not</u> map the boundary of J into itself. Embedding J as the middle third of a larger interval I, we can extend f_0 as illustrated in Figure 7.6 to a map f with the same kneading matrix which <u>does</u> satisfy the hypothesis $f(\dot{I}) \subset \dot{I}$. (Case Λ is illustrated, but the construction in Case N, V, or h is completely analogous.)

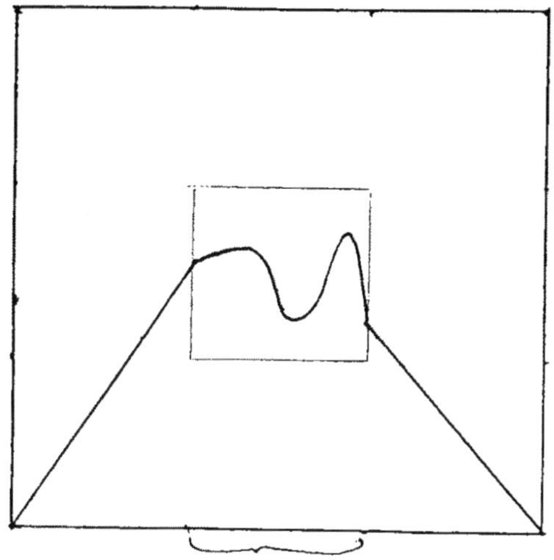

Figure 7.6.

<u>Lemma 7.7</u>. <u>The growth number</u> s <u>of this extended function</u> f <u>is precisely equal to the growth number</u> s_0 <u>of</u> f_0.

<u>Proof</u>. Clearly $\ell(f^n) \geq \ell(f_0^n)$; hence $s \geq s_0$. Suppose that s were strictly greater than s_0. Then $s > 1$, so Theorem 7.4 applies. Setting $r = 1/s$, since $r < 1/s_0$, the power series $L(J,r)$ converges. Hence the limit

$$\Lambda(J) = \lim_{t \to r} L(J,t)/L(I,t)$$

equals zero. Therefore, the function $\lambda(x) = \Lambda([c_0,x])$ maps the entire interval J into a single point λ_0.

The associated piecewise-linear function $F : \lambda(x) \longmapsto \lambda(f(x))$ must actually be linear, with $|\text{slope}| = s > 1$, throughout the interval $[0,\lambda_0]$, and also throughout the interval $[\lambda_0,1]$. Furthermore, since f maps the interval J into itself, we must have $F(\lambda_0) = \lambda_0$. Inspection shows that these conditions are impossible. Therefore, s must equal s_0. ∎

Proof of Theorem 6.3. The functions f and f_0 have the same kneading matrix and the same growth invariant. Since we have already proved 6.3 for the function f, it follows that it is true for f_0 also. ∎

7.8. Discussion of Uniqueness. It should be noted that the function λ constructed above may not be the only semiconjugacy from f to a piecewise-linear mapping with $|\text{slope}| = \text{constant}$. As an example, let us single out some index k, and consider the power series $\gamma_k(I,t)$ of §6.1. Evidently its radius of convergence r_k satisfies $r \le r_k \le 1$ (where $r_k = r$ for at least one value of k). We will need the following.

Hypothesis. This radius of convergence r_k should be strictly less than 1.

Defining $\Lambda_k(J) = \lim\limits_{t\to r_k} \gamma_k(J,t)/\gamma_k(I,t)$, the analogues of 7.1, 7.2, 7.3, and 7.4 are then easily verified. Thus the function

$$\lambda_k(x) = \Lambda_k([c_0,x])$$

has properties completely analogous to those of λ. In particular, $\lambda_k(f(x)) = F_k(\lambda_k(x))$, where F_k has slope $\pm r_k^{-1}$ everywhere.

As an example, for the function f illustrated in Figure 7.5, the function λ_3 will map the left half of the interval I to a point, and will provide a semiconjugacy from f to a piecewise-linear map F_3 which has $|\text{slope}|$ equal to a constant r_3^{-1} which is strictly less than 3.

This construction is, of course, closely related to the original construction of λ. Using 5.8 one sees that λ can be expressed as a sum

$$\lambda(x) = a_1\lambda_1(x) + \ldots + a_{\ell-1}\lambda_{\ell-1}(x),$$

where the coefficient

$$a_k = (r/(1-r))\lim_{t\to r} \gamma_k(I,t)/L(I,t)$$

is zero unless the function $\gamma_k(I,t)$ has a pole of precisely the right order, precisely at r.

In order to actually compute $\lambda(x)$, we proceed as follows.

Lemma 7.9. *This real valued function* $\lambda(x)$, *which semiconjugates* f *to a piecewise linear map, can be expressed as a linear combination*

$$\lambda(x) = a_1\theta_1(x,r) + \ldots + a_{\ell-1}\theta_{\ell-1}(x,r)$$

with constant coefficients.

The proof will provide an explicit procedure for determining the coefficients a_i. This formula is well adapted to numerical computations, since the power series $\theta_1(x,t)$ converges nicely at $t = r = 1/s$.

Proof of 7.9. Recall that the $(\ell-1)$ $(\ell-1)$ kneading matrix $[N_{ij}(t)]$ has an inverse matrix $M_{jk}(t)$, whenever $D(t) \neq 0$. Using 6.4 and 6.5, it is not difficult to show that the limit

$$\bar{M}_{jk} = \lim_{t\to r} M_{jk}(t)/L(I,t)$$

exists, and is finite. Note that the matrix product

$$\sum N_{ij}(r)\bar{M}_{jk} = \lim_{t\to r} \delta_{ij}/L(I,t)$$

is identically zero. Using 5.7 and 5.2, we can now compute the limit of the ratio $\gamma_k(J,t)/L(I,t)$, as $t\longrightarrow r$, for any interval $J = [a,b]$. We find that $\lim_{t\to r} \gamma_k(J,t)/L(I,t) = \sum_j (\theta_j(b,r) - \theta_j(a,r))\bar{M}_{jk}$.

Using 5.8, this implies that

$$\Lambda(J) = (r/(1-r))\sum_{j,k}(\Theta_j(b,r) - \Theta_j(a,r))\bar{M}_{j,k}.$$

Hence

$$\lambda(x) = (r/(1-r))\sum_{j,k}\Theta_j(x,r)\bar{M}_{jk} + (\text{constant}). \tag{7.10}$$

This is the required formula.

(It is conjectured that the functions $\lambda_k(x)$ of §7.8 can be expressed analogously as linear combinations of $\Theta_1(x,r_k),\ldots,\Theta_{\ell-1}(x,r_k)$.)

In the case $\ell = 2$, formula 7.10 takes the simple form

$$\lambda(x) = \Theta_2(x,r)$$

in Case Λ. Similarly, $\lambda(x) = 1-\Theta_1(x,r)$ in Case V.

As an example to illustrate this computational technique, the function $\lambda(x)$ associated with the mapping $f(x) = (x^2-7)/2$ is plotted in Figure 7.11.

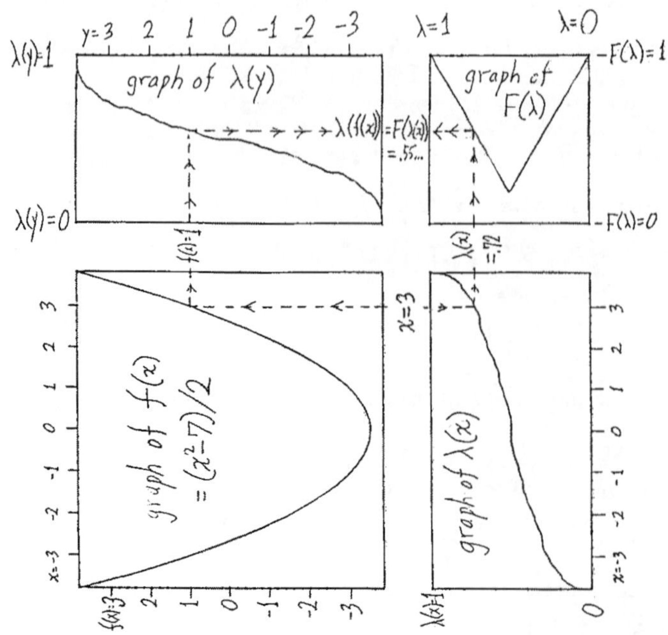

Figure 7.11. Piecewise-linearization of the function $f(x) = (x^2-7)/2$.

Note that λ has intervals of constancy, corresponding to points x whose forward images $f^*(x)$ converge towards the one-sided stable periodic orbit which was described in §2. The commutative diagram

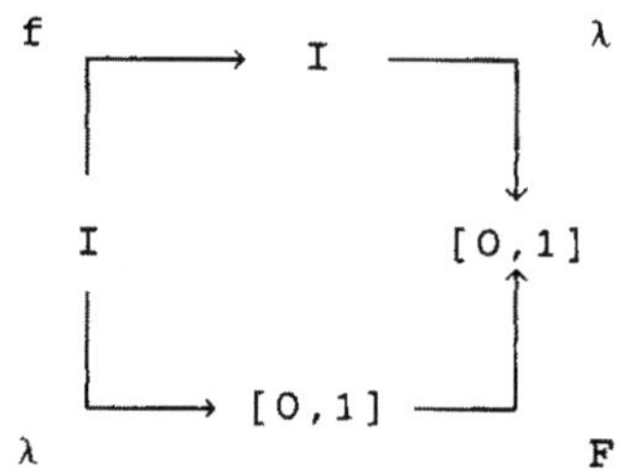

is illustrated for the particular value $x = 3$.

We conclude this section with three applications of 7.4. Let f be a function with just one turning point, having growth number s.

Corollary 7.12. *If s is not an algebraic unit satisfying a polynomial equation of the form $s^n \pm s^{n-1} \pm \ldots \pm 1 = 0$, then the kneading determinant of f is uniquely determined by s.*

We will see in §13.8 that some such hypothesis on s is definitely needed.

Proof. Evidently $s > 1$, so f is topologically semiconjugate to a map which can be put in the form

$$F(y) = s|y| - 1$$

by a suitable choice of variable. For any given s, the successive iterates $F^k(0) = \pm sF^{k-1}(0) - 1$ can easily be computed. Our hypothesis implies that these numbers are all nonzero. Hence, using 4.5, the successive coefficients of the kneading determinant can also be computed. ■

Now assume, to fix our ideas, that f takes its minimum at $c_1 = 0$. We continue to assume that there are no other turning points.

Corollary 7.13. *If the growth number satisfies $s > 1$, then $s \leq \sqrt{2}$ if and only if $f^k(0) < 0$ for all odd k, or if and only if the*

<u>kneading determinant of</u> f <u>takes the form</u> $(1-t)(1\pm t^2\pm t^4\pm t^6\pm\dots)$.

<u>Proof</u>. With F as above, let y_0 denote the fixed point $-1/(1+s)$. If $s \le \sqrt{2}$, then it is not difficult ot check that

$$F(0) < F^3(0) \le y_0 < 0 < F^2(0).$$

Hence F maps each of the two intervals $[f(0),y_0]$ and $[y_0,F^2(0)]$ into the other. Therefore $F^k(0) \le y_0 < 0$, and hence $f^k(0) < 0$, for all odd k. Using 4.5, this implies that the kneading determinant has the required form. Conversely, if $D(t)$ has this form, then the estimate $s \le \sqrt{2}$ follows easily from 6.3. ∎

To conclude this section, we provide a prelude to the study of periodic points. According to Stefan, any map of the interval which has a periodic point of odd period $p \ge 3$ must satisfy $s > \sqrt{2}$. We will prove a converse statement for maps with just one turning point, which we may assume is a minimum.

<u>Corollary 7.14</u>. <u>If</u> f <u>has just one turning point</u>, <u>then</u> $s > \sqrt{2}$ <u>if and only if</u> f <u>admits a periodic point of odd period</u> $p \ge 3$.

<u>Proof</u>. If $s > \sqrt{2}$, then the preceding corollary implies that $f^p(0) \ge 0$ for some odd p. If x_0 is the unique negative fixed point of f, note that $f^p(x) < x_0 < x$ for x just to the right of x_0. Hence, by the intermediate value theorem, the difference $f^p(x) - x$ must vanish at some point $\bar{x}$ of the interval $(x_0,0]$. Evidently the period of $\bar{x}$ is odd, and this period must be at least 3 since $f(\bar{x}) \le x_0 < \bar{x}$.

Conversely, if $s \le \sqrt{2}$, then according to Štefan every periodic point must have a period which is either even or equal to 1. (In the case $1 < s \le \sqrt{2}$, this could easily be proved by an argument similar to the proof of 7.13. However, for the case $s = 1$, this method would break down.) ∎

8. Periodic points: a review of the classics

This section will begin the study of periodic points by giving an outline of a number of important results from the literature.

First, some general definitions and notations. Given a function $f : X \to X$ from an arbitrary set to itself, a point x is periodic if $f^p(x) = x$ for some integer $p \geq 1$. The smallest such p is the period of x, and the finite set $o = \{x, f(x), \ldots, f^{p-1}(x)\}$ is a periodic orbit of period $p = p(o)$.

Suppose now that each iterate f^k has only finitely many fixed points. Following Artin and Mazur, the zeta function of f is defined to be the formal power series

$$\zeta(t) = \exp \sum_{k \geq 1} \underset{\sim}{n}(f^k) t^k / k,$$

where $\underset{\sim}{n}(f^k)$ is the number of fixed points of f^k and where $\exp \alpha = 1 + \alpha + \alpha^2/2! + \ldots$. (The idea of this formalism goes back to Weil. For a survey of applications, see Williams.)

In order to understand this definition, first consider the special case where X consists of a single periodic orbit of period p. Then evidently

$$\zeta(t) = \exp(t^p + t^{2p}/2 + \ldots) = \exp(-\log(1-t^p))$$

is equal to

$$(1-t^p)^{-1} = 1 + t^p + t^{2p} + \ldots .$$

In the general case, expressing the set of periodic points of f as a disjoint union of finite orbits o with periods $p(o)$, it follows that

$$\zeta(t) = \prod_o (1 + t^{p(o)} + t^{2p(o)} + \ldots).$$

Thus the zeta function of f, if defined at all, is always a formal power series with nonnegative integer coefficients.

In practice, it is often more convenient to work with the reciprocal series

$$\zeta(t)^{-1} = \prod_{o} (1-t^{p(o)}).$$

If the set of all periodic points is finite with d elements, note that $\zeta(t)^{-1}$ is a polynomial of degree d. Another alternate expression, which may be more convenient, is the logarithmic derivative multiplied by t. The identities

$$t\zeta'(t)/\zeta(t) = \sum_{k\geq 1} \underset{\sim}{n}(f^k)t^k = \sum_{o} p(o)(t^{p(o)}+t^{2p(o)}+\ldots)$$

are easily verified.

If X is a smooth compact manifold, then Artin and Mazur show that <u>the space of smooth maps from X to itself admits a dense subset consisting of maps whose associated zeta function has a positive radius of convergence</u>. In other words, for maps f in this subset, the numbers $\underset{\sim}{n}(f^k)$ grow at most exponentially with k. This statement applies also to maps of the unit interval

A most remarkable theorem concerning maps from the unit interval to itself was proved by A.N. Šarkovskii, in a paper which was published in 1964 but was little known in the west until much later. Define the <u>Šarkovskii ordering</u> of the positive integers by

$$3 < 5 < 7 < \ldots < 6 < 10 < 14 < \ldots < 12 < 20 < \ldots \quad \ldots < 4 < 2 < 1,$$

taking first the odd integers other than one in their natural order, then their doubles, then their quadruples,..., and finally the powers of two in descending order.

<u>Šarkovskii's Theorem</u>. <u>If a continuous map f from an open or closed interval to itself has a periodic point of period p, and if $p < q$ in the ordering described above, then f must have a periodic point of period q also</u>.

For example, the function $f(x) = (x^2-7)/2$, since it possesses a periodic point of period 3, must have periodic points of all possible periods (compare Li and Yorke).

For proofs of Šarkovskii's Theorem, see for example Targonski or Block, Guckenheimer, Misiurewicz and Young, as well as Štefan, Straffin, or Collet and Eckmann.

This theorem is related to the growth parameter s of §1, or to the topological entropy $\log s$, as follows. If f has a periodic point of some period p which is not a power of 2, then according to Bowen and Franks the topological entropy is strictly positive, hence $s > 1$. (More precisely, they show that $s > 2^{1/p}$.) For a converse result, and further discussion, see 9.6 below.

In studying periodic points of a map f, it is often important to single out these periodic orbits $o = \{x_0, f(x_0), \ldots, f^{p-1}(x_0)\}$ which are stable, in the sense that for any point x sufficiently close to o the successive images $f^k(x)$ converge to o, uniformly in x. If f is differentiable, then clearly f is stable whenever the inequality

$$-1 < df^p(x)/dx < 1$$

holds at x_0, and hence at every point of the orbit o. On the other hand, if the absolute value of the derivative of f^p at x_0 is strictly greater than one, then o is unstable; that is, there exists a neighborhood U of o and points x_i arbitrarily close to x_0 so that some forward image $f^k(x_i)$ does not belong to U.

If the derivative of f^p at x_0 is precisely equal to ± 1, then the orbit through x_0 may be stable or unstable or may exhibit some intermediate behavior. For example, there may be stability as x approaches x_0 from one side but unstability for the other side. As illustrative examples, the four functions $x - x^3$, $x + x^3$, $x + x^2$, and $-x/(2x+1)$ all exhibit different types of behavior at the origin.

In practice, it seems that "most" periodic orbits are unstable.

However, if f possesses stable periodic orbits, then these orbits tend to dominate the qualitative behavior of the successive iterates of f.

In the case of a polynomial function, a basic result was proved some 65 years ago. Adapting the hypotheses to our situation, it can be stated as follows.

Theorem of Julia and Fatou. *If $f : I \longrightarrow I$ is a polynomial function of degree $n \geq 2$, then f has at most $n-1$ periodic orbits which are stable or nearly stable in the sense that the associated derivative satisfies*

$$-1 \leq df^p(x)/dx \leq 1.$$

In fact, any such orbit can be obtained as the limit of the forward images $f^k(z_0)$ starting at some critical point of f in the complex plane, that is, at some complex number z_0 at which the derivative $f'(z_0)$ vanishes.

As an immediate corollary, *a quadratic function has at most one stable periodic orbit, which can always be located by starting at the critical point and iterating the function.*

A sketch of the proof is given at the end of this section. This theorem is proved with complex variable methods. If the derivative f' has non-real roots, then we definitely need to look at nonreal values of the variable, even though we are presumably interested only in real stable periodic orbits.

The Julia-Fatou proof applies also to rational functions of degree $n \geq 2$. In this case there are at most $2n-2$ critical points in the extended complex plane $\mathbb{C} \cup \infty$, hence at most $2n-2$ stable periodic orbits. The condition of rationality is essential. As an example, the transcendental function $f(x) = (e^x-1)/2$ has a stable fixed point, but has no critical points at all.

An extremely simple and useful alternative condition, which

eliminates the need for complex variable computations has recently been introduced by D.J. Allwright and by D. Singer. *Suppose first that* f *is smooth of class* C^3 *and satisfies the inequality*

$$f'(x)f'''(x) < \frac{3}{2}(f''(x))^2$$

everywhere in the interval I. For example, this condition will always be satisfied if f is a polynomial whose derivative f' has distinct real roots. (This can be proved, making use of the fact that the second derivative of $\log|f'|$ is strictly negative wherever $f' \neq 0$.) If this inequality is satisfied, then clearly the following is also satisfied.

Hypothesis. The map f is piecewise monotone, C^3-smooth, and the *Schwarzian derivative*

$$f'''/f' - \frac{3}{2}(f''/f')^2$$

is defined and strictly negative throughout the interior of each lap of f.

Theorem of Singer. *Let* f *be such a piecewise monotone map with negative Schwarzian derivative inside each lap, and let* $c_0, \ldots, c_\ell$ *be its local extrema. Then* f *has at most* $\ell + 1$ *periodic orbits* o *which are stable or nearly stable in the sense that*

$$-1 \le df^{p(o)}(x)/dx \le 1$$

for $x \in o$. *Any such orbit can be obtained as the limit of the successive images* $f^k(c_i)$ *as* $k \longrightarrow \infty$, *starting at one of the* $\ell + 1$ *local extreme points* c_i.

The proof can be outlined in three steps as follows.

Step 1. *If* f *satisfies this Hypothesis, then each iterate* $g = f^p$ *does also*. This is proved inductively by a straightforward computation.

Step 2. *The derivative* g' *cannot have any local minimum within an open interval where* $g' > 0$. For the inequalities $g' > 0$, $g'' = 0$, $g''' \geq 0$ would contradict the hypothesis of negative Schwarzian derivative.

Step 3. *If* x_0 *is a fixed point of* g *with* $0 < g'(x_0) \leq 1$, *lying in the interior of a lap* $[c', c'']$ *of* g, *then either the successive images* $g^k[c', x_0]$ *or the successive images* $g^k[x_0, c'']$ *converge uniformly to* x_0.

We must show either that $x < g(x) < x_0$ for all $x \in [c', x_0)$ or that $x_0 < g(x) < x$ for all $x \in (x_0, c'']$. But otherwise there would exist points x_1 between c' and x_0 and x_2 between x_0 and c'' with $g'(x_1) \geq 1$ and $g'(x_2) \geq 1$. Hence g' would attain a local minimum somewhere between x_1 and x_2, contradicting Step 2. Further details will be left to the reader (compare Singer, as well as Allwright and Collet-Eckmann). ∎

In the case $\ell = 2$, a similar argument proves the following statement. Suppose that $f(\dot{I}) \subset \dot{I}$, where the end fixed point is unstable, and suppose that the Hypothesis above is satisfied. *Then there exists a stable or one-sided stable periodic orbit if and only if the kneading determinant of* f *is periodic*.

To conclude this section, here is an outline proof of the Julia-Fatou Theorem. Let f be a rational function of degree ≥ 2 mapping the sphere $\mathbb{C} \cup \infty$ to itself, and let z_0 be a periodic point which we may assume finite. Let $g = f^p$, so that $g(z_0) = z_0$. First assume that the derivative $g'(z_0)$ has absolute value strictly less than 1. Then clearly there exists a small disk D_0 centered at z_0 so that the successive images

$$D_0 \supset g(D_0) \supset g^2(D_0 \supset \ldots$$

converge to z_0.

Let $g^{-n}(D_0)$ denote the full inverse image of D_0 under the iterate g^n. *We assert that some* $g^{-n}(D_0)$ *contains a critical point of* g. For if this statement were false, then we could derive a contradiction as follows.

The function g^n maps the compact set $g^{-n}(D_0)$ onto D_0 by a mapping which is assumed to be locally biholomorphic. It follows that each component of $g^{-n}(D_0)$ is a covering space of D_0, and hence maps biholomorphically onto D_0. In particular, this is true for that component D_n of $g^{-n}(D_0)$ which contains D_0. In this way we construct topological disks $D_0 \subset D_1 \subset D_2 \subset \ldots$, together with a single-valued branch $D_0 \xrightarrow{\cong} D_n$ of the many-valued function g^{-n}. These disks D_n cannot exhaust the sphere since g has degree ≥ 2, and hence is many-to-one. Therefore there must exist an open set disjoint from all of the D_n. After a change of coordinate we may assume that this open set contains the infinite point. In other words, we may assume that these single-valued holomorphic functions $g^{-n} : D_0 \xrightarrow{\cong} D_n$ are uniformly bounded. By a classical theorem, it follows that there exists a subsequence $\{g^{-n_i}\}$ which converges on the interior of D_0, uniformly on each compact interior subset, to a holomorphic limit (see for example Ahlfors, p. 216). It follows that the derivative $dg^{-n_i}(z)/dz$ must also converge on the interior of D_0 as $i \longrightarrow \infty$. But this is impossible since the derivative of g^{-n} at z_0 tends to infinity with n.

Now consider the more difficult case where $g'(z_0) = 1$. After a careful change of coordinate, we may assume that $z_0 = 0$ and that

$$g(z) = z - z^{k+1} + (\text{higher terms}),$$

with $k \geq 1$. The local behavior of the iterates of g can be indicated roughly by the sketch below.

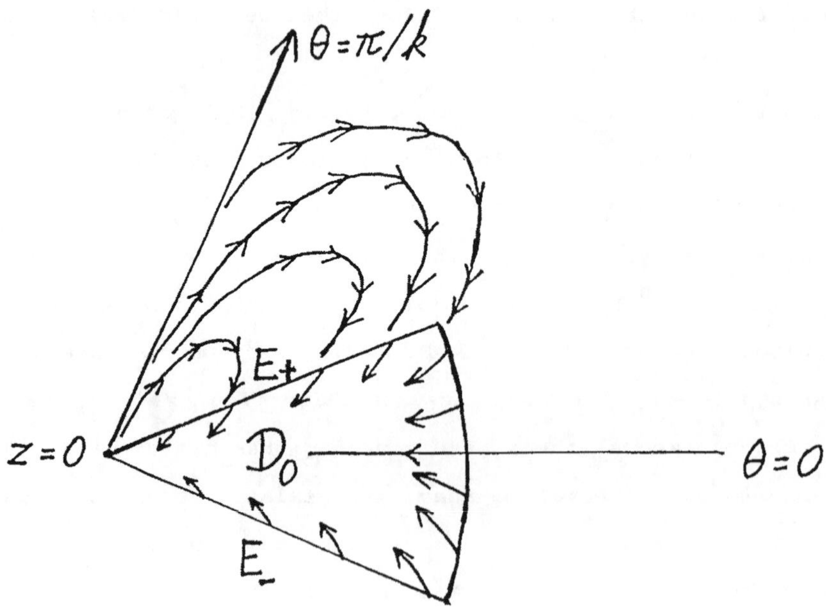

Figure 8.1.

Each arrow leads from some point $z = re^{i\theta}$ to $g(z) \approx z - z^{k+1}$.

In particular, if D_0 is a small "pie slice" consisting of all $z = re^{i\theta}$ with $r \le \varepsilon$ and $|\theta| \le \pi/3k$, then it can be shown that the successive images

$$D_0 \supset g(D_0) \supset g^2(D_0) \supset \ldots$$

converge uniformly to zero. As in the argument above, if the full inverse images $g^{-n}(D_0)$ do not contains any critical point of g, then, after a change of coordinate, we can construct a single-valued branch

$$g^{-n} : D_0 \xrightarrow{\cong} D_n \supset D_0,$$

and a subsequence $\{g^{-n_i}\}$ which converges uniformly on every compact interior subset of D_0 to a finite holomorphic limit.

Let E_+ and E_- be the two straight edges of the region D_0. A novel feature of the new situation is that *the successive images* $g^{-n}(E_{\pm})$ *converge uniformly to zero* (compare the figure). Therefore g^{-n_i} converges to zero on a set $g(E_+)$ containing an entire curve of interior points, so it follows that *the holomorphic limit of the sequence* $\{g^{-n_i}\}$ *is identically zero throughout the interior of* D_0.

Combining these two statements, we see that the image under g^{-n_i} of the entire boundary of the region $g(D_0)$ converges uniformly to zero. Hence $g^{-n_i}(g(D_0))$ converges to zero. But this is impossible, since the sets $g^{-n}(g(D_0))$ increase with n. ■

9. Computation of zeta functions.

This section will state two different forms of our main theorem, relating the Artin-Mazur zeta function of §8 to the kneading determinant of §4. It will also derive several applications.

First consider a piecewise monotone map $f : I \to I$ satisfying the following.

Instability Hypothesis. All but finitely many of the periodic points of f *are unstable.*

This is needed to avoid badly behaved examples such as $f(x) = 1 - x$. Recall from §8 that this hypothesis is satisfied whenever f is a polynomial or rational function of degree ≥ 2, or whenever f has negative Schwarzian derivative in the interior of each lap.

It will be convenient to use the phrase *cyclotomic polynomial* for any product of finitely many factors of the form $1 - t^n$.

Theorem 9.1. *If* f *satisfies this Instability Hypothesis, then the reciprocal* $\zeta(t)^{-1}$ *of its zeta function is equal to its kneading determinant* $D(t)$ *multiplied by a cyclotomic polynomial.*

The proof in §10.4 will provide an explicit recipe for computing this cyclotomic polynomial in terms of stable periodic orbits.

Here is an alternative version which does not require any special hypothesis, other than piecewise monotonicity. Let us say that two fixed points $x < y$ of some iterate f^k are *monotonely equivalent* if f^k maps the entire closed interval $[x,y]$ homeomorphically onto itself. It is easy to check that this is an equivalence relation. *Let* $\hat{n}(f^k)$ *be the number of monotone equivalence classes of fixed points of* f^k. These numbers are finite with $\hat{n}(f^k) \le \ell(f^k)$, since each lap of f^k clearly contains at most one monotone equivalence class of fixed points. The associated formal power series

$$\hat{\zeta}(t) = \exp \sum \hat{n}(f^k)t^k/k$$

will be called the *reduced zeta function* of f.

Theorem 9.2. The reciprocal $\hat{\zeta}(t)^{-1}$ *of this reduced zeta function is equal to* $D(t)$ *multiplied by a cyclotomic polynomial*.

The proof in §10.6 will provide an explicit way of computing this cyclotomic polynomial in terms of the kneading matrix.

The reduced zeta function satisfies a product formula, which can be written as

$$\hat{\zeta}(t) = \prod_{\hat{o}} (1-t^{p(\hat{o})})^{-1}, \tag{9.3}$$

where $\hat{o}$ varies over all monotone equivalence classes of periodic orbits. Here two periodic orbits o' and o'' are said to be *monotonely equivalent* if they contain representative points x' and x'' which are monotonely equivalent under some iterate of f.

Caution: It may happen that the period of one of these two monotonely equivalent orbits is twice the period of the other. (This happens, for example, for the function $f(x) = 1-x$.) *We define the period* $p(\hat{o})$ *to be the smaller of the periods* $p(o)$ *for orbits* o *in the monotone equivalence class* $\hat{o}$. With this understanding, the proof of the product formula 9.3 proceeds just as in §8.

The rest of this section will be devoted to applications of Theorem 9.1 or 9.2. To fix our ideas, we will concentrate on the reduced zeta function $\hat{\zeta}(t)$. However, similar arguments clearly apply to the original zeta function $\zeta(t)$, whenever the Instability Hypothesis is satisfied.

It follows immediately from 9.2 that the reciprocal $\hat{\zeta}(t)^{-1}$ extends as an analytic function of the complex variable t throughout the disk $|t| < 1$, and has zeros precisely at the zeros of $D(t)$.

Corollary 9.4. *The radius of convergence of the series* $\hat{\zeta}(t)$ *is precisely* $1/s$. *In fact, if* $s > 1$ *then* $\hat{\zeta}(t)$ *has a pole at* $t = 1/s$.

Proof. This follows from 6.3. ∎

Using the standard formula for the radius of convergence of a power series, this implies that

$$\limsup_{k\to\infty} \hat{\underset{\sim}{n}}(f^k)^{1/k} = s.$$

In other words, at least some of the numbers

$$\hat{\underset{\sim}{n}}(f^k) \le \ell(f^k)$$

grow almost as fast as the lap numbers $\ell(f^k)$ *as* $k\longrightarrow\infty$.

Here is a much more precise estimate for $\hat{\underset{\sim}{n}}(f^k)$. Note that the analytic function $D(t)$ can have only finitely many zeros within any disk of radius less than 1. (For examples illustrating the behavior of $D(t)$ near the boundary of the unit disk, see §14.) Choose some number $\varepsilon > 0$.

Corollary 9.5. *Suppose that* $D(t)$ *has zeros at the points* $a_1,\ldots,a_q$ *(listing an* m-*fold zero* m *times) within the disk* $|t| < (1+\varepsilon)^{-1}$. *Then* $\hat{\underset{\sim}{n}}(f^k) = a_1^{-k} + \ldots + a_q^{-k} + O((1+\varepsilon)^k)$.

Here the symbol $O((1+\varepsilon)^k)$ stands for an error term which has absolute value less than $(1+\varepsilon)^k$ times some constant independent of k.

Note: If the Instability Hypothesis of 9.1 is satisfied, then we can substitute $\underset{\sim}{n}(f^k)$ in place of $\hat{\underset{\sim}{n}}(f^k)$ in this estimate.

As an example, for the function $f(x) = (x^2-7)/2$, with $s = (1+\sqrt{5})/2$, the determinant $D(t) = (1-t-t^2)/(1-t^3)$ has just one simple zero in the unit disk, at the point $1/s$ (compare 6.3 and the discussion following 5.9). Hence in this case

$$\underset{\sim}{n}(f^k) = ((1+\sqrt{5})/2)^k + O((1+\varepsilon)^k)$$

for any $\varepsilon > 0$. It follows easily from this estimate that f has points of period p for all sufficiently large p. (Of course, Šarkovskii's theorem implies the sharper statement that f has points of all periods.)

On the other hand, for the function $f(x) = (x^2-5.9)/2$ computation shows that

$$D(t) = (1-t)(1-t^2-t^4)/(1-t^6),$$

with simple zeros at two points $1/s$ and $-1/s$ in the unit disk, where now $s = ((1+\sqrt{5})/2)^{1/2}$ (compare 14.4). Applying 9.5, it follows that

$$\underset{\sim}{n}(f^k) = 2s^k + O((1+\varepsilon)^k)$$

for k even and for any $\varepsilon > 0$, but

$$\underset{\sim}{n}(f^k) = O((1+\varepsilon)^k)$$

for k odd. In fact, it follows from 7.13 and 7.14 that f has no points at all of odd period, other than the two fixed points.

Proof of 9.5. If $D(t)$ has a zero of order m at a, then $\hat{\zeta}(t)$ has a pole of order m at a, and it follows easily that the logarithmic derivative of $\hat{\zeta}(t)$ has a simple pole of the form

$$\hat{\zeta}'(t)/\hat{\zeta}(t) = m/(a-t) + (\text{holomorphic function})$$

near a. Hence the function

$$t\hat{\zeta}'(t)/\hat{\zeta}(t) = \sum \hat{n}(f^k)t^k$$

is equal to

$$tm/(a-t) = m(t/a + t^2/a^2 + \ldots)$$

plus a function which is analytic near t. Subtracting off these correction terms for all zeros within a slightly larger circle, we obtain a power series which must converge throughout the circle $|t| \le (1+\varepsilon)^{-1}$. The conclusion follows easily. ∎

Next let us describe a relationship between the set of periods p of a mapping and its growth number s. According to Bowen and Franks, if there exists an orbit whose period p is not a power of 2, then

$$s > \sqrt[p]{2} > 1.$$

A somewhat sharper version of this inequality has been given by Štefan. See also Jonker and Rand, and Targonski. Conversely, using 9.5, we can prove the following.

Corollary 9.6. *If* $s > 1$, *then* f *possesses a periodic orbit whose period is not a power of* 2.

Proof. Choose $1+\varepsilon$ just a little smaller than s, so that all the zeros of $D(t)$ in the disk $|t| < (1+\varepsilon)^{-1}$ lie on the circle of radius s^{-1}. Let $a_1, \ldots, a_q$ be these zeros. (As usual, a zero of multiplicity m is to be listed m times.) Then

$$\hat{n}(f^k) = a_1^{-k} + \ldots + a_q^{-k} + O((1+\varepsilon)^k)$$

by 9.5. Since the q-tuple $(sa_1, \ldots sa_q)$ lies on the q-dimensional torus, which is a compact topological group, there exist arbitrarily large integers r satisfying the inequality

$$(*) \qquad |(sa_j)^{-r} - 1| < 1/10$$

for each j. It follows that the unit complex number $(sa_j)^{-3r}$ has real part greater than $1/2$, hence $\mathrm{Re}(a_j^{-3r}) > \frac{1}{2}s^{3r}$ for each j, and

$$\tilde{\hat{n}}(f^{3r}) > \frac{1}{2}qs^{3r} + O((1+\varepsilon)^{3r});$$

although

$$\tilde{\hat{n}}(f^{r}) \le qs^{r} + O((1+\varepsilon)^{r}).$$

Subtracting these two inequalities, it follows that

$$\tilde{\hat{n}}(f^{3r}) > \tilde{\hat{n}}(f^{r})$$

whenever r is large enough and satisfies (*). Hence f has a monotone-equivalence class of periodic points whose period divides $3r$ but not r. Evidently such a period cannot be a power of 2. ■

Note: This proof is quite nonconstructive. The following example shows that one cannot be too explicit as to which periods will occur. Start with a mapping f with growth number $s > 1$ such that all periods are divisible by some high power 2^N (compare 14.4 below). Then the iterate f^{2m+1} has growth number s^{2m+1} which (for suitable choice of m) can be arbitrarily large. Yet every period of f^{2m+1} is divisible by 2^N.

Here is a different but closely related consequence of 9.2.

Corollary 9.7. *The mapping f has only finitely many distinct points, or in other words only finitely many periodic points up to monotone-equivalence, if and only if the sequence of lap numbers $\ell(f^k)$ is bounded by a polynomial function of k.*

If these conditions are satisfied, then every period must be a power of 2 by Šarkovskii's thoeorem. Evidently the growth number s must be 1 (compare Block).

Proof. It will be convenient to say that a formal power series $A(t) = \sum a_k t^k$ satisfies a *polynomial growth condition of degree* d if its coefficients satisfy

$$|a_k| \le \varphi(k)$$

for all k, where φ is some polynomial of degree d. If $A(t)$ and $B(t)$ satisfy such conditions of degrees d and d', then evidently the product $A(t)B(t)$ satisfies a growth condition of degree $d+d'+1$. Given integers $k_0,\ldots,k_d \geq 1$, the product

$$(1-t^{k_0})^{-1} \ldots (1-t^{k_d})^{-1}$$

provides an example of a power series which satisfies a growth condition of degree d, but not of degree $d-1$.

Now consider the following possible conditions on f.

(1) The associated power series $\hat{\zeta}(t) = \prod_{\hat{o}}(1-t^{p(\hat{o})})^{-1}$ satisfies a polynomial growth condition;

(2) the power series $D(t)^{-1}$ does;

(3) the coefficients $M_{jk}(t)$ of the inverse kneading matrix (§5.6) all satisfy such conditions;

(4) the series $\gamma_i(J,t)$ of §5.4 all do;

(5) the series $L(t) = \sum \ell(f^k)t^{k-1}$ satisfies such a condition.

These conditions are all equivalent to each other. In fact (1) ⇒ (2) by 9.2; (2) ⇒ (3) since the entries of the kneading matrix $N_{ij}(t)$ all satisfy a polynomial growth condition; (3) ⇒ (4) by 5.7; (4) ⇒ (5) by 5.8; and (5) ⇒ (1) by the inequality $\hat{\underset{\sim}{n}}(f^k) \leq \ell(f^k)$. But, using the remarks above, we see that (1) is satisfied if and only if the product formula for $\hat{\zeta}(t)$ has only finitely many factors, or in other words if and only if there are only finitely many monotone equivalence classes $\hat{o}$ of periodic orbits. This completes the proof. ∎

10. Periodic points of negative type

This section will state a third and more precise form of our main theorem relating zeta functions and kneading determinant, and then derive Theorems 9.1 and 9.2 from it.

Definition. Let x be a fixed point lying in the interior of some lap of the piecewise monotone function f. We will say that s is of *positive type* or of *negative type* according as f is increasing or decreasing throughout a neighborhood of x. If x is an endpoint of some lap of f, then it will be called a fixed point of *critical type*.

Similarly, a periodic orbit $o = \{x, f(x), \ldots, f^{p-1}(x)\}$ will be called of *positive*, *negative*, or *critical type* according as x is of positive, negative, or critical type as a fixed point of f^p.

Let $N(f) = 2\underset{\sim}{n}_-(f) - 1$ *where* $\underset{\sim}{n}_-$ *is the number of fixed points of negative type*. Evidently $N(f) \le \hat{\underset{\sim}{n}}(f)$, since between any two fixed points of negative type there must be at least one other fixed point. Hence the numbers $N(f^k) \le \hat{\underset{\sim}{n}}(f^k) \le \ell(f^k)$ are all finite.

Definition. The formal power series

$$Z(t) = \exp \sum N(f^k) t^k / k$$

will be called the *modified zeta function* of f.

Our main theorem can now be stated as follows.

Theorem 10.1. *For any piecewise monotone function* f *we have* $Z(t) = D(t)^{-1}$.

The proof, in §11, will be quite indirect. (A more conceptual proof has recently been given in an unpublished manuscript by J.B. Rogers in Melbourne.) *For the remainder of this section, we simply assume* 10.1 *without proof*.

As is usual with zeta functions, we can also express $Z(t)$ as a product. If there are no periodic orbits at all of negative type, then $N(f) = N(f^2) = \ldots = -1$, so that

$$Z(t) = \exp(-t - t^2/2 - \ldots) = 1 - t.$$

Evidently each periodic orbit o of negative type with period $p = p(o)$ contributes a factor of

$$\exp 2(t^p + t^{3p}/3 + t^{5p}/5 + \ldots) = (1+t^p)/(1-t^p)$$

to the series $Z(t)$. Thus we obtain the following.

<u>Lemma 10.2</u>. <u>The modified zeta function</u> $Z(t)$ <u>is equal to</u>

$$(1-t)\Pi(1+t^{p(o)})/(1-t^{p(o)}) = (1-t)\Pi(1+2t^{p(o)}+2t^{2p(o)}+\ldots),$$

<u>taking the product over all periodic orbits</u> o <u>of negative type</u>.

One immediate consequence is that <u>all of the coefficients of the formal power series</u> $D(t)^{-1}(1-t)^{-1} = Z(t)/(1-t)$ <u>are nonnegative</u>. This would surely be difficult to check directly from the definition of the kneading determinant.

Next let us show how theorem 9.1 follows from 10.1. For this argument only, we make a special definition. Let us say that an isolated fixed point x_0 of f, not of negative type, is <u>formally left stable</u> if $f(x) > x$ throughout some interval $(x_0-\varepsilon, x_0)$, or also if x_0 is the leftmost point of the entire interval I. Similarly x_0 is <u>formally right stable</u> if $f(x) < x$ throughout an interval to the right of x_0, or if x_0 is the rightmost point of I. Fixed points of negative type must definitely be excluded in this definition.

<u>Remark</u>. For fixed points of positive type, this corresponds precisely to the intuitive notion of left or right stability. If f is smooth of class C^1, then the same is true for fixed points of critical type. (In fact, every turning point c_i with $f(c_i) = c_i$ is automatically both left and right stable in the C^1 case.) However, the non-differentiable function $f(x) = 2|x|$ provides an example of a critical fixed point which is <u>formally</u> left stable, but is not left stable in any ordinary sense.

<u>Lemma 10.3</u>. <u>If</u> f <u>is any piecewise monotone function with isolated fixed points, then the difference</u> $\underset{\sim}{n}(f) - N(f)$ <u>is equal to the number</u>

of formally left stable fixed points plus the number of formally right stable fixed points.

The proof is completely elementary. Let $x_1 < \ldots < x_{\underset{\sim}{n}}$ be the fixed points of f. Then x_1 is either formally left stable or of negative type and $x_{\underset{\sim}{n}}$ is either formally right stable or of negative type. For each intermediate interval (x_{i-1}, x_i) on which $f(x) > x$, we know that x_i is either formally left stable or of negative type. Similarly, if $f(x) < x$ in this interval then x_{i-1} is either formally right stable or of negative type. Thus the sum

number of formally left stable fixed points
+ number of formally right stable fixed points
+ twice number of fixed points of negative type

is equal to $\underset{\sim}{n} + 1$, which proves the lemma. ∎

Now suppose that each iterate f^k has isolated fixed points, so that the zeta function $\zeta(t)$ of f is defined.

Corollary 10.4. The ratio $Z(t)/\zeta(t) = D(t)^{-1}\zeta(t)^{-1}$ *is equal to a product of cyclotomic polynomials as follows. Each periodic orbit of period* p *contributes a factor of*

$(1-t^p)^2$ *if it is stable of positive or critical type*

$1 - t^{2p}$ *if it is stable of negative type, and*

$1 - t^p$ *if it is unstable of critical type, or if it is stable on one side but unstable on the other.*

There is no contribution at all from an unstable orbit of positive or negative type.

Note that a one-sided stable periodic orbit is necessarily of positive type. Thus the appropriate factors can be tabulated as follows.

	stable	one-sided stable	unstable
positive type	$(1-t^p)^2$	$1-t^p$	1
critical type	$(1-t^p)^2$		$1-t^p$
negative type	$1-t^{2p}$		1

If f is differentiable, note that an unstable orbit of critical type can only consist of one or both endpoints of the interval I, with p equal to 1 or 2. However, for nondifferentiable functions. like $2|x|$ near the origin, there can definitely be other unstable orbits of critical type.

Evidently theorem 9.1 is an immediate consequence of this corollary. For if all but finitely many periodic orbits are unstable, there can be only finitely many correction factors.

Here is an explicit example. As described in 2.2 and 4.6, the map $f(x) = (x^2-7)/2$ from the interval $|x| \le 1+\sqrt{8}$ to itself (or equivalently the map $g(y) = (1+\sqrt{8})\,y(1-y)$ from the unit interval to itself) has a one-sided stable orbit of period 3, and also an unstable fixed point $1+\sqrt{8}$ of critical type. By the Julia-Fatou Theorem or by Singer's Theorem, there are no other stable, one-sided stable, or critical periodic orbits. Since $D(t) = (1-t-t^2)/(1-t^3)$, it follows that

$$\zeta(t)^{-1} = (1-t)(1-t^3)D(t) = (1-t)(1-t-t^2)$$

is a polynomial of degree 3. On the other hand, for the map $f(x) = (x^2-(7+\varepsilon))/2$ on a corresponding maximal interval, where ε is small and positive, the kneading determinant is unchanged but the one-sided stable orbit of period 3 splits into one stable and one unstable orbit of period 3. Hence $\zeta(t)^{-1} = (1-t)(1-t^3)(1-t-t^2)$. This change in f corresponds to a perturbation in which the graph of f^3 passes through the diagonal as sketched schematically in Figure 10.5 (compare Jonker and Rand).

Figure 10.5.

<u>Proof of 10.4 (assuming 10.1)</u>. Let $\sigma(f^k)$ be the number of fixed points of f^k which are formally left stable plus the number which are formally right stable. Then

$$Z(t)/\zeta(t) = \exp \sum (-\sigma(f^k)t^k/k)$$

by 10.3. Separating out the contributions from the various periodic orbits of f, we can obtain a corresponding product formula for $Z(t)/\zeta(t)$.

Consider for example a one-sided stable orbit of period p. Evidently the points of this orbit contribute precisely p to the number $\sigma(f^k)$ of formal left or right stabilities of k is a multiple of p, and contribute nothing at all otherwise. Therefore such an orbit contributes a factor of

$$\exp(-t^p - t^{2p}/2 - \ldots) = 1 - t^p$$

to the ratio $Z(t)/\zeta(t)$. The contributions from other types of periodic orbits can be computed similarly. Details will be left to the reader. ∎

The proof of 9.2 (assuming 10.1) is completely analogous. We will say that a monotone equivalence class of fixed points of f, with leftmost point x_-, is <u>formally left stable</u> if x_- is not of negative type yet $f(x) > x$ for all x just to the left of x_-, or also if x_- is the leftmost point of the entire interval I. Formal right stability is defined similarly. In analogy with 10.3, the difference $\hat{\underset{\sim}{n}}(f) - N(f)$ is equal to the number of formally left stable monotone equivalence classes of fixed points plus the number of formally right stable ones. Expressing $Z(t)/\hat{\zeta}(t)$ as a product, in analogy with 10.4, the proof of 9.2 can easily be computed. ∎

In order to obtain a more explicit computation of this product, in terms of kneading invariants and the behavior of f on the boundary of the interval I, we proceed as follows.

Let $I_j = [c_{j-1}, c_j]$ be a lap of f. Recall that the invariant coordinate $\theta_*(c_{j-1}+)$ of §4 is a certain infinite sequence of formal symbols $\pm I_r$.

<u>Assertion</u>. <u>This infinite sequence</u> $\theta_*(c_{j-1}+)$ <u>is strictly periodic of period</u> 1, <u>being equal to</u> $(I_j, I_j, \ldots)$, <u>if and only if</u> f <u>is increasing on</u> I_j <u>and this lap contains a (necessarily unique) monotone equivalence class of fixed points of</u> f <u>which is formally left stable</u>. <u>Similarly the sequence</u> $\theta_*(c_{j-1}+)$ <u>has period dividing</u> p <u>if and only if the lap of</u> f^p <u>which is bounded on the left by</u> c_{j-1} <u>has the analogous property</u>. <u>Similar statements hold for</u> $\theta_*(c_j-)$ <u>and formal right stability</u>.

The proof is straightforward. However, in order to apply this assertion, we must take account of the fact that two different periodic invariant coordinates $\theta_*(c_j\pm)$ may correspond to the same monotone equivalence class of periodic orbits.

<u>Definition</u>. We will say that two different periodic infinite sequences θ_* and θ'_* are <u>cyclically equivalent</u> if $\theta'_i = \theta_{i+k}$ for

some constant $k \geq 0$ and for all $i \geq 0$.

The precise formula can now be stated as follows.

Theorem 10.6. *The ratio* $Z(t)/\hat{\zeta}(t) = D(t)^{-1}\hat{\zeta}(t)^{-1}$ *can be expressed as a product of at most* $\ell + 1$ *factors of the form* $1 - t^p$ *as follows. There is one factor of* $1 - t^p$ *corresponding to each cyclic equivalence class of sequences of period* p *among the* 2ℓ *infinite sequences* $\theta_*(c_{j-1}+)$ *and* $-\theta_*(c_j-)$ *for* $1 \leq j \leq \ell$.

The proof is not difficult, and will be omitted. ∎

The following special case is somewhat easier to state.

Corollary 10.7. *Suppose that* f *has just one turning point (so that* $\ell = 2$*), and suppose that* $f(\dot{I}) \subset \dot{I}$. *Then*

$$\hat{\zeta}(t)^{-1} = (1-t)D(t),$$

if the sequence of coefficients of the power series $D(t)$ *is not strictly periodic, while*

$$\hat{\zeta}(t)^{-1} = (1-t)(1-t^p)D(t)$$

if the sequence of coefficients is periodic of period p.

This follows immediately, since the fixed endpoint contributes a factor of $1 - t$, and it is easy to check that one of the two invariant coordinates $\pm\theta_*(c_1\pm)$ is periodic of period p if and only if the coefficients of D are periodic of period p. ∎

In the periodic case, it is interesting to note that $\hat{\zeta}(t)^{-1}$ is a polynomial of degree p.

In both 10.6 and 10.7 it is important to emphasize that the sequences must be precisely periodic. Eventual periodicity does not count. As an example, for the function $f(x) = (x^2-8)/2$ (or the equivalent function $g(y) = 4y(1-y)$) the coefficients of the kneading determinant

$$D(t) = (1-2t)(1-t) = 1 - t - t^2 - t^3 - \ldots$$

are not strictly periodic, hence there is no extra factor, and

$$\hat{\zeta}(t)^{-1} = (1-t)D(t) = 1 - 2t.$$

In fact

$$\hat{\zeta}(t) = \zeta(t) = (1-2t)^{-1} = 1 + 2t + 4t^2 + \ldots,$$

and

$$t\zeta'(t)/\zeta(t) = \sum n(f^k)t^k = 2t + 4t^2 + 8t^3 + \ldots$$

for this example.

11. Smooth deformations of f

This section will complete the arguments of the previous two sections by proving the equality $D(t) = Z(t)^{-1}$ of Theorem 10.1. The idea of the proof is first to check this equation for some particularly simple mapping f with given lap number ℓ and then to check that both sides change in precisely the same way as we deform f. We will first concentrate on changes in the kneading determinant as we change f.

Let $C^0(I,\mathbb{R})$ denote the Banach space consisting of all continuous functions f from the interval I to the real numbers $\mathbb{R}$, with norm

$$\|f\|_0 = \operatorname*{Max}_{x\in I} |f(x)|.$$

We will also make use of the Banach space $C^r(I,\mathbb{R})$ consisting of functions f which are smooth of class C^r, with norm

$$\|f\|_r = \operatorname{Max}(\|f\|_0, \|df/dx\|_0, \ldots, \|d^rf/dx^r\|_0).$$

We will be particularly interested in the closed convex subset $C^r(I,I)$ consisting of C^r-maps from I to itself. For our purposes, r will always be 0, 1, or 2.

Lemma 11.1. The map $(f,x) \longmapsto (f,f(x))$ from $C^r(I,I)\times I$ to itself is *smooth of class* C^r. *Hence its* n-*fold iterate* $(f,x) \longmapsto (f,f^n(x))$ *is also smooth of class* C^r.

For a proof, the reader is referred to Abraham and Robbin, p. 25.

∎

We must study piecewise-monotone mappings from I to itself. *To simplify the discussion, we will only consider the case where the set* $\{c_1,\ldots,c_{\ell-1}\}$ *of turning points is fixed*. For example, if I is the interval $[0,\ell]$ we could take $c_1 = 1,\ldots,c_{\ell-1} = \ell - 1$. We will assume that some such choice of turning points has been made. Also choose a sign $\varepsilon(I_1) = \pm 1$.

Definition. Let $P^r = P^r(I, c_1,\ldots,c_{\ell-1},\varepsilon(I_1))$ denote the convex subset of $C^r(I,I)$ consisting of piecewise-monotone maps f which are strictly monotone on each interval $I_j = [c_{j-1},c_j]$, being mono-none increasing or monotone decreasing according as $(-1)^{j+1}$ equals $\varepsilon(I_1)$ or $-\varepsilon(I_1)$.

We are interested in the continuity properties of the kneading increment $\nu_i(f)$ of §4 as f varies in P^r. Here is a first result. Choose some turning point c_i.

Lemma 11.2. If f *belongs to the space* P^0 *of piecewise-monotone maps and satisfies the condition that none of the forward images* $f(c_i), f^2(c_i), f^3(c_i),\ldots$ *is a turning point, then the correspondence*

$$g \longmapsto \nu_i(g)$$

from P^0 *to the module* $V[[t]]$ *of formal power series is continuous at* f.

In other words, for each N there exists $\varepsilon > 0$ so that

$$\nu_i(g) \equiv \nu_i(f) \bmod t^N$$

whenever g in P^0 satisfies $\|g - f\|_0 < \varepsilon$. The proof, using Lemma 11.1, is clear. ∎

Next, let us look at the simplest case where a discontinuity occurs. *Suppose that there exists one and only one integer* $p \geq 1$ *so that* $f^p(c_i)$ *is a turning point* c_j, *with* $i \neq j$. Let N be a preassigned large integer. Suppose, to fix our ideas, that f^p has a local minimum at c_i.

Lemma 11.3. *With these hypotheses, every* $g \in P^0$ *which is sufficiently close to* f *satisfies either*

$$\nu_i(g) \equiv \nu_i(f) \bmod t^N$$

or

$$\nu_i(g) \equiv \nu_i(f) - 2t^p\nu_j(f) \bmod t^N$$

according as $f^p(c_i) \geq c_j$ *or* $f^p(c_i) < c_j$.

The proof is straightforward. ∎

The case of a local maximum is completely analogous. The only difference is that the inequality symbols $\geq$ and $>$ must be replaced by $\leq$ and $<$ respectively.

The behavior of the kneading determinant $D(f)$ can now be described as follows.

Corollary 11.4. *For* $f \in P^0$ *if each of the turning points* $c_1, \ldots, c_{\ell-1}$ *satisfies the hypothesis of 11.2 or 11.3, then the correspondence*

$$g \longmapsto D(g)$$

from P^0 *to* $Z[[t]]$ *is continuous at* f.

Proof. This follows immediately, since the discontinuity of 11.3 corresponds to an elementary row operation in the kneading matrix $[N_{ij}]$, and hence does not change the determinant. ∎

A more complicated discontinuity occurs when $f^p(c_i) = c_i$. To get a reasonable result in this case we will need to make use of first derivatives.

Lemma 11.5. *Suppose that* $f^p(c_i) = c_i$, *where* f *is smooth of class* C^1. *Then, for every* g *which is sufficiently close to* f *in the space* P^1, *the real numbers*

$$g^p(c_i) - c_i,\ g^{2p}(c_i) - c_i,\ g^{3p}(c_i) - c_i, \ldots$$

are either all zero or all have the same sign. Furthermore, given $\varepsilon > 0$ *there exists* $\delta > 0$ *so that* $|g^m(c_i) - f^m(c_i)| < \varepsilon$ *uniformly for all* m *whenever* $\|g - f\|_1 < \delta$.

Proof. To simplify the notation, set $F = f^p$, $G = f^p$. Since the first derivative $\dot{F}$ of F must vanish at c_i, we can choose $\varepsilon > 0$ small enough so that $|\dot{F}(x)| < 1/4$ for $|x - c_i| \le \varepsilon$. Now choose g close enough to f so that $|\dot{G}(x)| < 1/2$ for $|x - c_i| \le \varepsilon$, and so that $|G(c_i) - c_i| < \varepsilon/2$. If $G(c_i) = c_i$, the proof is completely straightforward. But if $G(c_i) \ne c_i$, then it follows by an easy induction that

$$|G^{m+1}(c_i) - G^m(c_i)| < |G(c_i) - c_i|/2^m$$

and hence that

$$|G^m(c_i) - G(c_i)| < |G(c_i) - c_i| < \varepsilon/2$$

for all $m \ge 1$. The conclusion then follows easily. ∎

Now suppose that the forward images $f(c_i), f^2(c_i), \ldots, f^{p-1}(c_i)$ are not turning points, but that $f^p(c_i) = c_i$. Then it is not difficult to see that the i-th kneading increment takes the form

$$\nu_i(f) = \eta(t)/(1-t^p)$$

where $\eta(t)$ is a polynomial of degree p. More precisely,

$$\eta(t) = (I_{i+1} - I_i) \pm 2tI_{\alpha_1} \pm \ldots \pm 2t^{p-1}I_{\alpha_{p-1}} \pm t^p(I_{i+1} + I_i),$$

where the coefficient of t^m is positive or negative according as f^m has a local minimum or a local maximum at c_i, and where

$f^m(c_i) \in I_{\alpha_m}$. Suppose, to fix our ideas, that f^p has a local minimum at c_i.

Corollary 11.6. *With these hypotheses, for every g which is sufficiently close to f in the space P^1 we have either*

$$\nu_i(g) = \eta(t)/(1-t^p) \quad \underline{\text{or}} \quad \nu_i(g) = \eta(t)/(1+t^p)$$

according as $g^p(c_i) \geq c_i$ *or* $g^p(c_i) < c_i$.

The proof, using 11.5, is not difficult and will be left to the reader. ∎

If the other rows of the kneading matrix do not change, then it follows that the determinant $D(g,t)$ is equal to $D(f,t)$ in the first case and to $D(f,t)(1-t^p)/(1+t^p)$ in the second. If f^p has a local maximum at c_i, then the situation is completely analogous except that the inequalities are reversed.

Remark. The C^1-topology is essential here. For example, for the map $f(x) = x^2$ on the interval $[-1,1]$ we can choose a smooth map g, with $\ell = 2$, uniformly close to f so that g maps some small interval $[-\varepsilon,\varepsilon]$ to itself by a quadratic map having any growth number at all in the interval $1 \leq s \leq 2$. Thus there are uncountably many different possibilities for the kneading determinant $D(g)$.

We will next show that a "generic" homotopy between generic elements of the space P^2 involves only the discontinuities which have already been described in 11.4 and 11.6. We will need to work in an open subset of a Banach space, in order to make use of the inverse function theorem. To achieve this, we introduce second derivatives.

Definition. A mapping $f \in C^2(I,I)$ will be called a *Morse function* if the first derivative $f'(x)$ and the second derivative $f''(x)$ do not vanish simultaneously, and if f' does not vanish at the two boundary points. Let $M \subset P^2$ denote the dense open subset consisting of Morse functions in P^2 which satisfy $f(I) \subset$ interior (I).

In other words, M consists of all maps $f \in C^2$ (I, interior I) such that f' vanishes <u>only</u> at the turning points $c_1, \ldots, c_{\ell-1}$, and such that $f''(c_i) \neq 0$, this second derivative being alternately positive or negative according as $(-1)^i \varepsilon(I_1)$ is positive or negative.

It is easy to check that M is a convex open set in the Banach space B consisting of all elements $f \in C^2(I, \mathbb{R})$ which satisfy the linear equations $f'(c_i) = \ldots = f'(c_{\ell-1}) = 0$.

Define submanifolds $M_n(i,j) \subset M$ as follows. For any $n \geq 1$ and any turning points c_i and c_j, let $M_n(i,j)$ be the set of all maps $f \in M$ such that $f(c_i), f^2(c_i), \ldots, f^{n-1}(c_i)$ are not turning points, but $f^n(c_i) = c_j$.

<u>Lemma 11.7</u>. <u>Each</u> $M_n(i,j)$ <u>is a</u> C^2-<u>smooth codimension</u> 1 <u>submanifold of</u> M. <u>Furthermore, each intersection of two distinct submanifolds</u> $M_{n_1}(i_1,j_1)$ <u>and</u> $M_{n_2}(i_2,j_2)$ <u>is transverse, and hence is a codimension</u> 2 <u>submanifold of</u> M <u>whenever it is non-vacuous</u>.

By definition, a subset S of a Banach manifold M is a C^r-smooth <u>codimension</u> k <u>submanifold</u> if for every point s of S there exists a local chart mapping a neighborhood of s in the pair (M,S) by a C^r-diffeomorphism onto an open subset of some pair $(\mathbb{R}^k \times B', 0 \times B')$ where B' is a Banach space.

<u>Proof of 11.7</u>. We will make use of Lemma 11.1 together with the inverse function theorem. Fixing some point $x_0 \in I$, consider the C^2-map $\varphi : M \longrightarrow I$ defined by $\varphi(f) = f^n(x_0)$. If we set $x_k = f^k(x_0)$, then computation shows that the derivative $D\varphi(f)$ is the linear map

$$\Delta f \longmapsto \sum_{i=1}^{n} \Delta f(x_{i-1}) f'(x_i) f'(x_{i+1}) \ldots f'(x_{n-1})$$

from the ambient Banach space B to $\mathbb{R}$.

Now let us specialize to the case $x_0 = c_i$. If f belongs to the set $M_n(i,j) \subset M$, note that $D\varphi(f) \neq 0$. In fact, choosing

$\Delta f \in B$ so that $\Delta f(x)$ vanishes except for x in a small neighborhood of c_i, it follows easily that $D\varphi(f)(\Delta f) \neq 0$. Hence, using the inverse function theorem for Banach spaces, it follows that $M_n(i,j)$ is a codimension 1 submanifold (compare Lang, p. 15, Corollary 25).

The proof that intersections are transverse is completely analogous. ∎

<u>Remark</u>. The manifold $M_n(i,j)$ is not a closed subset of M when $n \geq 2$. However any limit point must belong to some 2-fold intersection of the form $M_p(i,h) \cap M_q(h,k)$ with $p+q \leq n$. Note that $i \neq h$. This follows almost by definition when $i \neq j$, and by 11.5 when $i = j$.

<u>Definition</u>. Let $M^{(1)}$ denote the union of all of these codimension 1 submanifolds $M_n(i,j)$, and let $M^{(2)}$ denote the union of the codimension 2 submanifolds $M_{n_1}(i_1,j_1) \cap M_{n_2}(i_2,j_2)$.

<u>Lemma 11.8</u>. <u>Any two points f_0 and f_1 in $M - M^{(2)}$ can be joined by a path in $M - M^{(2)}$ which intersects each manifold $M_n(i,j)$ only finitely often</u>.

<u>Proof</u>. Let H_0 denote the set consisting of all C^2-paths from f_0 to f_1 in the convex open set M, topologized as a subset of the Banach space $C^2(I \times I, \underset{\sim}{R})$. Define subsets $H_0 \supset H_1 \supset H_2 \supset \ldots$ inductively as follows. Let H_n be the set of all paths belonging to H_{n-1} which

(a) intersect the manifold $M_n(i,j)$ transversally for all i and j, and

(b) are disjoint from $M_n(i,j) \cap M_{n_1}(i_1,j_1)$ for all $n_1 \leq n$ and all $i \neq i_1$, j, and j_1.

It is not difficult to check inductively that each H_n is a dense open subset of the complete metric space $\bar{H}_0$. Hence the intersection of the H_n is non-vacuous, and contains the required homotopy from f_0 to f_1. ■

Finally we will need the following. Fix some large integer N.

<u>Lemma 11.9</u>. <u>Any map $f \in P^0$ can be uniformly approximated by a map in $M - M^{(1)}$ which has the same kneading matrix modulo t^N.</u>

<u>Proof</u>. First suppose that the various forward images $f^p(c_i)$ with $1 \le p < N$, $1 \le i \le \ell - 1$ are not turning points of f. Then the proof is straightforward: First uniformly approximate f by an element g belonging to the subspace $M \subset P^2$ of Morse functions. Then approximate g by an element h in the subspace $M - M^{(1)}$. If the approximations are sufficiently close, then it is easy to check that $\nu_i(f) \equiv \nu_i(g) \equiv \nu_i(h)$ modulo t^N (compare 11.2).

Now consider the more difficult case where, for example, $f^p(c_1) = c_j$. To fix our ideas, suppose that f has a local minimum at c_1. Let g coincide with f except in a very small neighborhood of c_1 and suppose that $g(x) > f(x)$ within this neighborhood. If the approximation is sufficiently close, then one can check that the forward images $g(c_1),\ldots,g^p(c_1),\ldots,g^{N-1}(c_1)$ are not turning points, and that the kneading invariants of g are congruent to those of f modulo t^N. Making analogous alterations at $c_2,\ldots,c_{\ell-1}$, and then proceeding as above, the proof can easily be completed. ■

We are now ready to prove our main theorem.

<u>Proof of 10.1</u>. Recall that the space M, consisting of all C^2-smooth Morse functions $f : I \longrightarrow \text{interior}(I)$ with prescribed turning points and prescribed sign $\varepsilon(I_1) = \pm 1$ for the derivative $f'(c_0)$ at the leftmost point, is a convex open subset of a suitable Banach space.

It is first necessary to check the formula $D(t) = Z(t)^{-1}$ for just one arbitrarily chosen function f_0 in M. For example, let f_0 map the entire interval I into the interior of a lap I_j with $\varepsilon(I_j) = +1$. Then all of the interesting part of the kneading matrix

$[N_{pq}(f_0)]$ is concentrated in the j-th column. Cancelling this j-th column, it follows easily that

$$D(f_0,t) = 1/(1-t).$$

On the other hand, there are no periodic orbits at all of negative type, so the formula 10.2 takes the form $Z(f_0,t) = 1 - t$, as required.

Now consider a generic homotopy f_u between maps f_0 and f_1, as constructed in 11.8. We will show that the product

$$Z(f_u,t)D(f_u,t) \in \underset{\sim}{Z}[[t]]$$

varies continuously with the parameter $u \in [0,1]$ and hence is constant.

Fixing some parameter value u, recall from 11.8 that there at most one turning point c_i such that some forward image $f_u^p(c_i)$ is a turning point c_j. There are two cases to consider.

<u>Case</u> 1. Suppose that no turning point is periodic, so that $f_u^k(c_i) \neq c_i$ for $k \geq 1$ and for all i. Then every fixed point of f_u is an interior point of a lap of f_u^k. Evidently the number of fixed points which occur in laps on which f_u^k is decreasing remains constant under small deformations of f_u. In other words the correspondence

$$g \longmapsto Z(g,t)$$

from P^0 to the ring of formal power series is continuous at f_u. Together with 11.4, it follows that the correspondence

$$g \longmapsto Z(g,t)D(g,t)$$

is continuous at f_u also.

<u>Case</u> 2. Suppose that just one turning point c_i is periodic, say $f_u^p(c_i) = c_i$ with p minimal. Suppose, to fix our ideas, that f_u^p has a local minimum at c_i as illustrated in Figure 11.10. Then

the graph

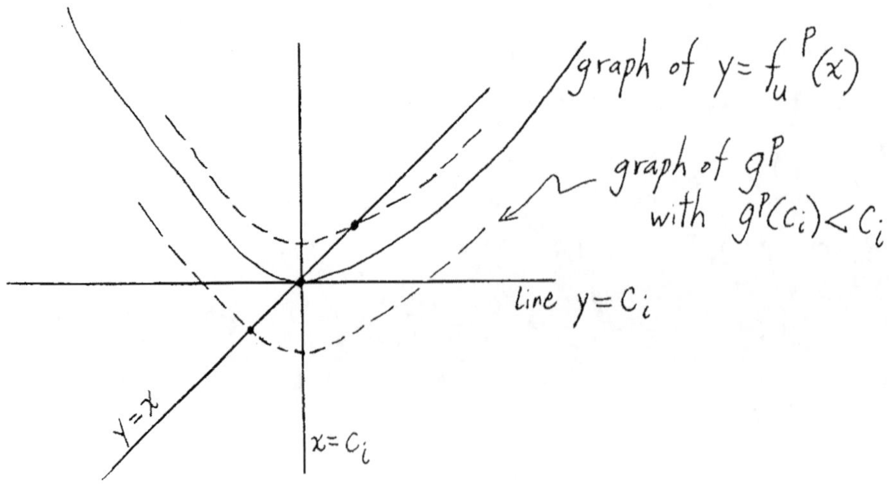

Figure 11.10

the graph of f_u^p intersects the diagonal transversally at c_i. For any g which is sufficiently close to f_u in the space P^1, it follows that the graph of g^p intersects the diagonal transversally somewhere near c_i. The resulting fixed point of g^p will be of negative type if and only if $g^p(c_i) < c_i$. In this case there is a new factor of $(1+t^p)/(1-t^p)$ in the zeta function $Z(g,t)$. There is no such new factor if $g^p(c_i) \geq c_i$.

Comparing Lemma 11.6, it follows that the correspondence

$$g \longmapsto Z(g,t)D(g,t)$$

from the space P^1 to $\mathbb{Z}[[t]]$ is continuous at f_u.

Since these are the only possibilities allowed by 11.8, this proves that the function $u \longmapsto Z(f_u,t)D(f_u,t)$ from $[0,1]$ to the totally disconnected space $\underset{\sim}{Z}[[t]]$ is continuous and hence constant. Therefore

$$Z(f_1,t)D(f_1,t) = Z(f_0,t)D(f_0,t) = 1.$$

This completes the proof of 10.1 for any function $f_1 \in M$ which does not belong to the exceptional subset $M^{(1)}$, so that 11.8 applies.

Now for any piecewise-monotone map $f \in P^0$, according to Lemma 11.9, we can approximate by a map $f_1 \in M - M^{(1)}$ which has the same kneading determinant modulo t^N. Here N can be any preassigned large integer. Inspecting the proof, it is not difficult to show that the congruence

$$Z(f_1,t) \equiv Z(f,t) \bmod t^N$$

will also be satisfied. Thus

$$Z(f,t) \equiv Z(f_1,t) = D(f_1,t)^{-1} \equiv D(f,t)^{-1} \bmod t^N$$

for every N, and hence $Z(f,t) = D(f,t)^{-1}$ as required. ∎

12. An intermediate value theorem

This section will continue to study the way in which various invariants of the mapping f change under smooth deformation of f. We concentrate on the case $\ell = 2$, but also study continuity properties of the growth number $s(f)$ for arbitrary fixed lap number.

Let us first ask which kneading determinants $D(t) = 1 \pm t \pm t^2 \pm \ldots$ can actually occur, for maps with $\ell = 2$. As in §3, we order the additive group $\underset{\sim}{Z}[[t]]$ of formal power series by setting $\alpha = a_0 + a_1 t + \ldots > 0$ whenever $a_0 = \ldots = a_{n-1} = 0$ but $a_n > 0$ for some $n \geq 0$. It is convenient to set $|\alpha| = \pm\alpha$, taking the plus sign if and only if $\alpha > 0$.

__Definition__. A formal power series $D(t) = 1 + D_1 t + D_2 t^2 + \ldots$ with coefficients $D_i = \pm 1$ will be called <u>admissible</u> if

$$D(t) \le |D_n + D_{n+1}t + D_{n+2}t^2 + \ldots|$$

for every $n \ge 1$.

To illustrate this definition, note that a series of the form $1 + t + (\text{higher terms})$ is admissible if and only if it is equal to $(1-t)^{-1} = 1 + t + t^2 + \ldots$.

<u>Theorem 12.1</u>. *A given power series $D(t) = 1 \pm t \pm t^2 \pm \ldots$ actually occurs as the kneading determinant of some map f with $\ell(f) = 2$ if and only if this series is admissible.*

The proof will show that every such admissible power series actually occurs as kneading determinant for some quadratic mapping

$$x \longmapsto bx(1-x)$$

from the unit interval to itself, with $1 \le b \le 4$. More generally, consider a one-parameter family of C^1-smooth maps $g_b : I \longrightarrow I$, depending C^1-smoothly on the parameter b for $b_0 \le b \le b_1$, all with the single turning point c_1.

<u>Intermediate Value Theorem 12.2</u>. *Any admissible power series $1 \pm t \pm t^2 \pm \ldots$ which lies between the kneading determinants of g_{b_0} and g_{b_1} must actually occur as the kneading determinant of g_b for some parameter value between b_0 and b_1.*

<u>Proof</u>. Let $D(b;t) \in \underset{\sim}{Z}[[t]]$ be the kneading determinant associated with the parameter value b, and let $A(t)$ be the given admissible series. Suppose, to fix our ideas, that $D(b_0;t) > A(t) > D(b_1;t)$. Let b' be the supremum of parameter values $b < b_1$ for which $D(b;t) \ge A(t)$. Thus there are parameter values b^- and b^+ arbitrarily close to b' with $D(b^-;t) \ge A(t) > D(b^+;t)$. We will show that $D(b';t)$ must be equal to $A(t)$.

To simplify notations, let $f = g_{b'}$. First suppose that none of the iterated images $f(c_1), f^2(c_1), \ldots$ is equal to the turning point

c_1. Then according to 11.2 or 11.4 the correspondence $b \longmapsto D(b;t)$ is continuous at b', and it follows immediately that $D(b';t) = A(t)$.

Now suppose that $f^p(c_1) = c_1$, with $p \ge 1$ minimal. Then the formal power series $D(b;t)$ is periodic of period p, having the form $\varphi(t)/(1-t^p)$ for some polynomial $\varphi(t) = 1 + D_1 t + \ldots + D_{p-1} t^{p-1}$ of degree $p-1$ (compare §4.5). Furthermore, for every b sufficiently close to b' we have

$$D(b;t) = \varphi(t)/(1 \pm t^p)$$

by §11.6. Thus

$$\varphi(t)/(1-t^p) \ge A(t) > \varphi(t)/(1+t^p)$$

by the definition of b'. However, it is not difficult to check that the only admissible series satisfying these inequalities is $\varphi(t)/(1-t^p)$ itself. This completes the proof of 12.2. ∎

Proof of 12.1. In one direction, this follows immediately by applying 12.2 to the one parameter family of maps $x \longmapsto bx(1-x)$. In fact, for $b = 1$ we obtain the largest possible admissible kneading determinant $1 + t + t^2 + \ldots$ while for $b = 4$ we obtain the smallest possible admissible kneading determinant $1 - t - t^2 - \ldots$. (For further discussion of such quadratic maps, see §13.)

For the proof in the other direction, we must study the kneading determinant $D(t) = \sum D_k t^k$ associated with an arbitrary map f with $\ell(f) = 2$. Suppose, to fix our ideas, that f attains its minimum at the unique turning point c_1. Then $f(c_1) \le f(x)$ for all x, and in particular,

$$f(c_1) \le f^{n+1}(c_1)$$

for all $n \ge 1$. Applying the invariant coordinate function θ of §3, it follows that $\theta(f(c_1)) \le \theta(f^{n+1}(c_1))$. Similarly, taking the limit as $x \longrightarrow c_1$, $x > c_1$, we have

$$\theta(f(c_1+)) \le \theta(f^{n+1}(c_1+)).$$

Recall that $\theta(x)$ is a formal power series in t, of the form $I_{i_0} \pm I_{i_1} + \dots$, provided that x is not a pre turning point. Since there are only two formal basis elements $I_1 < I_2$, corresponding to the two laps, we can simplify the situation by applying the linear transformation

$$I_1 \longmapsto -1, \quad I_2 \longmapsto +1$$

of §3.3. Then $\theta(x)$ will map to a formal power series in $\underset{\sim}{Z}[[t]]$ of the form

$$\bar{\theta}(x) = \pm 1 \pm t \pm t^2 \pm \dots ,$$

which still depends monotonely on x, so that

$$(*) \qquad \bar{\theta}(f(c_1+)) \le \bar{\theta}(f^{n+1}(c_1+)).$$

With this notation, it is not difficult to check that $\bar{\theta}(c_1+)$ is equal to the kneading determinant $1 + D_1 t + D_2 t^2 + \dots$. Similarly,

$$\bar{\theta}(f(c_1+)) = D_1 + D_2 t + D_3 t^2 + \dots,$$

and more generally

$$\bar{\theta}(f^{n+1}(c_1+)) = D_n(D_{n+1}+D_{n+2}t+\dots).$$

Now, multiplying the inequality (*) by t and adding $+1$, we obtain

$$1 + D_1 t + D_2 t^2 + \dots \le D_n(D_n+D_{n+1}t+\dots) = |D_n + D_{n+1}t + \dots|,$$

as required. ∎

Next let us study continuity properties of the growth number $s(f)$. We consider piecewise monotone C^1-maps from I to itself with fixed turning points $c_1,\dots,c_{\ell-1}$.

Lemma 12.3. *As g tends to f in the C^1-topology, keeping the lap number $\ell(g)$ and the turning points $c_1,\dots,c_{\ell-1}$ fixed, the growth number $s(g)$ tends to $s(f)$.*

Compare Guckenheimer, 1979.

Here it is essential to use the C^1-topology. Furthermore, it is essential to keep the lap number fixed, or at least bounded. For Misiurewicz and Szlenk, and independently Newhouse, have given examples to show that the correspondence $g \longmapsto s(g)$ is <u>not</u> continuous if the number of turning points is allowed to jump to an arbitrarily large value, even if we use the C^r-topology with r large.

<u>Proof of 12.3</u>. First let us prove continuity at f under the hypothesis that none of the turning points c_i is periodic under f.

We will use the notation $D(g,t)$ for the kneading determinant of g. Let t be a complex variable ranging over the disk $|t| \le r$ of some radius $r < 1$. Then we will show that the uniform norm

$$\sup_{|t|\le r} |D(g,t) - D(f,t)|$$

tends to zero as g tends to f in the space P^0 of §11, or, in other words, as g tends uniformly to f with fixed turning points.

Setting $D(g,t) = \sum D_n(g)t^n$, first note that the integer coefficients $D_n(g)$ are uniformly bounded,

$$|D_n(g)| \le A_n \quad \text{for all} \quad g \in P^0,$$

where A_n is some polynomial function of n. For example, one can take $A_n = 2^{\ell-1}(n+1)(n+2)\dots(n+\ell-1)$.

To verify this estimate, it is convenient to introduce two special notations. We will write $\sum a_n t^n \ll \sum b_n t^n$ whenever $a_n \le b_n$ for all n, and will use the notation $\|\sum a_n t^n\|$ for the power series $\sum |a_n| t^n$. Recall that the product $(1-t)D(g,t)$ is defined to be the determinant of a certain $(\ell-1)\times(\ell-1)$ kneading matrix. Each entry $\sum a_n t^n$ of this kneading matrix satisfies $|a_n| \le 2$; or, in other words, $\|\sum a_n t^n\| \ll 2/(1-t)$. Adding up $(\ell-1)!$ monomials to form the determinant, we obtain

$$\|(1-t)D(g,t)\| \ll (\ell-1)!(1/(1-t))^{\ell-1},$$

and therefore

$$\|D(g,t)\| \ll (\ell-1)!2^{\ell-1}(1-t)^{\ell} .$$

Expanding $(1-t)^{-\ell}$ by the binomial theorem, this yields the required upper bound for $|D_n(g)|$.

Since A_n is a polynomial in n, it follows that the comparison series $\sum A_n t^n$ converges uniformly throughout the disk $|t| \le r < 1$. Hence for each $\varepsilon < 0$ there exists an N so that the sum from N to infinity satisfies

$$(1) \qquad |\sum_N^\infty D_n(g)t^n| \le \sum_N^\infty A_n r^n < \varepsilon$$

uniformly throughout the disk $|t| \le r$, and uniformly for all g.

But if g is sufficiently close to f in the space P^0, then

$$(2) \qquad D_n(g) = D_n(f) \quad \text{for} \quad n < N.$$

This follows using 10.1, or can be verified directly by the methods of §11. Combining (1) and (2) it follows that

$$|D(g,t) - D(f,t)| < 2\varepsilon$$

for $|t| \le r$. In other words, the analytic function $D(g,t)$ tends uniformly to $D(f,t)$ throughout the disk of radius r as g tends to f in the C^0-topology.

Using 6.3 and the fact that the zeros of a complex analytic function vary continuously with the function, it follows that $s(g)$ tends to $s(f)$ as g tends to f.

Now consider the more general case where one or more of the turning points c_i may be periodic under f. We must now use the C^1-topology. If $f^p(c_i) = c_i$, then for g arbitrarily close to f in the space P^1 the function $D(g,t)$ may uniformly approximate not $D(f,t)$ but rather the product $D(f,t)(1-t^p)/(1+t^p)$ (compare §11.6).

If k of the $\ell - 1$ turning points are periodic, then there may be as many as k of these additional factors of the form $(1-t^p)/(1+t^p)$. However, these additional factors are all smooth non-zero analytic functions throughout the disk $|t| \le r$, so that they do not affect the location of zeros. The argument now proceeds just as before. ∎

One immediate corollary is that _every number in the interval_ $[1,\ell]$ _occurs as growth number_ $s(f)$ _for some map_ f _with lap number_ ℓ. For we can choose a smooth one parameter family of maps in the space P^1 leading from a map with $s = 1$ to a map with $s = \ell$.

For the case $\ell = 2$, it follows that _there are uncountably many distinct admissible kneading invariants_ $D(t)$.

13. _More on quadratic maps_

Let us take a further look at the archetypal one parameter family of maps

$$y \longmapsto by(1-y)$$

from the unit interval to itself, or the family of linearly conjugate maps $f_a(x) = (x^2-a)/2$, where $a = b^2 - 2b$. The growth number $s(f_a)$, as a function of the parameter a, is plotted in Figure 13.3, and tabulated in 13.5. We have noted earlier that the behavior of the iterates of the function f_a becomes more and more complicated as the parameter a increases. Here is a precise statement.

Theorem 13.1. _The kneading invariant_ $D(f_1) \in Z[[t]]$ _is monotone decreasing as a function of the parameter_ a.

Corollary 13.2. For each fixed n, _the number of laps_ $\ell(f_a^n)$ _is monotone increasing as a function of_ a. _Hence the growth number_ $s(f_a)$ _is also monotone increasing as a function of_ a.

These statements were first proved by Douady, Hubbard and

Sullivan (unpublished). The proofs given below were suggested by their argument. Another closely related statement, which follows from the Douady-Hubbard-Sullivan argument but will not be proved here, is the following. _For each fixed k, the number of periodic points of_ f_1 _of period k is monotone increasing as a function of_ a (see Jonker, 1982, for a weaker form of this result).

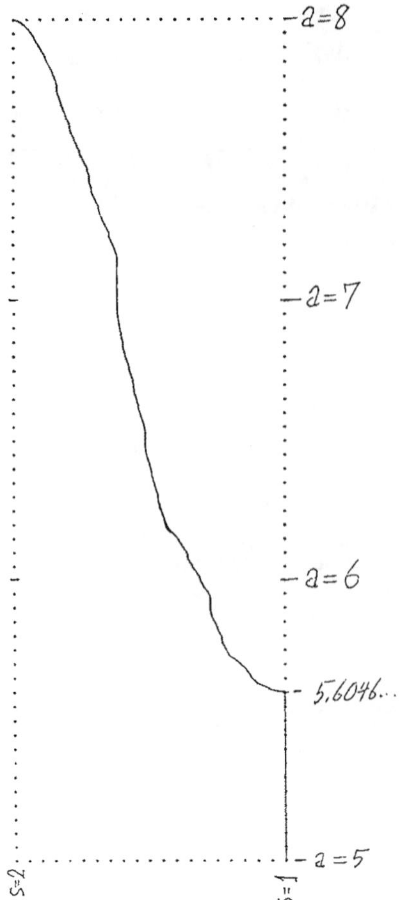

Figure 13.3. Graph of the growth number $s(f_a)$ as aa function of the parameter a, where $f_a(x) = (x^2-a)/2$. This function is continuous and monotone. Its intervals of constancy are everywhere dense. For example $s(f_a) = (1+\sqrt{5})/2$ for $7 \le a \le 7.161\ldots$ and $s(f_a) = \sqrt{(1+\sqrt{5})/2}$ for $5.89\ldots \le a \le 5.94$.

Before beginning of the proof of 13.1, let us briefly discuss other theorems and problems concerning quadratic maps. We have seen in Sections 11 and 12 that every strictly periodic admissible kneading invariant corresponds to an entire interval of parameter values a. However, it turns out that these open intervals become extremely small as the period increases. In fact, according to Jakobson [1981], we have the following remarkable statement: <u>The set of parameter values</u> a <u>for which the kneading invariant</u> $D(f_a)$ <u>is nonperiodic has positive Lebesgue measure</u>.

It is conjectured that this set of positive measure does not contain any interval, or equivalently that the complementary set, consisting of all parameter values for which $D(f_1)$ is strictly periodic, is everywhere dense in the interval $[-1,8]$. Still another equivalent conjecture would be that each admissible nonperiodic kneading invariant corresponds to a <u>unique</u> parameter value a. In the special case of an eventually periodic (but not periodic) kneading invariant, this would not be difficult to prove by the methods described below; but we have no idea how to handle the general case.

The proof of 13.1 will be based on the following statement.

<u>Lemma 13.4</u>. <u>Consider two different real quadratic mappings for which the critical point is contained in a periodic orbit of period</u> p. <u>If these two have the same kneading invariant, then they are linearly conjugate to each other</u>.

<u>Proof</u>. Since the cases $p = 1,2$ are easy to verify, we may assume that $p \geq 3$. The proof will be based on complex variable methods, and in particular, on Teichmüller theory. Let M_p be the complex $(p-2)$-dimensional manifold consisting of all p-tuples $(z_1,\ldots,z_p)$ of distinct complex numbers, well defined up to a simultaneous linear substitution

$$z_j \longmapsto \alpha z_j + \beta$$

with $\alpha \neq 0$. Then the universal covering $\tilde{M}_p$ can bé identified with the Teichmüller space consisting of all marked Riemann surfaces of genus zero with $p+1$ punctures (see for example Abikoff).

Consider any complex quadratic mapping q whose critical point c is periodic of period p. Then the finite orbit $\{z_j = q^j(c)\}$ determines a point $(z_1,\ldots,z_p)$ in the manifold M_p. It is not difficult to check that two different quadratic maps, each having such a superattractive periodic orbit, determine the same point of M_p if and only if they are linearly conjugate.

Let us define a many-valued, locally biholomorphic mapping F from M_p into itself by the formula

$$F(z_1,\ldots,z_p) = (\pm\sqrt{z_2-z_1},\ldots,\pm\sqrt{z_p-z_1},0).$$

Then the fixed points of F are precisely the points $(z_1,\ldots,z_p)$ associated with quadratic maps having a superattractive period p orbit. for if $z_j = q^j(c)$, then setting $q(z) = (z-c)^2/\alpha^2+z_1$ we will have $q^{-1}(w) = \pm\alpha\sqrt{w-z_1}+c$, hence

$$z_j = \pm\alpha\sqrt{z_{j+1}-z_1}+c \quad \text{for} \quad j < p,$$

and $z_p = \alpha\cdot 0 + c$, as required.

Now let us choose a basepoint b in the universal covering manifold $\tilde{M}_p$ which lies over such a fixed point of F. Evidently F lifts to a unique holomorphic mapping $\tilde{F}$ of $\tilde{M}_p$ into itself which fixes the point b. The key step of our proof is the following observation. <u>If we give the manifold $\tilde{M}_p$ the Teichmüller metric, then this map $\tilde{F}$ from $\tilde{M}_p$ into itself must be strictly distance decreasing. Hence $\tilde{F}$ can have no other fixed points</u>.

If we assume this last statement for a moment, then the proof of 13.4 can be easily completed as follows. Let q be a real quadratic map with a superattractive orbit of period p. Let $\Delta \subset M_p$ be the open $(p-2)$-simplex consisting of all p-tuples $(x_1,\ldots,x_p)$ of distinct real numbers, well defined up to a simultaneous linear substitution, which are ordered in the same way as the base p-tuple $(q(c),\ldots,q^p(c) = c)$. We may identify Δ with a subset of the Teichmüller space $\tilde{M}_p$, containing the basepoint b. Then $\tilde{F}$ maps Δ into itself. If there were another real quadratic map q' having the same kneading invariant as q, and also having a superattractive periodic orbit, then it would determine another fixed point of $\tilde{F}$ in Δ. Since there is only one such fixed point, it follows that q' and q must be linearly conjugate to each other.

In order to prove that $\tilde{F}$ reduces Teichmüller distance, we must first recall the definition of this distance. First consider a diffeomorphism $f : z \longmapsto z'$ between two Riemannian 2-manifolds. If we fix some tangent direction at a point of the first manifold, then the mapping stretches distances (at this point and in this direction) by a factor of $ds'/ds > 0$. The ratio $K(z) = \sup(ds'/ds)/\inf(ds'/ds)$, taking the sup and the inf over all directions through some given point z, is called the dilatation of f at z. A map f, which is a diffeomorphism except possibly at a finite number of points, is called admissible (or quasi-conformal) if K(z) is bounded, so that the number $K[f] = \sup_z K(z)$ is finite. The smallest possible value of $\log K[f]$ as f varies over a homotopy class of maps is called the Teichmüller distance associated with this homotopy class. In the hyperbolic case, a basic theorem of Teichmüller asserts that the minimum dilatation is always realized by one and only one mapping which has the following description (see for example Abikof, p. 59). There is an essentially unique quadratic differential $\varphi(z)dz^2$, which is holomorphic except possibly for simple poles at the puncture points, so that f stretches distances by a maximum factor in the positive directions associated with this quadratic differential, and by a minimum factor in the orthogonal negative directions, the ratio of these factors being exactly K[f] everywhere.

For our purposes, the two Riemann surfaces should be the complex plane $\underset{\sim}{C}$, punctured at the infinite point and either at the points $z_1,\ldots,z_p$ or at corresponding points $z'_1,\ldots,z'_p$. The mapping F assigns to each of these punctured Riemann surfaces a new punctured Riemann surface which can be described geometrically as follows. First form the 2-fold covering, branched only at the critical value z_1 (or z'_1) and at the infinite point. Each of the remaining puncture points has two distinct representatives in this 2-fold covering. We must fill in one of these two puncture points, but keep the other one. (There is a choice here, so that F is not a single-valued function.) Finally, we must renumber the finite puncture points cyclically.

Let us start with the unique admissible map f from $\underset{\sim}{C} - \{z_1,\ldots,z_p\}$ to $\underset{\sim}{C} - \{z'_1,\ldots,z'_p\}$, within a specified homotopy class, which minimized the maximum dilatation. Now pass to 2-fold branched covering surfaces, as described above. Evidently f lifts to an admissible map $\tilde{f}$ between these branched coverings, with the same constant dilatation K[f]. <u>This proves that $\tilde{F}$ can never increase distance</u>.

To see that $\tilde{F}$ must actually decrease Teichmüller distance, we must show that the lifted map $\tilde{f}$ cannot be the unique best possible map described in Teichmüller's theorem. Evidently the map $\tilde{f}$ is associated with a quadratic differential, obtained by lifting the given quadratic differential on $\underset{\sim}{C} - \{z_1,\ldots,z_p\}$ to the 2-fold covering. Recall that every quadratic differential on a surface of genus zero must have at least four poles. In general, these poles must be counted with the appropriate multiplicity, but in our case only simple poles can occur. Hence there must be at least two poles at the non-branch points $z_2,\ldots,z_p$. Each of these will be covered by two poles on the covering surface, one at a puncture point and one at an ordinary point. Thus this quadratic differential on the covering surface has poles at ordinary (nonpuncture) points, which proves that its associated mapping is not the best possible mapping which minimizes the maximal dilatation. Hence the Teichmüller distance between these two marked covering surfaces must be strictly less than the Teichmüller distance between the two original surfaces. ∎

Proof of Theorem 13.1. As the parameter a increases from -1 to 8, we must show that kneading invariant $D(f_a)$ decreases monotonely from $1+t+t^2+\ldots$ to $1-t-t^2-\ldots$. Suppose to the contrary that $D(f_a) < D(f_b)$ for some $a < b$. Since $D(f_a) \neq D(f_b)$, there must clearly exist some parameter value c between a and b so that f_c has a superattractive periodic orbit. Then either $D(f_c) < D(f_b)$ or $D(f_c) > D(f_a)$ or both. In the first case, according to the intermediate value theorem 12.2, there exists a parameter value $d > b > c$ with $D(f_d) = D(f_c)$. Furthermore, the proof constructs this parameter value d so that f_d also has a superattractive periodic orbit. But this is impossible by 13.4. Since the argument in the second case $D(f_c) > D(f_a)$ is completely analogous, this completes the proof of 13.1. ∎

Proof of Corollary 13.2. If two quadratic maps f and g satisfy $D(f) > D(g)$, using the usual total ordering of the formal power series ring $Z[[t]]$, then we assert that $\ell(f^n) \le \ell(g^n)$ for every n. For, as h varies continuously from f to g, the number $\ell(h^n)$ can change only as we pass through a map having a superattractive orbit of period less than n. Using 11.6 and 5.8, we see that $\ell(h^n)$ can only increase. Evidently, the corollary follows. ∎

To conclude this section, here is a table showing some actual kneading invariants, together with the associated parameter values a and growth numbers s. In order to make a short list, we have singled out those admissible kneading invariants which are periodic, with period $p \le 6$.

Table 13.5. <u>List of admissible kneading determinants of period</u> $p \le 6$.

parameter a in the interval	coefficients of periodic kneading determinant	period p	growth number s
[-1 , 0]	+ + + + + + + + ...	1	
(0 , 4]	+ - + - + - + - ...	2	1
(4 , 5.242..]	+ - - + + - - + ...	4	
***			
[5.898.., 5.904..]	+ - - + - + + - ...	6	1.272..
***			
[6.497.., 6.501..]	+ - - + - + - - ...	5	1.512..
***			
[7 , 7.019..]	+ - - + - - + - ...	3	1.618..
(7.019.., 7.091..]	+ - - - + + + - ...	6	
***			
[7.442.., 7.443..]	+ - - - + + - - ...	5	1.722..
***			
[7.629.., 7.629..]	+ - - - + - + - ...	6	1.792..
***			
[7.762.., 7.763..]	+ - - - + - - - ...	4	1.839..
***			
[7.867.., 7.867..]	+ - - - - + + - ...	6	1.883..
***			
[7.941.., 7.941..]	+ - - - - + - - ...	5	1.927..
***			
[7.985.., 7.985..]	+ - - - - - + - ...	6	1.965..
***			

This table is to be interpreted as follows: An entry of say "+ - - + - - + - ..." of period 3 stands for the periodic series $(1-t-t^2)/(1-t^3)$. Note that there are infinitely many other periodic admissible series with period $p > 6$ in every one of the starred gaps in this table. (The precise number of admissible series of period p is given by the expression $(2p)^{-1}\sum\mu(d)2^{p/d}$, to be summed over all squarefree odd divisors of p. Here $\mu(d) = \pm 1$ is the Moebius function, equal to +1 or -1 according as the number of primes dividing the squarefree number d is even or odd. This expression grows exponentially with p.)

Note that the lengths of the associated parameter intervals tend to zero very rapidly for large p. Thus, in several cases the three decimal places listed are not enough to distinguish the two endpoints of an interval.

14. Constructing new functions out of old

This section will briefly describe several ways of constructing interesting maps f from an interval to itself. Proofs will be largely left to the reader. We will use the notation D(f,t) to indicate the dependence of the determinant D(t) on the map f, and the notation L(f,t) for the power series $\sum\ell(f^k)t^{k-1}$.

Example 14.1 Juxtaposition. Suppose that $f : I \longrightarrow I$ maps the left half of I to itself by a map f_1, and maps the right half to itself by a map f_2 (compare Figure 7.5). Then the lap number $\ell(f^k)$ is evidently equal to $\ell(f_1^k) + \ell(f_2^k) - 1$, hence

$$L(f,t) = (L(f_1,t) + L(f_2,t) - (1-t)^{-1}.$$

It follows that the growth number s(f) is equal to the maximum of the two numbers $s(f_1)$ and $s(f_2)$. Similarly, inspection shows that $\hat{n}(f^k) = \hat{n}(f_1^k) + \hat{n}(f_2^k) - 1$ and $N(f^k) = N(f_1^k) + N(f_2^k) + 1$, hence

$$\hat{\zeta}(f,t) = \hat{\zeta}(t_1,t)\hat{\zeta}(f_2,t)(1-t),$$

$$Z(f,t) = Z(f_1,t)Z(f_2,t)/(1-t).$$

The corresponding formula

$$D(f,t) = D(f_1,t)D(f_2,t)(1-t)$$

for the kneading determinant then follows from 10.1. Alternatively, inspection shows that the kneading matrix of f takes the form

$$\left[\begin{array}{c|c|c} A & \begin{matrix} * \\ \vdots \\ * \end{matrix} & 0 \\ \hline 0 & \begin{matrix} * \\ \vdots \end{matrix} & B \end{array}\right]$$

Deleting the common column and applying 4.2, we obtain a different proof of the same formula for $D(f,t)$

<u>Example 14.2 Iteration</u>. Given any f, we can think of the 2-fold iterate $f^2 : I \longrightarrow I$ as a mapping in its own right. Evidently $s(f^2) = s(f)^2$. In order to describe the power series $tL(f^2,t) = \sum \ell(f^{2k})t^k$, it is convenient to substitute t^2 in place of t. Then inspection shows that

$$t^2L(f^2,t^2) = \sum \ell(f^{2k})t^{2k}$$

is equal to the average $(tL(f,t)-tL(f,-t))/2$. Similarly, a completely formal argument shows that

$$\hat{\zeta}(f^2,t^2) = \hat{\zeta}(f,t)\hat{\zeta}(f,-t), \quad \text{and} \quad Z(f^2,t^2) = Z(f,t)Z(f,-t).$$

Using 10.1, it follows that

$$D(f^2,t^2) = D(f,t)D(f,-t).$$

As an example, starting with the quadratic map $f(x) = (x^2-7)/2$, we obtain a quadratic map $f^2(x) = (x^4-14x^2+21)/8$. Since $D(f,t) = (1-t-t^2)/(1-t^3)$, it follows by multiplying out this formula that $D(f^2,t^2) = (1-3t^2+t^4)/(1-t^6)$, or in other words $D(f^2,t) = (1-3t+t^2)/(1-t^3)$.

Similarly, for any iterate f^k, an analogous argument based on 10.1 shows that

$$D(f^k,t^k) = D(f,\omega_1 t)\ldots D(f,\omega_k t),$$

where $\omega_1,\ldots,\omega_k$ are the k-th roots of unity. This computation can be expressed in a more perspicuous form if $D(f,t)$ happens to be a rational function of t. In this case we can write

$$D(f,t) = \Pi_j(1-a_j t)^{m_j},$$

where the a_j may be complex. It follows easily that

$$D(f^k,t) = \Pi_j(1-a_j^k t)^{m_j}. \tag{14.3}$$

As an example, the function f illustrated in Figure 1.5 has rational kneading determinant $D(f,t) = (1-3t)/(1-t^2)$. Using 14.3, it follows that

$$D(f^k,t) = \begin{cases} (1-3^k t)/(1-t^2) & \text{for } k \text{ odd} \\ (1-3^k t)/(1-t)^2 & \text{for } k \text{ even.} \end{cases}$$

<u>Example 14.4 Period Doubling</u>. Let us start with a function $f : [0,1]\longrightarrow[0,1]$ which maps both boundary points to zero. Define an associated map $g : [-2,2]\longrightarrow[-2,2]$, as illustrated in Figure 14.5, by setting

$$g(x) = \begin{cases} 2x+2 & \text{for } x \in [-2,-1] \\ f(-x) & \text{for } x \in [-1,0] \\ -x & \text{for } x \in [0,2] \end{cases}.$$

Inspection shows that $g^2(x) = f(x)$ for $x \in [0,1]$. The periodic points of g can be described very precisely as follows.

(1) There are just two fixed points, at -2 and 0.

(2) Each periodic point $x \neq 0$ of f with period p gives rise to periodic points $\pm x$ of g of period $2p$. There are no other periodic points.

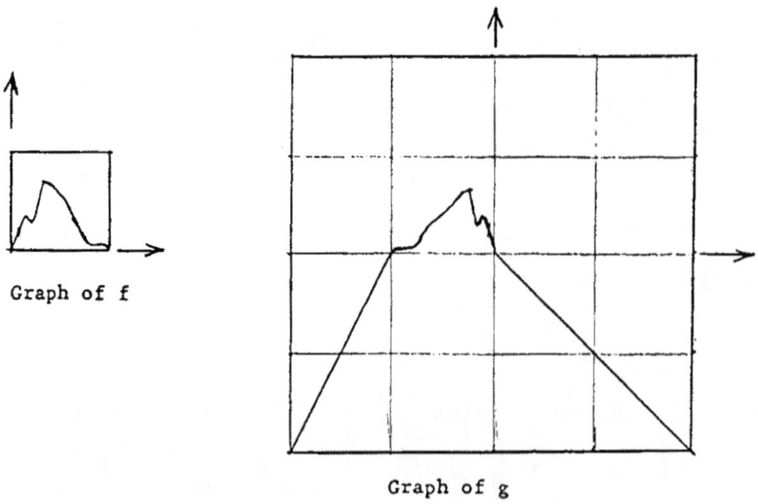

Figure 14.5

It follows easily that

$$\hat{\zeta}(g,t) = \hat{\zeta}(f,t^2)(1-t)/(1-t) \quad \text{and} \quad Z(g,t) = Z(f,t^2)/(1-t).$$

Using 10.1 or by direct computation, we find that

$$D(g,t) = (1-t)D(f,t^2).$$

It follows that $s(g) = s(f)^{1/2}$.

As an example, if $D(f,t) = (1-t-t^2)/(1-t^3)$, then $D(g,t) = (1-t)(1-t^2-t^4)/(1-t^6)$. (By §12, we can find a quadratic map with this kneading determinant.)

<u>Example 14.6</u>. After making a linear change of coordinates to replace the interval $[-2,2]$ by $[0,1]$, we can iterate this construction to obtain a sequence of maps $f_0 = f$, $f_1 = g$, f_2, $f_3, \ldots$ from $[0,1]$ to itself, with

$$D(f_1,t) = (1-t)D(f,t^2)$$

$$D(f_2,t) = (1-t)(1-t^2)D(f,t^4),$$

$$D(f_3,t) = (1-t)(1-t^2)(1-t^4)D(f,t^8),$$

and so on. Note that the growth numbers $s(f_i) = s(f)^{1/2^i}$ tend to 1 as $i \longrightarrow \infty$. The limit function f_∞, illustrated in Figure 14.7, is completely independent of f. Its kneading determinant is equal to infinite product

$$D(f_\infty,t) = (1-t)(1-t^2)(1-t^4)(1-t^8)\ldots\ .$$

Evidently this infinite product converges, and is nonzero, whenever $|t| < 1$. Hence $s(f_\infty) = 1$. Noteworthy properties of this limit function are the following (compare 1.5).

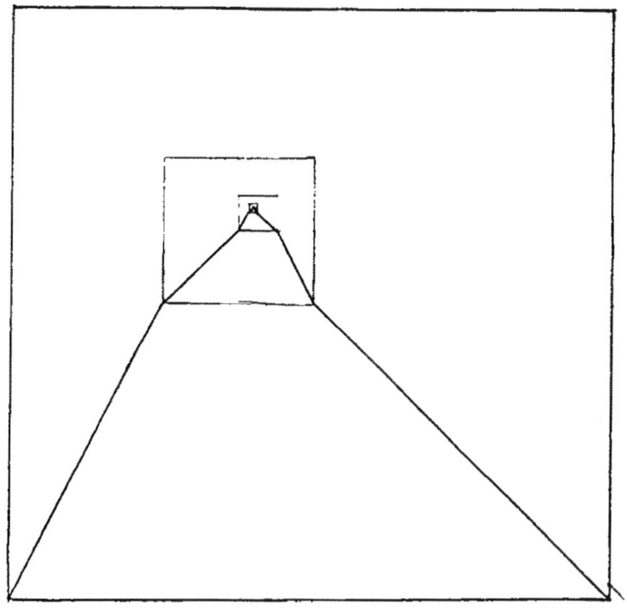

Figure 14.7. Graph of f_∞.

(1) The reciprocal zeta function $\zeta(f_\infty,t)^{-1} = \hat{\zeta}(f_\infty,t)^{-1}$ is equal to $(1-t)^2(1-t^2)(1-t^4)(1-t^8)\ldots$. In other words, there is just one periodic orbit 2^k for each $k \geq 1$, and there is no periodic point at all whose period is not a power of 2 (compare Šarkovskii's theorem in §8).

(2) From 9.7 or by direct computation we see that the sequence of lap numbers $\ell(f_\infty^k)$ grows faster than any polynomial function of k as $k \to \infty$. Yet this sequence does not have exponential growth since $s = 1$.

(3) Restricting t to the interval $[0,1]$, the function $D(t) = (1-t)(1-t^2)(1-t^4)\ldots$ clearly tends to zero faster than any power of $1-t$ as $t \to 1$. The same is true for $D(\omega t)$, where ω is any 2k-throot of unity. Thus the complex analytic function $D(t)$ has an essential singularity at every dyadic root of unity, and hence at evert point of the unit circle. In other words $D(t)$ cannot be extended as an analytic fucntion outside of the disk $|t| < 1$. Similarly, $\zeta(f_\infty,t)$ cannot be extended. The non-differentiability of the map f_∞ is of course not a necessary property, but is rather an artifact of this particular method of construction. According to 12, there exists a real number a so that the quadratic map $x \mapsto (x^2-a)/2$ has precisely the same kneading determinant $D(t) = (1-t)(1-t^2)(1-t^4)\ldots$ (compare Feigenbaum). This number $a = 5.6046\ldots$ is conjectured to be unique.

Example 14.8 Microimplantation of functions. The construction in 13.4 can be generalized as follows. Suppose that we are given $e : I \to I$ and $f : J \to J$ with the following properties. The map e should have just one turning point c_1, and this turning point should be periodic of period p, so that $e^p(c_1) = c_1$. Furthermore f should map the two and points of J to a single end point. Then we will construct a new map $g : I \to I$ by implanting a small copy of f into e.

Suppose, to simplify the notation, that $c_1 = 0$. Note that e^{p-1} maps a neighborhood of $e(0)$ homeomorphically onto a neighborhood of 0. It is not difficult to construct an auxiliary map E, also with lap number $\ell = 2$, which coincides with e outside of a small neighborhood of 0, but satisfies either $E^p(x) = |x|$ or $E^p(x) = -|x|$ for all x within some smaller neighborhood $-\delta \le x \le \delta$. Evidently E and e have the same kneading matrix.

Construct g as follows. Let $g(x) = E(x)$ for $|x| \ge \delta$, and let g^p map $[-\delta,\delta]$ to itself by a mapping which is linearly conjugate to f. Evidently g is uniquely determined by this description and has the same lap number as f. Each periodic orbit of E which does not intersect the interval $[-\delta,\delta]$ remains periodic for g. Similarly, each periodic orbit of f with period q gives rise to a periodic orbit of g with period pq. it is not difficult to verify that

$$N(g^k) = \begin{cases} N(e^k) & \text{if } k \not\equiv 0 \bmod p \\ N(e^k) + N(f^{k/p}) + 1 & \text{if } k \equiv 0 \bmod p. \end{cases}$$

Hence

$$Z(g,t) = Z(e,t)Z(f,t^p)/(1-t^p)$$

and

$$D(g,t) = (1-t^p)D(e,t)D(f,t^p). \tag{14.9}$$

(Note that the product $(1-t^p)D(e,t)$ is a polynomial of degree $p-1$ with all coefficients ± 1, by 4.5.)

<u>It follows that the growth number</u> $s(g)$ <u>is just the maximum of the two numbers</u> $s(e)$ <u>and</u> $s(f)^{1/p}$.

As an example, the quadratic mapping $e(x) = (x^2-r)/2$ satisfies $e^2(0) = 0$, with $p = 2$ and $(1-t^2)D(e,t) = 1-t$. In this case we recover the formula

$$D(g,t) = (1-t)D(f,t^2)$$

of 14.4, with $s(g) = \sqrt{s(f)}$.

On the other hand, if the period p is not a power of 2, and if $\ell(f) = \ell(g) = 2$, then according to Bowen and Franks we have

$$s(e) > 2^{1/p} \geq s(f)^{1/p},$$

hence $s(g) = s(e)$. Thus, in this case, we can construct uncountably many distinct admissible kneading determinants, all with the same growth number $s(e)$.

Here is an example with $p = 3$. Let $e(x) = (x^2-7.019\ldots)/2$ where the constant $7.019\ldots$ is to be chosen so that $e^3(0) = 0$. In this case, $(1-t^3)D(e,t) = 1 - t - t^2$, so the formula becomes

$$D(g,t) = (1-t-t^2)D(f,t^3),$$

with $s(g) = s(e) = (1+\sqrt{5})/2$. Just as in 14.6, this construction can be iterated infinitely often, thus constructing a map with this same growth number whose kneading determinant

$$D = (1-t-t^2)(1-t^3-t^6)(1-t^9-t^{18})\ldots$$

has infinitely many zeros within the unit disk. Again by §12, there exists a (presumably unique) real number $a = 7.145\ldots$ so that the quadratic map $(x^2-a)/2$ has this same kneading determinant.

References

Abikoff, W.: The real analytic theory of Teichmüller space, Lect. Notes Math. 820, Springer, 1980.

Abraham, R., Robbin, J.: Transversal Mappings and Flows, Benjamin, 1967.

Adler, R., Konheim, A., McAndrew, M.: Topological entropy, Trans. Amer. Math. Soc. 114, 309-319 (1965).

Ahlfors, L.: Complex Analysis, McGraw-Hill, 1966.

Allwright, D.J.: Hypergraphic functions and bifurcations in recurrence relations, SIAM J. Appl. Math. 34, 687-691 (1978).

Artin, M., Mazur, B.: On periodic points, Annals of Math. 81, 82-99 (1965).

Block, L.: Noncontinuity of topological entropy of maps of the Cantor set and of the interval, Proc. Amer. Math. Soc. 50, 388-393 (1975).

Block, L.: Mappings of the interval with finitely many periodic points have zero entropy, Proc. Amer. Math. Soc. 67, 357-360 (1977).

Block, L.: An example where topological entropy is continuous, Trans. Amer. Math. Soc. 231, 201-213 (1977).

Block, L.: Continuous maps of the interval with finite nonwandering set, Trans. Amer. Math. Soc. 240, 221-230 (1978).

Block, L., Guckenheimer, J., Misiurewicz, M., Young, L.-S.: Periodic points and topological entropy of one dimensional maps, pp. 18-34 of Global Theory of Dynamical Systems, ed. Nitecki and Robinson, Lecture Notes in Math. 819, Springer 1980.

Bowen, R.: Topological entropy and Axiom A, pp. 23-41 of Global Analysis, ed. Chern and Smale, Proc. Symp. Pure Math. 14, A.M.S., 1970.

Bowen, R.: Entropy for maps of the interval, Topology 16, 465-467 (1977).

Bowen, R.: On Axiom A Diffeomorphisms, Notes of NSF Regional Conference, North Dakota State Univ., Fargo, N.D., 1977.

Bowen, R.: Bernoulli maps of the interval, Israel J. Math. 28, 161-168 (1977).

Bowen, R., Franks, J.: The periodic points of maps of the disk and the interval, Topology 15, 337-442 (1976).

Bowen, R., Lanford, O.E., III: Zeta functions of restrictions of the shift transformation, pp. 43-49 of Global Analysis, ed. Chern and Smale, Proc. Symp. Pure Math. 14, A.M.S., 1970.

Collet, P., Eckmann, J.-P.: Iterated Maps of the Interval as Dynamical Systems, Progress in Physics 1, Birkhauser, 1980.

Fatou, M.P.: Sur les équations fonctionnelles, Bull Soc. Math. France 47, 161-271 (1919); 48, 33-94, 208-314 (1920).

Feigenbaum, M.: Quantitative universality for a class of nonlinear transformations, J. Statist. Phys. 19 22-52 (1978); 21, 669-706 (1979).

Feigenbaum, M.: The transition to aperiodic behavior in turbulent systems, Commun. Math. Phys. 77, 65-86 (1980).

Goodman, T.N.T.: Relating topological entropy and measure entropy, Bull. London Math. Soc. 3, 176-180 (1971).

Guckenheimer, J.: Endomorphisms of the Riemann sphere, pp. 95-123 of Global Analysis, ed. Chern and Smale, Proc. Symp. Pure Math. 14, A.M.S. (1970).

Guckenheimer, J.: On the bifurcation of maps of the interval, Invent. Math. 39, 165-178 (1977).

Guckenheimer, J.: Bifurcations of dynamical systems, C.I.M.E. Lectures, 1978.

Guckenheimer, J.: Sensitive dependence on initial conditions for one-dimensional maps, Comm. Math. Phys. 70, 133-160 (1979).

Guckenheimer, J.: The growth of topological entropy for one-dimensional maps, pp. 216-223 of Global Theory of Dynamical Systems, ed. Nitecki and Robinson, Lecture Notes in Math. 819, Springer 1980.

Jakobson, M.V.: Structure of polynomial mappings on a singular set, Mat. Sbornik 77, 105-124 (1968) (= Math. USSR Sb. 6, 97-114 (1968)).

Jakobson, M.V.: On smooth mappings of the circle into itself, Mat. Sb. 85, 163-188 (1971) (= Math. USSR Sb. 14, 161-185 (1971)).

Jakobson, M.V.: Absolutely continuous invariant measures for one-parameter families of one-dimensional maps, Com. Math. Phys. 81, 39-88 (1981).

Jonker, L.: Periodic orbits and kneading invariants, Proc. London Math. Soc. 39, 428-450 (1979).

Jonker, L.: A monotonicity theorem for the family $f_a(x) = a - x^2$, Proc. A.M.S. 85, 434-436 (1982).

Jonker, L., Rand, D.: A lower bound for the entropy of certain maps of the unit interval, preprint, Kingston, Ontario, and Univ. of Warwick (1978).

Jonker, L., Rand, D.: The periodic orbits and entropy of certain maps of the unit interval, J. London Math. Soc. (2), 22, 175-181 (1980).

Jonker, L., Rand, D.: Bifurcations in one dimension; I, The nonwandering set, Invent. Math. 62, 347-365 (1981); II, A versal model for bifurcations, Invent. Math. 63, 1-15 (1981).

Julia, G.: Mémoire sur l'iteration des fonctions rationelles, J. de Math. (Liouville), ser. 7, 4, 47-245 (1918).

Kolata, G.B.: Cascading bifurcations: the mathematics of chaos (news article), Science 189, 984-985 (1975).

Lanford, O.E.: Smooth transformations of intervals, Séminaire Bourbaki 563 (1980/81), Lecture Notes Math. 901, Springer 1981.

Lanford, O.E.: A computer-assisted proof of the Feigenbaum conjectures, Bull. A.M.S. 6, 427-434 (1982).

Lang, S.: Differential Manifolds, Addison-Wesley, 1972.

Li, T., Yorke, J.A.: Period three implies chaos, Amer. Math. Monthly 82, 985-992 (1975).

Lorenz, E.: On the prevalence of aperiodicity in simple systems, pp. 53-57 of Global Analysis, ed. Grmela and Marsden, Lecture Notes in Math. 755, Springer 1979.

May, R.M.: Biological populations obeying difference equations: stable points, stable cycles and chaos, J. Theor. Biol. 51, 511-524 (1975).

May, R.M.: Simple mathematical models with very complicated dynamics (review article), Nature 261, 459-467 (1976).

Metropolis, N., Stein, M.L., Stein, P.R.: On finite limit sets for transformations on the unit interval, Journal of Combinatorial Theory (A) 15, 25-44 (1973).

Misiurewicz, M., Szlenk, W.: Entropy of piecewise monotone mappings, Asterisque 50, 299-310 (1977) and Studia Math. 67, 45-63 (1980).

Oster, G., Guckenheimer, J.: Bifurcation phenomena in population models, pp. 327-353 of "The Hopf Bifurcation and its Applications," by Marsden and McCracken, Springer 1976.

Parry, W.: Symbolic dynamics and transformations of the unit interval, Trans. Amer. Math. Soc. 122, 368-378 (1966).

Rothschild, J.: On the computation of topological entropy, Thesis, CUNY, 1971.

Ruelle, D.: Applications conservant une mesure absolument continue par rapport a dx sur [0,1], Comm. Math. Phys. 55, 47-51 (1977).

Šarkovskii, A.N.: Coexistence of cycles of a continuous map of a line into itself (Russian), Ukr. mat. Ž. 16, 61-71 (1964).

Singer, D.: Stable orbits and bifurcation of maps of the interval, SIAM J. Appl. Math. 35, 260-267 (1978).

Smale, S., Williams, R.: The qualitative analysis of a difference equation of population growth, J. Mathematical Biology 3, 1-4 (1976).

Štefan, P.: A theorem of Šarkovskii on the existence of periodic orbits of continuous endomorphisms of the real line, Com. Math. Phys. 54, 237-248 (1977).

Straffin, P.D., Jr.: Periodic points of continuous functions, Math. Mag. 51, 99-105 (1978).

Targonski, G.: Topics in Iteration Theory (Studia Math.: Skript 6), Vandenhoeck u. Ruprecht, Göttingen 1981.

Weil, A.: Numbers of solutions of equations in finite fields, Bull. Amer. Math. Soc. 55, 497-508 (1949).

Williams, R.F.: The zeta function in global analysis, pp. 335-339 of Global Analysis, ed. Chern and Smale, Proc. Symp. Pure Math. 14, A.M.S. 1970.

On the dynamics of iterated rational maps.

William P. Thurston

1. Introduction

The dynamical behaviour of rational maps of the Riemann sphere under iteration is intricate and interesting. From the blossoming of this subject over the past few years, it is evident that it has many areas where an analysis can be pushed amazingly far — and yet, a complete understanding of iterated rational maps is nowhere in sight.

What is the mathematical interest of this subject, why is it worthwhile to study?

First, the internal test: this subject is beautiful, rich, and deep, and whether or not it has immediate applications to the outside, it brings in many outside theories for its own development. This is an important criterion, for very often a mathematical theory seems to bear edible fruit in proportion to its beauty, depth and richness.

Second, it can act to nurture our understanding of dynamical systems. Dynamical systems of many different sorts are everywhere, and many of them are quite important. Unfortunately, many of these dynamical systems are also very complicated, so that they are hard to understand in a systematic way and hard to predict. Here is a class of examples which are complicated, yet simple enough that we can develop many insights.

Third, there are many close parallels between the theory of iterated rational maps and the theory of Kleinian groups, which are intimately connected with three-dimensional topology. These parallels are aiding the development of both subjects, since some things which are known or which are easy to prove in one subject can point the way to an analysis in the other subject.

Fourth, there is a special (yet not so special) particular case of this theory which is iteration of Newton's method for finding a root of a polynomial. Newton's method is an extremely valuable and widely used numerical technique, but its use has been more of an art than a science — the question of how you set it up to find the root that you are interested in, rather than a random root or no root at all, is tricky. The intricate structure that is revealed by the systematic study of iterated rational maps should eventually lead to a sounder theory, and more efficient methods.

Finally, for many people, the major attraction of the subject is simply the visual and artistic appeal of the computer-generated pictures that can be, and have been, produced. There are pictures which map out the dynamical behavior of points in the complex plane, for individual rational maps. For instance, when iterating a polynomial, the most fundamental distinction among points is whether or not they tend to infinity; one kind of picture colors black those points which do not go to infinity, and white those which do. Another kind of picture maps the presence or absence of some property (or set of properties) as you range over the parameter space for a family of rational maps. These computer graphics studies have given rise to a number of interesting problems concerning computation, as well as many mathematical conjectures and problems. The availability of computers probably is what provided the impetus for the modern rebirth of interest in rational maps, after the subject remained mostly quiescent since the time of the original work by Julia and Fatou in 1918.

2. Preliminaries: uniformization and the Poincaré metric

In this section, we will summarize some of the fundamental background in complex analysis which will be useful for the theory of rational maps. There are many alternate sources for this material.

2.1 **Definition.** A *Riemann surface* is a (real) 2-manifold, equipped with a 1-dimensional complex structure. In other words, it is a space which is locally modeled on **C**, with holomorphic transition maps between local coordinate systems. A Riemann surface is the same as a complex curve, and often just called a curve; it is not to be confused with a complex surface, which in reality has four dimensions. The use of the words "curve" and "surface" is a common place for confusion in conversations between people with different background.

We will be dealing a lot with a very special Riemann surface, the Riemann sphere. There are several ways to think of the Riemann sphere:

(i) The Riemann sphere is the one-point compactification of **C** (the complex numbers), with the coordinate system $\frac{1}{z}$ defining the complex structure in a neighborhood of the added point, which is named ∞.

(ii) The Riemann sphere is an ordinary round sphere in euclidean 3-space $\mathbf{E}^3$. Rotations of the sphere are holomorphic. The complex structure is determined by a covering by a set of coordinate maps to **C**, defined by stereographic projection to the $(x-y)$-plane (which is identified with **C**), which may be preceded by an arbitrary rotation.

Riemann sphere

2.2 Figure. *It is often best to think of the Riemann sphere as the ordinary round sphere in Euclidean space. All rigid motions of the Riemann sphere are holomorphic.*

(iii) The Riemann sphere is complex projective 1-space, $\mathbf{CP}^1 = (\mathbf{C}^2 - 0)$ *mod* $\mathbf{C}_*$ (where $\mathbf{C}_*$ denotes the group of multiplicative complex numbers.)

The relation of (i) with (iii) is that the ratio $z = z_1/z_2$ is an invariant of an element $(z_1, z_2) \in \mathbf{C}^2$ up to scalar multiplication. This defines a holomorphic map of $\mathbf{CP}^1$ to the one point compactification of $\mathbf{C}$. Projective transformations of $\mathbf{CP}^1$ are the transformations inherited from linear transformations of $\mathbf{C}^2$; in terms of the coordinate z, the projective transformation inherited from the linear transformation

$$\begin{bmatrix} z_1 \\ z_2 \end{bmatrix} \longmapsto \begin{bmatrix} a & b \\ c & d \end{bmatrix} \begin{bmatrix} z_1 \\ z_2 \end{bmatrix}$$

is the fractional linear transformation or Moebius transformation

$$z \longmapsto \frac{az + b}{cz + d}$$

All holomorphic self-homeomorphisms of the Riemann sphere have the form of Moebius transformations.

Another very important Riemann surface is the open unit disk, D^2 in $\mathbf{C}$ (which corresponds to the southern hemisphere on the Riemann sphere). The subgroup of Moebius transformations which preserve the unit disk preserve a certain Riemannian metric, the Poincaré metric

$$ds^2 = \frac{4(dx^2 + dy^2)}{(1 - x^2 - y^2)^2}$$

This metric has constant negative curvature -1 , so that it is identified with the hyperbolic plane (or Lobachevskiian plane). Orientation-preserving hyperbolic congruences are the same as Moebius tranformations which preserve the unit disk. In order to actually calculate with this group, it is usually convenient to transform by a Moebius transformation sending the unit disk to the half-plane ($y>0$); the group of Moebius transformations preserving the upper half-plane are exactly those which can be written with real coefficients.

The Poincaré metric has an elementary but extremely convenient property:

2.3, Proposition: Self-maps of the disk shrink. *A holomorphic map f of the unit disk into itself cannot expand the Poincaré metric. In other words, for any points w and z,*

$$d(w,z) \leq d(f(w),f(z)) .$$

The inequality is strict unless f is a Moebius transformation.

Proof, 2.3, Self-maps of the disk shrink: This is proved on the infinitesimal level, by showing that the Poincaré length of a tangent vector V to the disk is at least as great as the Poincaré length of its image $f(V)$ by the derivative of f.

If it happens that V is a vector based at the origin, and its image is also based at the origin, this follows immediately from the Cauchy integral formula for the derivative of f — the derivative at the origin can never have modulus greater than 1. The derivative has modulus 1 only if f is a rotation of the disk, which is a special case of a Moebius transformation.

To take care of the general case, change coordinates in both the domain and range by Moebius tranformations, so that the target and source points are both at the origin.

— **2.3, Self-maps of the disk shrink**

2.4 **Definition.** A family of mappings from one metric space to another is *equicontinuous* if for each x in the domain and for each ϵ there exists a δ such that for any map f in the family,

$$d(x,y) < \delta \Rightarrow d(f(x),f(y)) < \epsilon.$$

This depends on the topology of the domain, not the particular metric, and if the range of the family is compact, it is independent of the metric in the range. If the domain is compact, the choice of $\delta(\epsilon)$ need not depend on x.

When the range is not compact, there is often more than one useful metric — for instance, in the case of the open disk, both the Poincaré metric and the Euclidean metric induced from **C** are useful. It is a very significant observation that equicontinuity with respect to the Poincaré metric implies equicontinuity with respect to the Euclidean metric, since Poincaré distances are greater than a fixed constant times Euclidean distances.

From proposition **Self-maps of the disk shrink, 2.3** it follows that the set of holomorphic maps of the disk to itself is equicontinuous with respect to the Poincaré metric. Of course, it is equicontinuous with respect to the Euclidean metric as well.

The concept of equicontinuity is useful because of the Ascoli theorem:

2.5, Proposition: Equicontinuity implies compactness. *Let X and Y be metric spaces with countable bases, and let F be an equicontinuous family of maps from X to Y such that for some point $x_0 \in X$, the closure of the set of image points of x_0 under maps in F is compact. Then the closure of F in (say) the topology of pointwise convergence among the set of all continuous maps $X \to Y$ is compact.*

There is a beautiful theory of "uniformization" of Riemann surfaces, which strongly influences the way that we see them — it basically says that any Riemann surface has a canonical geometric form, so that many sorts of questions about Riemann surfaces and holomorphic maps can be analyzed in terms of their topology, plus their geometry. The first step is the well-known

2.6, Riemann mapping theorem: Riemann mapping theorem. *Let U be any connected, simply-connected open subset of the Riemann sphere. If the complement of U has more than one point, then U is conformally equivalent to the open unit disk.*

The standard modern proof of this theorem is amazingly elementary: Choose some point $u \in U$, and consider the set of embeddings $g: U \to D^2$ which send u to the origin in the disk. Look for a holomorphic embedding which maximizes the derivative at u. The key point is that for any proper open subset V of the disk, there is a re-embedding of V in the disk which strictly expands the Poincaré metric (and which can be adjusted to send the origin to the origin). Such a map is easily constructed as a branch of the inverse of any map f which is proper, not 1-1 (in other words, a branched cover of the disk over itself) but does not have any critical value in V — traditionally, f is $z \longmapsto z^2$, normalized to move its critical value out of V. Since f shrinks the Poincaré metric, its inverse must expand it. A map $g: U \to D^2$ with maximal derivative must exist because of the compactness property for holomorphic mappings which follows from **Equicontinuity implies compactness, 2.5,** together with the fact that pointwise limits of holomorphic functions are holomorphic: and it must be an isomorphism, by the argument above.

Here is another case of the uniformization theorem which can be proven in much the same way. A lot of what we will do makes use of this statement, and does not require any uniformization theorem of greater generality.

2.7, Theorem: Uniformizing-domains. *Let U be any subset of the sphere whose complement contains at least three points. Then the universal covering is conformally equivalent to the unit disk.*

Proof, 2.7, Uniformizing-domains: This time, instead of considering embeddings $U \to D^2$, consider the set E of holomorphic embeddings of pointed covering spaces of U into D^2. *Pointed* means that a base point has been chosen). To make E into a manageable mathematical object, observe that there is a map of the universal cover $\tilde{U}$ of U into D^2 associated with any e. Therefore E is is in one-to-one correspondence with the subset of the space of maps $\tilde{U} \to D^2$ which are covering spaces of their images. We will henceforth think of E as the latter set. Clearly, E is an equicontinuous family.

A limit of maps which are covering spaces of their images is also a covering space of its image, provided it is not a constant map. (In other words, E is a closed subset of the space of all maps $\tilde{U} \to D^2$.) The reason is that, first, the limit map cannot have any critical points of finite order — because any perturbation of it would still have critical points — and second, any identifications $f(x)=f(y)$ made by the limit map f must be approximated by identifications by the functions in the sequence, so any such identifications must come from some deck transformation of $\tilde{U}$ over U.

Now one needs to check that E is non-empty — this can be done by explicit construction, doing it first for the three-punctured sphere with the help of classical functions.

Given any element of E which is not surjective, it can be lifted to a branched cover of the disk (using the square-root function, for instance). After normalizing so that the base-point goes to the origin, the new element of E has a greater first derivative at the origin.

Therefore, there is some element of E which takes the base point of E to the origin and maximizes the derivative among all elements of E which take the base point to the origin. It is surjective, and since the image is simply-connected, the map must be an isomorphism of the universal cover of U with the disk.

— **2.7, Uniformizing-domains**

Note that the universal covering of the sphere minus two points is conformally equivalent to the sphere minus one point, via the logarithm (or in the other direction, the exponential map).

Here is a stronger form of the theorem:

2.8, Theorem: Uniformization theorem.

i) *A Riemannian metric on a surface plus an orientation determines a Riemann surface structure (complex structure).*

ii) *Every Riemann surface (complex curve) has universal cover conformally equivalent either to the disk, the plane, or the sphere.*

If the Riemann surface has negative Euler number, its universal cover is equivalent to the disk.

A closed Riemann surface of zero Euler number has universal cover **C**. *(Closed means that it is compact and without boundary.)*

> *A closed Riemann surface of positive Euler number has universal cover the sphere.*
>
> *Any noncompact surface of non-negative Euler number has some Riemann surface structures whose universal covers are the disk, and other structures with universal cover* **C**.

We will not attempt to explain the proof of this general version. Note, however, that it is sufficient to find some cover of the Riemann surface which embeds in the Riemann sphere, and apply the previous theorem, **Uniformizing-domains, 2.7**

The Poincaré metric transfers to a canonical metric of constant curvature -1 on any Riemann surface S whose universal cover is the disk, and this transfered metric is called the *Poincaré metric* for S. Because of the uniformization theorem, this metric has great significance – often the knowledge that this metric exists helps a lot in analyzing properties of Riemann surfaces for which we may have only indirect descriptions.

A surface equipped with a metric of constant curvature -1 is called a *hyperbolic* surface.

If the universal cover of a Riemann surface is **C**, then it does not have a canonical metric – since an affine transformation of **C** (which is a similarity of the Euclidean plane) may change the scale factor for the Euclidean metric – but the Euclidean metric is defined up to change of scale by a constant. You can think of the Euclidean metric as a limiting case of the Poincaré metric. In other words, you can represent the universal cover of the Riemann surface as an increasing union of proper simply-connected open sets. Each of these sets has an associated Poincaré metric, and the limit of these Poincaré metrics, rescaled so that they converge to something non-zero, is the Euclidean metric.

Proposition **Self-maps of the disk shrink, 2.3** immediately extends to the more general case of maps between Riemann surfaces:

2.9, Proposition: Monotonicity-of-Poincare-metric.

(i) *A holomorphic map between two Riemann surfaces whose universal covers are the disk can never increase the Poincaré metric. Unless the map is a covering map, it decreases the metric.*

(ii) *The only holomorphic maps from a Riemann surface not covered by the disk to one which is covered by the disk are the constant maps.*

(iii) *The only holomorphic maps from the Riemann sphere to anything else are constant.*

Proof, 2.9, Monotonicity-of-Poincare-metric: A continuous map between two spaces lifts to a map between their universal covers. Apply **Self-maps of the disk shrink, 2.3** to deduce part (i). Part (ii) amounts to Liouville's theorem: it follows from the fact that there are maps of the disk to the plane and the sphere with arbitrarily high derivative at the origin. Composing with a non-constant map of the plane or sphere to the disk would give maps of arbitrarily high derivative from the disk to the disk, a contradiction. Part (iii) follows from the fact that topologically, a diffeomorphism from the sphere to

$\mathbf{R}^2$ has to reverse orientation at some points (since it must have degree 0) – this is impossible for a holomorphic map.

— 2.9, Monotonicity-of-Poincare-metric

2.10, Corollary: equicontinuity of holomorphic families. *Any family of holomorphic maps from any Riemann surface to the sphere minus at least three points is equicontinuous.*

Here are a couple of properties of holomorphic maps which will be useful:

2.11, Proposition: removable-smooth-arc. *If f is any continuous function defined in an open set U which is holomorphic except possibly on the intersection of a smooth arc with U, then it is holomorphic in all of U.*

Proof, removable-smooth-arc: This is a local proposition; for concreteness, we may suppose that the open set is a disk whose boundary is cut twice by the arc, and that f is continuous on the closed disk. Let α be the intersection of the arc with the closed disk. Let g be the function defined by the Cauchy integral formula in terms of f, that is,

$$g(z) = \frac{1}{2\pi i} \int_{\partial D^2} \frac{f(w)}{w-z} dw$$

For any continuous function f defined on the circle, this formula defines a holomorphic function g. The only trouble is that, g need not agree with f on the circle (unless f extends to a holomorphic function).

Let γ_1 and γ_2 be the two curves which go around the two halves of the D_2 cut by α. Since γ_1 and γ_2 traverse α in opposite senses, we have

$$\int_{\partial D^2} \frac{f(w)}{w-z} dw = \int_{\gamma_1} \frac{f(w)}{w-z} dw + \int_{\gamma_2} \frac{f(w)}{w-z} dw.$$

For z in the part of the disk enclosed by γ_1, the integral around γ_1 agrees with f, while the integral around γ_2 is 0. Thus the holomorphic function g agrees with f on the half of the disk enclosed by γ_1, and in the other half as well, by symmetry. Consequently, f is holomorphic.

— removable-smooth-arc

There are, of course, much more general statements about removable singularities of holomorphic maps, but this will meet our needs.

2.12, Corollary: no-interval-of-constancy. *Let f be a holomorphic function on D^2. If there is any interval along ∂D^2 where f converges to a constant limiting value, then f is constant.*

Proof, no-interval-of-constancy: If f converges to a constant value along any interval, one could tack on a little neighborhood on the outside of of this interval and extend f to be constant there. By the preceding proposition, the extended f is holomorphic, hence constant everywhere.

— **no-interval-of-constancy**

2.13 **Definition.** A *rational map* is a holomorphic map of the Riemann sphere to itself.

2.14, Proposition: rational maps are ratios of polynomials. *Any rational map* $z \mapsto R(z)$ *can be expressed as the ratio of two polynomials,* $R(z) = \frac{P(z)}{Q(z)}$. *Conversely, the ratio of two polynomials always extends to a holomorphic map of the Riemann sphere.*

Proof, 2.14, rational maps are ratios of polynomials: The second half of the assertion is just a matter of expanding a rational function in terms of partial fractions, and interpreting the resulting expression in terms of the coordinate $w = \frac{1}{z}$ near any pole.

The first half is also elementary. Given any holomorphic map R from the Riemann sphere to itself, look at the set of poles and zeros of R. Let $P(z)$ be a polynomial whose zeros match the finite zeros (including multiplicities) of R, and let $Q(z)$ be a polynomial whose zeros match the finite poles of R. The ratio $P(z)/Q(z)$ is now a rational function whose poles and zeros match those of R. (You can easily check that the multiplicities work out at ∞ as well as at the finite points). The quotient $\frac{R(z)}{P(z)/Q(z)}$ is a rational map which does not take the values 0 or ∞, hence it is a constant; after an adjustment we have an expression for R(z) as the ratio of two polynomials.

— **2.14, rational maps are ratios of polynomials**

The *degree* of a rational map can be thought of either algebraically or topologically. In the algebraic form, the degree of $P(z)/Q(z)$ is the maximum of the degree of P and the degree of Q, provide P and Q are relatively prime. In topological form, the degree is the topological degree of the map of the sphere to itself — and this can be computed as the number of preimages of any point on the sphere which is not a critical value, since any holomorphic map preserves local orientation, except at critical points where this does not make sense.

It is clear that these definitions are equivalent, since the equation $P(z)/Q(z) = const$ has the right number of solutions.

A Riemann surface S has *finite type* if it can be expressed as a closed Riemann surface $\bar{S}$ minus a finite set of points. $\bar{S}$ is called the *completion* of S. The completion $\bar{S}$ is determined by S, both topologically and holomorphically: topologically, it is the end-compactification of S; the analytic structure is unique around any added point because a continuous map which is holomorphic in the complement of a point is holomorphic at the point, as well. A point in $\bar{S}$ which is missing in S is called a *cusp* or *puncture* of S.

If S is hyperbolic, its Poincaré metric near a cusp can be estimated by comparing it to the Poincaré metric of a punctured disk embedded in the surface. The universal cover of the punctured disk is formed by taking log; it becomes a half-plane, whose Poincaré metric we know. Working backward, it follows that the metric in a neighborhood of the cusp is not bigger than the metric on the narrow end of a pseudo-sphere — with the narrow end of the pseudo-sphere pointing toward the puncture. A pseudo-sphere has finite area, so it follows that a Riemann surface of finite type has finite area. Conversely, every hyperbolic surface of finite area consists of the union of a compact part with a finite number of non-compact parts isometric to the thin ends of pseudo-spheres. (See, for example, [Thurston, 1984] for a discussion of the structure of hyperbolic manifolds of finite volume.) Therefore, the Riemann surface structure associated with a hyperbolic surface has finite type iff the hyperbolic surface has finite area.

Elements of the fundamental group of a Riemann surface which are represented by loops going some number of times around a cusp are called *parabolic* elements. They play a special role, about halfway between the role of the other non-trivial elements — the *hyperbolic* elements, and the role of the trivial element of the group. Parabolic elements are distinguished from hyperbolic elements by the property that a parabolic element can be represented (up to free homotopy, *i.e.*, homotopy which need not conserve base points) by an arbitrarily short loop in the Poincaré metric, while a hyperbolic element is represented by a unique shortest curve, a closed geodesic.

2.15, Proposition: punctures-are-removable. *Let R be a Riemann surface, $p \in R$ a point, and $f:(R-p)\to S$ a holomorphic map to a hyperbolic Riemann surface. Then either*

(a) *f extends to a holomorphic map defined on R, or*

(b) *S can be desribed as $S=T-q$, where T is another Riemann surface and q is a point, and f extends to a holomorphic map $\overline{f}:R\to T$ such that $f(p)=q$.*

This is a way of phrasing the "big Picard theorem", that a function f can omit at most two values near an essential singularity.

Proof, 2.15, punctures-are-removable: We need only consider the case that the original domain $R-p = D^2-0$. A small loop around the origin is also small in the Poincaré metric, so its image in S must have a small diameter. There are two cases:

Case (a). The image of a small loop around the origin is null-homotopic in S. Then the map can be lifted to a map into the universal cover of S, that is, the disk. Consider the image of any cylinder between two concentric loops. The boundary of its image must be contained in the image of its boundary (since a neighborhood of any interior point maps to an open set); therefore there can not be disjoint disks which contain the images of its two boundary components (since a cylinder is connected) — so the images of the boundary components are not only small, but they are close together. It follows that f extends to a continuous map $\overline{f}$ defined on R. Therefore, p is a removable singularity.

Case (b). The image of a small loop around the origin is not null-homotopic in S. Since it can be homotoped to be arbitrarily small, it follows that the element of the fundamental group of S must be parabolic, and the image of any representative of the loop which is sufficiently small must lie in a small neighborhood of a puncture. Complete S by adding the point for this puncture, and reduce to case (a).

— **2.15, punctures-are-removable**

2.16, Proposition: finiteness of self-maps. *Any Riemann surface of negative Euler number and finite type has only a finite number of non-constant holomorphic self-maps; each self-map is an isometry of finite order (under composition).*

Proof, 2.16, finiteness of self-maps: The degree of a map $f : R \to S$ between Riemann surfaces of finite type is well-defined, since such a map extends to their completions. Since the Poincaré metric can only be decreased, it follows that

$$Area(R) \geqslant deg(f) Area(S)$$

Therefore, the degree of a self-map cannot be greater than one. If the degree is one, the area element must be preserved, so the map is an isometry. The group of isometries of a hyperbolic manifold of finite volume is finite. If the degree is zero, the map must be a constant.

— **2.16, finiteness-of-self-maps**

What about nonconstant maps between different Riemann surfaces of finite type? In the hyperbolic case, if there is any holomorphic map, the area of the image surface must be less than the first. By the Gauss-Bonnet theorem, the area is S is $2\pi\chi(S) \times K$, where K is the curvature and $\chi(S)$ is its Euler number. The interpretation of this is that the image surface must be simpler than the domain. Here is a more precise statement:

2.17, Proposition: number-of-critical-points. *Let $f : S \to T$ be a nonconstant holomorphic map between Riemann surfaces of finite type. Then*

$$\chi(S) = deg(f) \times \chi(T) - c(f),$$

where $c(f)$ is the total multiplicity of the critical points of f. In particular, if S and T are both the sphere, the formula is

$$c(f) = 2 \times (deg(f) - 1)$$

Proof, 2.17, number-of-critical-points: One can see this just by counting cells. Make a triangulation of $\bar{T}$ which includes as vertices all the special points — the points added at the punctures, and the critical values of f. The inverse images of all the cells give a triangulation of $\bar{S}$. To compute Euler number, add up $(-1)^{dim}$ over all the cells in the uncompleted surfaces S and T. Each regular cell in T counts for $deg(f)$ cells in S, while the critical vertices count for $deg(f) - m$, where m is the multiplicity of the critical value. That is the formula. The formula for the sphere is obtained by substituting $\chi(S^2) = 2$.

— **2.17, number-of-critical-points**

3. The equicontinuity set and the Julia set

We will now begin the study of the dynamics of rational maps, with some classical material that was developed around 1918 independently by Fatou and Julia. A rational map of degree one is a Moebius transformation, and this case is not very interesting — we shall generally understand that our rational maps have degree greater than one, even when we forget to say it. Two rational maps f_1 and f_2 are *conjugate* if they are the same up to change of coordinates by a Moebius transformation, i.e., if there is a holomorphic 1-1 map g of the sphere such that $f_1 = g^{\circ -1} small \circ f_2 small \circ g$. (Since we are studying dynamics in $\mathbf{C}$, where arithmetic operations make sense, we use the notation $g^{\circ n}$ for the iterated composition $g small \circ g small \circ \cdots small \circ g$ of g with itself n times, rather than g^n, which we reserve for the meaning $g \times g \times \cdots \times g$ of g multiplied with itself n times). We are really interested in the properties of rational maps up to conjugacy, since a conjugating map transfers any dynamical properties of one map to properties of the other.

Here is a fundamental concept:

3.1 **Definition.** Let f be a rational map of the sphere to itself, and let $I(f) = \{f^{\circ i}\}$ be the sequence of its iterates, for $i \geq 0$. The *equicontinuity set* E_f for f is the set of points z on the sphere for which there exists some neighborhood $N(z)$ so that the sequence $I(f)$ restricted to $N(z)$ is equicontinuous. The *Julia set* J_f is the complement of the equicontinuity set.

3.2 **Definition.** A subset X of the sphere is *invariant* for f if the image of X by f is contained in X,

$$f(X) \subset X$$

The subset is *totally invariant* for f if

$$f^{\circ -1}(X) = X$$

This implies that $f(X) = X$.

3.3, Proposition: complementary invariance. *A set X is totally invariant*

$\Longleftrightarrow$ *it is invariant and also $f^{\circ -1}(X) \subset X$.*

$\Longleftrightarrow$ *its complement is totally invariant*

$\Longleftrightarrow$ *it and its complement are each invariant.*

$\Longrightarrow f(X)=X$.

Proof, 3.3, complementary invariance: These properties all follow from the definitions (with a little concentration).

— **3.3, complementary invariance**

3.4, Proposition: The Julia set is totally invariant. *The Julia set is totally invariant. So is the equicontinuity set.*

Proof, 3.4, The Julia set is totally invariant: It suffices to prove this for the equicontinuity set E_f. Let $z \in E_f$. The sequence of iterates of f at $f^{\circ -1}(z)$ is obtained by composing once more with f — clearly, this cannot destroy the property of equicontinuity, so $f^{\circ -1}(E_f) \subset E_f$. In the other direction, if $z \in E$ is a regular point for f, it is clear that $f(z) \in E$ as well, because the sequence of iterates of f in a small neighborhood of $f(z)$ is obtained from the sequence at z by composing with a branch of the inverse of f at $f(z)$. But even if z is a critical point of f, a similar argument works (although it uses the fact that f is holomorphic), because the inverse image of a small neighborhood of $f(z)$ is locally small — so epsilons which work for z transfer to epsilons which work for $f(z)$.

— **3.4, The Julia set is totally invariant**

Subsets of the sphere with fewer than three points are very different from closed sets with three or more points, since their complements do not have a Poincaré metric. We will give such a set a special name — an *elementary* set is a set with 0, 1 or 2 points. (This is in keeping with the terminology, in the theory of Kleinian groups, of an elementary group).

Sometimes a rational map can have a nonempty totally invariant set which is elementary. For instance, if f is a polynomial, the point at infinity is totally invariant.

3.5, Proposition: metis. *Let f be a rational map of degree greater than one. Any finite set which is totally invariant is elementary. There is a unique maximal elementary totally invariant set E for f — it consists of 0, 1, or two points. E has a neighborhood basis of invariant (in fact, contracting) neighborhoods.*

E has two points iff f is conjugate to a map of the form $z \longmapsto z^{\pm d}$, where d is the degree.

If E has a single point, f is conjugate to a polynomial.

Proof, 3.5, metis: That there cannot be a finite totally invariant set with more than two elements follows immediately from proposition **finiteness of self-maps** (using the fact that $d > 1$). Since the union of totally invariant sets is totally invariant, there is a unique maximal elementary totally invariant set E. If E has one element, then this point can be moved to the point at infinity (by conjugacy with a Moebius transformation), and f becomes a polynomial. If E has two elements, then you can assume they are zero and ∞. Either f or f *small* $\circ (z \longmapsto z^{\circ -1})$ fixes both zero and infinity. If f fixes zero and infinity, then it is a polynomial with a multiple zero of order d at zero, hence it is of the form $z \longmapsto \lambda z^d$. Conjugating by the linear transformation $z \longmapsto z \times \lambda^{(1/d-1)}$, we obtain the function $z \longmapsto z^d$. (Notice the use of the condition $d>1$. $\lambda+1/\lambda$ is a conjugacy invariant in the case $d=1$.) The other case also follows.

The set E clearly has a neighborhood basis of attracting neighborhoods.

— **3.5, metis**

We will call the maximal elementary totally invariant set by its acronym, the *metis* of f. Generically, of course, the metis is empty, but if we have a name for it anyway, we can avoid making lots of special exceptions when discussing rational maps.

The very definition of the Julia set says that f tends to expand, at least sporadically, near any point in the Julia set. Here is a crisper statement to the same effect.

3.6, Theorem: Julia-neighborhood-spreading. *Let U be any open set which intersects the Julia set. Then the sequence $f^{\circ n}(U)$ is eventually increasing, and it eventually envelopes the complement of the metis: more precisely, if V is any neighborhood of the metis, there is some n such that for all $m \geqslant n$, $f^{\circ m}(U) \bigcup V = S^2$.*

In the generic case when the metis is empty, this says that for some n, $f^{\circ n}(U)$ is the whole sphere.

Proof, 3.6, Julia-neighborhood-spreading: Let U be any open set which intersects the Julia set, and consider the sequence of images $f^{\circ n}(U)$. There cannot be an ϵ such that these images always omit three points whose distances (pairwise) are all at least ϵ, since the Poincaré metrics for the complements of uniformly separated triples are uniformly bounded below. (This is true for any particular triple, and the set of such triples is compact.)

We conclude that for every epsilon there is some n such that the complement of the image of $f^{\circ n}(U)$ is contained in the ϵ-neighborhood of an elementary set E.

Let F be the set of all elementary subsets $E \subset S^2$ such that for every ϵ, there is some n such that the complemnt of $f^{\circ n}(U)$ is contained in an ϵ neighborhood of E. F is non-empty; we will show that F contains the metis of f.

Define a transformation $c_f : A \longmapsto S^2 - f(S^2 - A)$ on subsets of the sphere. This describes how f transforms complements of sets.

... **3.7, Lemma: finite-complements-shrink.** *If A is a set with k elements, then $c_f^{\circ k+2}(A)$ is contained in the metis of f.*

Proof, 3.7, finite-complements-shrink: The transformation c_f always decreases the cardinality of a finite set A, except in the case that each $x \in A$, $f^{\circ -1}(f(x)) = x$. This condition says that element of A is a critical point of maximal multiplicity $d-1$. Since the total multiplicity of critical points is $2(d-1)$, there can be at most two critical points of maximal multiplicity.

Even when all elements of A do have maximal multiplicity, unless each such element maps to a critical point of maximal multiplicity, the cardinality decreases after one step. The only way A can avoid decay is to be a subset of the metis.

— **3.7, finite-complements-shrink**

The set F defined above is clearly invariant under the transformation c_f. It follows from the lemma that F contains the metis (since it contains any superset of any of its elements). This says that for some m, $f^{\circ m}(U) \bigcup V = S^2$. Since the metis of f is attracting, there is some n such that that this property is true for all $m > n$.

— **3.6, Julia-neighborhood-spreading**

3.8, Corollary: Julia-spreading. *For any subset U of the Julia set which is relatively open, there is some n such that*

$$f^{\circ n}(U) = J_f$$

Phrased another way, the Julia set is the closure of the backwards orbit of any point in the Julia set.

3.9, Corollary: Julia-perfection. *The Julia set is perfect (for degree bigger than one).*

Proof, 3.9, Julia-perfection: From the preceding corollary, this immediately reduces to showing that the Julia set has at least two points. If the whole sphere were the equicontinuity set, then $f^{\circ n}$ could expand the area element by only a bounded amount, for all n. This cannot happen when $deg(f)>1$, for then $deg(f^{\circ n}) \to \infty$. Thus, the Julia set contains at least one point. Therefore, the Julia set has infinitely many points (since it is not contained in the metis).

— **3.9, Julia-perfection**

3.10 **Definition.** A *periodic point* of f is a point x such that for some n, $f^{\circ n}(x) = x$. The minimal $n > 0$ for which this is true is the *period* of x.

Another invariant set canonically associated with a rational map f is the closure of the set of all periodic points for f. This set is not totally invariant; however, it is closely related to a familiar totally invariant set:

3.11, Theorem: Periodic-points-dense. *Periodic points are dense in the Julia set.*

Proof, 3.11, Periodic-points-dense:

Let x be any point in the Julia set. We can assume that $f^{\circ -1}(x)$ and $f^{\circ -2}(x)$ are disjoint from x, otherwise x is already a periodic point. Choose $x_0 = x$, x_1 and x_2 such that $f(x_{i+1}) = x_i [i = 0,1]$. The three x_i's are all in the Julia set. Consider an arbitrary neighborhood U of x; we will show that there is a periodic point for f in this neighborhood.

Let $U_0=U$, and let U_i+1 be the component of $f^{\circ -1}(U_i)$ containing x_{i+1}, for $i=0,1$. Consider the sequence $f^{\circ n}|U_2[n \geq 0]$, and their graphs in $U_2 \times S^2$. If any of these graphs intersect the graphs of the first three, it gives us a periodic point in U (since any periodic orbit intersecting U_2 or U_1 also travels through $U=U_0$.)

There is a holomorphic map of $U_2 \times S^2$ to itself, commuting with projection to U_2, and taking the graphs of id, f and $f^{\circ 2}$ to constant sections $U_2 \times 0, U_2 \times 1$, and $U_2 \times 2$. The graphs of the $f^{\circ n}$ transform to the graphs of some other sequence of functions. Since $\{f^{\circ n}\}$ is not equicontinuous, the new sequence is not equicontinuous either. Therefore, some of the functions take the value $0,1 or 2$. Points which hit 0, 1, or 2 become periodic points, when translated back to the original coordinate system.

— **3.11, Periodic-points-dense**

4. Some simple examples

It is time to look at a few examples.

4.1 Example. An invariant set with two points.

For a polynomial f of the form

$$z \longmapsto z^d$$

the Julia set is the unit circle. Every point inside the unit circle tends toward the origin, and everything outside the unit circle tends toward infinity; in either case, the derivative of f goes to zero and the point has a neighborhood whose radius gets squished to zero (in terms of the spherical metric.) Note that the derivative of $f^{\circ n}$ grows quickly on the unit circle -- it has norm d^n

4.2 Example. A nearly invariant set with two points.

Example **metis-2, 4.1** has a built in stability, near its Julia set: any polynomial g of the form $z^d+R(z)$ also has a circle as its Julia set, when R(z) is a small polynomial of lower degree. One way to prove this is to give a dynamical description of the topology of the circle. First we will describe the circle in terms of the dynamics of f, but the description will be robust enough that it carries over to g. Cover the circle by a finite number of neighborhoods $\{U_i\}$ – say, $2d$ – with the property that

(a) The inverse image $f^{\circ -1}(U_i)$ has d components, and the diameters of components of $f^{\circ -n}(U_i)$ go to zero with n. (For our particular example, the second half of this condition follows automatically from the first.)

(b) At most one component of the closure of $f^{\circ -1}(U_i)$ intersects the closure of U_j.

(c) Each component of the closure of $f^{\circ -1}(U_i)$ is contained in some U_j.

These properties imply in particular that the set of components of $f^{\circ -n}(U_i)$ form a basis for the topology of the circle. A good way to get a collection $\{U_i\}$ is to divide the circle into $2d$ equal segments, starting at the point 1, and take small neighborhoods of these segments for our U_i.

What condition (b) means is that any component of $f^{\circ -1}(U_i)$ is determined by naming any U_j which it intersects. Therefore, a component of $f^{\circ -2}(U_i)$ is determined by naming a component of a $f^{\circ -1}(U_j)$ which it intersects. Continuing on in this way and applying (a), it follows that a point x on the circle is completely determined by naming a sequence $\{i_n\}$ such that $f^{\circ n}(x) \in U_{i_n}$. A neighborhood basis for the x is formed by sets of points which are in the same U_{i_n}'s for a finite number of i's.

What happens when we perturb f a little, to get g? If there are any non-generic intersections for property (b), where closures of sets intersect but there interiors do not, modify the sets a little bit to make them generic. Similarly, we can make intersections of the U_i with each other generic. Then the set of intersections which occur are stable, so we can suppose that g is close enough to f that it has exactly the same qualitative structure of intersections for the closures of the inverse images of the U_i and the closures of the U_j. We can also assume that there are no critical points in the U_i.

Unfortunately, the intersections of $g^{\circ -n}(U_j)$ with the U_i probably are different for those with f. This is where condition (c) helps. Given a finite or infinite sequence of indices i_n, let us say that it is legitimate if $f(U_{i_{n+1}}) \subset f(U_{i_n})$ for each applicable n.

4.3, Lemma: legitimate-heir. *For each legitimate sequence, the set*

$$\{x \mid g^{\circ n}(x) \subset U_{i_n} \ [\text{applicable } n]\}$$

is non-empty.

Proof, legitimate-heir: For finite legitimate sequences, this is clear from the conditions on g. For infinite sequences, this set is the intersection of non-empty compact sets.

— **legitimate-heir**

Each infinite legitimate sequence corresponds to exactly one point on the circle. Define two legitimate sequences $\{i_n\}$ and $\{j_n\}$ to be adjacent if U_{i_n} intersects U_{j_n}, for each applicable n. There is exactly one legitimate sequence except for certain boundary cases, as in decimal expansions. Thus, a map h can be defined from the Julia set of f to some corresponding set for g. The sets of points admitting an infinite legitimate sequence beginning with a certain finite segment do not quite form a basis for the topology of the circle, any more than initial segments of decimal expansions define a basis for the topology of $\mathbf{R}$; but by taking unions of adjacent sets of this form, one obtains a basis. Therefore, h is continuous. Its image is closed and totally invariant by g, so the image $h(J_f)$ is the Julia set for g. Since h is 1-1, it is a homeomorphism between the two Julia sets.

The Julia set for the polynomials of these two examples is also the boundary between the set of points whose orbits tend toward infinity and the set of orbits which stay bounded. This is a general phenomenon:

4.4, Proposition: Julia-is-frontier. *The Julia set for a polynomial is the frontier of the set of points whose forward orbits are bounded.*

Proof, Julia-is-frontier: If a neighborhood consists of points whose orbits stay bounded, then the set of forward iterates of f on the neighborhood is clearly equicontinuous (by proposition **Self-maps, of**). If a point tends toward infinity, it is also clearly a point of equicontinuity. On the other hand, if the orbit of a point is bounded but points arbitrarily close to it go to infinity, it is clearly not a point of equicontinuity.

— **Julia-is-frontier**

4.5 Example. The Julia set can be a Cantor set.

Here is an example where the set of orbits which stay bounded has no interior. For concreteness, consider the map $f : z \longmapsto z + 4$. Let us analyze the set of points whose orbits are bounded. A point of modulus r goes to a point with modulus not less than $r^2 - 4$. If r is at least 3, then r increases, so the point tends toward infinity.

The disk of radius 3 maps to a disk of radius 9 centered about 4, as a two-fold branched cover. The set of points which do not escape the disk of radius 3 after one iterate is the inverse image of the disk of radius 3, which is two not-quite-round subdisks. The set of points which also do not escape the disk of radius 3 after 2 iterates is the union of two subdisks of each of these. To get the set of points which do not escape ever, continue this procedure, and form the infinite intersection.

Is the intersection a Cantor set, or something else? Note that for a point to remain in the disk of radius 3 after the first iterate, it must be outside the disk of radius 1, so the norm of the derivative of f is at least 2. Therefore, the diameters of the nested disks tend geometrically to zero, and the Julia set is indeed a Cantor set.

If you study examples of the form $z \longmapsto z^2 + a$, letting a vary, you can find many different types of behavior. Indeed, it should be clear that something else must happen to allow a transition from the case when a is small, when the Julia set is a circle and the case that a is large, when the Julia set is a Cantor set. In fact, many different things happen. It is worthwhile to experiment a little with a programmable calculator or a computer. Try, for example, the cases

$$z \longmapsto z^2 + i$$

where the critical point (0) goes to $i, i-1, -i, i-1, -i, \ldots$, and the case

$$z \longmapsto z^2 + (-.1 + .6i)???$$

where the critical point eventually spirals (at a fairly slow rate) toward a periodic orbit of period 3. By a little experimentation, you can see that in the first case almost no orbits remain forever bounded, while in the second, there is a definite open set of points whose orbits remain bounded, and they all spiral toward the same periodic orbit of period 3.

5. Periodic points

There are important distinctions between different kinds of periodic points. If x is a periodic point of period p for f, then for any neighborhood U of p, $f^{\circ p}$ maps U to another neighborhood V. This locally defined map is the *return map* for x; it is well-defined except for the ambiguity of domain.

5.1 **Definition.** The point x is called an *attracting* periodic point if there exists a neighborhood U of x such that the sequence of the iterates of the return map on this neighborhood tend uniformly toward the constant function $z \longmapsto x$. It is a *repelling* if there exists a neighborhood V of x such that the image of every neighborhood of x under high iterates of the return map contain V. If x is neither attracting nor repelling, it is *indifferent.*

5.2, Proposition: derivative-of-periodic. *Let x be a periodic point, and let g be its return map. Then x is*

attracting

$\Longleftrightarrow$ $|g'|<1$

$\Longleftrightarrow$ *there exists a neighborhood V of x with non-elementary complement mapped into a proper subset of itself by the return map;*

indifferent

$\Longleftrightarrow$ $|g'|=1$

repelling

$\Longleftrightarrow$ $|g'|>1$

$\Longleftrightarrow$ *there exists a neighborhood V of x which is mapped diffeomorphically by g to a proper superset of itself.*

Proof, derivative-of-periodic: By Taylor's theorem, it is clear that if $|g'|$ is less than or greater than one, then x is (respectively) attracting or repelling. If there exists any neighborhood V of x which is mapped into a proper subset of itself by the return map, then the return map must decrease the Poincaré metric of V, so that its derivative at x is less than one — provided, of course, that V is covered by the disk. Similarly, if any neighborhood is mapped to a strictly larger neighborhood, one can apply this reasoning to a locally defined inverse of g to conclude that the derivative must be strictly greater than one. When the derivative of g has modulus one, the only remaining possibility is that g is indifferent.

— **derivative-of-periodic**

In terms of derivatives, a further distinctions can be made:

5.3 **Definition.** The periodic point x is *super-attracting* of order k if the derivative of the return map vanishes to order k.

5.4 **Definition.** An indifferent periodic point x is *rationally indifferent* if the first derivative of its return map is a root of unity, and *irrationally indifferent* otherwise.

5.5, Proposition: periodic-conjugacy-classification. *Let x be a periodic point, and g be its return map. If x is not indifferent, then g is determined up to conjugacy, in any sufficiently small neighborhood, by its first derivative, or if this is zero, by the order of vanishing of its derivatives.*

Proof, periodic-conjugacy-classification:

Case (i): $0<|g'(x)|<1$

Let V be a neighborhood of x which the return map takes inside itself. Define a Riemann surface R by gluing together an infinite number of copies $V_{(0)}, V_{(1)}, V_{(2)}, \cdots$ of V, identifying each copy $V_{(n+1)}$ to the previous copy $V_{(n)}$ by a copy of g.

The maps $g_{(i)}$ of the ith copy to itself commute with the identifications, so there is a well-defined map of R to itself, which we also call g, extending the original definition of g on $V_{(0)}$. But on R, g has a globally-defined inverse. The Riemann surface R cannot be the disk (because of the canonical nature of the Poincaré metric), so it is **C**.

In the coordinate system of the **C**, g must be an affine map $g: z \longmapsto az+b$; by change of coordinates so that the fixed point x is at the origin, it has the form

$$z \longmapsto g'(x)^* z$$

This proof easily gives an actual formula for the linear coordinates. The uniformization for R is accomplished just by taking a sequence of embeddings of V in C, so that the image $g^{on}(V)$ (interpreted as the approximation for $V_{(0)}$ stays roughly constant in size, and passing to the limit. A formula for a linearizing map L is obtained by choosing a point $z_0 \neq x$ and setting

$$L(z)=\lim_{n\to\infty}\frac{g^{on}(z)}{g^{on}(z_0)}$$

Case (ii): $|g'(x)|>1$

Apply the argument of case (i) (which is strictly local in nature) to a local inverse of g.

Case (iii): $g'(x)=0$
We will make an argument similar to case (i), but we need to take into account the fact that g is not any longer a homeomorphism, but rather a branched covering.

Let V be a simply-connected neighborhood of x which is mapped into itself by g, and such that $g|V$ has a critical point only at x. Let k be the order of the first non-zero derivative of g at x, so that g has k-fold branching at x. Inductively define a sequence V_i by taking $V_0=V$, and V_{i+1} to be the k fold branched covering of V_i, branched over the point which maps to x. Let p_i be the covering projection $V_{i+1}\rightarrow V_i$. Thus, V_n is the k^n-fold cover of V, branched over x.

Since g is a k-fold branched covering, the map $g_0=g$ of V_0 to itself lifts to a diffeomorphism $h_0:V_0\rightarrow V_1$. ("Lifting" means that $p_0 small \circ h_0=g$). The composition $g_1=h_0 small \circ p_0:V_1\rightarrow V_1$ is a k-fold branched cover of V_1 into itself. g_1 corresponds to g via the branched covering, *i.e.*, $p_0 small \circ g_1=g small \circ p_0$. Having inductively defined g_i, define h_i and g_{i+1} in the same way.

Now we can form a Riemann surface R by gluing V_{i+1} to V_i by the embedding h_i. The maps g_i are compatible with the gluing, so they fit together to give a map g_∞ which is a k-fold branched covering of R over itself. By checking the two possibilities, one sees that R must be the disk.

There is only one k-fold branched cover of the disk branched only at the origin. One description of such cover is by the map $z\longmapsto z^k$. Since the only isometries of the Poincaré metric of the disk which fix the origin are multiplication by constants, the covering map at hand must agree with this one, up to multiplication by a constant in the domain and range. The constant, of course, can be pushed backward or forward to the domain or range.

Observe that if we conjugate $z\longmapsto z^k$ by the map $z\longmapsto az$, we get the map $z\longmapsto a^{k-1}z^k$. We can dispense with the constant (using the hypothesis that $k\neq 1$), and the proposition is proven.

Just as for the earlier cases, this proof translates into a formula for the conjugating map, although interpretation of the formula requires intelligence (or continuity) in deciding which k^nth roots to use. First, choose a local coordinate z so that the fixed point for the return map is at the origin and the kth derivative of the return map is 1. The formula for a local map h to the disk which conjugates g to $z\longmapsto z^k$ is

$$h(z) = \lim_{n\rightarrow\infty} (g^{\circ k}(z))^{(1/k^n)}$$

— **periodic-conjugacy-classification**

An attracting periodic point is always in the equicontinuity set, of course. Using the existence of the Poincaré metric, we can give deduce considerably more about how it sits in the equicontinuity set.

5.6, Proposition: attracting-gobble-critical.

All the attracting periodic points are in different components of the equicontinuity set.

Every point in a component of the equicontinuity set which contains an attracting periodic point p tends toward the periodic orbit of p in future time.

For each attracting periodic orbit, there is at least one critical point in the union of the components of the equicontinuity set which intersect the orbit.

Proof, attracting-gobble-critical: If x is an attracting periodic point of period p, then $f^{\circ p}$ actually decreases the Poincaré metric of its component in E_f, since its derivative at x is less than 1 in modulus. There cannot be another periodic point in the same component, for the Poincaré distance between the two would have to decrease.

The map of one component of E_f to its image is a branched cover; it is an isometry if it does not actually branch, *i.e.*, if it does not contain any critical points. Clearly, the component of at least one of the periodic points in any attracting periodic orbit must contain a critical point.

— **attracting-gobble-critical**

5.7, Corollary: attracting-bound. *There are at most $2(d-1)$ attracting periodic orbits for any rational map of degree d, and at most $d-1$ attracting periodic orbits in $\mathbf{C}$ for a polynomial of degree d.*

Proof, attracting-bound: Since each attracting periodic orbit consumes at least one critical point, there can be no more attracting periodic orbits than there are critical points.

— **attracting-bound**

Indifferent periodic points are considerably more subtle. What we really need to know is the qualitative description, especially for rationally indifferent periodic points. The local analysis of rationally indifferent periodic points with any given derivative, shows that there are countably many different possible qualitative types, and for each qualitative type there is still an uncountable family of different local conjugacy classes.

We begin with the simplest possible case, when the derivative of the return map g is one. We can change coordinates so that the origin is the fixed point, and for some k the Taylor series for g has the form

$$z \longmapsto z + a_k z^k + \cdots$$

By a linear change of coordinates, we may assume that $a_k = 1$. To get a rough picture of the qualitative behavior of g, we can consider the special case that g is the time-one map for a flow (that is, it is obtained by following a vector-field for a unit of time). An appropriate vector-field is $V = z^k e$, where e is the unit vector-field (often written $e = \frac{\partial}{\partial z}$) pointing in the x-direction in $\mathbf{C}$.

If $k=2$, then V is actually a holomorphic vector-field on the whole Riemann sphere: in terms of the coordinate $w=1/z$ with unit vector-field f, V is the constant vector-field $v=-f$. The orbits of V on the Riemann sphere are all asymptotic to 0, as time goes to $\pm\infty$. Each orbit (except the orbit of the fixed-point 0) is a circle, minus one point; all the circles have a common tangent at the origin.

If $k>2$, the vector-field V can be obtained by pulling back the vector-field for the case $k=2$ by a $(k-1)$-fold covering map $h: z \longmapsto (1/k-1)z^{k-1}$. In this case, V blows up on the sphere, at the point ∞, so it is not globally integrable. The time-one map for its flow is nonetheless defined for some neighborhood of the origin — in fact, the only problem is along rays passing through (k-1)st roots of unity. These rays are pushed outward along themselves, and a definite chunk goes out to the point ∞ in any positive time interval. The rays which pass through odd 2(k-1)st roots of unity get pulled inward, so that the image of an entire ray is a proper subset of itself, after any positive time interval. All other orbits of the flow exist for all time, and they form curves which are asymptotic to 0 as time goes to $\pm\infty$. All of these typical orbits are tangent to a ray through an odd 2(k-1)st root of unity in positive time, and to a ray through an even 2(k-1)st root of unity in negative time. The whole picture is like a flower with 2(k-1) petals.

For a general return map g with first derivative one, the local qualitative picture is similar. When normalized so that the next non-zero term in the Taylor series is z^k, there are regions around the rays through odd 2(k-1)st roots of unity which tend toward 0 in positive time. In fact, for any reasonably small open disk U, the set F of points which tend toward 0 without ever leaving the U has $k-1$ components, each of which is tangent on each side to a ray through a (k-1)st root of unity. F also looks like a flower (at least for k big enough), but with only (k-1) petals.

When the derivative of the return map g is a root of unity, but not one, then of course there is some iterate of g which has derivative one. Again, the set F consisting of points in a small disk which tend toward 0 without ever leaving the disk form a figure which is like a flower. Since F is invariant by g, the number of petals must be a multiple of the multiplicative order of the derivative of g.

5.8 **Definition.** For any rationally indifferent periodic point, an *attracting flower* is the set of points in a sufficiently small disk around the periodic point whose forward orbits always remain in the disk under powers of the return map. A component of the attracting flower is an *attracting petal*

We now have enough qualitative information to give a general discussion of the situation of rationally indifferent periodic points with respect to the Julia set and the equicontinuity set.

5.9, Proposition: rationally-indifferent-gobble-critical.

A rationally indifferent periodic point is in the Julia set, but each of its attracting petals is in the equicontinuity set.

Distinct attracting petals are in distinct components of the equicontinuity set; they are also distinct from any components of E_f which contain periodic orbits.

Every point in a component C of E_f containing an attracting petal tends toward the rationally indifferent periodic point in forward time. There is at least one critical point in $\bigcup_{i=1}^{pk} f^{\circ i}(C)$, where p is the period of the periodic point and k is the order of the first derivative of its return map.

Proof, rationally-indifferent-gobble-critical: It is obvious that any rationally indifferent periodic point x is in the Julia set, and that attracting petals are in E_f, but not so obvious that distinct attracting petals are in distinct components. The petals are somewhat artificial, depending on the choice of a small disk, and we do not have much control over what might happen to the material between petals once they leave this disk – why can't they be in E_f?

Consider any two points w and z in attracting flowers for rationally indifferent periodic points (possibly the same). Suppose that w and z are in the same component of E_f. Then there is a smooth path β joining w to z in E_f, with a certain Poinarè length L. Any image $f^{\circ n}(\beta)$ can only have shorter length.

Apply a large number N of iterates of f, where N is a multiple of the period of the periodic point x associated with w. Now $f^{\circ N}(w)$ is very close to x, so close that the disk of radius L in the Poincaré metric of E_f is contained in the flower-disk for x. We see that z and w are associated with the same periodic point x.

It also follows that $f^{\circ N}(\beta)$ is contained in the attracting flower of x: indeed, for all powers m of the return map g, $g^{\circ m}(f^{\circ N}(\beta)$ is contained in the flower-disk for x, by the choice of N. Consequently, its two endpoints are in the same attracting petal.

We will be able to deduce the existence of a critical point associated with each periodic cycle of attracting petals once we can prove that the Poincaré metrics of the components of E_f are not all mapped by isometries. This is not so obvious as it was for attracting periodic points – in fact, the local picture is topologically consistent with the hypothesis that the maps are all isometries.

Let F be the iterate of f which returns each petal to itself. Not all points in the petal are alike; some points have the property that the full orbit under F remain in the petal, while other points exit in backward time. Let z be any point whose full orbit is in the petal.

Let $h: D^2 \to E_f$ be a uniformizing map for the component of E_f containing the petal. Draw an arc β within the petal from the point z to $F(z)$, in such a way that β remains in the petal under all forward and backward iterates of F. F lifts to a map $\tilde{F}: D^2 \to D^2$ such that a certain lift $\tilde{z}$ of z extends to a lift $\tilde{\beta}$ of β connecting $\tilde{z}$ to $\tilde{F}(\tilde{z})$.

If F were an isometry, then $\tilde{F}$ would also be an isometry. The path $\bigcup_{n=-\infty}^{+\infty} \tilde{F}^{\circ n}(\beta tilde)$ would converge at both ends to a point in ∂D^2. The image $\bigcup_{n=-\infty}^{+\infty} \tilde{F}^{\circ n}(\beta)$ is closed; this forces one of the two non-compact components of $int(D^2)-\beta$ to map in such a way that the interval where it touches ∂D^2 maps to a point. By proposition **no-interval-of-constancy** it follows that this interval is a single point.

Now we could repeat the same analysis, but using a point z_1 and an arc β_1 which are on the other half of the attracting petal, so that their orbits move in the opposite direction. Make sure, though, that the lift $\tilde{F}$ of F is the same in both cases. The two points $\tilde{z}_1$ and $\tilde{z}$ are a finite distance away, so that their forward and backward limit points on ∂D^2 are the same; it would follow that ∂D^2 consists of a single point, an absurdity.

The conclusion is that F could not be an isometry after all, so at least one of the images of the component of E_f before F must contain a critical point.

— **rationally-indifferent-gobble-critical**

6. The limiting behavior of the Riemann map.

The Riemann mapping theorem constructs a holomorphic map from any simply-connected open subset $U \subset S^2$ to the disk, provided S^2-U has more than one point. What happens, though, near the frontier ∂U? In particular, suppose that $g: int(D^2) \to U$ is a 1-1 holomorphic mapping (in the inverse direction from the usual statement of the Riemann mapping theorem). Under what circumstances does g extend to a continuous map on all of D^2? Here is the definition which is most relevant.

6.1 **Definition.** A set X is *locally connected* if for every $x \in X$ and every neighborhood W of x, there is a smaller neighborhood $V \subset W$ such that if $A \subset W$ and $B \subset W$ are disjoint relatively closed subsets whose union is W, then one of them is disjoint from V. A set X is *locally path-connected* if for every $x \in X$ and for every neighborhood W of x there is a smaller neighborhood $V \subset W$ of x such that any two points in V are connected by a path in W. A *path* is a continuous map from the unit interval to the space.

Local path-connectedness easily implies local connectedness. For a set which is, say, locally compact, it is not too hard to prove the converse – that if it is local connected, it is also locally path-connected. To do this, first "join" any two nearby points by a sequence of points spaced no more than say 1/10th the distance apart – then join each adjacent pair of points in this sequence by a new sequence spaced no more than a 1/ 100th of the original distance apart, and not wandering too far from the pair they are joining – and continue on in this way, then pass to the limit.

Whenever the inverse Riemann mapping g extends continuously to the closed disk, then its frontier ∂U must be locally path-connected because of the following

6.2, Lemma: image(locally-path-connected-compact). *The continuous image of a compact, locally path-connected space is locally path-connected.*

Proof, 6.2, image(locally-path-connected-compact): Let $f : X \to Y$ be a continuous surjection, where X is compact and locally path-connected. Let $y \in W$ be a point and neighborhood in Y. For each element $x \in f^{\circ -1}(y)$, there is an open neighborhood V_x such that any two points of V_x are connected by a path in $f^{\circ -1}(W)$. Let U be the union of any collection of such open neighborhoods which cover $f^{\circ -1}(x)$. The image $V = f(U)$ is a neighborhood of x, since the image of $X - V$ is a compact set which does not contain x. Any point in V can be connected to x by a path in W. Therefore Y is locally path-connected.

— **6.2, image(locally-path-connected-compact)**

It turns out that local connectdness of $S^2 - U$ is, in fact, the precise condition for the inverse Riemann mapping to extend:

6.3, Theorem: Riemann-map-extension. *The inverse Riemann mapping g extends to a continuous map $\bar{g}$ defined on the closed disk D^2 if and only if the complement $S^2 - U$ is locally connected.*

This remarkable theorem is a consequence of Caratheodory's beautiful theory of prime ends. Rather than getting involved in this theory, we will give a direct proof.

Proof, 6.3, Riemann-map-extension: We have essentially taken care of the easy direction already, that if the inverse Riemann mapping extends, then $S^2 - U$ is locally connected. The lemma shows that if g extends continuously to ∂D^2, then the frontier ∂U is locally path-connected. It is also clear that if ∂U is locally path-connected, then $S^2 - U$ is also locally path-connected: draw a straight line between any two of its points; if it doesn't intersect U, it is a path connecting the two points, and if it does, replace a segment of it by a path in the frontier of U. Local connectivity follows.

Sufficiency of the condition that $S^2 - U$ be locally connected is somewhat trickier. We will prove the g extends without constructing the extension ahead of time, by showing that for every ϵ and every point $x \in \partial(D^2)$ there is a neighborhood V of x in the closed disk such that the image of its intersection with the open disk has diameter less than ϵ. This is really a form of Cauchy's condition for convergence.

We first construct a neighborhood of any point $x \in \partial(D^2)$ bounded in $int(D^2)$ by an arc whose image in U has an arbitrarily short length. This first step works for any simply-connected set U, even if its complement is not locally connected.

... **6.4, Lemma: small-area-short-cuts.** *Let* $f : Q \longmapsto S^2$ *be any holomorphic embedding of the unit square* $Q = (0,1) \times (0,1)$ *in the Riemann sphere. If* A *is the area of* $f(Q)$, *then at least one of the arcs of the form* $\{s\} \times (0,1)$ *and at least one of the arcs of the form* $(0,1) \times t$ *is rectifiable and has length less than* $(A)^{1/2}$.

Proof, 6.4, small-area-short-cuts:

$$A = \iint_Q |f'|^2 \, ds \, dt \geqslant \left(\iint_Q |f'| \, ds \, dt \right)^2$$

by the Schwarz inequality. The integral on the right is the average length of an arc, by Fubini's theorem.

— **6.4, small-area-short-cuts**

... **6.5, Corollary: short-dam-basis.** *For any point* $x \in \partial(D^2)$ *and any* $\epsilon > 0$, *there is a neighborhood* V *of* x *in* D^2 *whose intersection with* $int(D^2)$ *is bounded by an arc* α *such that* $g(\alpha)$ *is rectifiable and has length less than* ϵ.

Proof, 6.5, short-dam-basis: Map the unit square via the map $z \longmapsto z^2$ so that two of its sides cover the interval $(-1, 1)$, and then map it by a Moebius transformation to the disk, so that a vertex goes to x, and its image is the intersection of a neighborhood of x with the open disk. Compose with a contraction in the domain until the area of its image is tiny, and apply lemma to find a corner of the square bounded by two pieces of arcs whose images in U are short.

— **6.5, short-dam-basis**

Local connectedness comes in precisely through the following

... **6.6, Proposition: short-dam-small-lake.** *Let* $U \subset S^2$ *be any simply-connected open set whose complement is locally connected. Then for every* ϵ *there is a* δ *such that if* α *is any arc with interior in* U *and endpoints on* ∂U, *and if* α *has diameter less than* δ, *one or the other of the two components of* $U - \alpha$ *has diameter less than* ϵ.

Proof, 6.6, short-dam-small-lake: First, note that by a standard pattern of reasoning, local connectedness of a compact set K implies uniform local connectedness — for every ϵ there is a δ such that any δ-ball in K intersects only one component of the concentric ϵ-ball.

If α is an arc as in the proposition contained in a disk of radius δ, consider how the two components of $U - \alpha$ might intersect the concentric circle of radius ϵ. If both components were to intersect this circle, then there would be an arc β in the intersection of U with the disk of radius ϵ, beginning and ending on the boundary, and cutting across α exactly once. The arc β would cut the disk into two halves, and the two endpoints of α would be in opposite halves: this contradicts the choice of δ.

The only possible conclusion is that one or the other of the two components of $U - \alpha$ does not intersect the circle of radius ϵ, and hence it is contained in the disk of radius ϵ.

— **6.6, short-dam-small-lake**

— **6.3, Riemann-map-extension**

7. Identifications in the limit of the Riemann mapping

Consider a polynomial map P of degree d which satisfies the condition that the forward orbits of the finite critical points are bounded. We may assume that P has leading coefficient 1, after conjugating by a complex affine transformation.

The point at infinity is a super-attracting fixed point of degree d for P. Therefore, by proposition **periodic-conjugacy-classification** P is conjugate to the map $z \longmapsto z^n$ in a neighborhood of infinity. We may assume that the conjugacy is close to the identity near infinity. With this proviso, the conjugating homeomorphism g is unique. (Note, however, that there is non-uniqueness in the picture coming from the fact that the map $z \longmapsto \omega z$, where ω is a $(d-1)$st root of unity, conjugates any polynomial of degree d with leading coefficient 1 to another such polynomial.)

Suppose that a conjugating homeomorphism g has been defined for some disk D containing infinity with the property that $P(D) \subset D$. Every point inside D tends toward infinity with positive time. The inverse image $P^{-1}(D)$ is also a disk: it is obtained by taking a certain d-fold branched cover of the extended complex plane, branched over the critical values of P. Since none of the finite critical points tend toward infinity, none of the critical values lie in D, so the branched cover is also a disk. There is a unique way to extend g analytically to this disk $P^{\circ -1}(D)$. We can continue indefinitely until we have defined g on the set of all points which tend to infinity in forward time.

7.1, Proposition: polynomial-Riemann-map. *For any polynomial P such that no finite critical points tend to infinity in forward time, the set $E_\infty \subset S^2$ consisting of all points which are attracted to infinity is a simply-connected set, which has a Riemann mapping g to the disk $|z|>1$ which conjugates P in this region to the map $z \longmapsto z^d$. The conjugating Riemann mapping g is unique up to multiplication by a $d-1$st root of unity.*

Proof, polynomial-Riemann-map:

— **polynomial-Riemann-map**

7.2, Proposition: infinity-basin-envelopes. *For any polynomial map, every neighborhood of every point in the Julia set intersects the set E_∞*

Proof, infinity-basin-envelopes: This follows from the fact (Proposition **Julia-spreading**) that the union of the forward images of any neighborhood of any point in the Julia set is the sphere minus at most two points.

— **infinity-basin-envelopes**

This elementary proposition immediately suggests two questions as holding clues to the theory of polynomial self-maps of $\mathbb{C}$:

7.3 Under what circumstances does the inverse Riemann map $g^{\circ -1}$ extend to a continuous map of the disk to $\hat{\mathbb{C}}$? (By the results of section **Limiting-behavior-of-the-Riemann-map** this is equivalent to the question of whether or not the Julia set is locally connected.)

7.4 In case $g^{\circ -1}$ does extend to a continuous map on the disk, what identifications does g make on the boundary of the disk?

There are certainly big restrictions on what identifications can occur. First, there are restrictions which can be deduced from the separation properties of the sphere, conveniently described by this proposition:

7.5, Proposition: separation-properties. *Let $f: D^2 \to S^2$ be a continuous map of the disk to the sphere which is an embedding of the interior of the disk. Let K_1 and K_2 be inverse images of points in the image of the boundary of the disk. (In other words, they are equivalence classes in the identification.) Then the convex hull of K_1 is disjoint from the convex hull of K_2.*

Proof, separation-properties: Otherwise, there would be paths α_i beginning and ending on K_i which intersect transversely in the disk. In the image, they would become closed curves which cross exactly once — a contradiction.

— **separation-properties**

In view of the proposition, there is a convenient way to draw a picture of the set of identifications which occur on the boundary of the disk — as a collection of disjoint convex subsets of the disk which are the convex hulls of sets of points on the circle. It is worth emphasizing that the convex hulls must be disjoint in the closed disk, not just in the open disk — this condition is not always easy to discern from a sketch. The collection of convex hulls must be closed, in the sense that if a pair of points is the limit of a sequence of pairs of points such that each pair of the sequence belongs to some convex hull in the collection, then the limiting pair must also belong to a convex hull in the collection. Any equivalence relation satisfying these conditions can be shown to arise as the set of identifications created by some continuous map of the disk into the sphere which is an embedding of the open disk.

It is convenient to think of the picture of the convex hulls in the unit disk $D_1 \subset S^2$ as representing a set of identifications on the boundary of the complementary disk $D_2 = S^2 - \int(D_1)$. Then the image of D_2 is homeomorphic to the space obtained from $D_2 \bigcup(\bigcup(\textit{convex hulls}))$ by collapsing each convex hull to a point. The map of $D_2 \to S^2$ is surjective iff the union of the convex hulls is all of D_1.

More restrictions stem from the fact that the identifications must be compatible with the dynamics.

(a). If a set K of points on the circle is to be identified to a single point, then its image under $z \longmapsto z^d$ must also be identified to a single point.

(b). Suppose that the set K is an equivalence class under the identification. Let L be the set of d th roots of elements of K. Then L is divided into at most d equivalence classes $L_1 \cdots L_k$. Each L_i maps with a certain constant multiplicity m_i, *i.e.*, it contains exactly m_i of the d th roots of each element of K.

The assertion about multiplicities needs a little explanation. Think of points on the circle as endpoints of rays in the disk. Points on the circle are identified when rays of the disk, mapped via the inverse Riemann mapping to the complement of the Julia set, have the same endpoints. Any point p in the Julia set has some deleted neighborhood U so that the map $P: P^{\circ -1}(U) \to U$ is a covering map, and the components of $P^{\circ -1}(U)$ are in 1-1 correspondence with the elements of $P^{\circ -1}(p)$. Then for any element $q \in P^{\circ -1}(p)$ and for any ray converging to p, there are exactly $deg(V \to U)$ inverse images of the ray converging to q, where V is the component of $P^{\circ -1}(U)$ corresponding to q.

Every equivalence class therefore has a well-defined multiplicity under the map $z \longmapsto z^n$: the number of points equivalent to any point in the class which have the same image as that point. When the multiplicity is greater than one, the equivalence class is a *critical* equivalence class.

When we form the convex hulls of all the equivalence classes, there may or may not be regions of the disk left over, not filled by any convex hull. If R is any such region, then its boundary is made up of parts of the circle, and secants; no two secants can have an endpoint in common. The image of ∂R after the identification is homeomorphic to a circle. Since any circle on S^2 is the boundary of two disks, this implies that the equicontinuity set for P has a bounded component, bounded by this circle, corresponding to R.

The analysis will be greatly simplified by a theorem of Sullivan [1982a], which answered a longstanding problem:

7.6, No Wandering Domains Theorem (Sullivan): No wandering domains. *There are no wandering domains in the equicontinuity set. Every component of the equicontinuity set for a rational map is eventually periodic, i.e., its forward images eventually start repeating in some cycle.*

This means that any component of the complement of the union of the convex hulls of the equivalence classes is eventually periodic.

When any region returns to itself, there is an induced map on the quotient space of its boundary obtained by collapsing all secants to points. The return map of this quotient circle is a covering map.

8. Invariant laminations

In the preceding section (**Identifications**), we began the analysis of what the picture can look like when the Julia set of a polynomial is connected and locally connected. In this section, we will carry this line of investigation forward, but on a more abstract level. We will define the notion of an invariant lamination (which we secretly think of as determining a potential set of identifications for a polynomial map), and we will see how far we can go toward the goal of classifying all invariant laminations.

8.1 **Definition.** A *geodesic lamination* is a set L of chords in the unit disk, called *leaves* of L, satisfying

(a) Elements of L are disjoint, except possibly at their endpoints.

(b) The union of L is closed.

A *lamination* on the disk is a collection of embedded arcs, disjoint except possibly at their endpoints, whose intersection with any sufficiently small neighborhood in the closed disk is topologically equivalent to the intersection of a geodesic lamination. A *gap* of a lamination L is the closure of a component of the complement of $\bigcup L$. Any gap for a geodesic lamination is the convex hull of its intersection with the boundary of the disk. If G is a gap, we will denote this intersection G_0. When G_0 is finite G is a polygon.

8.2 **Definition.** The geodesic lamination L is *forward invariant* for the map

$$f : S^1 \to S^1$$
$$f : z \longmapsto z^d$$

if it satisfies

(c) Forward invariance: if any leaf $\overline{pq}$ is in L, then either $f(p) = f(p)$, or $\overline{f(p)f(q)}$ is in L.

The lamination L is *invariant* if it is forward invariant, and it also satisfies conditions (d) and (e):

(d) Backward invariance: if any leaf $\overline{pq}$ is in L, then there exists a collection of d disjoint leaves, each joining a preimage of p to a preimage of q.

(e) Gap invariance: For any gap G, the convex hull of the image of G_0 is either

  (i) a gap, or

  (ii) a leaf, or

  (iii) a single point.

  In case (i), the boundary of G must map as a covering map to the boundary of the image gap with positive orientation: *i.e.*, the image of a point moving clockwise around G must move clockwise around the image of G.

The notion of an invariant lamination is tricky, because the map $z \longmapsto z^d$ is not a homeomorphism. Condition (d) is not as strong as one might think: it is not specific about which of the preimages of p and q are joined by leaves, and there is not really any good rule to decide this in general. It does not say anything about additional leaves beyond the minimum Condition (e) helps control the possibilities, by guaranteeing that at least the image of a gap cannot cross leaves of the lamination. It sometimes happens, for instance, that there is a polygonal gap with $2k$ sides, whose vertices all map to the endpoints of a single leaf. If the sides of this polygon are isolated, then without condition (e), one could erase half of its sides (and all preimages of them) and still have a lamination satisfying the other conditions.

For any forward invariant geodesic lamination L, the map $z \longmapsto z^d$ of the unit circle to itself extends to a map of the circle union $\bigcup L$ to itself. This extension is formed by mapping each leaf linearly to the leaf (or point) joining the images of its endpoints. It is not quite so clear how to extend the map over the gaps, but it can still be done, provided the lamination is invariant. One way to define an actual map is to decree that the center of mass of any gap maps to the center of mass of its image, and that each ray from the center of mass to a point on the boundary of a gap maps linearly to the appropriate ray in the image.

8.3 **Definition.** Let L be an invariant lamination, and let G be a gap of L. The *degree* of G is defined to be the degree of the map of G to its image gap, provided its image is a gap. If the image of G is a leaf l, the degree of G is $1/2*$(the number of leaves on ∂G mapping to l). If its image is a point, then the degree is zero. A gap is *critical* if its degree is not one. A leaf $l \in L$ is a *critical leaf* if the images of its two endpoints are the same (so the leaf maps to a point).

The idea of an invariant lamination has a generalization which is often useful. Suppose we have two laminations L and M on two copies of the disk, and suppose that f is a map from the boundary circle of one disk to the boundary circle of the other, which is topologically a covering map. Then L and M *correspond* by f if conditions (c),(d) and (e) above are satisfied for f The notions of the degree of a gap, and the definition of a critical leaf go over to this more general situation.

8.4, Proposition: total critical degree of a lamination. *For any two laminations L and M which correspond by a map f of degree d, the total number of critical leaves of L with respect to f, plus the sum for critical gaps of their degree minus one, equals $d-1$.*

Proof, total critical degree of a lamination: If there are any critical leaves, then we can inductively reduce the proposition to cases with lower degree. Simply cut the circle for the domain lamination along any critical leaf l, and then glue the endpoints of the two resulting arcs, to form two circles. The lamination L produces a laminations L_1 and L_2 on the two pieces. The original map f splits into two new maps f_1 and f_2 of total degree d which make the two new new laminations correspond to M. Each critical leaf of the original lamination yields a critical leaf in one of the new laminations, except that if l is on a triangle made of three critical leaves, then the remaining two sides merge to give only one critical leaf – but there is also one less critical gap of degree 0. Any other kind of critical gap yields a critical gap of the same degree in one of the new laminations. By induction, this reduces the proposition to simpler cases.

If there are no critical leaves, then the extension of f over the disk, as discussed above, is a branched covering. The branch points correspond to critical gaps, and the degrees of branching $b_1 \cdots b_k$ are the degrees of the gaps. A simple Euler characteristic computation gives

$$d^* \chi(D^2) - \sum_{b_i} (b_i - 1) = \chi(D_2)$$

and the proposition follows, since the Euler characteristic $\chi(D^2)$ is one.

– **total critical degree of a lamination**

There is a basic and very useful construction for invariant geodesic laminations, which we will describe shortly. The idea is to start with a lamination L which is not invariant – only forward invariant – but which has a complete set of critical leaves and gaps. Often, it is possible to work backwards, successively producing a sequence of laminations $L_0 = L, L_1, \cdots$ such that $L_{(i+1)}$ corresponds to L_i by $f: z \longmapsto z^d$. Finally, one takes the closure L_∞ of the union of the L_i to obtain an invariant lamination. The original lamination might consist only of critical leaves. With a small amount of initial information, one constructs the lamination L_∞ which is essentially the only lamination having the given set of critical leaves. Here is the formal statement:

8.5, Proposition: constructing-invariant-laminations. *Let L be a lamination without critical gaps which satisfies the following conditions:*

(i) *L is forward invariant by the map $f: z \longmapsto z^d$, and*

(ii) *there are enough critical leaves – $d-1$ – to satisfy the formula of the preceding proposition.*

Then there is an invariant geodesic lamination, the f-saturation $S(L)$ of , which contains L.

Proof, constructing-invariant-laminations: We will construct a sequence of geodesic laminations $L_0 = L, L_1, L_2, \cdots$ such that $\underline{L}_{(i+1)}$ corresponds by f to L_i. To construct the sequence, consider a map $\overline{f}$ of D^2 to itself which extends the map f over the entire disk, mapping regular leaves linearly to leaves, mapping critical leaves to points on the boundary, and extending the map over gaps as a homeomorphism on the open gap if possible, otherwise by collapsing to the image leaf or point.

We can define a lamination M whose leaves are mainly the inverse images of leaves of L by $\overline{f}$, together with the critical leaves of L. This is closely related to the proof of the preceding proposition – we are thinking of the domain disk as being made up of d pieces, each of which maps 1-1 to its image, and just imposing a copy of L in each one. The only trouble with this definition is with leaves l of L which have an endpoint on the image (by $\overline{f}$) of a critical leaf of L. In this case, for each component of $f^{\circ -1}(l)$ which intersects a critical line, we put in two leaves, going to the two ends of the critical line or chain of contiguous critical lines; if the component of the preimage is a gap, use leaves on the boundary of the gap for this purpose. If both endpoints of l meet images of critical lines, we still put in at most only two leaves. Now straighten out all the leaves of M without changing their endpoints, to make a geodesic lamination L_1. The lamination L_1 contains the original lamination L, by the condition on surjectivity of f.

We repeat this construction a countable number of times, to obtain our sequence $\{L_i\}$ of geodesic laminations, first straightening out the map $\overline{f}$ at each step so that it really maps $L_{(i+1)}$ to L_i. We pass to the limit, forming the closure of the union of the L_i. Conditions (c), (d) and (e) are satisfied by the closure of the union because they are satisfied at each stage.

$-\square$

As an example of this construction, we can take the case that $d=2$, with a geodesic lamination having a single leaf l which is any diameter of the circle. A diameter is automatically a critical leaf for $z \longmapsto z^2$. There is a convenient notation for specifying a diameter: it is determined by the square of the two endpoints, which we represent as $exp(2\pi\alpha)$ for some $\alpha \in [0,1)$. We denote $L(\alpha)$ the lamination constructed from the α-diameter.

For simplicity, let us consider the case that neither endpoint of the diameter is a periodic point under $z \longmapsto z^2$, which is equivalent to the condition that α is not a rational number with odd denominator. Then L_1 has 3 leaves, the two new leaves being on the two sides of the chosen diameter; L_2 has 7 leaves, one new leaf in each of the four components of the complement of the existing three; L_3 has 15 leaves, and so forth.

As a further example, the geodesic lamination with a single leaf going from the 1/ 3 point of the circle to the 2/ 3 point is forward invariant by the map $z \longmapsto z^2$. Adjoin to this lamination any diameter which does not intersect this leaf — that is, a diameter with one of its endpoints between the 1/ 6 point and the 1/ 3 point. The f —saturation of this example contains the f - saturation of the corresponding example from the last paragraph. One would expect that, at least for a typical choice of diameter, the two laminations would be the same, since there is not usually much room left over to enlarge an invariant lamination. (We shall prove something to this effect below).

Similarly, there is a forward-invariant lamination with 3 leaves which make a triangle joining the 1/7 point, the 2/7 point, and the 4/7 point. The map $z \longmapsto z^2$ combinatorially rotates this triangle by 1/3. Any diameter of the circle with one endpoint between the 1/14 point and the 1/7 point is compatible with this triangle (doesn't intersect it), so that, applying the proposition, one obtains a whole interval of laminations, each containing this triangle.

As an illustration of what can happen when we begin with a critical line which has a periodic endpoint, let us consider the simplest case — we begin with the lamination L having a single leaf, joining the 0-point to the

1/2-point. Then L_1 has four new leaves – joining the 1/4-point and the 3/4-point to the ends of the original diameter. L_2 has eight new leaves, with two new leaves in each quarter-circle, joining the midpoint of the quarter-circle to its two ends. The process continues in the same way. This lamination is not good for the purpose of defining a set of identifications of the circle which might occur for a polynomial map of degree 2, because it would force all dyadic points on the circle, and hence all points, to be identified to a single point. However, it is still significant as a prototype for other laminations having critical leaves with periodic endpoints.

The construction raises two questions: first, what are the possible choices for critical leaves and critical gaps in an invariant lamination, and second, to what extent is the invariant lamination determined by these critical leaves and gaps?

9. Invariant laminations for the degree two case.

We shall now specialize to an analysis of invariant laminations for the degree two case. Much of what we shall do seems to have natural generalizations to the cases of higher degree, and I hope that someone will undertake a careful analysis of these cases. However, there is a particular simplicity and clarity in the phenomena for degree two, and it seems worthwhile to concentrate on this.

It is convenient to define the *length* of a leaf l of a lamination to be the distance between its endpoints, measured *along the circle* in units of radians/2π. Thus, the maximum length of a leaf is $1/2$; a leaf of length $1/2$ is a critical leaf.

Consider the sequence of lengths L_i of the sequence of forward images l_i of a leaf l of a forward invariant lamination (for $z \longmapsto z^2$). If the length L_i is less than $1/3$, then $L_{i+1} > L_i$, while if it is greater than $1/3$, $L_{i+1} < L_i$. Given a leaf l, let m be the (nonempty but possibly finite) sequence of indices m_i such that $L(l_{m_i})$ is not less than any preceding length, and $L(l_{m_i}) > 1/3$.

If l is any leaf, then $-l$ is a line which does not cross any leaves of the lamination (as you see by looking at the images under $z \longmapsto z^2$). Let $C(l)$ be the subset of the disk containing the origin, and bounded by $l \bigcup -l$.

9.1, Lemma: centrally-enlarging. *The leaf $l_{m_{i+1}}$ is the first leaf in the interval $l_{m_i+1} \cdots l_{m_{i+1}}$ to intersect the region $C(l_{m_i})$.*

Proof, centrally-enlarging: First, note that since l_{m_i}, $-l_{m_i}$ and $l_{m_{i+1}}$ all have length at least $1/3$, that $l_{m_{i+1}}$ must indeed be contained in $C(l_{m_i})$.

The length of the image of l_{m_i} is $1 - 2L_{m_i}$, while the two intervals in which $C(l_{m_i})$ touches the boundary of the disk are shorter, each having length $1/2 - L_{m_i}$. Any leaf which intersects $C(l_{m_i})$ must be shorter than $1/2 - L_{m_i}$, or longer than L_{m_i}. The leaf l_{m_i+1} is the shortest leaf until $l_{m_{i+1}+1}$, so the lemma follows.

— **centrally-enlarging**

9.2, Theorem: gaps-eventually-cycle. *Let L be an invariant lamination for* $f: z \longmapsto z^2$, *and let G be a gap of L. Then either*

G eventually collapses to a leaf: it is a triangle which eventually maps to a triangle with a critical leaf as one of its sides, or a quadrilateral which eventually maps to a quadrilateral containing the origin, or

G is eventually periodic: there is some integer s > 0 and some p > 0 such that $f^{\circ s+p}(G) = f^{\circ s}(G)$. *The first iterate* $f^{\circ s}$ *which maps G to a periodic gap may either be*

a homeomorphism,

a degree two covering, or

it may collapse one edge of G to a point, and map G otherwise homeomorphicly.

The cases of the eventually collapsing triangle or quadrilateral are closely related. When there is a collapsing quadrilateral, it can be subdivided into two triangles (in two ways) by putting in a critical leaf; a new invariant lamination with collapsing triangles is obtained by adding preimages for the new critical leaf. Any collapsing triangle has a symmetric mate. As long as the critical leaf has no preimages which are other sides of the collapsing triangle, this construction can be reversed by removing the critical leaf and all its preimages.

Similarly, the cases of a gap which maps as a degree two covering to a periodic gap and a gap which maps to a periodic gap after one of its edges is collapsed are closely related. If G maps as a degree two covering to another gap which is eventually periodic, then a critical leaf may be added to G which joins any two of its points. By adjoining all the preimages of the new leaf, an invariant lamination is obtained with a gap which eventually maps to a periodic gap, after first collapsing a side. Conversely, if G maps by collapsing one side to an eventually periodic gap, then there is a symmetric gap on the other side of its critical leaf. As long as the critical leaf has no other preimages which are sides of G, removal of the critical leaf and all its predecessors reverses the construction.

Proof, gaps-eventually-cycle:

First let us settle that if a gap eventually maps to a periodic gap, it maps in one of the three ways described. Among all the gaps in the lamination, all map homeomorphically to their images except for at most two: there might be one gap (containing the origin) which maps by a degree two covering, or else there might be two gaps which have a common side which is a critical leaf. If G eventually hits a gap of the first exceptional type, then it can never hit an exceptional gap again unless the exceptional gap is periodic. If G eventually hits a gap G_1 of the second exceptional type, then it might also eventually hit the symmetric exceptional gap G_2: but $f(G_1) = f(G_2)$ is then periodic, so that only one side of G is collapsed before it hits a periodic gap.

Let a, b, and c be three distinct points on $\partial D^2 \cap \overline{G}$ Let G_i be the sequence of images of G, and a_i, b_i, and c_i be the sequences of images of a, b and c, and let d_i be the minimum distance between any two of the three points a_i, b_i, c_i.

If for some i, d_i is zero, it means that the gap eventually maps to one which contains the origin, and two of the three points map to diametrically opposite points. There is at most one gap containing the origin, so if this ever happens again, the gap is eventually periodic, and we are done. If there are no more than four points on $\partial D^2 \cap \overline{G}$, we are also done, since G is a quadrilateral or triangle which eventually collapses to a leaf.

Otherwise, we can assume that a, b and c are chosen so that d_i is always greater than zero.

Each of these triples of points on the circle determines a point p_i in the three-torus. Since distinct gaps cannot overlap, the distance between p_i and p_j is at least $d_i + d_j$, as measured using the metric which is the maximum of the three coordinate metrics — unless $G_i = G_j$. The theorem will follow if we can show that lim sup $\{d_i\} > 0$.

In analogy to the proof of the lemma, let m_i be the sequence of indices such that the largest leaf L_{m_i} on the boundary of G_{m_i} is at least as large as any previous leaf on the boundary of any G_j, and also longer than $1/3$.

If $S = \sup \{L_{m_i}\} < 1/2$, we are done, since then the sequence d_i can never decrease, until it exceeds $1 - 2S$ — after which it can never decrease below $1 - 2S$.

If $L_{m_i} = 1/2$ for at least 3 of the i, then clearly G_i repeats. If this happens fewer than three times, we can throw out an initial sequence of G_i and begin the argument over again.

There can be at most two leaves of a lamination with a given length $L > 1/3$ — any set of three chords of length L has intersections. If it ever happens that $L_{m_{i+1}} = L_{m_i}$, then the longest leaf of $G_{m_{i+1}}$ is identified with the longest leaf of $G_{m_{i+2}}$ by one application of f, and it follows that $f(G_{m_{i+1}}) = f(G_{m_{i+2}})$. In this case, the sequence of gaps starts repeating, and we are done.

The only remaining possibility we have to consider is that the L_{m_i} form a strictly increasing sequence converging to $1/2$. We will show that this is absurd. (What we are really using about the sequence is that it is infinite.) In fact, whenever $L_{m_{i+1}} > L_{m_i}$, then there are exactly two leaves of $G_{m_{i+1}}$ longer than any leaf of G_{m_i}. If $L_{m_{i+2}} > L_{m_{i+1}}$, then the two longest leaves of $G_{m_{i+1}}$ must map to the two longest leaves of $L_{m_{i+2}}$, by the lemma.

In the meantime, d_j never decreases, unless it is greater than 1/ 3. But if ever $d_j > e = 1 - 2L_{m_{i+1}} < 1/3$, it cannot decrease below e again until after time $m_{i+2}+1$ — but this is impossible, since a leaf of length greater than e and less than $L_{m_{i+1}}$ cannot fit into the space between $G_{m_{i+1}}$ and its symmetric copy (rotated 180°).

This says that if, say, $a_{m_{i+1}}$ and $b_{m_{i+1}}$ are the two closest points, the distance between b_j and b_j can never exceed 1/ 3 for $m_{i+1} < j \leqslant m_{i+2}$, and by induction, for all time. The absurdity of this statement proves the theorem.

— **gaps-eventually-cycle**

This theorem is invaluable for classifying laminations in the degree two case. The question whether this generalizes to higher degrees, and if not what is the nature of the counterexamples, to me seems the key unresolved question about invariant laminations.

We are now in a position to make a qualitative analysis of gaps for degree two invariant laminations.

9.3, Theorem: gap-classification. *A gap G of a degree two invariant lamination intersects ∂D^2 in either a finite set of points, a Cantor set, or a countable set of points.*

If G is not periodic under f , then either

it maps by some iterate of f to a periodic gap (in one of the three ways mentioned in the preceding theorem), or

it is a triangle or quadrilateral which eventually collapses.

If G is periodic, then the return map g may have either

degree one, in which case G may intersect the circle in either

a finite number of points, in which case G is a finite-sided polygon on which g acts by a homeomorphism which permutes the sides transitively (as a rational rotation).

a Cantor set. In this case, g acts on ∂G as an irrational rotation blown up along the backward orbit of some point (see discussion below).

a countable set of points. For this to happen, there must be at least one point on the boundary of G which has finite order q under the return map. There is at most one side S of G which is fixed by the return map; except for S, all sides of G eventually collapse to points, and they are connected to each other in q one-sided-infinite chains.

degree two, in which case G intersects the circle in a Cantor set, where g acts as the shift map on two symbols (see discussion below).

Before proving the theorem, we will construct examples as mentioned in its statement.

The standard tertiary Cantor set in the unit interval may be defined as the set of numbers which possess base three expansions in which only the digits 0 and 2 occur. The *shift* map (or the one-sided Bernoulli shift on two symbols) is the homeomorphism which discards the first digit, shifting each of the others to the left. It is analogous to the squaring map of the circle, since it maps the bottom half and the top half of the Cantor set linearly to the whole. To say that g acts on a set as the shift map is to say that there is a homeomorphism of the set to the standard Cantor set conjugating the map to the shift.

For any irrational angle α (as measured in our units, radians/2π), let $R_\alpha : S^1 \to S^1$ be the rotation of the circle by α. Another map $h_\alpha : S^1 \to S^1$ can be constructed as follows. Construct a probability measure ν on the circle by picking an arbitrary point x_o (say 1), giving x_o measure 1/ 2, giving $R_\alpha^{\circ -1}(x_o)$ weight 1/ 4, $R_\alpha^{\circ -2}(x_o)$ weight 1/ 8, etc. It is easy to construct a continuous map $m : T^1 \to S^1$, where T^1 is a second copy of the circle such that m pushes the Lebesgue measure μ (normalized to have total mass one) of T^1 to the measure ν on S^1. (This process blows up the backward orbit of x_o). R_α comes from a unique map h_α of T_1 to itself with the property that the derivative of h_α is the constant 2. h_α has degree two. m is not quite a semi-conjugacy of h_α to R_α, but it acts as a semi-conjugacy everywhere except on the longest blown-up interval (which has length 1/ 2). This longest interval $m^{-1}(x_o)$ maps by h_α to the whole circle, so that its endpoints are mapped to a single point. Each of the other blown-up intervals maps to a linearly to a successor interval of twice the length.

It folows that h_α is conjugate to the map $z \longmapsto z^2$. There is a forward invariant lamination for h_α, consisting of all the lines which join endpoints the intervals which have been obtained by blowing up points. By the construction **constructing-invariant-laminations** there is an invariant lamination L_α containing this one. It has an invariant gap, which maps to itself by a blown-up rotation. This example is the general model: we will say that the return map for a gap is a blown up rotation of angle R_α if it is topologoically conjugate to the return map for this model.

The gaps which touch the boundary of the disk in a countable set of points are annoying special cases. The simplest example is comes by joining points in the sequence 0, *1/2*, *1/4*, *1/8*, *1/16*, $\cdots$. This forms an invariant gap, where the fixed leaf S_0 reduces to a point. By the technique of **constructing-invariant-laminations** it can be extended to an invariant lamination. Note that for the invariant lamination, there is a chain of leaves of the lamination coming arbitrarily near any pair of points: the lamination could not arise from a set of identifications.

As an example, where there is a side S fixed by the return map, form a gap G by joining points in the sequence

$$\frac{1}{3}, \frac{2}{3}, \frac{1}{6}, \frac{7}{24}, \frac{31}{96}, \cdots$$

G returns with period 2 ; the leaf *1/3— 2/3* flips under $z \longmapsto z^2$, so it is fixed by the return map. The leaf *2/3— 1/6* is a critical leaf; the other leaves form a never-ending line marching forward to the slaughter.

If you remove the leaf *1/3— 2/3* from the above example, you obtain an example of a gap which has period of return 1, but where there are two chains of sides which are interchanged by the return map.

Other examples can be constructed similarly with arbitrary rational rotation numbers p / q, by modifying the invariant laminations constructed above which have polygonal gaps which return with period 1 and rotation number p / q. Next to such a polygon is a central gap, which returns with period q and degree two. Transplant the example 0 — *1/2* — *1/4* — *1/8* — $\cdots$ into the central gap to obtain a gap which returns with period q, and rotation number 1, and for which there is one leaf (a side of the original polygon) fixed by the return map. Finally, remove each copy of the original polygon from the invariant lamination, to obtain an example with a gap which returns with period 1 and rotation number p / q.

Proof, gap-classification: This theorem reduces to the statements about the periodic cases, by the preceding theorem.

Among all the sides of the gaps in the future orbit of G, there is a side P which is largest. We may as well assume that P is a side of G.

There is at most one other side of G whose length is at least 1/3. We will call this other side Q, if it exists

Every leaf l of the lamination eventually lands on a leaf of length at least 1/3, by some forward iterate of f. If l is a side of G, the first time that a forward image of l has length greater than 1/3 need not be when G returns to itself. However, by lemma **centrally-enlarging**, it is easy to see by an inductive argument that any sides of the other gaps in the orbit of G which have length at least 1/3 map to the long sides of G upon their first return to G.

It follows that there are at most two orbits of sides of G under the return map g. We shall see that in most cases, there is only one orbit.

There are a variety of different situations, which we shall now analyze individually.

I. First we take up the case that g has degree one. This means that G does not contain the origin.

> **A.** Suppose that P is not a critical leaf. Then g is a homeomorphism on the boundary of G, so it acts as a bijection on the sides of G. P and Q (if it exists) must eventually return to P or Q, so

their future orbits are finite. It follows that G has a finite number of sides, because of the bijectivity and the fact that every side eventually hits P or Q.

g cannot act on the sides as the trivial permutation. Therefore, if Q exists, g maps it to a different side, which by lemma **centrally-enlarging** can only be P: hence, g permutes the sides transitively.

The sides of G must touch, since the backward orbit of P and Q is dense in the boundary of G. G is indeed a polygon, and g maps it as described.

B. If P is a critical leaf, the return map g maps the sides of G other than P bijectively to the sides of G. Therefore, G must have an infinite number of sides.

1. If there are any periodic points for g on the boundary of G, they all have the same period, since g preserves the circular order. Not all points on the boundary of G are periodic points, since g is not injective.

a. It may be that the periodic points for g are fixed points.

We claim that there is either exactly one fixed point or one fixed leaf on the boundary of G. Indeed, if there were more, we could obtain a new lamination by adjoining three leaves which join any three fixed points for g. By adding also all the preimages of these three leaves, we would obtain an invariant lamination with a gap whose return map is the identity, violating the conclusion from the case A.

If Q does not exist, there is only one orbit of sides of G under g. But even if Q exists and g does not fix it, then it can never return to itself (since there are no periodic points other than fixed points), so it eventually maps to P, and there is only one orbit of sides of G under g.

If x is an endpoint of a leaf and not a fixed point for g, then one of the two intervals between x and $g(x)$ does not contain a fixed point This interval must form exactly one leaf: otherwise there would be more than one orbit of sides of G which are not fixed. Thus, the non-fixed leaves form an infinite chain as described.

b. If the period q of the periodic points of g is not 1, then we will construct a new lamination, thereby reducing to case a when $q=1$.

Under this hypothesis, the lamination can be subdivided by adding leaves to form the boundary of the convex hull of all periodic points for g on G_o, and also adding all their preimages. If $q > 2$, then the convex hull of the periodic points is a new gap, which by case A must have exactly q sides; if $q = 2$, there can only be two periodic points, so their convex hull is a leaf. The new lamination has a gap with a countable number of sides and period q times as great; its return map has fixed points, reducing the statement to case b.

2. This is the case that the longest leaf P of G is a critical leaf, and there are no periodic points on the boundary of G.

As before, we conclude that there is only one orbit of the sides of G under g, since in the absence of periodic points no side can ever return to itself: hence, all sides must eventually land on P and collapse.

No two sides of G can touch, for otherwise two distinct iterates of their common point would hit the image of P. This would violate the assumption that the boundary of G has no periodic points.

Since the endpoints of the leaves for any invariant lamination are dense on the circle, it follows that G touches the circle in a set obtained by removing a countable dense sequence of open intervals, that is, a Cantor set.

If we form the quotient space of the boundary of G by the equivalence relation which identifies each leaf to a point, we obtain a circle, and g induces a homeomorphism h on the circle. It remains to show that h is topologically conjugate to a rotation.

It is not true that all homeomorphisms of the circle which have no periodic points are topologically conjugate to rotations: if you start with a rotation and blow up an orbit in both forward and backward time, you can obtain a counterexample. This construction is due to Denjoy, who showed that such homeomorphisms can even be constructed which are C^1 diffeomorphisms. What is true is that if a homeomorphism h of the circle has the property that the only minimal set (nonempty closed invariant subset) is the whole circle, then h is conjugate to an irrational rotation.

In our case, if there were any proper closed invariant subset K, for h, we could construct a new invariant lamination by adding leaves to form the boundary of the convex hull of its preimage K_1 on the boundary of G. Since the backwards orbit of P is dense in the boundary of G, the proper invariant subset K_1 could not contain P. The new lamination would have a periodic gap which does not have a side which eventually collapses, so that by case A it would have to be a polygon, which is absurd.

We conclude that h is actually conjugate to a rotation.

II. Now we consider the case that the return map $g: G \to G$ has degree two. We can suppose that G contains the origin.

In this case, Q exists and has the same length as P, and it is identified with it after one application of f. By lemma **centrally-enlarging**, it follows that either $g(P)=P$, or $g(Q)=Q$. Up to labeling, we can assume we have the first case.

The return map g acts on the boundary of G as a 2-fold covering. Since every side of G eventually lands on P, and sides are dense, it easily follows that the intersection of G with the circle is a Cantor set, and that the return map is topologically conjugate to the two-sided shift.

— **gap-classification**

10. The structure of the set of quadratic invariant laminations

In the preceding section, we gained some insight into the structure of particular quadratic invariant laminations. Now we will work toward an understanding of the relationships among different quadratic invariant laminations, and the structure of the set of quadratic invariant laminations.

We will need the following

10.1, Proposition: gap-density. *If L is an invariant lamination other than the lamination whose leaves are all vertical lines in D^2, then the gaps of L are dense in D^2.*

Proof, gap-density: Suppose that L is a lamination such that gaps are not dense. Then there is some one-parameter family l_t of distinct leaves of L. Let q_t and p_t be the moving endpoints of l_t.

There is at least one interval of time such that either p_t or q_t is not constant. Therefore, we have an interval I where there is a map τ defined such that for each x in the domain of τ, x is connected to $\tau(x)$. By forward invariance, the definition of τ can be extended to the whole of the circle, since I_1 eventually engulfs everything, and τ is locally determined by the leaves of L. In other words, there are no gaps at all in L. It follows that τ is a homeomorphism of the circle of order two.

Lifting τ to the universal cover of the circle, one obtains an involution of $\mathbf{R}$ which commutes with multiplication by two, and which conjugates translations by integers to translation by their negatives. Therefore, it takes every number of the form $p / 2^n$ to its negative, which implies that $\tau(z)=\bar{z}$.

— **gap-density**

Here is an important concept:

10.2 **Definition.** A leaf l of a forward-invariant lamination L (in the degree two case) is *principal* if l is at least as long as any of its forward images. For any invariant lamination L, let M be the longest length of any leaf of L. Any leaf of length M is a principal leaf. Unless M has length $1/2$, L has two leaves of length M. The longest leaf or leaves we will call *major* leaves of L. The images of the major leaves of L coincide; this image may be either a point or leaf. This image we will call the *minor* leaf of L (by abuse of notation, since it might not be a leaf at all). The minor leaf of an invariant lamination has the property that its length is the *infimum* of the *lim infs* of the sequence of lengths of the forward images of any leaf of L.

Our goal is to describe the set of laminations in terms of their minor leaves. Unfortunately, it is possible for two different laminations to have the same minor leaf: let's look at some examples.

In the first place, any lamination L having a gap with a critical leaf as one of its sides, one of whose endpoints is fixed by the return map for the gap, can be subdivided by putting in leaves joining this fixed point to all of the vertices of the gap. After adjoining all the backwards images of the new leaves, one obtains an invariant lamination L_1 with the same major leaf and hence the same minor leaf as the L.

Another class of examples for the non-unicity of an invarinat lamination with a given minor leaf is obtained from any lamination L with a gap G containing the origin (so that it returns with degree two). It has two major leaves, P and Q. Adjoin two new leaves connecting the ends of P with the ends of Q, forming a collapsing quadrilateral. The quadrilateral separates G into two additional components (besides itself), each of which maps by the return map for G homeomorphically to all of G. Subdivide each of these regions by putting in the backward image of the quadrilateral, and continue inductively. In this way, one can obtain an invariant lamination with a collapsing quadrilateral, having the same major leaves and minor leaf as the original.

These two constructions are related: there are four related laminations, joined by a cycle of four of the constructions we have discussed, as follows. You can begin with any lamination having a central gap which returns with degree two; it can be subdivided, yielding an invariant lamination having a gap with a critical side, a countable number of vertices and an edge which is fixed by the return map; the gap with a countable number of sides can be subdivided into eventually collapsing triangles; removing the critical edge along with all its preimages from the lamination yields an example with a eventually collapsing quadrilateral; and this lamination is the same that can be obtained by putting quadrilaterals into the original central gap.

Similarly, let L be any lamination with a central polygonal gap having 6 or more sides, which maps with degree two to a polygon with half the number of sides Then L can be subdivided by putting a quadrilateral in the middle. The polygons on the two sides of the quadrilateral map homeomorphically to their images; the central quadrilateral collapses to a leaf. The major and minor leaves for the new lamination are identical the major or minor leaves.

In all these cases, at least one of the two laminations with a given minor leaf has a shortcoming: it has leaves which intersect (on the circle) but are not sides of a single gap. Such laminations cannot arise from the process of outlining the convex hulls of the collections of points which are identified as limiting values of a conformal (or even continuous) mapping from the disk to the complement of a closed connected set in the sphere.

10.3 **Definition.** A lamination is *clean* if any two leaves with endpoints in common are sides of a common gap. Otherwise, it is *unclean*.

Only the clean laminations have potential direct applicability to the theory of rational maps, but unclean laminations are indirectly useful in helping to explain the structure of the set of laminations. They can arise as the limiting case of a sequence of clean laminations; in the original definition of a lamination, cleanness was not built in so that the condition for a set of lines to form a lamination would be a closed condition. In making constructions, for instance, it is convenient to be able to form an increasing sequence of laminations and then pass to the closure of the union. Afterwards, one can worry about whether the result is clean.

One way to separate clean from unclean laminations is by their gaps:

10.4, Proposition: clean-gaps. *If a lamination L is clean then*

a). *no side of any gap G collapses under f unless G is a periodic gap which returns by a blown up irrational rotation,*

b). *the sides of any eventually-collapsing quadrilateral are not isolated leaves of L, and*

c). *No vertex of a collapsing quadrilateral is a periodic point.*

Conversely, a lamination which satisfies a) and b) is clean.

Proof, clean-gaps:

If there is a gap G which has a side that is a critical leaf, and if then there is another gap on the other side of the critical leaf. If G does not return by a blown-up irrational rotation, then by examining the various cases in the previous theorem, we see that the critical leaf touches other leaves of each of the gaps, so L is not clean.

If there is a side of an eventually-collapsing quadrilateral which is an isolated leaf, then the gap on its other side must be a polygon (since a collapsing quadrilateral precludes collapsing leaves and periodic degree-two gaps), so L is not clean.

If there is a collapsing quadrilateral Q with a vertex v which is a periodic point, then some other quadrilateral having v as a vertex must map to Q when v returns. Therefore, L is not clean.

For the other direction, suppose that the lamination L is not clean, so that there are two leaves l_1 and l_2 which touch at there endpoints, but are not sides of a common gap. This means that there is at least one other leaf l_3 between them, with the same endpoint p. There must also be at least two gaps G_1 and G_2 which touch the circle at the same point, because gaps are dense in the disk.

Neither of the G_i can be eventual degree two periodic regions or eventual irrational rotation regions, since they each have two sides that touch at p — it is truly a vertex. They cannot both be eventual periodic polygons, for topologically, that would force the return map of one of the periodic polygons to fix a vertex. (Note that two gaps which touch at a vertex are never identified with each other, unless they touch along a common side which eventually collapses.) If no side of G_1 or G_2 eventually collapses under f, the only remaining possibility is that at least one of them is an eventually collapsing quadrilateral.

If G_1 and G_2 have a common face, then we are done. Otherwise, there must be at least one more gap with a vertex at p, and we may assume that it is an eventually collapsing quadrilateral, and that in fact there are an infinite number of eventually collapsing quadrilaterals meeting at p. It follows that at least one vertex of the collapsing quadrilateral is a periodic point.

To prove that b) holds as well as c), consider the set B of points on the circle which are endpoints of at least three leaves. If any point of B is joined to another point of B by a leaf, then this leaf is necessarily isolated, and by previous considerations we are done. Otherwise, let M consist of all diagonals of eventually collapsing quadrilaterals which do not have an endpoint in B. The union $L_1 = M \bigcup L$ is closed, so it is an invariant lamination. Let K be the set of leaves of L (or L_1) which have an endpoint in B. Let J be the interior of K relative to the space of leaves of L_1. J is an open, invariant set of leaves.

Now if we form $L_2 = L_1 - J$, we have a new invariant lamination. Each point of B is eventually periodic, and it is contained in a gap of L_2 with an infinite number of eventually collapsing sides. By the previous theorem, it follows that the sides of these gaps are connected in chains, which implies that the eventually-collapsing quadrilaterals of L had sides in common with other eventually-collapsing quadrilaterals.

— **clean-gaps**

Given a chord or boundary point m of the disk, when is it the minor leaf of a lamination?

There are four preimages of the endpoints of m (or two when m is degenerate), which can be joined in pairs in at most one way by leaves

M and M_1 of length at least 1/ 3. The first condition is that all of the forward images of M have disjoint interiors, and that they all have shorter length. If this is satisfied, then we immediately obtain a forward-invariant lamination $F(m)$ containing m.

There is also a fully invariant lamination $I(m)$ having m as a minor leaf. In the degenerate case, when m is a single point, we simply note that the lamination with a single leaf M is forward invariant and has enough critical leaves, so that the construction **constructing-invariant-laminations** applies to give us $I(m)$.

The construction in the case that m is a true leaf is similar, and goes as follows:

> The leaf m cuts the disk into two pieces a small piece D_0 and a large piece D_1. The two major leaves M and M_1 cut the disk into three pieces, E_0 in the middle, and E_1 and E_2 on the two sides. Let I_0 consist of all the forward images of M and M_1. (We do not take their closure at this stage, to avoid unnecessary problems with ambiguous liftings). Having constructed I_i, I_{i+1} is constructed inductively by adjoining preimages to leaves in I_i. Both E_1 and E_2 map to D_1; the map is a homeomorphism. We adjoin to E_1 and E_2 all leaves of I_i in D_1, transplanted by the inverse homeomorphism.
>
> The region E_0, which maps to D_0, is also straightforward, as long as we keep in mind that we must not create any leaves separating M from M_1. For any leaf l of D_0 of m, there are exactly four points on the boundary of E_0 which map to its endpoints, and there is only one way to join them in pairs without creating any leaves longer than M. Do it.
>
> By inductive hypothesis, I_i is forward invariant. This forces I_{i+1} to contain I_i, and it follows that I_{i+1} is forward invariant.
>
> Finally, we define $I(m)$ as the closure of the union of the I_i. $I(m)$ is fully invariant.

Here are three cogent question concerning $I(m)$:

(a) Under what conditions is $I(m)$ the unique lamination with m as minor leaf?

(b) Under what conditions is $I(m)$ clean?

(c) Under what conditions is $I(m)$ the smallest lamination having m as a minor leaf?

Property (c) is the easiest to characterize.

10.5, Proposition: invariant-of-chord. *If m is a (non-degenerate) chord in D^2 such that m and its forward images have disjoint interiors, and no forward image of m is shorter than m, then*

(a) *m is the minor leaf of an invariant lamination $I(m)$ (constructed above), which has the property that any other lamination having minor leaf m contains $I(m)$.*

(b) *m is the leaf of a forward invariant lamination $F(m)$, which has the property that any other forward invariant lamination containing m contains $F(m)$.*

Proof, invariant-of-chord: At each step of the construction, the existence of previous leaves forced all the new leaves.

— **invariant-of-chord**

10.6, Proposition: invariant-of-point. *Let m be any point on the circle. Then $I(m)$ is the minimal lamination with m as minor leaf iff m is not periodic under $z \longmapsto z^2$.*

Proof, invariant-of-point: The construction for $I(m)$ involves inductively adjoining preimages at each stage, beginning with M.

If m is periodic of period q, then there is a $q-1$st preimage of M which has an endpoint at m, and the construction for the next stage involves taking two lifts for any such leaf. This creates two triangles with a common side M. Continuing in the construction, one obtains an infinite collection of triangles forming a triangulation of a region in the center of the disk. By the construction we have already seen, some of the edges can be removed to leave a lamination still containing M.

If, on the other hand, m is not periodic, then no ambiguities ever arise in the construction, so that any invariant lamination containing M must contain $I(m)$.

— **invariant-of-point**

Here are some more observations concerning the three properties mentioned above:

10.7, Proposition: niceness-of-laminations.

(i) *Any invariant lamination which is clean is the minimal lamination with its given minor leaf.*

(ii) *If the minor leaf is nondegenerate and its endpoints are not eventually periodic, then $I(m)$ satisfies all three properties: it is clean, minimal and unique.*

(iii) *If the minor leaf is degenerate and not eventually periodic, then $I(m)$ is the unique invariant lamination with minor leaf m (but it is not necessarily clean)*

Proof, niceness-of-laminations: Let L be any clean lamination. Let m be its minor leaf. If L' is any other lamination with minor leaf m, then L' cannot be contained in L:

If it were, then L' would be clean as well. The additional leaves of L would have to fit in the gaps of L', and since L is clean, the only possible gaps would be those which intersect the circle as a Cantor set. The irrational rotation gaps have no invariant subdivisions at all, so those are not possible.

The only remaining possibility is that L' has a central gap which returns with degree two. Then, since L has the same major leaves, it must also have a central gap. It cannot be a degree two gap, since such a gap is determined by the major leaves; it could only be an eventually periodic polygon or a collapsing quadrilateral. But either of these would make L unclean.

To prove part (i), we need only show that L and L' both contain some invariant lamination with the same minor leaf.

Suppose first that m is degenerate.

Then m cannot be a periodic point: otherwise, preimages of M would intersect M, and because of the forced symmetry, L could not be clean. Therefore, L contains $I(m)$.

On the other hand, if m is nondegenerate, then $I(m)$ is contained in both L and L', so in either case we are done.

As for part (ii), suppose that L is an invariant lamination with a non-degenerate minor leaf m whose endpoints are not eventually periodic. Then L contains $I(m)$. The central gap of L in this case is a collapsing quadrilateral, and all other gaps are either eventually collapsing quadrilaterals or periodic polygons. None of the sides of the central quadrilateral can be sides of other gaps, because of the non-periodicity of the endpoints of m. Therefore, $I(m)$ is clean. The only gaps of $I(m)$ which can be subdivided to form another invariant lamination are eventually collapsing quadrilaterals, but this subdivision changes the minor leaf.

Part (iii) is proven similarly, by examining the gaps. We already know that $I(m)$ is minimal, so we just need to show that it cannot be subdivided. If there are gaps adjacent to the major leaf M, then they can only be irrational rotation domains or collapsing triangles. Gaps which do not eventually touch M can only be polygons. All three kinds of gaps are indivisible.

— **niceness-of-laminations**

Rather than to pursue all the intricacies of questions (a), (b) and (c), it is more important for us to investigate the set of all minor leaves. It turns out that the collection of all possible minor leaves has a beautiful structure, a structure which should by now be familiar:

10.8, Theorem: adolescents. *The set QML of all chords in D^2 which are minors of some invariant lamination for $z \longmapsto z^2$ forms a lamination.*

Proof, adolescents: Let m_1 and m_2 be any two minor leaves of laminations. Let $D(m_i)$ denote closure of the small region of the disk cut off by m_i. It suffices to prove that m_1 and m_2 have disjoint interiors under the hypothesis that the interior of $D(m_1)$ intersects the interior of $D(m_2)$.

If these two regions intersect, choose any point m in their intersection which is not eventually periodic, and let QML be the diameter joining its two preimages. Then $QML \bigcup F(m_i)$ is a forward-invariant lamination containing m_i, so it is contained in an invariant lamination L_i. By the preceding result,

$$L_1 = L_2 = I(m)$$

so that, as leaves of a common lamination, m_1 and m_2 are necessarily disjoint.

Since the condition to be a minor leaf of a lamination is closed, it follows that the set of all minors of all quadratic invariant laminations forms a lamination.

— **adolescents**

10.9 **Definition.** The lamination M of the disk by all minors of all invariant laminations is the *quadratic minor lamination,* which abbreviated is QML.

There is quite a bit of structure that can be found in the quadratic minor lamination. There is a convenient partial order which can be imposed on the set of leaves of this lamination:

10.10 **Definition.** If m_1 and m_2 are minor leaves for two quadratic invariant laminations, we will say that

$$m_1 < m_2$$

whenever

$$D(m_1) \supset D(m_2),$$

where $D(m)$ is the half-disk bounded by m, as above, when m is non-degenerate, the entire disk if m is the special degenerate minor 1, and the single point m in the general degenerate case.

Thus 1 is the minimum in the partial order: its lamination $I(1)$ is the empty lamination. Every other degenerate minor is a maximal element. Intuitively, larger minors correspond (roughly) to larger laminations although they have smaller half-disks $D(m)$.

The partial order $<$ has the property that if $m < n$ and $p < n$ then either $m < p$ or $p < m$ — that is to say, that the set of minors less than any given minor is linearly ordered.

Furthermore, there are greatest lower bounds for $<$ the intersection of all those half-disks of the form $D(m)$ which contain a given set of leaves S is a half-disk bounded by a leaf of QML which is the greatest lower bound. (The set of such half-disks is non-empty, in virtue of the fact that 1 is less than everything else.)

Every non-degenerate leaf l of QML is of course a leaf of some quadratic invariant lamination — in fact, it is a leaf for many such laminations, since it is contained in most laminations $I(m)$ whenever $m > l$. What about the gaps of QML?

10.11, Theorem: quadratic-minor-gaps. *The gaps of QML are of two types:*

(i) *A polygon P which occurs as a gap in some quadratic invariant lamination. If m is the side of P which is least in the partial order, then the associated lamination $I(m)$ is clean, and it has a central polygon of 6 or more sides which maps with degree two to P.*

Conversely, for any lamination having a central gap of 6 or more sides, the image of the central gap is a gap of QML.

and

(ii) *A copy of the circle, blown up at the rational points. The least side m of this gap (in the partial order) is a minor leaf with periodic endpoints. The associated lamination $I(m)$ is clean, and has a central gap which is periodic with degree two. The other sides of m are its immediate successors, and they also have periodic endpoints; each is the smallest leaf of a similar gap.*

There is one special case, in which m is the degenerate minor 1.

Conversely, every minor leaf with periodic endpoints is a side of a gap of QML whose other sides are the immediate successors of m.

Proof, quadratic-minor-gaps: If G be any gap of QML, there is some least side m of G, and all the other sides are immediate successors of m. Classifying the gaps of QML is equivalent to classifying leaves which have immediate successors in the partial order.

— **quadratic-minor-gaps**

11. Rational maps with preperiodic critical points

We will describe a procedure for constructing and classifying rational maps of the Riemann sphere which yield various combinatorial patterns of asymptotic periodic behavior of the critical points under iteration. For a point $z \in \hat{C}=CP^1=$ the Riemann sphere to be *asymptotically periodic* means that for some k the limit of "$f^{nk}(z)$" is a periodic point w. There is an important special case, when $f^{l+k}(z)=f^{l}(z)$ for some k and l; such a z is called an *eventually periodic* point. If l can be taken to be 0, then z is actually a periodic point; if z is not periodic, then it is a *preperiodic* point. Information about asymptotically periodic behavior of critical points can be derived by known techniques from information about eventually periodic behavior – see Sullivan, [1982]. The new idea we will describe works most simply for the eventually periodic case, although it can be adapted to work in the general asymptotically periodic case..

In practical terms, our method gives

(a) algorithms and computer methods for finding rational maps (or polynomials) with interesting behavior, such as Julia sets which are parametrized as sphere-filling curves.

(b) a mathematical classification of rational maps such that all the periodic points are asymptotically periodic, and

(c) a computer construction for making line drawings of the Julia set.

In the special case of polynomial maps of C, the information can be translated to a particularly nice form so that one can understand not just any individual case, but the set of all possible cases at once, using ideas of Douady and Hubbard [1982]. For a polynomial map such that every periodic point is asymptotically periodic, the Julia set is locally connected, and it is a quotient space of the circle with the dynamics coming from the degree d expanding map of the circle. If furthermore every critical point is eventually periodic but not periodic, then the complement of the Julia set is connected, and the Julia set is a dendrite (or very hairy tree). We shall give a classification of these cases in two equivalent forms: first in terms of a finite set of k-tuples of rational numbers that satisfy some simple conditions, in particular that each $k \geqslant 2$ and that $\sum(k_i\ minus\ 1)=$°, and second, in terms of certain kinds of self-maps of (finite) trees.

In the case of rational maps with real coefficients, the classification is also clean. In particular we will prove that for every possible type of eventually periodic kneading behavior of the critical points there is a unique (up to conjugacy by $PGL(2,\mathbf{R})$) rational map with all critical points real having that kneading behavior. More generally, real rational maps such that all critical points are real and asymptotically periodic are classified by the kneading behavior of the critical points together with the first derivatives of the return maps at each of the attractive or indifferent periodic orbits. (It is known from the classical theory that these periodic orbits are limit points of the critical points which are asymptotically periodic but not eventually periodic.)

For a general rational map for which all critical points are asymptotically periodic that does not have some special form such as in the polynomial case or the real case, I will show that it is determined by the homotopy class of the map rel the closure of the forward orbits of the critical points, together with the derivatives of the return maps at the attractive and rationally indifferent periodic orbits.

(Actually, there are families of exceptional cases, first described by Lattès [191], where this is information is not quite enough. In these cases, there are 1-complex-parameter families of inequivalent rational maps with the same topological behavior of the critical points and their forward orbits. Happily, the method picks out these special cases and shows that they are the only exceptions.)

I will give a necessary and sufficient topological condition on a branched covering of the sphere over itself such that branch points are asymptotically periodic that it be homotopic, rel the closure of the forward orbits of the branch point, to a rational map.

I will also describe procedures to generate large classes of examples of maps satisfying the necessary and sufficient condition, In certain cases, such as the polynomial case, the construction is exhaustive, but I do not know a uniform procedure to generate all examples in general.

12. Description of the method

The idea of the method is to construct a rational function by constructing its inverse. The inverse of a rational function of degree d is a somewhat more complicated entity than the function itself: it is a "multi-valued" function, which one can think of more properly as a d-sheeted branched cover of the sphere, together with a holomorphic map of this branched cover to the sphere. There is a geometrical compensation for the extra combinatorial complications of working with a multivalued function: it tends to decrease, rather than increase distances, provided one somehow gets around the ambiguity of the choice of branch. Intuitively, the extra branches give more material which can spread around and cover the whole sphere, without being stretched.

When we have a particular rational map such that all the critical points are asymptotically periodic, we will analyze it with the aid of a certain metric on the sphere which is canonically determined by the combinatorial data and which is, in fact, contracted by the inverse function. By techniques which are fairly standard, the combinatorial information together with the knowledge that there exists a contracting metric can be used to deduce a precise topological picture of the rational map.

If we do not have a rational map, but only a kind of combinatorial sketch of a potential rational map, how can we decide whether such a a rational map exists, and compute what it is (provided it exists)? For this question, we will make a related analysis, but on one higher level of abstraction — on the level of Teichmüller spaces. The combinatorial sketch of a potential rational map defines a certain map of an appropriate Teichmüller space to itself. Actual rational maps correspond to fixed points of the map of Teichmüller space. There is a certain metric on Teichmüller space — the Teichmüller metric — which is contracted by this map, or at least not expanded. This will provide a good structure to use to analyze existence and uniqueness of rational maps.

To describe the inverse function f ,as a start, one needs to know the locations of the critical values of f. In order to be able to describe *all* the iterates of the inverse function, (we think of the n th iterate of the inverse function as a map from a d^n-fold branched cover of the sphere to the sphere), one needs to know the critical values for *all* the iterates of f. The union of the critical values for all iterates of f is precisely the union of the future orbits $f^l(c)[l \geqslant 1]$ of critical points c. The hypothesis that all the critical points are eventually periodic is clearly suited to this purpose, since it means that the future orbits of the critical points are finite.

It is time for some definitions. First, the topological analogue of a rational map:

12.1 **Definition.** A continuous map $f: S^2 \rightarrow S^2$ is a *branched map* if for each $x \in S^2$ has a neighborhood U such that for some pair of (topological) coordinate systems for U and $f(U)$, the map has the form $z \longmapsto z^k$ for some k. The point x is a *branch point* or, synonomously, a *critical point* if f is not a local homeomorphism in any neighborhood of x, The image $f(x)$ of x will be called a *branch value* or a *critical value*. The *degree* of a branched map is the degree of the map of the sphere to itself (in the usual topological sense), i.e., the number of sheets (assuming that the map preserves orientation).

We will be constantly discussing maps for which all critical points are asymptotically periodic or eventually periodic, so let us give names to these conditions.

12.2 **Definition.** A rational map or a branched map of S^2 to itself is *critically stable* if all its critical points are asymptotically periodic. The map is *critically finite* if all critical points are eventually periodic.

12.3 **Definition.** A point on the forward orbit of any critical point is called a *postcritical* point.

Often, if we have a branched map which is critically finite, we do not really care about the precise map, but only about an equivalence class of maps which have the same structure for the critical and postcritical points, and the same combinatorial branching behavior.

12.4 **Definition.** The *critically finite combinatorial class* of a critically finite branched map f is the set of all critically finite branched maps g for which there exists an orientation-preserving homeomorphism h of S^2 such that

(a) The homeomorphism h conjugates f restricted to the postcritical points of f to g restricted to the postcritical points of g, *i.e.*, h matches the postcritical points of f 1-1 with the postcritical points of g, and on that set, *hsmall* $\circ f =$ *gsmall* $\circ h$.

(b) The composition *hsmall* $\circ f$ is isotopic, *rel* the postcritical points of f, to the composition *gsmall* $\circ h$. In other words, h conjugates f to g up to homotopy *rel* postcritical points.

We need to associate with any critically finite branched map a certain structure on the sphere so that we can describe the branching behavior for the inverse of the map and its iterates. This structure has variously been called a Riemann surface with signature, a V-manifold, or an orbifold. We will use the orbifold terminology.

12.5 **Definition.** An *orbifold* is a space which is locally modeled on $\mathbf{R}^n$ modulo the action of a finite group (which depends on the particular neighborhood being modeled).

Orbifolds are designed to describe the quotients spaces of discrete groups acting on manifolds, where some elements of the groups may have fixed points. The quotient space is described as a topological space, together with a description of how the group action looks in a neighborhood of the original space — that is, how elements of the group act which bring this neighborhood back to itself. For more information, the reader may refer to Thurston [1983], Scott [1983], Satake [???], Poincaré [???] or other source, although for present purposes it may be sufficient to absorb the concept intuitively. An oriented 2-orbifold Q can be described as a surface, together with a discrete collection of points $\{q_i\}$ to which integers $n(q_i) \geqslant 2$ are assigned. At a distinguished point q, the orbifold is modeled on $\mathbf{R}^2 mod\, \mathbf{Z}_{mod\ n(q)}$ acting by rotations; everywhere else it is modeled on $\mathbf{R}^2$. The underlying space for an orbifold is called X_Q.

In practice, what the integers mean is that when we form cover spaces, we allow ourselves to take branched covers $f: Q_1 \to Q$ as well, provided that only the distinguished points are branch values, and furthermore that the degree $deg_r(f)$ of branching at any branch point r (recall that the branch points are in the domain of a branched map) is a divisor of the integer $n(f(r))$ associated with the branch value. As an orbifold the cover Q_1 has as its distinguished points the collection of branch points r such that the integer $n(r)=\frac{n(f(r))}{deg_r}(f)$ is

An orbifold, just like a manifold, can have any of a variety of structures; in particular, there are topological orbifolds, differentiable orbifolds, and complex orbifolds. An orbifold can also have a more rigid kind of geometric structure, such as a Euclidean structure or a hyperbolic structure. For an orbifold to have a Euclidean structure means that it has a Riemannian metric of zero curvature away from the distinguished points, and at the distinguished points the metric is the metric of a cone which is obtained by taking an angle of $2\pi/n$ in the plane and gluing together its two bounding rays.

An orbifold Q is called *good* if it has some cover which has an empty collection of distinguished points, *i.e.*, a cover which is a manifold. If Q is not good, it is *bad*.

12.6, Theorem: orbifold-universal-cover. *An orbifold Q has a universal cover $h:\tilde{Q}\to Q$. In other words, if we keep track of base points in all our orbifolds, the universal cover is a connected orbifold satisfying the condition that for every connected covering orbifold $f:R\to Q$ there is a unique cover $g:\tilde{Q}\to R$ such that $h=f$ small $\circ g$ (where base points always go to base points). The orbifold Q is good iff its universal cover $\tilde{Q}$ is a manifold.* See Thurston, [1983] for a proof.

The *Euler number* $\chi(Q)$ of an oriented 2-orbifold Q is defined to be the Euler number of the complement of the set of distinguished points plus $SUM\frac{1}{n_i}$. In other words,

$$\chi(Q)=\chi(X_Q)-SUM(1-\frac{1}{n_i})$$

In particular, note that adding more distinguished points always reduces the Euler number. (This usually means that the Euler number becomes more negative.)

12.7, Proposition: covering-orbifold-index. *The Euler number of any finite-sheeted covering orbifold satisfies $f:\tilde{Q}\to Q$ is*

$$\chi(\tilde{Q})=\text{\textdegree}(f)\times\chi(Q)$$

12.8, Proposition: branched-map-indices. *A branched map f of degree d of the sphere to itself satisfies*

$$\underset{\text{critical points}q}{SUM}(deg_q(f)-1)=2(d-1)$$

12.9, Theorem: nonpositive-goodness. *Every closed 2-orbifold of non-positive Euler number is good. There are precisely five distinct closed oriented 2-orbifolds of zero Euler characteristic, and they all have Euclidean structures. Every 2-orbifold Q of negative Euler number has a hyperbolic structure, and its area is $2\pi|\chi(Q)$.* See Thurston, [1983] for a proof. The universal cover of a Euclidean orbifold is the Euclidean plane; the group of deck transformations acts as a two-dimensional crystallographic group, often called a wall-paper group. The reader may easily check the five examples:

(1). The torus. This is the quotient of the Euclidean plane by the group of translations which preserve a tiling by parallelograms.

(2). the sphere with four distinguished points of order 2. This is the quotient of the Euclidean plane by the full group of orientation-preserving symmetries of a tiling by parallelograms which are not squares. (Note that a parallelogram has an order two symmetry rotating it *180degrees*. The tiling by parallelograms has three more kinds of symmetries, rotating about the midpoint of either type of edge, or rotating about a vertex.) There are actually good physical models for these Euclidean orbifolds: any tetrahedron whose four sides are all congruent is a model. It is amusing to make a tetrahedron of this form out of cardboard and carefully roll it around on a large sheet of paper, tracing out the triangles where its base can rest.

(3) The (2,4,4)-orbifold: the sphere with three distinguished points, of orders 2, 4, and 4. The Euclidean plane can be tiled by triangles of angles $\pi/2, \pi/4$, and $\pi/4$ (called (2,4,4)-triangles) cutting up a tiling by squares by diagonal lines. The quotient is the (2,4,4)-orbifold. The group is in fact the same as the full group of orientation-preserving symmetries of the tiling by squares.

(4) The (3,3,3)-orbifold: the sphere with three distinguished points of order 3. The universal cover may be visualized as the Euclidean plane, tiled by equilateral triangles which are alternately black and white. The group of deck transformations is the full group of orientation-preserving symmetries of this pattern. There are three kinds of fixed points of order 3: at any vertex, in the middle of any white triangle, and in the middle of any black triangle.

(5) The (2,3,6)-orbifold: the sphere with three distinguished points of orders 2, 3, and 6. This is the quotient of the Euclidean plane by the full group of orientation-preserving symmetries of the tiling by equilateral triangles.

Given any combinatorially finite branched map f of the sphere, we will associate an orbifold Q_f. The orbifold Q_f is chosen to be the simplest orbifold that has a covering orbifold on which $f^{\circ -1}$ is well-defined. Here is the definition. First, if any of the critical points is actually periodic, we remove it and all of its images from the sphere. This surface is the underlying space X_{Q_f} of the orbifold. The set of distinguished points is the union of the remaining postcritical points. To each of these points q, we need to associate an integer which will accomodate any branching behavior which occurs in the branched covers described by f or any of its iterates. The appropriate integer is the least common multiple, among all powers f^k, of the degree of branching of f^k at p, among all points p such that $f^k(p)=q$. As you follow along on the orbit of p by successive iterates of f, checking the local degree of branching from p to $f^l(p)$, the local degree of branching changes only when the orbit hits a critical value, and in that case it multiplies by the degree of the critical point $f^{l-1}(p)$. Any particular orbit which does not hit a periodic critical point can only go through any given critical point once, so that the local degree at any postcritical point is always a divisor of the product of the degrees of the critical points. This guarantees that we really have finite integers associated with postcritical points.

The Euler number for an orbifold whose underlying space is a finitely-punctured surface is still well-defined. We can think of a punctured surface as the limiting case of an orbifold, where the integer associated with any puncture is infinity. The formulas for Euler number work in this context. and in fact

12.10, Theorem: orbifold-goodness. *Every orbifold with underlying space a closed surface minus a finite set of points (at least one) is good. It has a complete Euclidean structure iff its Euler number is zero, and it has a complete hyperbolic structure of finite area iff its Euler number is negative.* The Euclidean cases are

(6) The (∞,∞)-orbifold, a cylinder.

(7) The $(\infty,2,2)$-orbifold, a sphere minus one point with two distinguished points of order 2. Physically, one can think of this as an infinitely long pillowcase. This may be gotten as the quotient of a cylinder by an order two rotation about an axis perpendicular to the axis of the cylinder. It is the quotient of the Euclidean plane by an action of the infinite dihedral group; the map of the plane to the orbifold is the map $z \longmapsto \sin(z)$.

12.11, Proposition: non-positive Q sub f. *The orbifold Q_f associated with any branched map f of S^2 of degree $d \geq 2$ has non-positive Euler number.*

Proof, : The branched map does not (usually) become a covering map of Q_f to itself, but it satisfies the condition that for any point $p \in Q_f$, the product $n(p) \times deg_p(f)$ is a divisor of $n(f(p))$. This means that if we change the orbifold structure of Q_f in the domain of f to create a new orbifold structure R_f on the two-sphere by adding some new distinguished points and increasing integers associated with some existing distinguished points, then f will be a branched cover of R_f over Q_f. We have

$$\chi(Q_f) \geqslant \chi(R_f) = \chi(Q_f) \times d$$

so that

$$(d-1) \times \chi(Q_f) \leqslant 0$$

$-\square$

13. Holomorphic maps of orbifolds and the Poincaré metric.

The orbifold Q_f associated with a critically finite rational map f actually has a complex structure, not just a differentiable structure. The universal cover of Q_f is a simply-connected, non-compact Riemann surface, so therefore it is conformally equivalent to either **C** or to the unit disk in **C**. It is easy to tell which: the universal cover is **C** iff $\chi(Q)=0$, and the universal cover is the unit disk iff $\chi(Q) \leqslant 0$. This easily follows from the fact that Q is quasi-conformally equivalent to a Euclidean or hyperbolic orbifold, respectively.

13.1, Proposition: conformal orbifolds are geometric. *A complex 1-orbifold Q whose underlying space is a punctured surface has a canonical complete finite-area hyperbolic structure if its Euler number is negative, and a Euclidean structure which is well-defined up to similarity if its Euler number is 0.*

Proof, : In the case that the Euler number is negative, the statement follows from the fact that the group of conformal equivalences of the disk to itself is the same as the group of isometries of the hyperbolic metric. The quotient of the disk by the group of deck transformations over Q therefore inherits a hyperbolic structure. The group of conformal equivalences of **C** to itself is the complex affine group, or the group of Euclidean similarities. No deck tranformation of **C** over Q can actually expand or contract the Euclidean metric, for if it did, it would have an expanding or contracting fixed point. This is clearly impossible, so Q inherits a Euclidean structure. Uniqueness up to similarity is clear.

$-\square$

13.2 **Definition.** A holomorphic map from an orbifold Q to an orbifold R is a map which has local liftings to holomorphic mappings between the local models. In other words, if $p \in Q$ is a point with a neighborhood U which is modeled on $(\tilde{U} \subset \mathbf{C}) / \mathbf{Z}_{mod\, n}$ and $q \in R$ has a neighborhood V modeled on $(\tilde{V} \subset \mathbf{C}) / \mathbf{Z}_{mod\, m}$, and if $f(U) \subset V$, then there must be a holomorphic map $\tilde{f}: \tilde{U} \to \tilde{V}$ which is a lifting of f.

Note that the underlying space X_Q of a complex 1-orbifold Q itself has the structure of a complex 1-manifold. In fact, whenever there is a neighborhood $U = (\tilde{U} \subset C)/ Z_{mod\ n}$, just define a new coordinate $w = z^n$, where $z \in \tilde{U}$. The new variable w gives a holomorphic coordinate system for U. According to the definition above, the "identity" map of Q to X_Q is a holomorphic map. More generally, whenever the orbifold structure of Q is changed by removing certain distinguished points and by replacing integers associated to other distinguished points by divisors of those integers, the "identity" map is holomorphic. This is the general form for a holomorphic map which is 1-1 on the level of underlying spaces; such a map is called an *erasing* map.

The reader may readily verify that the general criterion for a map $f : Q \to R$ between two complex 1-orbifolds to be holomorphic is that

(a). The map between underlying spaces is holomorphic.

(b). For any point $q \in Q$ with associated integer $n(q)$ (defined to be 1 if q is not a distinguished point) the integer $n(f(q))$ divides the product $n(q) \times deg_q(f)$.

13.3, Proposition.: the inverse is holomorphic on the orbifold. *A critically finite rational map of S^2 is a holomorphic self-map of the orbifold Q_f if and only if Q_f is Euclidean. The inverse of f is always a holomorphic erasing map from the covering orbifold R_f of Q_f to Q_f.*

Proof, : If Q_f is Euclidean, then R_f must equal Q_f so that f is holomorphic.

Otherwise, some of the points for the orbifold structure R_f have higher integers than for Q_f and the map is not holomorphic. The map from the covering R_f to Q_f is clearly a holomorphic erasing map.

$-\square$

13.4, Proposition: holomorphic maps of orbifolds decrease distance. *A holomorphic map between two hyperbolic orbifolds can never increase distances as measured in the hyperbolic metric. Distances are strictly decreased unless the map is a covering map.*

Proof, : Let $f : Q \to R$ be a holomorphic map between two hyperbolic orbifolds. Let $\tilde{Q}$ and $\tilde{R}$ denote their universal covers (which are both conformally equivalent to the disk). Just as in the theory for manifolds, the composition $\tilde{Q} \to Q \to R$ lifts to a holomorphic map

$$\tilde{f} : \tilde{Q} \to \tilde{R}$$

between the universal covers. (It is easy to prove this: liftings exist locally, by the definition of holomorphic map, and the lifting is constructed globally by analytic continuation). But a holomorphic map from the disk to itself can never increase the Poincaré metric, and it decreases the metric at every point unless it is an isometry.

$-\square$

Keep in mind that even though the contraction factor is strictly less than one at any point, it need not be *uniformly* less than one unless the orbifold is compact.

13.5, Corollary: a rational map expands the orbifold metric. *Let f be any critically finite rational map of the sphere to itself. Then f expands the Euclidean or hyperbolic metric of Q (in the sense of a Riemannian metric). The expansion factor is uniformly bounded above one in any compact set which does not meet any periodic critical orbits.*

Proof, : In the hyperbolic case, this is an immediate consequence of the preceding proposition. In the Euclidean case, we know that f is a similarity of the Euclidean metric. It is easy to see that the expansion constant must be greater than one.

$-\square$

13.6, Corollary.: Julia set in the preperiodic case. *Let f be a critically finite rational map.*

(a). *If no critical points are periodic, then the Julia set is the whole sphere.*

(b). *If f is a polynomial, so that ∞ is a fixed critical point, and if there are no other periodic critical points, then the Julia set is a connected compact subset of $\mathbf{C}$ with a connected complement.*

Proof, : For a point z in the equicontinuity set (the complement of the Julia set), the sequence of derivatives of powers of f are uniformly bounded in some sufficiently small neighborhood of z. This is clearly impossible when the expansion constant of the metric is uniformly bounded above 1, so in case (a) the Julia set must be the entire sphere. For case (b), the equicontinuity set includes a neighborhood U of ∞ which is mapped into itself by f. Form the union $V = \bigcup_{i=0}^{\infty} f^{\circ -1}(U)$. Then V is also contained in the equicontinuity set. V is connected, as can easily be seen by induction in the union, because $f^{-(i+1)}(U)$ is a d-fold branched cover of $f^{-i}(U)$, connected because all d sheets intersect U.

On the other hand, V is the entire equicontinuity set because. in light of the preceding proposition, the only way that the expansion constant for the derivative of $f^n(z)$ can fail to tend to infinity along the orbit of z is that either z tends toward a periodic critical orbit, or it hits a critical point, so that the derivative of high iterates is zero. The latter case does not matter because the equicontinuity set is open.

$-\square$

From our list of the seven oriented Euclidean orbifolds whose underlying spaces are the sphere or a punctured sphere, it is easy to understand (at least on a first level) all possible examples of critically finite rational maps such that the associated orbifold is Euclidean.

14. The Teichmüller metric.

The Teichmüller space for a closed 2-dimensional surface is the space of complex structures on the surface, up to isotopy. When the surface is noncompact, however, this is not quite the correct definition: first one must choose a quasi-conformal structure on the surface, and then define the Teichmüller space to be the set of complex structures compatible with the quasi-conformal structure.

14.1 **Definition.** A *quasi-conformal* diffeomorphism between two complex 1-manifolds is a diffeomorphism such that a circle in the tangent space of the first is taken to an ellipse in the tangent space of the second with uniformly bounded eccentricity. A *quasi-conformal structure* on a surface is a maximal family of complex structures such that the identity map is a quasi-conformal map between any two of them.

When the surface is homeomorphic to a closed surface minus a finite number of points, the choice of a quasi-conformal class amounts to a choice, for each puncture, of whether the conformal structure near the point is locally biholomorphic to a neighborhood of the origin in $\mathbf{C}-0$, or whether it is locally biholomorphic to a neighborhood of the circle in $\mathbf{C}-D^2$. Each conformal structure is represented by a unique hyperbolic structure, as soon as the surface is sufficiently complicated. In terms of the hyperbolic structure, the choice is whether the homotopy class of a curve going around a puncture is represented by a geodesic, or by a cusp (which looks like a pseudo-sphere).

The Teichmüller space for an orbifold, similarly, is the space of complex structures for the orbifold up to isotopy. An isotopy of a 2-orbifold must fix any singular points, so that the Teichmüller space for a closed orbifold Q is the same as the Teichmüller space for the punctured surface obtained by removing the distinguished points of Q from X_Q.

Given a combinatorially finite branched map f of the sphere to itself, there is an associated map $T(f): T(Q_f) \to T(Q_f)$ of the Teichmüller space $T(Q_f)$ of the orbifold associated with f. We define this as follows. First, form the covering R_f of Q_f which is induced by f. A complex structure on Q_f certainly gives rise to a conformal structure on R_f, so that we have a map $T(Q_f) \to T(R_f)$. Second, we have an identification of the underlying spaces X_{R_f} with X_{Q_f}. This will transport a complex structure on R_f to a complex structure on Q_f. The composition of these two maps is $T(f)$.

14.2, Proposition: Formal setup for existence. *Fixed points of $T(f)$ are in one-to-one correspondence with rational maps m in the combinatorial class of f, up to isomorphisms isotopic to the identity (on the orbifold Q_f).*

Proof, : Any rational map m in the combinatorial class of f certainly gives rise to a fixed point of $T(f)$, by proposition **the inverse is holomorphic on the orbifold, .** Suppose that $x \in T(Q_f)$ is a fixed point of $T(f)$. This means that there is a holomorphic map from the covering R_f of the corresponding complex orbifold Q_f back to Q_f. In the hyperbolic case, the holomorphic map is unique, so that its inverse composed with the covering projection from R_f to Q_f is a rational map m uniquely determined by f together with the complex structure x. In all of the Euclidean cases when there are at least three distinguished points, the holomorphic identification of R_f with Q_f is uniquely determined also since a biholomorphic map of the sphere is determined by what it does to three points. The only remaining case is when Q_f is the once-punctured plane. The rational map in this case must be conjugate to the map $z \to z^d$ (where d is the degree). This case is slightly different, since (unlike the other cases) the conjugacy is not unique unless $n=2$ – if you multiply z by an $(n-1)th$-root of unity, you conjugate the map to itself.

Thus, a rational map determines a fixed point, and a fixed point determines a unique rational map in the combinatorial class.

–□

There is a metric on Teichmüller space, the Teichmüller metric, which has properties quite analogous to the Poincaré metric.

14.3 **Definition.** The *dilatation* of a real-linear self-map of **C** is the ratio of the length of the major axis of the image of a circle to the length of the minor axis. The *dilatation* $dil(f)$ of a quasi-conformal map f from one complex surface to another is the supremum of the dilation of its derivative at points in the domain surface. The *Teichmuller metric* on Teichmüller space assigns as distance $d(x,y)$ between two complex structures the infimum over quasi-conformal diffeomorphisms homotopic f to the identity of the quantity

$$\frac{1}{2}\log(dil(f))$$

It is clear that this definition of a distance function satisfies the triangle inequality, and it is not difficult to show that it assigns a positive distance to any pair of distinct complex structures. In fact, if $\{f_i\}$ is any sequence of quasi-conformal diffeomorphisms isotopic to the identity, it can be shown that there is a subsequence converging to a limit. (There are actually some exceptions here, in the trivial cases such as the punctured plane). The limit is not usually a diffeomorphism – it is a quasi-conformal homeomorphism (which we have not defined), however, and the theory of these is well developed. If the dilatations were converging to zero, then the limit would be holomorphic.

We will make use of Teichmüller's theorem:

14.4, Theorem: Teichmullers theorem. *In any isotopy class of quasi-conformal maps, there is a unique map which achieves a minimum dilatation. These extremal maps, called Teichm'*:uller maps, are characterized by the fact that the dilatation of the derivative of the map is constant over the surface, and the directions of maximal and minimal stretching are the tangents to the positive real foliation and the negative real foliation of a quadratic differential on the surface. For a punctured surface, the same statement holds, but with a restriction that the quadratic differential near a puncture can only be that of a simple pole or a regular point.*

A quadratic differential is something which, in local coordinates, can be written in the form $f(z)dz^2$, where $f(z)$ is holomorphic. At any point, it gives a homogeneous map of degree 2 from the tangent space of the Riemann surface to $\mathbf{C}$. If the quadratic differential is not zero at a point on the surface, it maps certain tangent vectors are mapped to the positive reals: these form a real subspace. It maps other vectors are mapped to the negative reals: these form a real subspace which is orthogonal to the first. At a generic zero of a quadratic differential (of order 1),

Quadratic differentials

14.5 Figure. *Sketches of the positive real and negative real foliations of quadratic differentials, at zeros and poles.*

the positive real foliation has a 3-prong singularity. (See figure) At a zero of order k, the quadratic differential has a singularity with $k+2$

prongs. At a simple pole, a quadratic differential has a 1-prong singularity

Teichmüller's theorem goes over for orbifolds – one reason is that the Teichmu'*:uller space for an orbifold is the same as that for a punctured surface. Note that the behavior that looks like that of a simple pole, on a punctured surface, is that of a regular point or a zero at a singular point of an orbifold – since when you pass to the branched cover which is the local model at the singular point, the single prong reproduces and forms 2 prongs if the associated integer is two – thus giving a regular point – and otherwise 3 or more prongs, thus giving a zero.

Teichmüller space is not just a space – it is a complex manifold. In the cases we are interested in, that is, punctured spheres, the complex structure is obvious – it just comes from the coordinates of a bunch of points on the Riemann sphere. In the general case, the complex structure is not at all obvious, and it was constructed and analyzed by Bers [???]. The Teichmüller metric has a property akin to that of the Poincaré metric for a domain in **C**:

14.6, Theorem: Teichmuller metric contracts. *A holomorphic map between Teichmüller spaces cannot increase the Teichmuller metric.* This theorem means that we are in a good position to analyze fixed points of the map $T(f)$ associated with a combinatorially finite branched map – if we can show that $T(f)$ strictly decreases the Teichmüller metric, that will prove uniqueness for a critically finite branched map in any combinatorial class, and if we can show that the contraction constant is uniform, that will prove existence for a rational map in the combinatorial class.

Unfortunately, $T(f)$ does not strictly decrease the Teichmüller distances. What will serve equally well, though, is that some finite iterate of $T(f)$ does strictly decrease the metric.

14.7, Theorem: eventual Teichmuller contraction. *For any combinatorially finite branched map f such that the associated orbifold O_f is hyperbolic, there is some integer n such that $T(f^{\circ n})$ strictly decreases the Teichmüller metric.*

Proof, : Let q_1 and q_2 be two complex structures on the orbifold O_f. There is a Teichmüller map g between these two conformal structures. This map g lifts to a map between the branched covers of q_1 and q_2 defined by f, to give a quasi-conformal map g_1 between $T(f)(q_i)$. This map g_1 in turn lifts to a map g_2 between $T(f^{\circ 2})(q_i)$. Each member g_i of this sequence has exactly the same dilatation. All we need to show is that after a while, they fail to be Teichmüller maps.

Let us think about the quadratic differential which describes the original Teichmüller map as a quadratic differential D
on the sphere (without its orbifold structure). It has a certain number of poles and a certain number of zeros, which must satisfy the formula

$$\#(poles) - (total\ multiplicity(zeros)) = 4$$

(by the index formula for the singularities of a line field, for instance). The distortion of the conformal structure by the derivative of g_i is described by the lift of D to the deg^i-fold branched cover of the sphere induced by f^{oi}.

When we lift a quadratic differential to a branched cover, mostly the zeros and the poles just multiply by the order of the cover – except for points whose preimages which are branch points A regular point for the differential which is a critical value has a lift which is a zero of order 2 or more. A pole which falls on a critical value lifts to a regular point or a zero, depending on the order of the branching. A zero which falls on a critical value lifts to a zero of order greater than the degree of branching times the order of the original zero. If the differential D has any zeros, the proof is easily completed – the total multiplicity of zeros in the nth branched cover must be at least d^n, hence by the index formula there must also be a lot of poles when n is sufficiently high, and not all of them can fall on singular points of the orbifold – hence, the map g^n is not a Teichmüller map.

15. The Teichmüller metric.

The Teichmüller space for a closed 2-dimensional surface is the space of complex structures on the surface, up to isotopy. When the surface is noncompact, however, this is not quite the correct definition: first one must choose a quasi-conformal structure on the surface, and then define the Teichmüller space to be the set of complex structures compatible with the quasi-conformal structure.

15.1 **Definition.** A *quasi-conformal* diffeomorphism between two complex 1-manifolds is a diffeomorphism such that a circle in the tangent space of the first is taken to an ellipse in the tangent space of the second with uniformly bounded eccentricity. A *quasi-conformal structure* on a surface is a maximal family of complex structures such that the identity map is a quasi-conformal map between any two of them.

When the surface is homeomorphic to a closed surface minus a finite number of points, the choice of a quasi-conformal class amounts to a choice, for each puncture, of whether the conformal structure near the point is locally biholomorphic to a neighborhood of the origin in $\mathbf{C}-0$, or whether it is locally biholomorphic to a neighborhood of the circle in $\mathbf{C}-D^2$. Each conformal structure is represented by a unique hyperbolic structure, as soon as the surface is sufficiently complicated. In terms of the hyperbolic structure, the choice is whether the homotopy class of a curve going around a puncture is represented by a geodesic, or by a cusp (which looks like a pseudo-sphere).

The Teichmüller space for an orbifold, similarly, is the space of complex structures for the orbifold up to isotopy. An isotopy of a 2-orbifold must fix any singular points, so that the Teichmüller space for a closed orbifold Q is the same as the Teichmüller space for the punctured surface obtained by removing the distinguished points of Q from X_Q.

Given a combinatorially finite branched map f of the sphere to itself, there is an associated map $T(f): T(Q_f) \to T(Q_f)$ of the Teichmüller space $T(Q_f)$ of the orbifold associated with f. We define this as follows. First, form the covering R_f of Q_f which is induced by f. A complex structure on Q_f certainly gives rise to a conformal structure on R_f, so that we have a map $T(Q_f) \to T(R_f)$. Second, we have an identification of the underlying spaces X_{R_f} with X_{Q_f}. This will transport a complex structure on R_f to a complex structure on Q_f. The composition of these two maps is $T(f)$.

15.2, Proposition: Formal setup for existence. *Fixed points of $T(f)$ are in one-to-one correspondence with rational maps m in the combinatorial class of f, up to isomorphisms isotopic to the identity (on the orbifold Q_f).*

Proof, : Any rational map m in the combinatorial class of f certainly gives rise to a fixed point of $T(f)$, by proposition **the inverse is holomorphic on the orbifold, .** Suppose that $x \in T(Q_f)$ is a fixed point of $T(f)$. This means that there is a holomorphic map from the covering R_f of the corresponding complex orbifold Q_f back to Q_f. In the hyperbolic case, the holomorphic map is unique, so that its inverse composed with the covering projection from R_f to Q_f is a rational map m uniquely determined by f together with the complex structure x. In all of the Euclidean cases when there are at least three distinguished points, the holomorphic identification of R_f with Q_f is uniquely determined also since a biholomorphic map of the sphere is determined by what it does to three points. The only remaining case is when Q_f is the once-punctured plane. The rational map in this case must be conjugate to the map $z \to z^d$ (where d is the degree). This case is slightly different, since (unlike the other cases) the conjugacy is not unique unless $n = 2$ — if you multiply z by an $(n-1)th$-root of unity, you conjugate the map to itself.

Thus, a rational map determines a fixed point, and a fixed point determines a unique rational map in the combinatorial class.

−□

There is a metric on Teichmüller space, the Teichmüller metric, which has properties quite analogous to the Poincaré metric.

15.3 **Definition.** The *dilatation* of a real-linear self-map of **C** is the ratio of the length of the major axis of the image of a circle to the length of the minor axis. The *dilatation* $dil(f)$ of a quasi-conformal map f from one complex surface to another is the supremum of the dilation of its derivative at points in the domain surface. The *Teichmuller metric* on Teichmüller space assigns as distance $d(x,y)$ between two complex structures the infimum over quasi-conformal diffeomorphisms homotopic f to the identity of the quantity

$$\frac{1}{2}\log(dil(f))$$

It is clear that this definition of a distance function satisfies the triangle inequality, and it is not difficult to show that it assigns a positive distance to any pair of distinct complex structures. In fact, if $\{f_i\}$ is any sequence of quasi-conformal diffeomorphisms isotopic to the identity, it can be shown that there is a subsequence converging to a limit. (There are actually some exceptions here, in the trivial cases such as the punctured plane). The limit is not usually a diffeomorphism -- it is a quasi-conformal homeomorphism (which we have not defined), however, and the theory of these is well developed. If the dilatations were converging to zero, then the limit would be holomorphic.

We will make use of Teichmüller's theorem:

15.4, Theorem: Teichmullers theorem. *In any isotopy class of quasi-conformal maps, there is a unique map which achieves a minimum dilatation. These extremal maps, called Teichm'*:uller maps, are characterized by the fact that the dilatation of the derivative of the map is constant over the surface, and the directions of maximal and minimal stretching are the tangents to the positive real foliation and the negative real foliation of a quadratic differential on the surface. For a punctured surface, the same statement holds, but with a restriction that the quadratic differential near a puncture can only be that of a simple pole or a regular point.*

A quadratic differential is something which, in local coordinates, can be written in the form $f(z)dz^2$, where $f(z)$ is holomorphic. At any point, it gives a homogeneous map of degree 2 from the tangent space of the Riemann surface to **C**. If the quadratic differential is not zero at a point on the surface, it maps certain tangent vectors are mapped to the positive reals: these form a real subspace. It maps other vectors are mapped to the negative reals: these form a real subspace which is orthogonal to the first. At a generic zero of a quadratic differential (of order 1),

the positive real foliation has a 3-prong singularity. (See figure) At a zero of order k, the quadratic differential has a singularity with $k+2$ prongs. At a simple pole, a quadratic differential has a 1-prong singularity

Quadratic differentials

15.5 Figure. *Sketches of the positive real and negative real foliations of quadratic differentials, at zeros and poles.*

Teichmüller's theorem goes over for orbifolds – one reason is that the Teichmu"*:uller space for an orbifold is the same as that for a punctured surface. Note that the behavior that looks like that of a simple pole, on a punctured surface, is that of a regular point or a zero at a singular point of an orbifold – since when you pass to the branched cover which is the local model at the singular point, the single prong reproduces and forms 2 prongs if the associated integer is two – thus giving a regular point – and otherwise 3 or more prongs, thus giving a zero.

Teichmüller space is not just a space — it is a complex manifold. In the cases we are interested in, that is, punctured spheres, the complex structure is obvious — it just comes from the coordinates of a bunch of points on the Riemann sphere. In the general case, the complex structure is not at all obvious, and it was constructed and analyzed by Bers [???]. The Teichmüller metric has a property akin to that of the Poincaré metric for a domain in **C**:

15.6, Theorem: Teichmuller metric contracts. *A holomorphic map between Teichmüller spaces cannot increase the Teichmuller metric.* This theorem means that we are in a good position to analyze fixed points of the map $T(f)$ associated with a combinatorially finite branched map – if we can show that $T(f)$ strictly decreases the Teichmüller metric, that will prove uniqueness for a critically finite branched map in any combinatorial class, and if we can show that the contraction constant is uniform, that will prove existence for a rational map in the combinatorial class.

Unfortunately, $T(f)$ does not strictly decrease the Teichmüller distances. What will serve equally well, though, is that some finite iterate of $T(f)$ does strictly decrease the metric.

15.7, Theorem: eventual Teichmuller contraction. *For any combinatorially finite branched map* f *such that the associated orbifold* Q_f *is hyperbolic, there is some integer* n *such that* $T(f^{\circ n})$ *strictly decreases the Teichmüller metric.*

Proof, : Let q_1 and q_2 be two complex structures on the orbifold Q_f. There is a Teichmüller map g between these two conformal structures. This map g lifts to a map between the branched covers of q_1 and q_2 defined by f, to give a quasi-conformal map g_1 between $T(f)(q_i)$. This map g_1 in turn lifts to a map g_2 between $T(f^{\circ 2})(q_i)$. Each member g_i of this sequence has exactly the same dilatation. All we need to show is that after a while, they fail to be Teichmüller maps.

Let us think about the quadratic differential which describes the original Teichmüller map as a quadratic differential D on the sphere (without its orbifold structure). It has a certain number of poles and a certain number of zeros, which must satisfy the formula

$$\#(poles) - (total\ multiplicity(zeros)) = 4$$

(by the index formula for the singularities of a line field, for instance). The distortion of the conformal structure by the derivative of g_i is described by the lift of D to the deg^i-fold branched cover of the sphere induced by $f^{\circ i}$.

When we lift a quadratic differential to a branched cover, mostly the zeros and the poles just multiply by the order of the cover — except for points whose preimages which are branch points A regular point for the differential which is a critical value has a lift which is a zero of order 2 or more. A pole which falls on a critical value lifts to a regular point or a zero, depending on the order of the branching. A zero which falls on a critical value lifts to a zero of order greater than the degree of branching times the order of the original zero. If the differential D has any zeros, the proof is easily completed — the total multiplicity of zeros in the *nth* branched cover must be at least d^n, hence by the index formula there must also be a lot of poles when n is sufficiently high, and not all of them can fall on singular points of the orbifold — hence, the map g^n is not a Teichmüller map.

Entropy in dimension one

William P. Thurston

Bill Thurston fell ill during the process of completing a draft version of this manuscript. During his illness, Sarah Koch assisted him in making revisions. Sadly, Bill passed away in August of 2012. The editors received help from several people in preparing this paper for publication, including Mladen Bestvina, Michael Handel, Wolf Jung, Sarah Koch, Doug Lind, Curt McMullen, John Milnor, Lee Mosher, and Tan Lei. Footnotes and a section of additional remarks (§13) have been added by John Milnor.

1 INTRODUCTION

The topological entropy $h(f)$ of a map from a compact topological space to itself, $f\colon X \to X$, is a numerical measure of the unpredictability of trajectories $x, f(x), f^2(x), \ldots$ of points under f: it is the limiting upper bound for exponential growth rate of ϵ-distinguishable orbits, as $\epsilon \to 0$. Here ϵ can be measured with respect to an arbitrary metric on X, or one can merely think of it as a neighborhood of the diagonal in $X \times X$, since all that matters is whether or not two points are within ϵ of each other. The number of ϵ-distinguishable orbits of length n is the maximum cardinality of a set of orbits of f such that no two are always within ϵ.

More formally, given a metric d on X, define the ϵ-count of X, $N(X,\epsilon,d)$ to be the maximum cardinality of a set $S \subset X$ such that no point is within ϵ of any other. Given a continuous function $f\colon X \to X$, define a metric $d_{f,n}$ on X as $d_{f,n}(x,y) = \sup_{0\le i<n}\left\{d(f^i(x), f^i(y))\right\}$. Then

$$h(f) = \lim_{\epsilon\to 0}\limsup_{n\to\infty}\frac{1}{n}\log\left(N(X,\epsilon,d_{f,n})\right).$$

If f is a Lipschitz self-map of a compact m-manifold with Lipschitz constant $K \ge 1$, then it's easy to see that

$$h(f) \le m\log(K) \tag{1.1}$$

(compare [Mis]). The upper bound is attained in cases such as $x \mapsto Kx$ acting on the torus $\mathbb{R}^m/\mathbb{Z}^m$, where K is an integer. On the other hand, for a continuous map f that is not Lipschitz, $h(f)$ need not be finite, even for simple situations such as homeomorphisms of S^2 or continuous maps of intervals.

A differentiable map f of an interval to itself is *postcritically finite* or *critically finite* if the union of forward orbits of the critical points is finite. In particular, the set of critical points for f must be finite.

For a map f that is not differentiable, we can define any interior point that is a local maximum or local minimum to be a turning point, or topological critical point. (Note that with this definition, not all smooth critical points are topological critical points.) Let $c(f)$ denote the *modality*, or number of turning points for such

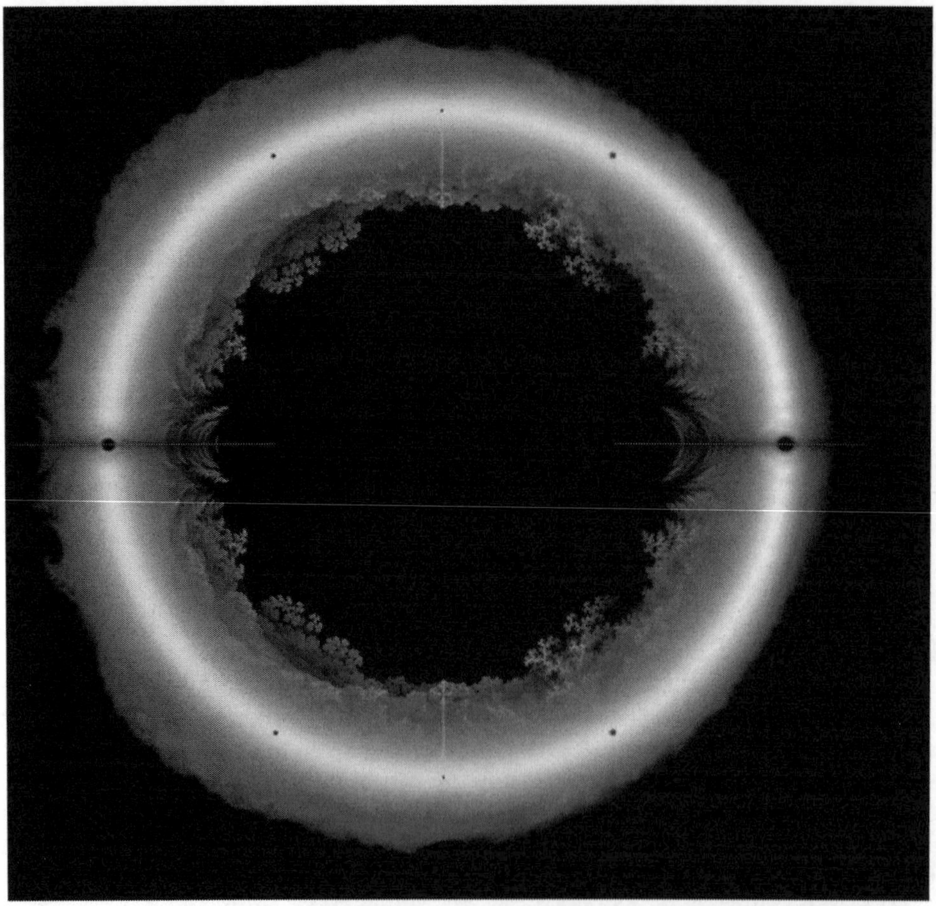

FIGURE 1.1: This is a plot of roughly 8×10^8 roots of defining polynomials for $\exp(h(f))$, where f ranges over a sample of about 10^7 postcritically finite quadratic maps of the interval with postcritical orbit of length ≤ 80. The brightness is proportional to the log of the density; the highest concentration is at the unit circle. (See also Plate 16.)

a map, and let $\mathrm{Var}(f)$ denote the total variation of f, i.e., its arclength considered as a path. For maps of the interval to itself with finitely many critical points, there are two simple ways to characterize the topological entropy.

THEOREM 1.2 (Misiurewicz-Szlenk, [MS1], [MS2]). *For a continuous map f of an interval to itself with finitely many turning points, the topological entropy equals*

$$h(f) = \lim_{n\to\infty} \frac{1}{n} \log(\mathrm{Var}(f^n)) = \lim_{n\to\infty} \frac{1}{n} \log c(f^n).$$

In particular, these are actual limits, not just limits of lim sups. There are good algorithms to actually compute the entropy [MT].

The first main goal of this paper is to characterize what values of entropy can occur for postcritically finite maps.

THEOREM 1.3. *A positive real number h is the topological entropy of a postcritically finite self-map of the unit interval if and only if* $\exp(h)$ *is an algebraic integer that is at least*

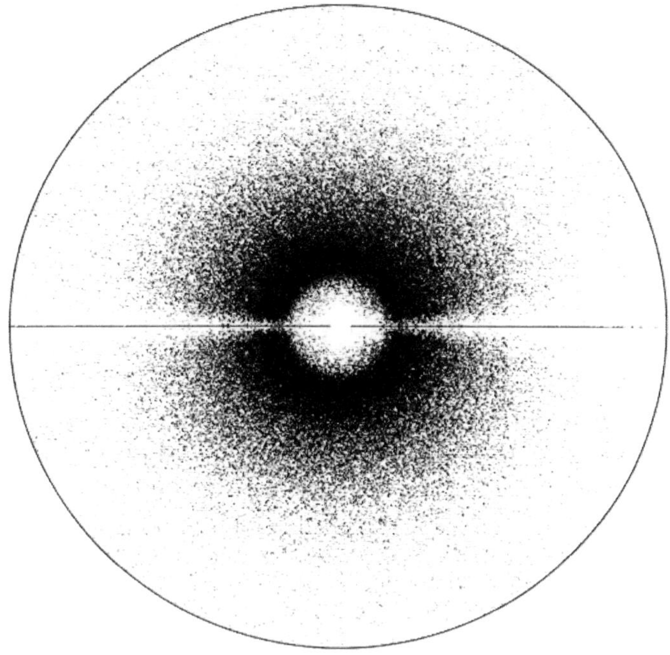

FIGURE 1.4: Shown are the roots of the minimal polynomials for 5,932 degree 21 Perron numbers, obtained by sampling 20,000 monic degree 21 polynomials with integer coefficients between 5 and -5, and keeping those that have a root in $[1,2]$ larger than all other roots. The indicated circle has radius 2.

as large as the absolute value of any conjugate of $\exp(h)$. *The map may be chosen to be a polynomial, all of whose critical points are in* $(0,1)$.

Two maps $f_1, f_2 \colon X \to X$ are *conjugate* if there is a homeomorphism g of X conjugating one to the other, i.e., $g \circ f_1 = f_2 \circ g$. They are *semiconjugate* if there is a map g satisfying the condition that is continuous and surjective but not necessarily a homeomorphism. Basically, a semiconjugacy can collapse out certain kinds of subsidiary behavior of the dynamics.

Many phenomena of 1-dimensional dynamics are irrelevant for the study of entropy; there is a relatively simple family of non-smooth examples that has central importance when f is PL (piecewise-linear) continuous and $|f'| = \lambda > 1$ is constant, wherever f' exists. We call such an f a *uniform expander*, or a *uniform* λ*-expander* if we want to be more specific. These maps are often called maps with *constant slope*. The importance of the uniform expanders is indicated by this theorem.

THEOREM 1.5 ([MT]). *Every continuous self-map g of an interval with finitely many turning points and with $h > 0$ is semi-conjugate to a uniform λ-expander* PL(g) *with the same topological entropy $h = \log(\lambda)$. If g is postcritically finite, so is* PL(g). *(But if* PL(g) *is postcritically finite, it does not imply that g is postcritically finite.)*

In [ALM], a more general version of Theorem 1.5 is proven, which applies in more circumstances, including, for instance, piecewise continuous maps and self-maps of graphs.

In other words, Theorem 1.3 reduces to the study of expansion constants for 1-dimensional uniform expanders.

In the case of a postcritically finite map f, a uniform expander model is easily computed. If necessary, first trim the domain interval until it maps to itself surjectively, by taking the intersection of its forward images. If this is a point, then the entropy is 0. Otherwise, the two endpoints are either images of an endpoint, or images of a turning point. In this case, conjugate by an affine transformation to make the interval $[0,1]$.

If we now subdivide $[0,1]$ by cutting at the union of postcritical orbits (including the critical points) into intervals J_i, each J_i maps homeomorphically to a finite union of other subintervals. If there is a uniform expander F with the same qualitative behavior, that is, having an isomorphic subdivision into subintervals each mapped homeomorphically to the corresponding union of other subintervals, then the lengths of the intervals satisfy a linear condition: The sum of the lengths of intervals J_j hits is λ times the length of J_j. In other words, the lengths of the intervals of the subdivision define a positive eigenvector for a non-negative matrix, with eigenvalue λ.

The Perron-Frobenius theorem gives necessary and sufficient conditions for this to exist. Here is some terminology: a non-negative matrix is *ergodic* if the sum of some number of its positive powers is strictly positive, and it is *mixing* if some power (and hence, all subsequent powers) is strictly positive. The incidence matrix for f is ergodic if and only if for each pair of intervals J_i and J_j, some $f^n(J_i)$ contains J_j. The incidence matrix is mixing if and only if for each J_i, some $f^n(J_i)$ image covers all intervals.

The Perron-Frobenius theorem says that any non-negative matrix has at least one non-negative eigenvector with non-negative eigenvalue $\geq$ the absolute value of any other eigenvalue. If the matrix is ergodic, there is a unique strictly positive eigenvector; its eigenvalue is automatically [strictly] positive.[1]

From any non-negative eigenvector for the incidence matrix, we can make a uniformly expanding model by subdividing the unit interval into subintervals whose lengths equal the corresponding coordinate of the eigenvector normalized to have L^1 norm $= 1$. We thus obtain a PL map; its topological entropy is the log of the eigenvalue.

If all the entries of the matrix are integers, then its characteristic polynomial has integer coefficients, so it's an immediate corollary that the expansion constant for a postcritically finite uniform λ-expander is at least as large as the absolute value of its Galois conjugates.

In [MT, Lemma 7.9] there is a more general formula (very quick on a computer) for a semiconjugacy to a uniform expander for a general map with finitely many critical points.

Doug Lind proved a converse to the integer Perron-Frobenius theorem.

THEOREM 1.6 ([Li]). *For any real algebraic integer $\lambda > 0$ that is strictly larger than its Galois conjugates in absolute value, there exists a non-negative integer matrix that is mixing (some power strictly positive) and has λ as an eigenvalue.*

We will prove Lind's theorem, on our way to other results, in Section 3.

[1]For an elegant description of the entire non-zero spectrum of a nonnegative mixing integer matrix, see [KOR].

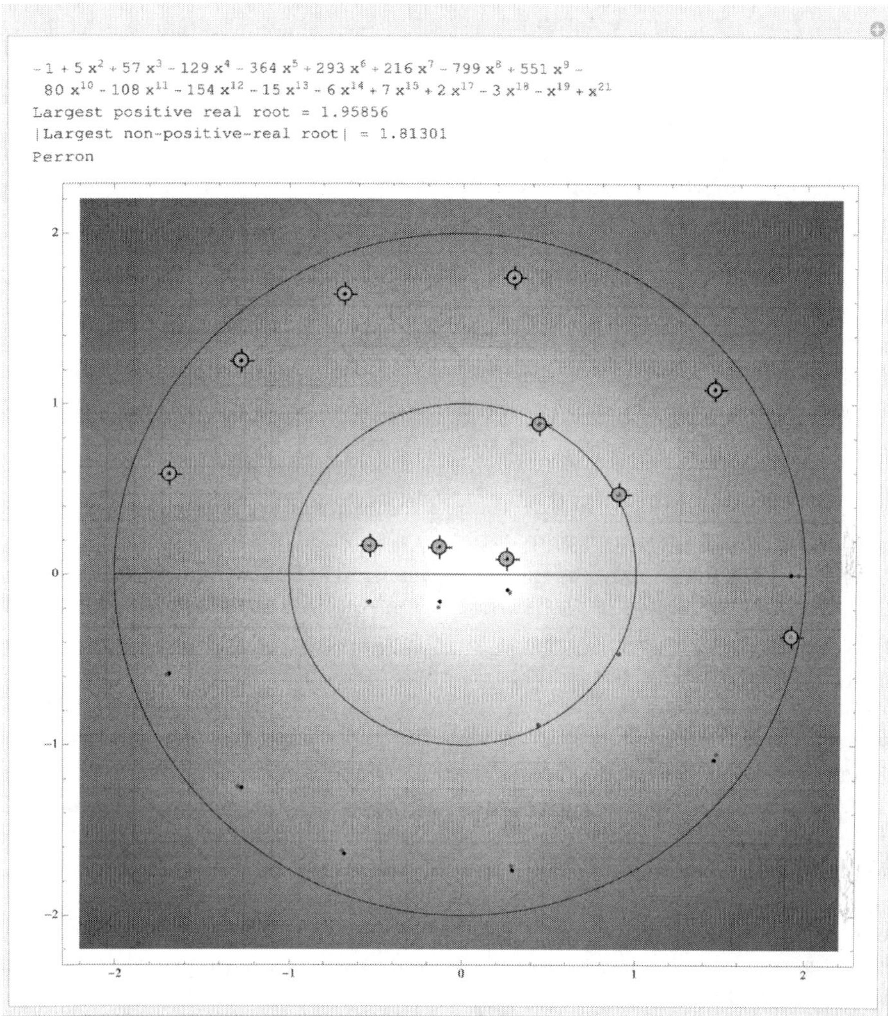

FIGURE 1.7: This is a plot of the roots of an irreducible polynomial P of degree 21, as printed above the picture, defining a Perron number, $\lambda \approx 1.95856$. Although λ is an algebraic unit < 2, the map $x \mapsto \lambda|x| - 1$ cannot be postcritically finite since P has (several) roots not in (far away from) the region shown in Figure 1.1. To construct P, first a monic real polynomial P_0 was defined by its roots with a mouse (the black dots). The coefficients of P_0 were rounded to define the integer polynomial P, whose roots are indicated by a lighter gray dot (typically close to a root of P_0), sometimes obscuring the roots of P_0. The constant term was made small by balancing roots inside and outside the unit circle. When roots for P_0 are chosen so that those away from the unit circle are spaced well apart (relative to the sizes of coefficients), they tend to be fairly stable under rounding. The shading, proportional to $\log(1 + |P|)$, is a guide to stability: clicking new roots into darker areas is typically stabilizing. (See also Note 13.1.)

In view of Lind's theorem together with the Perron-Frobenius theorem, these numbers are called *Perron* numbers. An algebraic integer λ that satisfies the weak inequality $\lambda \geq |\lambda^\alpha|$, where α ranges over the Galois group of λ, is a *weak Perron number*.

There are two important (and better-known) special cases of Perron numbers. A positive real algebraic integer is a *Pisot number*, or *Pisot-Vijayaraghavan number*, or *PV number* if all its Galois conjugates are in the open unit disk. Since the product of all Galois conjugates is a nonzero integer, λ is bigger than its conjugates. It turns out that the set of Pisot numbers is a closed subset of $\mathbb{R}$. If $\lambda > 1$ is a real algebraic integer that has at least one conjugate on the unit circle and all conjugates are in the closed unit disk, then λ is called a *Salem number*. Since the complex conjugate of a point on the unit circle is its inverse, every element of the Galois conjugacy class of a point on the unit circle is also Galois conjugate to its inverse; in particular λ is Galois conjugate to $1/\lambda$. Therefore, $1/\lambda$ is the only Galois conjugate of λ in the open unit disk, since the inverse of any other conjugate in the open unit disk would give another conjugate of λ outside the unit disk.

Note: The dimension of the matrix in Lind's theorem is not bounded by the degree of λ: for any integer $n > 0$, there are, in fact, cubic Perron numbers λ that are not eigenvalues for nonnegative matrices smaller than $n \times n$.[2]

The proof of Theorem 1.3 uses techniques motivated by Doug Lind's methods.

It is also interesting to investigate what happens for postcritically finite maps subject to a bound on the number of turning points, in particular, a single turning point (i.e., for quadratic polynomials). The situation is very different. Figure 1.1 shows the Galois conjugates of $\exp(h)$ for postcritically finite real quadratic maps. This is a path-connected set, with much structure visible.[3]

Most Perron numbers between 1 and 2 do not have roots in this set. For example, Figure 1.4 shows a sampling of degree 21 polynomials with coefficients between -5 and 5 that happen to define a Perron number between 1 and 2 (out of 20000 random polynomials, 5937 fit the condition). These however are not random Perron numbers of degree 21: more typically, many of the coefficients are much larger. Figure 1.7 shows a degree 21 example constructed by hand, first specifying a collection of real points and pairs of complex conjugate points in the disk of radius 2, expanding the monic polynomial with those points as roots, and rounding the coefficients to the nearest integers. With care in spacing and placement of roots, the integer polynomial has roots near the given choices. (When the points away from the unit circle cluster too much, their positions become unstable with respect to rounding).

More generally, if Γ is a finite graph and $f\colon \Gamma \to \Gamma$ is a continuous map which is an embedding when restricted to any edge, we will say that f is *postcritically finite* if the forward orbit of every vertex is finite. If f has the additional property that for all k, f^k restricted to any edge is an immersion (it is an embedding on a sufficiently small neighborhood of any point), then f is a *traintrack map*.

Entropy for graph maps behaves similarly to entropy for intervals.

[2]Lind points out an easy family of examples where the degree of the dominant eigenvalue λ for a non-negative $n \times n$ matrix must be strictly less than n. When the degree is equal to n, the trace of λ must equal the trace of the matrix, which is non-negative. But a Perron number may well have negative trace. The dominant root of $p(t) = t^3 + 3t^2 - 15t - 46$, with trace -3, provides an example.

[3]See Note 13.2, as well as the further discussion in §7.

THEOREM 1.8 (Alsedá-Llibre-Misiurewicz [ALM]). *Let $f\colon \Gamma \to \Gamma$ be a continuous self-map of a finite graph which has finitely many exceptional points $x(f)$, where f is not a local homeomorphism. Then*

$$h(f) = \lim_{n\to\infty} \frac{1}{n} \log \operatorname{Var}(f^n).$$

If f is a degree d covering map $S^1 \to S^1$, this yields $h(f) = \log(d)$; otherwise, the entropy also satisfies

$$h(f) = \lim_{n\to\infty} \frac{1}{n} \log(x(f^n)).$$

A self-map of a graph that is a homotopy equivalence defines an outer automorphism of its fundamental group, that is, an automorphism up to conjugacy (since we're not specifying a base point that must be preserved).

Handel and Bestvina [BH] showed that for any outer automorphism that is irreducible in the sense that no free factor is preserved up to conjugacy, there is a graph Γ and a traintrack map of Γ to itself representing the outer automorphism. They also developed a theory of relative traintracks that addresses outer automorphisms that are reducible. Traintrack theory is a powerful tool, parallel in many ways to pseudo-Anosov theory for self-homotopy-equivalences of surfaces. Algebraically, you can look at the action of an outer automorphism on conjugacy classes, represented by cyclically reduced words in a free group. The lengths of images of cyclically reduced words have a limiting exponential growth rate. There is a well-understood situation when traintrack automorphisms[4] can have conjugacy classes that are fixed under an automorphism: for instance, any automorphism of the free group on $\{a, b\}$ either fixes the conjugacy class of the commutator $[a, b]$ or carries it to its inverse. (See [LS, p. 44].) Apart from these, the lengths of the sequence of images of any conjugacy classes under iterates of a traintrack automorphism have exponent of growth equal to the topological entropy of the traintrack map, as measured in any generating set.

Peter Brinkmann wrote a very handy java application `Xtrain` that implements the Bestvina-Handel algorithm, `http://math.sci.ccny.cuny.edu/pages?name=XTrain`. I used this program extensively to work out and check examples for the next theorem, which is the second main goal of this paper.

THEOREM 1.9. *A positive real number h is the topological entropy for an ergodic traintrack representative of an outer automorphism of a free group if and only if its expansion constant* $\exp(h)$ *is an algebraic integer that is at least as large as the absolute value of any conjugate of* $\exp(h)$.

Note: Even though automorphisms are invertible, the expansion constant need not be an algebraic unit.

The relationship between the expansion constant of an automorphism and the expansion constant of its inverse is mysterious, but there is one special case where it's possible to control the expansion constant for both an automorphism ϕ and its inverse ϕ^{-1}. An automorphism ϕ is *positive* with respect to a set G of free generators if ϕ of any generator is a positive word in the generators, that is, it maps the semigroup they generate into itself. This implies that ϕ comes from a traintrack map of the bouquet of oriented circles defined by G.

[4]Presumably the term "traintrack automorphism" refers to the outer automorphism of a free group which is induced by some given traintrack map.

DEFINITION 1.10. *A linear transformation A is* bipositive[5] *with respect to a basis B if B can be expressed as the disjoint union $B = P \cup N$ such that A is non-negative with respect to B, and its inverse is non-negative with respect to the basis $P \cup (-N)$. An automorphism ϕ is* bipositive *if it is positive with respect to a set G of free generators, and its inverse is positive with respect to a set of generators obtained by replacing some subset of elements of G by their inverses.*

At one point, I hoped that the criteria in the following theorem would characterize all pairs of expansion constants for a free group automorphism and its inverse. This turned out to be false (Theorem 1.9), but the characterization of such pairs in this special case is still interesting. The proof of the following statement will be sketched in Section 12.

THEOREM 1.11. *A pair (λ_1, λ_2) of positive real numbers is the pair of expansion constants for ϕ and ϕ^{-1}, where ϕ is bipositive, if and only if it is the pair of positive eigenvalues for a bipositive element of $GL(n, \mathbb{Z})$ for some n, if and only if λ_1 and λ_2 are real algebraic units such that the Galois conjugates of λ_1 and λ_2^{-1} are contained in the closed annulus $\lambda_2^{-1} \le |z| \le \lambda_1$.*

Theorem 1.11 does not extend in an immediate way to the general case. Classification of the set of pairs of expansion constants that can occur in general remains mysterious. As already noted, these expansion constants need not be units. Moreover, there are examples of traintrack automorphisms where the Galois conjugates of λ_1 and λ_2^{-1} are *not* contained in the annulus $\lambda_2^{-1} \le |z| \le \lambda_1$. It is consistent with what I currently know that *every* pair of weak Perron numbers greater than 1 is the pair of expansion constants for ϕ and ϕ^{-1}.

2 SPECIAL CASE: PISOT NUMBERS

The special case that λ is a Pisot number has a particularly easy theory, so we will look at that first.

It is easy to see that the topological entropy of a map $f\colon I \to I$ with $d-1$ topological critical points can be at most $\log(d)$: each point has at most d preimages, so the total variation of f^n is at most d^n.

THEOREM 2.1. *For any integer $d > 1$ and any Pisot number $\lambda \le d$, there is a postcritically finite map $f\colon I \to I$ of degree d (that is, having $d-1$ critical points) with entropy $\log(\lambda)$.*

In fact, when λ is Pisot, every λ-uniformly expanding map whose critical points are in $\mathbb{Q}(\lambda)$ is postcritically finite.

PROOF. One way to construct uniform λ-expanders is to create their graphs by folding. Start with the graph of the linear function $x \mapsto \lambda x$ on the unit interval. Now reflect the portion of the graph above the line $y = 1$ through that line. Reflect the portion of the new function that is below the line $y = 0$ through that line. Continue until the entire graph is folded into the strip $0 \le y \le 1$.

[5]The following example was suggested by D. A. Lind: Let

$$A = \begin{pmatrix} 1 & 1 \\ 1 & 2 \end{pmatrix}, \quad N = \left\{ \begin{pmatrix} 1 \\ 0 \end{pmatrix} \right\} = \{\mathbf{e}_1\}, \quad \text{and} \quad P = \left\{ \begin{pmatrix} 0 \\ 1 \end{pmatrix} \right\} = \{\mathbf{e}_2\}.$$

Then A is non-negative with respect to $\{\mathbf{e}_1, \mathbf{e}_2\}$, and A^{-1} is non-negative with respect to $\{-\mathbf{e}_1, \mathbf{e}_2\}$.

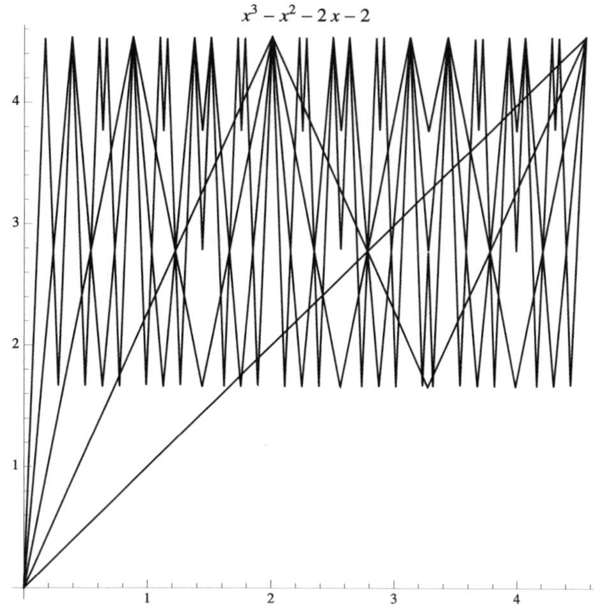

FIGURE 2.2: This is the graph of the first 4 iterates of the function described in the proof, for the Pisot number satisfying $x^3 - x^2 - 2x - 2 = 0$. The Pisot root is $\lambda = 2.2695308$, and the other roots are $-0.63476542 + 0.69160123\,i$ and its conjugate, of modulus 0.938743. The first critical point, 2, maps to the endpoint 2λ, which is fixed. The other critical point, $1 + \lambda$, maps to a fixed point on the third iterate.

This results in a function that may have fewer than $d - 1$ critical points, so repeatedly reflect segments of the graph through horizontal segments $y = \alpha$ at heights $\alpha \in \mathbb{Q}(\lambda)$ until the function has the desired number $d - 1$ of critical points.

For convenience, rescale by some integer n to clear all denominators, so that all critical points are algebraic integers. Therefore, the postcritical orbits are contained in $\mathbb{Z}[\lambda]$.

For each Galois conjugate λ^α of λ, there is an embedding of $\mathbb{Z}[\lambda]$ in $\mathbb{C}$. In each such embedding, the postcritical orbits remain bounded: the action of f_{λ^α} on any point is some composition of functions of the form $f_{\lambda^\alpha,i}(x) = \pm\lambda^\alpha(x) + a_i$, which act as contractions, so there is a compact subset $K \subset \mathbb{C}$ that the Galois conjugates of every piece, $f_{\lambda^\alpha,i}$ takes inside itself.

By construction, the orbit of the critical points under f_λ is also bounded, since f_λ is a map of an interval to itself.

For any fixed bound B, there is only a finite number of algebraic integers of $\mathbb{Q}(\lambda)$ satisfying the bound $|\lambda^\alpha| < B$, since in the embedding into the product of real embeddings and a selection of one from each pair of complex conjugate embeddings, the algebraic integers in $\mathbb{Q}(\lambda)$ form a lattice.[6] The postcritical set is contained in such a set, so it is finite. □

[6]This argument appears in the paper by K. Schmidt [Sch]. It was obtained independently by A. Bertrand (see [Ber] and [Ber1]). It actually goes back to an idea of Gelfond (see [Ge]). These papers concern the β-transformation $T_\beta : [0,1) \to [0,1)$, $T_\beta(x) = \beta x \mod 1$, which is not continuous, but the arguments are the same. In particular Schmidt's paper shows that if β is Pisot, then the set of points (pre-)periodic under T_β is exactly $\mathbb{Q}(\beta) \cap [0,1)$.

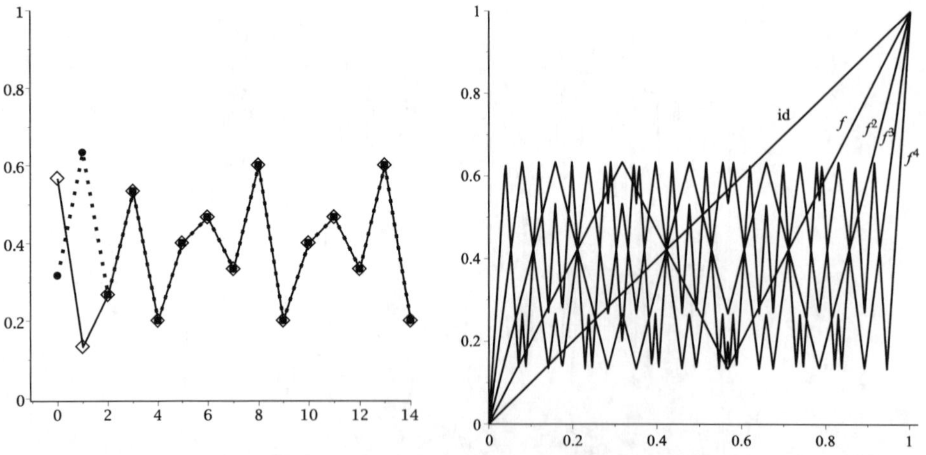

FIGURE 2.3: The diagram on the left shows the postcritical orbits for the Pisot construction where $\lambda = 2$, $d = 3$, with the critical points chosen as $19/60$ and $17/30$. Here the two critical orbits come together after two iterations, and land on a period five orbit after four iterations. On the right is the plot of the first four iterates of this piecewise linear function of entropy $\log 2$. (See also Plate 17.)

This phenomenon is closely related to why decimal representations of rational numbers are eventually periodic. There is a theory of β-expansions, similar to decimal expansions but with β a real number; when β is Pisot, the digits of the β-expansion of any element of $\mathbb{Q}(\lambda)$ are eventually periodic (by an almost identical proof).

The details of the preceding construction of f_λ are not important. *Any* λ-expander whose critical points are in $\mathbb{Q}(\lambda)$ will do. As long as $\lambda < d > 2$, there are infinitely many. To make the proof work as phrased, rescale the unit interval to clear all denominators, so that all critical points become algebraic integers in $\mathbb{Q}(\lambda)$.

Even in the case that λ is an integer less than d, this gives countably many different examples provided $d > 2$. For instance, Figure 2.3 shows the critical point orbits when $\lambda = 2$, $d = 3$ and the critical points are chosen as $19/60$ and $17/30$. On the fourth iterate, they settle into a single periodic orbit of period 4. There is a unique cubic polynomial, up to affine conjugacy, having the same order structure for the postcritical orbits, with entropy $\log(2)$.

In general, there is a non-empty convex $(d-2)$-dimensional space of λ-uniform-expanders for every $1 < \lambda < d$. If λ is Pisot, then postcritically finite examples are dense in this set.

There are many Pisot numbers: for any real algebraic number y, it is easy to see that there are infinitely many Pisot numbers in $\mathbb{Q}(y)$. In fact, the intersection of the lattice of algebraic integers with the unit cylinder centered around any line corresponding to an embedding of $\mathbb{Q}(y)$ in $\mathbb{R}$ consists of Pisot numbers, except for those in a closed ball containing the origin. However, in the geometric sense, Pisot numbers are rare: in [Sa] Salem proved that the set of Pisot number is a countable closed subset of $\mathbb{R}$, making use of a theorem of Pisot that a real number x is Pisot if and only if sequence of minimum differences of x^n from the nearest integer is square-summable. The golden ratio is the smallest accumulation point of Pisot numbers, and the *plastic number* 1.3247..., a root of $x^3 - x - 1$, is the smallest Pisot number.

It is elementary and well-known that postcritically finite maps are dense among uniform expanders, but the preceding construction raises a question that does not seem obvious for $d > 2$:

QUESTION 2.4. *For fixed d, is there a dense set of numbers $1 < \lambda < d$ for which the set of postcritically finite maps is dense among λ-uniform expanders? For which λ are there infinitely many non-affinely equivalent postcritically finite maps? For which λ are postcritically finite maps dense?*

One way to get infinite families of postcritically finite maps with the same λ is to take dynamical extensions of maps with fewer critical points (*c.f. Section 10*), taking care only to introduce new critical points that map to positions whose forward orbit is finite. But this construction cannot work when $\lambda > d - 1$.

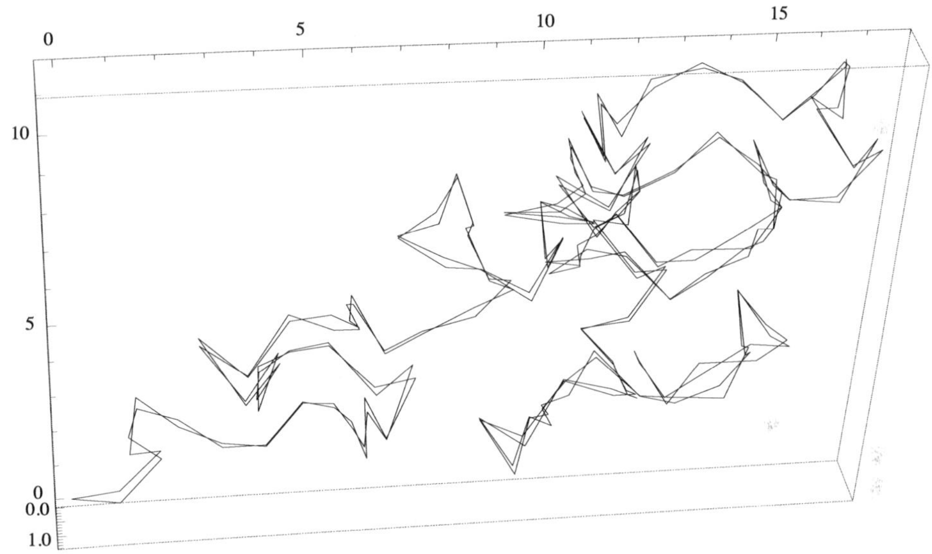

FIGURE 2.5: The tent map $x \mapsto \lambda|x| - 1$ for the degree 6 Salem number $\lambda = 1.4012683679\ldots$ that satisfies $\lambda^6 - \lambda^4 - \lambda^3 - \lambda^2 + 1 = 0$ is postcritically finite, with the critical point having period 270, quite large compared to the degree of λ. This figure shows the absolute value of the projection to the λ-line (the thin direction) as well as to the two complex places of λ. The trajectory resembles a random walk in the plane. Since random walks in $\mathbb{E}^2$ are recurrent (they have probability 1 of visiting any set of positive measure infinitely often), one would expect it to eventually return. It does, but as random walks often do (the mean return time is ∞), it takes a long time.

Salem numbers are closely related to Pisot numbers: some people conjecture that the union of Salem numbers and Pisot numbers is a closed subset of $\mathbb{R}$. However, the construction that worked for Pisot numbers is inadequate for Salem numbers. The Galois conjugates of the linear pieces of f_λ are isometries of $\mathbb{C}$ for any conjugate on the unit circle. Let n be the degree of the Salem number, so for each linear piece of f_λ there are $n/2 - 1$ Galois conjugate complex isometries, one for each complex place. If we take the product of these isometries over all complex places of $\mathbb{Q}(\lambda)$ we get an action by isometries on $\mathbb{C}^{n/2-1}$, where the first derivative of the action of each linear piece is $\pm U_\lambda$, where U_λ is a unitary transformation. In the unitary group, the orbit is dense on an $n/2-1$ torus, acting as an irrational rotation of the quotient of the torus by $\pm I$ (thus factoring out the complication of the variable sign of f_λ). The action of the sequence of linear functions on the $(n/2-1)$-tuple of moduli appears to behave like a random walk in $\mathbb{R}^{(n-2)/2}$. Sometimes they are periodic, sometimes with fairly large periods, but Brownian motion in dimension bigger than 2 is not recurrent, and a few experiments for $n \geq 8$ indicate that they typically drift slowly toward infinity and thus are not postcritically finite. However, there could be some reason (opaque to me) why they might not act randomly in the long run, and they could eventually cycle. At least it seems hard to prove any particular example is not postcritically finite.

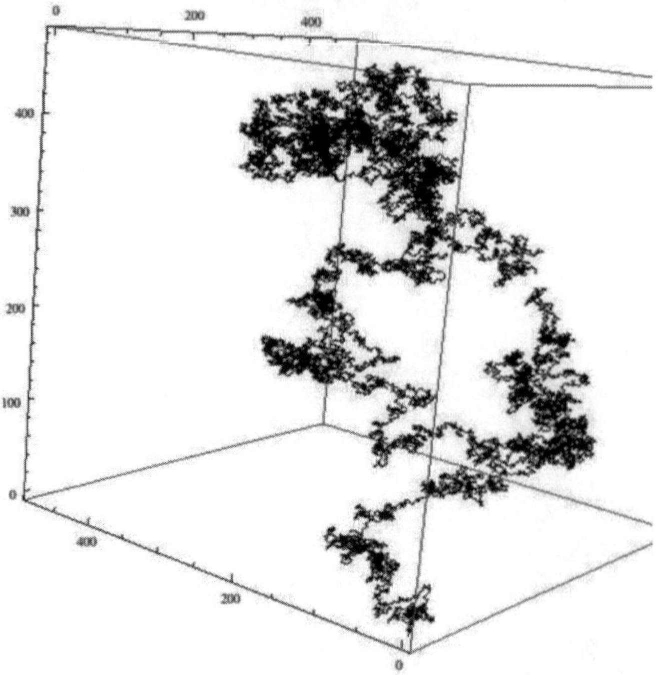

FIGURE 2.6: This diagram is a 3-dimensional projection of the first 200,000 iterates of the critical point for $x \mapsto \lambda|x| - 1$ where $\lambda = 1.17628\ldots$ is Lehmer's constant with minimal polynomial $1 + x - x^3 - x^4 - x^5 - x^6 - x^7 + x^9 + x^{10}$. There are 4 complex places; this shows a 3-dimensional projection for the quadruple of absolute values. It resembles a Brownian path in 3 dimensions. The map is postcritically finite if and only if the path closes. Its values are always algebraic integers, so if it comes sufficiently close to the start it's fairly likely to close (become eventually periodic). The quadruple of radii determines a 4-torus in $\mathbb{C}^4 \times \mathbb{R}^2$, with the current lattice point a bounded distance from that torus. When the radii are on the order of 200, this bounded neighborhood has volume on the order of $(2\pi \times 200)^4$, with roughly 10^{12} lattice points, so the chances of returning seem remote.

For example, for the Salem number $1.7220838\ldots$ of degree 4 which satisfies $x^4 - x^3 - x^2 - x + 1 = 0$, the critical point of the tent map is periodic of period 5.

For the Salem number 1.401268367939... satisfying $x^6 - x^4 - x^3 - x^2 + 1 = 0$, the critical point has period 270. Note that its square is also a Salem number, for which the period is $135 = 270/2$.

One of the most famous Salem numbers is the Lehmer constant, defined by the polynomial $1 + x - x^3 - x^4 - x^5 - x^6 - x^7 + x^9 + x^{10} = 0$. The single root outside the unit circle is approximately 1.17628. This is the smallest-known Salem number and, in fact, the smallest-known Mahler measure for any algebraic integer. (Mahler measure is the product of the absolute value of all Galois conjugates outside the unit circle.) Figure 2.6 is a diagram of the first 200,000 elements of the critical point in $x \mapsto \lambda|x| - 1$, projected from $\mathbb{C}^4 \times \mathbb{R}^2$ to the quadruple of radii in the complex factors and from there a projection into 3 dimensions. The map is postcritically finite if and only if the path closes. Its values are always algebraic integers, so if it comes sufficiently close to the start it's fairly likely to close. The quadruple of radii determines a 4-torus in $\mathbb{C}^4 \times \mathbb{R}^2$, with the current lattice point a bounded distance from that torus. When the radii are on the order of 200, as in this case, this bounded neighborhood has volume on the order of $(2\pi \times 200)^4$, with roughly 10^{12} lattice points, so the chances of looping appear remote unless the path wanders close to the origin, where the tori are smaller. For comparison, the variance of a random walk with step size 1 in $\mathbb{R}^n$ equals the number of steps, so for a random walk of length 200,000, the standard deviation is $\sqrt{200,000} \approx 447$; projected from 8 dimensions to 3, the standard deviation would be ≈ 274, in line with the picture. An experiment with a selection of small Salem numbers of moderate degree greater than 6 showed similar results, with none of them exhibiting a cycle within 500,000 iterates.[7]

3 CONSTRUCTING INTERVAL MAPS: FIRST STEPS

For a Perron number that is not Pisot, the situation is much more delicate. To develop a strategy, first we'll discuss incidence matrices. There are two reasonable versions for an incidence matrix that are transposes of each other. We will use the version whose columns each represent the image of a subinterval of the domain and whose rows represent a subinterval in the range, so that each entry counts how many times the subinterval that indexes its column crosses the subinterval that indexes its row.

Suppose the unit interval I is subdivided into n subintervals $J_1, \dots, J_n$ (in order). Let $V = \{0 = v_0, v_1, \dots, v_n = 1\}$ be the vertex set. Every function $g\colon V \to V$ that takes adjacent vertices to distinct vertices extends to a postcritically finite piecewise linear map. In this way, a large but finite and specialized set of $n \times n$ matrices can be realized as incidence matrices. Incidence matrices are easy to recognize. Each column consists of a consecutive sequence of 1s, and is otherwise 0. There is a matching between the ends of the blocks of consecutive 1s in adjacent columns, with every column except the first and last having one end of the block of 1s matched to the left and the other end of the block of 1s matched to the right.

Here is a generalization of this concept. Consider a map $f\colon I \to I$ with finitely many critical points such that $f(V) = V$ and f of the critical set is contained in V, that is, V contains all critical values. Under these conditions, an extended incidence matrix is still defined for f, whose entries a_{ij} count how many times $f(J_j)$ crosses interval J_i.

[7]For further discussion, see Note 13.3.

PROPOSITION 3.1. *An $n \times n$ non-negative integer matrix A is an extended incidence matrix if and only if*

1. *the nonzero entries in each column form a consecutive block, and*
2. *there is a map $\phi\colon \{0, 1, \ldots, n\} \to \{0, 1, \ldots, n\}$ such that in column i, the entries in rows greater than $\phi(i-1)$ and not greater than $\phi(i)$, and no other entries, are odd.*[8]

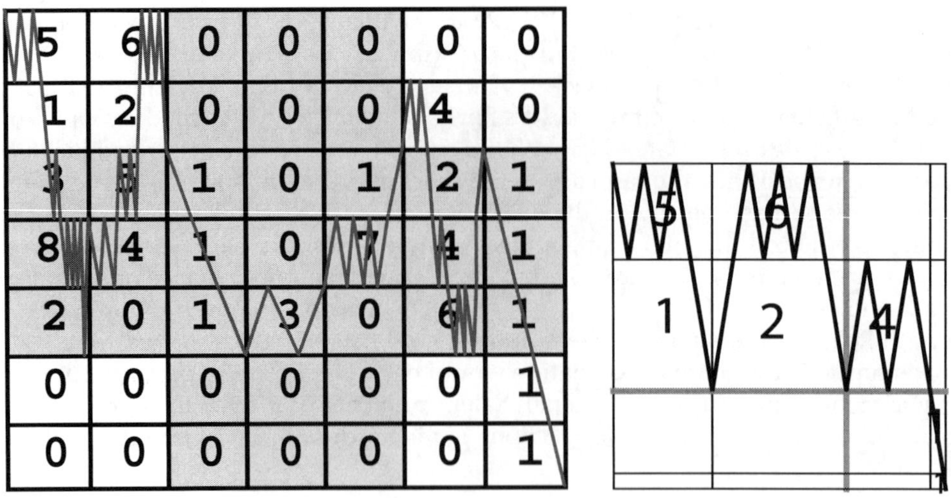

FIGURE 3.2: The matrix M on the left satisfies the necessary and sufficient conditions of proposition 3.1 to be an incidence matrix for a self-map of the interval: in each column, the positive entries in each column form a connected block as do the odd entries, and the odd blocks in the columns can be matched together end to end, from the left column to the right, to form a chain. The jagged line defines a PL function realizing the given matrix, since it crosses each square the specified number of times. If you reflect the line across the top edge of the matrix, it matches the usual convention for drawing a graph (since the convention for matrices that the row numbers increase going downward is opposite the convention for graphs of functions). On the right, the widths of rows and columns of the matrix have been adjusted in proportion to the positive eigenvector for the transpose M^t. The matrix M^t is not ergodic, entries 3,4,5 of the positive eigenvector are 0, and the shaded blocks at left have collapsed to the gray lines at right. Now the graph can be drawn with constant absolute slope.

PROOF. The necessity of the conditions is easy. The map ϕ represents the map f restricted to V. The image of any interval J_i is necessarily the union of a consecutive block of intervals. The subinterval between the images of the first and last endpoints, $[f(v_{i-1}), f(v_i)]$ (which could be a degenerate interval) is traversed an odd number of times, and the rest of the image is traversed an even number of times.

Sufficiency of the conditions is also easy. To map J_i, start from $v_{\phi(i-1)}$, go to the lowest vertex in the image, zigzag across the lowest interval until its degree is

[8]Condition (2) is not quite right as stated. See Note 13.4.

used up, then the next lowest, etc. until you get to $v_{\phi(i)}$, at which point proceed to the highest vertex in the image and work back. □

Note: Sufficiency can also be reduced to the familiar condition that a graph admits a Eulerian[9] path from vertex a to vertex b if and only if it is connected and either $a = b$ and all vertices have even valence, or a and b have odd valence and all other vertices have even valence. For each i, apply this to the graph Γ_i that has a_{ij} edges connecting v_{j-1} to v_j. The entire map f is really a Eulerian path in the graphs Γ_i connected in a chain by joining vertex $\phi(i)$ of Γ_i to that of Γ_{i+1}, followed by the natural projection to $[0,1]$.

PROPOSITION 3.3. *The topological entropy of any map with extended incidence matrix A is the log of the largest positive eigenvalue of A.*

PROOF. The total variation of f^n is a positive linear combination of matrix entries of A^n (if all intervals have equal length, it is their sum). By 1.2, $h(f)$ is the exponent of growth of total variation, so this equals the log of the largest positive eigenvalue of A. □

Suppose we are given a (strict) Perron number λ. Our strategy is to first construct a strictly positive extended incidence matrix that has positive eigenvalue λ^N, for a large power N of λ. We will promote this to an example with entropy λ by implanting it as the return map replacing a periodic cycle of a map with entropy less than λ.

Afterward, we will deal with additional issues involving questions of mixing and weak Perron numbers.

Given λ, let O_λ be the ring of algebraic integers in the field $\mathbb{Q}(\lambda)$, and let V_λ be the real vector space $V_\lambda = \mathbb{Q}(\lambda) \otimes_{\mathbb{Q}} \mathbb{R}$. Another way to think of it is that V_λ is the product of the real and complex places of $\mathbb{Q}(\lambda)$, that is, the product of a copy of $\mathbb{R}$ for each real root of the minimal polynomial for λ and a copy of $\mathbb{C}$ for each pair of complex conjugate roots. The ring operations of $\mathbb{Q}(\lambda)$ extend continuously to V_λ, but division is discontinuous where the projection to any of the places is 0. Yet another way to think of V_λ is in terms of the companion matrix C_λ for the minimal polynomial P_λ of λ. We can identify $\mathbb{Q}(\lambda)$ with the set of all polynomials in C_λ with rational coefficients, and V_λ with the vector space on which the companion matrix acts. The real and complex places of $\mathbb{Q}(\lambda)$ correspond to the minimal invariant subspaces of C_λ. When P_λ is factored into polynomials that are irreducible over $\mathbb{R}$, the terms are linear and positive quadratic; the subspaces are in one-to-one correspondence with these factors.

We are assuming that λ is a Perron number, so for multiplication of V_λ by λ, the λ-eigenvector is dominant. If we consider the convex cone K_λ consisting of all points where the projection to the λ eigenspace is larger than the projection to any of the other invariant subspaces, then $\lambda \times K_\lambda$ is contained in the interior of K_λ except at the origin.

PROPOSITION 3.4. *There is a rational polyhedral convex cone KR_λ contained in K_λ and containing $\lambda \times K_\lambda$.*

[9]An *Eulerian path* is one that passes through each edge exactly once. Thurston's manuscript uses the term *Hamiltonian path*, but this would be historically incorrect. Euler first studied such paths, while Hamilton studied paths which pass through each *vertex* exactly once.

PROOF. It is easiest to think of this projectively, in $\mathbb{P}(V_\lambda)$. The projective image $\mathbb{P}(K_\lambda)$ of K_λ is a convex set (specifically, a product of intervals and disks), with the image of $\mathbb{P}(\lambda \times K_\lambda)$ contained in its interior. Since rational points $\mathbb{P}(\mathbb{Q}(\lambda))$ are dense in $\mathbb{P}(V_\lambda)$, we can readily find a set of rational points in the interior of $\mathbb{P}(K_\lambda)$ whose convex hull contains $\mathbb{P}(\lambda \times K_\lambda)$. This gives us the desired rational polyhedral convex cone. □

Let S_λ be the additive semigroup $O_\lambda \cap KR_\lambda \setminus \{0\}$.

PROPOSITION 3.5. *S_λ is finitely generated as a semigroup.*

PROOF. This is a standard fact. Here's how to see it using elementary topology: We can complete S_λ by adding on the set of projective limits $\mathbb{P}(KR_\lambda)$. The completion is compact. It's easy to see that the set of closures U_s of the "ideals" $s + S_\lambda$ for $s \in S_\lambda$ form a basis for this topology.[10] By compactness, the cover by basis elements has a finite subcover $U_{s_1}, \ldots, U_{s_k}$. For any such cover, the set $G = \{s_1, \ldots, s_k\}$ form a generating set: given any element, write it as $s_i + s$, and continue until the remainder term s is 0. □

PROOF OF CONVERSE PERRON-FROBENIUS (THEOREM 1.6). Equipped with this picture, it is now easy to prove the converse Perron-Frobenius theorem of Lind. Consider the free abelian group $\mathbb{Z}^G$ on the set G of semigroup generators. The positive semigroup in $\mathbb{Z}^G$ maps surjectively to S_λ. Call this map p. We can lift the action of multiplication by λ to an endomorphism of the positive semigroup by sending each generator g to an arbitrary element of $p^{-1}(\lambda\, p(g))$. In coordinate form, this is described by a non-negative integer matrix. In S_λ, projection to the λ-space is a dual eigenvector, that is, a linear functional multiplied by λ under the transformation. It is strictly positive[11] on S_λ. Therefore, the pullback of this function is a strictly positive λ-eigenvector of the transpose matrix, proving Lind's theorem. □

Note that the minimum size of a generating set might be much larger than the dimension of V_λ. For instance, if λ is a quadratic algebraic integer, V_λ is the plane, and KR_λ could be bounded by any pair of rational rays that make slightly less than a 45° angle to the λ eigenvector. Minimal generating sets can be determined using continued fraction expansions of the slopes; they can be arbitrarily large.

As we shall presently see, it can happen in higher dimensions that the minimum size of a generating set for S_λ can be very large, no matter how we choose a semigroup S_λ on which multiplication by λ acts as an endomorphism.

[10] This is not quite right as stated, since these sets are not open and do not form a basis for the topology. However, each point in the limit set $\mathbb{P}(KR_\lambda) \subset \overline{S}_\lambda$ has an open neighborhood contained in some U_s. Choosing finitely many U_{s_j} which cover a neighborhood of the limit set, and adding the finitely many s_k which are not included in these U_{s_j}, we obtain a finite generating set.

[11] In the diagram

$$\begin{array}{ccccc} \mathbb{Z}_+^G & \xrightarrow{L} & \mathbb{Z}_+^G & & \\ p\downarrow & & p\downarrow & & \\ S_\lambda & \xrightarrow{\lambda\cdot} & S_\lambda & \overset{\iota}{\subset} & \mathbb{R} \end{array}$$

the composition $\iota \circ p \in \mathrm{Hom}(\mathbb{Z}_+^G, \mathbb{R})$ is a dual eigenvector which is strictly positive, in the sense that every basis element in $\mathbb{Z}_+^G$ maps to a positve real number.

4 SECOND STEP: CONSTRUCTING A MAP FOR λ^N

Now we need to address the special requirements for an incidence matrix for a map having a finite invariant set as the set of critical values. Given any Perron number λ of degree d, we will construct such a matrix for some power, probably large, of λ. From Section 3, we assume we have a set G of semigroup generators for a semigroup in $S_\lambda \subset V_\lambda = \mathbb{Q}(\lambda) \otimes_{\mathbb{Q}} \mathbb{R}$ invariant under multiplication by λ.

Choose a finite sequence of generators, including each generator at least once, such that the partial sums of the sequence contain all 2^d mod 2 congruence classes, that is, the partial sums map surjectively to $O_\lambda/(2O_\lambda)$. If necessary, adjoin additional elements to the sequence so that the sum T of the entire sequence is 0 (mod $2O_\lambda$).

The action of λ (by multiplication) on the projective completion of S_λ has a unique attracting fixed point. For any $s \in S_\lambda$, the closure of $s + S_\lambda$ contains[12] the fixed point, so there is some power N such that $\lambda^N \times S_\lambda \subset 3T + S_\lambda$.

Now mark off an interval of length T into segments whose lengths are given by the chosen sequence $g_1, \dots, g_k$ of generators (in the embedding of O_λ in $\mathbb{R}$ where λ goes to λ). We'll construct a λ^N-expander map by induction, going from left to right, starting with $0 \to 0$. There is at least one point $q = g_1 + g_2 + \cdots + g_k$ in this subdivision that has the same value mod $2O_\lambda$ as $\lambda^N \times g_1$. Since $\lambda^N \times g_1$ can be expressed as $3T$ plus a sum of generators, it can also be expressed as $q + 2T$ plus a sum of generators. Since $(\lambda^N \times g_1 - q - 2T)$ is in S_λ and congruent to zero mod 2, it can be expressed as 2α, where $\alpha \in S_\lambda$. We can write 2α as a linear combination of generators with even coefficients, so we can write $\lambda^N \times g_1$ as a strictly positive sum of generators, where each generator between 0 and q occurs an odd number of times, and each other generator occurs an even number of times.

We can continue in exactly the same way for λ^N times each of the points in the subdivision. We first pick where each point g_i goes based on the congruence class of $\lambda^N \times g_i$ mod 2, then express the difference as an even and strictly positive linear combination of all generators in the sequence. Finally, we end with T going to either 0 or T, as we choose. The incidence matrix satisfies the conditions of 3.1, so we have constructed a λ^N uniform expander.

5 POWERS AND ROOTS: COMPLETION OF PROOF OF THEOREM 1.3

Given a Perron number λ, we'll first construct a map of S^1 to itself that is a λ uniform expander, because the construction is a little nicer for S^1. From the preceding section, for some N we construct a λ^N uniform expanding map f_{λ^N} of an interval that takes each endpoint to itself. Let ρ be any rotation of the circle of order N. The circle can be subdivided into N intervals that are cyclically permuted by the rotation. Define a metric on the circle so that in this cyclic order, the N intervals have length $1, \lambda, \lambda^2, \dots, \lambda^{N-1}$. Now define $g_\lambda \colon S^1 \to S^1$ by mapping each of the first $N-1$ intervals affinely to the next and mapping the last interval to the first using f_{λ^N} with affine adjustments in the domain and range to send the last interval exactly to the first.[13] Since the first interval has length $1/\lambda^{N-1}$ times the last, with this affine adjustment f_{λ^N} also expands uniformly by λ.

[12] In fact this closure is a neighborhood of the fixed point. For any generator s_j of S_λ, the succesive products $\lambda^n \times s_j$ converge to the fixed point, and it follows that $\lambda^N \times S_\lambda \subset s + S_\lambda$ for large N.

[13] For a figure to illustrate this construction, see Note 13.5.

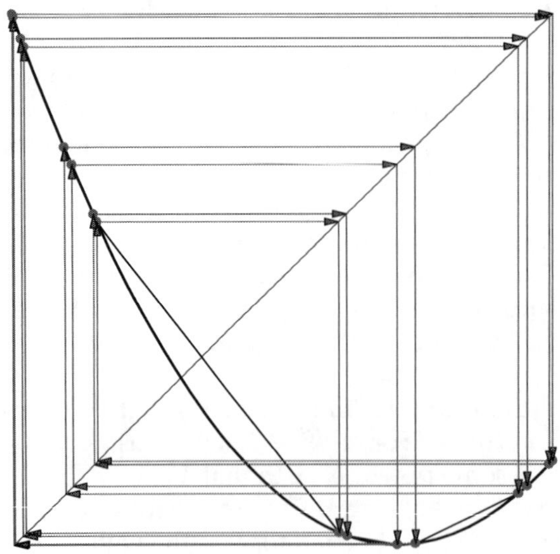

FIGURE 5.1: This is the graph of a quadratic map in the initial period-doubling cascade, with critical point of period 16. This, and all other quadratic maps in this family, have entropy 0. If the critical orbit is blown up and replaced with a sequence of intervals where the return map has is a λ^{16}-uniform expander, then the resulting map is semiconjugate to a λ-uniform expander.

In the case of the circle, we can easily modify the construction of g_λ to make the incidence matrix mixing: this will happen if we change any small piece of f_{λ^N} to stray into a neighboring interval and back, when it gets to the upper endpoint. If the rotation is chosen as a rotation by $1/N$ and if the first generator is chosen to be a "small" element in S_λ, these intervals are small, and straying is probably possible with ease, but in any case, by taking N to be a somewhat higher power, it can be readily guaranteed.

In the case of the unit interval, we need a substitute for a rotation of the circle. For this, we can use the well-known period-doubling cascade for quadratic self-maps of an interval (see Figure 5.1 for an illustrative example). In this period-doubling family, there is a quadratic map q_n with entropy 0 in which the critical point has period 2^n. We can blow up the forward and backward orbit of the critical point, replacing each point x in the orbit by a small interval I_x of any length l_x such that the set of lengths is summable. Extend the map to these intervals by affine homeomorphisms, with the exception of the interval for the critical point; for that, we can use any affine map that takes both endpoints to the right hand endpoint.

Since the entropy of the quadratic map is 0, the number of critical points of q_n^k grows subexponentially in k, so if we assign length λ^{-k} to each interval for a point that is critical for q_n^k but not for q_n^{k-1}, the set of lengths is summable. After blowing up the full orbit of the critical point in this way, we can define a pseudo-metric on the interval where the length of an interval is the measure of its intersection with the blown-up orbit; everything else collapses to measure 0.

By the preceding section, we can find an f_{λ^N} of the form $N = 2^n$ that maps the unit interval to itself, taking both endpoints to 0. Implant this, using affine adjustments in the domain and the range, for the map from the interval for the critical point of q_n to its image. The image of the critical interval has length λ^{2^n-1} since the original critical point had period 2^n, so the implanted map is a uniform λ-expander.

We now have a map which is a local homeomorphism with Radon derivative λ in the complement of the critical interval, where it is a uniform λ-expander. Therefore, the entire map is a piecewise-linear uniform λ-expander, with entropy $\log(\lambda)$. This completes the proof of Theorem 1.3 for strong Perron numbers.[14]

Now we'll address weak Perron numbers.

PROPOSITION 5.2. *A positive real number λ is a weak Perron number if and only if some power of λ is a [strong] Perron number.*

PROOF. In one direction this is pretty obvious: the nth roots of any algebraic integer are algebraic integers, and their ratios to each other are nth roots of unity. Thus the positive real nth root of a Perron number is a weak Perron number.

In the other direction, suppose λ is a weak Perron number. Let B be the set of Galois conjugates of λ with absolute value λ, and let b be their product. This product b is a real number equal to $\lambda^{\#(B)}$. The Galois conjugates of b are products of $\#(B)$ Galois conjugates of λ; for any subset of conjugates of this cardinality other than B, the product is strictly smaller, so b is a Perron number. □

(Since $\lambda^n = b$, the other elements of B also satisfy this equation,[15] so they are just the products of λ with an nth root of unity.)

Now given a weak Perron λ, we first find a power k so that λ^k is a Perron number. In the family of degree 2 uniform expanders, functions with periodic critical point are dense, and functions with critical point having period a multiple of k are also dense. Choose such a function g whose entropy is less than λ, where the critical point has period a sufficiently high multiple kn of k. Blow up the full (backwards and forward) orbit of the critical point, and implant a map of the form $f_{\lambda^{kn}}$, as constructed in Section 4, in the interval replacing the critical point. Since the growth rate of critical points for powers of g is less than the growth rate of powers of λ, we can construct a metric, just as before, that is uniformly expanded by λ.

Note that the λ-uniform expanders we have constructed are very far from mixing. This is, of course, impossible if λ is only a weak Perron number, by the Perron-Frobenius theorem, but for a Perron number λ it is tempting to try to generalize the straying technique that worked earlier.

There are two difficulties. The first is an essential problem.

PROPOSITION 5.3. *For any self-map f of the interval with entropy strictly between* 0 *and* $\log\sqrt{2}$, *there are two disjoint subintervals that are interchanged by the map, and, in particular, f is not mixing.*

---

[14]For a simplified version of this argument, see Note 13.6; and for a proof that the required map can be realized as a polynomial, see [dMvS, Theorem 6.4].

[15]In fact, any Galois automorphism α which takes λ to an element of B must permute the elememts of B, since otherwise the element $(\lambda^\alpha)^n = (\lambda^n)^\alpha = (\prod_{\beta\in B}\lambda^\beta)^\alpha$ would have a smaller absolute value. It follows that $(\lambda^\alpha)^n = \lambda^n$.

PROOF.[16] Let $D(f)$ be the set of points that have more than one preimage. Note that $D(f) \subset D(f^2)$ and also $f(D(f)) \subset D(f^2)$; in fact, $D(f^2) = D(f) \cup f(D(f))$. Let $a = \inf(D(f^2))$, and $b = \sup(D(f^2))$. Then $[a, b]$ is mapped into itself and has the same entropy as f. □

In particular, they never mix. By induction, if the entropy is less than $2^{-n} \log 2$, there are 2^n disjoint intervals cyclically permuted by the map.

The second problem is that the piecewise-linear map we constructed before has coefficients in $\mathbb{Q}(\lambda)$ (either because the formulas from kneading theory for the infinite sums of intervals are rational functions of λ or because they are determined by linear functions with coefficients in O_λ), but it is not obvious how to get coordinates to be in O_λ. Maybe it's possible to analyze and control the algebra, but if so it's beyond the scope of this paper.

6 MAPS OF ASTERISKS

Postcritically finite self-maps of graphs, including in particular Hubbard trees, give another interesting collection of one-dimensional postcritically finite maps.

REMARK 6.1. Any expanding self-map of a tree can be promoted to a self-map of a Hubbard tree by adding extra information as to a planar embedding and choosing a branched covering map for the planar neighborhood of each vertex that acts on its link in the given way. Each such promotion is the Hubbard tree for a unique polynomial up to affine automorphism.

For use later in constructing automorphisms of free groups, we will look at a special case, the *asterisks* $*$. An *n-pointed asterisk* is the cone on a set of n points, which we'll refer to as the *tips* of the asterisk.

THEOREM 6.2. *For every Perron number λ there is a postcritically finite λ-uniform expanding self-map f of some asterisk such that*

- *f fixes the center vertex,*
- *f permutes the n tips,*[17] *and*
- *the incidence matrix for f is mixing.*

PROOF. This is similar to the proof for self-maps of the interval. In principal it is easier because there is no order information to worry about, but we will use a very similar method.

An incidence matrix for a self-map of an asterisk of the given form is a non-negative integer matrix that has exactly one odd entry in each row and each column. Given any such matrix, we can construct a corresponding self-map of an asterisk by permuting the tips according to the matrix mod 2, which is a permutation matrix, and running each edge out and back various edges to generate the even part of the matrix.

As before, find a subsemigroup S_λ of O_λ that excludes 0 and is invariant under multiplication by λ.

[16] This proof is seriously incomplete. Compare Note 13.7.
[17] This condition has been simplified by the editors

Let G be a finite set of generators for S_λ that maps surjectively to $O_\lambda/(2O_\lambda)$. Let $T = \sum_{g \in G} g$.

Let n be an integer such that $\lambda^n - 1$ is congruent to 0 mod 2, and for each $g \in G$, $\lambda^n \times g$ is contained in $g + 2(T + \lambda \times g) + S_\lambda$.

Now make an asterisk whose points are indexed by $G \times \{1, 2, \ldots, n-1\}$. Map the edge to point (g, i) homeomorphically to the edge to point $(g, i+1)$ when $i < n - 2$.

The final set of tips $(g, n-1)$ will map to tips $(g, 1)$. For each g, $\lambda^n \times g - g$ is congruent to 0 mod $2O_\lambda$ and so can be expressed as a strictly positive even linear combination of $G \cup \{\lambda \times g\}$.

Use these linear combinations to construct an asterisk map. Each edge to the first set of tips maps homeomorphically to a new edge for the first $n - 1$ iterates. On the nth iterate, it maps to a path that makes at least one round trip to all the first-layer points as well as one second-layer point, finally ending back where it started. We can easily arrange the order of traversal so that every edge is represented in the first segment of one of these edge paths.

If the asterisk is given a metric where edge (g, i) has length equal to the value of $\lambda^i g$ in the standard embedding of O_λ in $\mathbb{R}$, where λ is the Perron number, this map is a λ-uniform expander. It is mixing: by the nth iterate, the image of an edge with index (g, i) contains all generators of the form $(*, i)$; by the $2n$th iterate, the image of an edge contains all generators of the form $(*, i)$ and $(*, i+1)$. After n^2 iterates, the edge maps surjectively to the entire asterisk. □

This construction was intended to avoid the need for complicated conditions and bookkeeping. It's clear that a more careful construction could produce suitable asterisk maps for a typical Perron number λ that are much smaller (but still might be quite large).

7 ENTROPY IN BOUNDED DEGREE

The constructions for maps of given entropy have been very uneconomical with the complexity of the maps. First, there is a potentially dramatic (but sometimes unavoidable) blowup in going from a Perron number λ to a finite set of generators for a subsemigroup S_λ of algebraic integers invariant by multiplication by λ. Even then, there is another possibly large blowup in finding a power of λ such that $\lambda^N \times S_\lambda$ is sufficiently deep inside S_λ to guarantee an easy construction of a suitable incidence matrix. In other words: unlike Pisot numbers, a typical Perron number is probably unlikely to be the growth constant for a typical postcritically finite map of an interval to itself.

In some sense, expansion constants for bounded degree systems are almost Pisot: most of their Galois conjugates don't seem to wander very far outside the unit circle. Figure 1.1 illustrate this point: most of the Galois conjugates of $\exp(h(f))$ for postcritically finite quadratic maps are in or near the unit circle. Since their minimal polynomials have constant term ± 1 or ± 2, the inside and outside roots are approximately balanced. If they don't wander outside the circle, they can't wander very far inside the circle, and most roots are near the unit circle. In contrast, Figure 1.4 illustrates that Perron numbers less than 2 can have roots spread the disk of the their radius.

DEFINITION 7.1. *To get some understanding of what's going on, we'll translate questions about the distribution of roots into questions about dynamics of a semigroup of affine maps, elaborating on the point of view taken in the discussion of Pisot numbers in Section 2. We'll consider the sets of piecewise linear uniform expander functions $F(d,\epsilon)$ that take ∂I onto ∂I, where $d > 1$ is the number intervals on which the function is linear and $\epsilon = \pm 1$ determines the sequence of slopes. $\epsilon(-1)^i\lambda$ is the sequence of slopes in subintervals $i = 0$ through $i = d-1$. The constant terms for the linear function f_i in the first and last interval are determined by the condition that $f(\partial I) = \partial I$, implying that $f_0(0) = 0$ if $\epsilon = 1$ and $f_0(0) = 1$ when $\epsilon = -1$, with a similar equation for the last interval. In all other subintervals, the constant term is a free variable C_i subject to linear inequalities. That is, the ith critical point c_i, determined by $\epsilon(-1)^{i-1}\lambda c_i + C_{i-1} = \epsilon(-1)^i\lambda c_i + C_i$ must be inside the unit interval, so we have the inequalities*[18]

$$0 \le \epsilon(-1)^{i-1}\frac{C_i - C_{i-1}}{2\lambda} \le 1.$$

Now for any other field embedding $\sigma\colon \mathbb{Q}(\lambda, C_1, \dots, C_{i-1}) \to \mathbb{C}$, we can look at the collection of image functions f_i^σ. Since the choice of which f_i to apply is determined by inequalities, we will look at the product action. Define $f^{1+\sigma}$ to act on $I \times \mathbb{C}_\sigma$ by $f^{1+\sigma}\colon (x,z) \mapsto (f(x), f_i^\sigma(z))$ where f_i is a linear piece that f applies to x. In the ambiguous case where x is one of the critical points, this definition is discontinuous, so we will look at both images, which are the limit from the left and the limit from the right.

Define the *boundedness set* $B(f^{1+\sigma})$ to be the set of (x,z) such that its (forward) orbit stays bounded. If f is postcritically finite, then the critical points[19] in particular have bounded orbits, so in addition the full orbit of the critical points (under taking inverse images as well as forward images) are contained in the boundedness locus. Define the limit set $L(f^{1+\sigma})$ to be the smallest closed set containing all ω-limit sets for (x,z).

There are three qualitatively different cases, depending on whether $|\sigma(\lambda)|$ is less than 1, equal to 1, or greater than 1.

In the first case, $|\sigma(\lambda)| < 1$, every orbit remains bounded in the $\mathbb{C}_\sigma$ direction, since the map is the composition of a sequence of contractions. Everything outside a certain radius is contracted, so the limit set[20] is compact. Some special examples of this are illustrated in Figures 7.2, 7.3, and 7.4.

The second case, when $|\sigma(\lambda)| = 1$, seems hardest to understand, since the f_i^σ act as isometries. Perhaps the limit set in these cases is all of $\mathbb{C}$.

In the third case, when $|\sigma(\lambda)| > 1$, the maps f_i^σ are expansions, so there is a radius R such that everything outside the disk $D_R(0)$ of radius R centered at the origin escapes to ∞. Given $x \in [0,1]$, the only z such that $f^{1+\sigma}(x,z)$ has second coordinate inside $D_R(0)$ are inside a disk of radius $R/\sigma(\lambda)$ about the preimage of 0. Starting far along in the sequence of iterates and working backward to the beginning, we find a sequence of disks, shrinking geometrically by the factor $1/\sigma(\lambda)$, that contain all bounded orbits. In the end, there's a unique point $b(x) \in \mathbb{C}$ such that $(x, b(x))$ remains bounded. The point $b(x)$ can also be expressed as the sum of a power series in $\sigma(\lambda)^{-1}$ with bounded coefficients that depend on the C_i and the kneading data for x, matching with formulas from [MT].

[18] Actually the C_i must satisfy the stronger inequalities $0 \le f(c_i) = (C_{i-1} + C_i)/2 \le 1$; and also $C_0 > C_2 > C_4 > \cdots$ and $C_1 < C_3 < \cdots$ if $\epsilon = +1$, or the opposite inequalites if $\epsilon = -1$.

[19] By a critical point for $f^{1+\sigma}$ is meant a pair $\big(x, \sigma(x)\big)$, where x is a critical point for f.

[20] The original manuscript stated incorrectly that the boundedness set is compact.

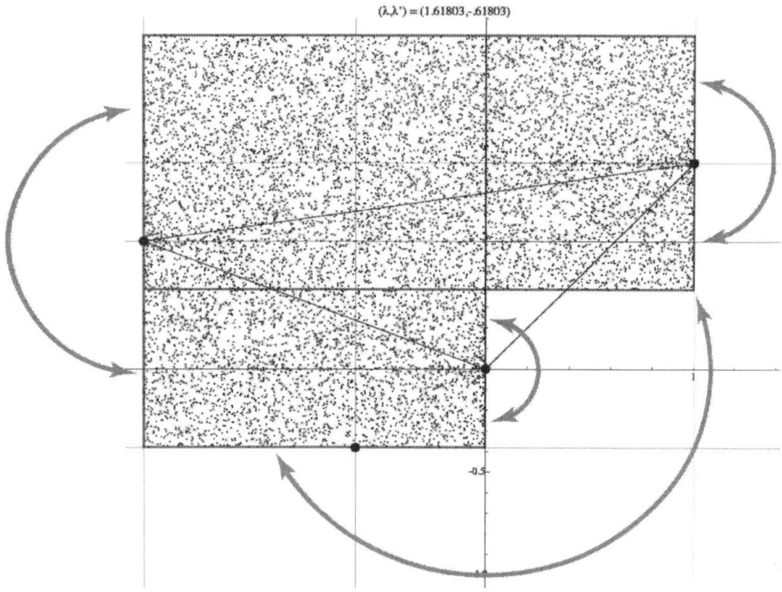

FIGURE 7.2: This diagram shows $L(f^{1+\sigma})$, where $\lambda = 1.61803\ldots$, the golden ratio, is a Pisot number, and $\sigma(\lambda) = -1/\lambda$. It was drawn by taking a random point near the origin, first iterating it 15,000 times and then plotting the next 30,000 images. The dynamics is hyperbolic and ergodic, so almost any point would give a very similar picture. The critical point is periodic of period 3, and the limit set is a finite union of rectangles with a critical point on each vertical side. The dynamics reflects the rectangle on the upper right in a horizontal line and arranges it as the rectangle on the lower left. The big rectangle formed by the two stacked rectangles at left is reflected in a vertical line, stretched horizontally and squeezed vertically, and arranged as the two side-by-side rectangles on the top. Although the dynamics is discontinuous, the sides of the figure can be identified, as indicated (partially) by the arrows to form a tetrahedron to make it continuous: see Figure 7.3. (See also Plate 18.) For further discussion, see Notes 13.8 and 13.9.

FIGURE 7.3: This is the diagram from Figure 7.2 taped together into a tetrahedron, where the dynamics acts continuously (but reverses orientation). The map $f^{1+\sigma}$ acts on the tetrahedron as an Anosov diffeomorphism. There are coordinates in which it becomes the Fibonacci recursion $(s,t) \mapsto (t, s+t)$, modulo a $(2,2,2,2)$ symmetry group generated by $180°$ rotations about lattice points. Compare Note 13.9. (See also Plate 19.)

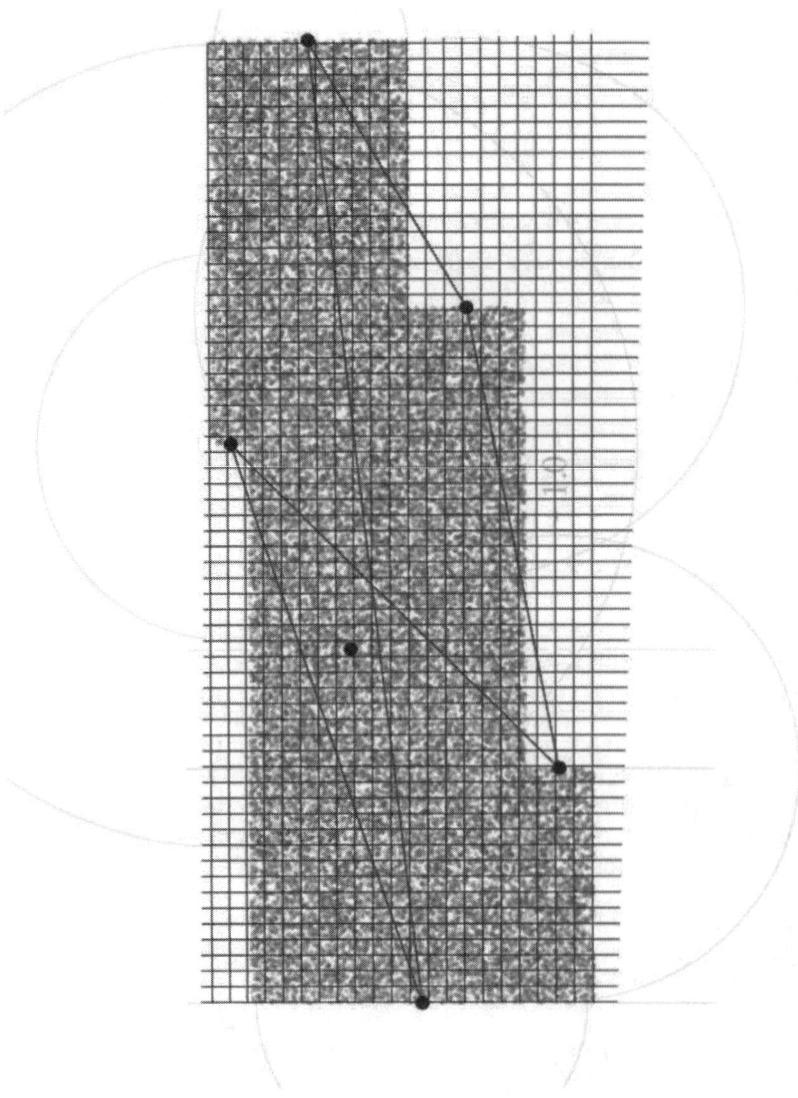

FIGURE 7.4: This diagram shows $L(f^{1+\sigma})$ where $\lambda = 1.7220838\ldots$ is a Salem number of degree 4 satisfying $\lambda^4 - \lambda^3 - \lambda^2 - \lambda + 1 = 0$, and $\sigma(\lambda) = 1/\lambda$. The critical point is periodic of period 5, and the limit set is a finite union of rectangles with a critical point on each vertical side. The dynamics multiplies the left half (to the left of the vertical line through the uppermost black dot) by the diagonal matrix with entries $(-\lambda, -1/\lambda)$, and then translates until the lowermost black dot goes to the uppermost black dot. The right half of the figure is multiplied by the diagonal matrix $(\lambda, 1/\lambda)$, and translated to fit in the lower left. As with the example in Figure 7.2, it can be folded up, starting by folding the vertical sides at the black dots, to form (topologically) an S^2 on which the dynamics acts continuously, a pseudo-Anosov map of the (2,2,2,2,2)-orbifold. This phenomenon has been explored, in greater generality, by André de Carvalho and Toby Hall [CH], and it is also related to the concept of the dual of a hyperbolic groupoid developed by Nekrashevych [Ne]. (See also Plate 21.)

PROPOSITION 7.5. *If f is postcritically finite and if $|\sigma(\lambda)| > 1$, then $b(x)$ depends continuously on x.*

PROOF. When f is postcritically finite, the preperiodicity of f at any critical point c_i is equivalent to an identity among compositions of the f_i. Therefore, $f^{1+\sigma}$ satisfies the same identity when applied to $(c_i, \sigma(c_i))$; therefore, its orbit is bounded so $b(c_i) = \sigma(c_i)$.

Since $b(x)$ is the sum of a geometrically convergent series, the value depends continuously on the coefficients. The coefficients change continuously except for where x is a precritical point; but the limits from the two sides coincide at precritical points, since they coincide for critical points. Therefore, $b(x)$ is continuous. □

When f is not postcritically finite, $b(x)$ might not be continuous; there could well be different limits from the left and from the right at critical points and, therefore, at precritical points: the natural domain for $b(x)$ in general is a Cantor set obtained by cutting the interval at the countable dense set of precritical points. For example, if λ is transcendental, then σ could send it to any other transcendental number, and it's obvious that the generically $b(x)$ would not be continuous.

This phenomenon points to the inadequacy of considering only the algebraic properties of σ when our real goal is to control the geometry. Here is a formulation of the appropriate condition, in a slightly bigger context.

DEFINITION 7.6. *Suppose that $f \in F(d, \epsilon)$ (see Definition 7.1). Let V be a complex vector space, and let $g\colon [0,1] \times V \to [0,1] \times V$ have the form*

$$g\colon (x, v) \mapsto (f(x), \epsilon(-1)^i A(v) + C_i)$$

for $c_i < x < c_{i+1}$, where $A\colon V \to V$ is a linear expanding map, and each C_i is a constant vector. Then g is a friendly extension *of f if there is a (necessarily unique) continuous map $x \mapsto b(x)$ such that the orbit of $g(x, b(x))$ is bounded.*

Since postcritically finite systems are dense in $F(d, \epsilon)$, there are many postcritically finite examples, and many have Galois conjugates of λ outside the unit circle. Since boundedness depends continuously on the kneading data, the coefficient λ and the constant terms C_i, we can pass to limits of postcritically finite systems; if we take a convergent sequence of postcritically finite systems having a convergent sequence of friendly extensions, then the limit is also a friendly extension.

In the case $d = 2$, these correspond to the kneading roots which were collected in Figure 1.1. Note that even for a λ for which $x \mapsto \lambda|x| - 1$ is postcritically finite, it is a limit of λ_i of much higher degree for which $x \mapsto \lambda_i|x| - 1$ is also postcritically finite; in fact, it is a limit of cases where the critical point is periodic. The minimal polynomials for these nearby polynomials can be of much higher degree, so there can be many more friendly extensions than just the ones associated with the roots of the minimal polynomial for λ. Figure 7.7 is a 3-dimensional figure of friendly extensions, where the vertical axis is the λ direction, and the horizontal direction is $\mathbb{C}$; the plotted points outside the unit cylinder are expansion factors for friendly extensions.[21]

Figure 7.8 is a very thin slice of the set depicted in Figure 7.7, halfway up and of thickness 10^{-9}. A movie made from frames of this thickness, at 30 frames per second, would last a year. This slab is thin enough to freeze the motion of 7 isolated

[21] Here Thurston is considering the case $V = \mathbb{C}$ with $A(v) = uv$, where the coefficient u is called the expansion factor.

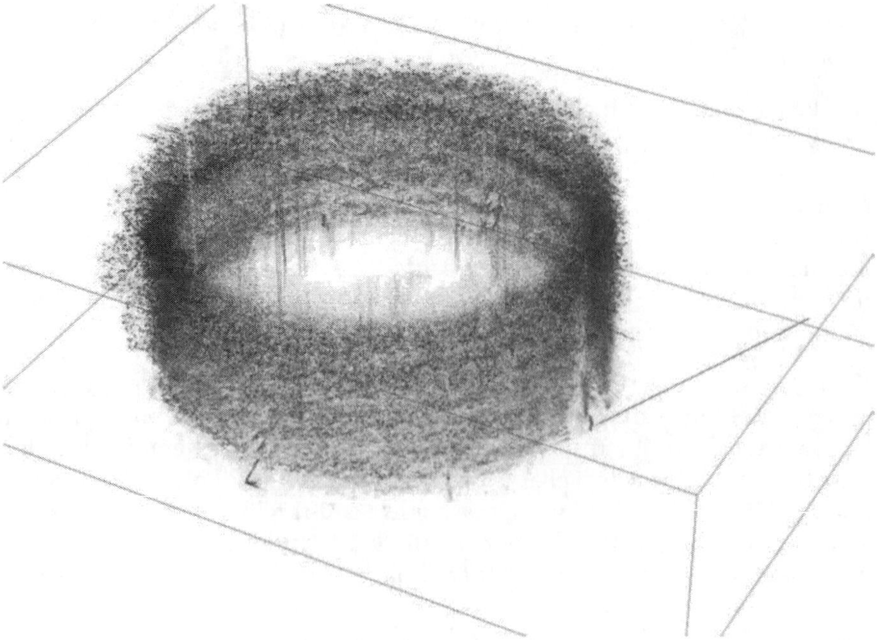

FIGURE 7.7: This picture shows the parameter values for friendly extensions of tent maps (the space $F(2,+1)$). The vertical direction is the λ-axis. The horizontal plane is $\mathbb{C}$, and the points shown are roots of minimal polynomials for postcritically finite examples. With a finer resolution and expanded vertical scale, the friendly extensions would appear as a network of very frizzy hairs (usually of Hausdorff dimension greater than 1), sometimes joining and splitting, but always transverse to the horizontal planes. The diagonal line on the right is λ itself. (See also Plate 22.)

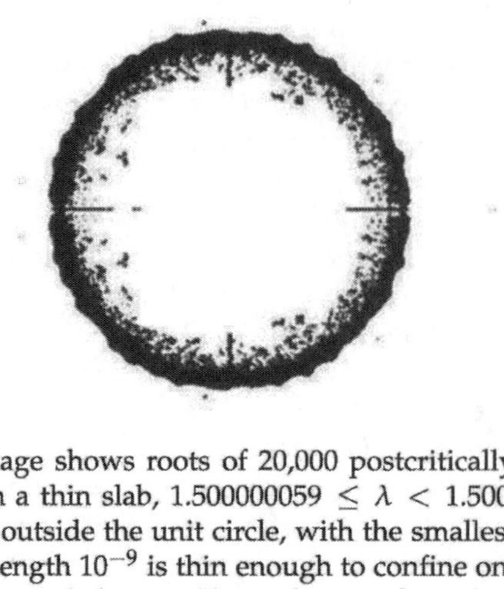

FIGURE 7.8: This image shows roots of 20,000 postcritically finite polynomials of degree about 80 in a thin slab, $1.500000059 \leq \lambda < 1.500000060$. There are 7 isolated spots visible outside the unit circle, with the smallest at 1.5. This slice of expansion factors of length 10^{-9} is thin enough to confine one friendly extension to stay within each spot as λ changes. Roots closer to the unit circle move quite fast with λ, so that even in this short interval the closer roots wander over large areas that overlap and sometimes perhaps collide and split. Roots in the closed unit disk do not depend continuously on λ, but they are confined to (and dense in) closed sets that include the unit circle and increases monotonically with λ, converging at $\lambda = 2$ to the inside portion of Figure 1.1.

friendly parameters, which you can see around the periphery. One of the 7, at position 1.5, is the original controlling expansion constant λ. Closer to the unit circle, the friendly parameters are packed closer together, and they move so quickly that they blur together into a big cloud.

REMARK 7.9. The "set of friends" changes continuously, regarded as an atomic measure on the parameter space for candidate systems, in the weak topology.

8 TRAINTRACKS

We have defined a traintrack map of a graph to be a map such that each edge is mapped by a local embedding under all iterates; now we'll define a traintrack structure.

DEFINITION 8.1. *A traintrack structure τ for a graph Γ is a collection of 2-element subsets of the link of each vertex, called the set of legal turns.*

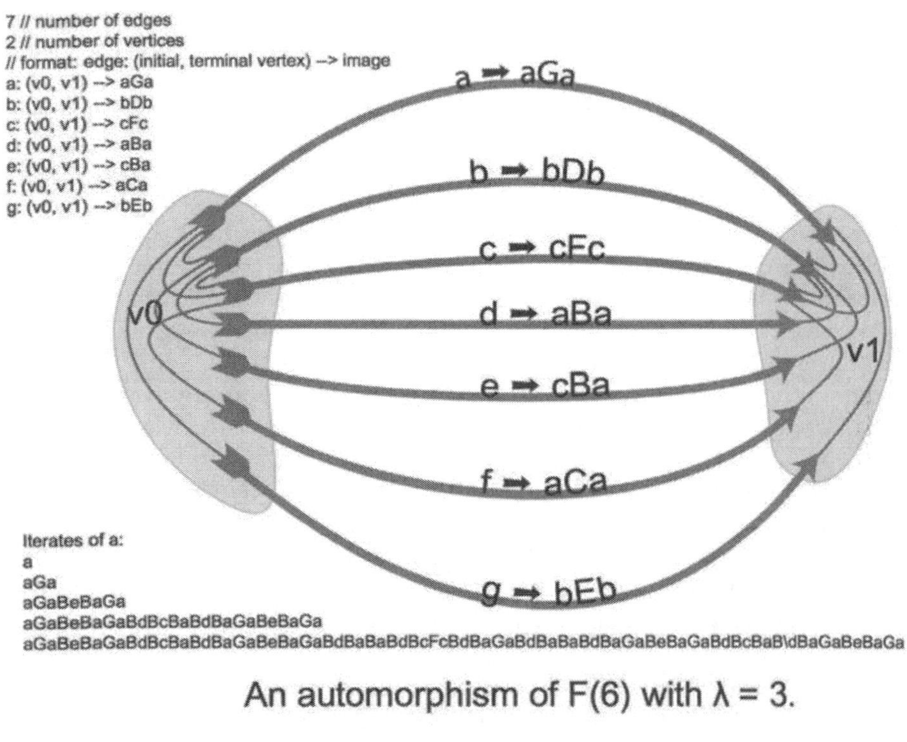

FIGURE 8.2: This is a traintrack structure preserved by an automorphism of the free group of rank 6 with expansion constant 3. The edges are labeled with letters, where a capital letter indicates the reverse direction. One set of generators is[22] $\{bA, cA, dA, eA, fA, gA\}$. The two vertices v_0 and v_1 are "exploded" to show legal turns, depicted as thin lines connecting the heads or tails of the arrows. This example will be used in §9 as a template for constructing traintrack maps with any possible Perron number as expansion constant.

The mental image is that of a railroad switch, or more generally a switchyard, where for each incoming direction there is a set of possible outgoing directions where trains can be diverted without reversing course. A path on Γ is *legal* if it is a local embedding, and at each vertex it takes a legal turn.

In describing edge paths, we must first choose an orientation for each edge. We use the convention that a lowercase letter denotes the forward direction on the edge, and the corresponding capital letter to denote the backward direction on the edge. This convention also applies to generators of groups (which we can think of as edges in a cell complex having the given group as its fundamental group).

A map $f\colon \Gamma \to \Gamma$ *preserves* τ if f maps every τ-legal path to a τ-legal path. It follows that f is a traintrack map: since the forward images of edges must always be legal paths, in particular, they are mapped by local embeddings.

For any traintrack map f, there is always at least one traintrack structure that f preserves. The *minimal invariant traintrack structure* allows turns only if they are ever taken by the forward image of some edge of the graph. The *maximal invariant traintrack structure* allows all turns that are never folded by iterates of f, that is, it allows any turn that is always mapped to be locally embedded.

There is an interesting special case of traintrack structure for a bouquet of circles: the positive traintrack structure, where the legal paths correspond to paths that are positive words in the generators of the free group. These paths follow a consistent orientation along the circles. More generally, an *orientation* for a traintrack structure is a choice of direction for each edge such that every legal path maintains a consistent orientation, either always positive or always negative. A traintrack structure is *orientable* it admits such an orientation. This implies the weaker condition that we can choose a consistent orientation along any legal path.[23]

An oriented traintrack on a graph Γ defines a convex cone C in $H^1(\Gamma, \mathbb{Z})$. For an automorphism ϕ that preserves such a structure, $H^1(\phi, \mathbb{R})$ takes C to C or to $-C$. Therefore, $\pm\phi$ has an eigenvector inside C. Just as in the Perron-Frobenius theorem, an eigenvector strictly inside C has the largest eigenvalue, and in any case, there is an eigenvector in the cone (possibly on its boundary) whose eigenvalue is largest. This eigenvalue is the same as the expansion constant for ϕ, since the incidence matrix for ϕ is the matrix for $\pm\phi$ acting on (simplicial) 1-chains of Γ. As an eigenvector for an element of $\mathrm{GL}(n, \mathbb{Z})$, such an eigenvalue is always an algebraic unit.

As illustrated by the example of Figure 8.2, the expansion constant for a traintrack map need not be an algebraic unit: the figure describes a map ϕ_3 which clearly has expansion constant 3. In fact, if we assign each edge the length one, then every edge maps to a path of length three.

To check that this map induces a group automorphism, first write down the images of the generators $\{bA, cA, dA, eA, fA, gA\}$ and collapse the a edge to a point[24] (thus striking out all a's):

---

[22]Note that in discussing the fundamental group of a graph, there is no need to consider only legal paths.

[23]This definition has been changed by the editors. Let us call the weaker condition "weak orientability." We are grateful to Lee Mosher for the following example of a traintrack structure which is weakly orientable but not orientable. Let Γ be the graph with four vertices v_0, v_1, v_2, v_3, together with a loop at v_i for $i > 0$ and an edge joining v_i to v_0 for $i > 0$; and let τ be the traintrack structure which contains all three turns at v_0, together with exactly two of the three turns at v_i including the turn of the loop at v_i.

[24]As an example, we must first expand, $bA \mapsto (bDb)(AgA)$ and only then delete each a or A to obtain $b \mapsto bDbg$.

$$\begin{aligned}\phi_3\colon b &\mapsto bDbg\\ c &\mapsto cFcg\\ d &\mapsto Bg\\ e &\mapsto cBg\\ f &\mapsto Cg\\ g &\mapsto bEbg\end{aligned}\tag{8.1}$$

Every surjective selfmap of a free group is an automorphism,[25] so it's enough to check that all 6 generators are in the image. We get c from $eD \mapsto c$, and given c we get g from $f \mapsto Cg$. Given c and g, we get b from the image of E and f from the image of C. Given these four generators we get the remaining two, d from the image of B and e from the image of G. Therefore, ϕ_3 is a self-homotopy-equivalence of the graph, so it gives a traintrack map. To see that the map preserves the traintrack structure, first note that every edge maps to a legal path. Now note that the three edges a, b, c each map to a path starting and ending in the same way, with a, b or c, respectively. All words involving only these three edges are legal for this traintrack. The other four edges map to words beginning and ending with a, b or c, so it is easy to check that legal turns are mapped to legal turns.

Since the expansion constant is not an algebraic unit, there can be no orientable traintrack structure invariant by the automorphism up to homotopy. For this particular example, the structure has an anti-orientation, such that every legal path reverses orientation at every vertex.

For $m \geq 0$, a map ϕ_{3+2m} with expansion factor $3 + 2m$ can be defined, preserving the same traintrack:

$$\begin{aligned}\phi_{3+2m}\colon a &\mapsto aG(aB)^m a\\ b &\mapsto bD(bC)^m b\\ c &\mapsto cF(cA)^m c\\ d &\mapsto aB(aB)^m a\\ e &\mapsto cB(aB)^m a\\ f &\mapsto aC(aB)^m a\\ g &\mapsto bE(bA)^m b\end{aligned}\tag{8.2}$$

A sequence of steps identical to that for ϕ_3 shows that ϕ_{3+2m} is a traintrack map. For completeness, we can define ϕ_1 as the identity map of this traintrack; ϕ_1 is also a traintrack map. There are similar constructions for even expansion factors, but we do not need them.

9 SPLITTING HAIRS

Let $f\colon \Gamma_0 \to \Gamma_0$ be any purely expanding self-map of a tree. Every tree is bipartite; that is, we can partition the vertices into to sets V_0 and V_1 such that every edge of Γ_0 has one endpoint in each of the two sets.

[25] See, for example, [LS, p. 14].

The map f may not respect this partition; if not, let Γ be the barycentric subdivision of Γ_0, with one new vertex $v(e)$ in the middle of each edge e of Γ_0. For each edge e of the original Γ_0, choose one of the edges e' in its image, and adjust f so that $v(e) \mapsto v(e')$. The new map still has a non-negative incidence matrix. Any positive linear function that is a dual eigenvector for its transpose pulls back to a linear function that is an eigenvalue for the original f, so its eigenvalue is the same.[26] Therefore, there are positions for the new vertices $v(e)$ so all vertices of Γ are mapped (by the original f) to vertices of Γ. The graph Γ has a bipartite partition, where V_0 consists of original vertices and V_1 consists of new vertices, where each partition element is invariant by f.

Given any graph Γ and an integer n, define a new graph $\mathrm{Split}_n(\Gamma)$ to be obtained by replacing each edge of Γ with n edges having its same pair of endpoints. For each edge e, label the new edges with subscripts $e_a, e_b, \dots$.

When Γ has a bipartite structure $\{V_0, V_1\}$, define a traintrack structure on $\mathrm{Split}_7(\Gamma)$ using the prototype of Figure 8.2: that is, a path is legal if and only if the sequence of subscripts defines a legal path in the prototype.

When f is a map of Γ to itself that preserves the partition elements and maps each edge to an edge-path of length at least one, define

$$\mathrm{Split}_7(f)\colon \mathrm{Split}_7(\Gamma) \to \mathrm{Split}_7(\Gamma)$$

using the prototypes ϕ_i defined in Section 8. That is, for an edge e_x of $\mathrm{Split}_7(\Gamma)$, if $f(e_x)$ has combinatorial length k, lift f to the map $\mathrm{Split}_7(f)$ by applying the sequence of subscripts $\phi_k(x)$ to the sequence of edges $f(e_x)$.

Note that the ϕ_{2m+1} themselves are defined by this process, starting with the self-maps of the unit interval that fold it over itself an odd number of times.

PROPOSITION 9.1. *For any bipartite structure $\{V_0, V_1\}$ on a graph Γ and any map $f\colon \Gamma \to \Gamma$ that preserves the partition elements and maps each edge to an edge-path of length at least one,* $\mathrm{Split}_7(f)$ *is a traintrack map.*

PROOF. Notice that in the prototype traintrack, every legal turn has at least one of its ends among the edges a, b, c. Any turns among these three edges are legal, and every ϕ_{2m+1} preserves their beginnings and end. Every ϕ_{2m+1} with $m > 1$ maps beginnings and ends in the same way, so a turn between edges whose combinatorial image length is more than 1 maps to a legal turn. Similarly, a legal turn between edges one or both of which have combinatorial image length 1 maps to a legal turn. □

PROPOSITION 9.2. *If Γ is a tree with bipartite structure $\{V_0, V_1\}$, and if $f\colon \Gamma \to \Gamma$ is a self-map such that*

- *f preserves V_0 and V_1, and*
- *f is a local embedding on each edge,*
- *the map $f_1\colon e \mapsto f(e)_1$, that is, e goes to the first element of its image edgepath, is a permutation, and*

[26]The incidence matrix M' for the subdivided traintrack map is made up out of 2×2 blocks $\begin{bmatrix} a & b \\ c & d \end{bmatrix}$ such that $a + b = c + d$ is equal to the corresponding entry for the original incidence matrix M. If $(x_1, x_2, \dots, x_{2n})$ is a left eigenvector for M', then $(x_1 + x_2, \dots, x_{2n-1} + x_{2n})$ is a left eigenvector for M with the same eigenvalue.

- *the map* $f_2 \colon e \mapsto f(e)_2$ *if* $f(e)$ *has length more than 1, and* $e \mapsto f(e)_1$ *otherwise, is also a permutation,*

then $\mathrm{Split}_7(\Gamma)$ *is a homotopy equivalence.*

PROOF. The set of edges with subscript a forms a spanning tree for $\mathrm{Split}_7(\Gamma)$. If we collapse the spanning tree to a point, we obtain a bouquet of circles, so the remaining edges give a free set of generators for the fundamental group.

Pick a basepoint $*$ on the graph $\mathrm{Split}_7(\Gamma)$. As an edge path, the generator corresponding to e_x is obtained by prefixing e_x by the subscript a path from $*$ to its first vertex, and appending the A path from its end vertex back to $*$, then striking out all as and As.

As before, we just need to show that all generators are in the image of $\mathrm{Split}_7(f)$. Consider the 6 generators lying over any particular edge x of Γ. Let x' be the first edge in the edgepath $f(x)$. We will follow the same outline that showed ϕ_3 is a homotopy equivalence. The image of $f(x_e X_D)$ is x'_c.[27] Applying this to all edges x of Γ, we obtain all generators x_c in the image. Modulo the x_c generators, from $\mathrm{Split}_7(f)(x_f)$ we get all generators of the form $f_2(x)_g$. Since f_2 is a permutation, this gives all subscript g generators. From $\mathrm{Split}_7(f)(X_D)$ modulo the g generators, we get the $f_1(x)_b$ generators. Continue in the same sequence that was used to show ϕ_3 is surjective: modulo previous generators, each edge has a payload generator in the first or second slot of its image, so we get all generators. □

Given any Perron number λ, we can now apply Proposition 9.2 to the asterisk map f_λ constructed by Theorem 6.2, where we take V_0 to consist of the center vertex and V_1 to consist of all the tips. We obtain a traintrack map $\mathrm{Split}_7(f_\lambda)$. The maps were constructed so that every edge occurs as the first segment of some image edge-path. The same edge occurs as the second segment, if the path has length more than 1, so the maps $f_{\lambda,1}$ and $f_{\lambda,2}$ are both permutations. Therefore, $\mathrm{Split}_7(f_\lambda)$ is traintrack homotopy equivalence of $\mathrm{Split}_7(\Gamma)$, which uniformly expands with expansion factor λ.

The maps ϕ_{2m+1} are mixing when $m > 1$, from which it follows that $\mathrm{Split}_7(f_\lambda)$ is mixing.

If λ is a weak Perron number, let n be the least integer such that λ^n is a Perron number. Make an asterisk from n copies of an asterisk for λ^n, and map it to itself by permuting the factors except for the final map, which is a copy of f_{λ^n}. From this asterisk map f_λ, we obtain a traintrack automorphism $\mathrm{Split}_7(f_\lambda)$ with expansion constant λ. This completes the proof of Theorem 1.9.

10 DYNAMIC EXTENSIONS

If $f \colon X \to X$ is a continuous map, then an *extension* of f is a space Y with a continuous surjective map $F \colon Y \to Y$ and a semiconjugacy $p \colon Y \to X$ of F to f, that is, it satisfies $p \circ F = f \circ p$.

The hair-splitting construction of section 9 is a special case, which we'll call a *graph extension*, where X and Y are graphs and edges of Y are mapped to nontrivial edge-paths of X. If f is a λ-uniform expander, then so is F. Given two graph extensions F_1 and F_2, they have a fiber product F_3, consisting of all pairs of points in Γ_1 and Γ_2 that map to the same point in Γ. We can pick a connected component

[27] The image of $x_e X_D$ under $\mathrm{Split}_7(f)$ is $x'_c X'_A$, which becomes x'_c after the collapse.

of the fiber product to obtain a connected graph that is a common extension of both; thus the set of connected graph extensions is a partially ordered set where any two elements have an upper bound (but not necessarily a least upper bound).

It seems natural to ask which uniformly expanding self-maps of graphs have graph extensions that are traintrack self-homotopy-equivalences. It would also be interesting to strengthen the condition, to require that the fundamental group of the extension graph maps surjectively to the fundamental group of the base; in that case, a necessary condition is that f itself be a homotopy equivalence, otherwise neither it nor its extension would be surjective on π_1. If necessary, we could also weaken the condition by allowing a subdvision of Γ before taking the extension.

Although Proposition 9.2 required that the first and second elements in the edgepaths for f are partitions, this condition does not seem essential. One trick would be to modify the formula of 8.2 to lift f, we could use lifts in patterns such as $b \mapsto b(Cb)^{m1}D(bC)^{m2}g$ and $e \mapsto (aB)^{m1}c(Ba)^{m2}$ to adjust the payload edges to be somewhere else along the edgepath, taking care that the payload edges map surjectively, and that they are chosen so that there is an order that will unlock them inductively.

It seems plausible that by a combination of duplicating edges, subdividing edges, and lifting with adjustable payloads, any expanding self-homotopy-equivalence of a graph would have an extension that is a traintrack map of a graph whose fundamental group maps surjectively to the base. However, it is beyond the scope of the current paper to pursue this.

It would also seem interesting to understand a theory of minimal uniformly expanding maps, ones that cannot be expressed as non-trivial extensions. Perhaps extensions that merely identify vertices should be factored out. This should be related to a theory of finitely-generated abelian semigroups that exclude 0 and are invariant by a transformation and which, tensored with $\mathbb{R}$, have λ as dominant eigenvalue. *They can be thought of as positive λ-modules.*

11 BIPOSITIVE MATRICES

DEFINITION 11.1. *A pair of real numbers (a, b) is a* conjugate pinching pair *if a and b^{-1} are algebraic units whose Galois conjugates are contained in the open annulus of inner radius b^{-1} and outer radius a. They are a* weak conjugate pinching pair *if their conjugates are contained in the closed annulus.*

Here is an elementary fact.

PROPOSITION 11.2. *A real number is the expansion constant for an element of* $\mathrm{GL}(n, \mathbb{Z})$ *for some n if and only if it is a weak Perron number.*

A pair (a, b) of algebraic units occurs as the pair of expansion constants for A and A^{-1} if and only if the pair is a weak conjugate pinching pair.

PROOF. Sufficiency is easy. Given a weak Perron number a, the companion matrix for the minimal polynomial for a has expansion constant a. Given a conjugate pinching pair (a, b), take the sum of the companion matrix for a minimal polynomial for a with the inverse of a companion matrix for the minimal polynomial for b.

Necessity is pretty obvious: a must be the maximum absolute value, and b the minimum, among all roots of χ_A. □

DEFINITION 11.3. *An invertible matrix A is* bipositive *with respect to a basis B if B admits a partition into two parts P and Q such that A maps the orthant spanned by $P \cup Q$ to itself, and A^{-1} maps the orthant spanned by $P \cup -Q$ to itself.*

If either of the two positive matrices associated with a bipositive matrix is mixing, the other is as well, so in this case we also say the bipositive matrix is mixing. Similarly, we will call a bipositive matrix *mixing* if all sufficiently high powers have all non-zero entries. More geometrically, given a positive matrix, we can make a graph whose vertices are the basis elements and whose edges correspond to nonzero entries. The matrix is primitive if there are edge-paths of every sufficiently long length from each vertex to each others. The set of lengths of edge-paths from one vertex to another is a semigroup, and it is easy to see that its complement is finite if and only if the elements have a common divisor larger than one. It follows that an irreducible positive or bipositive matrix A is mixing if and only if every positive power of A is ergodic.

THEOREM 11.4. *A pair (a, b) is the pair of positive eigenvalues for an invertible bipositive integral matrix of some dimension if and only if it is a conjugate pinching pair, or a weak conjugate pinching pair such that all conjugates of a and b^{-1} of modulus a or b^{-1} are a or b^{-1} times roots of unity.*

REMARK 11.5. The clause about roots of unity addresses the issue that although all Galois conjugates of a weak Perron number a that have *maximum* modulus are at angles that are roots of unity, the Galois conjugates of *minimum* modulus need not be.

For instance, if a is the plastic number $1.32472\ldots$ which is a root of $x^3 - x - 1$, its other two conjugates, $-0.662359\ldots \pm i0.56228\ldots$, have modulus $1/\sqrt{a}$, so $(a, \sqrt{a})$ is a weak conjugate pinching pair. However, the two complex Galois conjugates of a are at angles that are irrational multiples of π. Therefore, $(a, \sqrt{a})$ is not the pair of expansion constants for any bipositive matrix.

PROOF. We will first establish the theorem in the case (a, b) is a [strict] conjugate pinching pair.

First, let $A \in \mathrm{GL}(n, \mathbb{Z})$ have eigenvectors S and T of eigenvalues a and b^{-1}. By restricting to the smallest rationally defined subspace of $\mathbb{R}^n$ containing S and T, we may assume that all other characteristic roots of A are in the interior of the annulus $b^{-1} < |x| < a$. Let H be the real subspace spanned by S and T and $p\colon \mathbb{R}^n \to H$ be the projection that commutes with A.

Let L be the line through the origin of slope $a \times b$ in the ST-plane. The linear transformation A multiplies slopes by $1/a \times b$, so $A(L)$ has slope 1. Let L' be the reflection of L in the S-axis; thus $A(L')$ is the reflection of $A(L)$.

Choose a set P_0 of lattice points such that $p(P_0)$ lies in the interior of the first quadrant of the ST-plane, P_0 generates the lattice $\mathbb{Z}^n$ as a group, and the convex cone generated by P_0 contains the angle between L and $A(L)$, except 0, in its interior. Similarly, choose Q_0 to be a set of lattice vectors that generate $\mathbb{Z}^n$ as a group, that are mapped by p to the interior of the fourth quadrant, generates $\mathbb{Z}^n$ as a group, and contains the angle between L' and $A(L')$.

Let σ_P be the semigroup generated by P_0 and let σ_Q be the semigroup generated by Q_0.

We claim that for sufficiently large $k > 0$ and for any two elements $p_1, p_2 \in P_0$, there is a $j > 0$ such that

$$A^j(p_2) - A^{-k}(p_1) \in \sigma_Q,$$

and, similarly, for any $q_1, q_2 \in Q_0$, there is a $j > 0$ such that

$$A^j(q_2) - A^{-k}(q_1) \in \sigma_P.$$

To see this, we will make use of a basic fact about the semigroups of lattice elements:[28]

LEMMA 11.6. *Let U be a finite set of elements of the lattice $\mathbb{Z}^n \subset \mathbb{R}^n$, and let C_U be the convex cone generated by U. There is some constant R such that any element g of the group G_U generated by U whose radius R ball $B_R(g)$ is contained in C_U is also contained in the semigroup S_U generated by U.*

PROOF. The convex cone C_U is the set of all non-negative real linear combinations of elements of U. If we express an element of $G_U \cap C_U$ as a non-negative real linear combination of elements of U and round the coefficients to the nearest integer, we see that g is within a bounded distance R_1 of some element of S_U.

Let F be the set $G_U \cap B_{R_1}(0)$. Express each element of $f \in F$ as an integer linear combination $\sum_{u \in U} k_{u,f} u$. Let R_2 be the maximum, among these linear combinations, of the norm of the set of negative coefficients, and set $R = R_1 + R_2$.

Now for any element $g \in G_U$ such that $B_R(g) \subset C_U$, there are elements $s \in S_U$ and $f \in F$ such that $g = s + f$. □

Now for k large, the image in projective space of $A^{-k}(p_1)$ is close to the image of the eigenspace T. The images $A^j(p_2)$ march in the direction of the S axis, projectively converging to the S eigenspace, with the slope in the ST-plane of the line from $A^{-k}(p_1)$ to $A^j(p_2)$ decreasing by a factor of $a \times b$ at each iterate. If k is suitably large, at least one of these iterates is captured in the interior of the cone $A^{-k}(p_1) + C_Q$ and so, by the lemma, in the semigroup σ_Q.[29]

Now we can describe an irreducible bipositive matrix. Take the free abelian group FA generated by

$$\bigcup_{g \in P_0 \cup Q_0} \left\{A^{-k}(g), \ldots, A^{j(g)-1}(g)\right\}.$$

We will choose a bipositive map $\tilde{A}$ of FA to itself that commutes with evaluation in $\mathbb{Z}^n$. The generators $A^h(g)$ pass off from one to the next until $A^{j(g)-1}(g)$. Choose a cyclic permutation of p of P_0 and a cyclic permutation q of Q_0. For each $g \in P_0$ choose an expression $A^{j(g)}(g) = A^{-k}(p(g)) + sq$ where sq is an element of σ_Q, and similarly for $g \in Q_0$ express $A^{j(g)}(g) = A^{-k}(q(g)) + sp$, with $sp \in \sigma_P$. Use this to give the final links in the chain, to define $\tilde{A}$.

The matrix for the linear transformation can be expressed as an upper triangular matrix followed by a permutation; hence, it is invertible. In block form, the

[28] See for example Exercise 7.15 of [MSt].

[29] If k is suitably large, at least one of these iterates is the center of a large ball captured in the interior of the cone $A^{-k}(p_1) + C_Q$ and so, by the lemma, in the semigroup σ_Q.

P generators are each expressed as another P generator plus an element of the σ_Q semigroup, and vice versa. The inverse has the same form, but with semigroup elements subtracted; reversing the sign of the Q generators turns the inverse into a positive matrix.

The matrix for FA is irreducible because of the cyclic permutations: the images of each generator eventually involves each other generator, and (except in the trivial case $a = b = 1$) the powers of the matrix are eventually strictly positive. The positive eigenvalue for FA is a, since projection to the S-axis gives a linear function, positive on the positive orthant for FA, that is a dual eigenvector of eigenvalue a. Similarly, the positive eigenvalue for FA^{-1} is b.

If (a, b) is a weak conjugate pinching pair such that all conjugates of a or b^{-1} on the circle of maximal or minimal modulus have arguments that are rational multiples of 2π, let h be a common multiple of the denominators. Then (a^h, b^h) is a strict conjugate pinching pair; let A_0 be a matrix realizing this pair of pinching constants. Take the direct sum of h copies of the underlying vector space, permute them cyclically, with return map A_0. This is a bipositive matrix realizing (a, b).

For the converse: consider any bipositive matrix A with positive eigenvalue a and positive eigenvalue b for A^{-1}, where one or both are not the unique characteristic roots of maximum modulus *say there are k roots of modulus a*. Then there is a k-dimensional subspace with a metric where A acts as a similarity, expanding by a factor of a. Hence, the projective image of this subspace is mapped isometrically. Its intersection with the image of the positive orthant is a polyhedron mapped isometrically to itself; hence, its vertices are permuted. It follows that all characteristic roots of modulus a have the form of a times an mth root of unity. □

A simple example of the construction for a weak conjugate pinching pair is the Fibonacci transformation $x \mapsto y$ and $y \mapsto x + y$, with eigenvalues $\phi = (1 + \sqrt{5})/2$, the golden ratio and $-1/\phi$. The square of this transformation is $x \mapsto x + y$ and $y \mapsto x + 2y$, which is bipositive: its inverse maps the second quadrant to itself. This translates into a bipositive map for $(\phi, 1/\phi)$ in dimension 4, that expresses a bipositive linear recurrence for four successive terms of the Fibonacci sequence,

$$x_1 \mapsto x_2, \quad x_2 \mapsto x_3, \quad x_3 \mapsto x_2 + x_3, \quad x_3 \mapsto x_1 + 2x_2.$$

As another example, $a = (1 + \sqrt{2})^{1/3}$ and $b = (\sqrt{2} - 1)^{1/3})$ can be realized by a bipositive transformation in dimension 12.

QUESTION 11.7 (suggested by Martin Kassabov). *Suppose $A \in \mathrm{GL}(n, \mathbb{Z})$ has dominant eigenvalue a and A^{-1} has dominant eigenvalue b, where (a, b) is a [strict] conjugate pinching pair. Is there a basis for which some power of A is bipositive?*

REMARK 11.8. Although elementary matrices generate $\mathrm{SL}(n, \mathbb{Z})$, and together with permutations generate $\mathrm{GL}(n, \mathbb{Z})$, the semigroup is a very different matter: for $n \geq 3$, the elementary positive semigroup is not even finitely generated.

To see the gap between the elementary positive semigroup and the full positive semigroup of $\mathrm{GL}(n, \mathbb{Z})$, let's focus on the case $n = 3$ (the embedding in $\mathrm{GL}(3 + n, \mathbb{Z})$ that fixes all but the first 3 basis elements gives examples, albeit atypical, for arbitrary dimension ≥ 3).

Let's look at the action of these semigroups on the basis triangle B in $\mathbb{RP}^2$. The image of the triangle by a word in the generators gives a sequence of subtriangles of this triangle, where at each step you bisect one of the sides and throw one half away. In particular, each possible proper image is contained in one of six half-triangles of B.

Now consider a general positive element of $GL(3, \mathbb{Z})$. It maps the tetrahedron spanned by 0, together with the basis elements to a "clean" lattice tetrahedron, that intersects lattice points only at its vertices. This property (clean) characterizes the possible images. Furthermore, any clean triangle in the positive orthant with one vertex at the origin can be extended (in many ways) to a clean tetrahedron: just add any vertex in one of the two lattice planes neighboring the plane containing the triangle.

But there are many clean triangles; in fact, the set of lattice points in the positive orthant that are primitive (i.e., the 1-simplex from the origin to the point is clean) have density $(1-2^{-3}) \times (1-3^{-3}) \times (1-5^{-3}) \cdots = 1/\zeta(3) \approx 0.831907\ldots$, and given a primitive lattice point p, the density of lattice points q such that the triangle $\Delta(0,p,q)$ is clean is $1/\zeta(2) = 6/\pi^2 \approx 0.607927\ldots$. It follows that every line segment contained in B can be approximated in the Hausdorff topology by an image of B under the positive semigroup of $GL(3, \mathbb{Z})$. Many such line segments cross all 3 altitudes of B, so a positive element of $GL(3, \mathbb{Z})$ that maps B to a nearby triangle is not in the positive elementary semigroup.

Nonetheless, images of B under the positive semigroup of $GL(3, \mathbb{Z})$ are quite restricted. For instance, it's easy to see that the centroid of B, corresponding to the line $x = y = z$, cannot be in the interior of any image of B. For any finite collection of triangles not containing the centroid in their interior, most lines through the centroid are not contained in any one of them. Therefore, no finite set of positive $GL(3, \mathbb{Z})$ images of B cover all possible images.

Note that the proof of Theorem 11.4 actually gave something stronger. If we have a basis B that is partitioned into two parts P and Q, then any elementary transform that replaces an element of P by its sum with an element of Q, or an element of Q by its sum with an element of P is bipositive. The *elementary bipositive semigroup* [with respect to (P, Q)] is the semigroup generated by these cross-type elementary transformations, together with permutations that preserve P and Q.

THEOREM 11.9. *A pair of real algebraic units (a, b) is the pair of expansion constants for an elementary bipositive matrix and its inverse if and only if it is a conjugate pinching pair such that all Galois conjugates of a or b^{-1} of maximal or minimal modulus have arguments that are roots of unity.*

PROOF. From the proof of Theorem 11.4, the condition on arguments is an equivalent form of the hypothesis in the case that (a, b) does not strictly pinch. □

12 TRACKS, DOUBLETRACKS, ZIPPING AND A SKETCH OF THE PROOF OF THEOREM 1.11

Continuous maps are often inconvenient for representing homotopy equivalences between graphs, because a self homotopy equivalence cannot be made into a self-homeomorphism of any graph in the homotopy class unless it has finite order up to homotopy.

Continuous maps of graphs can be inconvenient as geometric representatives of group automorphisms of the free group, since they are usually not invertible. As we have seen, it is not easy to see at a glance whether a given map of graphs is a homotopy equivalence. There are algorithms to check, but they can be tedious.

As an alternative, we can represent self-homotopy equivalences by continuous 1-parameter families of graphs. These have the advantage of being reversible. If we restrict to graphs that have no vertices of valence 1, these can be locally described by moving attachment points of edges along paths in the complement of the edge.

A *zipping* of a traintrack τ to a traintrack σ is a 1-parameter family of structures $(\Gamma_t, \tau_t)|\tau \in [0,1]$ that may be thought of as squeezing together legal paths. It's elementary to see that for any traintrack map f, there is a zipping that yields the homotopy class of f: just progressively and locally zip together the identifications that will be made by the map.

A zipping can be translated into a sequence of reversible steps, consisting of a motion of an attachment point of one edge along a legal path on its complement, starting in a direction in its linkgroup. When (Γ, τ) is zipped to (Γ', τ'), every bi-infinite τ-legal path becomes a bi-infinite τ'-legal path. The inverse of a zipping is an *unzipping*, or *splitting*.

A *doubletrack* structure for Γ is a pair (σ, τ) of traintrack structures on Γ. A graph Γ equipped with a pair of traintrack structures is a *doubletrack*.

A *bizipping* between doubletracks is a 1-parameter family of doubletracks that is a zipping of the first traintrack structure and an unzipping of the second. This yields a traintrack map of one structure whose inverse is a traintrack map for the other structure.

REMARK 12.1. It seems likely that invariant foliations could provide a good alternative to traintracks. Bestvina and Handel introduced a concept of traintracks relative to an invariant filtration of a graph and showed that relative traintrack maps exist for every outer automorphism of a free group (not just in the irreducible case). Instead, one could look at foliations of finite depth on a manifolds of sufficiently high dimension (as a function of the rank of the free group), with singularities having links based on polyhedra, and satisfying the condition (used to great effect by Novikov) that there are no null-homotopic closed transversals. Such a foliation picks out a class of bi-infinite words in the free group. Bestvina-Handel's theorem on existence of relative traintracks would appear to translate to the existence of a homeomorphism of some open manifold homotopy-equivalent to a bouquet of circles that preserves such foliation. Pairs of foliations could substitute for doubletracks.

We will not take the detour of trying to develop this point of view here.

We now sketch the proof of Theorem 1.11.

PROOF. We will now analyze the [main] case when there are strict inequalities: let (a, b) be a strictly pinching pair of algebraic units.

From Theorem 11.4, let A be a bipositive matrix with positive eigenvectors S for A and T for A^{-1} having eigenvalues a and b. Let p be the invariant projection of the vector space for A to the plane spanned by S and T. The basis is partitioned into two sets, P and Q, with $p(P)$ in the first quadrant and $p(Q)$ in the fourth quadrant. The negations of these two sets give vectors that are mapped into the other two quadrants.

We claim that, for a suitable choice of P and Q, there is a permutation α of the basis such that for any $g \in P \cup Q$ the difference $A(\alpha g) - g$ is a non-negative linear combination of $\pm$ basis vectors that map to the quadrant neighboring the quadrant of $p(g)$ across the S axis. (To help with visualization: for most basis elements, α will be chosen so that $A(g) = \alpha^{-1}(g)$.)

Suppose this claim is true. Let Γ be a graph with a single vertex whose edges correspond to $P \cup Q$. Let E be the homomorphism $\pi_1(\Gamma) \to \mathbb{R}^n$ that maps the loop of an edge labeled by a vector to that vector, and let $\tilde{\Gamma}$ be the corresponding covering space. (Γ can be visualized as an embedded graph in the torus $\mathbb{R}^n/\mathbb{Z}^n$, and $\tilde{\Gamma}$ is lift to a graph in $\mathbb{R}^n$). Define a doubletrack structure (σ, τ) on Γ where a σ-legal path lifted to $\tilde{\Gamma}$ has monotone projection to the S-axis and a τ-legal path lifts to have monotone projection to the T-axis. The linear transformation A maps Γ to a new graph $A(\Gamma)$ mapped to $\mathbb{R}^n$. Using the preceding claim we can construct a bizipping that slides the image edges back to the originals.[30] □

13 SUPPLEMENTARY NOTES (MOSTLY BY JOHN MILNOR)

NOTE 13.1 (See Figure 1.7). The following email exchange between Curt McMullen and Thurston took place in February 2012.

> *CTM: What proof do you have in mind that this Perron number does not give a critically finite tent map?*
>
> *WPT: I check these things with a simple computer program that computes in the field, represented as polynomials modulo the defining polynomial for the field. The field can be mapped to $\mathbb{C}$ according to various roots, but it's computationally simpler just to look at several of the coefficients. As you iterate the self-map of the interval, if the image in one of the real or complex places gets outside some easily computed bound, it escapes to infinity. This becomes very obvious (without worrying about precise bounds) when you actually do it, because in the non-Perron cases, if they escape, you quickly get numbers on the order of 10^4 or 10^6 or so which are way out of bounds. Anyway, with the computer, it's immediate (for non-Perron cases).*

Presumably, in the Perron case it takes a bit longer, but still becomes obvious. Note that this procedure makes sense only if the polynomial is known to be irreducible. However, this is easy to check using Maple or Mathematica.

NOTE 13.2. Thurston states on page 344:

> *Figure 1.1 shows the Galois conjugates of* $\exp(h)$ *for postcritically finite real quadratic maps. This is a path-connected*[31] *set, with much structure visible.*

The same message from February 2012 clarifies this statement. But first let us provide some necessary background from kneading theory. Let f be a unimodal map with topological entropy h. According to [MT], the kneading data for f gives rise to a function of the form

$$k_f(t) = 1 \pm t \pm t^2 \pm \dots ,$$

[30] The inverse of the permutation α is implicit in the description of the basis for the free abelian group FA in the proof of Lemma 11.6.

[31] Or, more precisely, the topological closure of this set is path-connected.

which is defined and holomorphic throughout the open unit disk, which has no roots within the disk $|t| < e^{-h}$, and which has e^{-h} as a root provided that $h > 0$. In the postcritically finite case, the coefficients of this power series are eventually periodic, say, of period p. It follows that k_f extends to a rational function with poles only at pth roots of unity. If the expansion constant $\lambda = e^h$ has defining polynomial $P(t)$ of degree n, it follows that the numerator of the rational function $k_f(t)$ is a multiple of the "reversed" polynomial $t^n P(1/t)$.

Here is Thurston's commentary (mildly edited).

> *For the portion of Figure 1.1 outside the unit circle, the roots depend continuously on the entropy. This comes from a general principle — first of all, among power series with ± 1 coefficients, there are a bounded number of roots inside a circle of radius $r < 1$. (See the original paper that Milnor and I wrote.) Further, the roots change continuously with the coefficients. Since the set of possible coefficients is totally disconnected, this doesn't yet logically say they move continuously, but in fact they do change continuously with entropy via the maps*
>
> $$\textit{entropy} \mapsto \textit{power series (from kneading theory)} \mapsto \textit{roots}.$$
>
> *However, the roots move very fast with entropy.*
>
> *For the roots inside the unit circle, if you look at the case when they are roots of a finite power series, they are approximately determined by the last bunch of terms of the series (the reciprocal is a root of the reverse series) — they don't depend continuously on entropy. What actually happens is that in a small interval of entropy, there is a set of length N tails that can occur; in the usual topology on power series they are traced over in a discontinuous, dense way. The portion "blacked out" inside the unit disk gradually increases, and it always remains connected.*[32] *The unit circle remains in the set as well.*

The statement that the roots of k_t within the unit disk depend continuously on entropy depends on the observation that the only discontinities of the correspondence $f \mapsto k_f(t)$ occur when f is critically periodic. (Here we assume that f varies within a well-behaved family, for example, quadratic maps, or uniformly expanding maps.) As we pass through a critically periodic f, there is a jump discontinuity in which $k_f(t)$ is multiplied by a function which has no zeros in the unit disk. Thus the roots of k_t within the unit disk always depend continuously on entropy. (More precisely, in order to keep track of roots near the boundary, it would be clearer to say that the correspondence $h \mapsto \{\text{roots in } \mathbb{D}\} \cup \partial\mathbb{D}$ is continuous, using the Hausdorff topology.)

Note that in an example such as that of Figure 1.7, it would be perfectly possible to compute the roots of $k_f(t)$ within any disk $|t| < 1 - \epsilon$, and to compare them with the roots of the defining polynomial for $1/\lambda$. There will always be at least one root $1/\lambda$ in common; but in the postcritically finite case every Galois conjugate of $1/\lambda$ must also be a root of k_f.

[32] Here he does not say path-connected. We are not sure whether this is just an oversight or whether there is a problem with path-connectedness. See [Ti] for a proof of path connectedness for the closure of the possibly larger set consisting of reciprocals of zeros of the kneading determinant for critically periodic unimodal maps.

Evidently the conjugates of λ inside the unit disk (or the roots of $k_f(t)$ outside the unit disk), form a much more difficult object of study, well worth a more detailed exposition and further investigation. As one example, it would be curious to compare this with Oylyzko and Poonen study [OP] of polynomials with 0, 1 coefficients. For other similar studies, see `http://math.ucr.edu/home/baez/roots`.

NOTE 13.3 (D. Lind). Here are further remarks on the discussion in §2 (page 351). Let $T_\beta : [0,1) \to [0,1)$ defined by $T_\beta(x) = \beta x \mod 1$ be the β-transformation. A key property is whether 1 has a finite orbit under T_β. This is the case iff the corresponding subshift has finite type (a result due to Parry). Such a β is called a *Parry number* by some. Every Pisot number is Parry, and Schmidt suggested that every Salem number is Parry.

David Boyd (see [Bo1]) confirmed that Salem numbers of degree 4 are Parry. In [Bo2], he investigated the degree 6 case. In this paper Boyd describes in detail the heuristic based on random walks discussed in this paper, and in particular the predictions that *almost every* degree 6 Salem number should be Parry, while in degree 8 and higher both a positive proportion of Salem numbers should be Parry and a positive proportion should be *not* Parry.

He considers the 11,836 Salem numbers of degree 6 and trace at most 15. For all but a handful, he computes a (finite) orbit of 1. However, there are few examples ($x^6 - 3x^5 - x^4 - 7x^3 - x^2 - 3x + 1$ being one of them) for which the orbit size is at least 10^9 and possibly infinite. This is left unresolved in Boyd's paper.

NOTE 13.4 (See page 352). Given an arbitrary function $\phi\colon \{0,\ldots,n\} \to \{0,\ldots,n\}$, we can extend uniquely to a function $f\colon [0, n] \to [0, n]$ which is linear on each subinterval $I_j = [j-1, j]$. The associated incidence matrix $A = [a_{i,j}]$ is defined by

$$a_{i,j} = \begin{cases} 1 & \text{if} \quad f(I_i) \supset I_j\,, \\ 0 & \text{otherwise}\,. \end{cases}$$

It follows easily that $a_{i,j} = 1$ if and only if

$$\min\left(\phi(i-1), \phi(i)\right) < j \le \max\left(\phi(i-1), \phi(i)\right).$$

If the leading eigenvalue λ of this matrix is non-zero, then the topological entropy of f is given by $h = \log(\lambda)$.

More generally, consider a piecewise monotone map $f\colon [0, n] \to [0, n]$ which maps the finite set $\{0, \ldots, n\}$ into itself, and such that every critical value lies in this finite set. We can then define $\widehat{a}_{i,j} \ge 0$ to be the number of times that the image of I_i covers I_j. The topological entropy of f is the logarithm of the leading eigenvalue $\lambda \ge 1$.

Let ϕ be the restriction of f to $\{0, \ldots, n\}$, and let $[a_{i,j}]$ be the matrix of zeros and ones associated with this function ϕ. Then it is not hard to see that $\widehat{a}_{i,j} \equiv a_{i,j} \pmod 2$. Furthermore, for each j, the set of i for which $\widehat{a}_{i,j} > 0$ must form a connected block. Conversely, given a matrix $\widehat{a}_{i,j}$ satisfying these conditions for an appropriately chosen ϕ, it is not hard to construct a corresponding piecewise montone map.

NOTE 13.5 (See page 355). The figure below illustrates Thurston's procedure for passing from a postcritically finite λ^N-expanding map of the interval to a postcritically finite λ-expanding map of the circle as discussed at the beginning of Section 5. In this example, $N = 3$ and $\lambda^3 = 2$, so that the subintervals on the right have lengths $1, 2^{1/3}, 2^{2/3}$. Critical orbits have been marked.

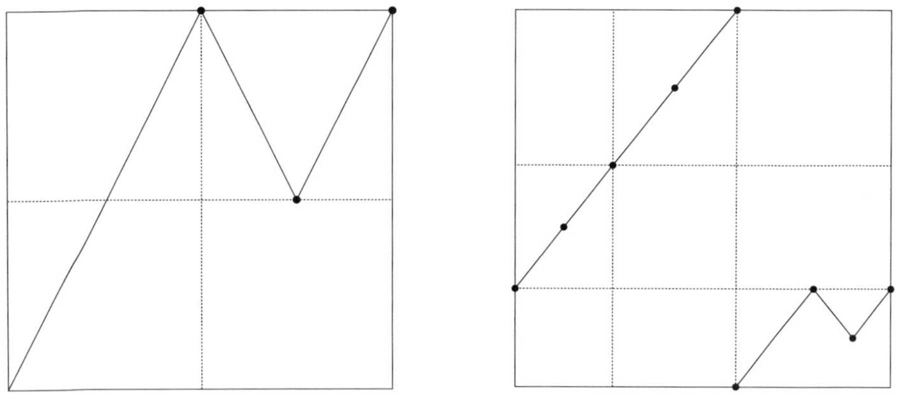

NOTE 13.6 (See page 357). The proof that existence of a postcritically finite λ^N-expanding interval map implies the existence of a postcritically finite λ-expanding interval map can be simplified as follows. First consider the special case $N = 2$. Divide the unit interval into four subintervals by the points

$$b = \lambda a, \quad c = \lambda b, \quad \text{and} \quad b = \frac{1}{1+\lambda}.$$

The following figure illustrates the case $\lambda^2 = 5$. A uniform 5-expander is graphed on the left. Reflecting in the y-axis and squeezing the unit square onto the rectangle $[b,c] \times [a,b]$ we obtain the graph on the right, with slope $\pm\sqrt{5}$ everywhere.

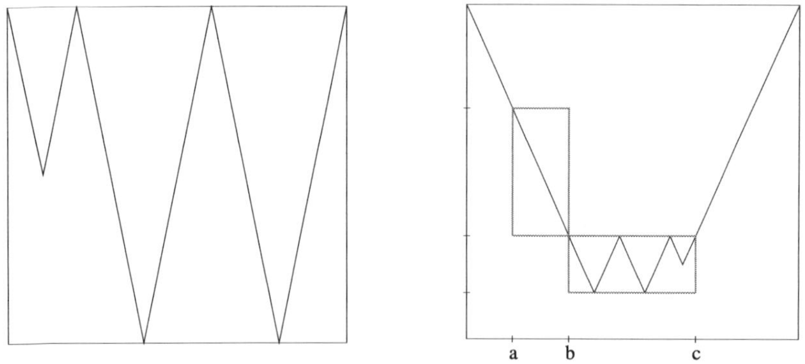

A similar construction works for any $\lambda > 1$ and any uniform λ^2-expander which maps both endpoints of the unit interval to 1. Note that the number of critical points is preserved. In the critically periodic case, the period is doubled.

As an example, consider the family of inverted tent maps

$$V_\mu(x) = \begin{cases} 1 - \mu x & \text{for } 0 \le x \le \frac{1}{2}, \\ \mu x + 1 - \mu & \text{for } \frac{1}{2} \le x \le 1, \end{cases}$$

with $1 < \mu \leq 2$. This construction applied to V_μ yields the map $V_{\sqrt{\mu}}$. If we choose μ_0 so that V_{μ_0} is critically periodic of period p, then by iterating we obtain a sequence of maps V_{μ_k} of critical period $2^k p$, where the expansion constant $\mu_k = \sqrt[2^k]{\mu_0} > 1$ converges to 1 as $k \to \infty$.

Now let λ be a Perron number or weak Perron number, and let λ^N be a power such that there exists an uniformly λ^N-expanding map f which is postcritically finite. After replacing N with some multiple if necessary, we can construct an inverted tent map V_μ which is critically periodic of period N, with $\mu < \lambda$. Let

$$\frac{1}{2} \mapsto x_1 \mapsto \cdots \mapsto x_N = \frac{1}{2}$$

be the critical orbit for V_μ. For each $1 \leq i \leq N$ let L_i be a short interval of length $\epsilon\lambda^i$ centered at x_i. Construct a new piecewise linear map g as follows. Each L_i with $i < N$ is to map linearly onto L_{i+1}, with slope either λ or $-\lambda$ according as $x_i > 1/2$ or $x_i < 1/2$. On the other hand, map L_N into L_1 by a composition $a' \circ f \circ a''$, where a' and a'' are carefully chosen affine maps, so the the iterate $g^{\circ N}$ maps L_N onto itself by a map which is affinely conjugate to f. Finally, extend g over the $N+1$ complementary intervals by affine maps. If ϵ is sufficiently small, then the slope of this map g will have absolute value at most λ everywhere. Hence it will be a Lipschitz map with Lipschitz constant λ. Hence it follows from the inequality (1.1) that $h(g) \leq \log(\lambda)$. Since the inequality $h(g) \geq \log(\lambda)$ can be checked by looking at $g^{\circ n}|_{L_n}$ and since g is postcritically finite, this completes the proof. □

NOTE 13.7 (See Proposition 5.3). For any interval map f with entropy

$$0 < h(f) < \sqrt{2},$$

there exists a pair of disjoint subintervals which are mapped to each other by f. Thurston's argument is incomplete. However, the statement can be proved as follows. It suffices to consider the case of a uniform λ-expander. The proof will be by induction on the number of critical points. If f has only one critical point, the argument is straightforward. In fact, if $c_0 \mapsto c_1 \mapsto \cdots$ is the critical orbit, then the two intervals $[c_2, c_4] \leftrightarrow [c_3, c_1]$ are disjoint whenever $\lambda < \sqrt{2}$. For the general case, replacing I by a smaller interval if necessary, we may assume that $f\colon I \to I$ is onto.

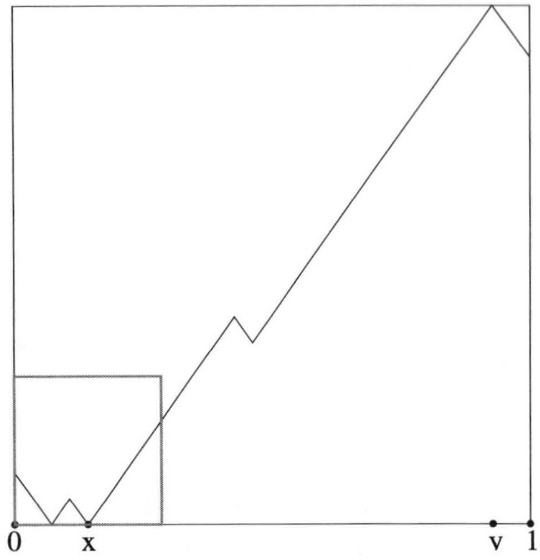

Let $[x, y] \subset I$ be the unique shortest subinterval which maps onto $I = [0,1]$. If $f(x) = 0$ with $x > 0$, as in the diagram, then the interval $[0, 2x]$ maps into itself and has fewer critical points; hence, the conclusion follows by induction. If $x = f(x) = 0$, then the interval $[\epsilon, 1]$ maps into itself for small $\epsilon > 0$. As ϵ increases, we must either reach a value where the map $[\epsilon, 1] \to [\epsilon, 1]$ has ϵ as a critical value, so that the preceding argument applies, or a value such that f has only one critical point in $[\epsilon, 1]$. In either case we are finished. Now suppose that $f(x) = 1$. In this case consider the map $f^2 = f \circ f$ from $[0, 1]$ to itself. Then there is a shortest interval $[x', y']$ which maps onto $[0, 1]$ under f^2, where now $f^2(x')$ is necessarily equal to 0. If $x' > 0$, then the interval $[0, 2x']$ is mapped into itself by f^2. (Here the inequality $\lambda^2 < 2$ is essential.) There are now two possibilities. If the intervals $[0, 2x']$ and $f[0, 2x']$ intersect in at most a single endpoint, then (replacing $[0, 2x']$ by $[0, 2x' - \epsilon]$ if necessary), we have found the required pair of disjoint intervals. On the other hand, if these two intervals overlap, then their intersection is f-invariant and has fewer critical points, and the conclusion follows by induction. Finally, if $x' = 0$, then again we can pass to the interval $[\epsilon, 1]$ for carefully chosen $\epsilon > 0$, and continue as before. □

NOTE 13.8 (See Figure 7.2 and Plates 18–20). It will be convenient to work with the map $f(x) = \lambda|x| - 1$, where $\lambda = (1 + \sqrt{5})/2$ with conjugate

$$\lambda^\sigma = 1 - \lambda = -1/\lambda\,.$$

The corresponding map $F = f^{1+\sigma}$ then has the form

$$F(x, y) = \begin{cases} (\lambda x - 1,\ \lambda^\sigma y - 1) & \text{if } x > 0, \\ (-\lambda x - 1,\ -\lambda^\sigma y - 1) & \text{if } x < 0. \end{cases}$$

This is piecewise linear (with a discontinuity along the line $x = 0$), and with Jacobian determinant -1. If we place three boxes

$$A = [-1,0] \times [1 - \lambda,\ \lambda - 1], \quad B = [-1,0] \times [-1 - \lambda,\ 1 - \lambda]\,,$$
$$C = [0,\ \lambda - 1] \times [-1 - \lambda,\ 1 - \lambda]$$

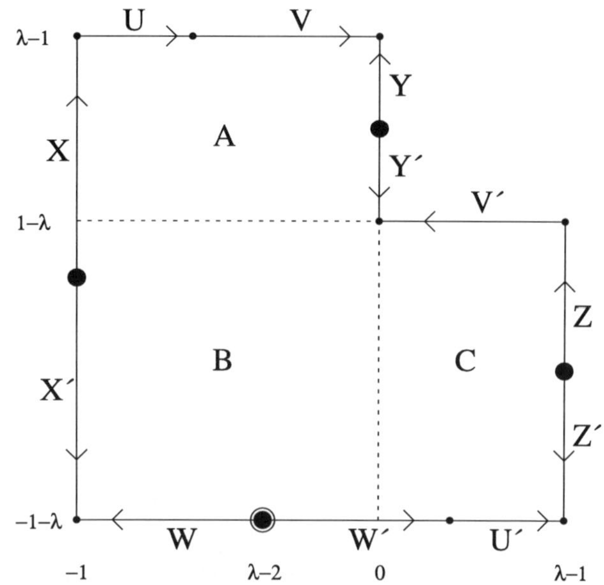

in the (x, y)-plane, as shown in the figure, then it is not hard to check that

$$A \cup B \cup C$$

maps bijectively onto itself with

$$C \xrightarrow{\cong} A \qquad \text{and} \qquad A \cup B \xrightarrow{\cong} B \cup C\,.$$

Here the fundamental period three orbit has been indicated by heavy dots, and the fixed point by a circled dot. Every point in $[-1,\ \lambda - 1] \times \mathbb{R}$ eventually maps into this union. (Here the horizontal scale has been exagerated by a factor of two. To compare this with Thurston's picture, it is necessary to reflect, interchanging top and bottom.)

NOTE 13.9 (See Figure 7.3 and Plate 19). Thurston points out that the discontinuity along the line $x = 0$ (and even the discontinuity in derivatives) can be eliminated by identifying the top and bottom halves of each of the outer vertical lines. This will yield a topological 2-disk bounded by the four top and bottom edges. In fact, he further identifies boundary points so as to obtain a closed surface of genus zero with a flat Euclidean structure except at four cone points. (The cone points correspond to the the large dots in the figures: the period three orbit together with the fixed point.) The boundary of $A \cup B \cup C$ is divided into twelve intervals which are identified in pairs, as indicated in the figure of Note 13.8. The dynamics remains well defined and continuous under these identifications, with

$$U \xrightarrow{\cong} V' \xrightarrow{\cong} A \cap B\,, \qquad U' \xrightarrow{\cong} V \xrightarrow{\cong} A \cap B\,,$$

and with

$$W \xrightarrow{\cong} W' \cup U', \qquad W' \xrightarrow{\cong} W \cup U\,.$$

The map is a linear area preserving Anosov map which preserves the horizontal and vertical foliations, except at the cone points.

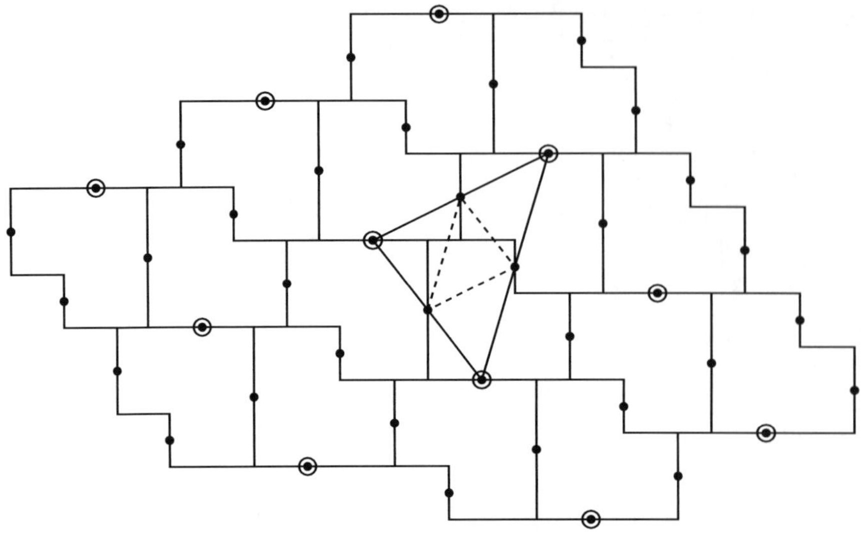

For a different way of describing this construction, consider the group of Euclidian motions of the plane generated by 180° rotation about the four marked periodic points. The images of $A \cup B \cup C$ under the action of this group forms a tiling of the plane, as shown in the figure. (See also Plate 20.) The map F on $A \cup B$ now extends to a linear automorphism of the plane. If we translate coordinates so that the fixed point $W \cap W'$ is at the origin, then this extended map is given simply by $(x, y) \mapsto (-\lambda x,\ (\lambda - 1)y)$.

The triangle in the figure forms an alternate fundamental domain for this action. If we fold this triangle along the dotted lines, then we obtain the tetrahedron which Thurston described as a model for the quotient space. (In fact there is a one parameter family of such examples, since the ratio between horizontal scale and vertical scale can be chosen arbitrarily—the ratio 2 was used for convenience in these figures.)

14 Bibliography

[ALM] L. Alsedà, J. Llibre, and M. Misiurewicz. *Combinatorial dynamics and entropy in dimension one*, in *Advanced Series in Nonlinear Dynamics*. **5** World Scientific Publishing Co. Inc., River Edge, NJ, second edition, 2000.

[Ber] A. Bertrand. Développements en base de Pisot et répartition modulo 1. *C. R. Acad. Sci. Paris Sér. A-B*, **285**(6):A419–A421, 1977.[33]

[Ber1] A. Bertrand-Mathis. Développement en base θ, répartition modulo un de la suite $(x\theta^n)_n > 0$, langages, codes et θ-shift. *Bull. Soc. Math. France* **114** (1986), 271–323, 1986.[33]

[BH] M. Bestvina and M. Handel. Train tracks and automorphisms of free groups. *Ann. of Math. (2)*, **135**(1):1–51, 1992.[33]

[Bo1] D. Boyd. Salem numbers of degree four have periodic expansions. In *Théorie des nombres (Quebec, PQ, 1987)*, pages 57–64. de Gruyter, Berlin, 1989.[33]

[Bo2] D. Boyd. On the beta expansion for Salem numbers of degree 6. *Math. Comp.*, **65**(214):861–875, S29–S31, 1996.[33]

[CH] A. de Carvalho and T. Hall. Unimodal generalized pseudo-Anosov maps. *Geometry & Topology*, **8**:1127–1188, 2004.

[Ge] A. O. Gelfond. A common property of number systems. *Izv. Akad. Nauk SSSR. Ser. Mat.*, **23**:809–814, 1959.[33]

[KOR] K. Kim, N. Ormes and F. Rousch. The spectra of nonnegative integer matrices via formal power series. *J. AMS* **13**:773–806, 2000.[33]

[Li] D. A. Lind. The entropies of topological Markov shifts and a related class of algebraic integers. *Ergodic Theory Dynam. Systems*, **4**(2):283–300, 1984.

[LS] R. Lyndon and P. Schupp *Combinatorial Group Theory.* Springer, 1977, 2001.[33]

[dMvS] W. de Melo and S. van Strien, *One-Dimensional Dynamics.* Springer, 1993.[33]

[33]These references have been added by the editors.

[MSt] E. Miller and B. Sturmfels. *Combinatorial Commutative Algebra.* Graduate Texts in Mathematics. Springer, New York; 2005 edition.[33]

[MT] J. Milnor and W. Thurston. On iterated maps of the interval. In *Dynamical systems (College Park, MD, 1986–87)*, **1342** of *Lecture Notes in Math.*, pages 465–563. Springer, Berlin, 1988.

[Mis] M. Misiurewicz. On Bowen's definition of topological entropy. *Discrete and Continuous Dynamical Systems* **10**(3):827–833, 2004.[33]

[MS1] M. Misiurewicz and W. Szlenk. Entropy of piecewise monotone mappings. In *Dynamical systems, Vol. II—Warsaw*, pages 299–310. Astérisque, **50** Soc. Math. France, Paris, 1977.

[MS2] M. Misiurewicz and W. Szlenk. Entropy of piecewise monotone mappings. *Studia Math.*, **67**(1):45–63, 1980.

[Ne] V. Nekrashevych. Hyperbolic groupoids: Definitions and duality. 2011.

[OP] A. M. Odylyzko and B. Poonen. Zeros of polynomials with 0, 1 coefficients. *L'Enseign. Math.* **39**: 317–348, 1993.[33]

[Pa] W. Parry On the β-expansions of real numbers, *Acta Math. Acad. Sci. Hungar.*, **11**: 269–278 (1960).[33]

[Sa] R. Salem. A remarkable class of algebraic numbers. Proof of a conjecture of Vijayaraghavan. *Duke Math. Journal*, **11**:103–108, 1944.[33]

[Sch] K. Schmidt. On periodic expansions of Pisot numbers and Salem numbers. *Bull. London Math. Soc.*, **12**(4):269–278, 1980.[33]

[Ti] G. Tiozzo. On Thurston's entropy spectrum for unimodal maps. Preprint, Harvard University, 2013

Degree-d-invariant Laminations

William P. Thurston, Hyungryul Baik, Gao Yan,
John H. Hubbard, Kathryn A. Lindsey,
Tan Lei and Dylan P. Thurston

PREFACE

During the last year of his life, William P. Thurston developed a theory of *degree-d-invariant laminations*, a tool that he hoped would lead to what he called "a qualitative picture of [the dynamics of] degree d polynomials." Thurston discussed his research on this topic in his seminar at Cornell University and was in the process of writing an article on this topic, but he passed away before completing the manuscript. Part I of this document consists of Thurston's unfinished manuscript. As it stands, the manuscript is beautifully written and contains a lot of his new ideas. However, he discussed ideas that are not in the unfinished manuscript and some details are missing. Part II consists of supplementary material written by the other authors based on what they learned from him throughout his seminar and email exchanges with him. Tan Lei also passed away during the preparation of Part II. William Thurston's vision was far beyond what we could write here, but, hopefully, this paper will serve as a starting point for future researchers.

Each semester since moving to Cornell University in 2003, William Thurston taught a seminar course titled "Topics in Topology," which was familiarly (and perhaps more accurately) referred to by participants as "Thurston seminar." On the first day of each semester, Thurston asked the audience what mathematical topics they would like to hear about, and he tailored the direction of the seminar according to the interests of the participants. Although the course was nominally a seminar in topology, he discussed other topics as well, including combinatorics, mathematical logic, and complex dynamics. Between 2010 and 2012, a high percentage of the seminar participants were dynamicists, and so (with the exception of one semester) Thurston's seminar during this period primarily focused on topics in complex dynamics. He discussed his topological characterization of rational maps on the Riemann sphere, as well as how to understand complex polynomials via topological entropy and laminations on the circle. During this time, Thurston developed many beautiful ideas, motivated in part by discussions with people in his seminar and in part by email exchanges with others who were at a distance.

His seminar was not an organized lecture series; it was much more than that. He talked about ideas that he was developing at that very moment, as opposed to previously known findings. Thurston invited members of the seminar to be actively involved in the exploration. He often demonstrated computer experiments in class, and he encouraged seminar participants to experiment with his codes. Seminar participants frequently received drafts of Thurston's manuscript as his thinking evolved. Those fortunate enough to learn from Thurston observed how his understanding of the subject gradually transformed into a beautiful theory.

This paper was the work of many people, written by different people at different times. We have done our best to make it a unified whole, and apologize for remaining inconsistencies of notation.

Part I

Degree-d-invariant Laminations

William P. Thurston
February 22, 2012

1 INTRODUCTION

Despite years of strong effort by an impressive group of insightful and hardworking mathematicians and many advances, our overall understanding and global picture of the dynamics of degree d rational maps and even degree d complex polynomials has remained sketchy and unsatisfying.

The purpose of this paper is to develop at least a sketch for a skeletal qualitative picture of degree d polynomials. There are good theorems characterizing and describing examples individually or in small-dimensional families, but that is not our focus. We hope instead to contribute toward developing and clarifying the global picture of the connectedness locus for degree d polynomials, that is, the higher-dimensional analogues of the Mandelbrot set.

To do this, the main tool will be the theory of degree-d-invariant laminations.

We hope that by developing a better picture for degree d polynomials, we will develop insights that will carry on to better understand degree d rational maps, whose global description is even more of a mystery.

2 SOME DEFINITIONS AND BASIC PROPERTIES

A degree d polynomial map $z \mapsto P(z) : \mathbb{C} \to \mathbb{C}$ always "looks like" $z \mapsto z^d$ near ∞. More precisely, it is known that P is conjugate to $z \mapsto z^d$ in some neighborhood of ∞.

We may as well specialize to monic polynomials such that the center of mass of the roots is at the origin (so that the coefficient of z^{d-1} is 0), since any polynomial can be conjugated into that form. In that case, we can choose the conjugating map to converge to the identity near ∞; this uniquely determines the map. As Douady and Hubbard noted, we can use the dynamics to extend the conjugacy near ∞ inward toward 0 step by step. If the Julia set is connected, we obtain in this way a Riemann mapping of the complement of the Julia set to the complement of the closed unit disk in $\hat{\mathbb{C}}$ that conjugates the dynamics outside the Julia set (which is the attracting basin of ∞) to the standard form $z \mapsto z^d$.

It has been known since the time of Fatou and Julia (and easy to show) that the Julia set is the boundary of the attracting basin of ∞. We want to investigate the topology of the Julia set, and how this topology varies among polynomial maps of

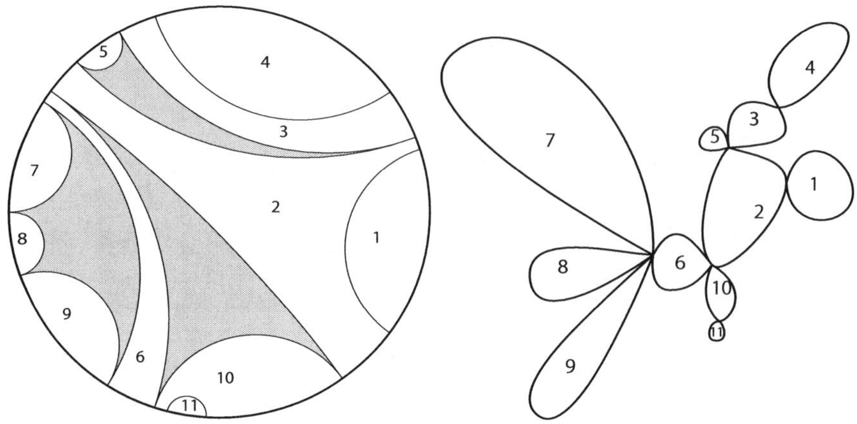

Figure 1: On the left is a finite lamination associated with a treelike equivalence relation with finitely many non-trivial equivalences: the two ends of any leaf are equivalent, and (as a consequence) the vertices of the polygons are all equivalent. The leaves are drawn as geodesics in the Poincaré disk model of the hyperbolic plane. The leaves and shaded areas can be shrunk to points by a pseudo-isotopy of the plane to obtain (topologically) the figure on the right.

degree d. As long as the Julia set is locally connected, the Julia set is the continuous image of the unit circle that is the boundary of the Riemann map around ∞; the key question is to understand the identifications on the circle made by these maps, and the way the identifications vary as the polynomial varies.

Define a *treelike* equivalence relation on the unit circle to be a closed equivalence relation such that for any two distinct equivalence classes, their convex hulls in the unit disk are disjoint. (A relation $R \subset X \times X$ on a topological space X is *closed* if it is a closed subset of $X \times X$.) The condition that the convex hulls of equivalence classes be disjoint comes from the topology of the plane: it translates into the condition that if we take the complement of the open unit disk and make the given identifications on the circle, two simple closed curves that cross the circle quotient using different equivalence classes cannot have intersection number 1, otherwise the quotient space would not embed in the plane.

To develop a geometric understanding of the possible equivalence relations, it helps to translate the concept into the language of laminations. Given a treelike equivalence relation R, there is an associated lamination $\mathrm{Lam}(R)$ of the open disk, where the leaves of $\mathrm{Lam}(R)$ consist of boundaries of convex hulls of equivalence classes intersected with the open disk. The regions bounded by leaves are called *gaps.* Some of the gaps of $\mathrm{Lam}(R)$ are *collapsed gaps*, which touch the circle in a single equivalence class, while other gaps are *intact gaps.* An intact gap necessarily touches the circle in an uncountable set, and its boundary mod R is homeomorphic to a circle.

Conversely, given a lamination λ there is an equivalence relation $\mathrm{Rel}(\lambda)$, usually obtained by taking the transitive closure of the relation that equates endpoints of leaves. However, this relation may not be topologically closed. One can take the closure, which may not be an equivalence relation. It is possible to alternate the operations of transitive closure and topological closure by transfinite induction until it stabilizes to a closed equivalence relation $\mathrm{Rel}(\lambda)$, or simply define $\mathrm{Rel}(\lambda)$ as the intersection of all closed equivalence relations that identify endpoints of leaves of λ.

Often, but not always, the operations Lam and Rel are inverse to each other. If R has equivalence classes that are Cantor sets, then $\mathrm{Rel}(\mathrm{Lam}(R))$ will collapse these only to a circle, not to a point. If λ has chains of leaves with common endpoints, then any leaves in the interior of the convex hull of the chain of vertices will disappear, and new edges may also appear.

A circular order on a set intuitively means an embedding of the set on a circle, up to orientation-preserving homeomorphism. In combinatorial terms, a circular order can be given by a function C from triples of elements to the set $\{-1,0,1\}$ (interpreted as giving the sign of the area of the triangle formed by three elements in the plane) such that $C(a,b,c)=0$ if and only if the elements are not distinct, and C is a 2-cocycle, meaning for any four elements a, b, c and d, the sum of the values for the boundary of a 3-simplex labeled with these elements is 0. (This implies, in particular, that $C(a,b,c)=-C(b,a,c)$, etc.) With this definition, it can be shown that any countable circularly ordered set can be embedded in the circle in such a way that $C(a,b,c)$ is the sign of the area of the triangle formed by the images of a, b and c. One way to prove this assertion is to first observe that a circular order can be "cut" at a point a into a linear order, defined by $b<c \iff C(a,b,c)=1$. It is reasonably well-known that a countable linear order on a set is induced by an embedding of that set in the interval, from which it follows that a circular ordering is induced by an embedding in S^1.

An *interval* J of a circularly ordered set is a subset with a linear order (=total order) satisfying the condition that

i. for $a,b,c\in J$, $a<b<c \iff C(a,b,c)=1$, and
ii. for any x and any $a,b\in J$, if $C(a,x,b)=1$ then $x\in J$.

Usually the linear order is determined by the subset, but there is an exception if the subset is the entire circularly ordered set. In that case the definition is like cutting a circle somewhere to form an interval. Just as for linear orders, any two elements a,b in a circularly ordered set determine a closed interval $[a,b]$, as well as an open interval (a,b), etc.

The *degree* of a map $f:X\to Y$ between two circularly ordered sets is the minimum size of a partition of X into intervals such that on each interval, the circular order is preserved. If there is no such finite partition, the degree is ∞.

A treelike equivalence relation R is *degree-d-invariant*

i. if sRt then s^dRt^d, and
ii. if for any equivalence class C, the total degree of the restrictions of $z\mapsto z^d$ to the equivalence classes on the set $C^{1/d}$ is d.

Similarly, a lamination λ is *degree-d-invariant*

i. if there is a leaf with endpoints x and y, then either $x^d=y^d$ or there is a leaf with endpoints x^d and y^d, and
ii. if there is a leaf with endpoints x and y, there is a set of d disjoint leaves with one endpoint in $x^{1/d}$ and the other endpoint in $y^{1/d}$.

These conditions on leaves imply related conditions for gaps, from the behavior of the boundary of the gap. In particular, an equivalence relation R is degree-d-invariant if and only if $\mathrm{Lam}(R)$ is degree-d-invariant. As customary, we will use the word "quadratic" as a synonym for degree 2, "cubic" for degree 3, "quartic" for degree 4, etc.

It is worth pointing out that the map $f_d : z \mapsto z^d$ is centralized by a dihedral group of symmetries of order $2(d-1)$, generated by reflection $z \mapsto \overline{z}$ together with rotation $z \mapsto \zeta z$ where ζ is a primitive $(d-1)$th root of unity. Therefore, the set of all degree-d-invariant laminations and the set of all degree-d-invariant relations has the same symmetry group.

If R is a degree-d-invariant equivalence relation, a *critical class* C is an equivalence class that maps with degree greater than 1. Similarly, a critical gap in $\mathrm{Lam}(R)$ is a gap whose intersection with the circle is mapped with degree greater than 1. The *criticality* of a class or a gap C is its degree minus 1. A critical class may correspond to either a critical leaf, which must have criticality 1, or a critical gap. A critical gap may be a collapsed gap that corresponds to a critical class, or an intact gap.

Proposition 2.1. *For any degree-d-invariant equivalence relation R, the total criticality of equivalence classes of R together with intact critical gaps of* $\mathrm{Lam}(R)$ *(i.e., the sum of the criticality of the critical classes and the intact critical gaps) equals $d-1$.*

Proof. First, let's establish that if $d > 1$ there is at least one critical leaf or critical gap. We can do this by extending the map $z \mapsto z^d$ to the disk: first extend linearly on each edge of the lamination, then foliate each gap by vertical lines and extend linearly to each of the leaves of this foliation. If there were neither collapsing leaves nor collapsing gaps, this would be a homeomorphism on each leaf and on each gap, so the map would be a covering map, impossible since the disk is simply-connected.

Now consider any critical leaf $\overline{xy}$, the circle with x and y identified is the union of two circles mapping with total degrees d_1 and d_2 where $d_1 + d_2 = d$. Proceed by induction: if the total criticality in these two circles is $d_1 - 1$ and $d_2 - 1$, then the total criticality in the whole circle is $d-1$.

Now consider any critical gap G. Let X be the union of S^1 with the boundary of G, and X' be its image under $z \mapsto z^d$, extended linearly on each edge. Any leaf in the image of the boundary of G has d preimages, from which it follows readily that the total criticality of the map on the complementary regions of X is $d-1$. □

Definition 2.2. *The* major *of a degree-d-invariant lamination is the set of critical leaves and critical gaps. The* major *of a degree-d-invariant equivalence relation R is the set of equivalence classes corresponding to critical leaves and critical gaps of* $\mathrm{Lam}(R)$.

3 MAJORS

What are the possibilities for the major of a degree-d-invariant lamination?

We'll say that the major of a lamination is *primitive* if each critical gap is a polygon whose vertices are all identified by $z \mapsto z^d$. We will see later that any degree-d-invariant lamination is contained in a degree-d-invariant lamination whose major is primitive; there are sometimes several possible invariant enlargements, and sometimes uncountably many. The invariant enlargements might not come from invariant equivalence relations, but they are useful nonetheless for understanding the global structure of invariant laminations, invariant equivalence relations, and Julia sets for degree d polynomials.

Even without a predefined lamination, we can define a *primitive degree-d major* to be a collection of disjoint leaves and polygons each of whose vertices are identified

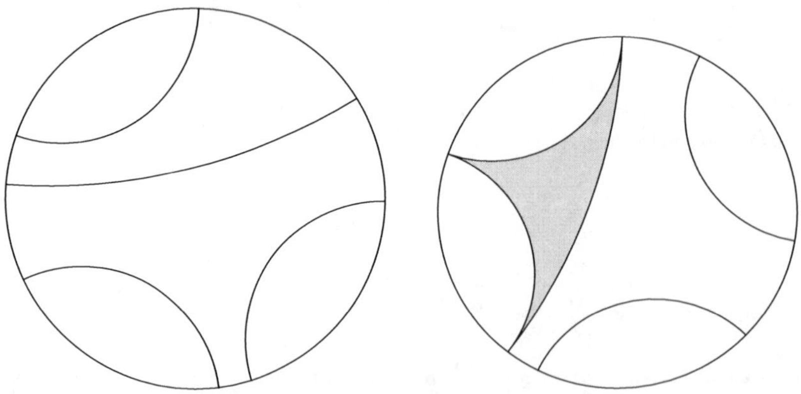

Figure 2: Here are two examples of primitive majors for a quintic-invariant lamination. Each intact gap that touches S^1 touches with total length $2\pi/5$. As a consequence, the two endpoints of any leaf are collapsed to a single point under $z \mapsto z^5$. Each isolated leaf has criticality 1; the three vertices of the shaded triangular gap in the example at right map to a single point, so it has criticality 2. If the leaves of the lamination at left move around so that two of them touch, then a third leaf joining the non-touching endpoints of the touching leaves is implicit, as in the figure at right. The example at right can be perturbed so that any one of the three leaves forming the ideal triangle disappears and the other two become disjoint.

under $z \mapsto z^d$, with total criticality $d-1$. In Section 4 we will show how to construct a degree-d-invariant lamination whose major is the given major. First though, we will analyze the set $\mathrm{PM}(d)$ of all primitive degree-d majors.

An element $m \in \mathrm{PM}(d)$ determines a quotient graph $\gamma(m)$ obtained by identifying each equivalence class to a point. The path-metric of S^1 defines a path-metric on $\gamma(m)$. In addition, $\gamma(m)$ has the structure of a planar graph, that is, an embedding in the plane well-defined up to isotopy, obtained by shrinking each leaf and each ideal polygon of the lamination to a point. These graphs have the property that $H^1(\gamma(m))$ has rank d, and every cycle has length a multiple of $2\pi/d$. Every edge must be accessible from the infinite component of the complement, so the metric and the planar embedding together with the starting point, that is, the image of $1 \in \mathbb{C}$, is enough to define the major.

The metric $\mathrm{met}(m)$ on the circle induced by the path-metric on $\gamma(m)$ determines a metric md on $\mathrm{PM}(d)$, defined as the sup difference of metrics,

$$\mathrm{md}(m, m') = \sup_{x,y \in S^1} |\mathrm{met}(m)(x, y) - \mathrm{met}(m')(x, y)| .$$

In particular, there is a well-defined topology on $\mathrm{PM}(d)$.

There is a recursive method one can use to construct and analyze primitive degree-d majors, as follows. Consider any isolated leaf l of a major $m \in \mathrm{PM}(d)$. The endpoints of l split the circle into two intervals, A of length k/d and B of length $(d-k)/d$, for some integer k. If we identify the endpoints of A and make an affine reparametrization to stretch it by a factor of d/k so it fits around the unit circle and place the identified endpoints at 0, the leaves of m with endpoints in A become a primitive degree k major. Similarly, we get a primitive degree $(d-k)$ major from B.

If p is any ideal j-gon of an element $m \in \mathrm{PM}(d)$, we similarly can derive a sequence of j primitive majors whose total degree is d.

Theorem 3.1. *The space* $\mathrm{PM}(d)$ *can be embedded in the space of monic degree d polynomials as a spine for the set of polynomials with distinct roots, that is, the complement of the discriminant locus. The spine (image of the embedding) consists of monic polynomials whose critical values are all on the unit circle and such that the center of mass of the roots is at the origin.*

A *spine* is a lower-dimensional subset of a manifold such that the manifold deformation-retracts to the spine.

Proof. We'll start by defining a map from the space of polynomials with distinct zeroes to $\mathrm{PM}(d)$. Given a polynomial $p(z)$, we look at the gradient vector field for $|p(z)|$, or better, the gradient of the smooth function $|p(z)|^2$. Any simple critical point for p is a saddle point for this flow. Near infinity, the flow lines align very closely to the flow lines for the gradient of $|z|^{2d}$, so each flow line that doesn't tend toward a critical point goes to infinity with some asymptotic argument (angle). Define a lamination whose leaves join the pairs of outgoing separatrices for the critical points. If there are critical points of higher multiplicity or if the unstable manifolds from some critical points coincide with stable manifolds for others (separatrices collide), then define an equivalence relation that decrees that for any flow line, if the limit of asymptotic angles on the left is not the same as the limit of asymptotic angles on the right, then the two limits are equivalent. For each such equivalence class, adjoin the ideal polygon that is the boundary of its convex hull.

We claim that the lamination so defined is a primitive degree-d major. To see that, notice that the level sets of $|p(z)|$ are mapped under $p(z)$ to concentric circles, so the flow lines for the gradient flow map to the perpendicular rays emanating from the origin. Each equivalence class coming from discontinuities in the asymptotic angles therefore maps to a single point, since there are no critical points for the image foliation except at the origin in the range of p. The total criticality is $d-1$ since the degree of $p'(z)$ is $d-1$.

There are many different polynomials that give any particular major. In the first place, note that for any polynomial, if we translate in the domain by any constant (which amounts to translating all the roots in some direction, keeping the vectors between them constant), the asymptotic angles of the upward flow lines from the critical points do not change, so then lamination does not change. We may as well keep the roots centered at the origin. Algebraically, this is equivalent to saying that the critical points are centered at the origin (by considering the derivative of the defining polynomial).

You can think of the spine this way. In the domain of p, cut $\mathbb{C}$ along each upward separatrix of each critical point. This cuts $\mathbb{C}$ into d pieces, each containing one root of p (that you arrive at by flowing downhill; whenever there is a path from z_1 to z_2 whose downhill flows don't ever meet a critical point, they end up at the same root). For any region, if you glue the two sides of each edge together, starting at the lowest critical point and matching points with equal values for $p(z)$, you obtain a copy of $\mathbb{C}$ which is really just a copy of the image plane; the seams have joined together to become rays.

You can construct other polynomials associated with the same major by varying the length of the cuts. (These cuts are classical branch cuts). Assuming for now that we are in the generic case with no critical gaps, for each leaf choose a positive real number. Take one copy of $\mathbb{C}$ for each region of the complement of the lamination, and for each leaf l $\overline{xy}$ on its boundary, pick a positive real number r_l and make a slit on the ray at angle $x^d = y^d$ from ∞ to the point $r_l x^d$. Now glue the slit copies of $\mathbb{C}$ together so as to be compatible with the parametrization by $\mathbb{C}$. (This is equivalent to forming a branched cover of $\mathbb{C}$, branched over the various critical values, with given combinatorial information or *Hurwitz data* describing the branching.) By uniformization theory, the Riemann surface obtained by gluing the copies of $\mathbb{C}$ together is analytically equivalent to $\mathbb{C}$, and the induced map is a polynomial.

In the nongeneric case (in which there are critical gaps), the set of possibilities branches: it is not parametrized by a product of Euclidean spaces, because of the different possibilities for the combinatorics of saddle connections. Let's look more closely at the possible structure of saddle connections for the gradient flow of $|p(z)|$. A small regular neighborhood of the union of the upward separatrices for the critical points has boundary consisting of $k+1$ lines (here k is the multiplicity of the critical point); these map to the leaves of the boundary of the corresponding gap of the major. The union of upward separatrices is a tree whose leaves are the vertices of the gap polygon. The height function $\log|p(z)|$ has the property that induces a function with exactly one local minimum on each boundary component of the regular neighborhood projected to the graph.

We can bypass the enumeration of all such structures by observing that there is a direct way to define a retraction to the case when all critical values equal 1. We think of the graph as a metric graph where the height function $log|p|$ has speed 1. Multiply the height function on the compact edges of the graph by $(1-t)$ while adding the constant times t to the height function on each unbounded component that makes it continuous at the vertices. When $t=1$, the compact edges all collapse, only the unbounded edges remain, and the polygonal gap contains a single multiple critical point with critical value 1. □

Corollary 3.2. PM(d) *is a* $K(B_d, 1)$ *where* B_d *is the d-strand braid group. In other words,* $\pi_1(\mathrm{PM}(d)) = B_d$ *and all higher homotopy groups are trivial.*

This follows from the well-known fact that the complement of the discriminant locus is a $K(B_d, 1)$. (See, e.g., [FM12, Section 9.12].) A loop in the space of primitive degree-d majors can be thought of as braiding the d regions in the complement of the union of its leaves and critical gaps.

A primitive quadratic major is just a diameter of the unit circle, so PM(2) is itself a circle. This corresponds to the fact that the 2-strand braid group is $\mathbb{Z}$.

A primitive cubic major is either an equilateral triangle inscribed in the circle, or a pair of chords that each cut off a segment of angle $2\pi/3$. There is a circle's worth of equilateral triangles. To each primitive cubic that consists of a pair of leaves, there is associated a unique diameter that bisects the central region. Thus, the space of two-leaf primitive cubic majors fibers over S^1, with fiber an interval. When the diameter is turned around by an angle of π, the interval maps to itself by reversing the orientation, so this is a Moebius band. The boundary of the Moebius band is attached to the circle of equilateral triangle configurations, wrapping around 3 times, since, given an equilateral triangle, you can remove any of its three edges to get a limit of two-leaf majors.

The space of cubic polynomials can be normalized to make the leading coefficient 1 (monic) and, by changing coordinates by a translation, make the second

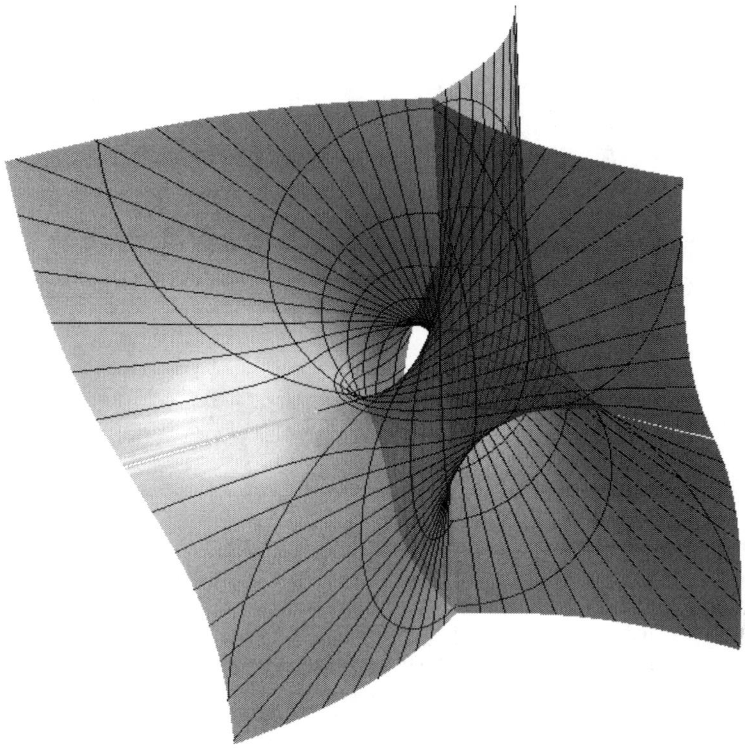

Figure 3: This is the space PM(3) embedded in S^3. It is formed by a 3/2-twisted Moebius band centered on a horizontal circle that grows wider and wider until its boundary comes together in the vertical line, which closes into a circle passing through the point at infinity, and wraps around this circle 3 times. The complement of PM(3) consists of 3 chambers which are connected by tunnels, each tunnel burrowing through one chamber and arriving 2/3 of the way around the post in the middle. A journey through all three tunnels ties a trefoil knot: such a trefoil knot is the locus where the discriminant of a cubic polynomial is 0, and PM(3) is a spine for the complement of the trefoil.

coefficient (the sum of the roots) equal to 0. By multiplying by a positive real constant, one can then normalize so that the other two coordinates, as an element of $\mathbb{C}^2$, are on the unit sphere. The discriminant locus intersects the sphere in a trefoil knot. The spine can be embedded in S^3 as follows: start with a 3/2-twisted Moebius band centered around a great circle in S^3. This great circle can be visualized via stereographic projection of S^3 to $\mathbb{R}^3$ as the unit circle in the xy-plane. The Moebius band can be arranged so that it is generated by geodesics perpendicular to the horizontal great circle, in a way that is invariant by a circle action that rotates the horizontal great circle at a speed of 2 while rotating the z-axis, completed to a great circle by adding the point at infinity, at a speed of 3. Extend the perpendicular geodesics all the way to the z-axis; this attaches the boundary of the Moebius band by wrapping it three times around the vertical great circle, as shown in Figure 3.

The spine gives a graphic description for one presentation of the 3-strand braid group,

$$B_3 = \langle a, b | a^2 = b^3 \rangle ,$$

where a is represented by the core circle of the Moebius band, and b is the circle of equilateral triangles, the two loops being connected via a short arc to a common base point. The presentation is an amalgamated free product of two copies of $\mathbb{Z}$ over subgroups of index 2 and 3. As braids, a is a 180 degree flip of three strands in a line, while b is a 120 degree rotation of 3 strands forming a triangle; the square of a and the cube of b is a 360 degree rotation of all strands, which generates the center of the 3-strand braid group.

4 GENERATING INVARIANT LAMINATIONS FROM MAJORS

Let m be a degree-d primitive major. How can we construct a degree-d-invariant lamination having m for its major?

A leaf of a lamination is defined by an unordered pair of distinct points on the unit circle. The space of possible leaves is topologically an open Moebius band. To see this, consider that any leaf divides the circle into two intervals. The line connecting the midpoints of these two intervals is the unique diameter perpendicular to the leaf. For each diameter, there is an interval's worth of leaves, parametrized by the point at which a leaf intersects the perpendicular diameter. When the interval is rotated 180 degrees, the parameter is reversed, so the space is an open Moebius band. One way to graphically represent the Moebius band is as the region outside the unit disk in $\mathbb{RP}^2$. The pair of tangents to the unit circle at the endpoints of a leaf intersect somewhere in this region, which is homeomorphic to a Moebius band.

Another way to represent the information is by passing to the double cover, the set of ordered pairs of distinct points on a circle, that is, the torus $S^1 \times S^1$ minus the diagonal, which is a $(1,1)$-circle on the torus, going once around each axis. Here $S^1 = \mathbb{R}/\mathbb{Z}$. For $x, y \in S^1$, we use $\overline{xy}$ to represent the geodesic in the unit disc linking $e^{2\pi ix}$ and $e^{2\pi iy}$. Each leaf $\overline{xy}$ is represented twice on the torus, as (x,y) and (y,x).

If a lamination contains a leaf $l = \overline{xy}$, then a certain set $X(l)$ of other leaves are excluded from the lamination because they cross $\overline{xy}$. On the torus, if you draw the horizontal and vertical circles through the two points (x,y) and (y,x), they subdivide the torus into four rectangles having the same vertex set; the remaining two common vertices are (x,x) and (y,y). The leaves represented by points in the interior of two of the rectangles constitute $X(l)$, while the leaves represented by the closures of the other two rectangles are all compatible with the given leaf $l = \overline{xy}$. We will call this good, compatible region $G(l)$, as shown in Figure 4. The compatible rectangles are actually squares, of side lengths $a - b \mod 1$ and $b - a \mod 1$. They form a checkerboard pattern, where the two squares of $G(l)$ are bisected by the diagonal. Another way to express it is that two points p and q define compatible leaves if and only if you can travel on the torus, without crossing the diagonal, from one point to either the other point or the other point reflected

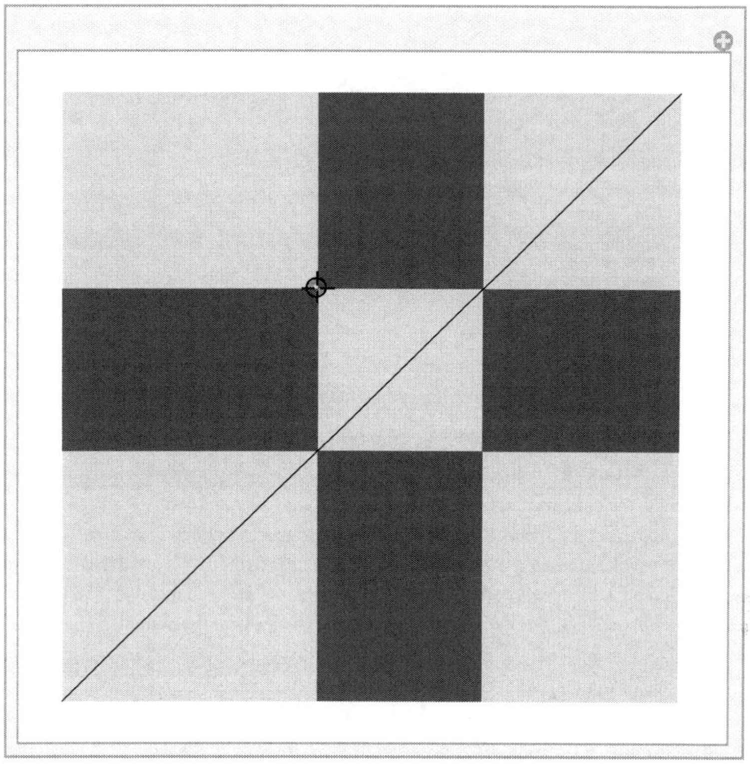

Figure 4: A leaf $\overline{xy}$ of a lamination can be represented by a pair of points $\{(x,y),(y,x)\}$ on a torus. Leaves that are excluded by $\overline{xy}$ because they intersect it are represented in two darker shaded rectangles, and leaves compatible with it are represented in two squares of side length $b-a \mod 1$ and $a-b \mod 1$.

in the diagonal, heading in a direction between north and west or between south and east (using the same conventions as on maps, where up is north, left is west, etc.).

Given a set S of leaves, the excluded region $X(S)$ is the union of the excluded regions $X(l)$ for $l \in S$, and the good region $G(S)$ is the intersection of the good regions $G(l)$ for $l \in S$. If S is a finite lamination, then $G(S)$ is a finite union of rectangles that are disjoint except for corners.

In the particular case of a primitive major lamination $m \in \mathrm{PM}(d)$, each region of the disk minus m touches S^1 in a union of one or more intervals $J_1 \cup \cdots \cup J_k$ of total length $1/d$. This determines a finite union of rectangles $(J_1 \cup \cdots \cup J_k) \times (J_1 \cup \cdots \cup J_k)$ of $G(m)$ whose total area is $1/d^2$ that maps under the degree d^2 covering map $(x,y) \mapsto d \cdot (x,y)$ to the entire torus, as seen in Figures 5 and 6.

Consequently we have (here we use $\mathbb{D}^2$ to denote the unit disc):

Proposition 4.1. *For any primitive degree-d major m, the total area of $G(m)$ is $1/d$. Almost every point $p \in T^2$ has exactly d preimages in $G(m)$ by the degree d^2*

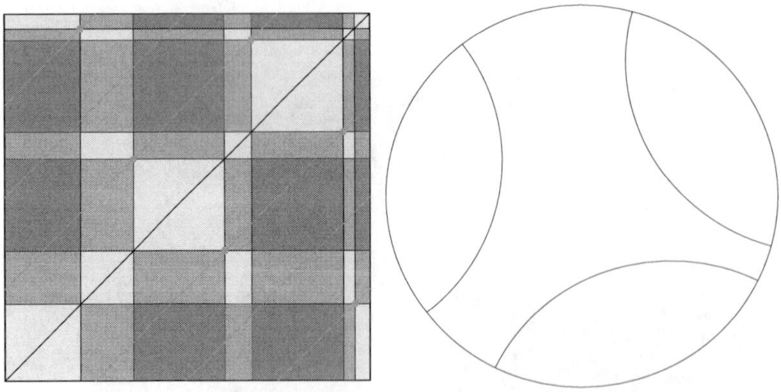

Figure 5: On the left is a plot showing the excluded region $X(m)$, shaded, together with the compatible region $G(m)$ on the torus, where $m \in \mathrm{PM}(4)$ is a primitive degree-4 major. The figure is symmetric by reflection in the diagonal. The quotient of the torus by this symmetry is a Moebius band. Note that $G(m)$ is made up of three $1/4 \times 1/4$ squares (one of them wrapped around) corresponding to the regions that touch the circle in only one edge, together with $3 \cdot 3 = 9$ additional rectangles whose total area is $1/4^2$, corresponding to the region that touches S^1 in 3 intervals. The 6 green dots represent the leaves of the major, one dot for each orientation of the leaf.

covering map $f_d : (x, y) \mapsto d \cdot (x, y)$ with one preimage representing a leaf in each of the d regions of $\mathbb{D}^2 \setminus m$, and all points have at least d preimages in $G(m)$ with at least one preimage representing a leaf in each region of $\mathbb{D}^2 \setminus m$.

For $m \in \mathrm{PM}(d)$ we can now define a sequence of backward-image laminations $b_i(m)$. Let $b_0(m) = m$ and inductively define $b_{i+1}(m)$ to be the union of m with the preimages under f_d of $b_i(m)$ that are in $G(m)$.

Proposition 4.2. *For each i, $b_i(m)$ is a lamination.*[1]

Proof. We need to check that the good preimages under f_d of two leaves of b_i are compatible. If they are in different regions of $\mathbb{D}^2 \setminus m$ they are obviously compatible. For two leaves in a single region of $\mathbb{D}^2 \setminus m$, note that when the boundary of the region is collapsed by collapsing m, it becomes a circle of length $1/d$ that is mapped homeomorphically to S^1. By the inductive hypothesis, the two image leaves are compatible; since the compatibility condition is identical on the small circle and its homeomorphic image, the leaves are compatible. □

[1]This proposition from the original text is not correct as stated. Two counter-examples are the cubic lamination $m = \{(0, \frac{1}{3}), (\frac{1}{3}, \frac{2}{3}), (\frac{2}{3}, 0)\}$ (for which $b_1(m)$ fails to be a lamination) and the quadratic lamination $m = \{(0, \frac{1}{2})\}$ (for which $b_2(m)$ fails to be a lamination). Section 14 has a corrected algorithm in the quadratic case.

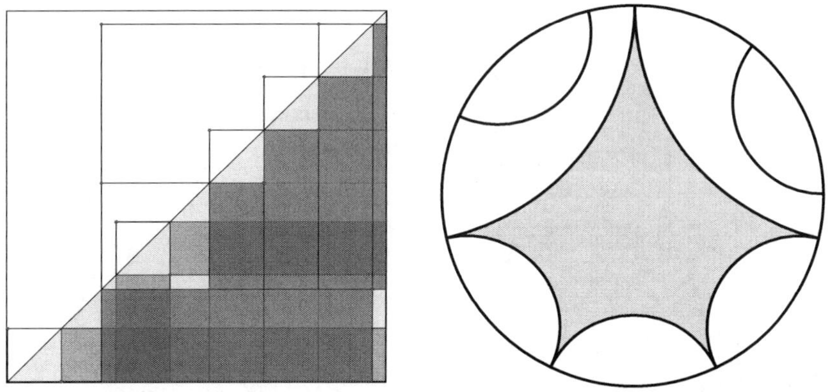

Figure 6: Here is a primitive heptic (degree-7) major with a pentagonal gap, shown in a variation of the torus plot, along with the standard Poincaré disk picture. On the left, half of the torus has been replaced by a drawing that indicates each leaf of the lamination by a path made up of a horizontal segment and a vertical segment. The upper left triangular picture transforms to the Poincaré disk picture by collapsing the horizontal and vertical edges of the triangle to a point, bending the collapsed triangle so that it goes to the unit disk with the collapsed edges going to $1 \in \mathbb{C}$, then straightening each rectilinear path into the hyperbolic geodesic with the same endpoints. Notice how the ideal pentagon corresponds to a rectilinear 10-gon, where two pairs of edges of the 10-gon overlap. The lower right triangle is a fundamental domain for a group Γ that is the covering group of the torus together with lifts of reflection through the diagonal of the torus. The horizontal edge of the triangle is glued by an element $\gamma \in \Gamma$ to the vertical edge. The action of γ on the edge is the same as a 90 degree rotation through the center of the square; γ itself follows this by reflection through the image edge. The quotient space $\mathbb{E}^2/\Gamma$ is topologically a Moebius band. As an orbifold it is the Moebius band with mirrored boundary.

Note that this is an increasing sequence, $b_i(m) \subset b_{i+1}(m)$. By induction, the good region $G(b_i(m))$ has area $1/d^m$.

It follows readily that:

Theorem 4.3. *The closure $b_\infty(m)$ of the union of all $b_i(m)$ is a degree-d-invariant lamination having m as its major.*

The lamination $b_\infty(m)$ may have various issues concerning its quality. In particular, it does not always happen that $b_\infty = \mathrm{Lam}(\mathrm{Rel}(b_\infty(m)))$. We will study the quality of these laminations later, and develop tools for studying more general degree-d-invariant laminations by embedding them in $b_\infty(m)$ for some m. The process is illustrated in Figure 7.

Note that as a finite lamination λ varies, a compatible rectangle can become thinner and thinner as two endpoints approach each other, and disappear in the limit. Thus $G(\lambda)$ is not continuous in the Hausdorff topology.

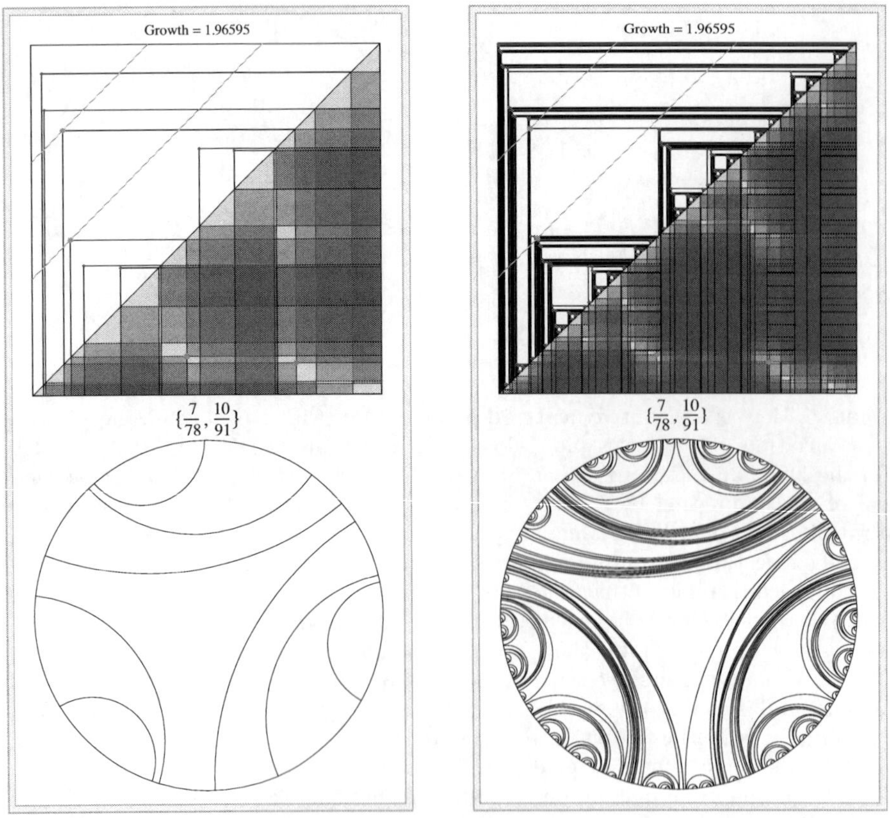

Figure 7: On the left is stage 1 ($b_1(m)$) in building a cubic-invariant lamination for $m = \{(\frac{10}{91}, \frac{121}{273}), (\frac{7}{78}, \frac{59}{78})\} \in \mathrm{PM}(3)$. The two longer leaves of m subdivide the disk into 3 regions, each with two new leaves induced by the map f_3. On the right is a later stage that gives a reasonable approximation of $b_\infty(m)$.

On the other hand,

Proposition 4.4. *The map from* $\mathrm{PM}(d)$ *to the space of compact subsets of* T *endowed with the Hausdorff topology defined by* $m \mapsto X(m)$ *is a homeomorphism onto its image, i.e., the topology of* $\mathrm{PM}(d)$ *coincides with Hausdorff topology on the set* $\{X(m) : m \in \mathrm{PM}(d)\}$.

Proof. Perhaps the main point is that $X(m)$ is a union of fat rectangles, with height and width at least $1/d$, so pieces of $X(m)$ can't shrink and suddenly disappear.

Suppose m and m' are majors that are within ϵ in the metric md, and suppose $p \in X(m)$, so there is some leaf l_m of m intersecting the leaf l_p represented by p. The endpoints of l_m have distance 0 in the quotient graph S^1/m, so there must be a path of length no greater than ϵ on S^1/m' connecting these two points. In particular, assuming $\epsilon < 1/d$, some leaf of m' comes within at most ϵ of intersecting l_p, therefore a leaf of m' intersects a leaf $l_{p'}$ near l_p. Then p' is in $X(m')$. Since this works symmetrically between m and m', it follows that they are close in the Hausdorff metric.

Conversely, given $m \in \mathrm{PM}(d)$ and $\epsilon > 0$, we will show that there exists $\delta > 0$ such that for any $m'' \in \mathrm{PM}(d)$ with the Hausdorff distance between $X(m)$ and $X(m')$ less than δ, the distance $\mathrm{md}(m, m') < \epsilon$. We will do this by induction on d. It is obvious for $d = 2$, since a major is a diameter and $X(m)$ is a union of two squares that intersect at two corners that are the representatives of the single leaf of m.

When $d > 2$, we choose a region of $\mathbb{D}^2 \setminus m$ that touches S^1 in a single interval A of length $1/d$, bounded by a leaf l. Suppose $X(m')$ is Hausdorff-near $X(m)$. Then most of the square $A \times A$ of $G(m)$ is in $G(m')$, and most of the rectangles $A \times (S^1 \setminus A)$ and $(S^1 \setminus A) \times A$ of $X(m)$ is also in $X(m')$. This implies that m' has a nearby l' spanning an interval A' of length $1/d$. Now we can look at the complementary regions, with l or l' collapsed, normalized by the affine transformation that makes the restriction of m or m' a primitive degree-$(d-1)$ major having the collapsed point at 0. Call these new majors m_1 and m_1'. The excluded region $X(m_1)$ is obtained from $X(m)$ intersected with a $(d-1) \times (d-1)$ square on the torus, and similarly for $X(m_1')$; hence if the Hausdorff distance between $X(m)$ and $X(m')$ is small, so is that between $X(m_1)$ and $X(m_1')$. By induction, we can conclude that the pseudo-metrics $\mathrm{met}(m)$ and $\mathrm{met}(m')$ on S^1 induced from the quotient graphs are close. □

5 CLEANING LAMINATIONS: QUALITY AND COMPATIBILITY

We will use the parametrization of the circle by *turns*, that is, numbers interpreted as fractions of the way around the circle. Thus $\tau \in [0, 1]$ corresponds to the point $\exp(2\pi i \tau)$ in the unit circle in the complex plane.

Let's look at the quadratic major $\overline{0, \frac{1}{2}}$. We'll use a variation of the process of backward lifting, depicted in Figure 8, which is really the limit as $\epsilon \to 0$ of the standard process applied to $\{-\epsilon, \frac{1}{2} - \epsilon\}$. The first backward lift adds 2 new leaves, $\overline{0, \frac{1}{4}}$ and $\overline{\frac{1}{2}, \frac{3}{4}}$, and at each successive stage, a new leaf is added joining the midpoint of each interval between endpoints to the last clockwise endpoint of the interval. (The full construction would also join each of these midpoints to the counterclockwise endpoint of the interval.) In the limit, there are only a countable set of leaves, but they are joined in a single tree. The closed equivalence relation they generate collapses the entire circle to a point!

Each finite-stage lamination λ_k can be cleaned up into the standard form $\mathrm{Lam}(\mathrm{Rel}(\lambda_k))$.

6 PROMOTING FORWARD-INVARIANT LAMINATIONS

A degree-d primitive major is a special case of a lamination that is forward invariant. We have seen how to construct a fully invariant degree-d lamination containing it. How generally can forward-invariant laminations be promoted to fully invariant laminations?

Figure 8: The top four laminations are successive steps of building an invariant quadratic lamination for the major $\{0, \frac{1}{2}\}$. The bottom four laminations are obtained by performing Lam $\circ$ Rel.

7 HAUSDORFF DIMENSION AND GROWTH RATE

Figure 9: The top nine laminations are successive steps of building an invariant quadratic lamination for the major $\{\frac{1}{7}, \frac{9}{14}\}$. The bottom nine laminations are obtained by performing Lam $\circ$ Rel (at every third step). It gives the lamination of Douady's rabbit.

Part II

Part II consists of supplementary sections expanding on the material in Part I, which were written by the authors other than W. Thurston. We start by recalling related work on laminations in Section 8. Section 9 gives various dynamical interpretations of PM(d). Section 10 looks at several different parametrizations of PM(d) and describes PM(2). Section 11 gives a complete description of PM(3), and Section 12 discusses PM(4) to some extent.

The remaining sections are devoted to issues around entropy. Section 13 explains and proves correctness of a simple algorithm for computing core entropy extracted from computer code by W. Thurston. Sections 14 and 15 examine different types of laminations related to a primitive major (presumably related to the material intended for Section 6), while Section 16 shows equivalence of several different entropies and Hausdorff dimensions related to a polynomial (presumably related to Section 7).

8 RELATED WORK ON LAMINATIONS

Invariant laminations were introduced as a tool in the study of complex dynamics by William Thurston [Thu09]. Thurston's theory suggests using spaces of invariant laminations as models for parameter spaces of dynamical systems defined by complex polynomials. In degree $d=2$, Thurston uses *QML* (quadratic minor lamination) to model the space of 2-invariant laminations. Underpinning this approach in degree 2 is Thurston's No Wandering Triangle Theorem. Thurston conjectured that the boundary of the Mandelbrot set is essentially the quotient of the circle by the equivalence relation induced by QML. The precise relationship between invariant laminations and complex polynomials is even less clear for higher degree.

Not all degree-d-invariant laminations correspond to degree d polynomials. One necessary condition for a lamination to be directly associated to a polynomial is that it be generated by a tree-like equivalence relation on S^1 (see, for example, [BL02, BO04a, Kiw04, Mim10] and Section 2 of Part I). To distinguish between arbitrary invariant laminations and those associated to polynomials, we call the invariant laminations defined by Thurston *geometric invariant laminations*, and the smaller class of laminations defined by tree-like equivalence relations *combinatorial invariant laminations.* (In [BMOV13], such laminations are called q-laminations).

The difference between these two notions can be explained via the following examples. Let a, b, c be three distinct points on S^1 which form an equivalence class under a tree-like equivalence relation. Then in the combinatorial lamination obtained from the given tree-like equivalence, all the leaves $(a,b), (b,c), (a,c)$ are contained. But as a geometric lamination, it may have leaves $(a,b), (b,c)$ without having (a,c) as a leaf. Here is another important case to consider. From the "primitive major" construction, we can end up with leaves $(a,b), (b,d)$, and (c,d), where a,b,c,d are four distinct points on S^1 so that (a,b,c,d) appear in cyclic order. On the other hand, if this were a combinatorial lamination, it would only include the ones around the outside, namely $(a,b), (b,c), (c,d)$, and (d,a). One way to understand both of these examples geometrically is to identify the circle with the ideal boundary of the hyperbolic plane $\mathbb{H}^2$. Then a combinatorial lamination can be understood as the one consisting of the boundary leaves of the convex hulls of finitely many points on the ideal boundary of $\mathbb{H}^2$.

A fundamental global result in the theory of combinatorial invariant laminations is the existence of *locally connected models for connected Julia sets*, obtained by Kiwi [Kiw04]. He associates a combinatorial invariant lamination $\lambda(f)$ to each polynomial f that has no irrationally neutral cycles and whose Julia set is connected. Then the topological Julia set $J_{\sim_f} := S^1/_{\sim_f}$ is a locally connected continuum, where $\sim_f$ is the equivalence relation generated by $x \sim_f y$ if x and y are connected by a leaf of $\lambda(f)$, and $f|_{J_f}$ is semi-conjugate to the induced map $f_{\sim_f}$ on $J_{\sim_f}$ via a monotone map ϕ: $J_f \to J_{\sim_f}$ (by monotone we mean a map whose point preimages are connected). Kiwi characterizes the set of combinatorial invariant laminations that can be realized by polynomials that have no irrationally neutral cycles and whose Julia sets are connected. In [BCO11], Blokh, Curry and Oversteegen present a different approach, one based upon continuum theory, to the problem of constructing locally connected dynamical models for connected polynomial Julia sets J_f; their approach works regardless of whether or not f has irrational neutral cycles. These locally connected models yield nice combinatorial interpretations of connected quadratic Julia sets that themselves may or may not be locally connected.

The No Wandering Triangle Theorem is a key ingredient in Thurston's construction [Thu09] of a locally connected model $\mathcal{M}_c^2$ of the Mandelbrot set. The theorem asserts the non-existence of wandering non-(pre)critical branch points of induced maps on quadratic topological Julia sets. Branch points of $\mathcal{M}_c$ correspond to topological Julia sets whose critical points are periodic or preperiodic. Thurston posed the problem of extending the No Wandering Triangle Theorem to the higher-degree case. Levin showed [Lev98] that for "unicritical" invariant laminations, wandering polygons do not exist. Kiwi proved [Kiw02] that for a combinatorial invariant lamination of degree d, a wandering polygon has at most d edges. Blokh and Levin obtained more precise estimates on the number of edges of wandering polygons [BL02]. Soon after, Blokh and Oversteegen discovered that some combinatorial invariant laminations of higher-degree ($d \geq 3$) do admit wandering polygons [BO04b, BO09].

Extending Thurston's technique of using invariant laminations to construct a combinatorial model $\mathcal{C}_d$ of the connectedness locus for polynomials of degree $d > 2$ remains an area of inquiry. In [BOPT14] and [Pta13], A. Blokh, L. Oversteegen, R. Ptacek and V. Timorin make progress in this direction. They establish two necessary conditions of laminations from the polynomials in the *Main Cubioid CU*, i.e., the boundary of the principal hyperbolic component of the cubic connectedness locus $\mathcal{M}_3$; CU is the analogue of the main cardioid in the quadratic case. They propose this set of laminations as the *Combinatorial Main Cubioid* CU^c, a model for CU.

9 INTERPRETATIONS OF THE SPACE OF DEGREE d PRIMITIVE MAJORS

9.1 Recalling definitions

We begin by recalling some concepts from Part I. A *critical class* of a degree-d-invariant equivalent relation is subset of $S^1 = \partial\mathbb{D} \subset \mathbb{C}$ that consists of all elements of an equivalence class and that maps under $z \mapsto z^d$ with degree greater than 1; the associated subsets of $\mathbb{D}$ are the *critical leaves* and *critical gaps* of the lamination. The *criticality* of a critical class is defined to be one less than the degree of the restriction of the map to that subset. Per Definition 2.2, the *major* of a degree-d-invariant lamination (resp. equivalence relation) is the set of critical leaves and critical gaps (resp. equivalence classes corresponding to critical leaves and critical gaps). Such a major is said to be *primitive* if every critical gap is a "collapsed polygon" whose vertices are identified under the map $z \mapsto z^d$. (The restriction of $z \mapsto z^d$ to an intact gap of a primitive degree-d-invariant lamination is necessarily injective.) A critical leaf may be thought of as a critical gap defined by a polygon with precisely two vertices, and those vertices are identified by $z \mapsto z^d$. By Proposition 2.1, the sum over the critical classes of their criticalities equals $d-1$.

9.2 Metrizability of PM(d)

As described in Part I, an element $m \in \mathrm{PM}(d)$ determines a quotient graph $\gamma(m)$ obtained by identifying each equivalence class to a point. The path-metric of S^1 defines a path-metric on $\gamma(m)$. In addition, $\gamma(m)$ has the structure of a planar graph, that is, an embedding in the plane well-defined up to isotopy, obtained by shrinking each leaf and each ideal polygon of the lamination to a point. These graphs have the property that $H^1(\gamma(m))$ has rank d, and every cycle has length a multiple of $1/d$. Every edge must be accessible from the infinite component of the complement, so the metric and the planar embedding together with the starting point, that is, the image of $1 \in \mathbb{C}$, is enough to define the major.

The pseudo-metric $\mathrm{met}(m)$ on the circle induced by the path-metric on $\gamma(m)$ determines a continuous function on $S^1 \times S^1$. The sup-norm on the space of continuous functions on $S^1 \times S^1$ induces a metric md on PM(d):

$$\mathrm{md}(m,m') = \sup_{(x,y)\in S^1\times S^1} |\mathrm{met}(m)(x,y) - \mathrm{met}(m')(x,y)|\,.$$

For the sake of completeness, we give a proof of the fact that md is indeed a metric.

Lemma 9.1. *The function* md *is a metric on* PM(d).

Proof. Non-negativity and symmetry follow automatically from the definition. The rest of the proof is also pretty straightforward.

Suppose $\mathrm{md}(m,m')=0$ for some $m, m'' \in \mathrm{PM}(d)$. Since $\sup_{x,y} | \mathrm{met}(m)(x,y) - \mathrm{met}(m')(x,y)| = 0$, this means $\mathrm{met}(m)(x,y) = \mathrm{met}(m')(x,y)$ for all $x, y \in S^1$. This implies indeed $m = m'$. In order to see this, suppose $m = \{W_1, \ldots, W_n\}$ and $m'' = \{W_1', \ldots, W_k'\}$ are different. Then one can pick two distinct points p, q such that $p, q \in W_i$ for some i but there is no W_j' which contains both p and q. Clearly one has $\mathrm{met}(m)(p,q) = 0$ while $\mathrm{met}(m')(p,q) > 0$. This proves that $\mathrm{md}(m,m') = 0$ if and only if $m = m'$.

It remains to prove the triangular inequality. Let m_1, m_2, m_3 be three elements of PM(d). Then

$$\begin{aligned}
&\mathrm{md}(m_1, m_3)\\
&\quad = \sup_{x,y} |\mathrm{met}(m_1)(x,y) - \mathrm{met}(m_3)(x,y)|\\
&\quad \le \sup_{x,y} \left(|\mathrm{met}(m_1)(x,y) - \mathrm{met}(m_2)(x,y)| + |\mathrm{met}(m_2)(x,y) - \mathrm{met}(m_3)(x,y)|\right)
\end{aligned}$$

$$\leq \sup_{x,y} |\mathrm{met}(m_1)(x,y) - \mathrm{met}(m_2)(x,y)| + \sup_{x,y} |\mathrm{met}(m_2)(x,y) - \mathrm{met}(m_3)(x,y)|$$
$$= \mathrm{md}(m_1, m_2) + \mathrm{md}(m_2, m_3). \qquad \square$$

9.3 A spine for the complement of the discriminant locus

Let $\mathcal{P}_d$ be the space of monic centered polynomials of degree d, and $\mathcal{P}_d^0 \subset \mathcal{P}_d$ be the space of polynomials with distinct roots.

There is a natural map $f: \mathcal{P}_d^0 \to \mathrm{PM}(d)$ defined as follows: for $p \in P_d^0$ consider the meromorphic 1-form

$$\frac{1}{2d\pi i}\frac{d}{dz}\log p(z) = \frac{1}{2d\pi i}\frac{p'(z)}{p(z)}dz.$$

Denote by $Z(p)$ the set of roots of p. This 1-form gives $\mathbb{C} - Z(p)$ a Euclidean structure. Near infinity, we see a semi-infinite cylinder of circumference 1 (∞ is a simple pole of $d \log p$ with residue d) and near all the zeroes of p we see a semi-infinite cylinder with circumference $1/d$.

We restate Theorem 3.1, and give another proof here.

Theorem 9.2. *The map $f: \mathcal{P}_d^0 \to \mathrm{PM}(d)$ is a homotopy equivalence. More specifically, there exists a section $\sigma: \mathrm{PM}(d) \to \mathcal{P}_d^0$ which is a deformation retract.*

Proof. For $m \in \mathrm{PM}(d)$ consider the half-infinite cylinder $X'_m = (S^1 \times [0,\infty))/\sim$ where

$$(\theta_1, t_1) \sim (\theta_2, t_2) \iff (\theta_1, t_1) = (\theta_2, t_2) \quad \text{or} \quad t_1 = t_2 = 0 \text{ and } \theta_1 \sim_m \theta_2.$$

The graph quotient of $S^1 \times \{0\}$ by $\sim_m$ consists of d closed curves of length $1/d$. Glue to each of these closed curves a copy of $(\mathbb{R}/\frac{1}{d}\mathbb{Z}) \times (-\infty, 0)$, to construct X_m, which is a Riemann surface carrying a holomorphic 1-form ϕ_m given by $d\theta + idt$ on the upper cylinder. The integral of this 1-form around any of the d punctures at $-\infty$ is $1/d$, so we can define a function $p_m: X_m \to \mathbb{C}$

$$p_m(x) = e^{2\pi i d \int_{x_0}^{x} \phi_m},$$

well-defined up to post-multiplication by a constant depending on x_0. (Since the integral of ϕ_m along a loop around one of the lower cylinders $(\mathbb{R}/\frac{1}{d}\mathbb{Z}) \times (-\infty, 0)$ is $1/d$, p_m is well-defined.) But the endpoint compactification $\overline{X}_m$ of X_m is homeomorphic to the 2-sphere, and the complex structure extends to the endpoints, so $\overline{X}_m$ is analytically isomorphic to $\overline{\mathbb{C}}$. With this structure we see that p_m is a polynomial of degree d with distinct roots at the finite punctures. We can normalize p_m to be centered and monic by requiring that $1 \times [0,\infty]$ is mapped to a curve asymptotic to the positive real axis. The map $m \mapsto p_m$ gives the inclusion $\sigma: \mathrm{PM}(d) \to \mathcal{P}_d^0$.

We need to see that σ is a deformation retract. For each $p \in \mathcal{P}_d^0$, we consider the manifold with 1-form $(\mathbb{C} - Z(p), \phi_p)$, and adjust the heights of the critical values until they are all 0. $\square$

9.4 Polynomials in the escape locus

Again let $\mathcal{P}_d$ be the space of monic centered polynomials of degree d; this time they will be viewed as dynamical systems. For $p\in\mathcal{P}_d$ let G_p be the *Green's function* for the filled Julia set K_p.

Let $Y_d(r)\subset\mathcal{P}_d$ be the set of polynomials p such that $G_p(c)=r$ for all critical points of p. The set $Y_d(0)$ is the degree d *connectedness locus*; it is still poorly understood for all $d>2$.

For $r>0$, there is a natural map $Y_d(r)\to \mathrm{PM}(d)$ that associates to each polynomial $p\in\mathcal{P}_d$ the equivalence relation m_p on S^1 where two angles θ_1 and θ_2 are equivalent if the external rays at angles θ_1 and θ_2 land at the same critical point of p.

Theorem 9.3. *For $r>0$ and $p\in Y_d(r)$, the equivalence relation m_p is in* $\mathrm{PM}(d)$, *and the map $p\mapsto m_p$ is a homeomorphism $Y_d(r)\to$* $\mathrm{PM}(d)$.

The above theorem is a combination of a theorem of L. Goldberg [Gol94] and a theorem of Kiwi [Kiw05]. To see a more recent proof using quasiconformal surgery, the readers are referred to [Zen14].

9.5 Polynomials in the connectedness locus

For $m\in\mathrm{PM}(d)$, we geometrically identify it as the unions of convex hulls within $\overline{\mathbb{D}}$ of non-trivial equivalence classes of m. Let us denote by $J_1(m),\ldots,J_d(m)$ the open subsets of S^1 that are intersections with S^1 of the components of $\mathbb{D}\setminus m$; each $J_i(m)$ is some finite union of open intervals in S^1. An *attribution* will be a way of attributing each non-empty intersection $\overline{J_i(m)}\cap\overline{J_j(m)}$ to $J_i(m)$, or to $J_j(m)$, or to both. Call A such an attribution, and denote by J_i^A the interval J_i together with all the points attributed to it by A. Define the equivalence relation $\sim_{(m,A)}$ to be

$$\theta_1\sim_{m,A}\theta_2 \iff d^k\theta_1 \text{ and } d^k\theta_2 \text{ belong to the same } J_i^A \text{ for all } k\geq 0.$$

Suppose that some $p\in\mathcal{P}_d$ belongs to the connectedness locus without Siegel disks, and that K_p is locally connected, so that there is a Carathéodory loop γ_p: $\mathbb{R}/\mathbb{Z}\to\mathbb{C}$. Then γ_p induces the equivalence relation $\sim_p$ on $\mathbb{R}/\mathbb{Z}$ by $\theta_1\sim_p\theta_2$ if and only if $\gamma_p(\theta_1)=\gamma_p(\theta_2)$.

Theorem 9.4. *There exists $m\in\mathrm{PM}(d)$ and an attribution A such that the equivalence relation $\sim_p$ is precisely $\sim_{m,A}$.*

This theorem is essentially proved in [Zen15, Theorem 1.2]. Understanding when different $\sim_{m,A}$ correspond to the same polynomial is a difficult problem, even for quadratic polynomials.

Proof. (Sketch) Choose an external ray landing at each critical value in J_p, and an external ray landing at the root of each component of $\overset{\circ}{K}_p$ containing a critical value if the critical value is attracted to an attracting or parabolic cycle. Then for each critical point $c\in J_p$, the angles of the inverse images of the chosen rays landing at c form an equivalence class for (m,A).

For each critical point $c \in \overset{\circ}{K}_p$, the angles of the inverse images of the chosen rays landing on the component of $\overset{\circ}{K}_p$ containing c form the other equivalence classes. These are the ones that need to be attributed carefully. □

9.6 Shilov boundary of the connectedness locus

Conjecture 9.5. *All stretching rays through $Y_d(r)$ land on the Shilov boundary of the connectedness locus. Furthermore, if $r \to 0$, $Y_d(r)$ accumulates to the Shilov boundary.*

If this is true, it gives a description of the Shilov boundary of $Y_d(0)$, which is probably the best description of the connectedness locus we can hope for.

10 PARAMETRIZING PRIMITIVE MAJORS

As part of his investigations into core entropy, William P. Thurston wrote numerous Mathematica programs. This section presents an algorithm found in W. Thurston's computer code which cleverly parametrizes primitive majors using starting angles.

Denote by I the circle of unit length. We will interpret I as the fundamental domain $[0,1)$ in $\mathbb{R}/\mathbb{Z}$ with the standard ordering on $[0,1)$. Simultaneously, we will think of I as the space of angles of points in the boundary of the unit disk.

Throughout this section, we will let $m = \{W_1, \cdots, W_s\} \in \mathrm{PM}(d)$ denote a generic primitive major. By "generic," we mean that the associated lamination $\mathcal{M}$ consists of $d-1$ leaves and has no critical gaps. Each leaf ℓ_i in $\mathcal{M}$ has two distinct endpoints in I. We will call the lesser of these two points the *starting point* of ℓ_i and denote it s_i, and we will call the greater the *terminal point* of ℓ_i and denote it t_i. We will adopt the labelling convention that the labels of the leaves are ordered so that

$$s_1 < s_2 < \cdots < s_{d-1}.$$

Since, for each i, $d \cdot s_i \pmod 1 = d \cdot t_i \pmod 1$, there exists a unique natural number $k_i \in \{1, ..., d-1\}$ such that $t_i = s_i + \frac{k_i}{d}$; we will show how to find the k_i.

Each leaf $\ell_i \in \mathcal{M}$ determines two open arcs of I: $I_i = (s_i, t_i)$ and $I_i^0 = I \setminus [s_i, t_i]$. The complement in $\overline{\mathbb{D}}$ of the lamination $\mathcal{M}$ consists of d connected sets. We will adopt the notation that C_i is the connected component whose boundary contains ℓ_i and has non-empty intersection with the arc I_i, for $1 \le i \le d$, and C_0 is the connected set whose boundary contains an arbitrarily small interval $(1-\epsilon, 1) \subset I$. For each connected set C_i, denote by $\mu(C_i)$ the Lebesgue measure of the boundary of C_i in $I = \partial\overline{\mathbb{D}}$.

Lemma 10.1. *Let $m \in \mathrm{PM}(d)$ be a generic primitive major. Then $\mu(C_i) = 1/d$ for all i.*

(This lemma is used implicitly several times in Part I.)

Proof. Since for every leaf ℓ_i the lengths of the arcs I_i and I_i^0 are both integer multiples of $1/d$, $\mu(C_i)$ is also an integer multiple of $1/d$. Thus, we can write $\mu(C_i) = m_i/d$

for a unique natural number m_i. Then, since the C_i are pairwise disjoint,

$$1=\sum_{i=0}^{d-1}\mu(C_i)=\frac{1}{d}\sum_{i=0}^{d-1}m_i.$$

Consequently, $m_i=1$ for all i. □

Lemma 10.2. *Let $m\in \mathrm{PM}(d)$ be a generic primitive major. Then $s_{d-1}<\frac{d-1}{d}$ and $t_{d-1}=s_{d-1}+\frac{1}{d}$.*

Proof. First, we will prove that $s_{d-1}<\frac{d-1}{d}$. To see this, suppose $s_{d-1}\in[\frac{d-1}{d},1)$. We know t_{d-1} is the fractional part of $(s_{d-1}+\frac{k_{d-1}}{d}) \pmod 1$ for some $k_{d-1}\in\mathbb{N}$. Consequently, $t_{d-1}\notin(\frac{d-1}{d},1)$, contradicting the fact that $s_{d-1}<t_{d-1}$.

Now, suppose $s_{d-1}+\frac{1}{d}<t_{d-1}$. Since s_{d-1} is the biggest of the s's, there is no leaf in $\mathcal{M}$ whose starting point lies in the arc I_{d-1}. Therefore the boundary of C_{d-1} contains the entire arc I_{d-1}, and so $\mu(C_{d-1})>1/d$, a contradiction. □

Definition 10.3. *Let $m\in\mathrm{PM}(d)$ be a generic primitive major. The* derived primitive major *m' is an equivalence relation on I that is the image of m under the following process: collapse the interval $[s_{d-1},t_{d-1}]$ in I to a point, and then affinely reparametrize the quotient circle so that it has unit length, keeping the point 0 fixed.*

Lemma 10.4. *For any generic primitive major $m\in\mathrm{PM}(d)$, the derived primitive major m' is in $\mathrm{PM}(d-1)$.*

Proof. By Lemma 10.2, the arc $[s_{d-1},t_{d-1}]$ has length $\frac{1}{d}$, so the reparametrization affinely stretches the quotient circle by a factor of $\frac{d}{d-1}$. For $i\le d-2$, denote the image of s_i and t_i in $\mathcal{M}'$ by s_i' and t_i'. If $t_i-s_i=\frac{k_i}{d}$, then

$$t_i'-s_i'=\frac{k_i'}{d}\cdot\frac{d}{d-1}=\frac{k_i'}{d-1},$$

where $k_i'=k_i-1$ if $[s_{d-1},t_{d-1}]\subset[s_i,t_i]$ and $k_i'=k_i$ otherwise. In either case, $(d-1)\cdot(t_i'-s_i')=0 \pmod 1$. Hence $m'=\{(s_i',t_i')\,|\,i=1,\cdots,d-2\}$ is in $\mathrm{PM}(d-1)$. □

Lemma 10.5. *Let $m\in\mathrm{PM}(d)$ be a generic primitive major. Then $s_i<\frac{i}{d}$ for all i.*

Proof. Repeatedly deriving the major m yields a sequence of primitive majors

$$m^{(0)}(:=m),m^{(1)}(:=m'),m^{(2)},\dots,m^{(d-3)}.$$

The major $m^{(j)}$ consists of $d-1-j$ leaves, $\ell_1^{(j)},\dots,\ell_{d-1-j}^{(j)}$, that are the images under j derivations of the leaves $\ell_1^{(0)},\dots,\ell_{d-1-j}^{(0)}$ of the original major m. We will denote the starting point of the leaf $\ell_k^{(j)}$ by $s_k^{(j)}$.

We wish to show, for any fixed i, that $s_i^{(0)}<\frac{i}{d}$. The major $m^{(d-1-i)}$ consists of i leaves, of which $\ell_i^{(d-1-i)}$ has the largest starting point. Hence by Lemma 10.2, we have

$$s_i^{(d-1-i)} < \frac{i}{i+1}.$$

When deriving $m^{(j-1)}$ (which is in PM$(d-j+1)$) to form $m^{(j)}$, for any j, we collapse an interval of length $\frac{1}{d-j+1}$ and rescale by a factor of $\frac{d-j+1}{d-j}$. Thus,

$$\begin{aligned} s_i^{(d-1-i)} &= s_i^{(0)} \cdot \frac{d}{d-1} \cdot \frac{d-1}{d-2} \cdot \ldots \cdot \frac{d-(d-1-i)+1}{d-(d-1-i)} \\ &= s_i^{(0)} \cdot \frac{d}{i+1}. \end{aligned}$$

Hence

$$s_i^{(0)} \cdot \frac{d}{i+1} < \frac{i}{i+1},$$

implying

$$s_i^{(0)} < \frac{i}{d}.$$

□

Theorem 10.6. *Given any increasing sequence $0 \le s_1 < s_2 < \ldots < s_{d-1} < 1$ such that $s_i < \frac{i}{d}$ for all i, there is a unique degree-d-invariant primitive major m whose starting points are $s_1, \ldots, s_{d-1}$. There is an algorithm to find m.*

Proof. Let $m \in \mathrm{PM}(d)$ be any primitive major whose leaves have starting points $s_1, \ldots, s_{d-1}$; we will find possible values for the terminal points of each leaf of m are uniquely determined.

When we derive m n times, we collapse a union of arcs, namely $\bigcup_{i=1}^{n} cl(I_{d-i})$. The starting point $s_{d-1-n}^{(n)}$ is the biggest starting point of the resulting major, $m^{(n)}$. By Lemma 10.2, $t_{d-1-n}^{(n)} - s_{d-1-n}^{(n)} = \frac{1}{d-n}$. Reversing the rescaling process which at each derivation rescales the quotient circle to have unit length, an interval of length $\frac{1}{d-n}$ in $\mathcal{M}^{(n)}$ corresponds to an interval of length $\frac{1}{d}$ in the original major $\mathcal{M}$ when we measure the arcs which get collapsed as having length 0.

Thus, for each natural number $n < d-1$, we can choose t_{d-1-n} to be the smallest number in $[0,1)$ such that

$$m\left([s_{d-1-n}, t_{d-1-n}] \setminus \bigcup_{i=1}^{n} I_{d-i}\right) = \frac{1}{d}, \tag{1}$$

where m is Lebesgue measure. For $n=0$, we have $t_{d-1} = s_{d-1} + \frac{1}{d}$, and thus $I_{d-1} = (s_{d-1}, s_{d-1} + \frac{1}{d})$, by Lemma 10.2. Inductively, if $t_{d-1}, \ldots, t_{d-n}$ are known, Equation (1) gives t_{d-n-1}.

The resulting union of leaves might not be a primitive major, since in non-generic cases it might not come from an equivalence relation. Once we add in the missing leaves as necessary to make ideal polygons, the result is unique. (There might be more than one value for t_{d-1-n} satisfying Equation (1), but they are identified by the rest of the lamination.) □

Theorem 10.6 describes an algorithm which associates a primitive major to any ordered sequence of points $\{s_i\}_{i=1}^{i=d-1}$ such that $s_i < \frac{i}{d}$ and $s_i < s_{i+1}$ for all i. We now describe an algorithm used in W. Thurston's code for constructing such sequences of points from arbitrary collections of points.

Definition 10.7. *For any sequence $X = \{x_1, \ldots, x_{d-1}\}$ of $d-1$ distinct points x_i in I, define A to be the map*

$$A : X \mapsto Y$$

where $Y = \{y_1, \ldots, y_{d-1}\}$ is the sequence of $d-1$ points y_i in I defined by the following process:

1. Reorder and relabel (if necessary) the elements of the sequence X so that

$$x_1 < x_2 < \cdots < x_{d-1}.$$

2. Set

$$y_i = \begin{cases} x_i & \text{if } x_i < \frac{i}{d} \\ x_i - \frac{1}{d} & \text{if } x_i \geq \frac{i}{d}. \end{cases}$$

Theorem 10.8. *Let $X = \{x_1, \ldots, x_{d-1}\}$ be any sequence of $d-1$ points x_i in I. Then there exists $n \in \mathbb{N}$ such that $A^n(X) = A^m(X)$ for all $m \geq n$. For such an n, $A^n(X) = \{a_1, \ldots, a_{d-1}\}$ is a sequence of $d-1$ distinct points a_i in I such that*

$$a_1 < a_2 < \cdots < a_{d-1}$$

and $a_i < \frac{i}{d}$ for all i.

Proof. For convenience, denote by $B^j = \{b_1^j, \ldots, b_{d-1}^j\}$ the result of applying Step 1 of the definition of the map A to $A^j(X)$. Thus, B^j is just the sequence $A^j(X)$ reordered and relabeled so that $b_1^j < \cdots < b_{d-1}^j$. Notice that when passing from $A^j(X)$ to $A^{j+1}(X)$ (with B^j as an intermediate step), if we do not subtract $1/d$ from at least one of the elements b_i^j, then $B^j = A^m(X)$ for all $m > j$.

Since we only subtract $1/d$ from b_i^j if $b_i^j \geq i/d \geq 1/d$, every b_i^j is non-negative. Since $\sum_i b_i^0 < \infty$, we can only subtract $1/d$ finitely many times from elements of B^0 without some b_i^j becoming negative. Hence, there are only finitely many integers j such that we subtract $1/d$ from at least one element of B^j when passing from $A^j(X)$ to $A^{j+1}(X)$. Hence, there exists $n_0 \in \mathbb{N}$ such that $B^{n_0} = A^m(X)$ for all $m > n_0$. Set $n = n_0 + 1$. Then $B^{n_0} = A^n(X) = A^m(X)$ for all $m \geq n$.

By assumption, $b_1^{n_0} < \cdots < b_{d-1}^{n_0}$. Since we do not subtract $1/d$ from any $b_i^{n_0}$ when passing from $A^{n_0}(X)$ to $A^n(X)$, this means $b_i^{n_0} < i/d$ for all i. □

11 UNDERSTANDING PM(3)

11.1 Topology of PM(3)

We now explicitly describe PM(3). For brevity, we will say simply *a major* to denote a cubic primitive major throughout this section. One can represent points on the circle by their angle from the positive real axis. This angle is measured as the number

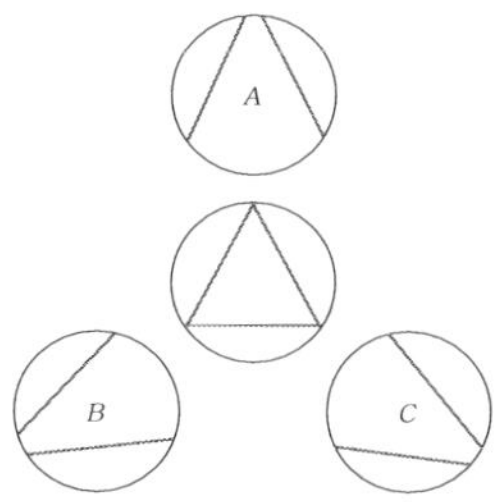

Figure 10: The three generic majors A, B and C are close to the degenerate major m shown in the center.

of turns we need to get to that point, i.e., as a number between 0 and 1. Let $m \in \mathrm{PM}(3)$. In the generic case, m has two leaves, each of which bounds one third of the circle. Assume that we start with a generic choice of a major and rotate it counterclockwise. We get a new major at each angle until we make one full turn. But there are two types of special cases to look at.

One case is when two major leaves share an endpoint. Putting an extra leaf connecting the non-shared endpoints of the major leaves, we get a regular triangle. In fact, which two sides of this regular triangle you choose as the major does not matter. Therefore, we get an extra symmetry in this case; if we rotate such a major, then it does not take a full turn to see the same major again: only a 1/3-turn is enough. Let's call the set of all majors of this type the *degeneracy locus*. One can think of this space as the space of regular triangles inscribed in the unit circle.

Another special case is when two major leaves are parallel (if drawn as straight lines). We call the set of all majors of this type the *parallel locus*. In this case, after making a half-turn, we see the same major again.

Starting from these two types of singular loci allows us to understand the topology of the space PM(3). Note that both the degeneracy locus and the parallel locus are topological circles. Let's see what a neighborhood of a point on the degeneracy locus looks like. From a major m on the degeneracy locus, there are three different ways to move into the complement: remove one of the sides of the regular triangle and open up the shared endpoint of the remaining two, as illustrated in Figure 10. They are all nearby, but there is no short path connecting two of A, B, C without crossing the degeneracy locus.

On the other hand, if you move along a locus consisting of majors that are distance ϵ away from the degeneracy locus for a small enough positive number ϵ and look at the closest degenerate major at each moment, then you will see each degenerate major exactly three times, as in Figure 11.

We can thus obtain a neighborhood of the degeneracy locus from a tripod cross an interval by gluing the tripod ends with a 2/3-turn. (We choose to rotate by 2/3 to agree with the other part of PM(3) below.) The boundary is again a topological circle which is embedded in $\mathbb{R}^3$ as a trefoil knot, as on the right side of Figure 12.

Now, we move on to the local picture around the parallel locus. Almost the same argument works, except now the situation is somewhat simpler and the neighborhood is homeomorphic to a Moebius band. One can embed this space into $\mathbb{R}^3$ so that the boundary is again a trefoil knot. (See the image on the left side of Figure 12.) Now the whole space PM(3) is obtained from these two spaces by gluing along the boundary. Figure 3 illustrates what the space looks like after this gluing.

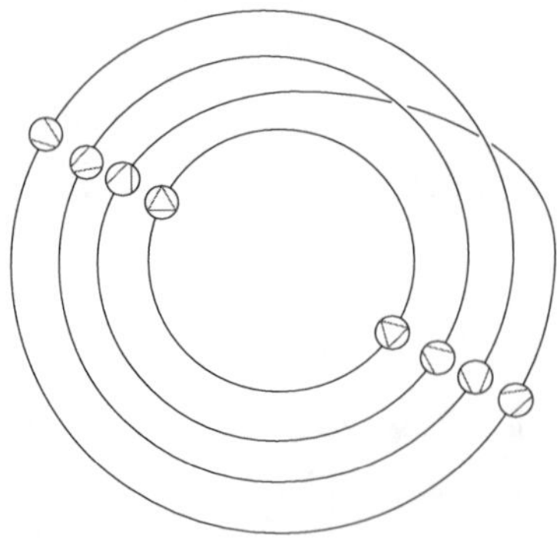

Figure 11: The degeneracy locus (the inner-most circle) and the locus of majors which are distance ϵ away from it.

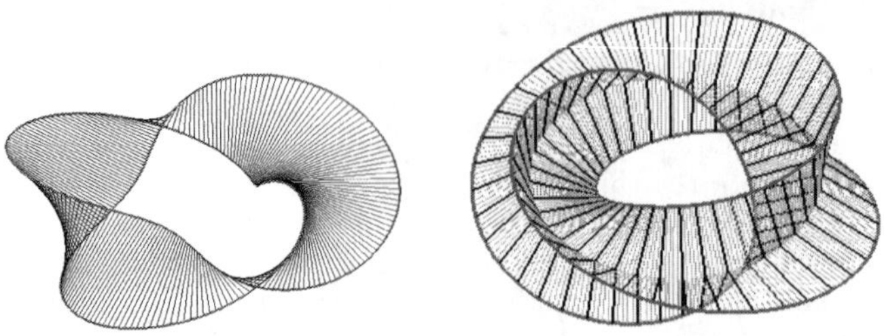

Figure 12: Neighborhoods of the parallel locus and the degeneracy locus.

Visualizing PM(3) by dividing into neighborhoods of two singular loci in this way allows one to see that PM(3) is a $K(B_3, 1)$-space, as follows.

We first construct the universal cover of PM(3). Write PM(3) as $A \cup B$, where A is the closure of the neighborhood of the degeneracy locus and B is the closure of the neighborhood of the parallel locus. We glue them in a way that $A \cap B = \partial A = \partial B$.

Now it is easy to see that $\widetilde{A}$ is just the product of a tripod with $\mathbb{R}$, and $\widetilde{B}$ is simply an infinite strip. $\widetilde{A}$ has three boundary lines each of which is glued to a copy of $\widetilde{B}$, and each end of each copy of $\widetilde{B}$, one needs to glue a copy of $\widetilde{A}$, and so on. To get $\widetilde{\mathrm{PM}(3)}$, we need to do this infinitely many times and finally get the product of an infinite trivalent tree with $\mathbb{R}$, which is obviously contractible. Therefore, all the higher homotopy groups of PM(3) vanish.

On the other hand, the Seifert–van Kampen theorem says that

$$\begin{aligned}\pi_1(\mathrm{PM}(3)) &= \pi_1(A) *_{\mathbb{Z}} \pi_1(B)\\ &= \langle \alpha, \beta \mid \alpha^3 = \beta^2 \rangle\end{aligned}$$

which is one presentation for B_3.

11.2 Parametrization of PM(3) using the angle bisector

Let m be a non-degenerate cubic major. The endpoints of leaves of m divide the circle into four arcs, two of them with length $1/3$ and the other two have length

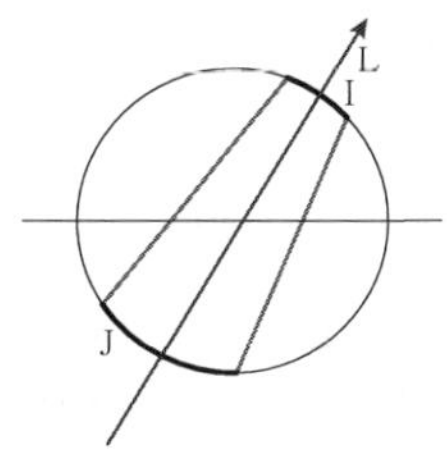

Figure 13: A cubic major (red) and its angle bisector (blue).

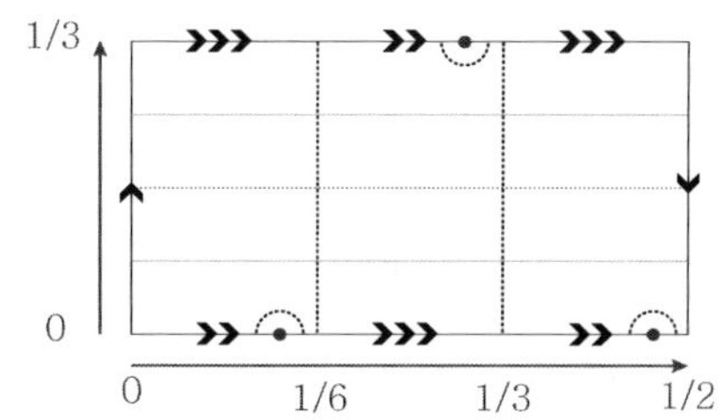

Figure 14: The parameter space for cubic majors using the angle bisector.

between 0 and 1/3. Call these other two arcs I and J. Draw a line L passing through the midpoint of I and the midpoint of J, as in Figure 13. Let θ be the angle from the positive real axis to L. Relabeling I and J if necessary, let I be the interval that L meets at the angle θ, and let a be the length of the interval I. Our parameters are a and θ. Note that we can choose θ from $[0, 1/2]$, since (a, θ) represents the same major as $(1/3 - a, \theta - 1/2 \mod 1)$. Also note that a runs from 0 to 1/3. Hence, the set $\{(a, \theta) : 0 \leq a \leq 1/3, 0 \leq \theta \leq 1/2\}$ with appropriate identifications on the boundary gives a parameter space of PM(3) (see Figure 14).

When a is either 0 or 1/3, either I or J becomes a single point, and this corresponds to the degeneracy locus. The locus where $a = 1/6$ is the parallel locus (the red line in Figure 14).

It is easy to see that $(a, 0)$ and $(1/3 - a, 1/2)$ represent the same major. Also observe that, for any $0 \leq \theta \leq 1/6$, the pairs $(0, \theta)$, $(1/3, \theta + 1/6)$, and $(0, \theta + 1/3)$ are just different choices of two edges of the regular triangle whose vertices are at $(\theta, \theta + 1/6, \theta + 1/3)$. Hence they must be identified. Similarly, for $1/6 \leq \theta \leq 1/3$, $(1/3, \theta - 1/6)$, $(0, \theta)$, and $(1/3, \theta + 1/6)$ represent the same point. For instance, the three large dots in Figure 14 should be identified.

11.3 Embedding of PM(3) into S^3

We can visualize how PM(3) embeds into S^3. Consider the decomposition of S^3 into two solid tori glued along the boundary. Put the degeneracy and parallel locus as central circles of the solid tori. Seeing S^3 as $\mathbb{R}^3$ with a point at infinity, one may assume that the parallel locus coincides with the unit circle on xy-plane and the degeneracy locus is the z-axis with the point at infinity. Then one can view PM(3) with one point removed as a 2-complex in $\mathbb{R}^3$.

We already know that PM(d) embeds as a spine of the complement of the discriminant locus, so it would be instructive to see the discriminant locus in this picture. A monic centered cubic polynomial is written as z^3+az+b for some complex numbers a, b. Hence the space of all such polynomials can be seen as $\mathbb{C}^2$. The unit sphere is the locus $|a|^2+|b|^2=1$, and the discriminant locus is $4a^3+27b^2=0$. The intersection of these two loci is a trefoil knot. We will embed PM(3) into S^3 so that it forms a spine of the complement of this trefoil knot, and thus also as a spine of the complement of the discriminant locus in $\mathbb{C}^2$.

Consider the stereographic projection $\phi: S^3\setminus\{(0,0,0,1)\}\to\mathbb{R}^3$ defined by

$$\phi(x_1,x_2,x_3,x_4)=\left(\frac{x_1}{1-x_4},\frac{x_2}{1-x_4},\frac{x_3}{1-x_4}\right).$$

Instead of taking the line segment connecting a point on the unit circle of xy-plane and the z-axis, we first take the preimages of these two points under ϕ and consider the great circle passing through them in S^3. Then we take the image of this great circle under ϕ. While one wraps up the parallel locus twice and the degeneracy locus three times, we construct a surface as the trajectory of the image of the great circle passing through the preimages of the points on the parallel and the degeneracy locus. Now it is guaranteed to be an embedded 2-complex (not a manifold, since the degeneracy locus is singular) by construction.

Let's return to the space of normalized cubic polynomials

$$\{\, z^3+az+b \mid a,b\in\mathbb{C}, |a|^2+|b|^2=1\,\}.$$

We can identify the degeneracy locus inside this space as the subset cut out by $a=0$, and the parallel locus as the subset cut out by $b=0$. To connect the degeneracy locus to the parallel locus by spherical geodesics, running three times around the degeneracy locus while running twice around the parallel locus, we look at the subset

$$\begin{aligned}\mathrm{PM}(3)&=\{\, z^3+s\theta^2+t\theta^3 \mid s,t\in\mathbb{R}_{\geq 0}, \theta\in S^1, |s\theta^2|^2+|t\theta^3|^2=1\,\}\\&=\left\{\, z^3+az^2+bz^3 \,\middle|\, a,b\in\mathbb{C}, |a|^2+|b|^2=1, \frac{a^3}{b^2}\in[0,\infty]\,\right\}.\end{aligned}$$

This is clearly disjoint from the discriminant locus.

Recall that in the proof of Theorem 9.2, we constructed a section $\sigma:\mathrm{PM}(d)\to\mathcal{P}_d^0$ where $\mathcal{P}_d^0$ is the space of all monic centered polynomials of degree d with distinct roots. The above embedding of PM(3) gives another section into $\mathcal{P}_3^0$, but it is not exactly the same as σ in Theorem 9.2. For instance, σ has the property that for each polynomial f in the image of PM(3) under σ, all the critical values of f have the same modulus, while the above embedding does not have this property.

11.4 Other parametrizations

It is not a priori clear that the angle bisector parametrization of PM(3) can be generalized to the parametrization of PM(d) for higher d. In this section, we briefly survey several different ways of parametrizing PM(3) and see the advantages and disadvantages of each method.

11.4.1 Starting point method

Here we start with the most naive way to parametrize the space of primitive majors, from Section 10. Start from angle 0 and walk around the circle until you meet an end of a major leaf, and say the angle is x_1. Keep walking until you meet an end of different major leaf, say the angle is x_2. The numbers x_1, x_2 are regarded as the starting points of the major leaves of given cubic major. We have two cases: either $x_2 - x_1 < 1/3$ and the leaves are $\{(x_1, x_1 - 1/3 \mod 1), (x_2, x_2 + 1/3)\}$ or $x_2 - x_1 \geq 1/3$ and the leaves are $\{(x_1, x_1 + 1/3), (x_2, x_2 + 1/3)\}$. As you see, it is fairly easy to get a neat formula for leaves. x_1 has the range from 0 to 1/3 and x_2 has the range from 0 to 2/3. But not every point in the rectangle $[0, 1/3] \times [0, 2/3]$ is allowed. First of all, there is a restriction $x_2 \geq x_1$, and sometimes even $x_2 \geq x_1 + 1/3$. So, this parametrization method is not as neat as the rectangular parameter space obtained by the angle bisector method. Another issue is that there are many more combinatorial possibilities in higher degrees. We will see this in more detail while we discuss the next method.

11.4.2 Sum and difference of the turning number

Given a cubic lamination, start from angle 0 and walk around the circle until the first time you meet two consecutive ends $x < y$ belonging to distinct leaves. We call these numbers x, y turning numbers of the given lamination. Let $S = y + x$ and $D = y - x$ (S stands for the "sum" and D stands for the "difference"). Then one can easily get $x = (S - D)/2$ and $y = (S + D)/2$, and the leaves are $\{x, x - 1/3 \mod 1\}, \{y, y + 1/3 \mod 1\}$. Note that D runs from 0 to 1/3 and for a given D, S runs from D to $4/3 - D$. Hence we get a trapezoid shape domain for the parameter space and one can figure out which points on the boundary are identified as we did for the other models.

It is pretty clear what each parameter means and one gets a neat formula for the leaves. On the other hand, it has some drawbacks when one tries to generalize to higher-degree cases. Even for degree 4, the complement of the degeneracy locus in PM(4) is not connected. (This is what we postponed discussing in the last subsection. See Figure 15.) Hence, however one defines the turning numbers, it is hard to determine which configuration one has. One can divide the domain into pieces, each of which represents one combinatorial configuration, and give a different formula for each such piece. This requires understanding the different possible configurations. In particular, one can start with counting the number of connected components of the complement of the degeneracy locus in PM(d). Tomasini counted the number of components in his thesis (see Theorems 4.3.1 and 4.3.2, pp. 118–121, in [Tom14]).

11.4.3 Avoiding the Moebius band

One way to avoid the Moebius band is this: if you take the quotient of the set of majors by the symmetry $z \leftrightarrow -z$ that conjugates a cubic polynomial to a dynamically isomorphic polynomial, then the Moebius band folds in half to an annulus. One boundary component of the annulus is wrapped three times around a circle (laminations with a central lamination), and the other boundary component consists of laminations whose majors are parallel. The parameter transverse to the annulus is the shortest distance between endpoints of the majors, in the interval [0,1/6]. The other parameter is the most clockwise endpoint of this shortest distance interval.

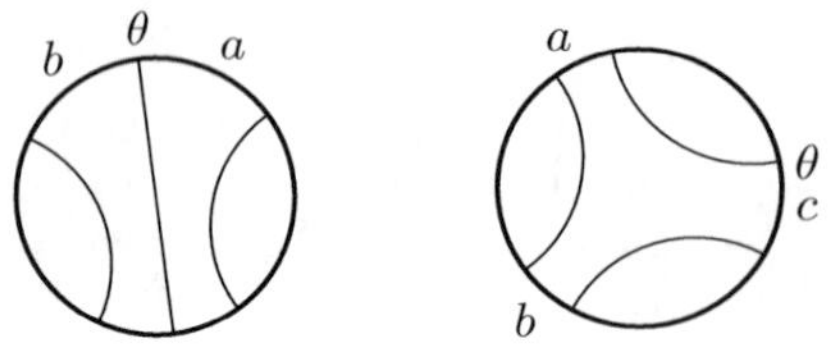

Figure 15: The two topological types of generic primitive majors in PM(4), with the parameters marked.

12 UNDERSTANDING PM(4)

We now give some brief comments on the shape of PM(4). Unlike in PM(2) and PM(3), there is more than one way that a generic primitive major can be arranged topologically: the complement of the degeneracy locus is not connected. The two possibilities are illustrated in Figure 15.

PM(4) is a 3-complex, and each of these topological types of generic majors contributes a top-dimensional stratum of the 3-complex. The *parallel stratum* is the piece corresponding to the stratum on the left, containing the part where the three leaves are parallel. Topologically, the parallel stratum may be parametrized by triples (θ, a, b) with θ in the circle and $a, b \in (0, 1/4)$. Here θ is the angle to one endpoint of the central leaf and $\theta - a$ and $\theta + b$ are the angles to adjacent endpoints of the other two leaves, so that the three leaves have endpoints

$$(\theta - a, \theta - a - 1/4), \quad (\theta, \theta + 1/2), \quad \text{and} \quad (\theta + b, \theta + b + 1/4)$$

(with coordinates interpreted modulo 1). There is an equivalence relation:

$$(\theta, a, b) \equiv (\theta + 1/2, 1/4 - b, 1/4 - a).$$

Therefore, the parallel stratum topologically is a square cross an interval, with the top glued to the bottom by a half-twist. As a manifold with corners, this stratum has 2 codimension-1 faces, each an annulus.

The other stratum, the *triangle stratum*, can be parametrized by quadruples (θ, a, b, c) with $a, b, c > 0$, $a + b + c = 1/4$, and θ in the circle. Here a, b, and c are the lengths of the three intervals on the boundary of the central gap, and θ is the angle to the start of one leaf, so that the three leaves have endpoints

$$(\theta, \theta + 1/4), \quad (\theta + 1/4 + a, \theta + 1/2 + a), \quad \text{and} \quad (\theta - 1/4 - c, \theta - c).$$

(The last leaf is also $(\theta + 1/2 + a + b, \theta + 3/4 + a + b)$.) Again, there is an equivalence relation:

$$(\theta, a, b, c) \equiv (\theta + 1/4 + a, b, c, a) \equiv (\theta + 1/2 + a + b, c, a, b).$$

This stratum is therefore topologically a triangle cross an interval, with the top glued to the bottom by a 1/3 twist. As a manifold with corners, it has only one codimension-1 face, an annulus.

We next turn to the codimension-1 degeneracy locus. Here there is only one topological type (the stratum is connected), consisting of a triangle and another leaf. As illustrated in Figure 16, the major can be uniquely parametrized by an angle θ,

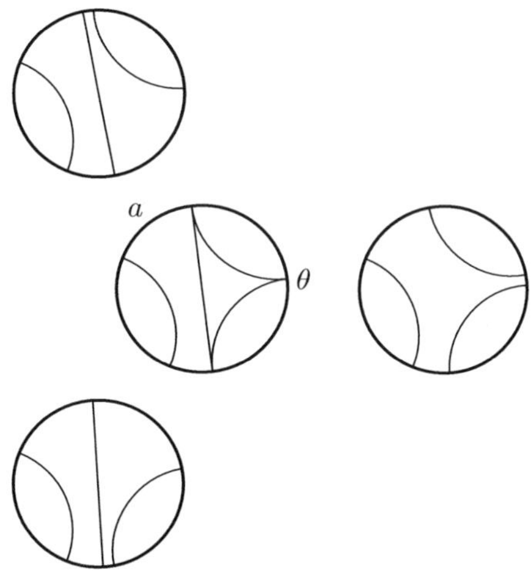

Figure 16: A codimension-1 primitive major in PM(4) (in the center) and three perturbations to generic majors.

the angle to the vertex of the triangle opposite the leaf, and a number $a \in (0, 1/4)$, the length of one of the intervals on the boundary of the gap between the triangle and the leaf. The codimension-1 degeneracy locus is therefore an annulus.

There are three ways to perturb a codimension-1 degenerate major into generic majors. Note that two of the generic majors are in the parallel stratum and one is in the triangle stratum. Thus, all three annuli that we found on the boundary of the top-dimensional strata are glued together.

13 THURSTON'S ENTROPY ALGORITHM AND ENTROPY ON THE HUBBARD TREE

To any rational angle θ (mod 1), Douady-Hubbard [DH84] associated a unique postcritically finite quadratic polynomial $f_\theta : z \mapsto z^2 + c_\theta$. This polynomial induces a Markov action on its Hubbard tree. The topological entropy [AKM65] of the polynomial f_θ on its Hubbard tree is called the *core entropy* of f_θ.

In order to combinatorially encode and effectively compute the core entropy, W. Thurston developed an algorithm that takes θ as its input, constructs a non-negative matrix A_θ (bypassing f_θ), and outputs its Perron-Frobenius leading eigenvalue $\rho(A_\theta)$. We will prove that $\log(\rho(A_\theta))$ is the core entropy of f_θ.

More precisely, we will define the notions in the following diagram and establish the equality on the right column (Theorem 13.10):

$$\begin{array}{ccc} \text{rational } \theta & \xrightarrow{\text{Thurston's entropy algorithm}} & \log \rho(A_\theta) \\ {\scriptstyle (DH)}\downarrow & & \Big\| \ {\scriptstyle \text{equality by Thm. 13.10}} \\ f_\theta & \xrightarrow[\text{entropy on Hubbard tree}]{} & h(\mathcal{H}_\theta, f_\theta). \end{array}$$

13.1 Thurston's entropy algorithm

Set $\mathbb{S}=\mathbb{R}/\mathbb{Z}$. All angles in this section are considered to be mod 1, i.e., elements of $\mathbb{S}$.

Let $\tau:\mathbb{S}\to\mathbb{S}$ denote the angle doubling map. An angle θ is *periodic* under the action of τ if and only if it is rational with odd denominator, and (strictly) *preperiodic* if and only if it is rational with even denominator.

Fix a rational angle $\theta\in\mathbb{S}\setminus\{0\}$. If θ is periodic, exactly one of $(\theta+1)/2$ and $\theta/2$ is periodic and the other is preperiodic. If θ is preperiodic, both $(\theta+1)/2$ and $\theta/2$ are preperiodic. Set $2^{-1}\theta$ to be the periodic angle, if it exists, among $(\theta+1)/2$ and $\theta/2$, and otherwise set it to be $\theta/2$. Define the set

$$O_\theta:=\left\{\{2^n\theta,2^l\theta\}\,\middle|\, l,n\geq -1 \text{ and } 2^n\theta\neq 2^l\theta\right\},$$

with the convention that the pairs $\{2^n\theta,2^l\theta\}$ that constitute O_θ are unordered sets. We divide the circle $\mathbb{S}$ at the points $\{\theta/2,(\theta+1)/2\}$, forming two closed half circles, with the boundary points belonging to both halves.

Define Σ_θ to be the abstract linear space over $\mathbb{R}$ generated by the elements of O_θ. Define a linear map $\mathcal{A}_\theta:\Sigma_\theta\to\Sigma_\theta$ as follows. For any basis vector $\{a,b\}\in O_\theta$, if a and b are in a common closed half circle, set $\mathcal{A}_\theta(\{a,b\})=\{2a,2b\}$; otherwise set $\mathcal{A}_\theta(\{a,b\})=\{2a,\theta\}+\{\theta,2b\}$. Denote by A_θ the matrix of $\mathcal{A}_\theta$ in the basis O_θ; it is a non-negative matrix. Denote its leading eigenvalue, which exists by the Perron-Frobenius theorem, by $\rho(A_\theta)$. It is easy to see that A_θ is not nilpotent, so $\rho(A_\theta)\geq 1$.

Definition 13.1. Thurston's entropy algorithm *is the map*

$$(\mathbb{Q}\cap\mathbb{S}\setminus\{0\})\ni\theta\mapsto\log\rho(A_\theta).$$

We will relate the output of Thurston's entropy algorithm, $\log\rho(A_\theta)$, to quadratic polynomials in Subsection 13.3.

Example 13.2. Set $\theta=\frac{1}{5}$. The abstract linear space Σ_θ has basis

$$O_{\frac{1}{5}}=\{\{\tfrac{1}{5},\tfrac{2}{5}\},\{\tfrac{1}{5},\tfrac{3}{5}\},\{\tfrac{1}{5},\tfrac{4}{5}\},\{\tfrac{2}{5},\tfrac{3}{5}\},\{\tfrac{2}{5},\tfrac{4}{5}\},\{\tfrac{3}{5},\tfrac{4}{5}\}\}.$$

We divide the circle $\mathbb{S}$ by the pair $\{\frac{1}{10},\frac{3}{5}\}$. The linear map $\mathcal{A}_{\frac{1}{5}}$ acts on the basis vectors as follows:

$$\left\{\frac{1}{5},\frac{2}{5}\right\}\mapsto\left\{\frac{2}{5},\frac{4}{5}\right\},\quad \left\{\frac{1}{5},\frac{3}{5}\right\}\mapsto\left\{\frac{2}{5},\frac{1}{5}\right\},\quad \left\{\frac{1}{5},\frac{4}{5}\right\}\mapsto\left\{\frac{1}{5},\frac{2}{5}\right\}+\left\{\frac{1}{5},\frac{3}{5}\right\},$$
$$\left\{\frac{2}{5},\frac{3}{5}\right\}\mapsto\left\{\frac{4}{5},\frac{1}{5}\right\},\quad \left\{\frac{2}{5},\frac{4}{5}\right\}\mapsto\left\{\frac{4}{5},\frac{1}{5}\right\}+\left\{\frac{1}{5},\frac{3}{5}\right\},\quad \left\{\frac{3}{5},\frac{4}{5}\right\}\mapsto\left\{\frac{1}{5},\frac{3}{5}\right\}.$$

We then compute $\log\rho(A_\theta)=0.3331$.

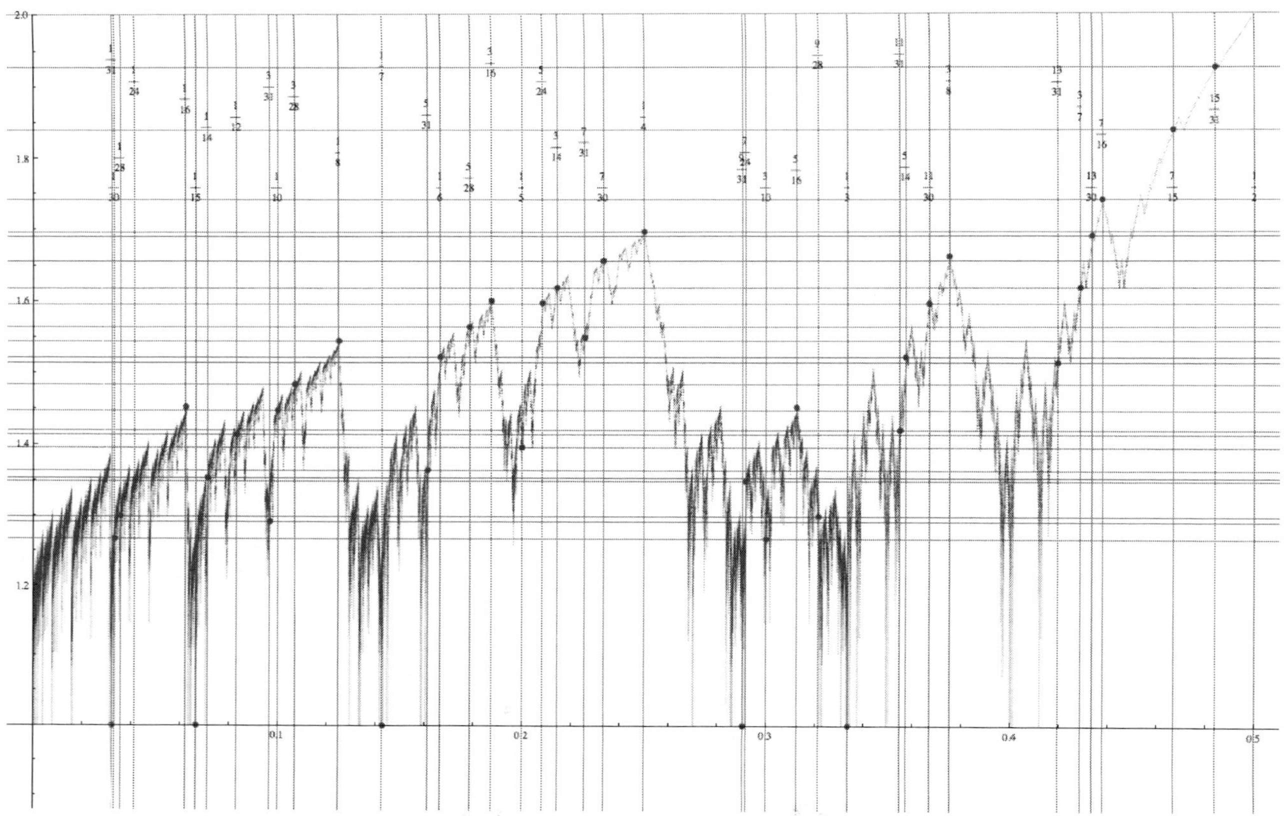

Figure 17: W. Thurston's first plot of core entropy. The horizontal axis is the θ-axis, for rational $\theta \in [0, 1/2]$ (half of the unit circle), and the vertical axis is core entropy, or $\log \rho(A_\theta)$.

13.2 Hubbard trees following Douady and Hubbard

We now recall background material about Hubbard trees, used in later sections to justify Thurston's entropy algorithm as computing core entropy. (See, for example, [DH84, Poi09, Poi10] for additional information about Hubbard trees.)

Let f be a postcritically finite polynomial, i.e., a polynomial all of whose critical points have a finite (and hence periodic or preperiodic) orbit under f. By classical results of Fatou, Julia, Douady, and Hubbard, the *filled Julia set*

$$\mathcal{K}_f = \{z \in \mathbb{C} \mid f^n(z) \not\to \infty\}$$

is compact, connected, locally connected and locally arc-connected. These conditions also hold for the *Julia set* $\mathcal{J}_f := \partial \mathcal{K}_f$. The Fatou set $\mathcal{F}_f := \overline{\mathbb{C}} \setminus \mathcal{J}_f$ consists of one component $U(\infty)$ which is the basin of attraction of ∞, and at most countably many bounded components constituting the interior of $\mathcal{K}_f$. Each of the sets $\mathcal{K}_f, \mathcal{J}_f, \mathcal{F}_f$ and $U(\infty)$ is fully invariant by f; each Fatou component is (pre)periodic (by Sullivan's no-wandering domain theorem, or by hyperbolicity of the map); and each periodic cycle of Fatou components contains at least one critical point of f (counting ∞).

There is a system of Riemann mappings

$$\left\{\phi_U : \mathbb{D} \to U \,\middle|\, U \text{ Fatou component}\right\}$$

each extending to a continuous map on the closure $\overline{\mathbb{D}}$, so that for all U and some d_U, the following diagram commutes:

$$\begin{array}{ccc} \overline{\mathbb{D}} & \xrightarrow{z \mapsto z^{d_U}} & \overline{\mathbb{D}} \\ \phi_U \downarrow & & \downarrow \phi_{f(U)} \\ \overline{U} & \xrightarrow[f]{} & \overline{f(U)}. \end{array}$$

In particular, on every periodic Fatou component U, including $U(\infty)$, the map ϕ_U realizes a conjugacy between a power map and the first return map on U. The image in U under ϕ_U of radial lines in $\mathbb{D}$ are, by definition, *internal rays* on U if U is bounded and *external rays* if $U = U(\infty)$. Since a power map sends a radial line to a radial line, the polynomial f sends an internal/external ray to an internal/external ray.

If U is a bounded Fatou component, then $\phi_U : \overline{\mathbb{D}} \to \overline{U}$ is a homeomorphism, and thus every boundary point of U receives exactly one internal ray from U. This is in general not true for $U(\infty)$, where several external rays may land at a common boundary point.

Definition 13.3 (Supporting rays). *We say that an external ray R* supports *a bounded Fatou component U if*

1. *the ray lands at a boundary point q of U, and*
2. *there is a sector based at q delimited by R and the internal ray of U landing at q which does not contain other external rays landing at q.*

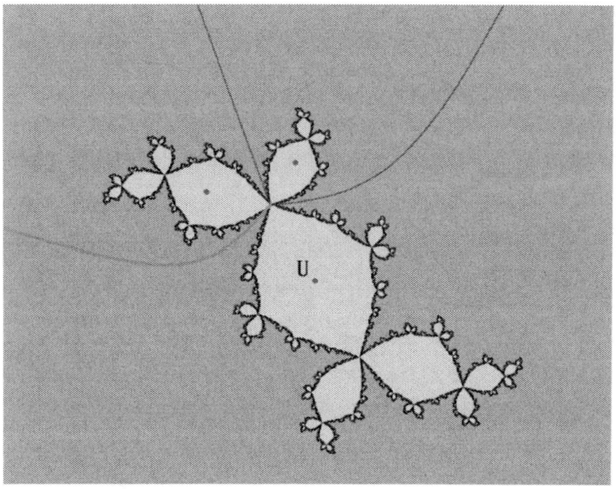

Figure 18: The green (left) and red (right) rays are supporting rays of the Fatou component U, but the blue (middle) one is not.

It follows from Definition 13.3 that for any bounded Fatou component U and point $z \in \partial U$ there are at most two external rays which support U and land at z. Start from the internal ray in U which lands at z and turn in the counterclockwise direction centered at z. The first (resp. last) encountered external ray landing at z is called the *right-supporting ray* (resp. *left-supporting ray*) of U at z. See Figure 18.

The system of internal/external rays does not depend on the possible choices of ϕ_U. If $f: z \mapsto z^2 + c$ and f is postcritically finite, there is actually a unique choice of ϕ_U for each Fatou component U. In particular, $\phi_{U(\infty)}$ conjugates z^2 to f and ϕ_U conjugates z^2 to f^p if U is a bounded periodic Fatou component and p is the minimal integer such that $f^p(0) = 0$ (if no such p exists, $\mathcal{K}_f = \mathcal{J}_f$). In this case, for any $x \in \mathbb{S}$ we use $\mathcal{R}_f(x)$ or simply $\mathcal{R}(x)$ to denote the image under $\phi_{U(\infty)}$ of the ray $\{re^{2\pi i x}, 0 < r < 1\}$ and will call it the *external ray of angle* x. Angles of internal rays can be defined similarly. We also use $\gamma(x) = \phi_{U(\infty)}(e^{2\pi i x})$ to denote the *landing point* of the ray $\mathcal{R}(x)$.

Any pair of points in the closure of a bounded Fatou component can be joined in a unique way by a Jordan arc consisting of (at most two) segments of internal rays. We call such arcs *regulated* (following Douady and Hubbard). Since $\mathcal{K}_f$ is arc-connected, given two points $z_1, z_2 \in \mathcal{K}_f$, there is an arc $\gamma: [0,1] \to \mathcal{K}_f$ such that $\gamma(0) = z_1$ and $\gamma(1) = z_2$. In general, we will not distinguish between the map γ and its image. It is proved in [DH84] that such arcs can be chosen in a unique way so that the intersection with the closure of every Fatou component is regulated. We still call such arcs regulated and denote them by $[z_1, z_2]$. We say that a subset $X \subset \mathcal{K}_f$ is *allowably connected* if for every $z_1, z_2 \in X$ we have $[z_1, z_2] \subset X$.

Definition 13.4. *We define the* regulated hull *of a subset X of $\mathcal{K}_f$ to be the minimal closed allowably connected subset of $\mathcal{K}_f$ containing X.*

Proposition 13.5. *For a collection of $z_1, \dots, z_n$ finitely many points in $\mathcal{K}_f$, their regulated hull is a finite tree.*

Definition 13.6. *Let f be a postcritically finite polynomial. The* postcritical set *$\mathcal{P}_f$ is defined to be*[2]

$$\mathcal{P}_f = \left\{ f^n(c) \,\middle|\, f'(c) = 0, n \geq 0 \right\}.$$

The Hubbard tree *$\mathcal{H}_f$ is defined to be the regulated hull of the finite set $\mathcal{P}_f$.*

The *vertex set* $V(\mathcal{H}_f)$ of $\mathcal{H}_f$ is the union of $\mathcal{P}_f$ together with the branching points of $\mathcal{H}_f$, namely the points p such that $\mathcal{H}_f \setminus \{p\}$ has at least three connected components. The closure of a connected component of $\mathcal{H}_f \setminus V(\mathcal{H}_f)$ is called an *edge.*

Lemma 13.7. *For a postcritically finite polynomial f, the set $\mathcal{H}_f$ is a tree with finitely many edges. Moreover $f(\mathcal{H}_f) \subset \mathcal{H}_f$ and $f : \mathcal{H}_f \to \mathcal{H}_f$ is a Markov map (as defined in Appendix A).*

Definition 13.8. *For a polynomial f such that its Hubbard tree exists and is a finite tree, the* core entropy *of f is the topological entropy of the restriction of f to its Hubbard tree, $h(\mathcal{H}_f, f)$.*

Using Proposition A.6, we may relate the topological entropy of f on $\mathcal{H}_f$ to the spectral radius of a transition matrix D_f constructed from f by $h(\mathcal{H}_f, f) = \log \rho(D_f)$.

We remark that if a polynomial f has a connected and locally connected filled Julia set such that the postcritical set lies on a finite topological tree, then its Hubbard tree exists. This condition is called "topologically finite" by Tiozzo [Tio15], and is more general than postcritically finite.

13.3 Relating Thurston's entropy algorithm to polynomials

Thurston's entropy algorithm effectively computes the topological entropy $h(\mathcal{H}_f, f)$ for any postcritically finite polynomial without actually computing the Hubbard tree. We will see how to relate the quadratic version of the algorithm given in Section 13.1 to quadratic polynomials.

On one hand, Thurston's entropy algorithm produces a quantity, $\log \rho(A_\theta)$, from any given rational angle θ. On the other hand, Douady-Hubbard defined a finite-to-one map $\mathbb{Q} \ni \theta \mapsto c_\theta$ so that the quadratic polynomial $z \mapsto z^2 + c_\theta$ is postcritically finite. More precisely:

Theorem 13.9 (Douady-Hubbard). *If $\theta \in \mathbb{Q}$ is preperiodic (resp. p-periodic) under the angle doubling map, there is a unique parameter c_θ such that for $f : z \mapsto z^2 + c_\theta$ both external rays $\mathcal{R}(\frac{\theta}{2})$ and $\mathcal{R}(\frac{\theta+1}{2})$ land at 0, and 0 is preperiodic (resp. support the Fatou component containing 0, and 0 is p-periodic). Furthermore, every postcritically finite quadratic polynomial arises in this way.*

Our objective is to establish:

Theorem 13.10. *For θ a rational angle, $\log \rho(A_\theta) = h(\mathcal{H}_f, f)$ for $f : z \mapsto z^2 + c_\theta$.*

[2]For the purpose of this section, we include critical points in the postcritical set.

Proof. The idea of the proof is the following, inspired by [Gao13, Jun14]:

1. Construct a topological graph G and a Markov action $L: G \to G$ such that the spectral radius of the transition matrix $D_{(G,L)}$ is equal to the spectral radius of A_θ.
2. Construct a continuous, finite-to-one, and surjective semi-conjugacy Φ from $L: G \to G$ to $f: \mathcal{H}_f \to \mathcal{H}_f$.

Then we may conclude that

$$\log \rho(A_\theta) \overset{\text{same matrix}}{=} \log \rho(D_{(G,L)}) \overset{\text{Prop. A.6}}{=} h(G,L) \overset{\text{Prop. A.3}}{=} h(\mathcal{H}_f, f).$$

Let G be a topological complete graph whose vertex set is the forward orbit after identifying the diagonal angles $\frac{\theta}{2}$ and $\frac{\theta+1}{2}$:

$$V_G = \{2^n\theta, n \geq -1\} \Big/ \left(\frac{\theta}{2} \sim \frac{\theta+1}{2}\right).$$

The set of edges of G is

$$E_G = \{\, e(x,y) |\; x \neq y \in V_G\}$$

(with $e(x,y) = e(y,x)$). Being a topological graph means that G is a topological space and each edge $e(x,y)$ is homeomorphic to a closed interval with ends x and y. Thus E_G is in bijection with the set O_θ indexing A_θ.

Mimicking the action of the linear map A_θ, we can define a piecewise monotone map $L: G \to G$ as follows. On vertices, $L = \tau$. Let x, y be two distinct vertices in V_G. If x, y belong to the same closed half circle, i.e., the closure of a complete component of $\frac{\theta}{2}, \frac{\theta+1}{2}$, then $\tau(x) \neq \tau(y) \in V_G$. In this case, let L map the edge $e(x,y)$ homeomorphically onto the edge $e(\tau(x), \tau(y))$. If x, y belong to distinct open half circles, subdivide the edge $e(x,y)$ into non-trivial arcs $e(x,z)$ and $e(z,y)$, and let L map the arc $e(x,z)$ (resp. $e(z,y)$) homeomorphically onto $e(\tau(x), \theta)$ (resp. $e(\theta, \tau(y))$).

It is easy to see that the transition matrix of (G,L) is exactly A_θ. By Proposition A.6, the topological entropy $h(G,L)$ is always equal to $\log \rho(A_\theta)$, regardless of the precise choices of L as homeomorphisms on the edges.

However, to relate $h(G,L)$ to $h(\mathcal{H}_f, f)$, we will redefine the homeomorphic action of L on each edge by lifting corresponding actions of f via suitably defined conjugacies.

We will treat the periodic and preperiodic cases separately.

Case 1: θ is periodic

In this case the rays of angles $\frac{\theta}{2}$ and $\frac{\theta+1}{2}$ both land at the boundary of the Fatou component U containing 0 and support that component. If $\frac{\theta}{2}$ is periodic then the rays are right-supporting rays, and if it is $\frac{\theta+1}{2}$, that is, periodic then the rays are left-supporting rays. For simplicity we will only treat the former case and thus right-supporting rays.

The angles in V_G form a periodic cycle. For each $x \in V_G$, the external ray $\mathcal{R}(x)$ of angle x may support one or two Fatou components, but it right-supports a unique periodic Fatou component, denoted by U_x. Define a map $\Phi: V_G \to \mathcal{P}_f$ such that $\Phi(x)$

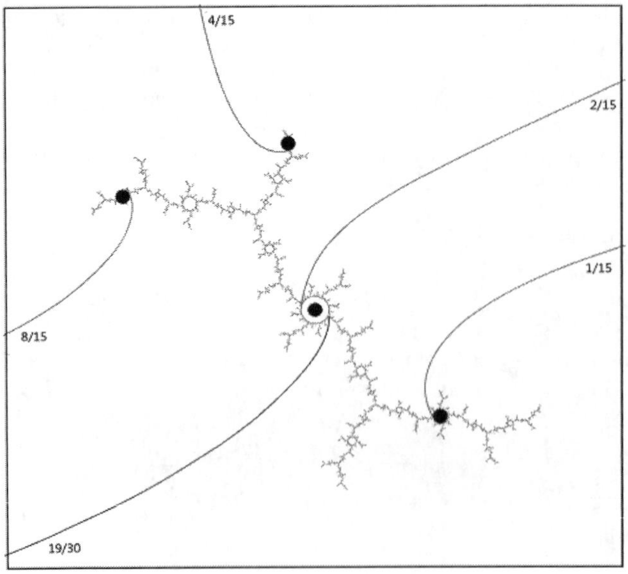

Figure 19: In this periodic example, $\theta = 4/15$ and $\theta/2 = 2/15$. All the rays in this figure are right-supporting rays.

is the center of the Fatou component U_x. Since external rays with distinct angles of V_G right-support distinct Fatou components, we have $U_x \neq U_y$ if $x \neq y \in V_G$. Thus the map $\Phi : V_G \to \mathcal{P}_f$ is bijective.

Extend Φ to a map, also denoted by Φ, from G to $\mathcal{H}_f$ such that Φ maps the edge $e(x,y)$ homeomorphically to the regulated arc $[\Phi(x), \Phi(y)] \subset \mathcal{H}_f$. We now assert and justify several facts about Φ:

1. Φ is finite-to-one.
   It follows directly from the fact that $\Phi : V_G \to \mathcal{P}_f$ is a bijection.
2. Φ is surjective.
   Since $\Phi(V_G) = \mathcal{P}_f$ and G is a complete graph, for any $p, q \in \mathcal{P}_f$, $[p,q] \subset \Phi(G)$. So we only need to evoke the fact that each edge of $\mathcal{H}_f$ is contained in a regulated arc $[p,q]$ with $p, q \in \mathcal{P}_f$.
3. $f \circ \Phi = \Phi \circ L$ after suitable modification of L on each edge.
   If two distinct vertices $x, y \in V_G$ belong to the same closed half circle, the interior of the regulated arc $[\Phi(x), \Phi(y)]$ does not contain the critical point of f. Its f-image is

$$\begin{aligned} f([\Phi(x), \Phi(y)]) &= [f(\Phi(x)), f(\Phi(y))] = [\Phi(2x), \Phi(2y)] \\ &= \Phi(e(2x, 2y)) = \Phi(L(e(x,y))). \end{aligned}$$

   So we can redefine L on the edge by lifting f, i.e., by setting

$$L = \Phi^{-1} \circ f \circ \Phi$$

   on the edge of $e(x,y)$.

If two vertices $x, y \in V_G$ belong to distinct open half circles, the interior of the regulated arc $[\Phi(x), \Phi(y)]$ contains the critical point 0. Its f-image is

$$\begin{aligned} f([\Phi(x), \Phi(y)]) &= [f(\Phi(x)), c_\theta] \cup [c_\theta, f(\Phi(y))] \\ &= [\Phi(2x), \Phi(\theta)] \cup [\Phi(\theta), \Phi(2y)] \\ &= \Phi(e(2x, \theta) \cup e(\theta, 2y)) = \Phi(L(e(x, y))). \end{aligned}$$

So we can redefine L on $e(x, y)$ such that $\Phi \circ L = f \circ \Phi$ on each of the two segments of $e(x, y)$ subdivided by $\Phi^{-1}(0)$.

Thus the maps L, Φ and f have been shown to satisfy the properties of Proposition A.3, so the equation

$$h(G, L) = h(\mathcal{H}_f, f)$$

holds, as desired.

Case 2: θ is preperiodic

In this case, the filled Julia set is equal to the Julia set. We can also define a map $\Phi : V_G \to \mathcal{P}_f$ such that $\Phi(x)$ is the landing point of $\mathcal{R}(x)$. It is easy to see that Φ is surjective. However, if we extend Φ piecewise monotonically on the edge of G as we did in the periodic case, the map Φ will lose the property of being finite-to-one, because some rays with distinct angles in V_G may land at the same point, which means that Φ collapses some edges of G to points. So we may no longer apply Proposition A.3 directly. To overcome this difficulty, let us define a subgraph Γ of G as follows:

$$V_\Gamma := V_G \quad \text{and} \quad E_\Gamma := \{e(x, y) \in E_G | \ \Phi(x) \neq \Phi(y)\}.$$

We will check that the graph Γ satisfies the following properties.

1. Γ is connected.
   This is because for any $x \in V_\Gamma$, the edge $e(\theta/2, x)$ belongs to Γ.
2. Γ is L-invariant.
   First, we observe that for any two distinct vertices $x, y \in V_G$ belonging to the same closed half circle, if $\mathcal{R}(x)$ and $\mathcal{R}(y)$ land at distinct points, then $\mathcal{R}(2x)$ and $\mathcal{R}(2y)$ also land at distinct points.
   Let $e(x, y)$ be an edge of Γ. Then the rays $\mathcal{R}(x)$ and $\mathcal{R}(y)$ land at distinct points $\Phi(x)$ and $\Phi(y)$ respectively. If x, y belong to the same closed half circle, by this observation the image edge

$$L(e(x, y)) = e(2x, 2y)$$

   belongs to Γ. If x, y belong to distinct open half circles, then $(x, \theta/2)$ and $(y, \theta/2)$ belong to the same closed half circle and the image

$$L(e(x, y)) = e(2x, \theta) \cup e(\theta, 2y)$$

   belongs to Γ.

3. $h(G, L) = h(\Gamma, L)$.

We claim that the set $G \setminus \Gamma$ is L-invariant. In fact, an edge $e(x, y)$ belongs to $E_G \setminus E_\Gamma \overset{\text{definition}}{\Longleftrightarrow} \Phi(x) = \Phi(y) \Longleftrightarrow$ the rays $\mathcal{R}(x), \mathcal{R}(y)$ land at a common periodic point in $\mathcal{P}_f$. To see the last implication, we only need to show the "$\Rightarrow$" part. Since x, y are both in the forward orbit of θ, we may assume $y = 2^k x, k \geq 1$. Then

$$\Phi(y) = \Phi(2^k x) = f^k(\Phi(x)).$$

So $\Phi(x)$ is a periodic point, receives both rays $\mathcal{R}(x), \mathcal{R}(y)$, and belongs to $\mathcal{P}_f$. In this case the angles x, y must belong to the same half circle and $\Phi(2x) = \Phi(2y)$. It follows that

$$L(e(x, y)) = e(2x, 2y)$$

also belongs to $E_G \setminus E_\Gamma$.

The argument above also shows that L maps an edge in $E_G \setminus E_\Gamma$ homeomorphically onto an edge in $E_G \setminus E_\Gamma$. According to the definition of the topological entropy, we have

$$h(G \setminus \Gamma, L) = 0.$$

So by Proposition A.1, we obtain

$$h(G, L) = \max \left\{ h(\Gamma, L), h(G \setminus \Gamma, L) \right\} = h(\Gamma, L).$$

By these properties, it is enough to prove that $h(\Gamma, L) = h(\mathcal{H}_f, f)$. In this case, the map $\Phi\big|_\Gamma : \Gamma \to \mathcal{H}_f$ is finite-to-one. By the same argument as in the periodic case, we also obtain that $\Phi\big|_\Gamma$ is surjective and

$$f \circ \Phi = \Phi \circ L$$

on Γ. So the equality

$$h(\Gamma, L) = h(\mathcal{H}_f, f)$$

holds.

This finishes the proof of Theorem 13.10. □

Thurston's entropy algorithm and Theorem 13.10 are generalized to higher-degree maps in [Gao].

14 COMBINATORIAL LAMINATIONS AND POLYNOMIAL LAMINATIONS

The action of a degree d postcritically finite polynomial on its Julia set can be combinatorially encoded by the action of a degree d expanding map on an invariant lamination on the circle $\mathbb{S} = \mathbb{R}/\mathbb{Z}$ or on the torus $\mathbb{T} = \mathbb{S} \times \mathbb{S}$. Here we illustrate this connection.

14.1 Combinatorial laminations

We denote the diagonal Δ of $\mathbb{T}$ by

$$\Delta = \{(x,x), x \in \mathbb{S}\}.$$

For $x \neq y \in \mathbb{S}$, we use $\overline{xy}$ to denote the closure in $\overline{\mathbb{D}}$ of the hyperbolic chord in $\mathbb{D}$ connecting $e^{2\pi ix}$ and $e^{2\pi iy}$, which is called a *leaf.* This leaf is represented (twice) on the torus $\mathbb{T} \setminus \Delta$ as (x,y) and (y,x).

Two leaves $\overline{xy}$ and $\overline{x'y'}$ are said to be *compatible* if they are either equal or do not cross inside $\mathbb{D}$. We say also that the two points (x,y) and (x',y') on the torus are compatible. In this case the four points $(x,y),(y,x),(x',y'),(y',x')$ are pairwise compatible.

A *lamination in* $\mathbb{D}$ is a collection of leaves which are pairwise compatible and whose union is closed in $\mathbb{D}$. A *lamination on the torus* is a subset L closed in $\mathbb{T} \setminus \Delta$ and symmetric with respect to the diagonal Δ so that points in L are pairwise compatible.

14.2 Polynomial laminations

Let f be a monic degree d polynomial with a connected and locally connected Julia set. There is a unique Riemann mapping $\phi : \overline{\mathbb{C}} \setminus \overline{\mathbb{D}} \to \overline{\mathbb{C}} \setminus \mathcal{K}_f$ tangent to the identity at ∞; it extends continuously to the closure by the Carathéodory theorem. External rays are parametrized by external angles. The set of pairs of external rays landing at a common point gives another combinatorial characterization of the polynomial dynamics.

Two distinct external rays $\mathcal{R}(x)$ and $\mathcal{R}(y)$ landing at the same point are said to be a *ray-pair of* f. We also say that $\{x,y\}$ is an *angle-pair of* f. If the rays land at z, we say that $\{x,y\}$ is an angle-pair at z.

Let $\{x,y\}$ be an angle-pair at z. Then the union of the two rays $\mathcal{R}(x)$ and $\mathcal{R}(y)$ together with $\{z\}$ divides the plane into two regions. If at least one of the two regions does not contain other rays landing at z, we say that $\{x,y\}$ is an *adjacent angle-pair.*

A polynomial f induces a lamination $L(f)$ whose leaves (i.e., points in $\mathbb{T}$) consist of geodesics connecting adjacent angle-pairs. $L(f)$ is said to be the *polynomial lamination* of f.

14.3 Good and excluded regions of a lamination

We recall here some notions introduced in Section 4. If a lamination on $\mathbb{T}$ contains a point $l = (x,y)$, then a certain set $X(l)$ of other points are excluded from the lamination because they are not compatible with (x,y). If you draw the horizontal and vertical circles through the two points (x,y) and (y,x), they divide the torus into four rectangles having the same vertex set; the remaining two common vertices are (x,x) and (y,y). The two rectangles bisected by the diagonal are actually squares, of side lengths $a - b \mod 1$ and $b - a \mod 1$. Together, they form the compatible region $G(l)$. The leaves represented by points in the interior of the remaining two rectangles constitute the excluded region $X(l)$. See Figure 4, where the blue region is $X(l)$ and the tan region is $G(l)$.

Given a set S of leaves, the *excluded region* $X(S)$ is the union of the excluded regions $X(l)$ for $l \in S$, and the *good region* $G(S)$ is the intersection of the good regions $G(l)$ for $l \in S$. If S is a finite lamination, then $G(S)$ is a finite union of closed rectangles that are disjoint except for corners.

14.4 Invariant laminations from majors

Define the map $F : \mathbb{T} \to \mathbb{T}$ by

$$(x,y) \mapsto (\tau(x), \tau(y)),$$

where τ is the angle-doubling map defined in Section 1.1. A lamination $L \subset \mathbb{T} \setminus \Delta$ is said to be *F-invariant* if, for every $(x,y) \in L$,

1. either $F(x,y) \in \Delta$ or $F(x,y) \in L$; and
2. there exist two preimages of (x,y) in L, which have different x components and different y components.

The polynomial lamination $L(f)$ for a postcritically finite quadratic polynomial $f : z \mapsto z^2 + c$ is an example of an F-invariant lamination.

Given an angle $\theta \in \mathbb{S}$, there are several more or less natural ways to define an F-invariant lamination, mimicking the polynomial lamination of $f : z \mapsto z^2 + c_\theta$. And as we shall see, they differ by at most a countable set. Here we choose one that mimics the fact that preimages of the critical point of f accumulate on every Julia point. Other definitions will be given later (see Section 16).

Set, inductively,

$$b_0 := \left\{ \left(\frac{\theta}{2}, \frac{\theta+1}{2}\right), \left(\frac{\theta+1}{2}, \frac{\theta}{2}\right) \right\} \quad \text{the major leaves}$$

$$b_{i+1} := \left(F^{-1}(b_i) \cap G(b_i)\right) \cup b_i$$

$$\mathrm{Pre}_\theta := \bigcup_{i \geq 0} b_i \quad \text{the set of pre-major leaves}$$

$$\mathrm{cluster}(\mathrm{Pre}_\theta) := \text{the set of cluster points in } \mathbb{T} \text{ of the pre-major leaves.}$$

$$L_\theta := \text{the set of cluster points in } \mathbb{T} \setminus \Delta \text{ of the pre-major leaves.}$$

(Cluster points are also known as accumulation points.)

14.5 Relating combinatorial laminations to polynomial laminations

Proposition 14.1. *For a rational angle θ, the set L_θ is equal to the polynomial lamination of $f : z \to z^2 + c_\theta$. In particular, L_θ is an F-invariant lamination.*

Proof. This says that every accumulation point (x,y), $x \neq y$, of Pre_θ is an adjacent angle-pair for $f : z \mapsto z^2 + c_\theta$, and conversely every adjacent angle-pair is the accumulation point of a sequence $(x_n, y_n) \in \mathrm{Pre}_\theta$.

If θ is preperiodic, the ray-pair of angles $\frac{\theta}{2}$ and $\frac{\theta+1}{2}$ land at the critical point 0. So each point in Pre_θ is a (not necessarily adjacent) angle-pair at a precritical point. If a sequence of distinct leaves $(x_n, y_n) \in \mathrm{Pre}_\theta$ converges to (x,y), one can find a subsequence converging from one side of the limit leaf (x,y). It follows that (x,y) is an angle-pair, and is an adjacent angle-pair. Conversely, one just needs to consider an adjacent angle-pair at a point z of the Hubbard tree, and then use expansions on the tree to see that z is approximated by preimages of 0 on the tree. Details are presented in Appendix B.

The case that θ is periodic is considerably more complicated. The ends of a leaf in Pre_θ are not an angle-pair, but correspond to a pair of rays supporting a common Fatou component with dyadic internal angles, as in Figure 19. They would form a cutting line if we add the two internal rays. See details in Appendix B. □

15 COMBINATORIAL HUBBARD TREE FOR A RATIONAL ANGLE θ

We have now seen the combinatorial encodings of Julia sets in $\mathbb{S}$ and $\mathbb{T}$. Let us turn now to the corresponding encodings of Hubbard trees. There are actually two characterizations of a Hubbard tree: one by looking at ray-pairs separating the postcritical set, the other by looking at forward images of ray-pair landing points.

15.1 Combinatorial Hubbard tree

Note that the Julia points of the Hubbard tree $\mathcal{H}_f$ can be considered as the set of Julia points that "separate" the postcritical set $\mathcal{P}_f$. Such Julia points receive necessarily at least two external rays. One can therefore define a combinatorial counterpart of the postcritical set and the Hubbard tree for quadratic maps as follows.

Let $\theta \in \mathbb{S}$ be a rational angle. Define the post-major angle set to be:

$$P_\theta^{\mathbb{S}} = \{\tau^n\theta \mid n \geq -1\} \quad \text{and} \quad P_\theta = \{(\alpha,\alpha) \in \mathbb{T} \mid \alpha \in P_\theta^{\mathbb{S}}\}.$$

As θ is rational, the set P_θ is a finite F-forward-invariant set in the diagonal of the torus (also $P_\theta^{\mathbb{S}}$ is a finite τ-forward-invariant set).[3]

For $(x,y) \in \mathbb{T}$, we say that (x,y) *separates* P_θ if $x \neq y$ and both components of $\Delta \setminus \{(x,x),(y,y)\}$ contain points of P_θ, or equivalently, both components of $\mathbb{S} \setminus \{x,y\}$ contain points of $P_\theta^{\mathbb{S}}$. Visually, this is equivalent to assert that both open tan squares in $\mathbb{T}$ in Figure 4 generated by (x,y) contain points of P_θ in the interior. We say that (x,y) *intersects* P_θ if either (x,x) or (y,y) is in P_θ.

Recall that

$$L_\theta = \text{cluster}(\text{Pre}_\theta) \setminus \Delta = \text{the cluster set in } \mathbb{T} \setminus \Delta \text{ of the pre-major leaves.}$$

Set

$$X_\theta = \{\, (x,y) \in L_\theta \mid (x,y) \text{ intersects or separates } P_\theta\}.$$

Definition 15.1 (Combinatorial Hubbard tree). *The set $H_\theta := X_\theta \cup P_\theta$ is called the combinatorial Hubbard tree of θ.*

[3] More formally, P_θ should be written $P_\theta^{\mathbb{T}}$. Since we work more often on the torus model, we omit the superscript $\mathbb{T}$.

It is known that the Hubbard tree of a postcritically finite polynomial is an attracting core of the landing points of ray-pairs, in the sense that some forward iterate of any such point will be in the Hubbard tree and stay there under further iterations. Here is the combinatorial counterpart.

Proposition 15.2. *Let $\theta \in \mathbb{S}$ be a rational angle. The set X_θ is an attracting core of L_θ, in the sense that $X_\theta \subset L_\theta$ and for any $(x,y) \in L_\theta$, there is $n \geq 0$ so that $F^n(x,y) \in X_\theta$.*

Proof. By definition, $X_\theta \subset L_\theta$.

For a pair of points $x, y \in \mathbb{S}$, let us define the distance of x, y, denoted by $|x-y|$, to be the arc-length of the shortest of the two arcs in $\mathbb{S}$ determined by x, y. Let $(x,y) \in L_\theta$. Note that if $|x-y| \leq \frac{1}{4}$, then

$$|\tau(x) - \tau(y)| = 2|x-y|.$$

So there exists a minimal $n \geq 0$ such that $|\tau^n(x) - \tau^n(y)| > \frac{1}{4}$. In this case,

$$\frac{1}{2} \geq |\tau^n(x) - \tau^n(y)| > \frac{1}{4}.$$

It follows that the shorter closed arc between $\tau^{n+1}(x)$ and $\tau^{n+1}(y)$ must contain the point θ. Since the leaf $\overline{\tau^{n+1}(x)\tau^{n+1}(y)}$ belongs to the closure of a component of

$$\overline{\mathbb{D}} \setminus \overline{\tfrac{\theta}{2}\tfrac{\theta+1}{2}},$$

it follows that the leaf $\overline{\tau^{n+1}(x)\tau^{n+1}(y)}$ separates or intersects

$$P_\theta^{\mathbb{D}} := \{e^{2\pi i t} \mid t \in P_\theta^{\mathbb{S}}\}. \qquad \square$$

15.2 Relation between combinatorial trees and pre-major leaves

Fix a rational angle $\theta \in \mathbb{S}$. It is relatively easy to find in the countable set Pre_θ of points those that separate or intersect the post-major angle set. We will see in Proposition 15.4 that, starting from these points, we can recover the combinatorial Hubbard tree. Set

$$\begin{aligned} S_\theta &:= \{(x,y) \in \mathrm{Pre}_\theta \mid (x,y) \text{ separates or intersects } P_\theta\} \\ &= \{\text{pre-major leaves separating or intersecting post-major angles}\} \\ \mathrm{cluster}(S_\theta) &:= \text{the set of cluster points of } S_\theta \text{ in } \mathbb{T}. \end{aligned}$$

Lemma 15.3. *Any $(x,y) \in$ cluster(S_θ) intersects or separates P_θ.*

Proof. This is an easy consequence of the fact that P_θ is finite. $\square$

Proposition 15.4. *The combinatorial Hubbard tree H_θ and the set cluster(S_θ) are both compact and F-forward-invariant. Moreover the two sets differ by a finite set.*

Proof. The compactness of cluster(S_θ) is obvious since $\mathbb{T}$ is compact. We proceed to prove that it is F-forward-invariant.

We show first that the union of the post-major set P_θ with the set S_θ is F-forward-invariant. As Pre_θ is the set of pre-major leaves, if $(x,y)\in \mathrm{Pre}_\theta$, either $F(x,y)\in \mathrm{Pre}_\theta$ or $F(x,y)\in P_\theta$. So we are left to prove that if (x,y) separates or intersects P_θ, so does $(\tau(x),\tau(y))$.

Denote, as before, the set

$$P_\theta^{\mathbb{D}} = \{e^{2\pi i t} \mid t \in P_\theta^{\mathbb{S}}\},$$

which is the post-major angle set in the unit disc model. Note that (x,y) separates or intersects P_θ if and only if $\overline{xy}$ separates or intersects $P_\theta^{\mathbb{D}}$ in $\overline{\mathbb{D}}$. If $\overline{xy}$ intersects $P_\theta^{\mathbb{D}}$, the leaf $\overline{\tau(x)\tau(y)}$ must intersect $P_\theta^{\mathbb{D}}$. If $\overline{xy}$ separates (in $\overline{\mathbb{D}}$) but does not intersect $P_\theta^{\mathbb{D}}$, the leaf $\overline{xy}$ belongs to a component W of

$$\overline{\mathbb{D}} \setminus \overline{\tfrac{\theta}{2}\tfrac{\theta+1}{2}}.$$

Then, $\overline{xy}$ must separate a point $e^{2\pi i t}$ of $P_\theta^{\mathbb{D}}$ and the point $e^{2\pi i \frac{\theta}{2}}$. It follows that $\overline{\tau(x)\tau(y)}$ separates the two points $e^{2\pi i\tau(t)}, e^{2\pi i\theta} \in P_\theta^{\mathbb{D}}$. This shows that $S_\theta \cup P_\theta$ is F-forward-invariant.

Second, we will show that cluster(S_θ) is F-forward-invariant. Take $\{(x_n,y_n)\}$ a sequence of distinct points in S_θ such that

$$\lim_{n\to\infty}(x_n,y_n) = (x,y) \in \mathrm{cluster}(S_\theta).$$

According to the F-invariant property of $S_\theta\cup P_\theta$ proved above, the sequence of points $\{F(x_n,y_n)\}$ belong to $S_\theta \cup P_\theta$. As P_θ has only finitely many F-preimages, countably many of $\{F(x_n,y_n)\}$ must belong to S_θ. Since

$$\lim_{n\to\infty} F(x_n,y_n) = F(x,y),$$

we have $F(x,y)\in \mathrm{cluster}(S_\theta)$.

The proof that H_θ is compact and F-forward-invariant is similar and left as an exercise.

By Lemma 15.3 we have cluster$(S_\theta)\subset H_\theta$.

Let us show that $H_\theta \setminus \mathrm{cluster}(S_\theta)$ is also finite. Note that if a point $(x,y)\in X_\theta$ separates P_θ, then $(x,y)\in L_\theta$, which is the cluster set in $\mathbb{T}\setminus\Delta$ of Pre_θ. One can thus find a sequence $\{(x_n,y_n)\}$ in Pre_θ converging to (x,y). So for all large n, the point (x_n,y_n) must separate the finite post-major set P_θ. It follows by definition that $(x_n,y_n)\in S_\theta$, and therefore $(x,y)\in \mathrm{cluster}(S_\theta)$. So a point of $X_\theta\setminus \mathrm{cluster}(S_\theta)$ must intersect but not separate P_θ.

On the other hand, since L_θ is the polynomial lamination of f_θ (by Proposition 14.1), each angle belongs to at most two adjacent angle-pairs. In particular, for any point $(p,p)\in P_\theta$, there are at most two points of L_θ intersecting (p,p). So $X_\theta\setminus\mathrm{cluster}(S_\theta)$, and therefore $H_\theta\setminus \mathrm{cluster}(S_\theta)$, is a finite set. □

16 ENTROPIES

The objective is to prove that several quantities in the combinatorial and dynamical worlds all encode the core entropy of a quadratic polynomial.

16.1 Various entropies and Hausdorff dimensions

Let $\theta \in \mathbb{S}$ be a rational angle. Recall that for the polynomial $f_\theta : z \mapsto z^2 + c_\theta$, we use $\gamma(\eta)$ to denote the landing point of the external ray of angle η, and $\mathcal{H}_\theta$ to denote its Hubbard tree.

We have several natural actions (τ on the circle $\mathbb{S}$, F on the torus $\mathbb{T}$, and f_θ on the Hubbard tree $\mathcal{H}_\theta$). We have also defined the polynomial lamination L_θ and the combinatorial Hubbard tree H_θ, both in $\mathbb{T}$.

The set H_θ is compact and F-forward-invariant, and thus has a topological entropy $h(H_\theta, F)$.

The polynomial f_θ induces two more combinatorial sets[4]

- the angle-pair set

$$\mathrm{B}_\theta = \{(\beta, \eta) \in \mathbb{T} \mid \beta \neq \eta \text{ and } \gamma(\beta) = \gamma(\eta)\}$$

- and its projection to the circle

$$\mathrm{B}_\theta^{\mathbb{S}} = \{\beta \in \mathbb{S} \mid \exists\ \eta \neq \beta \text{ s.t } \gamma(\beta) = \gamma(\eta)\}.$$

There is also a non-escaping set NE, defined later.

We can then consider entropies or Hausdorff dimensions of these objects.

The Actions	$F : (\eta, \zeta) \mapsto (2\eta, 2\zeta)$	f_θ	$\tau : \eta \mapsto 2\eta$	On a Graph
Entropies On Trees	$h(H_\theta)$	$h(\mathcal{H}_\theta)$		
Dimensions On trees	$\mathrm{HD}(H_\theta)$			
Dimensions On Bi-acc Sets	$\mathrm{HD}(L_\theta)$ $\mathrm{HD}(NE \setminus \Delta)$ $\mathrm{HD}(\mathrm{B}_\theta)$		$\mathrm{HD}(\mathrm{B}_\theta^{\mathbb{S}})$	
Algorithm				$\rho(A_\theta)$

Theorem 16.1. *For a rational angle $\theta \in \mathbb{S}$, all the quantities in the above table (as well as several other quantities detailed in the proof) are related by*

$$\log \rho(A_\theta) = h(\cdot) = HD(*) \cdot \log 2.$$

[4] These sets are sometimes referred to in the literature as the set of bi-accessible angles.

Proof. We will establish the equalities following the schema:

$$\begin{array}{ccc} & \log\rho(A_\theta) & \\ & \Big\| \ \text{by Thm 13.10} & \\ h(H_\theta,F)= & h(\mathcal{H}_\theta,f_\theta) & \text{(entropies)}\\ \Big\uparrow \cdot\log 2 & \Big\uparrow\cdot\log 2 & \\ \mathrm{HD}(H_\theta) & \mathrm{HD}(\mathrm{B}^{\mathbb{S}}_\theta) & \text{(dimensions)}\\ \Big\| & & \\ \mathrm{HD}(L_\theta) & =\mathrm{HD}(NE\setminus\Delta)=\mathrm{HD}(\mathrm{B}_\theta) & \end{array}$$

(by Prop. 16.5 below)

For simplicity set $f=f_\theta$ and $\mathcal{H}=\mathcal{H}_\theta$.

- $h(H_\theta,F)=h(\mathcal{H},f)$.
  By Proposition 15.4, the set H_θ is compact F-invariant. By Proposition 14.1 the set L_θ is equal to the polynomial lamination of f. By the definition of H_θ, there is a continuous surjective and finite-to-one map $\varphi\colon H_\theta\to\mathcal{H}\cap\mathcal{J}$ realizing a semi-conjugacy: $f\circ\varphi=\varphi\circ F$. By Proposition A.3, we have that

$$h(H_\theta,F)=h(\mathcal{H}\cap\mathcal{J},f).$$

  If θ is preperiodic, $\mathcal{H}\cap\mathcal{J}=\mathcal{H}$ so

$$h(H_\theta,F)=h(\mathcal{H},f).$$

  If θ is periodic, the points in $\mathcal{H}\setminus\mathcal{J}$ are attracted to the critical orbit $\mathcal{P}_f$, on which f has 0 entropy. So

$$\begin{aligned} h(\mathcal{H},f) &= h\Big((\mathcal{H}\cap\mathcal{J})\cup\mathcal{P}_f,f\Big) \ \text{ by Proposition A.2}\\ &=\max\Big(h(\mathcal{H}\cap\mathcal{J},f),h(\mathcal{P}_f,f)\Big) \ \text{ by Proposition A.1}\\ &=h(\mathcal{H}\cap\mathcal{J},f)=h(H_\theta,F) \ \text{ by Proposition A.3.}\end{aligned}$$

  Note that this entropy is also equal to

$$h(\text{cluster}(S_\theta),F),$$

  as cluster(S_θ) and H_θ differ by a finite set (Proposition 15.4).
- $\mathrm{HD}(H_\theta)\cdot\log 2=h(H_\theta,F)$.
  This follows directly from Proposition 15.4 and Lemma A.4.
- $\mathrm{HD}(\mathrm{B}^{\mathbb{S}}_\theta)\cdot\log 2=h(\mathcal{H},f)$.
  This fact was first noticed by Tiozzo [Tio15, Tio16]. It is based on the fact that the set $\mathrm{B}^{\mathbb{S}}_\theta$ has an attracting core which is the set $H^{\mathbb{S}}_\theta$ of angles whose rays land at the Hubbard tree, so has the same dimension. One can then relate the dimension of $H^{\mathbb{S}}_\theta$ to entropy of τ on $H^{\mathbb{S}}_\theta$ by Lemma A.4, and then relates to

the entropy on the Hubbard tree via the ray-landing semi-conjugacy which is finite-to-one (so one can apply Proposition A.3).

- $\mathrm{HD}(L_\theta)=\mathrm{HD}(X_\theta)$ since X_θ is an attracting core of L_θ (Proposition 15.2).
- $\mathrm{HD}(L_\theta)=\mathrm{HD}(NE\setminus\Delta)=\mathrm{HD}(\mathrm{B}_\theta)$ by Proposition 16.5 below. □

16.2 Non-escaping sets on the torus

Let us fix an angle $\theta\in\mathbb{S}$ (not necessarily rational). We define two invariant subsets mimicking filled Julia sets as follows:

Definition (Non-escaping sets). Set inductively

$$\begin{aligned}\Omega_0 &:= \text{the interior of the good region } G\Big(\frac{\theta}{2},\frac{\theta+1}{2}\Big)\\ \Omega_{i+1} &:= F^{-1}(\Omega_i)\cap\Omega_0\\ NE_1(\theta) &:= \bigcap_{n\geq 0}\Omega_n\\ NE_2(\theta) &:= \bigcap_{n\geq 0}\overline{\Omega_n}.\end{aligned}$$

In other words, $NE_1(\theta)$ consists of the points in $\mathbb{T}$ whose orbits never escape Ω_0.

Lemma 16.2. *We have $\Delta\subset\overline{NE_1}(\theta)\subset NE_2(\theta)$.*

Proof. We omit θ for simplicity.

For each $n\geq 0$, the diagonal Δ is contained in $\overline{\Omega_n}$, and the set $\Delta\setminus\Omega_n$ is finite. It follows that $\Delta\subset NE_2$ and $\Delta\setminus NE_1$ is a countable set and hence $\Delta\subset\overline{NE_1}$. The inclusion $\overline{NE_1}\subset NE_2$ is obvious. □

For $\theta\neq 0$, the diameter $\overline{\frac{\theta}{2}\frac{\theta+1}{2}}$ subdivides the unit circle into two half open halves. Denote by S_0 the half circle containing the angle 0 and by S_1 the other one. Set

$$\mathrm{Pre}^{\mathbb{S}}_\theta=\bigcup_{n\geq 1}\tau^{-n}(\theta).$$

For any angle $\alpha\in\mathbb{R}/\mathbb{Z}\setminus\mathrm{Pre}^{\mathbb{S}}_\theta$, we can assign a sequence $\epsilon_0\epsilon_1\ldots\in\{0,1\}^{\mathbb{N}}$ such that $\epsilon_i=\delta$ if and only if $\tau^i(\alpha)\in S_\delta$ ($\delta=0$ or 1). This sequence is said to be the *itinerary of α relative to θ*, denoted by $\iota_\theta(\alpha)$.

Let θ be a rational angle. There is a parallel description in the dynamical plane of $f:z\mapsto z^2+c_\theta$.

If θ is preperiodic, the external rays $\mathcal{R}\left(\frac{\theta}{2}\right)$ and $\mathcal{R}\left(\frac{\theta+1}{2}\right)$ both land at the critical point 0. In this case, we set

$$\mathcal{R}\left(\frac{\theta}{2},\frac{\theta+1}{2}\right):=\mathcal{R}\left(\frac{\theta}{2}\right)\cup\{0\}\cup\mathcal{R}\left(\frac{\theta+1}{2}\right).$$

This set is called the *major cutting line.*

If θ is periodic, say of period p, the rays $\mathcal{R}\left(\frac{\theta}{2}\right)$ and $\mathcal{R}\left(\frac{\theta+1}{2}\right)$ support the Fatou component U that contains 0. The period of U is also p. The first return map

$f^p : \overline{U} \to \overline{U}$ is conjugated to the action of z^2 on $\overline{\mathbb{D}}$, and thus admits a unique fixed point on the boundary, called the root of U. This root point has internal angle 0, and its unique f^p-preimage on ∂U has internal angle 1/2. Denote by $r_U(0)$ and $r_U(1/2)$ the corresponding internal rays in U. They land at $\gamma\left(\frac{\theta}{2}\right)$, $\gamma\left(\frac{\theta+1}{2}\right)$ respectively (or vice versa, depending on which of $\frac{\theta}{2}$, and $\frac{\theta+1}{2}$ is periodic). In this case, we define the *major cutting line* to be

$$\mathcal{R}\left(\frac{\theta}{2}, \frac{\theta+1}{2}\right) := \mathcal{R}\left(\frac{\theta}{2}\right) \cup \overline{r_U(0) \cup r_U(1/2)} \cup \mathcal{R}\left(\frac{\theta+1}{2}\right).$$

In both cases, $\mathcal{R}\left(\frac{\theta}{2}, \frac{\theta+1}{2}\right)$ is a simple curve which subdivides the complex plane into two open regions. Assume $\theta \neq 0$. Denote by V_0 the region containing the ray $\mathcal{R}(0)$ and by V_1 the other region. For an angle

$$\alpha \in \mathbb{R}/\mathbb{Z} \setminus \mathrm{Pre}_\theta^{\mathbb{S}},$$

it is easy to see that $\iota_\theta(\alpha) = \epsilon_0\epsilon_1 \ldots$ if and only if

$$f_\theta^i(\gamma_{c_\theta}(\alpha)) \subset V_{\epsilon_i},$$

for all $i \geq 0$.

Set

$$\text{Pre-cr} = \bigcup_{n \geq 1} f^{-n}(\gamma(\theta)).$$

For $z \in \mathcal{J} \setminus \text{Pre-cr}$, the orbit of z does not meet the major cutting line. So we can define its itinerary relative to this cutting line:

$$\iota_\theta(z) = \epsilon_0\epsilon_1\epsilon_2 \ldots \qquad \text{if} \quad f^i(z) \in V_{\epsilon_i}.$$

It is easy to see that if $z \in \mathcal{J} \setminus \text{Pre-cr}$ has an external angle α, then $\iota_\theta(\alpha) = \iota_\theta(z)$.

The following known result will be useful for us (see Appendix C for a proof):

Lemma 16.3. *Let θ be a rational angle, and let $\alpha \neq \beta \in \mathbb{R}/\mathbb{Z} \setminus \mathit{Pre}_\theta^{\mathbb{S}}$. Then the rays $\mathcal{R}(\alpha)$ and $\mathcal{R}(\beta)$ land together if and only if $\iota_\theta(\alpha) = \iota_\theta(\beta)$.*

Corollary 16.4. *For $z_1, z_2 \in \mathcal{J} \setminus$ Pre-cr, then $z_1 \neq z_2$ if and only if $\iota_\theta(z_1) \neq \iota_\theta(z_2)$.*

Proposition 16.5. *For θ a rational angle, the four sets L_θ, $NE_1(\theta) \setminus \Delta$, $NE_2(\theta) \setminus \Delta$ and B_θ are pairwise different by a set that is at most countable. Consequently, they have the same dimension.*

Proof. Fix a rational angle θ. Let us introduce a new set

$$Z = \{(x, y) \in \mathbb{T} \mid \{x, y\} \text{ is an angle-pair (not necessarily adjacent)} \text{ at a point in } \mathcal{J} \setminus \text{Pre-cr}\}.$$

We will compare this set with the four sets L_θ, $NE_1 \setminus \Delta$, $NE_2 \setminus \Delta$ and B_θ. The notation "a.e" below on top of the equal sign means that the symmetric difference between the two sets has measure zero.

1. $NE_1 \setminus \Delta = Z$.

   By the definition of NE_1, a point $(x,y) \in NE_1 \setminus \Delta \iff x \neq y$ and the orbit of (x,y) stays in $\Omega_0 \iff x \neq y$ and $\iota_\theta(x) = \iota_\theta(y) \overset{\text{Lemma 16.3}}{\iff}$ the rays $\mathcal{R}(x)$ and $\mathcal{R}(y)$ land at a common point $z \in \mathcal{J} \setminus \text{Pre-cr} \iff (x,y) \in Z$.
2. $NE_1 \setminus \Delta \overset{a.e}{=} NE_2 \setminus \Delta$.

   We already know $NE_1 \setminus \Delta \subset NE_2 \setminus \Delta$. Let us fix a point (x,y) in $NE_2 \setminus (NE_1 \cup \Delta)$. By the definition of NE_2 and NE_1, we have that

$$(x,y) \in \left(\bigcap_{n \geq 0} \overline{\Omega}_n \right) \setminus \left(\bigcap_{k \geq 0} \Omega_k \right)$$

so

$$(x,y) \in \overline{\Omega}_n \ \ \forall n \ \text{ and } \ (x,y) \notin \Omega_N \text{ for some } N \geq 0.$$

But Ω_k is decreasing with respect to k (it is the set of points that remain in Ω_0 up to k iterates). So for every $n \geq N$,

$$(x,y) \in \overline{\Omega}_n \setminus \Omega_n = \partial \Omega_n.$$

We can then select a point $(x_n, y_n) \in \Omega_n$ such that $\{(x_n, y_n), n \geq 1\}$ converges to (x,y) as n goes to infinity. Set

$$z_n = \gamma(x_n), \quad w_n = \gamma(y_n), \quad z = \gamma(x), \quad w = \gamma(y).$$

We obtain that

$$\lim_{n \to \infty} z_n = z \quad \text{and} \quad \lim_{n \to \infty} w_n = w.$$

It is known that f_θ is uniformly expanding on a neighborhood of $\mathcal{J}$ with respect to the orbifold metric of f_θ (see [DH84, McM94, Mil06]). Since $(x_n, y_n) \in \Omega_n$, the first n-entries of $\iota_\theta(x_n)$ and $\iota_\theta(y_n)$ are identical. By the argument in the proof of Lemma 16.3, the orbifold distance of z_n and w_n is less than $C\lambda^{-n}$ for some constant C and $\lambda > 1$. It follows that $z = w$ and thus (x,y) is an angle-pair.

On the other hand, by an inductive argument, one can see that for a point $(p,q) \in \partial \Omega_n$, either p or q belongs to the set $\bigcup_{i=1}^{n+1} \tau^{-i}(\theta)$. So, in our case, either x or y belongs to $\mathrm{Pre}_\theta^{\mathbb{S}}$. It follows that

$$\gamma(x) = \gamma(y) \in \text{Pre-cr}.$$

It is known that the angle-pairs at points in Pre-cr form a countable set, so we obtain that $NE_1 \setminus \Delta \overset{a.e}{=} NE_2 \setminus \Delta$.

3. $L_\theta \overset{a.e}{=} Z$.

   By Proposition 14.1, the set L_θ is the lamination of f_θ. Then a point $(x,y) \in L_\theta \setminus Z$ represents an angle-pair at a common point of Pre-cr, so

$$\text{cluster}(\mathrm{Pre}_\theta) \setminus (\Delta \cup \setminus Z)$$

is a countable set. On the other hand, the set $Z \setminus L_\theta$ consists of angle-pairs that are not adjacent and that land at non-precritical points. Then the landing points each receive at least 4 external rays and the number of rays remain constant under forward iteration. But each point receiving at least two rays will be mapped eventually into the Hubbard tree and stay there. So a landing point of angle-pairs in $Z \setminus L_\theta$ is mapped eventually to a branching point of the Hubbard tree. As the tree is finite, there are only finitely many branching points forming finitely many forward-invariant orbits. So the set of such landing points is countable, and each point receives finitely many ray-pairs. It follows that the set $Z \setminus L_\theta$ is countable.

4. Finally, $\mathrm{B}_\theta \supset Z$ and $\mathrm{B}_\theta \setminus Z$ is countable as it concerns only angle-pairs whose rays land at the countable set Pre-cr. □

APPENDIX A. BASIC RESULTS ABOUT ENTROPY

The following can be found in [Dou95, dMvS93]. Let X be a compact topological space, $f : X \to X$ a continuous map.

Proposition A.1. *If $X = X_1 \cup X_2$, with X_1 and X_2 compact, $f(X_1) \subset X_1$ and $f(X_2) \subset X_2$, then*

$$h(X, f) = \sup\big(h(X_1, f), h(X_2, f)\big).$$

Proposition A.2. *Let Z be a closed subset of X such that $f(Z) \subset Z$. Suppose that for any $x \in X$, the distance of $f^n(x)$ to Z tends to 0 uniformly on any compact set in $X \setminus Z$. Then $h(X, f) = h(Z, f)$.*

In the setting of Proposition A.2, we say that Z is an *attracting core* of f.

Proposition A.3. *Assume that π is a surjective semi-conjugacy*

$$\begin{array}{ccc} Y & \xrightarrow{q} & Y \\ {\scriptstyle\pi}\downarrow & & \downarrow{\scriptstyle\pi} \\ X & \xrightarrow{f} & X. \end{array}$$

We have $h(X, f) \le h(Y, q)$. Furthermore, if $\sup_{x \in X} \#\pi^{-1}(x) < \infty$, then $h(X, f) = h(Y, q)$.

The following can be found in [Fur67].

Lemma A.4. *Let A be a compact τ-invariant subset of $\mathbb{S}$. Then*

$$e^{h(A,\tau)} = 2^{HD(A)}.$$

Similarly, if A is a compact F-invariant subset of $\mathbb{T}$,

$$e^{h(A,F)} = 2^{HD(A)}.$$

Definition A.5. (Finite connected graph). *A finite connected graph is a connected topological space G consisting of the union of the finite vertex set V_G and the edge set E_G with the properties that*

- *each edge $e \in E_G$ is homeomorphic to a closed interval and connects two points of V_G; these two vertices are called the ends of e;*
- *the interior of an edge does not contain the vertices and two edges can intersect only at their ends.*

A *finite tree* is a finite connected graph without cycles.

Let T be a finite tree. A point p in T is called an *endpoint* if $T \setminus \{p\}$ is connected and is called a *branching point* if $T \setminus \{p\}$ has at least 3 connected components. An important property of the finite tree is that for any two points $p, q \in T$, there is a unique arc in T that connects p and q. This arc is denoted by $[p,q]_T$. For a subset $X \subset T$, denote by $[X]_T$ the connected hull of X in T.

Let X be a finite connected graph. We call $f: X \to X$ a *Markov map* if each edge of X admits a finite subdivision into segments and f maps each segment continuously monotonically onto some edge of X.

Enumerating the edges of X by γ_i , $i = 1, \ldots, k$, we obtain a *transition matrix* $D_f = (a_{ij})_{k \times k}$ of (X, f) such that $a_{ij} = \ell$ ($\ell \geq 1$) if $f(\gamma_i)$ covers ℓ times γ_i and $a_{ij} = 0$ otherwise. Note that for different enumerations of the edges, the obtained transition matrices are pairwise similar. Denote by $\rho(D_f)$ the leading eigenvalue of D_f. By the Perron-Frobenius theorem, $\rho(D_f)$ is a non-negative real number and it is also the growth rate of $\|D_f^n\|$ for any matrix norm.

Since the matrix D_f has integer coefficients, we have $\rho(D_f) \geq 1$ unless there exists n such that $D_f^n = 0$ (nilpotent). But the latter case cannot happen because every edge is mapped onto at least one edge. It follows that we have always $\rho(D_f) \geq 1$.

The following is classical:

Proposition A.6. *The topological entropy $h(X, f)$ is equal to* $\log \rho(D_f)$.

APPENDIX B. PROOF OF PROPOSITION 14.1

For a rational angle $\theta \neq 0$, we want to prove that L_θ, the cluster set in $\mathbb{T} \setminus \Delta$ of the pre-major lamination, is equal to the polynomial lamination of $f: z \mapsto z^2 + c_\theta$.

Proof. To get a good geometric intuition of an invariant lamination, we choose to use the unit disk model. Let

$$\wp : \mathbb{T} \to \{\text{closures of hyperbolic geodesics in } \mathbb{D}\}$$

map a point (x, y) to $\overline{xy}$, which is called a *leaf* (or a point in the case $x = y$). Denote the image of Pre_θ and L_θ under $\wp$ by $\mathrm{Pre}_\theta^{\mathbb{D}}$ and $L_\theta^{\mathbb{D}}$, respectively. For each $i \geq 0$, define

$$L_0 = \wp(b_0) \qquad \text{and} \qquad L_i = \wp(b_i \setminus b_{i-1}), \quad i \geq 1.$$

Then we have

$$L_0 = \left\{ \overline{\tfrac{\theta}{2}\tfrac{\theta+1}{2}} \right\}$$

and $\tau : L_{i+1} \to L_i$ is at least 2 to 1. (Here by abuse of notation we set $\tau(\overline{xy}) = \overline{\tau(x)\tau(y)}$.)

The leaf $\overline{\frac{\theta}{2}\frac{\theta+1}{2}}$ corresponds in the dynamical plane to the major cutting line $\mathcal{R}\left(\frac{\theta}{2}, \frac{\theta+1}{2}\right)$.

For any $n \geq 1$, let $\overline{x_n y_n}$ be a leaf of L_n. Set

$$(x_{n-i}, y_{n-i}) := F^i(x_n, y_n), \quad i \in [0, n].$$

Then $\overline{x_i y_i}$ is a leaf of L_i. For each i, the set $\{x_i, y_i\}$ belongs to $\overline{S_0}$ or $\overline{S_1}$. (Recall that S_0 denotes the half open circle containing 0 bounded by $\frac{\theta}{2}, \frac{\theta+1}{2}$.) The ray-pair $\mathcal{R}(x_i) \cup \mathcal{R}(y_i)$ belongs to the corresponding component of $\overline{V_0}$, $\overline{V_1}$. Inductively, this follows:

- If θ is preperiodic, each $\{x_i, y_i\}$ is an angle-pair at a point z_i with $f_\theta(z_i) = z_{i-1}$. In this case, we define a cutting curve corresponding to $\overline{x_i y_i}$ by

$$\mathcal{R}(x_i, y_i) := \mathcal{R}(x_i) \cup \{z_i\} \cup \mathcal{R}(y_i)$$

- If θ is periodic, each $\{x_i, y_i\}$ is a pair of angles whose rays support a common Fatou component U_i, and $f(U_i) = U_{i-1}$. In this case, we define a cutting curve corresponding to $\overline{x_i y_i}$ by

$$\mathcal{R}(x_i, y_i) := \mathcal{R}(x_i) \cup \overline{r_{U_i}(\alpha_i) \cup r_{U_i}(\beta_i)} \cup \mathcal{R}(y_i)$$

where $r_{U_i}(\alpha_i)$ (resp. $r_{U_i}(\beta_i)$) is the internal ray in U_i landing at $\gamma(x_i)$ (resp. $\gamma(y_i)$).

For $i \geq 0$, set

$$\Gamma_i = \left\{ \mathcal{R}(x, y) \;\middle|\; \overline{xy} \in L_i \right\} \quad \text{and} \quad \Gamma_\infty = \bigcup_{i \geq 0} \Gamma_i.$$

Then there is a natural bijection $l_i : L_i \to \Gamma_i$ which maps a leaf $\overline{xy} \in L_i$ to the curve $\mathcal{R}(x, y)$ and satisfies the commutative diagram:

$$\begin{array}{ccc} L_i & \xrightarrow{\tau} & L_{i-1} \\ l_i \downarrow & & \downarrow l_{i-1} \\ \Gamma_i & \xrightarrow{f_\theta} & \Gamma_{i-1}. \end{array}$$

We will talk a little more about the curves in Γ_∞ for a periodic angle θ. In this case, there is the unique Fatou component cycle of degree 2 which contains the critical point 0. Then the internal angle for each Fatou component can be uniquely defined. Using the fact that the two internal rays contained in $\mathcal{R}\left(\frac{\theta}{2}, \frac{\theta+1}{2}\right)$ have angles 0 and 1/2, together with the construction of L_i $(i \geq 1)$, the correspondence between L_i and Γ_i, and the definition of internal angles of a Fatou component, it is not difficult to check the following results:

- Every curve $\mathcal{R}(x,y) \in \Gamma_\infty$ passes through a single Fatou component U, and the rays $\mathcal{R}(x)$, $\mathcal{R}(y)$ support the component U.
- The two internal rays of U contained in $\mathcal{R}(x,y)$ have angles $\frac{j}{2^n}, \frac{j+1}{2^n}$ for some $n \geq 1$, $0 \leq j \leq 2^n - 1$.
- For any Fatou component U, any integer $n \geq 1$ and $0 \leq j \leq 2^n - 1$, there exists a unique curve $\mathcal{R}(x,y) \in \Gamma_\infty$ such that it contains the two internal rays $r_U\left(\frac{j}{2^n}\right)$ and $r_U\left(\frac{j+1}{2^n}\right)$. In fact, there exists a minimal k such that f_θ^k sends $r_U(\frac{j}{2^n})$ and $r_U(\frac{j+1}{2^n})$ to the internal rays contained in $\mathcal{R}(\frac{\theta}{2}, \frac{\theta+1}{2})$. Suppose that $f_\theta^k(r_U(\frac{j}{2^n}))$ is the periodic one. Then $\mathcal{R}(x,y)$ is a lift by f_θ^k of $\mathcal{R}(\frac{\theta}{2}, \frac{\theta+1}{2})$ based at $r_U(\frac{j}{2^n})$.

With these preparations, we are ready to prove the proposition. First, we will show that each leaf in $L_\theta^{\mathbb{D}}$ represents an adjacent angle-pair.

Let $\overline{xy}$ be a leaf in $L_\theta^{\mathbb{D}}$. Then there exists a sequence of leaves $\overline{x_i y_i} \in L_i$ such that $\{\overline{x_i y_i}, i \geq 1\}$ converges to $\overline{xy}$ in the Hausdorff topology. Without loss of generality, we may assume that

$$\lim_{n\to\infty} x_n = x, \qquad \lim_{n\to\infty} y_n = y.$$

To see the dynamical explanation of the point (x,y), we need to discuss the set of accumulation points R of the corresponding curves $\mathcal{R}(x_i, y_i)$. At first, we only consider the case that $\overline{x_i y_i}$ converges to $\overline{xy}$ from one side. We distinguish 3 basic cases.

1. No curves in $\{\mathcal{R}(x_i, y_i) \mid i \geq 1\}$ pass through Fatou components. In this case, the external rays $\mathcal{R}(x_i)$ converge to $\mathcal{R}(x)$ and the rays $\mathcal{R}(y_i)$ converge to $\mathcal{R}(y)$. Consequently, the common landing point z_i of $\mathcal{R}(x_i)$ and $\mathcal{R}(y_i)$ converge to a point $z \in \mathcal{J}_f$, which must be the common landing point of $\mathcal{R}(x)$ and $\mathcal{R}(y)$. Then we have

$$R := \lim_{i\to\infty} \mathcal{R}(x_i, y_i) = \mathcal{R}(x) \cup \{z\} \cup \mathcal{R}(y).$$

We claim that $\{x,y\}$ is an adjacent angle-pair. Otherwise, each component of $\mathbb{C} \setminus \mathcal{R}(x,y)$ contains at least two components of $\mathcal{K}_f \setminus z$. Since $\lim_{i\to\infty} (x_i, y_i) = (x,y)$, for sufficiently large i, the points z_i belong to a common component of $\mathcal{K}_f \setminus z$. Set

$$\{z_i\}_{i\geq 1} \subset K \subset U,$$

where K is a component of $\mathcal{K}_f \setminus z$ and U is a component of $\mathbb{C} \setminus \mathcal{R}(x,y)$. Let $\{v,w\}$ be the angle-pair at z bounding K. On one hand, we know

$$[v,w] \subsetneq [x,y],$$

since, by assumption, the pair $\{x,y\}$ is not adjacent. On the other hand, the angles x_i, y_i $(i \geq 1)$ belong to $[v,w]$ and hence the interval $[x,y]$ is a subset of $[v,w]$. This leads to a contradiction.

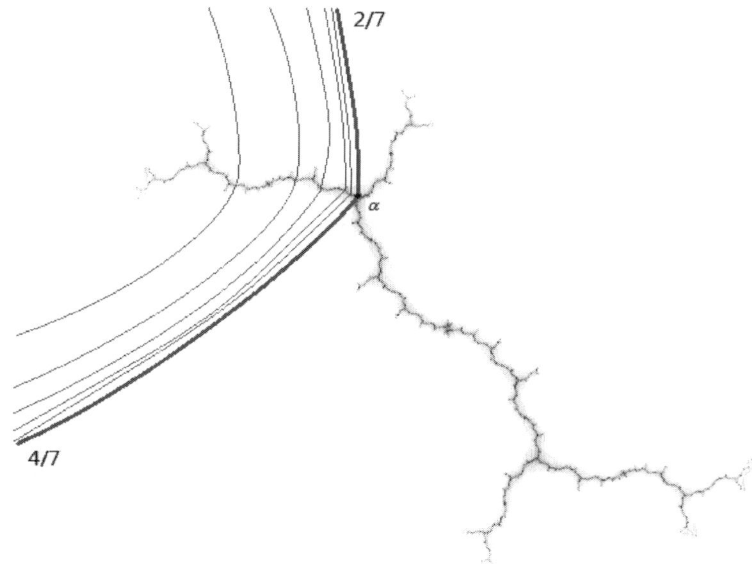

Figure 20: The case of $\theta = 1/4$.

Example B.1. Set $\theta = 1/4$. The red curves in Figure 20 are cutting curves $\{\mathcal{R}(x_i, y_i)\}$ satisfying case (1) such that $x_i \to 2/7, y_i \to 4/7$ as $n \to \infty$. It is seen that $\{\mathcal{R}(2/7), \mathcal{R}(4/7)\}$ is an adjacent ray-pair landing at the α-fixed point of $f_{1/4}$ and $R = \mathcal{R}(2/7) \cup \{\alpha\} \cup \mathcal{R}(4/7)$.

2. For each $i \geq 1$, the curve $\mathcal{R}(x_i, y_i)$ passes through a Fatou component U_i and the diameters of U_i $(i \geq 1)$ converge to 0. In this case, the external rays $\mathcal{R}(x_i)$ converge to $\mathcal{R}(x)$, the rays $\mathcal{R}(y_i)$ converge to $\mathcal{R}(y)$ and the Fatou components U_i converge to a point $z \in \mathcal{J}_f$. This point must be the common landing point of $\mathcal{R}(x)$ and $\mathcal{R}(y)$. Then we have

$$R = \lim_{i \to \infty} \mathcal{R}(x_i, y_i) = \mathcal{R}(x) \cup \{z\} \cup \mathcal{R}(y).$$

   Using the same argument as in (1) above, we see that the angle-pair $\{x, y\}$ is an adjacent angle-pair.
3. For each $i \geq 1$, the curve $\mathcal{R}(x_i, y_i)$ passes through a Fatou component U_i and the infimum of the diameters of U_i is bounded from 0. In this case, for sufficiently large i, and passing to a subsequence if necessary, all U_i coincide, and hence may be denoted by U. It follows that the two internal rays of U contained in the curves $\mathcal{R}(x_i, y_i)$ tend to a single internal ray as $i \to \infty$. So the two limit external rays $\mathcal{R}(x)$ and $\mathcal{R}(y)$ must land together. In fact,

$$R = \lim_{i \to \infty} \mathcal{R}(x_i, y_i) = \mathcal{R}(x) \cup \mathcal{R}(y) \cup r_U(\alpha) \cup \{z\}$$

   where $z = \gamma(x) = \gamma(y)$ and $r_U(\alpha)$ is the internal ray in U landing at z. Since each pair of rays $\mathcal{R}(x_i)$, $\mathcal{R}(y_i)$ support the component U, the angle-pair $\{x, y\}$ must be an adjacent angle-pair.

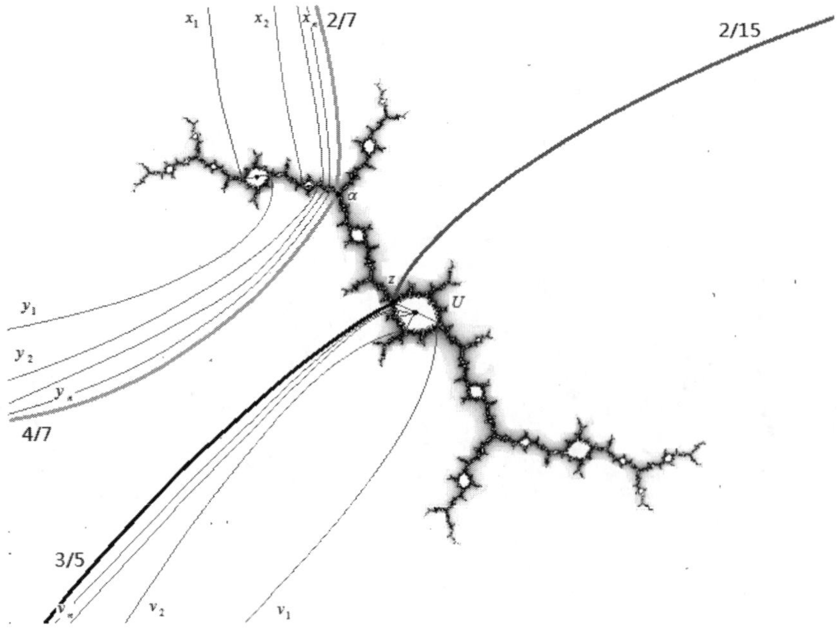

Figure 21: The case of $\theta = 4/15$.

Example B.2. Set $\theta = 4/15$. The blue curves in Figure 21 are cutting curves $\{\mathcal{R}(x_i, y_i)\}$ satisfying case (2) such that $x_i \to 2/7, y_i \to 4/7$ as $n \to \infty$. It is seen that $\{\mathcal{R}(2/7), \mathcal{R}(4/7)\}$ is an adjacent ray-pair landing at the α-fixed point of $f_{4/15}$ and $R = \mathcal{R}(2/7) \cup \{\alpha\} \cup \mathcal{R}(4/7)$. The red curves in Figure 21 are cutting curves $\{\mathcal{R}(2/15, v_i)\}$ satisfying case (3) such that $v_i \to 3/5$ as $n \to \infty$. It is seen that $\{\mathcal{R}(2/15), \mathcal{R}(3/5)\}$ is an adjacent ray-pair landing at z and $R = \mathcal{R}(2/15) \cup r_U(0) \cup \mathcal{R}(3/5)$.

Generally, we allow the leaves $\overline{x_i y_i}$ to converge to $\overline{xy}$ from two sides. Then the sequence $\{(x_i, y_i)\}_{i \geq 1}$ may contain subsequences satisfying 1, 2 or 3 of the basic cases described above from either side of $\overline{xy}$. So the set of accumulation points R is formed by the possible combinations of the set of accumulation points in the basic cases from two sides. That is, R consists of an adjacent ray-pair $\{\mathcal{R}(x), \mathcal{R}(y)\}$ together with eventual 0, 1 or 2 internal rays. Anyway, a leaf $\overline{xy} \in L_\theta^{\mathbb{D}}$ corresponds to the unique adjacent angle-pair $\{x, y\}$.

Next, we will show the opposite implication. That is, any adjacent angle-pair $\{x, y\}$ is contained in $L_\theta^{\mathbb{D}}$ as a leaf $\overline{xy}$.

Let $\{x, y\}$ be an adjacent angle-pair at z. Let V be one of the regions bounded by $\mathcal{R}(x)$ and $\mathcal{R}(y)$ without other rays landing at z, set $K = V \cap \mathcal{K}$. Choose any point $w \in K$, denote by $[w, z]_K$ the regulated arc-connecting z and w in K. We distinguish two cases:

1. $[w, z]_K \cap \mathcal{J}$ clusters at z. Then two further cases may happen.
   - $[w, z]_K$ passes through an ordered infinite sequence of Fatou components $\{U_i, i \geq 1\}$ from w to z so that U_i converges to z as $i \to \infty$. We claim that

for any $i \geq 2$, we can pick up a curve $\mathcal{R}(x_i, y_i) \in \Gamma_\infty$ such that it separates U_{i-1} and z (in Figure 21, we have $\{x, y\} = \{2/7, 4/7\}$ and the blue curves are what we want). By this claim, the sequence of curves $\mathcal{R}(x_i, y_i)$ $(i \geq 2)$ converge to the ray-pair of $\{x, y\}$ in the Hausdorff topology, so $\overline{xy} \in L_\theta^{\mathbb{D}}$.

Proof of this Claim: Fix U_i. The intersection of $U_i \cap [w, z]_K$ consists of two internal rays $r_{U_i}(\alpha)$ and $r_{U_i}(\beta)$ of U_i. We assume that $r_{U_i}(\alpha)$ lies in between z and $r_{U_i}(\beta)$. If α is not equal to $\frac{j}{2^n}$ for any $n \geq 1, j \geq 0$, we can choose integers n_0 and j_0 so that

$$\alpha \in \left(\frac{j_0}{2^{n_0}}, \frac{j_0+1}{2^{n_0}}\right) \quad \text{and} \quad \beta \notin \left[\frac{j_0}{2^{n_0}}, \frac{j_0+1}{2^{n_0}}\right].$$

Then the curve $\mathcal{R}(x_i, y_i) \in \Gamma_\infty$ that contains $r_{U_i}\left(\frac{j_0}{2^{n_0}}\right)$ and $r_{U_i}\left(\frac{j_0+1}{2^{n_0}}\right)$ separates U_{i-1} and z. If $\alpha = \frac{j_0}{2^{n_0}}$ for some integers n_0, j_0, we can choose a sufficiently large n and a suitable integer j such that

$$\alpha = j/2^n \quad \text{and} \quad \beta \notin \left[\frac{j-1}{2^n}, \frac{j+1}{2^n}\right].$$

In this case, either the curve (in Γ_∞) containing $r_{U_i}\left(\frac{j-1}{2^n}\right), r_{U_i}\left(\frac{j}{2^n}\right)$ or the curve (in Γ_∞) containing $r_{U_i}\left(\frac{j-1}{2^n}\right), r_{U_i}\left(\frac{j}{2^n}\right)$ separates U_{i-1} and z. Denote this curve by $\mathcal{R}(x_i, y_i)$.

- No subsequence of the Fatou components (if any) passed through by $[w, z]_K$ converges to z. Then, replacing w by a point closer to z if necessary, we may assume $[w, z]_K \subset \mathcal{J}$. In this case, we can pick up a sequence of points

$$\{z_i, \ i \geq 1\} \subset [w, z]_K \cap (\mathcal{J} \setminus \text{Pre-cr})$$

from w to z such that z_i $(i \geq 1)$ converges to z as $n \to \infty$. We claim that for any $i \geq 1$, we can pick up a curve $\mathcal{R}(x_i, y_i) \in \Gamma_\infty$ so that it separates z_i and z. Thus, we also obtain that $\overline{xy} \in L_\theta^{\mathbb{D}}$ (in Figure 20, we have $\{x, y\} = \{2/7, 4/7\}$ and the red curves are what we want).

Proof of this Claim: We only need to prove that each $\mathcal{R}(x_i, y_i)$ separates z_i and z_{i+1}. Fix $i \geq 1$. By Corollary 16.4, $\iota_\theta(z_i) \neq \iota_\theta(z_{i+1})$. Denote by p_i the index of the first distinct entries of $\iota_\theta(z_i)$ and $\iota_\theta(z_{i+1})$. Then the points $f^{p_i}(z_i)$ and $f^{p_i}(z_{i+1})$ are separated by the curve $\mathcal{R}\left(\frac{\theta}{2}, \frac{\theta+1}{2}\right)$. Lifting this curve along the orbit of the pair $\{z_i, z_{i+1}\}$, we obtain a curve $\mathcal{R}(x_i, y_i) \in \Gamma_\infty$ that separates z_i and z_{i+1}.

2. The set $[w, z]_K \cap \mathcal{J}$ does not cluster z. In this case, the point z is on the boundary of a Fatou component $U \subset V$ and the angle-pair $\{x, y\}$ bounds U. Denote by $r_U(\alpha)$ the internal ray of U landing at z. If α is not equal to $\frac{j}{2^n}$ for any $n \geq 1, j \geq 0$, we can choose a sequence of integers $\{j_n, n \geq 1\}$ such that $\alpha \in \left(\frac{j_n}{2^n}, \frac{j_n+1}{2^n}\right)$ for any $n \geq 1$. Let $\mathcal{R}(x_n, y_n)$ be the curve in Γ_∞ that contains $r_U\left(\frac{j_n}{2^n}\right)$ and $r_U\left(\frac{j_n+1}{2^n}\right)$. Then this sequence of curves converge to the ray-pair of $\{x, y\}$ as $n \to \infty$. If $\alpha = j_0/2^{n_0}$ for some integers n_0, j_0, then α can be expressed as $j_n/2^n$ for every $n \geq n_0$. Note that one of x and y, say x, belongs to the set $\text{Pre}_\theta^{\mathbb{S}}$. Then we obtain

two sequences of curves,

$$\{\mathcal{R}(x,y_n)\}_{n\geq n_0} \quad \text{and} \quad \{\mathcal{R}(x,s_n)\}_{n\geq n_0},$$

belonging to Γ_∞ such that each ray $\mathcal{R}(x,y_n)$ contains the internal rays $r_U(\frac{j_n-1}{2^n})$, $r_U\left(\frac{j_n}{2^n}\right)$ and each ray $\mathcal{R}(x,s_n)$ contains the internal rays $r_U\left(\frac{j_n+1}{2^n}\right), r_U\left(\frac{j_n}{2^n}\right)$. By the construction of the curves in Γ_∞, we can see that one sequence of curves converge to a single ray $\mathcal{R}(x)$ and the other sequence converges to the ray-pair of $\{x,y\}$. So $\overline{xy}\in L_\theta^{\mathbb{D}}$ (in Figure 21, we have $x=2/15, y=3/5$ and the red curves are what we want). □

APPENDIX C. PROOF OF LEMMA 16.3

For a rational angle $\theta\in\mathbb{S}\setminus\{0\}$, and $f:z\mapsto z^2+c_\theta$, this lemma claims that two non pre-major angles α and β form an angle-pair (i.e., the external rays $\mathcal{R}(\alpha)$ and $\mathcal{R}(\beta)$ land together) if and only if they have the same itinerary relative to the major leaf.

The necessity is obvious because α and β have the same itineraries as that of z, where z is the common landing point of $\mathcal{R}(\alpha)$ and $\mathcal{R}(\beta)$.

For the sufficiency, we only need to prove the following result: if the first n-th entries of $\iota_\theta(\alpha)$ and $\iota_\theta(\beta)$ are identical (see Subsection 16.2 for the definition of $\iota_\theta(\alpha)$), then the distance of z,w is less than $C\cdot\lambda^{-n}$ with some metric, where $\lambda>1$, C are constants and z,w are the landing points of $\mathcal{R}(\alpha),\mathcal{R}(\beta)$ respectively.

We will prove this by distinguishing between the following two cases: i) the case that c_θ is (strictly) periodic and ii) the case that c_θ is (strictly) preperiodic.

i) Case c_θ is periodic:

In this case, $\mathcal{P}_f$ consists of the orbit of c_θ. For a point $p\in\mathcal{P}_f$, set

$$B_p=\left\{z\in U_p\,\middle|\,|\phi_{U_p}(z)|<e^{-1}\right\}$$

and set

$$B_\infty=\left\{z\in U_\infty\,\middle|\,|\phi_{U_\infty}(z)|>e\right\}.$$

Define $\mathcal{L}$ the complement of the union

$$B_\infty\cup\left(\bigcup_{p\in\mathcal{P}_f}B_p\right)$$

in $\mathbb{C}$. Then $\mathcal{L}$ is a compact connected neighborhood of the Julia set $\mathcal{J}$ on which f is uniformly expanding with respect to the hyperbolic metric ρ in a neighborhood U of $\mathcal{L}$, i.e., $\exists\lambda>1$ such that if $\gamma\subset\mathcal{L}$ is an arc and $f:\gamma\to f(\gamma)$ is a homeomorphism, then

$$\mathrm{diam}_\rho(f(\gamma))>\lambda\cdot\mathrm{diam}_\rho(\gamma).$$

Define

$$\mathcal{R}(P_\theta):=\bigcup_{n\geq1}f^n\left(\mathcal{R}\left(\frac{\theta}{2},\frac{\theta+1}{2}\right)\right).$$

For any $p, q \in \mathcal{L}$, we can find an arc in $\mathcal{L}$ connecting p, q such that it doesn't cross the rays in $\mathcal{R}(P_\theta)$. So we can define a number

$$M_{p,q} = \inf \Big\{ \text{length}_\rho(\gamma) \mid \gamma \subset \mathcal{L} \text{ is an arc connecting } p \text{ and } q \text{ but not crossing the rays in } \mathcal{R}(P_\theta) \Big\}.$$

It is not difficult to check that $M_{p,q}$ is uniformly bounded for $p, q \in \mathcal{L}$. That is, there is a $C > 0$ so that $M_{p,q} < C$ for any $p, q \in \mathcal{L}$.

Now, suppose the first n-th entries of $\iota_\theta(\alpha)$ and $\iota_\theta(\beta)$ are identical. Set z, w to be the landing points of $\mathcal{R}(\alpha), \mathcal{R}(\beta)$, respectively, and z_i, w_i the i-th iteration of z, w by f. Choose an arc $\gamma_n \in \mathcal{L}$ with

$$\text{length}_\rho(\gamma_n) < C$$

such that it connects z_n, w_n and doesn't cross the rays in $\mathcal{R}(P_\theta)$. Lift γ_n to the arc γ_{n-1} with starting point z_{n-1}. Since γ_n doesn't cross the rays in $\mathcal{R}(P_\theta)$, then so doesn't γ_{n-1} and it implies that γ_{n-1} is contained in the same component of

$$\mathcal{L} \setminus \mathcal{R}\left(\frac{\theta}{2}, \frac{\theta+1}{2}\right).$$

Note that the i-th entries of $\iota_\theta(\alpha)$ and $\iota_\theta(\beta)$ are equal if and only if z_i, w_i belongs to the same component of $\mathcal{L} \setminus \mathcal{R}\left(\frac{\theta}{2}, \frac{\theta+1}{2}\right)$. Then w_{n-1} is the unique preimage of w_n contained in the component of $\mathcal{L} \setminus \mathcal{R}\left(\frac{\theta}{2}, \frac{\theta+1}{2}\right)$ that contains z_{n-1}. It follows that γ_{n-1} connects w_{n-1} and z_{n-1}. As

$$\gamma_{n-1} \subset f^{-1}(\mathcal{L}) \subset \mathcal{L},$$

by the uniform expansion of f in $\mathcal{L}$, we have

$$\text{length}_\rho(\gamma_{n-1}) < \lambda^{-1} \text{length}_\rho(\gamma_n).$$

Repeating this process n times, we obtain an arc γ_0 which is the n-th lift of γ_n with the starting point z. With the same argument as before, the ending point of γ_0 is w and

$$\text{dist}_\rho(z, w) \leq \text{length}_\rho(\gamma_0) < \lambda^{-n} \text{length}_\rho(\gamma_n) < C\lambda^{-n}.$$

ii) Case c_θ is preperiodic:

In this case, $\mathcal{J} = \mathcal{K}$ and f is uniformly expanding with respect to an *admissible metric (orbifold metric)* ρ in a compact neighborhood of $\mathcal{J}$ (see [DH84, McM94, Mil06]).

Since $\mathcal{J}$ is compact, the length of the regulated arc $[z, w]$ (with respect to the metric ρ) is uniformly bounded by a constant C for any $z, w \in \mathcal{J}$. Using the same notions as the periodic case, if the first n-th entries of $\iota_\theta(\alpha)$ and $\iota_\theta(\beta)$ are equal, then for $0 \leq i \leq n-1$, the polynomial f maps the regulated arc $[z_i, w_i]$ homeomorphically to the regulated arc $[z_{i+1}, w_{i+1}]$. By the uniform expansion of f on $\mathcal{J}$, we have that

$$\mathrm{dist}_\rho(z,w) \le \mathrm{length}_\rho([z,w]) \le \lambda^{-n}\mathrm{length}_\rho([z_n,w_n]) < C\cdot\lambda^{-n}.$$

We note that the result of Lemma 16.3 is covered by a more general theorem in [Zen15].

APPENDIX D. ADDITIONAL IMAGES

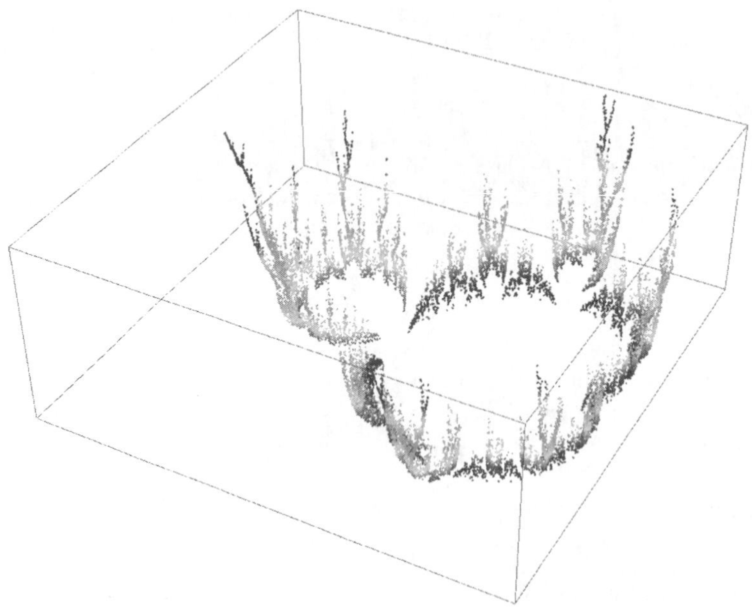

Figure 22: A plot of core entropy, using data computed by Wolf Jung. The horizontal plane is $\mathbb{C}$ and the vertical axis is the interval $[0, \log 2] \subset \mathbb{R}$. A plotted point $(c,h) \in \mathbb{C}\times\mathbb{R}$ represents the data that the core entropy of the polynomial $z \mapsto z^2 + c$ is h. Data is shown for a selection of parameters c in the boundary of the Mandelbrot set with rational external angle. Points are color-coded according to the h-coordinate, core entropy.

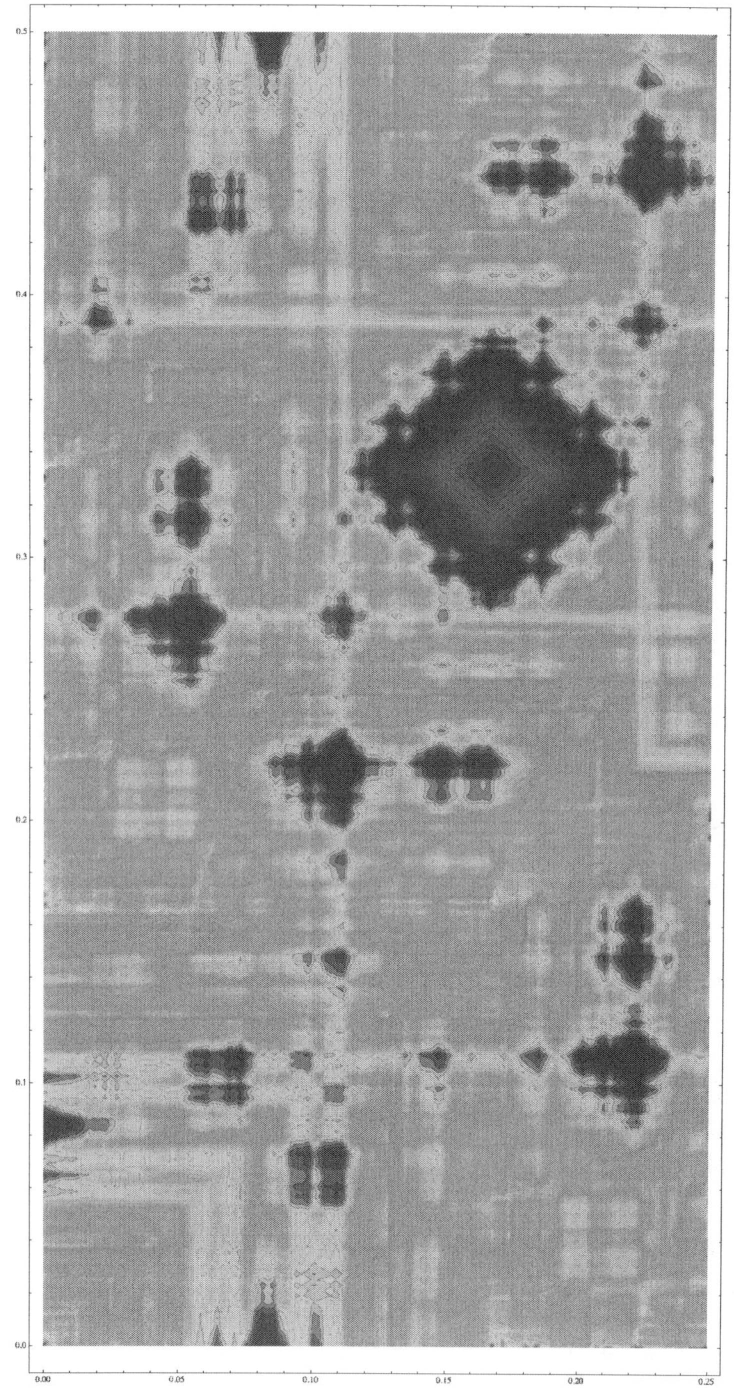

Figure 23: A contour plot by W. Thurston of core entropy as a function on PM(3), using the starting point parametrization of Section 11.4.1.

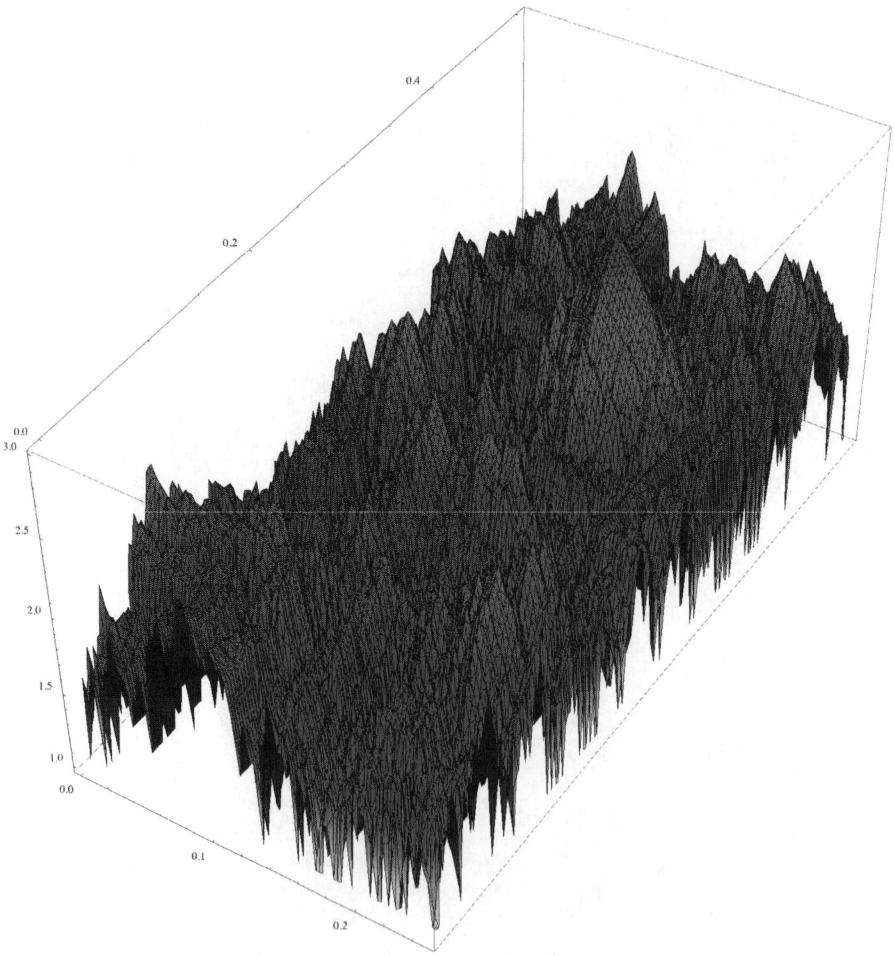

Figure 24: A plot by W. Thurston of the exponential of core entropy as a function on PM(3), using the starting point parametrization of Section 11.4.1.

Acknowledgments

The authors thank the anonymous referee for their careful reading and helpful comments. During the preparation of this manuscript, Kathryn Lindsey received support from the NSF via a Graduate Research Fellowship and, later, a Postdoctoral Research Fellowship. Hyungryul Baik was partially supported by Samsung Science & Technology Foundation grant No. SSTF-BA1702-01. Yan Gao was partially supported by NSFC grant No. 11871354. Dylan Thurston was partially supported by NSF Grant Number DMS-1507244.

REFERENCES

[AKM65] R. L. Adler, A. G. Konheim, and M. H. McAndrew, *Topological entropy*, Trans. Amer. Math. Soc. **114** (1965), 309–319.

[BCO11] Alexander M. Blokh, Clinton P. Curry, and Lex G. Oversteegen, *Locally connected models for Julia sets*, Adv. Math. **226** (2011), no. 2, 1621–1661. MR 2737795 (2012d:37106).

[BL02] A. Blokh and G. Levin, *An inequality for laminations, Julia sets and "growing trees,"* Ergodic Theory Dynam. Systems **22** (2002), no. 1, 63–97. MR 1889565 (2003i:37045).

[BMOV13] Alexander M. Blokh, Debra Mimbs, Lex G. Oversteegen, and Kirsten I. S. Valkenburg, *Laminations in the language of leaves*, Trans. Amer. Math. Soc. **365** (2013), no. 10, 5367–5391. MR 3074377.

[BO04a] Alexander Blokh and Lex Oversteegen, *Backward stability for polynomial maps with locally connected Julia sets*, Trans. Amer. Math. Soc. **356** (2004), no. 1, 119–133 (electronic). MR 2020026 (2005c:37081).

[BO04b] ———, *Wandering triangles exist*, C. R. Math. Acad. Sci. Paris **339** (2004), no. 5, 365–370. MR 2092465 (2005g:37081).

[BO09] ———, *Wandering gaps for weakly hyperbolic polynomials*, Complex Dynamics: Families and Friends, A. K. Peters, Wellesley, MA, 2009.

[BOPT14] Alexander Blokh, Lex Oversteegen, Ross Ptacek, and Vladlen Timorin, *The main cubioid*, Nonlinearity **27** (2014), no. 8, 1879–1897. MR 3246159.

[DH84] Andrien Douady and John H. Hubbard, *Étude dynamique des polynômes complexes*, Publications Mathématiques d'Orsay, vol. 84–2, Université de Paris-Sud, Département de Mathématiques, Orsay, 1984, available in English as *Exploring the Mandelbrot set. The Orsay notes* from http://math.cornell.edu/~hubbard/OrsayEnglish.pdf.

[dMvS93] Welington de Melo and Sebastian van Strien, *One-dimensional dynamics*, Ergebnisse der Mathematik und ihrer Grenzgebiete (3) [Results in Mathematics and Related Areas (3)], vol. 25, Springer-Verlag, Berlin, 1993. MR 1239171.

[Dou95] A. Douady, *Topological entropy of unimodal maps: monotonicity for quadratic polynomials*, Real and complex dynamical systems (Hillerød, 1993), NATO Adv. Sci. Inst. Ser. C Math. Phys. Sci., vol. 464, Kluwer Acad. Publ., Dordrecht, 1995, pp. 65–87.

[FM12] Benson Farb and Dan Margalit, *A primer on mapping class groups*, Princeton Mathematical Series, vol. 49, Princeton University Press, Princeton, NJ, 2012.

[Fur67] Harry Furstenberg, *Disjointness in ergodic theory, minimal sets, and a problem in Diophantine approximation*, Math. Systems Theory **1** (1967), 1–49.

[Gao] Gao Yan, *On Thurston's core entropy algorithm*, To appear in Transactions of the AMS.

[Gao13] ———, *Dynatomic curve and core entropy for iteration of polynomials*, Ph.D. thesis, Université d'Angers, France, April 2013.

[Gol94] L. R. Goldberg, *On the multiplier of a repelling fixed point*, Invent. Math. **118** (1994), 85–108.

[Jun14] Wolf Jung, *Core entropy and biaccessibility of quadratic polynomials*, Preprint, 2014, arXiv:1401.4792.

[Kiw02] Jan Kiwi, *Wandering orbit portraits*, Trans. Amer. Math. Soc. **354** (2002), no. 4, 1473–1485. MR 1873015 (2002h:37070).

[Kiw04] ———, *$\mathbb{R}$eal laminations and the topological dynamics of complex polynomials*, Adv. Math. **184** (2004), no. 2, 207–267. MR 2054016 (2005b:37094).

[Kiw05] J. Kiwi, *Combinatorial continuity in complex polynomial dynamics*, Proc. London Math. Soc. **91** (2005), no. 3, 215–248.

[Lev98] G. Levin, *On backward stability of holomorphic dynamical systems*, Fund. Math. **158** (1998), no. 2, 97–107. MR 1656942 (99j:58171).

[McM94] Curtis T. McMullen, *Complex dynamics and renormalization*, Annals of Mathematics Studies, vol. 135, Princeton University Press, Princeton, NJ, 1994. MR 1312365.

[Mil06] John Milnor, *Dynamics in one complex variable*, third ed., Annals of Mathematics Studies, vol. 160, Princeton University Press, Princeton, NJ, 2006. MR 2193309.

[Mim10] Debra L. Mimbs, *Laminations: A topological approach*, ProQuest LLC, Ann Arbor, MI, 2010, Thesis (Ph.D.), University of Alabama at Birmingham. MR 2801683.

[Poi09] Alfredo Poirier, *Critical portraits for postcritically finite polynomials*, Fund. Math. **203** (2009), no. 2, 107–163.

[Poi10] ———, *Hubbard trees*, Fund. Math. **208** (2010), no. 3, 193–248.

[Pta13] Ross M. Ptacek, *Laminations and the dynamics of iterated cubic polynomials*, ProQuest LLC, Ann Arbor, MI, 2013, Thesis (Ph.D.), University of Alabama at Birmingham. MR 3187504.

[Thu09] William P. Thurston, *Polynomial dynamics from combinatorics to topology*, Complex Dynamics: Families and Friends, A. K. Peters, Wellesley, MA, 2009, pp. 1–109.

[Tio15] Giulio Tiozzo, *Topological entropy of quadratic polynomials and dimension of sections of the Mandelbrot set*, Adv. Math. **273** (2015), 651–715.

[Tio16] ———, *Continuity of core entropy of quadratic polynomials*, Invent. Math. **203** (2016), no. 3, 891–921.

[Tom14] J. Tomasini, *Géométrie combinatoire des fractions rationnelles*, Ph.D. thesis, Université d'Angers, 2014.

[Zen14] Jinsong Zeng, *On the existence of shift locus for given critical portrait*, Preprint, 2014.

[Zen15] ———, *Criterion for rays landing together*, Preprint online at https://arxiv.org/abs/1503.05931, March 2015.

ISRAEL JOURNAL OF MATHEMATICS **130** (2002), 21–27

THE SYMBOLIC DYNAMICS OF TILING THE INTEGERS*

BY

ETHAN M. COVEN

Department of Mathematics, Wesleyan University
Middletown, CT 06459-0128, USA
e-mail: ecoven@wesleyan.edu

AND

WILLIAM GELLER

Department of Mathematics, Indiana University-Purdue University at Indianapolis
Indianapolis, IN 46202-3216, USA
e-mail: wgeller@math.iupui.edu

AND

SYLVIA SILBERGER

Department of Mathematics, Hofstra University
Hempstead, NY 11549, USA
e-mail: matsbs@hofstra.edu

AND

WILLIAM THURSTON

Department of Mathematics, University of California Davis
Davis, CA 95616, USA
e-mail: wpt@math.ucdavis.edu

ABSTRACT

A finite collection $\mathcal{P}$ of finite sets tiles the integers iff the integers can be expressed as a disjoint union of translates of members of $\mathcal{P}$. We associate with such a tiling a doubly infinite sequence with entries from $\mathcal{P}$. The set of all such sequences is a sofic system, called a tiling system. We show that, up to powers of the shift, every shift of finite type can be realized as a tiling system.

* Some of this work was done at the Mathematical Sciences Research Institute (MSRI), where research is supported in part by NSF grant DMS-9701755. The first two authors thank K. Schmidt for useful conversations and ideas.
Received August 25, 1999 and in revised form April 18, 2001

1. Introduction

For notation, terminology, and basic results of symbolic dynamics, see the book by D. Lind and B. Marcus [LM].

Let $\mathcal{P} = \{P_1, \ldots, P_K\}$ be a finite collection of finite subsets of the integers $\mathbb{Z} = \{\ldots, -1, 0, 1, \ldots\}$, called **prototiles**. We normalize the prototiles so that each has minimum 0. A **tile** is a translate of a prototile. If the integers can be expressed as a disjoint union of tiles, $\mathbb{Z} = \bigcup(t_j + P_{k_j})$, we say that $\mathcal{P}$ **tiles the integers.** The corresponding **tiling of the integers** by $\mathcal{P}$ is the point $x = (x_i) \in \prod_{i=-\infty}^{\infty}\{1, 2, \ldots, K\}$ defined by $x_i = k$ if and only if there exists j such that $i \in t_j + P_{k_j}$ and $k_j = k$. Thus we can think of a tiling as being given by a bi-infinite sequence of colors, where the colors are in one-to-one correspondence with the prototiles.

Let σ denote the **shift**, $(\sigma(x))_i = x_{i+1}$. The collection of points corresponding to tilings of the integers by $\mathcal{P}$, denoted $T(\mathcal{P})$, is closed and shift-invariant. We call $\sigma\colon T(\mathcal{P}) \to T(\mathcal{P})$ a **tiling system**. We first show that every tiling system is sofic. We then prove our main result: up to powers of the shift, every shift of finite type can be realized as a tiling system.

MAIN THEOREM: *Let $\sigma\colon \Sigma \to \Sigma$ be a shift of finite type. Then there is a positive integer m and a tiling system $\sigma\colon T \to T$ such that*

(1) *$T = T_0 \cup T_1 \cup \cdots \cup T_{m-1}$, where the T_i are closed and cyclically permuted by the shift σ;*

(2) *$\sigma^m\colon \Sigma \to \Sigma$ is topologically conjugate to every $\sigma^m\colon T_i \to T_i$.*

Recall that a **Perron number** is a positive real algebraic integer that dominates its algebraic conjugates.

COROLLARY: *The set of topological entropies of tiling systems is the same as that of shifts of finite type, i.e., the set of logarithms of roots of Perron numbers.*

Remark: In the sequel we will sometimes, as is common in symbolic dynamics, call the space T a tiling system, the space Σ a shift of finite type, etc.

We note that partial results on nonemptiness of a tiling system generated by a single prototile are known. D. Newman [N] gave a simple number-theoretic criterion for determining whether or not a single prototile of prime power cardinality can be used to tile the integers. The first author and A. Meyerowitz [CM] found a related criterion which they showed to be sufficient for an arbitrary single prototile to tile the integers, and necessary if the cardinality is the product of two prime powers; its necessity for an arbitrary single prototile is unknown.

We are not aware of any results in this direction for tiling systems arising from multiple prototiles.

2. Tiling systems are sofic

Consider the following examples.

(1) $\mathcal{P} = \{\{0\},\{0,1\}\}$. It is more convenient to think of $\mathcal{P}$ as $\{R, BB\}$, (R=red, B = blue). Then $T(\mathcal{P})$ is the set of all bi-infinite indexed concatenations of R and B such that between any two consecutive occurrences of R there is an even number of B's, the well-known even system. In this case $T(\mathcal{P})$ is also a renewal system, although we do not use that fact here. Recall that a **renewal system** is the collection of indexed bi-infinite sequences which are concatenations of a finite set of finite words from some alphabet. In the sequel, we shall abuse notation and write $T(R, BB)$ in place of $T(\{\{0\},\{0,1\}\})$.

(2) $\mathcal{P} = \{\{0\},\{0,2\}\}$, which we replace by $\{R, B\_B\}$. Then $T(\mathcal{P})$ is the renewal system generated by words R, BRB, and $BBBB$.

The next example shows that not every subshift that is topologically conjugate to a tiling system is itself a tiling system.

(3) $\mathcal{P} = \{\{0\},\{0,2,3\}\}$, or $\{R, B\_BB\}$. $T(\mathcal{P})$ is the renewal system generated by R and $BRBB$. It is topologically conjugate to the renewal system generated by R and $BRBY$, but the latter is not a tiling system.

(4) $\mathcal{P} = \{R, BB\_B, Y\_\_Y\}$, i.e., $\{\{0\},\{0,1,3\},\{0,3\}\}$. In this case $T(\mathcal{P})$ is not a renewal system. (Otherwise, consider the renewal word W which includes the first B in $\dots RRRBBYBYYY\dots \in T(\mathcal{P})$. But no such W can exist, since some R^n must be a renewal word and $\dots RRRWRRR\dots \notin T(\mathcal{P})$.)

To show that every tiling system is sofic, recall the proof that the even system $\sigma\colon T(R, BB) \to T(R, BB)$ is sofic — it is the image of the shift of finite type $\sigma\colon \widetilde{T}(R_1, B_1B_2) \to \widetilde{T}(R_1, B_1B_2)$ under the "drop the subscripts" map. Here $\widetilde{T}(R_1, B_1B_2)$ is the set of all bi-infinite indexed concatenations of R_1 and B_1B_2. We show that every "subscripted tiling system" is a shift of finite type. Clearly every tiling system can be obtained from a subscripted tiling system by dropping the subscripts.

Formally, let $\mathcal{P} = \{P_1, \dots, P_K\}$ be a finite collection of prototiles. Write

$$P_k = \{0 = p_{k,1} < p_{k,2} < \cdots < p_{k,\ell_k}\}$$

and define $\widetilde{T} = \widetilde{T}(\mathcal{P})$, on alphabet $\{(k,\ell) : 1 \le k \le K, 1 \le \ell \le \ell_k\}$, by $x \in \widetilde{T}$ iff there is a tiling of the integers by members of $\mathcal{P}$, $\mathbb{Z} = \bigcup(t_j + P_{k_j})$, such that for every i, there exist $j = j(i)$ and $\ell = \ell(i)$ such that $i \in t_j + P_{k_j}$ and $x_i = (k_j, \ell)$.

Equivalently, $x \in \prod\{(k,\ell) : 1 \le k \le K, 1 \le \ell \le \ell_k\}$ is in $\widetilde{T}$ if and only if for every i, $x_i = (k,\ell)$ and $1 \le \ell' \le \ell_k$ imply $x_{i+p_{k,\ell'}-p_{k,\ell}} = (k,\ell')$. Informally, if x_i is an element of a tile, then the other elements of that tile appear in the appropriate places of x.

The following result was proved in conversations with K. Schmidt in Warwick in 1994.

THEOREM: *Every "subscripted tiling system" is a shift of finite type.*

Proof: Let L be the length of a longest prototile in $\mathcal{P}$. (For example, $B \_ B$ has length 3.) We show that $\widetilde{T} = \widetilde{T}(\mathcal{P})$ is a shift of finite type by showing that if $x \in \prod\{(k,\ell)\}$ and every solid L-word which appears in x appears in some point of $\widetilde{T}$, then $x \in \widetilde{T}$.

Suppose that every solid L-word which appears in x appears in some $y \in \widetilde{T}$. Let $x_i = (k,\ell)$. Since $p_{k,\ell_k} + 1 \le L$, there exists $y \in \widetilde{T}$ such that

$$y_{i-p_{k,\ell}}, \ldots, y_{i-p_{k,\ell}+p_{k,\ell_k}} = x_{i-p_{k,\ell}}, \ldots, x_{i-p_{k,\ell}+p_{k,\ell_k}}.$$

But $y \in \widetilde{T}$ and $y_i = (k,\ell)$, so $y_{i-p_{k,\ell}+p_{k,\ell'}} = (k,\ell')$ for $1 \le \ell' \le \ell_k$. Hence $x \in \widetilde{T}$.

Informally, suppose that every solid L-word which appears in x appears in a subscripted tiling. Since no tile is longer than L, if x_i is an element of a tile, then the other elements of that tile appear in the appropriate places of x. Therefore $x \in \widetilde{T}$. ∎

COROLLARY: *Every tiling system is sofic.*

Remark: We cannot use $L-1$ in the proof of the theorem. Again let $\mathcal{P} = \{R, BB\}$, so $\widetilde{T} = \widetilde{T}(R_1, B_1B_2)$ and $L = 2$. Every 1-word appearing in $x = \ldots B_1B_1B_1\ldots$ appears in some point of $\widetilde{T}$, but $x \notin \widetilde{T}$.

Not every sofic system can be realized as a tiling system, as is shown by the following

PROPOSITION: *A tiling system which has a point of period 2 must have at least two fixed points.*

By period here we will mean least period.

Proof: The point of period 2 is $\ldots abab\ldots$, so there are two prototiles, each consisting entirely of even integers. Each tiles the integers, so both $\ldots aaa\ldots$ and $\ldots bbb\ldots$ are in the tiling system. ∎

Similarly, if a tiling system has a point of period 3 or one of period 4, then it must have at least one fixed point. The existence of a point of period greater than 4 does not imply the existence of a fixed point.

We can also use the preceding proposition to show that powers of tiling systems need not even be conjugate to tiling systems:

(5) Consider $T = T(\mathcal{P})$ where $\mathcal{P} = \{RR\_R, B\_\_\_B\}$, i.e., $\{\{0,1,3\},\{0,4\}\}$. The tiling system $\sigma: T \to T$ has exactly one orbit of period 4 – $(RRBR)^\infty$, no points of period 2, and exactly one fixed point – B^∞. Its square, $\sigma^2: T \to T$, has a point of period 2 – $[(RR)(BR)]^\infty$, but only one fixed point – $(BB)^\infty$. So it is not topologically conjugate to a tiling system.

3. The main theorem

Main Theorem: *Let $\sigma: \Sigma \to \Sigma$ be a shift of finite type. Then there is a positive integer m and a tiling system $\sigma: T \to T$ such that*

(1) $T = T_0 \cup T_1 \cup \cdots \cup T_{m-1}$, *where the T_i are closed and cyclically permuted by the shift σ;*

(2) $\sigma^m: \Sigma \to \Sigma$ *is topologically conjugate to every* $\sigma^m: T_i \to T_i$.

Proof: We may assume that $\Sigma = \Sigma_A$, the edge shift determined by a matrix A with nonnegative integer entries. A is the adjacency matrix of a directed graph G. The alphabet of Σ_A is the set of arcs (directed edges) of G, and $(x_i) \in \Sigma_A$ if and only if for every i, the terminal vertex of x_i is the initial vertex of x_{i+1}.

For every positive integer m, $\sigma^m: \Sigma_A \to \Sigma_A$ is topologically conjugate to $\sigma: \Sigma_{A^m} \to \Sigma_{A^m}$. We find a tiling system $\sigma: T \to T$ and a positive integer m such that $T = T_0 \cup T_1 \cup \cdots \cup T_{m-1}$, the T_i are closed and cyclically permuted by the shift, and $\sigma: \Sigma_{A^m} \to \Sigma_{A^m}$ is topologically conjugate to $\sigma^m: T_0 \to T_0$. Hence $\sigma^m: \Sigma_A \to \Sigma_A$ is topologically conjugate to every $\sigma^m: T_i \to T_i$.

Suppose that A is $V \times V$. Choose $n > V$ so that

$$(V \max A_{ij})^{13n} < (n+1)!.$$

Let $m = 13n$. Then every entry of $A^m = A^{13n}$ can be written (uniquely) as

$$c_1(1!) + c_2(2!) + \cdots + c_n(n!),$$

where $0 \le c_k \le k$ for $1 \le k \le n$.

We now construct the tiling system. The prototiles will be of two types: **barbells** and **racks** (to hold barbells). We will use the same terms for the corresponding tiles. In the sequel we will use colors to label prototiles. The symbols a, a' will stand for generic colors.

The barbells are the broken words of the form

$$a^2 \overset{\leftarrow 2r+1 \rightarrow}{\ldots} a^2$$

for $0 \le r \le 2n-2$.

The racks are chosen from the broken words of the form HCT of length $13n + 2J$, $1 \le J \le V$, where the **head** is

$$H = (a\,\_)^I a^{2n-2I}$$

for some I, $1 \le I \le V$; the **tail** is

$$T = (\_\,a)^J$$

for some J, $1 \le J \le V$; and the **center** is

$$C = a^{3n+i} \overset{\leftarrow 2k \rightarrow}{\ldots} a \overset{\leftarrow 2k \rightarrow}{\ldots} a^{8n-4k-1-i}$$

for some i and k, $0 \le i \le k-1$ and $1 \le k \le n$.

Given I, J with $1 \le I, J \le V$, write

$$(A^{13n})_{IJ} = c_1(1!) + c_2(2!) + \cdots + c_n(n!),$$

where $0 \le c_k \le k$ for $1 \le k \le n$. If $c_k \ne 0$, choose the racks to be the $c_k = c_k(I,J)$ broken words of the form

$$\left[(a\,\_)^I a^{2n-2I}\right] \left[a^{3n+i} \overset{\leftarrow 2k \rightarrow}{\ldots} a \overset{\leftarrow 2k \rightarrow}{\ldots} a^{8n-4k-1-i}\right] \left[(\_\,a)^J\right]$$

for $0 \le i \le c_k - 1$.

The barbells and racks have the following properties.

- The head $H = (a\,\_)^I a^{2n-2I}$ of a rack can be filled by the tail $T = (\_\,a')^J$ of a rack in a tiling if and only if $I = J$.
- Label the blanks in the center

$$C = a^{3n+i} \overset{\leftarrow 2k \rightarrow}{\ldots} a \overset{\leftarrow 2k \rightarrow}{\ldots} a^{8n-4k-1-i}$$

of a rack by $\{1, 2, \ldots, 4k\}$. Barbells can appear in a tiling only in the gaps in the centers of racks, starting only in odd places and straddling the a. Furthermore, the blanks in this center can be tiled by barbells in exactly $k!$ ways. To see this, define a permutation π of $\{1, 2, \ldots, k\}$ by $\pi(j) = \ell$ if and only if a barbell $a'a' \overset{\leftarrow 2r+1 \rightarrow}{\ldots} a'a'$ occupies places labelled $2j-1, 2j, 2(k+\ell)-1, 2(k+\ell)$.
- The heads of racks can appear in a tiling starting only at places which differ by multiples of $13n$.

Let T be the tiling system with prototiles the barbells and racks chosen above. Then $T = T_0 \cup T_1 \cup \ldots \cup T_{13n-1}$, where T_i is the set of indexed bi-infinite sequences in T in which the heads appear starting at places congruent to i modulo $13n$. Thus T_0 consists of all indexed bi-infinite concatenations of words of length $13n$, of the form $\bar{H}\bar{C}$, starting at multiples of $13n$, where H and C are the head and center of a rack, and $\bar{H}$ and $\bar{C}$ are the solid words resulting from filling them in a tiling. Recall that if $H = (a\ \_)^I a^{2n-2I}$, then it must be filled by a tail $T = (\_\ a')^I$. C can be filled only by barbells.

Define an edge shift as follows. Let G' be the directed graph with vertices $1, 2, \ldots, V$, and an arc from I to J for each rack with head $(a\ \_)^I a^{2n-2I}$, tail $(\_\ a)^J$, and center tiled by barbells. There are $(A^{13n})_{IJ}$ arcs from I to J, and an arc with head $(a\ \_)^I a^{2n-2I}$ can follow an arc with tail $(\_\ a')^J$ if and only if $I = J$. Therefore since $m = 13n$, the adjacency matrix of G' is A^m and $\sigma^m \colon T_0 \to T_0$ is topologically conjugate to $\sigma \colon \Sigma_{A^m} \to \Sigma_{A^m}$, which in turn is topologically conjugate to $\sigma^m \colon \Sigma_A \to \Sigma_A$. ∎

It follows immediately from the Main Theorem that every topological entropy of a shift of finite type occurs as the entropy of a tiling system. Since the topological entropies of shifts of finite type and also of sofic systems are both equal to the logarithms of positive spectral radii of nonnegative integral matrices, or equivalently, the logs of roots of Perron numbers, we have

COROLLARY: *The set of topological entropies of tiling systems is the same as that of shifts of finite type, i.e., the set of logarithms of roots of Perron numbers.*

The main question we have considered has a natural analogue in $d > 1$ dimensions: can every $\mathbb{Z}^d$ shift of finite type be realized, up to a power, as a $\mathbb{Z}^d$ tiling system? Forthcoming work of the first author, A. Johnson, N. Jonoska, and K. Madden adapts the methods used here to address the higher-dimensional case.

References

[CM] E. M. Coven and A. Meyerowitz, *Tiling the integers with translates of one finite set*, Journal of Algebra **212** (1999), 161–174.

[LM] D. Lind and B. Marcus, *An Introduction to Symbolic Dynamics and Coding*, Cambridge University Press, Cambridge, 1995.

[N] D. Newman, *Tesselation of integers*, Journal of Number Theory **9** (1977), 107–111.

Extending holomorphic motions

by

DENNIS P. SULLIVAN and WILLIAM P. THURSTON

I.H.E.S., Bures-sur-Yvette
France

Institute for Advanced Study
Princeton, NJ, U.S.A.

1. Introduction

The notion of an isotopy of one set within another set is one of the key concepts of topology. Here is one way this concept can be generalized to a holomorphic context:

Definition. If X is a subset of $\mathbf{C}$, a *holomorphic motion* of X in $\mathbf{C}$ is a map

$$f: T \times X \to \mathbf{C}$$

defined for some connected open subset $T \subset \mathbf{C}$ containing 0 such that

(a) for any fixed $x \in X$, $f_t(x) = f(t, x)$ is a holomorphic mapping of T to $\mathbf{C}$,

(b) for any fixed $t \in T$, f_t is an embedding, and

(c) f_0 is the identity map of X.

We think of t as a kind of complex time parameter. Note that in the definition, there is no requirement of holomorphy in the X-direction. X should be thought of with just its topological structure or its quasiconformal structure, although even continuity doesn't directly enter into the definition; the only restriction is in the t direction. We will see that continuity is a consequence of the hypotheses, by the lambda lemma of Mañé, Sad and Sullivan ([2], Theorem 2).

This definition is applicable in a number of interesting situations. For instance, the limit sets of Kleinian groups often move holomorphically as parameters are varied. Similarly, the Julia sets for iterated rational maps often move holomorphically with the parameters.

In topology, it is important to know whether an isotopy of one space within another can be extended to an ambient isotopy, that is, to an isotopy of the big space which restricts to the given isotopy of the small space. Without additional conditions, an extension is not generally possible. For example, consider a loosely knotted piece of string within $\mathbf{R}^3$. Physically, one can take hold of the ends and pull it tight. Mathematically, this corresponds to an isotopy of the knot in $\mathbf{R}^3$ which shrinks the knotted part to an arbitrarily small size—and finally, makes it disappear. Such an isotopy certainly cannot be extended to all of $\mathbf{R}^3$, since the fundamental group of the complement of the circle is different at the beginning of the isotopy and the end.

It is not hard to arrange the isotopy so that the motion of any point is differentiable in time. The isotopy can even be chosen so that it is real-analytic in the time parameter, if the knot vanishes and then instantly reappears.

There are also examples in two dimensions: there are isotopies of closed sets in the plane which cannot be extended to ambient isotopies. The simplest example is for X a countable set with one accumulation point. If the set is represented as say $\{x_i = 1/i | i \in \mathbf{Z}_+\}$ together with $x_\infty = 0$, the isotopy can be choreographed by making each point x_i swing around its two neighbors, two turns about x_{i+1} for each turn about x_{i-1}, faster and faster as $i \to \infty$. Any extension of the isotopy restricted to the first k points sends the arc $\overline{x_{k-1} x_k}$ at the beginning of the isotopy to an arc which winds around x_i by the end of the isotopy. To do this for all i would violate continuity at 0.

The question of extending an isotopy to an ambient isotopy also has significance in the holomorphic context. Our main result addresses this problem, as follows:

THEOREM 1. *There is some universal constant $a>0$ such that any holomorphic motion of any set $X \subset \mathbf{C}$ parametrized by the disk $T=D_1$ of radius* 1 *about* 0 *can be extended to a holomorphic motion of* $\mathbf{C}$ *with time parameter the disk D_a of radius a about* 0.

An important correlate of an open set moving holomorphically in $\mathbf{C}$ is quasiconformality:

PROPOSITION. *For any holomorphic motion f of* $\mathbf{C}$, *each map f_t is quasicon formal.*

Proof. The cross-ratio of any quadruple of points in $\mathbf{C}$ during the motion is a holomorphic function which omits the values 0, 1, and ∞. The derivative of a holomorphic function sends a tangent vector in its domain to a vector in its range which cannot be longer, when lengths are measured by the Poincaré metrics for the domain and the range. Therefore, for any particular value of t and for all quadruples of points whose initial cross-ratios are in some compact set K of cross-ratios, the cross-ratios of f_t of the quadruples are uniformly bounded. It readily follows that f_t is quasiconformal.

By the theory of quasiconformal maps, the converse is also true: any quasiconformal orientation-preserving map of **C** can be connected to the identity by a holomorphic motion. To find such a holomorphic motion, one simply takes the Beltrami coefficient for the quasiconformal map and multiplies it by a complex variable t.

COROLLARY. *If f is a holormorphic motion of a set X, then for each time t the map f_t extends to a quasiconformal map of* **C**.

Proof. By the preceding two assertions, it follows that the corollary is true when t is sufficiently small. To obtain the result for a general t, walk from 0 to t in small steps and compose the quasiconformal maps for each step to obtain the big quasiconformal map.

In a companion paper, Bers and Royden [1] will give another proof which shows that a can be taken to be 1/3. We do not know if the constant a of the theorem can be taken to be 1.

In section 6 we will state and prove another theorem which does work globally. This result is obtained from our main statement by changing all holomorphic motions to quasiconformal motions. The definition of a quasiconformal motion of an arbitrary set will be found in that section.

There is a simple but very useful special case of the main theorem which was proven by Mañé, Sad and Sullivan [2], which they dubbed the λ-lemma.

THEOREM 2 (λ-lemma). *A holomorphic motion of a set $X \subset \mathbf{C}$ can be extended to a holomorphic motion of the closure of X, with the same time parameter set T.*

Proof. The idea is that the motion of the points of the set must be equicontinuous because they remain disjoint.

This can be seen more explicitly by first choosing two finite points of X together with infinity, then normalizing by changing coordinates via the unique family of affine transformations sending these three points to $\{0, 1, \infty\}$. The Möbius transformations depend holomorphically on T, so the problem of extending the motion of the new family is the same as that for the old family. (We ignore here the exceptional case that there is only one point of X; the problem in this case is trivial.)

The points in the new family—except those which are fixed—move in the complement of $\{0, 1, \infty\}$. The map from the parameter space T to the three-punctured sphere takes vectors of unit length in the Poincaré metric of T to vectors of length not exceeding one in the Poincaré metric of the three-punctured sphere.

The motion of points with respect to T is equicontinuous, so the closure of the set of all the maps $t \mapsto f_t(x)$ is compact in the topology of uniform convergence on compact sets. Consider the closure, which consists of holomorphic functions from T to $\mathbf{C}$.

If g and h are any two distinct holomorphic functions on an open set T, their graphs intersect iff $g-h$ has a zero of some order at some point in T. This property is stable under perturbation. Therefore, the property of a pair of holomorphic functions that their graphs intersect is open. The opposite property, that the graphs of g and h are disjoint, is closed. The set of graphs of functons in the closure of our family are disjoint, and therefore there is exactly one through each element of X. This defines an extension of the holomorphic motion of X to a holomorphic motion of its closure.

It is easy to extend Theorem 1 to the case that the parameter space T has arbitrary dimension.

It is also easy to extend it to the case of the motion of a set X in an arbitrary Riemann surface, or even a family of Riemann surfaces depending on the parameter space T. For the latter case, one uses the Bers embedding of Teichmüller space to translate the problem into a motion of a set in $\mathbf{C}$.

A problem closely related to the problem of extending holomorphic motions is the holomorphic axiom of choice: Given a holomorphically moving set, is it possible to choose a holomorphically moving point in the complement of the set? If we strengthen this slightly so that the point for $t=0$ is chosen in advance, then a positive answer, even in the case that X consists of only a finite set of points implies the general extendibility of holomorphic motions to all of $\mathbf{C}$.

To see this, first observe that the ability to choose an additional point with arbitrary initial value in the complement of a holomorphically moving finite set implies the same ability in the complement of a general holomorphically moving set: if $F \subset X$ is a large finite set of points which comes very close to every point in X, then a choice which starts out not too close to X and remains in the complement of F must also avoid X for most of the time parameter (since the distance to X remains large in the Poincaré metric of the complement of F). Using equicontinuity, one can pass to the limit as F becomes dense in X and obtain a choice, through the given point, which misses X entirely. Thus the ability to choose a moving point in the complement of finite sets implies this ability for general sets. By adding more and more moving points, after a countable number of steps one has a dense set of moving points, so that the motion extends by continuity to a motion of the complex plane.

2. Framework for the proof

The strategy of the proof is to divide and conquer. We may assume that X is closed. First, we will show that for any point p in the complement of X the motion of X can be extended to a motion of some neighborhood of p for some subset of the time parameter space T. We will do this so that the subset T_0 is independent of p. We must do this carefully, so that it will be possible to piece together the choices from different neighborhoods, using a partition of unity for $\mathbf{C}-X$.

The local choices near some point p are fairly easy to make in the case that the set $X\cup\{\infty\}$ is connected, so we will take care of that case first, in section 3. In fact, the local choices are made by using a solution for the case that X consists of two points. This solution can be transplanted (for a limited subset of $T\times X$) if for each p, we choose two reasonably spaced points from the actual set X. A reasonable spacing is achieved if the distance between the two points and the distance of the two points from p is comparable to the distance of p from X. The hypothesis that $X\cup\{\infty\}$ is connected is used to guarantee the existence of two such points.

To do the general case, we will need to analyze the complement of X in terms of the thick-thin decomposition for its Poincaré metric. The thick part of the complement of X is handled like the connected case. The thin part divides into two new cases: cusps and short geodesics. The analysis of cusps again reduces to the case that X has only two points. We will construct a solution for the case of a thin part of X which contains a short geodesic by transplanting a solution for the case that X has exactly three points. This corresponds to the quadruply-punctured sphere, the simplest Riemann surface which can have short geodesics.

To see that local solutions can be pieced together, it helps to picture the trace or graph of the motion, in $T\times\mathbf{C}$. We begin with the motion of X, which gives a foliation in a subset of $T\times\mathbf{C}$ above T. For each p, we find some rule to fill in a foliation in some neighborhood of p which is disjoint from the leaves of X. We must do this so that each leaf projects to a uniform size neighborhood in T.

To pass between choices made for different neighborhoods which overlap (at time $t=0$), we average them (considered as maps of T to $\mathbf{C}$) using a partition of unity. To see what happens in any region of overlap, we make a t-dependent affine change of coordinates to fix two points of X at 0 and 1. These two points are picked so that they are near to the region in question, compared to their distance from each other and the distance of the region from X. There is no overlapping of different local choices in the region of influence of a short geodesic or a cusp—such a region is contained in a single coordinate patch, so it is always possible to find two points as above. Note that the process of averaging is unaffected by affine changes of coordinates.

In this coordinate system, no matter what the overlapping choices of holomorphic motions, they can move at speed at most 1 in the Poincaré metric for the thrice-punctured sphere. Since the region at time $t=0$ has a distance from X (following the affine renormalization) which is bounded above zero uniformly for all problems under consideration, convex combinations of the moving choices remain disjoint from X for a uniformly bounded time.

When we understand how to make local extensions of a holomorphic motion, we must still make sure that after averaging by a partition of unity, the motions of different points remain disjoint from each other.

There are two effects to take care of. First, if the diffeomorphisms from one fiber to another are not near each other in the C^1 topology, then convex combinations of them might have derivative zero. In order to take care of this, we will make sure that (after the affine renormalization) the maps from one copy of $\mathbf{C}$ to another all have derivatives near the identity. This precaution would suffice if we were taking convex combinations of the different choices with constant coefficients.

The second problem is that if the derivatives of the partitioning functions (with respect to the fiber variable) are too large, then convex combinations made with them will oscillate wildly. To take care of this problem, we will make sure that we define local motions on neighborhoods of uniform size (with respect to the coordinates as above) and that these neighborhoods have ample overlap, so that the fiber partitioning functions can be chosen with uniformly bounded derivatives.

To convince oneself that these two conditions are exactly what is needed without actually writing down the formula, one can think of the foliation picture: picture a small disk in T, and enlarge it until it has size 1. Then the various local choices automatically become uniformly near to horizontal, so that any effect due to averaging by a controlled partition of unity is absorbed.

3. Extending the motion of a connected set

Let X be a holomorphically moving set in $\mathbf{C}$ such that $X \cup \{\infty\}$ is connected. Consider any point $p \in \mathbf{C}$, disjoint from X at time $t=0$. We will extend the motion of X to include a motion of a neighborhood of p.

Let r be the distance of p from X. Let x_0 be a point in X which has distance r from p. The circle of radius $2r$ around p intersects X, since X is connected and unbounded. Let x_1 be a point of intersection of this circle with X.

The points x_0 and x_1 are separated by at least r and at most $3r$. Normalize the motion of X by t-dependent affine transformations so that the points x_0 and x_1 remain fixed at 0 and 1. After this normalization, the distance of p from 0 at time $t=0$ is at least 1/3, and the distance from the point 1 is at least 2/3. Define a holomorphic motions of the neighborhood U of radius 1/6 about p by fixing it, in the new coordinates.

At time 0, the distance of X from U as measured in the Poincaré metric of the thrice-punctured sphere $\mathbf{C}-\{0,1\}$ is bounded from zero. Since the set X moves with speed at most 1 in the Poincaré metric of the thrice-punctured sphere, this prescription gives a holomorphic motion of the neighborhood for a uniform disk in the time parameter T.

These choices can be combined, using a partition of unity, to give a motion of $\mathbf{C}$ defined for a disk in T of a size which is uniform with respect to different problems of this form. In fact, consider the motion of another neighborhood V defined in this way where V and U intersect at time 0. In the coordinates for which points in U are fixed, the motion of V is given by a t-dependent affine transformation. The motion of V is governed by two points x_3 and x_4 of X, which at time $t=0$ are uniformly bounded away from each other, from infinity, from U and from V.

The points x_3 and x_4 move at speed at most 1 in the Poincaré metric of the thrice-punctured sphere $\mathbf{C}-\{0,1\}$. In any bounded subset of $\mathbf{C}$, the unit circles for this Poincaré metric have bounded diameter. Consequently, x_3 and x_4 are moving at a bounded speed also in the Euclidean metric. Therefore we can find a uniform disk in T such that the derivatives of the comparison between the U motion and the V motion are close to the identity.

Since neighborhoods of the form of U can be chosen to overlap uniformly, this completes the proof in the case that X is connected.

4. Solution for the quadruply-punctured sphere

The method above does not carry over to the case that X is not connected, even if it contains only three points.

For example, let C be any constant, and define a holomorphic motion of a set $x_0=0$, $x_1=1$ and $x_2=e^{-2C}$ by the formula

$$f_t(x_0) = 0,$$

$$f_t(x_1) = 1,$$

$$f_t(x_2) = e^{C(t-2)}.$$

We can restrict attention to the unit disk. Suppose that C is very large. The motion of x_2 wraps it many times around the origin, while keeping it quite close to $x_0=0$. If we use a local prescriptions for extending this motion based on choosing subsets of X consisting of two points, we see that the only reasonable choice for x very near 0 is based on x_0 and x_2, while the only reasonable choice for x near, say, -7 is based on x_1 and either of the other two points. These choices clash fiercely with each other wherever they overlap. There is no reasonable way to average them and get a choice which works over the entire quadruply-punctured sphere.

Luckily, there is an alternate method which gives a global solution in the case of the quadruply-punctured sphere, provided the set T is contractible. When this is the case, there is a well-defined homotopy class of maps from the complement of X at time 0 to the complement of X at time t. Given this homotopy class, there is a well-defined element of the Teichmüller space for a quadruply-punctured sphere associated with each t, and the element of Teichmüller space depends holomorphically on t.

To extend the motion of X, we simply use the Teichmüller maps between the surfaces. The Teichmüller space for a quadruply-punctured sphere is the Poincaré disk, and the Teichmüller maps depend holomorphically on Teichmüller space for the quadruply-punctured sphere. This holomorphic dependence of the Teichmüller maps is a special feature of Teichmüller spaces of complex dimension 1. The holomorphic motion can be seen more explicitly by using the fact that the universal 2-fold-branching cover of the quadruply-punctured sphere is isomorphic to $\mathbf{C}$, with the isomorphism classically given by elliptic integrals. A holomorphically moving configuration of the three points together with ∞ transforms to a holomorphically moving lattice in $\mathbf{C}$, which has an extension to a motion of all of $\mathbf{C}$ in the form of t-dependent real affine transformations.

5. The general case

Now we are prepared to consider a general holomorphically moving subset $X \subset \mathbf{C}$. We will use local solutions usually based on the thrice-punctured sphere, but we will switch to local solutions based on the quadruply-punctured sphere in neighborhoods where X appears to be split into two small but nontrivial pieces.

One way to express the choice between the two kinds of local solutions is based on the thick-thin decomposition of the Riemann surface, considered with its hyperbolic metric. In the thick part of the surface (where the injectivity radius is greater than some fixed constant $\varepsilon>0$), we can use the thrice-punctured sphere. Each component of the thin part of the surface is either parabolic (coming from a cusp) or hyperbolic. In the hyperbolic thin case, we use the quadruply-punctured sphere, while in the parabolic case we again use the thrice-punctured sphere.

For simplicity, instead of this hyperbolic description of choices for local solutions, we will use an approximately equivalent but more direct description in terms of the Euclidean geometry of $\mathbf{C}$.

Let p be any point in $\mathbf{C}-X$. Choose a point x_0 which is at least as close to p as any other point of X. Change coordinates so that p is at 1, x_0 is at 0, and normalize the motion (via a t-dependent translation of $\mathbf{C}$) to keep x_0 fixed at the origin. Choose a small number ε (fixed throughout this case).

Case (i). (The thick case.) Suppose that there is some other point q in X such that $\varepsilon \leq |q| \leq 1/\varepsilon$. Then choose x_1 to be an element of X which minimizes $|\log |x_1||$. Define a local motion around p to be the affine motion governed by x_0 and x_1.

Case (ii). (The cusp case.) Suppose that (i) fails, and either there is no point in X with modulus greater than that of p, or there is no point except 0 in X with modulus less than p. By inverting if necessary, we can assume that there is no point of X inside the unit disk except for the origin. Let x_1 be a point of X with minimal modulus m. Define a motion in the disk of radius $2\varepsilon m$ about 0 to be the affine motion governed by x_0 and x_1. We use only one such choice for this entire disk, governed by x_0 and some point of X of minimal modulus.

Case (iii). (The short geodesic case.) Suppose that (i) and (ii) fail, so that p is contained in an annulus bounded by circles of radii $0<r<\varepsilon<1/\varepsilon<R<\infty$ which touches X on its two boundary components, but is otherwise disjoint from X. Let x_1 be an element of X of modulus r, and let x_2 be an element of modulus R. Define a motion of the subannulus of inner and outer radii $s=r/2\varepsilon<1<2R\varepsilon=S$ by restricting the solution for the quadruply-punctured sphere. Make only one such choice for this annulus.

These solutions have been chosen so that they are defined on sets which have ample overlaps. Overlapping can occur only

between cases (i) and (i),
between cases (i) and (ii), or
between cases (i) and (iii).

Overlapping between a pair of thin cases ((ii) or (iii)) is not possible.

The first kind of overlap is exactly like the kind we have already considered for X connected; only the constants are different.

The second kind of overlap is just like the first kind of overlap—it is an overlap of two thrice-punctured sphere coordinates in a region where both choices are reasonable.

The third kind of overlap is somewhat different. To understand it, we must first understand the behaviour of the solution for a quadruply-punctured sphere which has a short geodesic, in the vicinity of its thick part.

Any quadruply-punctured sphere with a short geodesic can be arranged so that its puncture are $x_0=0$, x_1, x_2, and $x_3=\infty$, where the ratio x_2/x_1 is very large. The short geodesic separates x_0 and x_1 from x_2 and x_3. By symmetry, we can focus on the thick part of the quadruply-punctured sphere which is in the half containing x_0 and x_1. To get a convenient coordinate system, normalize so that x_1 becomes 1.

LEMMA. *In the coordinates above, the derivatives of the maps from fiber to fiber given by thin quadruply-punctured sphere solutions are uniformly equicontiuous in the region* $P\subset \mathbf{C}$ *obtained by removing* ε*-disks about* 0, 1, *and* ∞. (*An* ε*-disk about* ∞ *means the complement of a disk of radius* $1/\varepsilon$ *about* 0.)

Proof. As long as the shape of the quadruply-punctured sphere remains in a compact region of the modular space of the quadruply-punctured sphere, the derivatives automatically remain equicontinuous in P. Thus, the problem is to show that as the quadruply-punctured sphere gets thinner and thinner, these derivatives remain reasonable.

The universal 2-fold branching cover of any sphere over four points is $\mathbf{C}$. We will transfer the question to $\mathbf{C}$ by taking these universal two-fold branching covers.

The preimage of the branch points forms a lattice in $\mathbf{C}$. We can normalize the lattices in $\mathbf{C}$ so that the origin is a preimage of 0, and so that is a preimage of 1 nearest to the origin. Thus, the lattices all contain the integers.

The Poincaré metric of the plane minus the lattice maps isometrically to the Poincaré metric of the quadruply-punctured sphere, so as the length of the short geodesic goes to zero, the distance between the two thick parts of the quadruply-punctured sphere tends to infinity, and the distance from the row of lattice points which are integers the next row also tend to infinity. In the limit, the branched covering from $\mathbf{C}$ to $\hat{\mathbf{C}}$ converges to a branched covering map of $\mathbf{C}$ to $\mathbf{C}$, with 2-fold branching above $0=x_0$ and $1=x_1$.

The advantage of passing to the branched covers, $\mathbf{C}$, is that the Teichmüller maps between quadruply-punctured spheres lift to $\mathbf{R}$-linear maps of $\mathbf{C}$ to itself. They are represented by matrices of the form

$$A = \begin{pmatrix} 1 & x \\ 0 & y \end{pmatrix}.$$

The distance between two of the lattices in Teichmüller space depends only on the matrix A which transforms one into the other, not on particular lattices. Therefore, a sequence of Teichmüller maps between pairs of quadruply-punctured spheres with bounded Teichmüller separation always has a geometrically convergent subsequence, as one passes to the limit, which is a thrice-punctured sphere. The limiting thrice-punctured sphere is identified with the thrice-punctured sphere defined by $0=x_0$, $1=x_1$, and $\infty=x_3$. The geometric limit map, lifted to the branched coverings $\mathbf{C}$ of the thrice-punctured sphere, is also a $\mathbf{R}$-linear map, described by a matrix A of exactly the same form.

Since the Teichmüller space for the quadruply-punctured sphere is equivalent to unit disk, and the Teichmüller metric is its Poincaré metric it follows that, in a holomorphically moving family, the Teichmüller distance between quadruply-punctured spheres defined by the motion of four points does not exceed the Poincaré distance in the parameter space.

The lemma follows.

With the aid of the lemma, we can take care of overlaps between coordinate neighborhoods of types (i) and type (iii), thus completing the proof of the main theorem.

6. Extending quasiconformal motions

What is the right definition for a quasiconformal map between two arbitrary subsets of $\hat{\mathbf{C}}$? It does not work to study only what happens to the sphere of a given radius about a point in the set: such a condition would say absolutely nothing about any of the geometric Cantor sets contained on the line which are obtained by iteratively removing the middle α subinterval, when $\alpha<1/3$, since the spheres about points in such a Cantor set intersect the Cantor set in at most one point! Similarly, it does not work to look just at what happens to quadruplets of points whose cross-ratios lie in any given compact set, because one can construct Cantor sets such that the cross-ratios of points in that set avoid the given compact set of cross-ratios.

Definition. A *quasiconformal homeomorphism* $f: X \to Y$ between subsets of $\hat{\mathbf{C}}$ is a homeomorphism such that the cross-ratio of any quadruple of points in X has a bounded distance from the cross-ratio of the image points in Y. The distance between cross-ratios is measured using the Poincaré metric on the space of values of the cross-ratio function, namely the thrice-punctured sphere $\mathbf{C}-\{0,1,\infty\}$. A *quasiconformal motion* of a subset $X\subset\mathbf{C}$ is a map from $T\times X$ to $\hat{\mathbf{C}}$, where T is a connected space (frequently an interval of time), such that

(i) for some basepoint $t_0\in T$, $f_{t_0}=\text{id}$, and

(ii) for any $t\in T$ and for any $\varepsilon>0$ there is a neighborhood of t such for any quadruple of points in X, cross-ratios of the various images of the quadruple for time parameters in the neighborhood all lie within an ε-ball in the Poincaré metric of the thrice-punctured sphere.

This definition precisely captures the property of holomorphic motions which we used to make the definitions for local holomorphic motions. The definitions immediately suggest the following statement:

THEOREM 3. *For any quasiconformal motion of any subset X in $\mathbf{C}$ whose time parameter space is an interval I, there is an extension of the motion to all of $\mathbf{C}$ defined over all of I.*

Proof. We may suppose that two finite points of X are 0 and 1, and that these two points do not move.

First we extend the motion to the closure of X, by passing to limits of sequences of motions of points. The motion of other points is equicontinuous in the Poincaré metric of $\mathbf{C}-\{0,1\}$ (since the cross-ratio of 0, 1, ∞ and x changes at a controlled rate) so the closure of the set of functions $f_*(x)$ which define the motion of points is compact. Two distinct functions g and h in the closure of this set of functions have disjoint graphs, since the cross-ratio of 0, 1, $g(x)$, $h(x)$ changes at a controlled rate, so it can never reach 0, 1, or ∞ in a finite time.

For small neighborhoods in the T plane, the extension of the motion from the closure of X to all of $\mathbf{C}$ is done just as for the extension of holomorphic motions. These extensions can be pieced together using a partition of unity of I, to obtain a global quasiconformal extension.

7. Naturality

It would be good if we could define an extension of a holomorphic motion which would be completely canonical—in particular, a canonical extension would entail that

(a) it would be invariant under any change of coordinates by a Möbius transformation, and

(b) it would be independent of choice of origin in the T plane.

Condition (b) seems quite hard. In particular, it would entail that the motion could be extended globally, over the entire T plane. In the companion paper by Bers and Royden, they give a construction of an extension of a holomorphic motion which depends on the choice of origin in the T plane but is otherwise canonical, so that property (a) is satisfied.

Condition (a) does not come as easily from our point of view, since the construction we made involved arbitrary choices of local holomorphic motions, subject to certain inequalities. To get around the fact that our choices are arbitrary, we can average different choices, but another difficulty crops up: the averaging procedure we have used itself depends on an arbitrary choice, the choice of the point at infinity. A probability measure on $\mathbf{C}$ has a well-defined mean, provided the measure is not too dense near ∞. When the measure is transformed by a Möbius transformation, its mean is not usually the image of its old mean.

To circumvent this problem, we can define a more general mean: if μ and ν are probability measures on the Riemann sphere, we will define the mean of μ with respect to ν.

First, if ν is a measure concentrated on a single point Q, then the mean of μ with respect to ν is the obvious mean, calculated by sending Q to infinity, forming the mean of μ, and transforming back to our original coordinates.

For arbitrary ν and μ, the mean will not always exist, but if there are disjoint round disks containing the support of ν and the support of μ, that is sufficient to guarantee existence. The idea in the general case is that the measure ν defines a transformation T_ν from probability measures on the Riemann sphere to probability measures on the Riemann sphere. The transformation is defined by taking the mean of mean of μ with respect to the various points in the support of ν, and weighting them according to ν.

LEMMA. *If the support of μ is contained in a disk D_1 and the support of ν is contained in a disjoint disk D_2, then the support of $T_\nu(\mu)$ is contained in a proper subdisk D_1' of D_1.*

Proof. This is transparent.

COROLLARY. *If the support of μ and the support of μ are contained in disjoint disks, then the sequence of measures $T_\nu^{\circ n}(\mu)$ converge to a measure concentrated at a point.*

17−868286 *Acta Mathematica* 157. Imprimé le 12 novembre 1986

Proof. The sequence of images of μ by T_ν is supported in a shrinking sequence of disks. If minimal support disks were to converge to disks of finite radius, then there would be a subsequence of the measures which would converge weakly to a measure such that T_ν would violate the previous lemma.

Definition. The mean of one probability measure μ on the Riemann sphere with respect to another measure ν is defined to be the support of the limit of $T_\nu^{\circ n}(\mu)$, provided this limit exists and is supported on a single point.

Now we are prepared to construct an extension of a holomorphic motions of a set X which is equivariant with respect to any group Γ of Möbius transformations which preserves the original motion. The extension to the closure of X is immediate from the lambda lemma, so we may assume that X is closed. We may also suppose that the Γ is a closed subgroup of the group of Möbius transformations—otherwise, form its closure.

To extend the motion equivariantly to the complement of a closed set, the idea is that we pick a collection of triples and quadruples of points in X together with a measure μ_p on this collection for any point p, to govern the motion of p. (The measure here is a generalization of a partition of unity.) The collections of triples and quadruples, and the measures on them can easily be made equivariant. Since thin parts of the complement of X are disjoint, we can also make sure that for each thin part there is exactly one triple or quadruple associated, which will govern the motion on the bulk of this thin part. We can make sure that the measures vary slowly, so that the total measure of any set of triples and quadruples has a bounded derivative with respect to p as measured with the Poincaré metric of the complement of X.

In addition, we can choose for each point $p \in S^2 - X$ a measure ν_p on X, in such a way that ν_p depends equivariantly on p and has bounded derivative with respect to p.

Now we can define the motion of p to be the mean with respect to ν_p of the motions of p weighted according to μ_p. This mean exists for a definite neighborhood of 0 in T, by the corollary. It depends holomorphically on t, since it is obtained by a procedure of averaging holomorphic motions. The motions of different points are disjoint for a definite neighborhood in T, by the same considerations as before.

References

[1] BERS, L. & ROYDEN, H. L., Holomorphic families of injections. *Acta Math.*, 157 (1986), 259–286.
[2] MAÑÉ, R., SAD, P. & SULLIVAN, D., On the dynamics of rational maps *Ann. Sci. Ecole Norm. Sup.*, 16 (1983), 193–217.

Received August 20, 1984
Received in revised form August 14, 1985

Zippers and Univalent Functions

WILLIAM P. THURSTON

1. Introduction. Every simply connected domain **C**, with the exception of **C** itself, can be parametrized by a conformal map from the unit disk, that is, by a univalent function. Families of domains which vary continuously with parameters arise in several mathematical contexts, particularly in the theory of Kleinian groups and in the theory of other 1-dimensional complex dynamical systems. A family of domains gives rise to a family of univalent functions. There are various topologies on the set of univalent functions. One topology which has proven valuable for the theory of Kleinian groups is the topology of uniform convergence of the Schwarzian derivative (to be described presently), uniform as measured using the hyperbolic metric on the disk. In this note, we will construct simple examples of domains whose corresponding univalent functions are isolated points in this topology. These domains have complements which are arcs, in fact, quasi-intervals, and they are reminiscent of zippers.

The construction and result here are related to that of Gehring [**1**].

2. Preliminaries. For the benefit of people to whom the Schwarzian derivative may seem a mystery, we will set the stage by discussing the Schwarzian derivative. Actually, the Schwarzian derivative is not really essential to the current discussion. We will use it only for the definition of a topology on the set of univalent maps, and logically we really use an equivalent description of the topology (to be explained later) which makes no mention of Schwarzian derivatives. This is not meant to downplay the significance of the Schwarzian derivative; by analogy, many qualitative constructions in differential geometry can be done without ever mentioning curvature, even though curvature is central to differential geometry.

The Schwarzian derivative is very much like a kind of curvature: the various kinds of curvature in differential geometry measure deviation of curves or manifolds from being flat, while the Schwarzian derivative measures the deviation of a conformal map from being a Mœbius transformation. A Mœbius transformation is determined by its two-jet (its value and first two derivatives) at any point. The two-jet is arbitrary, except for the proviso that the first derivative be nonsingular. Thus, for any locally univalent f and any point w in the domain of f there is associated a unique Mœbius transformation $M(f, w)$ which agrees with f through second order. Note that if A is a Mœbius transformation, then

$$M(A \circ f, w) = A \circ M(f, w).$$

More generally, if f and g are two locally univalent functions, then

$$M(g \circ f, w) = M(g, f(w)) \circ M(f, w).$$

Conformal mappings up to postcomposition (composition on the left) with Mœbius transformations are therefore conveniently described by their three-jets, up to the action of Mœbius transformations on three-jets. The quotient space of this action, at any point z in the domain of the mapping, is isomorphic to $\mathbf{C}$. To see the isomorphism, normalize a map f by postcomposing with the Mœbius transformation $M^{\circ^{-1}}(f, w)$ to obtain a new map having the two-jet of the identity at w. The information that remains in the three-jet is the third derivative, which is an arbitrary element of $\mathbf{C}$. Note that if the original map is a Mœbius transformation, it will have been normalized to be the identity, so its third derivative will be 0.

As z varies, this determines a complex vector bundle over the domain of the map whose fiber consists of the bundle of three-jets modulo the chain rule action of Mœbius transformations on three-jets. The linear structure in the fiber arises not because the action is linear (which it isn't), but because a linear structure is inherent in the complex structure of $\mathbf{C}$, once the origin is determined.

What vector bundle is it? A third derivative is a cubic mapping from the tangent space of the domain to the tangent space of the range. However, for a nonsingular mapping, the tangent space of the domain is identified with the tangent space of the range via the first derivative, so we may describe the third derivative in terms of a cubic map P from the tangent space of the domain to itself. We can form the quotient $P(w)/w$ to replace this cubic map by a quadratic map from the tangent bundle of the domain to $\mathbf{C}$ which contains the same information.

It follows that this invariant for locally univalent functions up to Mœbius transformations, called the Schwarzian derivative, is a holomorphic quadratic differential form, usually referred to as a quadratic differential.

Sketching the picture of a conformal map which has the two-jet of the identity at a point (Figure 2.1: jets of locally univalent mappings), one can see that the third derivative tends to make the image of a small circle around a point in the domain look elliptic in the range. Since Mœbius transformations preserve circles, the Schwarzian derivative may be interpreted as measuring the third-order distortion of the shape of the image of a small circle.

A formula for the Schwarzian derivative can be readily determined from the information above, or it may be looked up—someplace else. Like the formula for the curvature of a curve in the plane, the formula looks somewhat mystical at first, and in a qualitative discussion, the formula tends to be a distraction from the real issues.

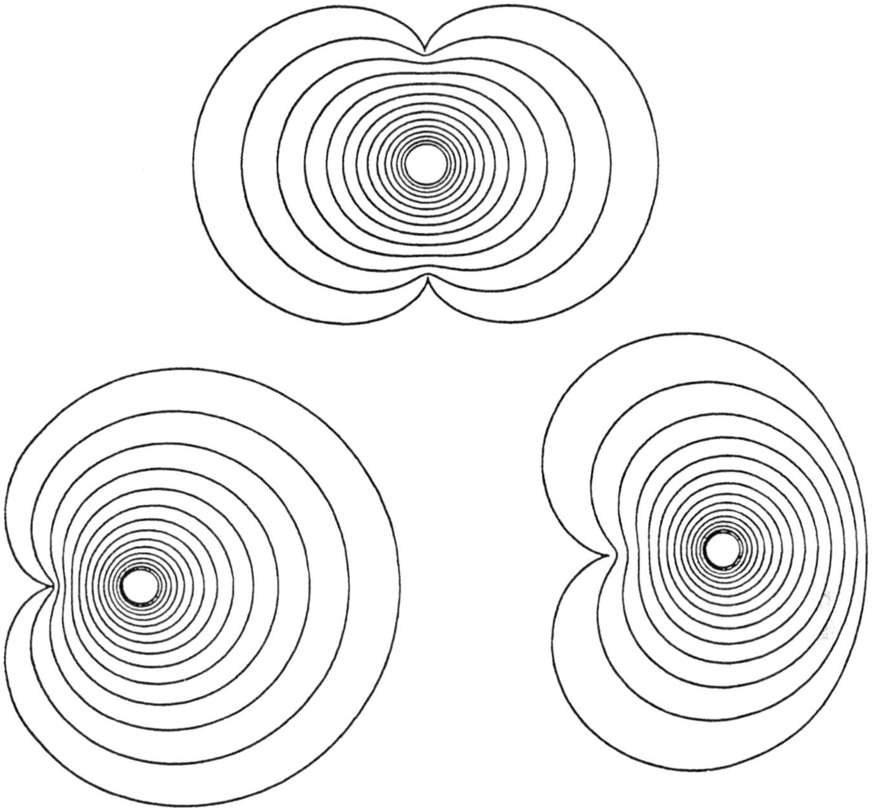

2.1. FIGURE: JETS OF LOCALLY UNIVALENT FUNCTIONS. Here are three univalent mappings illustrating the effects of second and third derivatives of a locally univalent mapping on a family of concentric circles. In (a) the mapping is $z \mapsto z + z^3/3$, and in (b), $z \mapsto z + z^2/2$. Thus, mapping (a) has the two-jet of the identity, and the third-order effect on the eccentricity of circles, as measured by the Schwarzian derivative, can be seen. Mapping (b) has the 1-jet but not the 2-jet of the identity, so even though its third derivative is 0, its Schwarzian derivative is not 0. The second derivative can be seen in the way that the centers of the images of circles are displaced toward the right. The 2-jet can be corrected by composing with the Mœbius transformation $z \mapsto z/(1 - .5z)$. The composition is illustrated in (c). The third-order term in the Taylor expansion of the composition is $-z^3/4$, and again it can be seen in the third-order effect on eccentricity.

How does the Schwarzian derivative behave under composition? If f and g are two locally univalent mappings which have order two contact with the identity at a point z, then the third derivative of the composition is the sum of their third derivatives, so the Schwarzian derivative of the composition is the sum of the Schwarzian derivatives, at least in this case. The same formula holds in general, provided we are careful to treat the Schwarzian derivative as a quadratic form. This easily follows from the fact that the Schwarzian derivative is invariant under postcomposition by a Mœbius transformation, and it transforms as a quadratic form under precomposition by a Mœbius tranformation. More explicitly, for the Schwarzian derivative of an arbitrary composition $f \circ g$ evaluated at a point z, one can replace it by $(f \circ M_1) \circ (M_1^{\circ^{-1}} \circ g \circ M_2)$, where M_1 and M_2 are the Mœbius transformations chosen so that the terms in parentheses have the 2-jet of the identity at $f \circ g(z)$.

The Schwarzian derivative together with the 2-jet of a locally univalent mapping determines the 3-jet, which is to say that the mapping satisfies a third-order differential equation in terms of its Schwarzian derivative. Since it is an ordinary differential equation, the equation can be locally integrated no matter what the holomorphic quadratic differential. Thus, every quadratic differential occurs, at least locally, as the Schwarzian derivative of a conformal map. Since Mœbius tranformations act transitively on nonsingular 2-jets, the three degrees of freedom for the solution are accounted for by compositions with Mœbius transformations. If the quadratic differential is defined over a simply connected domain, local solutions can be pasted together using Mœbius transformations to give a locally univalent mapping defined from the entire domain to the Riemann sphere $\hat{\mathbf{C}}$ and having the prescribed Schwarzian derivative.

Here is a slightly different way to think of the Schwarzian derivative. As a point w varies in the domain of f, the approximating Mœbius transformation $M(f, w)$ varies. The derivative of $M(f, w)$ with respect to w associates to any tangent vector V at w an infinitesimal Mœbius transformation, which is a holomorphic vector field on the Riemann sphere

$$X(f, w, V) = \frac{\partial}{\partial t} M^{\circ^{-1}}(f, w) \circ M(f, w + tV).$$

The order of composition has been chosen so that $X(A \circ f, w, V) = X(f, w, V)$, so we may normalize f to have order 2 contact with the identity at w without changing X. Then it is clear that $X(f, w, V)$ vanishes through order 1 at w. X is thus determined by its second derivative, which is an element of $\hom(T \otimes T, T)$, where T is the tangent space to $\hat{\mathbf{C}}$ at w. Therefore we have as an invariant of f at w an element of $\hom(T, (T \otimes T, T))$, which is canonically isomorphic to $\hom(T \otimes T \otimes T, T)$, which in turn is isomorphic to $\hom(T \otimes T, \mathbf{C})$ (taking into account the fact that T is 1-dimensional).

This invariant is, of course, the same as the Schwarzian derivative of f, but it has perhaps a clearer interpretation as measuring the rate of change of the best approximating Mœbius transformation.

Consider a univalent mapping f, that is, a conformal embedding of the open unit disk in $\mathbf{C}$. We can normalize f by postcomposition with an affine transformation until it has the 1-jet of the identity at the origin. Once this is done, it is easy to see that each derivative of f is bounded in a way which is independent of f. It follows that the Schwarzian derivatives of univalent functions are uniformly bounded with respect to the hyperbolic metric on the domain disk.

One useful topology on the set of univalent functions is the topology of uniform convergence on compact subsets of the open disk; this topology is used in the theory of normal families, and in particular, in the standard proof of the Riemann mapping theorem. The group of complex affine transformations acts on univalent functions, and the quotient space is contractible with respect to this topology: if f is normalized to have derivative 1 at the origin, then the isotopy

$$f_t(z) = \frac{1}{t} f(tz) \qquad [0 < t \leq 1]$$

connects f to the identity map (which is the limit as t approaches 0).

This topology is much too weak for many other purposes, however. Since Schwarzian derivatives are uniformly bounded, another natural topology comes from uniform convergence on quadratic differentials in the hyperbolic plane. This works best on the set of univalent maps up to postcomposition with Mœbius transformations. Denote this space by S. An element of S may be represented by any conformal embedding of the unit disk on the Riemann sphere $\hat{\mathbf{C}}$—up to Mœbius transformations; it is pointless to require that infinity is in the complement of the image.

Because of the chain rule above for Schwarzian derivatives, an ε neighborhood of a univalent mapping f consists precisely of mappings which have the form $g \circ f$, where g is a univalent mapping defined on the image of f whose Schwarzian derivative has norm less than ε as measured in the Poincaré metric of the image of f.

Here is a macroscopic (rather than infinitesimal) description of the topology of S. If f_1 and f_2 are two univalent mappings, we can compare them by considering the disk of radius R with respect to the hyperbolic metric about a point z in the domain of the mappings. Change coordinates to put z at the origin, and normalize the mappings to have the 2-jet of the identity at 0, i.e., let B be a conformal automorphism of the unit disk sending the origin to z, and let the normalized transformation be $g_i = M^{\circ^{-1}}(f_i \circ B, 0) \circ f_i \circ B$. Then define $d_{z,R}(f_1, f_2)$ to be the maximum distance between g_1 and g_2 in the hyperbolic disk of radius R about the origin, measured with respect to the spherical metric in the range. We can define f_1 and f_2 to be (ε, R) close if $d_{z,R} < \varepsilon$ for all z in the open unit disk. This determines a neighborhood basis for a topology. Actually, we would arrive at a neighborhood basis for the same topology if we picked any particular R and held it fixed.

In fact, the resulting topology would be the same if the normalized maps were prescribed to be not only pointwise close, but C^r close, since derivatives of holomorphic maps are estimated in terms of values, on account of the Cauchy integral formula. Clearly, this implies the Schwarzian derivatives are close. On the other hand, solutions of differential equations depend continuously on the data, so if the Schwarzian derivatives of f_1 and f_2 are uniformly close, then f_1 and f_2 are (ε, R) close. Thus, the topology is the same as that of S.

What is the macroscopic analogue of the description of neighborhoods of univalent functions in terms of compositions with functions having small Schwarzian derivatives? To construct such a description, we must modify the definition of $d_{z,R}$ so that it can be applied to functions defined on a domain U other than the unit disk. The difficulty is that we need a way to measure the distance between the values of two mappings which is not affected by postcomposition with a Mœbius transformation. In the infinitesimal setting, we used the Poincaré metric of the domain, but this will not do in the macroscopic setting since the mapping is unlikely to have image contained in U.

The solution is to measure using the unique round metric $R(z, U)$ on $\hat{\mathbf{C}}$ which has first-order contact to the Poincaré metric at a point z. By round, we mean the metric is equivalent to the standard spherical metric by a Mœbius transformation. If B is a conformal mapping of the unit disk to U which sends 0 to z, then this round metric is the image of the standard spherical metric under $M(B, 0)$. More generally, the round metric associated to a point z in a domain U transforms by $M(g, z)$ if g is a conformal embedding of U in $\hat{\mathbf{C}}$. (Note that the transformation of the 1-jet of a metric depends on the 2-jet of a mapping.)

One way to visualize the associated round metric is to observe that a choice of a round metric is equvialent to a choice of a point p in the $\mathbf{H}^3$ bounded by $\hat{\mathbf{C}}$. The round metric is the visual metric at p, that is, it is induced from the metric on the unit tangent space to $\mathbf{H}^3$ at p via the homeomorphism which sends a tangent vector to the endpoint of its geodesic on the sphere at infinity. Thus, to a point z in a domain U there is canonically associated a point $p(z, U)$ in $\mathbf{H}^3$. In the case of a round disk, the associated point is the perpendicular projection to the hyperbolic plane with the same bounding circle. The point $p(z, U)$ is transformed by the formula $p(f(z), f(U)) = M(f, z) \circ p(z, U)$ whenever U is mapped to another domain by an embedding (or even a covering map).

Now it is clear that another neighborhood basis for an element f of S consists of the sets of maps of the form $g \circ f$, where g is a conformal embedding of the image U of the unit disk which has the property that for each $z \in U$, $M^{\circ^{-1}}(g, z) \circ g$ is within ε of the identity when restricted to the Poincaré ball of radius R about z and measured using $R(z, U)$.

We shall prove that S has uncountably many isolated points. This contrasts strongly with the contractibility of the set S when it is equipped with the topology of uniform convergence on compact subsets.

In passing, we point out that there is a close connection developed by Lipman Bers between the space S and Teichmüller spaces. In fact, denote by U the space of locally univalent functions with the topology of uniform convergence of

Schwarzian derivatives, so that $S \subset U$ has the topology induced from U. The interior T of S is called the universal Teichmüller space. All ordinary Teichmüller spaces are embedded in the universal Teichmüller space in a natural way. T is also characterized as consisting of those univalent functions whose image is bounded by a quasicircle (see definition below). The main result of Gehring [**1**] is that S is not the closure of T.

3. Zippers. The image of a univalent mapping determines the mapping, up to precomposition with an isometry of the domain hyperbolic plane. The isometries of $\mathbf{H}^2$ do not act continously on S. Still, we will analyze univalent functions in terms of the shapes of their images.

We will say that an open set U is *rigid* if there is some ε such that any conformal embedding U in $\hat{\mathbf{C}}$ with Schwarzian derivative less than ε is a Mœbius transformation. When U is simply connected, this is equivalent to the condition that *every* Riemann mapping from the open disk to the domain represents an isolated point in S. Note that the condition of rigidity makes sense even when the domain is not simply connected.

A *quasi-interval* in $\hat{\mathbf{C}}$ is an arc α in $\hat{\mathbf{C}}$ with the property that there is a constant K such that for any two points $a, b \in \alpha$, the subarc between a and b lies in a disk of radius $K\, d(a, b)$ about a. Ahlfors showed that quasi-intervals are exactly the quasiconformal images of a unit interval in the real line. (Note: this follows from a similar characterization of the quasiconformal images of circles, using a square root construction.) The main point of this paper will be to prove

3.1. THEOREM: RIGID QUASI-INTERVALS. *There are many quasi-intervals with rigid complement.*

A more precise formulation of the statement will be given in Theorem 4.3: incorrigible implies rigid, where “many” will be replaced by a specific definition.

The proof of this theorem may make it sound like S has such a strong topology that almost no domains should be nonrigid, except for cases such as Jordan domains whose complement contains an open set, and geometrically simple cases such as domains with piecewise-smooth frontier. Perhaps so. On the other hand, there are many interesting nonrigid examples which arise in the theory of Kleinian groups: the complements of the limit sets of all known degenerate groups are nonrigid. Maybe the technique here can be turned around to help give information about the geometry and topology of these limit sets.

We will now give an intuitive description of a construction for rigid quasi-intervals. Rather than making the specific estimates needed to demonstrate the particular example rigorously, we will in the next section give a general proof that a general class of quasi-intervals (including the ones we construct) have rigid complements.

The idea is based on zippers. Zippers have the property that they cannot be pulled apart in the middle without a lot of distortion. In the same way, the

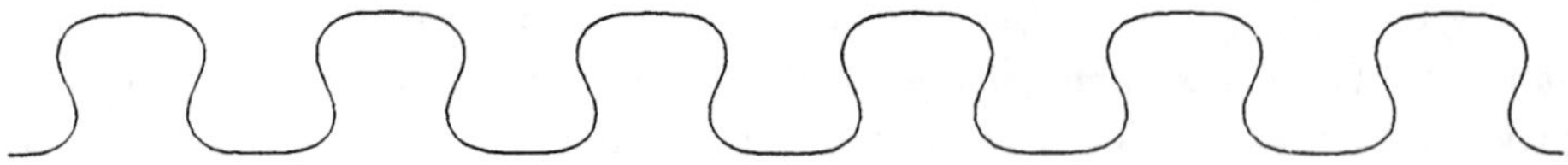

3.2. FIGURE. A simple zipper curve.

simple zipper curve of Figure 3.2 has the property that it separates the plane into two components which cannot be moved apart disjointly without first making a definite amount of distortion, because the zipper teeth on the two sides of the curve interlock. The usual custom in the physical world is to make zippers with rectilinear teeth, but these smooth teeth will serve the same purpose while being more convenient for constructing higher-order zippers.

The curve of Figure 3.2 is part of an embedded line in $\mathbf{C}$ invariant by a translation, which is a parabolic Mœbius transformation. Similarly, we can construct a simple zipper arc Z_1 which is invariant by a hyperbolic Mœbius transformation. Any conformal embedding of the complement of Z_1 with small Schwarzian derivative (measured as a quadratic differential using the Poincaré metric of the complement of Z_1) will have zipper teeth of approximately the same shape. If ε is small enough that a few of the teeth are interlocked, there is a cascade effect: all the other teeth will be forced to be interlocked.

The complement of Z_1 is not rigid. For one thing, the Schwarzian derivative varies continuously if the curve is wiggled in the C^1 topology. To take care of this, we iterate the zipper construction. Inductively replace the curve Z_k by a new curve Z_{k+1} lying in a thin band about the original and having teeth of a smaller scale. There is a limit curve Z_∞.

Any conformal embedding of the complement of Z_∞ with Schwarzian derivative uniformly near 0 must have the largest of the top level teeth interlocked. Because of the cascade effect, all the top level teeth are interlocked. Also, this forces the level 2 teeth to be approximately interlocked; hence they too are interlocked. The cascade effect proceeds to finer and finer levels, forcing all teeth at all levels to be interlocked.

This implies that the new conformal embedding extends continuously over Z_∞, giving a homeomorphism of $\hat{\mathbf{C}}$ which is conformal at least in the complement of Z_∞.

3.3. PROPOSITION: EXTENSION CONFORMAL. *Any homeomorphism of* $\hat{\mathbf{C}}$ *which is conformal in the complement of a quasi-interval* α *is a Mœbius transformation.*

PROOF. By the theorem of Ahlfors mentioned above, α can be straightened to a line segment by a quasiconformal homeomorphism. The proposition is easy to prove in the case of a line segment, using the Cauchy integral formula. This implies that the original map f is at least quasiconformal. Its quasiconformal distortion is zero almost everywhere; therefore f is conformal. Consequently, it is a Mœbius transformation. $\square$

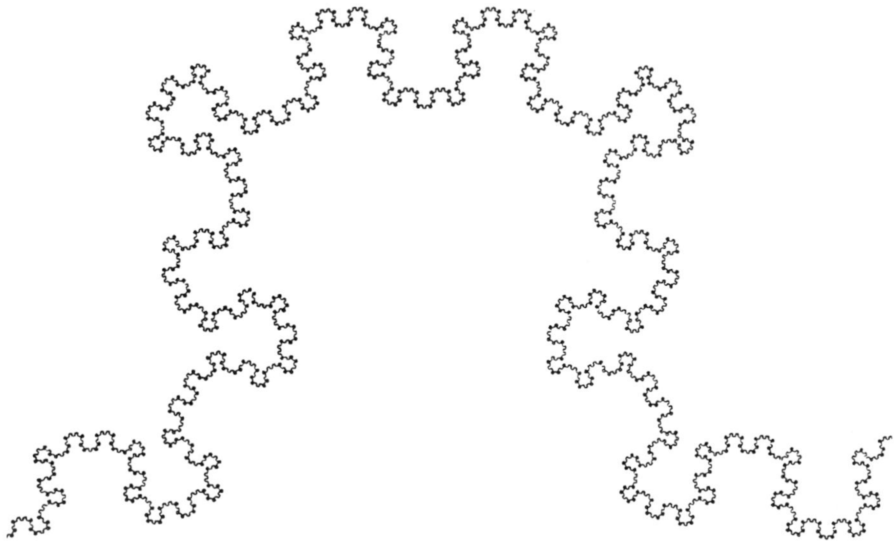

3.4. FIGURE: ITERATED ZIPPER CURVE. A detail of the iterated zipper arc Z_∞. The picture will look much the same if any small tooth is enlarged to the size of this large tooth. On a large scale, the entire zipper arc fits inside a banana-shaped region of the $\mathbf{C}$, and it is invariant by a hyperbolic transformation.

This gives a demonstration, at least on the intuitive level, that the complement of Z_∞ is rigid.

4. Compactness. We will make a more rigorous and more general analysis by considering all possible quasi-intervals microscopically. For this, it is convenient to consider quasi-intervals in $\mathbf{C}$, so that a K-quasi-interval when magnified by any constant is still a K-quasi-interval. (On the 2-sphere, the magnification would bring the two endpoints close together, so that the constant would have to go to infinity.) As a limiting case, there are also quasilines and quasihalf-lines, which are properly embedded copies of $(-\infty, \infty)$ or $[0, \infty)$.

The Hausdorff metric is a metric on the set of closed subsets of a metric space which says that two sets are within ε if every point in either set is within ε of some point of the other set. Since $\mathbf{C}$ is noncompact, it is convenient to extend closed subsets of $\mathbf{C}$ to subsets of $\hat{\mathbf{C}}$ by forming the closure, and to measure the Hausdorff distance of the resulting closed subsets of $\hat{\mathbf{C}}$ using the spherical metric. The set of closed subsets of $\mathbf{C}$ with this topology forms a compact metric space.

Let $Q(K)$ be the set of closed sets consisting of the empty set and all singletons, together with all K-quasi-intervals, K-quasilines, and K-quasihalf-lines.

4.1. PROPOSITION: CLOSURE QUASIARCS. *The set $Q(K)$ is closed and hence compact.*

PROOF. This is left as a nice exercise for the reader. □

Given any K-quasi-interval α, let $E(\alpha)$ consist of all enlargements of α, that is, the set of all images of α under transformations $z \mapsto az + b$, where $|a| \geq 1$.

4.2. DEFINITION. A K-quasi-interval α is *incorrigible* if the closure of $E(\alpha)$ contains no straight lines.

It is easy to see that incorrigibility is equivalent to the condition that the closure of $E(\alpha)$ contains no circles.

4.3. THEOREM: INCORRIGIBLE IMPLIES RIGID. *The complement of any incorrigible K-quasiarc is rigid.*

4.4. COROLLARY: RIGID DOMAINS EXIST. *There are domains which are rigid.*

PROOF. For one example, use the complement of Z_∞. There will be no lines in the closure of $E(Z_\infty)$ provided the construction is carried out in such a way that the ratio of the size of the teeth of the kth level to the teeth of the $(k+1)$st level is bounded.

A more interesting example comes from the limit set of any quasi-Fuchsian group which has no parabolic elements and which is not Fuchsian. Let α be any arc on such a limit set. The limit set has the property that the set of subsets of $\mathbf{C}$ obtained by conjugating the limit set by any Mœbius transformation which makes it pass through 0 and ∞ is already a closed subset of $Q(K)$, for some K. In fact, the set of such conjugating transformations, up to the action of the quasi-Fuchsian group, is compact. This implies that $E(\alpha)$ contains no lines. □

Note that if a quasi-Fuchsian group contains parabolics, the closure of $E(\alpha)$ in fact does contain lines, although a refinement of the proof of the theorem would show that the complement of any subarc of such a limit set is rigid anyway.

PROOF OF THEOREM 4.3. We will prove the theorem by proving a more positive form of the statement: If α is a K-quasi-interval and f_i a sequence of conformal embeddings which are not Mœbius transformations but whose Schwarzian derivatives converge uniformly to 0, then there are lines (in fact, many lines) in the closure of $E(\alpha)$.

Note that the closure of $E(\alpha)$ is invariant under the process of enlargement. What we will do is pass to a limit of enlargements of α, then a limit of enlargements of this, and so on until we find a line.

We may assume that f_i is normalized at some base point on $\hat{\mathbf{C}}$, to have the 2-jet of the identity transformation at that point. It is convenient to take as base point the point at infinity. The f_i converge uniformly on compact sets of the domain $U = \hat{\mathbf{C}} - \alpha$ to the identity.

We need a bit of information about the qualitative properties of a quasi-interval and the behavior of maps with small Schwarzian derivatives.

4.5. LEMMA. *Let t be any point in the interior of α, and let D be the disk of radius $3Kr$ about t. There must be points within D which are at distance exactly r from α.*

PROOF. Indeed, t separates α into two subarcs α_1 and α_2. Let A be the annulus about t with inner radius $2Kr$ and outer radius $3Kr$. The r-neighborhoods of the α_i intersected with A must be disjoint, since α is a K-quasiarc.

In the typical case that these intersections are nonempty, since the annulus is connected we can find points in the frontier of their union, which is the r-neighborhood of α, as claimed. If each α_i intersects the circle of radius $3Kr$, we can even find a point as claimed on both homological sides of the intersection of α with the $3Kr$-disk.

Even near the ends of α, where only one of the pieces (say α_1) of α intersects the annulus, a related argument still works. To see this, consider the double cover of $\mathbf{C}$ branched at t. The inverse image of the r-neighborhood of α_1 intersect A must have two components, so there are points in their frontier. □

The lemma implies that the collection of disks centered about points in the complement of α whose radius is $3K$ times the distance to α cover $\mathbf{C}$. In fact, the subcollection consisting only of those disks of any fixed radius cover some neighborhood of α.

We will define a quantity $D(f,z)$ which is a measure of how far f is distorted from a Mœbius transformation in the vicinity of $z \in U$. D will measure distortion roughly on the scale of the distance $r(z)$ of z from α. The definition is that $D(f,z)$ is the maximum distance of $M^{\circ^{-1}}(f,z) \circ M(f,w)$ from the identity, measured with respect to the round metric $R(z,U)$ on $\hat{\mathbf{C}}$, as w ranges over the set of points a distance at least $r(z)/2$ from α but distance at most $10Kr(z)$ from z.

Note that $D(f,z)$ goes to 0 as z goes to infinity. What happens for z near α?

4.6. PROPOSITION: SMALL DISTORTION IMPLIES CONTINUOUS. *There is a constant $a > 0$ such that if $D(f,z)$ is less than a in some neighborhood of α, then f extends continuously to α, and hence f is a Mœbius transformation.*

PROOF. The condition that a point w is within the disk of radius $10Kr(z)$ about z but has distance more than $r(z)/2$ from α implies that the round metrics $R(z,U)$ and $R(w,U)$ are roughly compatible, or in other words, there is an a priori bound for $d(p(z,U),p(w,U))$. In fact, among points z such that $r(z)$ is less than a bounded fraction of the diameter of α, the metric $R(z,U)$ deviates by a bounded amount from the metric obtained by enlarging the picture until $r(z)=1$, then mapping to a sphere via the stereographic projection which sends the unit circle about z to the equator of the sphere. To see this, consider the case that $r(z)=1$. Make a parabolic transformation fixing z so that α goes through ∞. The first two derivatives of a univalent function which sends the origin to z are bounded, which means that the 1-jet of the Poincaré metric deviates by a bounded amount from the 1-jet of the Euclidean metric, which is the same as the 1-jet of the round metric defined by stereographic projection.

The proof of the proposition has to do with the nesting properties of the disks of radius $10Kr(z)$ about points z of distance $r(z)$ from α.

For any point z' with $r(z') = r(z)/2$ and such that the $3Kr$ disks for z and z' intersect, then they are at distance no more than $3Kr(z) + 3Kr(z') = 4.5Kr(z)$ from each other. Then the $10Kr(z') = 5Kr(z)$ disk about z' is contained in the $10Kr(z)$ disk about z, leaving a $r(z)/2$ margin of safety.

Under the hypothesis that $D(f, z)$ is small, the transformation $M^{\circ^{-1}}(f, z) \circ M(f, z')$ sends the $10Kr(z')$ disk safely inside the $9.9Kr(z)$ disk about z.

Consider a pair $A \supset B$ of nested disks and all possible images of them under Mœbius transformations taking A to the finite complex plane. There is a maximum for the ratio of the radii of these images; this upper bound is attained when the two disks are concentric. It follows that there is an upper bound $\beta < 1$ to the ratio of the radius of the image of the $10Kr(z')$ disk about z' by $M(f, z')$ to that of the $10Kr(z)$ disk about z under $M(f, z)$, where z and z' range over pairs of points chosen as above. In fact, β can be found as a function only of $D(f, z)$; in the circumstances above, certainly $\beta < .99$.

Renormalize f for convenience to make sure that $M(f, z)$ maps the $10Kr(z)$ disk about z to a disk E in the finite complex plane. Let R be the radius of E. The image of the $10Kr(z')$ disk about z' by $M(f, z')$ is contained in E and has a radius less than βR.

Inductively, it follows that for any point w such that $r(w) = 2^{-k} r(z)$ whose $3Kr(w)$ disk intersects the $3Kr(z)$ disk about z, the diameter of its image under $M(f, w)$ is a disk of radius less than $R\beta^k$ contained in the original disk E. In particular, the image by f of the entire $3Kr(z)$ disk about z is contained in the image by $M(f, z)$ of the $10Kr(z)$ disk.

Consequently, for any point t on α, there are neighborhoods of t in $\mathbf{C}$ whose images under f (where f is defined) have arbitrarily small diameter. Therefore f converges along α to a continuous map. □

It is interesting to remark here that when f is a mapping with small Schwarzian derivative, the hypothesis of the proposition would be satisfied if the measure of distortion $D(f, z)$ were modified so that only points w are considered which are on the same side of α as z. The proof of the proposition therefore shows that f has a continuous limit on each side of α—but in general, the limits from the two sides will be different.

PROOF OF THEOREM 4.3 (CONTINUED). Now back to our sequence f_i. By assumption, the f_i are not Mœbius transformations, so there is a constant a such that $D(f_i, z) > a$ for points z arbitrarily close to α.

By the intermediate value theorem, for each sufficiently large integer $n > 0$, there is a point $x_{i,n}$ where D_i takes the value $1/n$. Let $f_{i,n}$ be a rescaling of f_i so that $x_{i,n}$ is at the origin, and α becomes a curve $\alpha_{i,n}$ at distance one from the origin. We renormalize $f_{i,n}$ to have the 2-jet of the identity at the origin.

If we fix n and let i tend to infinity, there is a subsequence of the i's so that $\alpha_{i,n}$ converges to a quasiline or quasihalf-line $\alpha_{\infty,n}$, and so that the $f_{i,n}$ converge to a transformation $f_{\infty,n}$ which is the identity near 0 and is locally Mœbius (since its Schwarzian derivative is 0) but not Mœbius, since $D(f_{i,n}, 0)$ is constant. Obviously, the complement of $\alpha_{\infty,n}$ cannot be connected. Therefore, $\alpha_{\infty,n}$ must be a quasiline which separates $\mathbf{C}$ into two components, U_0 which contains the origin and U_1 which does not. The transformation $f_{\infty,n}$ is the identity on U_0, and on U_1 it is a nontrivial Mœbius transformation taking U_1 into itself.

Now we let n go to infinity. There is some subsequence such that the $\alpha_{\infty,n}$ converge to a quasiline α_∞. The transformations $f_{\infty,n}$ converge to the identity on both sides of the limiting quasiline, since $D(f_{\infty,n}, 0)$ converges to 0.

Let X be any vector field which represents a limiting direction for the Mœbius transformations $f_{\infty,n}$. By "direction," we refer to the direction from the identity in the Lie group of Mœbius transformations. More explicitly, let

$$X(z) = \left[\lim a_{i_j}(f_{\infty,i_j}(z) - z)\right] \frac{\partial}{\partial z},$$

where the subsequence $\{i_j\}$ and sequence of constants a_{i_j} are chosen to make the limit exist and be nonzero.

We claim that X carries U_1 into itself. To see this, consider any flow line of X through a point x in U_1. For any point y downstream from x on the flow line, there is some $f_{\infty,n}$ and some integer q such that its qth iterate carries x approximately to y. It follows that y is at least in the closure of U_1. Since the image of an open set by a time t map of a flow is open, y must be in U_1 itself.

Choose a point t on α where the flow is not zero, and conjugate by a sequence of enlargements centered about t so that in the limit X becomes an infinitesimal translation in the i direction. Passing to the limit of a subsequence, we obtain a new quasiline β, which intersects each line parallel to the y-axis in an interval or a point.

If any of these intersections are intervals, we immediately obtain a line in the closure of $E(\alpha)$. In fact, if the slopes of secant lines connecting pairs of points of β are unbounded (either above or below), there is a sequence of enlargements carrying such pairs of points to pairs at distance n from each other. The limit is a vertical line, so we are done in this case also.

If the slopes of the secant lines are bounded, then β is the graph of a Lipschitz function. A Lipschitz function is differentiable almost everywhere. A sequence of enlargements centered at any point where the function is differentiable converges to a line whose slope is the derivative. □

References

F. W. Gehring, *Spirals and the universal Teichmüller space*, Acta Math. **141** (1978), 99–113.

Part 2.

Computer Science

COMPUTER SCIENCE

Bill Thurston not only developed revolutionary and transformative ideas within pure mathematics, he exported his perspective into other areas. We discuss below two examples within theoretical computer science. Consistent with his holistic world view, computer science in turn influenced his work on geometric group theory. For example, the paper *Groups, tilings and finite state automata* [T] and the book *Word Processing in Groups* (with Epstein, Cannon, Holt, Levy and Paterson) [ECHLPT] introducing the theory of automatic groups.

Papers [MT] (with Miller) and [MTTV1], [MTTV2], [MTTV3] (with Miller, Teng and Vavasis) introduce, for the first time, geometric methods to find fast algorithms to partition a large class of meshes in two and three-dimensional space into subsets of approximately equal size. This method is applicable to planar graphs thanks to the Andreev-Thurston circle packing theorem. This is important because [MTTV1] "One approach to achieving the large memory and computation power requirements for large-scale computational problems is to use massively parallel distributed-memory machines. In such an approach, the underlying computational mesh is divided into submeshes, inducing a subproblem to be stored on each processor in the parallel system and boundary information to [be] communicated [SBFGGMOST]. To fully utilize a massively parallel machine, we need a subdivision in which subproblems have approximately equal size and the amount of communication between subproblems is relatively small. This approach will decrease the time spent per iteration."

Binary trees are fundamental in computer science. A natural operation called *rotation* collapses a non-root edge and expands it in the other direction. Quoting [STT2] "In a common computer-related application of binary trees the tree is used to store an ordered collection of pieces of information (called items). Each internal node of the tree is labeled with an item, and the order of the items is represented by the symmetric order of the nodes. ... The rotation corresponding to a node changes the structure of the tree near that node, but leaves the structure elsewhere intact. A rotation maintains the symmetric order of the nodes but changes the depths of some of them. Rotations are primitives used by most schemes that maintain 'balance' in binary trees [ST], [STT1]."

A basic problem is to compute the maximal distance between two binary trees with the same number n of nodes. In [STT1] Bill (with Sleater and Tarjan) give the exact solution for $n \leq 16$ and when n is sufficiently large. An astonishingly elementary observation equates the problem to one of understanding elementary moves on triangulations of polygons with $n + 2$ edges. An equally elementary argument improved the known upper bound to $2n - 6$ when $n \geq 11$. A deep and original argument, involving volumes of hyperbolic 3-manifolds shows that, for n sufficiently large $2n - 6$ is the exactly realized maximum. This solved using transcendental methods a purely combinatorially stated problem for which good asymptotic estimates were not even known. With computer assistance they computed the exact answer for $n \leq 16$.

References

[ECHLPT] D.Epstein, J. Cannon, D. Holt, S. Levy, M. Paterson and W. Thurston, Word processing in groups, Jones & Bartlett, 1992, xi + 330.

[MT] G. L Miller and W. Thurston, Separators in two and three dimensions, Proceedings of the 22th Annual ACM Symposium on Theory of Computing, 300–309, Maryland, May 1990. ACM.

[MTTV1] G. Miller, S-H Teng, W. Thurston and S. Vavasis, Automatic mesh partitioning. Graph theory and sparse matrix computation, IMA Vol. Math. Appl., **56**, Springer, New York, 1993.

[MTTV2] G. Miller, S-H Teng, W. Thurston and S. Vavasis, Separators for sphere-packings and nearest neighbor graphs, *J. ACM* 44 (1997), 1–29.

[MTTV3] G. Miller, S-H Teng, W. Thurston and S. Vavasis, Geometric separators for finite-element meshes, *SIAM J. Sci. Comput.* 19 (1998), no. 2, 364–386.

[SBFGGMOST] E. Schwabe, G. Blelloch, A. Feldmann, O. Ghattas, J. Gilbert, G. Miller, D. O'Hallaran, J. Schewchuk and S.-H. Teng. A separator-based framework for automated partitioning and mapping ofparallel algorithms in scientific computing, First Annual Dartmouth Summer Institute on Issues and Obstacles in the Practical Implementation of Parallel Algorithms and the use of Parallel Machines, 1992.

[ST] D. Sleater and R. Tarjan, Self-adjusting binary search trees, *J. Assoc. Comput. Mach.* 32 (1985), 652–686.

[STT1] D. Sleater, R. Tarjan and W. Thurston Rotation distance, triangulations, and hyperbolic geometry, Proc. 18th Annual ACM Symposium on Theory of Computing, May 28-30, 1986, Berkeley, CA, Association for Computing Machinery, NY 122–135.

[STT2] D. Sleater, R. Tarjan and W. Thurston, Rotation distance, triangulations, and hyperbolic geometry, *J. of Amer. Math. Soc.* **1** (1988), 647–681.

[T] W. Thurston, Groups, tilings and finite state automata, 1989 AMS Colloquium Lectures, Research Report GCG 1.

Physica 19D (1986) 423–432
North-Holland, Amsterdam

SOLITON-LIKE BEHAVIOR IN AUTOMATA†

James K. PARK‡ and Kenneth STEIGLITZ
Dept. of Computer Science, Princeton University, Princeton, NJ 08544, USA

and

William P. THURSTON
Dept. of Mathematics, Princeton University, Princeton, NJ 08544, USA

Received 13 May 1985
Revised 25 October 1985

We propose a new kind of automaton that uses newly computed site values as soon as they are available. We call them *Filter Automata* (FA); they are analogous to Infinite Impulse Response (IIR) digital filters, whereas the usual Cellular Automata (CA) correspond to Finite Impulse Response (FIR) digital filters. It is shown that as a class the FA's are equivalent to CA's, in the sense that the same array of space-generation values can be produced; they must be generated in a different order, however.

A particular class of irreversible, totalistic FA's are described that support a profusion of persistent structures that move at different speeds, and these particle-like patterns collide in nondestructive ways. They often pass through one another with nothing more than a phase jump, much like the solitons that arise in the solution of certain nonlinear differential equations.

Histograms of speed, displacement, and period are given for neighborhood radii from 2 to 6 and particles with generators up to 16 bits wide. We then present statistics, for neighborhood radii 2 to 9, which show that collisions which preserve the identity of particles are very common.

1. Introduction

Cellular automata have attracted attention recently as non-numerical models for nonlinear physical phenomena [1]. Vichniac [2] points out that they "exhibit behaviors and illustrate concepts that are unmistakably physical...", and he goes on to mention "relaxation to chaos through period doublings," "a conspicuous arrow of time in reversible microscopic dynamics," "causality and light-cone," and others. The purpose of this paper is to describe a new kind of automaton that supports soliton-like structures in a strikingly clear way.

Scott *et al.* [3] propose as a working definition that "a soliton $\phi_s(x - ut)$ is a solitary wave solution of a wave equation which asymptotically preserves its shape and velocity upon collision with other solitary waves." The simplest examples are provided by solutions to the dispersionless linear wave equation. What is remarkable about solitons is that they can be supported by nonlinear equations with dispersion.

The notion of a solitary wave can be carried over in a natural way to the context of automata (either CA or FA) as follows: The term *solitary wave* or *particle* in an automaton will be taken to mean a periodic pattern of non-zero cell values that propagates with fixed finite velocity. A collision between two particles will be said to be a

†This work was supported in part by NSF Grant ECS-8307955, U. S. Army Research–Durham Grant DAAG29-82-K-0095, DARPA Contract N00014-82-K-0549, and ONR Grant N00014-83-K-0275.

‡Now with the Laboratory for Computer Science, Massachusetts Institute of Technology, Cambridge, MA 02139, USA.

0167-2789/86/$03.50 © Elsevier Science Publishers B.V.
(North-Holland Physics Publishing Division)

soliton collision if the particles retain their identities after the collision. (For an example see Fig. 3.)

The particular class of filter automata described here, which we call *parity-rule* FA's, support thousands of particles of relatively small size, are irreversible, and totalistic – that is, they belong to the simple class of automata that depend for their next state only on the number of 1's in the argument field of the next-state function (see below). Furthermore, as we will see in what follows, soliton collisions are quite common, occurring for some next-rule radii and ranges of particle widths 99% of the time. In contrast, in the totalistic one-dimensional CA's studied extensively by Wolfram [5] particles are relatively rare and non-destructive collisions extremely rare. There are also two-dimensional CA's that support particles – for example the gliders in the game of Life [7], or the billiard-ball models in the reversible CA's described by Margolus [4]. But the ease with which particles are supported by parity-rule FA's, and their propensity for passing through one another, appear to be unknown in the study of CA's.

2. Filter automata

We will restrict ourselves here to one-dimensional automata with k-valued site values a_i^t, where the subscript i refers to the space variable ($-\infty \le i \le +\infty$), and the superscript t refers to time ($0 \le t \le +\infty$). In the usual CA [5], the evolution of the automation is determined by a fixed rule F of the form

$$a_i^{t+1} = F\left(a_{i-r}^t, a_{i-r+1}^t, \ldots, a_i^t, \ldots, a_{i+r}^t\right), \tag{1}$$

with

$$F(0,0,\ldots,0) = 0.$$

The next value of site i is a function of the previous values in a neighborhood of size $2r+1$ that extends from $i-r$ to $i+r$. Given initial states at all the sites, which we assume run from $-\infty$ to $+\infty$, repeated application of the rule F determines the time evolution of the automaton.

In an FA, the next-state rule is of the form

$$a_i^{t+1} = F\left(a_{i-r}^{t+1}, a_{i-r+1}^{t+1}, \ldots, a_{i-1}^{t+1}, a_i^t, \ldots, a_{i+r}^t\right). \tag{2}$$

Now the next state is computed using the newly updated values $a_{i-r}^{t+1}, a_{i-r+1}^{t+1}, \ldots, a_{i-1}^{t+1}$, instead of $a_{i-r}^t, a_{i-r+1}^t, \ldots, a_{i-1}^t$. This is precisely analogous to the operation of an IIR digital filter, whereas a CA corresponds to an FIR digital filter (see, [6], for example).

Although we allow the sites in an FA to extend from $-\infty$ to $+\infty$, we must assume that to the left, anyway, there are only a finite number of sites containing non-zero values. This will then give us an unambiguous way to compute the evolution of the FA, using a left-to-right scan. We always start with an initial configuration that has only a finite number of non-zero site values.

Following Wolfram's terminology [5], when the next-state function F depends only on the sum

$$S(i) = \sum_{j=-r}^{r} a_{i+j} \tag{3}$$

we say an automation is *totalistic*. This class, although small and easy to specify, appears to exhibit all the interesting kinds of behavior found in general automata, and the particular class of FA's described here will be totalistic.

We will focus attention on the class of filter automata with binary-valued sites ($k=2$) defined by the following next-state rule. If $S(i)$ is the number of 1's in the $i-r$ to $i+r$ window at time t, then the new value of site i is

$$a_i^{t+1} = \begin{cases} 1, & S \text{ even but not } 0, \\ 0, & \text{otherwise.} \end{cases} \tag{4}$$

These we will call the *parity-rule* filter automata, and we will think of them as parameterized by the single integer $r = 2, 3, \ldots$, the *radius*.

3. Some examples

Fig. 1 shows a typical particle, one that occurs in the $r = 3$ parity-rule FA. The first line corresponds to generation 0 – it indicates the initial values assigned to the array of sites (a black square indicates a site with value 1, and the absence of such a square indicates a site with value 0). Subsequent lines correspond to subsequent generations. It is apparent from this figure that a particle has a well-defined *period* (the number of generations needed for the bit pattern to repeat), and *displacement* (the number of sites moved during a period, with plus measured to the left). In this case the period is 3 and the displacement 1.

We will refer to the succession of states passed through by a particle as its *orbit*. A convenient way to identify a particle uniquely is to view each orbital state as a binary number, and to take the smallest binary number in its orbit as the *canonical code* or *generator* of the particle.

It is not hard to see that the position of the rightmost 1 of a particle can never move right. For this to happen, we must be in the situation where the values of $a_i^t, \ldots, a_{i+r}^t$ are all 0, and there are an even nonzero number of 1's among the values $a_{i-r}^{t+1}, \ldots, a_{i-1}^{t+1}$. As the window slides right, this situation must be repeated, and so an infinite number of 1's would be generated, a contradiction. In fact it has been proved [11] that such situations can never be reached from an initial condition with a finite number of 1's; that is, that the parity-rule FA's are *stable* in this sense.

From the previous observation, we know that particles are either stationary or move left. It is also not hard to see that the maximum speed of a particle is $r - 1$, and this is realized by the particle consisting of $r+1$ consecutive 1's, which has period 1.

As Hirota and Suzuki [10] describe, one characteristic of solitons is that "A wave packet at any given position dissolves into many solitons each of which travels at its own velocity." Fig. 2 shows a typical evolution from a disordered state for the $r = 3$ parity-rule FA. Exactly the same kind of dissolution into several particles with different velocities can be observed.

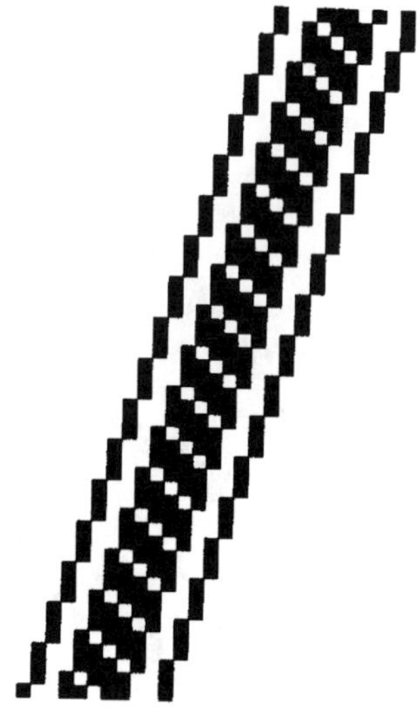

Fig. 1. A typical particle supported by a parity-rule filter automaton, illustrating period and displacement. This example is for $r = 3$, has canonical code 629, displacement 1, and period 3. The code sequence in its orbit is 629, 697, 1241.

Fig. 3 illustrates pairwise collisions that we call *soliton* – those in which the identity of both particles is preserved, while fig. 4 shows examples of non-soliton collisions. Extensive empirical evidence suggests that in a soliton collision the fast particle cannot be shifted to the right, and the slow particle cannot be shifted to the left. That is, the fast particle may only be pushed forward, and the slow only retarded. The latest collision in fig. 4 results in 2 particles moving in parallel and is particularly interesting because it shows that particle collisions are not always reversible. By *reversible* here, we mean that the picture rotated 180° would be a valid evolution of this or some other automaton. If we turn this picture upside down, it becomes clear that spontaneous splitting would be required for reversibility.

We next want to present some statistics for pairwise particle collisions, but first we need to study the maximum possible number of different

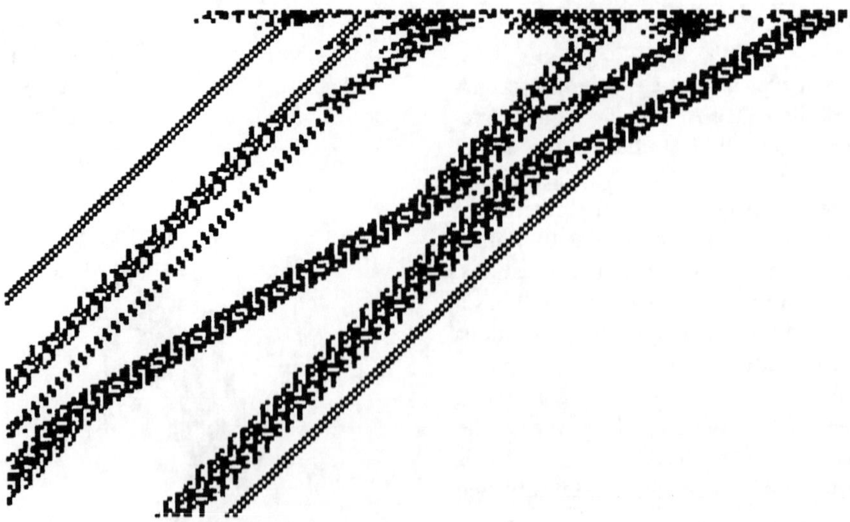

Fig. 2. Typical evolution from a disordered state, for the $r = 3$ parity-rule filter automaton, showing dissolution into several particles with different speeds.

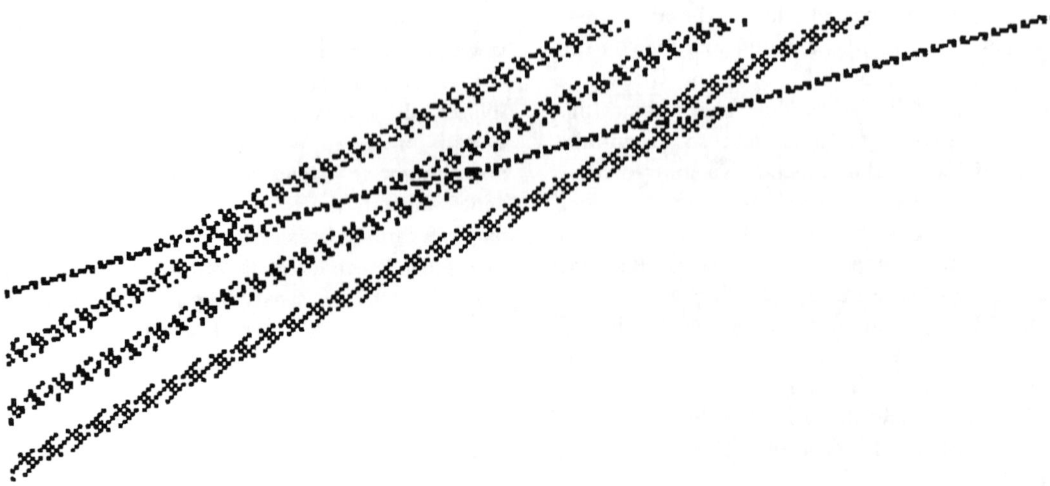

Fig. 3. Some typical soliton collisions, for the $r = 5$ parity-rule filter automaton. The initial particle canonical codes and displacement/periods are, from left to right: 145 (12/6), 201 (12/6), 273 (12/6), and 27 (7/2).

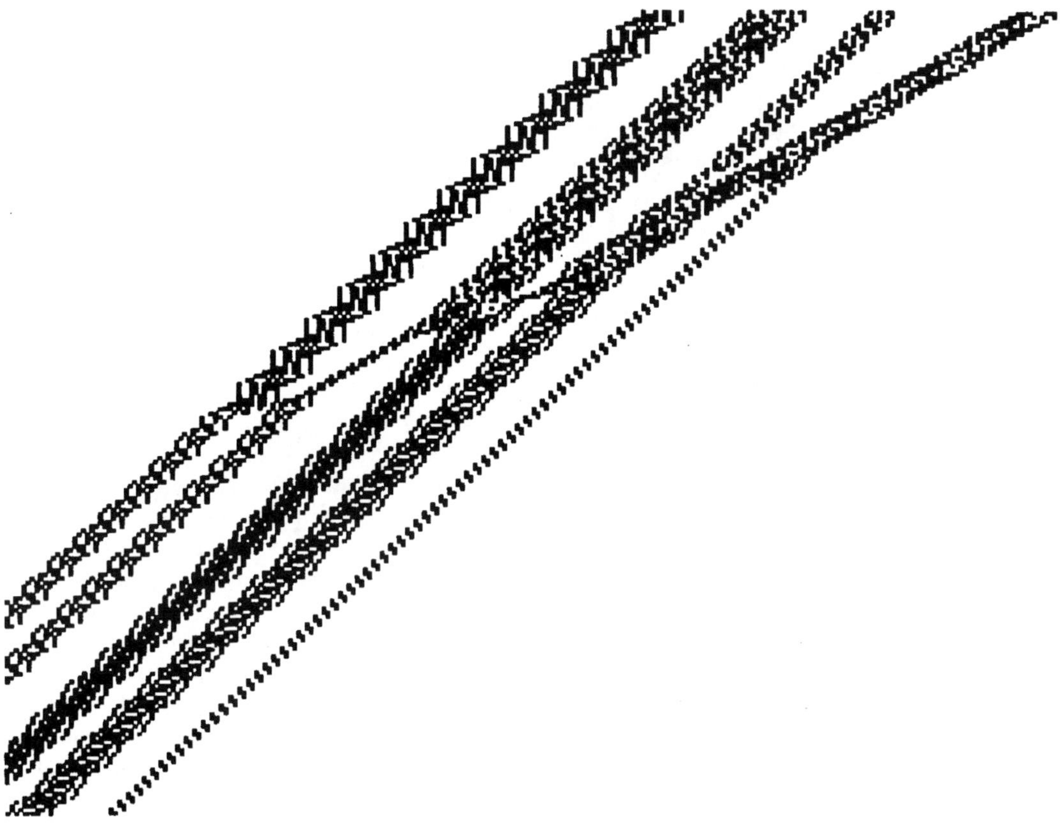

Fig. 4. Some typical collisions in which the identities of both particles are changed, for the $r = 3$ parity-rule filter automaton. The initial particle canonical codes and displacement/periods are, from left to right: 601 (8/8), 9451 (10/10), 43 (6/6), and 967 (20/12).

ways two particles can collide, given their periods and displacements.

4. The determinant of a particle pair

Given any two particles that move at different speeds, it is clear that we can arrange a collision between them by choosing an initial configuration with the faster particle to the right of the slower. If the two particles start close enough together, it may happen that they interact in a way that is impossible when they start far apart. In such cases we say the collision is *improper*; otherwise we say it is *proper*. We will restrict our attention to proper collisions, because we will always allow an initial spacing adequate for typical interactions.

We will say that two collisions are the *same* if the bit patterns of their history can be put in concordance by shifts in space and time, and *different* otherwise. We can now calculate a strict limit on the number of different proper collisions possible between two particles.

Theorem 1. Let the two particles have displacements d_1, d_2, periods p_1, p_2, and speeds $d_1/p_1 < d_2/p_2$, so that particle 2 hits 1. Let $q = \text{lcm}(p_1, p_2)$, and the difference in speeds be $\Delta s = d_2/p_2 - d_1/p_1$. Then the number of different proper collisions is no larger than

$$\mathrm{DET} = p_1 p_2 \cdot \Delta s = p_1 d_2 - p_2 d_1.$$

Proof. After q generations, the same relative configuration of bits in the particles' orbits repeats, so we need only cycle through $q \cdot \Delta s$ initial separations to obtain all possible proper collisions.

Case 1: p_1, p_2 relatively prime. In this case those $q \cdot \Delta s = p_1 p_2 \cdot \Delta s$ situations include all combinations of orbital states.

Case 2: p_1, p_2 not relatively prime. Then in $q \cdot \Delta s$ situations we have covered only $q/(p_1 p_2)$ of the total possible. Therefore there are at most $q p_1 p_2 \cdot \Delta s / q = p_1 p_2 \cdot \Delta s$ possible situations, as before.

We call DET the *determinant* of the collision, because it is precisely the 2×2 determinant with rows $p_1 p_2$ and $d_1 d_2$.

5. Particle and collision statistics

Dictionaries of particles were compiled in the following simple way. All bit patterns with width up to 16 were used as starting configurations of the parity-rule FA's, and the automata played forward in time long enough for particles to be generated and separated. The resulting bit strings were than analyzed to catalog the canonical codes, speeds, displacements, and periods of the particles so generated. Table I shows histograms of the results, sorted by speed for each value of r, for all distinct displacement-period pairs, and for all canonical codes of width ≤ 16.

There is a somewhat hazy distinction between two particles with the same speed traveling in parallel, and one particle. For our purposes we insist that a single particle not have a gap of $2r$ or more consecutive 0's in its canonical code.

The number of particle generators increases sharply with r, growing from only 8 for $r = 2$ to 13109 out of a possible 32768 – the number of odd integers up to 2^{16} – for $r = 6$. Another striking fact is the tendency for certain displacement/period pairs to be preferred. For example, for $r = 6$ the two pairs $d/p = 44/12$ and $56/14$ account for 65% of the total.

These particle dictionaries were used to study the question of how often a 2-particle collision is *soliton*; that is, how often the two particles pass through one another without change in their identities. Given a set of particles, all possible pairwise collisions were sampled uniformly as follows: a random pair (a, b) was chosen, with b faster than a and all such pairs equally likely. Then the fast particle was played a random number g generations forward in time, where g was chosen uniformly between 0 and the period of the fast particle less 1; that is, the fast particle was put in a random orbital state. The slow particle was then placed in its canonical orbital state with its right end a random number x spaces to the left of the left end of the fast particle, where x was chosen uniformly between k and $k + \mathrm{DET} - 1$, k being large enough to ensure a proper collision. Finally, the result of the ensuing collision was weighted by its corresponding DET to make the sample uniform over all possible collisions, rather than over all possible pairs of particles. This was done for 2000 collisions, for various values of radius r and for particle dictionaries with code-widths up to 10, 14, and 16. The results are shown in table II.

The general trend is that the estimated probability of soliton collisions increases with r for fixed code-widths, and decreases with code-width, although there are exceptions. What is perhaps most striking is the high level of the observed frequencies, reaching 99% for code-widths up to 10, and $r = 8$ and 9. It was also the case that mutual annihilation was never observed; every collision resulted in at least one particle.

6. Quasi-equivalence of CA's and FA's

A natural question is whether FA's are essentially different from CA's, or whether any FA can be simulated in some sense by a CA. In this

Table I

Speed, displacement, period, and frequency of occurrence for all particles with canonical code width ≤ 16, radius 2 to 6.

Speed	Disp.	Per.	Freq.
Radius = 2, no. of pars. = 8			
0.000	0	1	1
0.500	1	2	1
0.500	2	4	1
0.500	3	6	1
0.500	8	16	1
1.000	1	1	3
Radius = 3, no. of pars. = 198			
0.333	1	3	1
0.500	1	2	1
0.500	4	8	4
1.000	1	1	2
1.000	2	2	1
1.000	4	4	4
1.000	5	5	1
1.000	6	6	3
1.000	8	8	11
1.000	10	10	41
1.000	16	16	8
1.333	8	6	2
1.333	16	12	63
1.400	7	5	3
1.400	14	10	17
1.500	3	2	1
1.500	6	4	1
1.500	12	8	4
1.571	22	14	16
1.667	5	3	4
1.667	10	6	3
1.667	20	12	3
1.750	14	8	1
2.000	2	1	3
Radius = 4, no. of pars. = 682			
0.667	2	3	2
1.500	3	2	4
1.500	6	4	10
1.500	9	6	6
1.500	12	8	44
1.500	15	10	2
2.000	2	1	1
2.000	4	2	1
2.000	10	5	14
2.000	20	10	219
2.125	17	8	19
2.333	7	3	18
2.333	14	6	19
2.333	28	12	216
Radius = 4, no. of pars. = 682			
2.333	35	15	2
2.500	5	2	1
2.500	25	10	14
2.571	18	7	8
2.571	36	14	68
2.750	11	4	5
2.750	22	8	4
2.750	44	16	4
3.000	3	1	1
Radius = 5, no. of pars. = 6534			
1.000	1	1	1
1.000	3	3	3
2.000	2	1	3
2.000	4	2	6
2.000	8	4	20
2.000	10	5	2
2.000	12	6	10
2.000	16	8	110
2.000	20	10	4
2.600	13	5	41
2.600	26	10	926
2.750	11	4	3
2.750	22	8	46
3.000	3	1	2
3.000	6	2	1
3.000	9	3	51
3.000	12	4	4
3.000	18	6	69
3.000	36	12	2251
3.000	45	15	12
3.200	32	10	71
3.286	23	7	53
3.286	46	14	1929
3.500	7	2	2
3.500	14	4	35
3.500	21	6	3
3.500	28	8	32
3.500	35	10	1
3.500	42	12	37
3.500	56	16	693
3.500	70	20	4
3.667	33	9	7
3.667	66	18	96
3.800	19	5	2
3.800	38	10	1
3.800	76	20	2
4.000	4	1	1
Radius = 6, no. of pars. = 13109			
1.333	4	3	4
2.500	5	2	3
2.500	10	4	19
2.500	15	6	15
2.500	20	8	195
2.500	25	10	8
3.200	16	5	30
3.200	32	10	1635
3.375	27	8	95
3.667	11	3	5
3.667	22	6	35
3.667	44	12	4157
3.900	39	10	213
4.000	28	7	21
4.000	56	14	4353
4.250	17	4	5
4.250	34	8	7
4.250	51	12	231
4.250	68	16	1742
4.444	40	9	1
4.444	80	18	227
4.500	9	2	1
4.500	63	14	102
4.600	23	5	3
4.833	29	6	1
5.000	5	1	1

Table II
Frequency of collisions that are soliton, in *per cent*. Based on samples of 2000 collisions.

Radius	Width ≤ 10	Width ≤ 14	Width ≤ 16
2	38.74	38.74	38.74
3	24.63	11.02	9.55
4	65.63	30.50	35.38
5	73.97	45.13	45.20
6	80.35	80.76	52.67
7	80.62	84.51	
8	99.78	83.63	
9	99.42	78.16	

section we will show that the latter is true; in particular, we will show that the space-time array generated by any particular FA can also be generated by some CA, and vice-versa.

First consider the constraints imposed on the order in which site values in FA and CA space-time arrays can be generated. Any site value a_i^t in an FA space-time array is determined by $a_{i-r}^t, \ldots, a_{i-1}^t, a_i^{t-1}, \ldots, a_{i+r}^{t-1}$. These site values are in turn determined by the values of other sites, but all sites that could possibly affect a_i^t lie within the region $\{a_{i+j}^{t-k}\colon\ k \geq 0,\ j \leq kr\}$ (this is the shaded region in fig. 5). For the space-time array of a CA, the corresponding region of sites whose values can affect a site value a_i^t is depicted in fig. 6 – it is just the region $\{a_{i+j}^{t-k}\colon\ k \geq 0,\ -kr \leq j \leq kr\}$.

It is clear from figs. 5 and 6 that there exists no 1-to-1 mapping from site values in an FA space-time array to site values in a CA space-time array, with the property that the orientation of the space axis is preserved. Changing the value of some site to the left of a_i^t (but in the same generation) in an FA space-time array can affect this site value, but it cannot in a CA space-time array. Thus, if we wish to simulate an FA with a CA, we must somehow change the orientation of the space axis. This motivates a "tilting" mapping from the FA space-time array to the CA space-time array, a mapping that rotates the region in fig 5 so that it looks more like the region of fig. 6.

Suppose then that we are given an FA, with parameters r and k, and we want to generate the

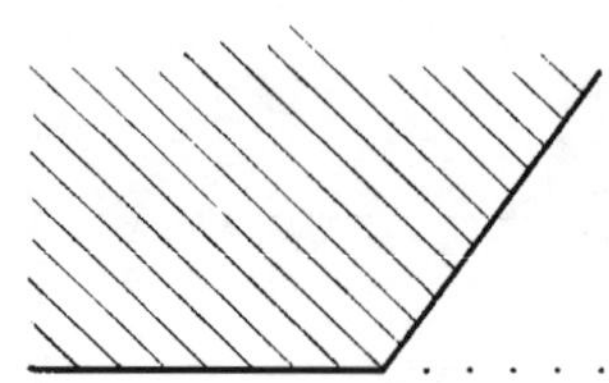

Fig. 5. The region that can affect a point in an FA.

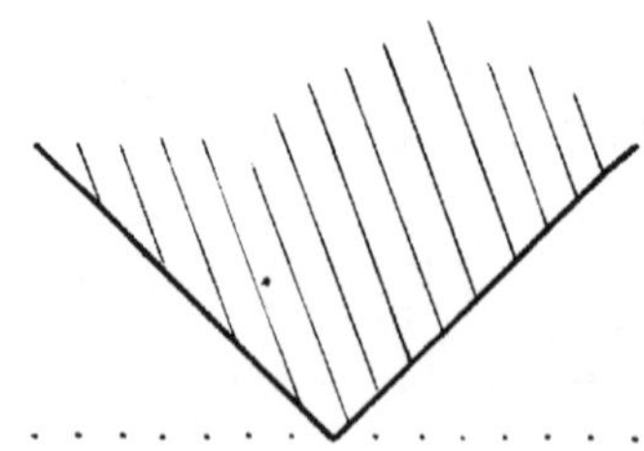

Fig. 6. The region that can affect a point in a CA.

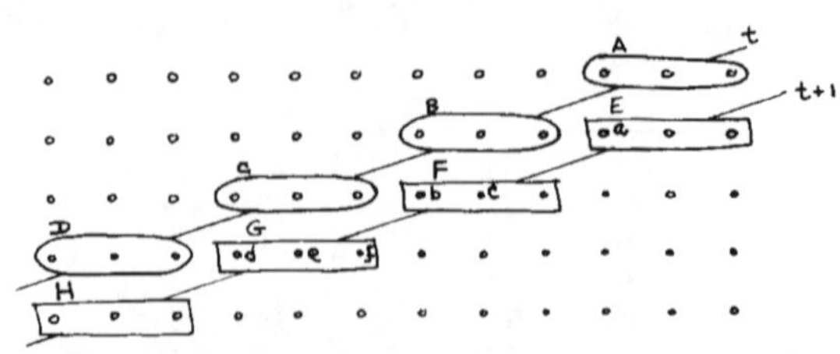

Fig. 7. Illustration of the simulation of an FA by a CA, for the case $r = 2$. The space axis of the CA is tilted, and each cell of the CA is a conglomerate of $(r+1)$ cells of the FA. To compute the value of CA cell G, we compute in turn the value of FA cell a, b, c, d, e and f.

same array of space-time values using a CA, using parameters r' and k'. The mapping we will use is illustrated in fig. 7 for the case $r = 2$. First, groups of three consecutive cell values in a particular generation of the FA array are coalesced to form one cell in the corresponding CA, so that the CA has $k' = 2^{r+1}$. Next, we take the space axis (constant-time axis) of the CA to be tilted so that cells in the same CA generation correspond to cells along a line rotated counterclockwise from horizontal in the FA space-time array.

Using capital letters to represent the cells in the CA, and lower case for the FA, we next show how the value of the cell G in generation $t+1$ can be computed from the values in the neighboring cells A, B, C, D in generation t. First, find the value of the FA cell a from the FA components in A and B. Next find the values of cells b and c from the value of a and the components of B and C. Finally, find the values of FA cells d, e, and f from the values of b, c and the components of C and D. This then gives the value of the CA cell G. We can summarize this in the following

Theorem 2. Every space-time array that is generated by an r, k – FA can also be generated by a CA with $r' = r$ and $k' = 2^{r+1}$, provided that we allow the values to be computed in a different order. The FA cell values are also coded in groups of $r+1$ alphabet symbols.

Next consider the problem of simulating a CA with an FA. A glance at fig. 8 shows that this is easy, provided that we allow a slippage to the left: just compute the value at site $x + r$ at the point x, and choose the FA rule to depend only on the values in the preceding generation. Each generation will therefore be shifted r cells to the left with respect to the CA array, but will otherwise be identical; in this case the time axis is tilted. We summarize this as

Theorem 3. Every space-time array that is generated by a r, k' – CA can also be generated with an FA with $r = 2r'$ and $k = k'$, provided that we allow each successive row in the FA to be displaced r units to the left with respect to the corresponding row in the CA.

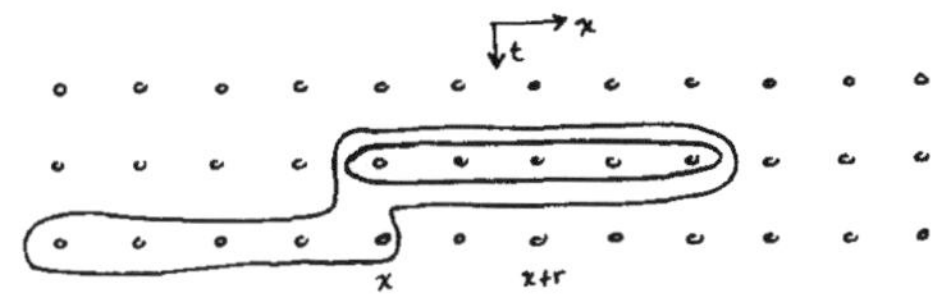

Fig. 8. Illustration of the simulation of a CA by an FA. The value at cell $x+r$ is computed at point x in the FA, so that the time axis of the FA is shifted to the left r cells every generation.

7. Discussion

Many questions about the class of filter automata remain unexplored. Some rules other than the parity rule appear to support particles the way the parity rule does. For example, the following variation of the parity rule seems to support particles and soliton collisions, and the particles are on the average slower:

$$a_i^{t+1} = \begin{cases} 1, & S \text{ even but not 0 or 2,} \\ 0, & \text{otherwise.} \end{cases} \tag{5}$$

Other rules are unstable in the sense described in section 3. While the question of stability in linear Infinite Impulse Response digital filters is settled by the criterion of characteristic values being inside the unit circle, an analogous general technique for FA's is unknown.

One of the motivations for studying cellular automata in general is to gain insight into the nonlinear phenomena that occur in the solution of differential equations, and in the physical systems they are used to model. We have seen that certain simple one-dimensional automata give rise to solitary waves that very often pass through one another non-destructively. Whether such automata capture the mechanism of soliton generation in differential equations is uncertain, but we feel worthy of further study.

Another reason for studying CA's is the possibility of embedding useful computation within such regular and simple structures. One-dimensional cellular automata can be implemented in VLSI in a highly pipelined and efficient way [8], resulting in what amounts to a cellular automaton machine with an almost unlimited degree of parallelism. Even if the embedding of a useful computation is very inefficient, it may still be more than compensated for in certain applications by the parallelism and efficiency of the VLSI implementation. There exist complex CA's that simulate a universal Turing machine, but we still do not know how to construct a simple one-dimensional CA that does useful computation. It appears, however, that it will help to have particles that can pass through one another, because that will make possible communication between different elements of the automaton.

Carter [9] describes the transmission of information in molecules using physically supported solitons. The general notion of processing information in simple, homogeneous media via solitons opens the way for speculation about such computation at the level of the molecule or the biological cell.

Acknowledgements

We thank Irfan Kamal and Arthur Watson, who helped in many ways, especially with the development and testing of programs. Doug West and Stephen Wolfram took the time to offer useful suggestions.

References

[1] D. Farmer, T. Toffoli and S. Wolfram, Cellular Automata (North-Holland, Amsterdam, 1984).

[2] G.Y. Vichniac, "Simulating Physics with Cellular Automata," pp. 96–116 in [1].

[3] A.C. Scott, F.Y.F. Chu, D.W. McLaughlin, "The Soliton: A New Concept in Applied Science," Proc. IEEE, vol. 61, no. 10, pp. 1443–1483, October 1973.

[4] N. Margolus, "Physics-Like Models of Computation," pp. 81–95 in [1].

[5] S. Wolfram, "Universality and Complexity in Cellular Automata," pp. 1–35 in [1].

[6] A.V. Oppenheim, R.W. Schafer, Digital Signal Processing (Prentice-Hall, Englewood Cliffs, NJ, 1975).

[7] E.R. Berlekamp, J.H. Conway and R.K. Guy, Winning Ways for your Mathematical Plays, Vol. 2: Games in Particular, (Chapter 25, "What is Life?") (Academic Press, New York, NY, 1982).

[8] K. Steiglitz and R.R. Morita, "A Multi-Processor Cellular Automaton Chip," Proc. 1985 IEEE Int. Conf. on Acoustics, Speech, and Signal processing, Tampa, Florida, March 1985.

[9] F.L. Carter, "The Molecular Device Computer: Point of Departure for Large Scale Cellular Automata," pp. 175–194 in [1].

[10] R. Hirota and K. Suzuki, "Theoretical and Experimental Studies of Lattice Solitons on Nonlinear Lumped Networks," Proc. IEEE, vol. 61, no. 10, pp. 1483–1491, October 1973.

[11] C.H. Goldberg, manuscript in preparation.

JOURNAL OF THE
AMERICAN MATHEMATICAL SOCIETY
Volume 1, Number 3, July 1988

ROTATION DISTANCE, TRIANGULATIONS, AND HYPERBOLIC GEOMETRY

DANIEL D. SLEATOR, ROBERT E. TARJAN, AND WILLIAM P. THURSTON

1. Introduction

A *rotation* in a binary tree is a local restructuring of the tree that changes it into another tree. One can execute a rotation by collapsing an internal edge of the tree to a point, thereby obtaining a node with three children, and then re-expanding the node of order three in the alternative way into two nodes of order 2. The *rotation distance* between a pair of trees is the minimum number of rotations needed to convert one tree into the other. The problem addressed in this paper is: what is the maximum rotation distance between any pair of n-node binary trees? We show that for all $n \geq 11$ this distance is at most $2n - 6$ and that for all sufficiently large n this bound is tight. Culik and Wood [2] showed that the maximum rotation distance is at most $2n-2$. Tom Leighton (private communication) showed that there exist trees whose rotation distance is at least $7n/4 - O(1)$. Pallo [7] proposed a heuristic search algorithm to compute the rotation distance between two given trees.

Our interest in this problem stems from our attempt to solve the dynamic optimality conjecture concerning the performance of splaying [8, 10]. Splaying is a heuristic for modifying the structure of a binary search tree in such a way that repeatedly accessing and updating the information in the tree is efficient. Although our solution to the problem of maximum rotation distance did not resolve the conjecture about splaying, the results in this paper are interesting for at least two other reasons. First, the combinatorial system of trees and their rotations is a fundamental one that is isomorphic to other natural combinatorial

Received by the editors January 29, 1988.

1980 *Mathematics Subject Classification* (1985 *Revision*). Primary 51M20, 51M10, 05C05, 68P05.

This is a revised and expanded version of a paper that appeared in the 18th Annual ACM Symposium on Theory of Computing [9], Berkeley, CA, May 28–30, 1986.

Partial support provided for the first author by DARPA, ARPA order 4976, amendment 19, monitored by the Air Force Avionics Laboratory under contract F33615-87-C-1499, and by the National Science Foundation under grant CCR-8658139.

Partial support provided for the second author by the National Science Foundation under grant DCR-8605962.

Partial support provided for the third author by the National Science Foundation under grants DMR-8504984 and DCR-8505517.

0894-0347/88 $1.00 + $.25 per page

systems. Results concerning this system are of interest from a purely mathematical point of view. Second, the method we use to solve the problem is novel and interesting in its own right and can potentially be applied to related problems.

A system that is isomorphic to binary trees related by rotations is that of triangulations of a polygon related by the *diagonal flip* operation. This is the operation that converts one triangulation of a polygon into another by removing a diagonal in the triangulation and adding the diagonal that subdivides the resulting quadrilateral in the opposite way. This type of move was studied by Wagner [14] in the context of arbitrary triangulated planar graphs and by Dewdney [3] in the case of graphs of genus one. They showed that any such graph can be transformed to any other by diagonal flips, but did not try to accurately estimate how many flips are necessary.

Our approach to solving the rotation distance problem is based on the observation that any sequence of diagonal flips converting one triangulation of a polygon into another gives a way to dissect (into tetrahedra) a polyhedron formed from the two triangulations. Using hyperbolic geometry, we construct polyhedra that require many tetrahedra to triangulate them. (Here and hereafter we use the word "triangulation" in a general sense meaning a dissection into simplices of appropriate dimension. A more rigorous definition is given in §2.4.) These polyhedra can be used to exhibit pairs of n-node trees (for all sufficiently large n) such that the rotation distance between them is $2n-6$.

In §2 we define the problem on trees, make the connection between trees and triangulations of a polygon, and show that sequences of diagonal flips are related to triangulations of polyhedra. In §3 we show how to use hyperbolic geometry to obtain a lower bound on the number of tetrahedra required to triangulate any polyhedron. We then construct particular polyhedra that require many tetrahedra to triangulate them. §4 contains remarks and some open problems.

2. Definitions and equivalences

2.1. Binary trees. A binary tree is a collection of nodes of two types, external and internal, and three relations among these nodes: parent, left child, and right child. Every node except a special one called the root has a parent, and every internal node has a left and a right child. External nodes have no children. A tree is said to be of size n if it has n internal nodes. A tree of size n has $n+1$ external nodes. (See [5,10] for a more complete description of binary trees and tree terminology.) The number of steps required to walk from the root of the tree to a node is the depth of that node. (Each step moves from a node to one of its children.)

A *symmetric order* traversal of the tree visits all of the nodes exactly once. This order can be described by a recursive algorithm as follows: If the node is an internal node, traverse its left subtree in symmetric order, visit the node itself, then traverse its right subtree in symmetric order. If the node is an external node, merely visit it. The order in which the nodes are visited is called the

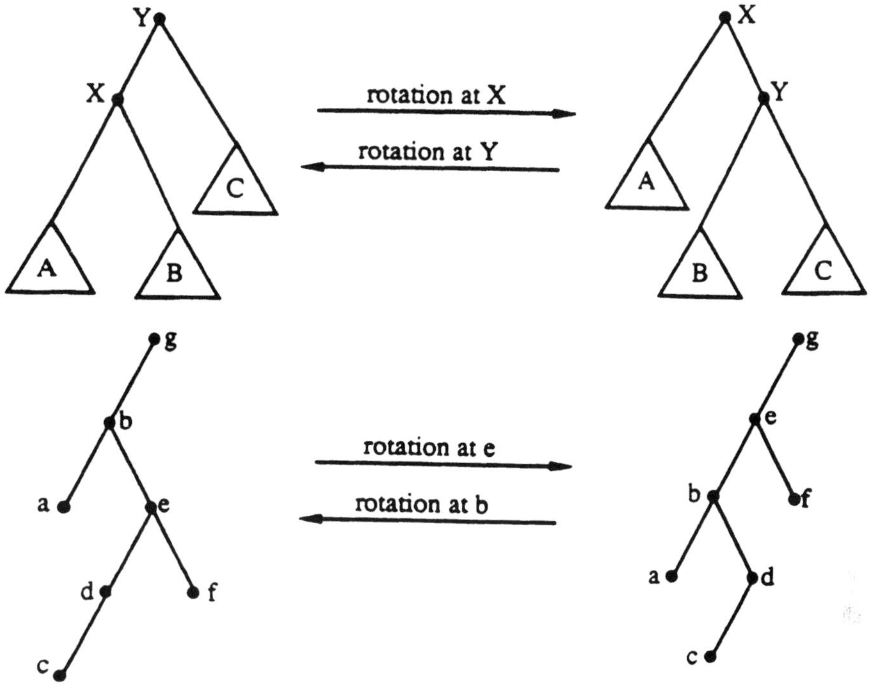

FIGURE 1. (a) The general definition of a rotation. Triangles denote subtrees. The tree shown could be part of a larger tree. (b) A rotation in a seven node tree. External nodes are not shown.

symmetric order permutation of the nodes (or simply the symmetric order of the nodes).

In a common computer-related application of binary trees the tree is used to store an ordered collection of pieces of information (called items). Each internal node of the tree is labeled with an item, and the order of the items is represented by the symmetric order of the nodes.

A *rotation* is an operation that changes one binary tree into another. In a tree of size n there are $n-1$ possible rotations, one corresponding to each nonroot internal node. Figure 1 shows the general rotation rule and the effect of a particular rotation on a particular tree. The rotation corresponding to a node changes the structure of the tree near that node, but leaves the structure elsewhere intact. A rotation maintains the symmetric order of the nodes, but changes the depths of some of them. Rotations are the primitives used by most schemes that maintain "balance" in binary trees [5, 10].

A rotation is an invertible operation; that is, if tree T can be changed into T' by a rotation, then T' can be changed back into T by a rotation. The *rotation graph* for trees of size n (denoted $RG(n)$) is an undirected graph with

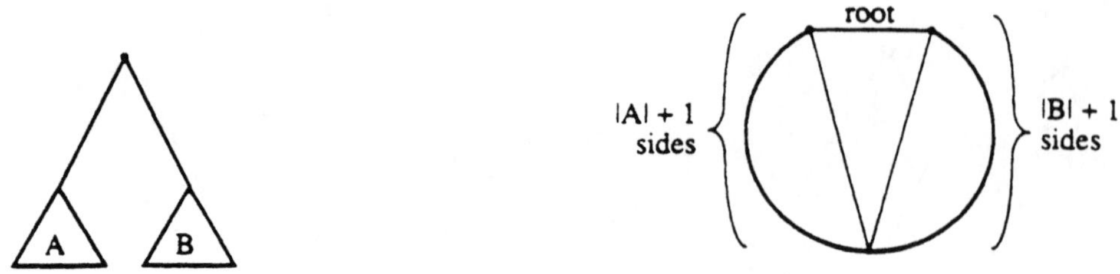

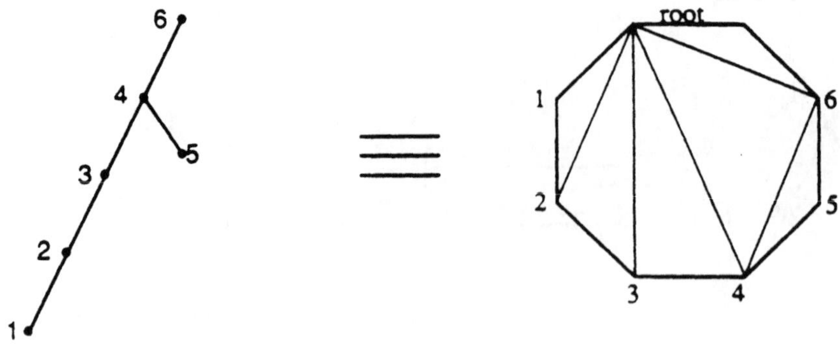

FIGURE 2. An example of a tree and its corresponding triangulation.

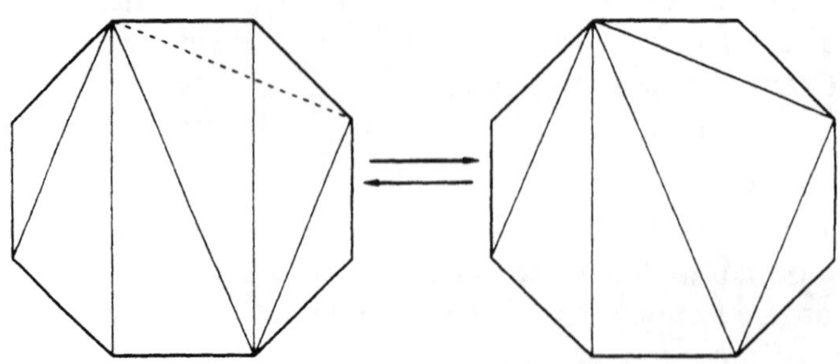

FIGURE 3. A diagonal flip in a triangulation of an octagon.

one vertex for each tree of size n and an edge between vertices T and T' if there is a rotation that changes T into T'.

Any binary tree of size n can be converted into any other by performing an appropriate sequence of rotations. Therefore the rotation graph is connected. We can define the *rotation distance* between two trees as the length of the shortest path in the rotation graph between the two trees, i.e. the minimum number of rotations required to convert one tree into the other. The main problem we address in this paper is that of estimating the diameter of $RG(n)$, i.e., the maximum rotation distance between any two n-node binary trees.

2.2. Polygon triangulations. Problems concerning rotation distance can be formulated with respect to a different system of combinatorial objects and their transformations. This alternative formulation is perhaps more natural and also seems to supply more insight.

Suppose we are given a binary tree T of size n. Take a collection of triangles indexed by the internal nodes of T. Now glue the triangles together along their edges according to the pattern of the tree, i.e., according to the parent-child relation. The resulting surface is homeomorphic to a disk. In fact, we can choose a standard convex $(n+2)$-gon and choose one of its edges to be the "root" edge. Label the n vertices of the $(n+2)$-gon that are not endpoints of the root edge by the n internal nodes of T, counterclockwise in symmetric order. Now the triangles can be inductively mapped into the $(n+2)$-gon by gluing one edge of each triangle to the appropriate edge of its parent triangle and sending the remaining vertex to the vertex labeled by its node. As a special case, the root triangle is attached to the root edge and the root vertex. See Figure 2.

In this way, we obtain a 1-1 correspondence between binary trees and triangulations of the $(n+2)$-gon with no interior vertices. We refer to the $n+2$ sides of the polygon as *edges* and the chords that divide it into triangles as *diagonals.* Any triangulation of the $(n+2)$-gon has $n-1$ diagonals and n triangles. We regard the polygon as having a distinguished edge and orientation.

A *diagonal flip* is an operation that transforms one triangulation of a polygon into another. The effect of a diagonal flip is shown in Figure 3 and can be described as follows: A diagonal inside the polygon is removed, creating a face with four sides. The opposite diagonal of this quadrilateral is inserted in place of the one removed, restoring the diagram to a triangulation of the polygon.

Let $TG(n+2)$ be a graph with one node for each triangulation of an $(n+2)$-gon and an edge between two nodes if the two nodes are related by a diagonal flip. We see that

Lemma 1. *The graph $TG(n+2)$ is isomorphic to the rotation graph $RG(n)$.*

The proof of this lemma is a straightforward application of the correspondence described above. A more detailed discussion of the relationships between trees, triangulations, and Catalan numbers can be found in Chapter 20 of [4].

2.3. Results of polygon triangulations. As we saw in §2.2, a study of the rotation distance between trees can be formulated as a study of the distance between triangulations under the diagonal flip operation. Let $d(\tau_1, \tau_2)$ be the minimum number of diagonal flips needed to transform triangulation τ_1 into triangulation τ_2. For convenience, we shall now change our use of the variable "n." We consider triangulations of an n-gon and let $d(n)$ be the maximum distance between any pair of such triangulations. That is, $d(n)$ is the diameter of $TG(n)$ or equivalently of $RG(n-2)$. Figure 4 shows $TG(6)$, whose vertices are the fourteen triangulations of a hexagon. The greatest distance between a pair of triangulations is four; there are several pairs that achieve this distance.

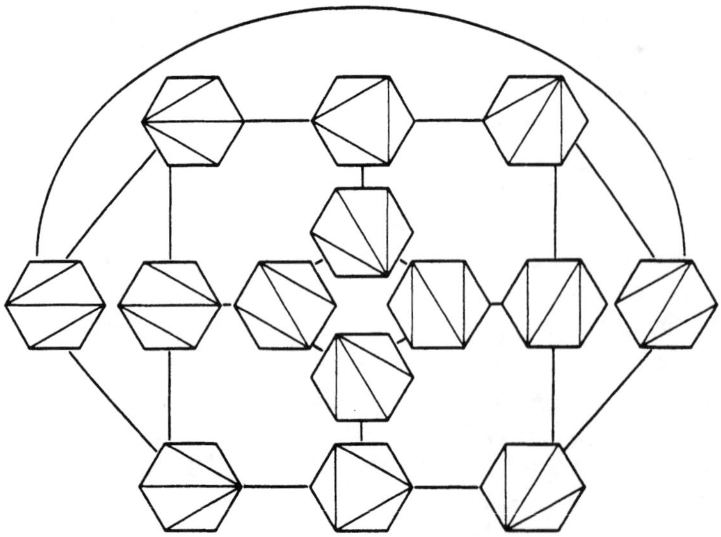

FIGURE 4. The rotation graph of a hexagon, $RG(6)$.

The added symmetry revealed in the triangulation system that is hidden in the binary tree system enables us to improve Culik and Wood's upper bound on $d(n)$ from $2n-6$ to $2n-10$.

Lemma 2. $d(n) \leq 2n-10$ *for all* $n > 12$.

Proof. Any triangulation of an n-gon has $n-3$ diagonals. Given any vertex x of degree $\deg(x) < n-3$, we can increase $\deg(x)$ by one by a suitable flip. Thus in $n-3-\deg(x)$ flips we can produce the unique triangulation all of whose diagonals have one end at x. It follows that given any two triangulations τ_1 and τ_2 we can convert τ_1 into τ_2 in $2n-6-\deg_1(x)-\deg_2(x)$ flips, where x is any vertex and the degree of x is $\deg_1(x)$ in τ_1 and $\deg_2(x)$ in τ_2. The average over vertices x of $\deg_1(x)$ is $2-6/n$, and of $\deg_1(x)+\deg_2(x)$ is $4-12/n$. It follows that if $n > 12$, there is a vertex x such that $\deg_1(x)+\deg_2(x) \geq 4$. □

The following lemma about sequences of diagonal flips shows that in some situations it is easy to find the first flip in an optimal sequence of flips.

Lemma 3. (a) *If it is possible to flip one diagonal of* τ_1 *creating* τ_1' *so that* τ_1' *has one more diagonal in common with* τ_2 *than does* τ_1, *then there exists a shortest path from* τ_1 *to* τ_2 *in which the first flip creates* τ_1'. (b) *If* τ_1 *and* τ_2 *have a diagonal in common, then a shortest path from* τ_1 *to* τ_2 *never flips this diagonal. In fact, any path that flips this diagonal is at least two flips longer than a shortest path.*

Proof. Let S be a sequence of adjacent triangulations connecting τ_1 to τ_2.

$$S = t_0(=\tau_1), t_1, t_2, \ldots, t_k(=\tau_2).$$

Assume that $t_1 \neq \tau_1'$. We shall construct a new sequence of adjacent triangulations S' also connecting τ_1 and τ_2 whose length is no longer than the length of S and in which the first flip creates τ_1'. This will suffice to prove (a) of the lemma.

Let l and r be the endpoints of the diagonal that τ_1' and τ_2 have in common but τ_1 and τ_2 do not have in common. Any triangulation τ can be *normalized* with respect to the diagonal (l, r) to create a new triangulation $N(\tau)$. The diagonals of $N(\tau)$ are of three types: (1) $N(\tau)$ contains the diagonal (l, r), (2) $N(\tau)$ contains every diagonal of τ that does not cross the diagonal (l, r) (two diagonals with an endpoint in common are not said to cross), (3) if τ contains a diagonal (a, b) that crosses the diagonal (l, r) then $N(\tau)$ contains the diagonals (a, r) and (b, r). (See Figure 5.)

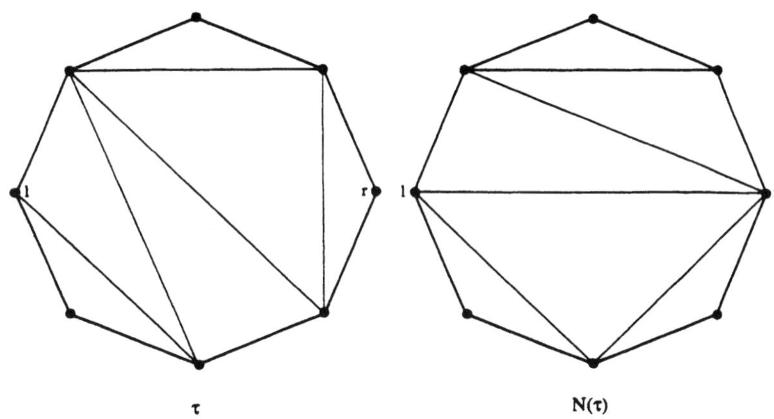

FIGURE 5. A triangulation τ and its normalized version $N(\tau)$.

Consider the sequence of triangulations

$$N = t_0, N(t_0), N(t_1), \dots, N(t_k).$$

A straightforward case analysis shows that successive triangulations of this sequence are either identical or adjacent. Eliminating all but one of each group of identical consecutive triangulations in this sequence gives the desired sequence S'. A priori S' might contain $k+2$ triangulations, but this cannot be the case for the following reason. Consider the triangulations t_i and t_{i+1} in S with the property that t_i does not contain diagonal (l, r) and t_{i+1} does. (There must be such a pair since the final triangulation contains the diagonal (l, r) and the initial one does not.) It is easily verified that the triangulations $N(t_i)$ and $N(t_{i+1})$ must be equal, and therefore occur only once in S'. Thus S' contains at most $k + 1$ triangulations. Verifying that S' starts and ends with τ_1 and τ_2, and that its second triangulation is τ_1', is straightforward. This completes the proof of (a) of the lemma.

The same technique serves to prove (b) of the lemma. Let $S = t_0(= \tau_1)$, $t_1, \ldots, t_k(= \tau_2)$ be a sequence that transforms τ_1 into τ_2 in which the first move is one that flips a diagonal (l, r) common to both τ_1 and τ_2. Normalize this sequence with respect to the diagonal (l, r) and eliminate redundancies to create a sequence S'. Then S' transforms τ_1 into τ_2 in two fewer flips than does S. The reason is that neither the first flip of S, which misaligns (l, r), nor a later flip that aligns (l, r), occurs in S'. □

A refinement of the lower bound proof in Lemma 2 for small values of n and a computer search have produced the exact values of $d(n)$ for $n \leq 18$, which appear below.

n	3	4	5	6	7	8	9	10	11	12	13	14	15	16	17	18
$d(n)$	0	1	2	4	5	7	9	11	12	15	16	18	20	22	24	26

2.4. Simplicial complexes. This section is a digression into the notation and terminology of simplicial complexes. Although we will apply these concepts in at most three dimensions, we have developed them in full n-dimensional generality. The sole purpose of this and the next section is to develop tools required to prove Lemma 7. At the end of §2.5 is an example that illustrates many of the concepts about to be introduced.

An *n-simplex* is the n-dimensional generalization of a triangle. It has $n+1$ vertices and $n+1$ faces each of which is an $n-1$-simplex. The simplex can be thought of as a list of the names of its vertices (in sorted order by vertex name).

A *simplicial complex* K is the union of a collection of simplices of assorted dimensions, called *cells*. Whenever α is an n-simplex in the collection and β is a k-face of α, then β is also in the collection. The intersection of any two simplices in the collection is also a simplex in the collection. The *dimension* of K is the maximum dimension of a simplex of K. The *k-skeleton* K_k of K is the union of the simplices of dimension not exceeding k, with its inherited structure as a simplicial complex.

A *triangulation* of a space X is a simplicial complex K and a map f from simplices onto X such that the images of the simplices in X intersect exactly as they do in K.

An *orientation* for a k-simplex is an ordering of its vertices. An oriented simplex thus consists of a sorted list of the names of its vertices, in addition to a permutation of its vertex names. The orientation is said to be positive if the permutation of the vertices is an even permutation, and negative otherwise.

Associated with a simplicial complex K of dimension n is a collection of vector spaces $C_k(K)$, for $0 \leq k \leq n$, where C_k is defined to consist of finite formal linear combinations of oriented k-simplices of K with real coefficients, subject to the relation that two oriented k-simplices whose orientations differ by a single inversion are negatives of each other. Elements of C_k are called

k-chains of K. Using the above relation we can reduce a k-chain to an equivalent one in which each simplex of K occurs at most once.

The *boundary map*

$$\partial_k : C_k \to C_{k-1},$$

is a linear map from k-chains to $(k-1)$-chains. Because it is linear, its value on an arbitrary k-chain is a linear combination of its values on each simplex of the given k-chain. The boundary map is defined on an oriented simplex to be its oriented boundary: to determine the orientation on a $(k-1)$-face β of a k-simplex α, choose an ordering of the vertices of α compatible with the orientation and such that the vertices of β come last. The induced ordering of the vertices of β is the correct orientation. As a special case, we define ∂_0 as a trivial map to a trivial vector space consisting of 0 alone.

Elements of the kernel of ∂_k (that is, chains whose boundary is 0) are called *k-cycles*. The image of ∂_{k+1} automatically is contained in the kernel of ∂_k because the boundary of a simplex is a cycle. (Elements of this image are called *k-boundaries*. The quotient space of k-cycles by k-boundaries is called the kth homology of K, $H_k(K;R)$. This does not depend on the structure of K as a simplicial complex, but only on the underlying space of K.)

A k-simplex is said to be *geodesic* if it is contained in an affine k-dimensional subspace of R^n (but not in any $(k-1)$-dimensional subspace of R^n), and all of its $(k-1)$-faces are also geodesic. A map f is said to be *affine* on a k-simplex if the image of the simplex under f is geodesic.

Let f be an affine map that maps an n-simplex s into an n-simplex s' in R^n. We may assume that the domain of f is simplex s, placed in R^n in the canonical fashion. (The canonical placement of an n-simplex in R^n is obtained by taking the convex body formed by convex combinations of the origin and the n unit vectors. To label the canonical placement, place the labels in sorted order on the origin, then on each of the unit vectors of R^n respectively.) Each vertex of s' may be labeled according to which vertex of s is mapped to it by f. The map f is said to be *orientation preserving* if the canonical placement of s in R^n can be transformed to s' via a continuous family of affine embeddings of s in R^n so that the labels on the vertices match. If this is not possible, then f is said to be *orientation inverting.*

2.5. The volume of an $(n-1)$-cycle. We shall now develop the tools necessary to define the volume of an $(n-1)$-cycle embedded in R^n.

Suppose that we have a simplicial complex K, an n-chain c in $C_n(K)$, and a map f of K into R^n that is affine on each simplex. Choose a basis α_1, $\alpha_2, \ldots$ of $C_n(K)$ such that the orientation of α_i is positive if f is orientation preserving on α_i, and negative otherwise. (This definition of orientation has the property that all the basis simplices have the "same" orientation. For example, if we are working in two dimensions, then the orientation of each triangle is either clockwise or counterclockwise, and in this case our basis has the property that all the basis triangles are counterclockwise.)

This data determines a certain step function *degree* (c, f) on R^n, as follows. Write $c = \sum a_i \alpha_i$. First suppose that x is any point in R^n which is not in the image of the $n-1$ skeleton of K. To compute the degree at x, we add the coefficients of the simplices whose image contains the point x. If x is in the image of the $n-1$ skeleton, we define *degree* $(c, f)(x)$ to be the maximum value of the degree that occurs in arbitrarily small neighborhoods of x.

The degree map we have defined satisfies the following lemma, which makes it useful in defining volumes.

Lemma 4. *If c is an n-cycle in a simplicial complex K, then degree $(c, f) = 0$. Furthermore if d and e are any chains such that $\partial d = \partial e$, then degree $(d, f) =$ degree(e, f).*

Proof. The degree function *degree* (c, f) is constant everywhere except possibly in the image of the union of the boundaries of the α_i. We shall first show that crossing these boundaries also leaves this function constant. Let α and α' be two simplices from the basis that have a boundary simplex β in common. If α and α' are mapped by f to the same side of β then the coefficient of β in $\partial\alpha$ is the same as the coefficient of β in $\partial\alpha'$. If α and α' are mapped to opposite sides of β, then the coefficients are negatives of each other. Since c is a cycle, the boundary vanishes, so the coefficient of β in the boundary is zero. Therefore the sum of the coefficients in c over those n-simplices that map to one side of β equals the sum of the coefficients in c over those that map to the other. This shows that the degree is the same on both sides of β.

If f is a generic map (where f does not have degeneracies, such as mapping several $(n-1)$-simplices to the same place) this is enough to show that *degree* (c, f) is continuous everywhere (and therefore constant). This is because in this case one can get from any generic point to any other generic point only crossing the image of a single $(n-1)$-face at a time. Since the degree is zero near infinity, the degree must be zero everywhere. In the general case, when f might not be generic, f can be perturbed a little, without changing the degree at most points. The degree for the perturbations must be identically zero, so *degree* $(c, f) = 0$ in this case as well.

As for the second assertion of the lemma, if d and e are chains such that $\partial d = \partial e$, then $d - e$ is a cycle, so by the first part of the lemma *degree* $(d - e, f) = 0$. Because the degree operator is linear, *degree* $(d, f)-$ *degree* $(e, f) =$ *degree*$(d - e, f)$, which gives the result. □

If z is an $n-1$ cycle in a simplicial complex K, and if f is a map of K into R^n, it is always possible to enlarge K to a simplicial complex K' in which there are chains c for which $z = \partial c$. The easiest way is to define K' as the cone on K; that is, K' has one more vertex v than K, and for each simplex α of dimension k there is an additional $(k+1)$-simplex with one face on α and its extra vertex at v. The map of K into R^n easily extends to a map of K' determined by choosing where to send v. Now consider the map *degree* (c, f). Lemma 4 tells us that this step function only depends on z, not on the way z is expressed as a boundary. Thus we can rewrite it as $w(z, f)(x) = degree(c, f)(x)$ and call it the *wrapping number* of z about x. The *algebraic volume* enclosed by $f(z)$ is defined to be the integral of the wrapping number.

A special case that will become important later is the volume of a three-dimensional polyhedron. A triangulation σ of the sphere is a way to dissect the sphere into curvilinear triangles. Each such σ has a *fundamental 2-cycle*, defined to be the sum of the triangles of σ with orientation coming from the counterclockwise ordering of the vertices (when looking at the sphere from the outside). If f maps the sphere into R^3 in such a way that each simplex of σ is geodesic, then we define the volume enclosed by f to be the volume enclosed by the fundamental 2-cycle of σ.

Figure 6 illustrates a simplicial complex K with 15 simplices: $\langle 123\rangle$, $\langle 234\rangle$, $\langle 235\rangle$, $\langle 12\rangle$, $\langle 13\rangle$, $\langle 23\rangle$, $\langle 24\rangle$, $\langle 34\rangle$, $\langle 35\rangle$, $\langle 25\rangle$, $\langle 1\rangle$, $\langle 2\rangle$, $\langle 3\rangle$, $\langle 4\rangle$, $\langle 5\rangle$. (Here the numbers between brackets are the vertices of the simplex in sorted order.) The way these are mapped into R^2 by a map f is shown in (a) of the figure. The canonical imbedding of simplices $\langle 123\rangle$, $\langle 234\rangle$, and $\langle 235\rangle$ into R^2 is shown in (b). The map f is orientation inverting on simplices $\langle 123\rangle$ and $\langle 234\rangle$, and orientation preserving on simplex $\langle 235\rangle$.

To define the degree of f, we could use the following basis for $C_2(K)$: $\alpha_1 = \langle 123\rangle(132)$, $\alpha_2 = \langle 234\rangle(243)$, $\alpha_3 = \langle 235\rangle(235)$. (The permutation for an oriented simplex is shown in parentheses after the sorted vertex list of the simplex.) Other bases are possible; for example, we could have chosen α_1 to be $\langle 123\rangle(321)$.

Let c be the chain $-\langle 123\rangle(132) + \langle 234\rangle(243) + 2\langle 235\rangle(235)$. Part (c) of Figure 6 shows the value of degree(c, f) in R^2. Note that

$$\partial\langle 123\rangle(132) = \langle 12\rangle(21) + \langle 13\rangle(13) + \langle 23\rangle(32),$$

and

$$\partial c = \langle 12\rangle(21)+\langle 13\rangle(13)+\langle 34\rangle(34)+\langle 24\rangle(42)+2\langle 23\rangle(23)+2\langle 35\rangle(35)+2\langle 25\rangle(52).$$

2.6. Triangulations of the sphere and the ball. In this section we show that the quantity $d(n)$ is related to the number of tetrahedra that are required to triangulate certain polyhedra.

Let σ be a triangulation of the sphere and let z be a fundamental 2-cycle of σ. Then T is an *exposed triangulation of the ball extending* σ if (1) T is a triangulation of the ball, (2) there exists a 3-chain c in $C_3(T)$ such that $z = \partial c$ and all the coefficients of c are ± 1, and (3) all of the vertices of T occur as vertices in σ. The 3-chain c satisfying this definition in unique.

An exposed triangulation of the ball extending σ is the three dimensional analogue of a triangulation of an n-gon described in §2.2. In contrast to the situation in two dimensions, there are generally triangulations extending σ containing different numbers of tetrahedra. Although it happens to be true that

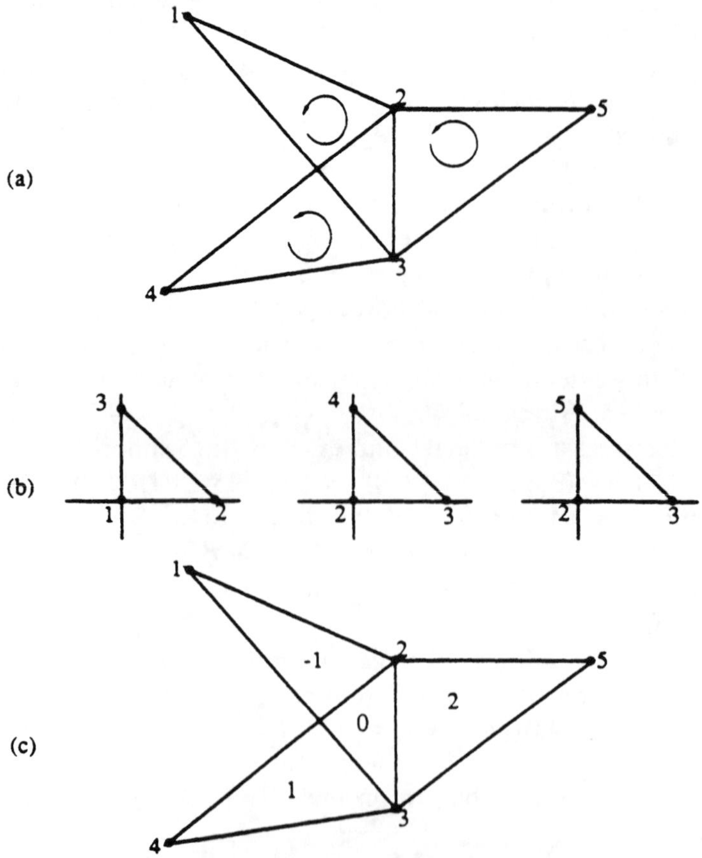

FIGURE 6. (a) shows the way a simplicial complex with fifteen simplices is mapped by a map f into R^2. (b) shows the canonical imbedding of simplices $\langle 123 \rangle$, $\langle 234 \rangle$, and $\langle 235 \rangle$ into R^2. (The map f is orientation inverting on simplices $\langle 123 \rangle$ and $\langle 234 \rangle$, and orientation preserving on simplex $\langle 235 \rangle$.) (c) shows the value of degree (c, f) in R^2.

every triangulation of S^2 is homeomorphic to a triangulation with geodesic faces, the situation is quite different from exposed triangulations of the ball: it is even possible to construct examples with knotted edges, so that they are forcibly curvilinear.

The union of two triangulations τ_1 and τ_2 of an n-gon, glued along their boundary, usually gives a triangulation of the sphere, which we denote by $U(\tau_1, \tau_2)$. (There are certain degenerate cases in which the union does not satisfy the definition of a simplicial complex. This happens if τ_1 and τ_2 have a diagonal in common. Nonetheless, $U(\tau_1, \tau_2)$ describes a triangulation of a slightly more general sort, which we shall not discuss.)

For any σ there is an exposed triangulation of the ball extending σ. This allows us to define $t(\sigma)$ to be the minimum number of tetrahedra in any exposed triangulation of the ball extending σ. The following lemma relates triangulations to rotation distance.

Lemma 5. *If τ_1 and τ_2 have no diagonal in common, then*

$$t(U(\tau_1, \tau_2)) \leq d(\tau_1, \tau_2).$$

Proof. There exists a sequence of $d(\tau_1, \tau_2)$ diagonal flips that changes τ_1 into τ_2. We shall describe how to extract from this sequence an exposed triangulation of the ball extending $U(\tau_1, \tau_2)$ containing $d(\tau_1, \tau_2)$ tetrahedra.

Imagine that there is a planar base with triangulation τ_1 drawn on it. Suppose the first diagonal flip replaces diagonal (a, c) with diagonal (b, d). Create a flat quadrilateral that is the same shape as quadrilateral (a, b, c, d). On the back side of the quadrilateral draw diagonal (a, c). On the front draw diagonal (c, d). Now place the quadrilateral onto the base in the appropriate place with diagonal (a, c) down and (b, d) up. Looking at the base we see a picture of a triangulation which is the result of making the first diagonal flip. For each successive move we create another quadrilateral and place it onto the base. After placing $d(\tau_1, \tau_2)$ such quadrilaterals we will see τ_2 when we view the base.

The triangulation of the ball that we construct has one tetrahedron for each quadrilateral. The tetrahedra are glued together according to the way the quadrilaterals are stacked. Two triangles are identified with each other if they face each other in the stack of quadrilaterals.

To finish the proof we need only verify that the resulting triangulation is an exposed triangulation of the ball extending $U(\tau_1, \tau_2)$. The fact that it is a triangulation of the ball is made clear by inflating each quadrilateral so that it turns into a tetrahedron. The resulting collection of tetrahedra is homeomorphic to a ball. (This is where we use the assumption that τ_1 and τ_2 have no diagonal in common.) The fact that the triangulation extends $U(\tau_1, \tau_2)$ is obvious because the boundary triangles are exactly those of τ_1 and τ_2. □

What the proof of Lemma 5 tells us is that for every sequence of diagonal flips from τ_1 to τ_2 there is an exposed triangulation of the ball extending $U(\tau_1, \tau_2)$. In fact, the same triangulation of the ball may result from many different sequences of moves from τ_1 to τ_2. It is not the case that every exposed triangulation of the ball extending $U(\tau_1, \tau_2)$ comes from a sequence of diagonal flips. In fact, it is possible to construct exposed triangulations of the ball with the property that no tetrahedron touches the boundary on more than one face, whereas in a triangulation obtained by the construction in the proof of Lemma 5 some tetrahedra touch the boundary on at least two faces.

Let $t(n)$ be the maximum of the quantity $t(\sigma)$ over all n-vertex four-connected triangulations σ of the sphere. (A triangulation is said to be

k-connected if its 1-skeleton, regarded as a graph, is k-connected. An undirected graph is k-connected if deletion of any $k-1$ vertices leaves the graph connected.)

Lemma 6. $t(n) \leq d(n)$.

Proof. Let σ be an n-vertex four-connected triangulation of the sphere such that $t(\sigma)$ is maximized. By a theorem of Hassler Whitney [15] or its generalization, Tutte's Theorem [13], any four-connected triangulated planar graph must have a Hamiltonian circuit. Draw the triangulation on a sphere. Cut the sphere along the edges of the Hamiltonian circuit. This separates the sphere into two disks, each of which is triangulated. Let these two triangulations be τ_1 and τ_2. Now $\sigma = U(\tau_1, \tau_2)$. By the preceding discussion and Lemma 5,

$$t(n) = t(\sigma) = t(U(\tau_1, \tau_2)) \leq d(\tau_1, \tau_2) \leq d(n). \quad \square$$

To make these concepts more concrete, consider the two triangulations τ_1 and τ_2 of a hexagon whose diagonals form a triangle. (See Figure 7.) The triangulation obtained by gluing τ_1 and τ_2 together is the boundary of the octahedron. There are six paths of length four between τ_1 and τ_2 (see Figure 4). Each of these paths gives rise to a triangulation of the octahedron. Three different triangulations of the octahedron are obtained in this way. (Each is produced by two different paths from τ_1 to τ_2.) These triangulations are the ones in which a single edge has been added between a pair of opposite vertices. The octahedron cannot be triangulated with fewer than four tetrahedra because no tetrahedron can contact more than two faces of the boundary.

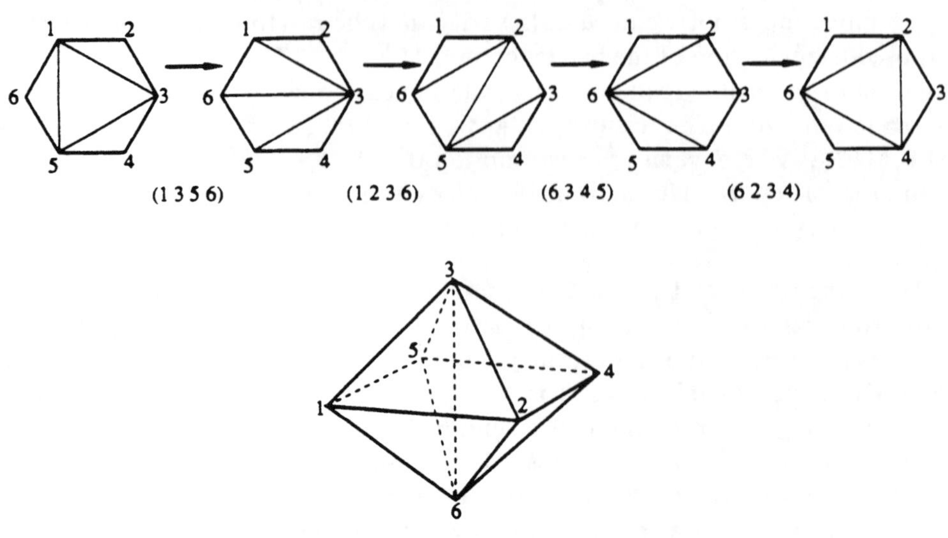

FIGURE 7. A sequence of diagonal flips and the corresponding triangulation.

By Lemma 6, an upper bound on $d(n)$ is an upper bound on $t(n)$, and a lower bound on $t(n)$ is a lower bound on $d(n)$. In the remainder of this paper we show that $2n - 10 \leq t(n)$ for all sufficiently large n. Combining this with Lemma 2 gives $t(n) = d(n) = 2n - 10$ for all sufficiently large n.

3. Lower bounds on $t(n)$

Our approach to deriving accurate lower bounds on $t(n)$ is geometric rather than combinatorial. We convert the combinatorial objects described in the previous section into geometric objects. We then infer properties of the combinatorial objects from the properties of the geometric objects. The following two paragraphs summarize our approach.

Let σ be an n-vertex triangulation of the sphere that is four-connected. Suppose σ is the boundary of a polyhedron P in R^3 such that the vertices of P are on the unit sphere. Let T be an exposed triangulation of the ball extending σ. For each tetrahedron Δ of T there is a geodesic tetrahedron Δ' whose vertices are the appropriate vertices of P. The sums of the volumes of these geodesic tetrahedra must be at least the algebraic volume enclosed by P. Let V_Δ be the volume of the largest tetrahedron that can be inscribed in a sphere. Let vol(P) be the volume of P. We conclude that at least vol$(P)/V_\Delta$ tetrahedra are required to cover P. In other words:

$$\frac{\mathrm{vol}(P)}{V_\Delta} \leq t(n).$$

In Euclidean space, this inequality does not give interesting bounds because the ratio of the volume of a sphere to the volume of the largest tetrahedron inscribed in the sphere is a small constant. In hyperbolic space, however, this method does lead to useful results. This is because in hyperbolic space the volume of a tetrahedron is bounded above by a constant V_0, while the volume of a polyhedron can grow linearly as a function of the number of vertices. Our problem is thus reduced to finding a polyhedron P with n vertices in hyperbolic space that has large volume.

First we present the necessary fundamentals of hyperbolic geometry. These ideas are described in more detail in Coxeter's book [1], Milnor's paper [6], and an expository article by Thurston and Weeks [11].

3.1. Hyperbolic geometry. In hyperbolic geomerty there are many lines through a given point parallel to a given line, the sum of the angles of a triangle is less than 180 degrees, and the circumference of a circle is greater than π times the diameter. There are various ways of mapping hyperbolic space into Euclidean space. These mappings enable us to draw pictures on Euclidean paper of hyperbolic polygons, but these pictures are distorted; two congruent hyperbolic triangles may not look congruent when mapped into Euclidean space.

One mapping of two-dimensional hyperbolic space into the Euclidean plane is called the *upper half-plane model*. In this model all of hyperbolic space is mapped into the upper half of the complex plane (the points with positive imaginary parts). This mapping is conformal, which means that angles are preserved. The geodesics (straight lines) in hyperbolic space are mapped into the semicircles with centers on the real axis and the vertical half lines with ends on the real axis. Most of the area of hyperbolic space is mapped into the region near the real axis. See Figure 8.

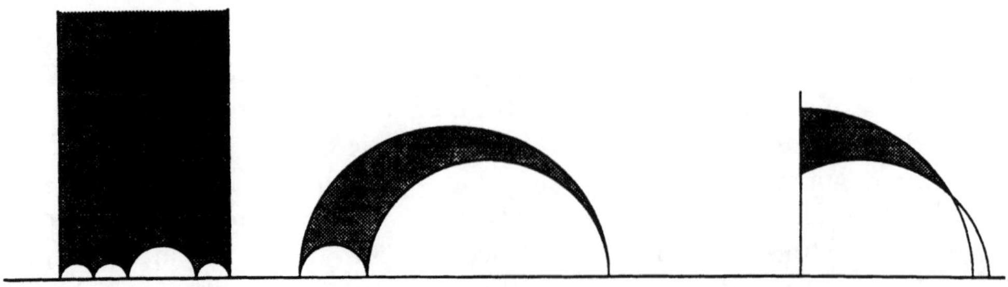

FIGURE 8. The upper half-plane model of two-dimensional hyperbolic space.

The area of a triangle in hyperbolic space is $\pi - \Sigma$, where Σ is the sum of the internal angles. An *ideal triangle* is one with three distinct vertices on the real axis or at infinity. All ideal triangles have area π. In fact all ideal triangles are congruent, that is, any ideal triangle can be transformed to any other by a rigid motion. (The rigid motions of the space form a group known as the group of orientation-preserving isometries.)

The upper half-space model of three-dimensional hyperbolic space consists of the complex plane plus all the points above the plane in Euclidean three-space plus a point at infinity. The complex plane plus the point at infinity is sometimes called the *sphere at infinity*. A geodesic in hyperbolic three-space is mapped to a semicircle perpendicular to the complex plane or a straight half line perpendicular to the complex plane going to infinity. The geodesic surfaces are mapped to hemispheres with centers in the complex plane and half planes orthogonal to the complex plane.

An *ideal hyperbolic tetrahedron* is a tetrahedron in which all the vertices are distinct and on the sphere at infinity. Any hyperbolic tetrahedron can be transformed by a rigid motion to one in which three of the vertices are at 0, 1, and ∞ and the other vertex is at a point z in the complex plane. (This motion is possible because all four of the triangles of the tetrahedron are ideal and any ideal triangle can be moved to any other. Note that despite this fact, not all ideal hyperbolic tetrahedra are congruent.) The tetrahedron then looks like three vertical flat walls above the Euclidean triangle $(0, 1, z)$, bounded below by part of a hemispherical bubble.

The hyperbolic cross section of the vertical chimney, in the hyperbolic metric, scales in a way that decreases with increasing height. (Most of the volume of hyperbolic space is near the complex plane.) It can be seen by integrating that the volume of an ideal tetrahedron is finite. Let this volume be denoted by $v(z)$. There are explicit formulas for $v(z)$, from which it can be seen that the maximum is attained at the point $z = \omega$, where ω is defined as $\omega = e^{2\pi i/6}$. (See [6] for a discussion of how to compute hyperbolic volumes.) The tetrahedron of maximum volume is the most symmetrical one. Its base triangle $(0, 1, \omega)$ is equilateral, its dihedral angles are all 60 degrees, and its volume is $v(\omega) = V_0 = 1.014941606\cdots$

3.2. The volume of hyperbolic polyhedra. Let σ be a triangulation of the sphere. For concreteness think of σ as though it is embedded in the sphere in some particular way. For any mapping of the vertices of σ to distinct points in three-dimensional hyperbolic space (or Euclidean space for the purposes of this discussion), there is a continuous map f from the sphere into hyperbolic space that (1) maps the vertices of σ to the appropriate places, and (2) maps every triangle of σ one-to-one onto a geodesic triangle in hyperbolic space. (Just as in Euclidean space, a simplex is *geodesic* in hyperbolic space if it and all of its subsimplices are in geodesic surfaces of appropriate dimension.) In other words, f maps the surface of the sphere into the surface of some hyperbolic polyhedron.

Lemma 7. *Let σ be a triangulation of the sphere. Let z be the fundamental 2-cycle of σ (the sum of all its triangles with counterclockwise orientation). Let f be a map from σ into hyperbolic 3-space that maps each triangle of σ into a geodesic triangle. Let P be the hyperbolic polyhedron defined by $f(\sigma)$. Let $\mathrm{vol}(P)$ be the hyperbolic algebraic volume enclosed by $f(z)$. Then*

$$\frac{\mathrm{vol}(P)}{V_0} \leq t(\sigma).$$

Proof. Let T be an exposed triangulation of the ball extending σ containing $t(\sigma)$ tetrahedra. Let f be a map from the vertices of σ to the vertices of P as described above, with the additional property that f maps all of the triangles of all of the tetrahedra of T to geodesic triangles in hyperbolic space.

By the definition of T, there is a 3-chain c on T such that $\partial c = z$, and the coefficients of c are ± 1. The algebraic volume of P is the integral of the wrapping number $w(z, f) = \mathit{degree}(c, f)$. Since the degree map is a linear operator on c and integration is a linear operator, we may separate this calculation into the sum of several terms, one for each tetrahedron of T. The contribution to this sum of each tetrahedron of T is its coefficient times the volume of the tetrahedron (degree is ± 1 inside, 0 outside). No term exceeds V_0, hence the lemma is true. □

We have now reduced the problem of finding lower bounds on $t(n)$ to that of finding n-vertex hyperbolic polyhedra with large volumes. The remainder of this section is devoted to constructing such polyhedra.

3.3. A preliminary bound: $2n - O(n^{1/2})$. There is a tessellation of hyperbolic space consisting of copies of the simplex of maximal volume. This tessellation can be constructed by starting with some maximal simplex, reflecting it through its faces, reflecting these through their faces, and so on. For any finite union of these tetrahedra whose boundary is a sphere, we obtain a polyhedron. The triangulation we have is automatically a minimal extension of its boundary, since all the simplices are disjoint and have maximal volume. Indeed, it is the unique minimal extension: in any minimal triangulation, every tetrahedron must have maximal volume, hence it is determined by any of its faces. This determines, one-by-one, where the tetrahedra have to be.

Consider the special case in which all the simplices have a common vertex. (In general, we call a triangulation that is obtainable by coning to some vertex a *cone-type triangulation.*) We may assume that the common point is the point at infinity in the upper half space model. When drawn in this way, each of the tetrahedra lies above an equilateral triangle in the tessellation of the complex plane by equilateral triangles. Any set of triangles whose union is homeomorphic to a disk will define such a polyhedron. Consider the case when the bounding polygon is hexagonal, with k edges on a side. The hexagon contains $6k^2$ triangles; hence the polyhedron contains $6k^2$ tetrahedra. (See Figure 9.) The hexagon has $3k^2 + 3k + 1$ vertices, so the polyhedron has $3k^2 + 3k + 2$ vertices (including the one at infinity). In particular we obtain

$$t(n) \geq 6k^2 = 2(3k^2 + 3k + 2) - O(k) = 2n - O(n^{1/2}).$$

Note that by using other triangulations we can actually get explicit lower bounds for each n, not just those n of the form $3k^2 + 3k + 2$.

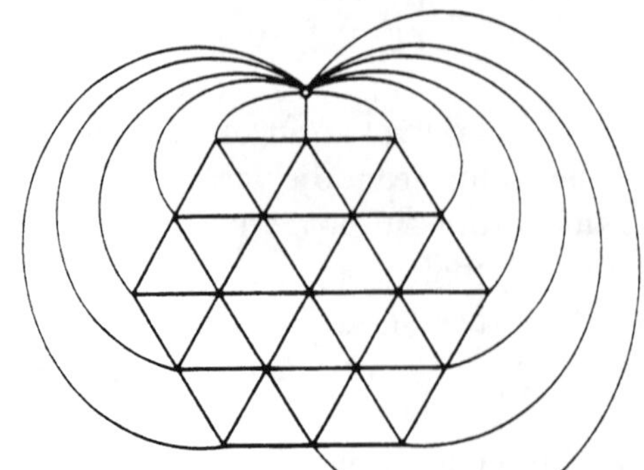

FIGURE 9. The boundary of the polyhedron used to show that $t(n) \geq 2n - O(n^{1/2})$.

3.4. A better lower bound: $2n - O(\log(n))$. To construct polyhedra that require more simplices for a given number of vertices we must eliminate the vertex of high degree. (Roughly speaking, polyhedra in which the vertices are spread over the sphere at infinity as uniformly as possible have the largest volumes.)

A natural sequence of triangulations for this purpose can be derived from a regular icosahedron. Divide each face of the icosahedron into k^2 equilateral triangles, giving $20k^2$ triangles in all and $n = 10k^2 + 2$ vertices.

We need to map the vertices of the triangulation of the sphere defined above into hyperbolic space in such a way that the resulting polyhedron has a large volume. The Riemann mapping theorem gives a way to do this. Corresponding to the icosahedron, there is a subdivision of the sphere into triangles bounded by segments of great circles, obtained by projecting the edges of the icosahedron out to the sphere. The Riemann mapping theorem implies that there is a unique conformal map of the faces of the icosahedron to the spherical triangles, sending vertices of the icosahedron to the corresponding vertices of the spherical triangles. By symmetry the maps determined on individual triangles piece together to give a map h of the entire surface of the icosahedron to the sphere. This map is conformal everywhere except at the vertices of the icosahedron. Note that it is conformal even on the edges of the icosahedron because they can be flattened out (locally) in the plane. Define the ideal hyperbolic polyhedron $P(k)$ to have its vertices at those places on the sphere at infinity in hyperbolic space to which h maps the vertices of the subdivided icosahedron.

Lemma 8. *The volume of* $P(k)$ *is* $2nV_0 - O(\log(n))$, *where* $n = 10k^2 + 2$.

Proof. To make the estimate of volume, pick a vertex of $P(k)$ of degree six that is as far as possible from vertices of degree five and arrange this vertex to be at infinity in the upper half-space model. (This is a rigid rotation of $P(k)$ in hyperbolic space.) Triangulate $P(k)$ as the union of cones from the vertex at infinity to the triangles with all vertices finite. (This is a cone-type triangulation.) Call a vertex of $P(k)$ "bad" if it is a vertex of the icosahedron or if it is the vertex mapped to infinity. Now h can be thought of as a map from the icosahedron to $\mathbf{C}$ (the complex plane) that is conformal everywhere except at the bad vertices. For large k the triangles far away from bad vertices get mapped by h to triangles that are nearly equilateral (because h is conformal). Figure 10 shows how the vertices of $P(k)$ near a vertex of the icosahedron get mapped to the complex plane by h. If the vertices were vertices of true equilateral triangles then the tetrahedra formed by coning them to infinity would all be congruent to the tetrahedron of maximal volume. We must show that the deficit caused by the fact that the triangles are not quite equilateral is small.

The shape of a triangle Δ with vertices p, q, r is conveniently described by a complex number $s(\Delta) = (r - p)/(q - p)$, its *shape parameter*. (The triangle $(0, 1, s(\Delta))$ is congruent to Δ.) The volume of the hyperbolic tetrahedron $C(\Delta)$ formed by coning Δ to the point at infinity is a function of the shape parameter of Δ, $\mathrm{vol}(C(\Delta)) = v(s(\Delta))$. Note that z, $1/(1-z)$, and $(z-1)/z$ describe similar triangles, so v takes the same value at these three points. Since v attains its maximum at ω, the first derivative of v at ω is 0, and the second derivative is the same in every direction because of symmetry. Thus, by Taylor's theorem, the volume deficit of $C(\Delta)$, defined as $V_0 - \mathrm{vol}(C(\Delta))$ (where V_0 is

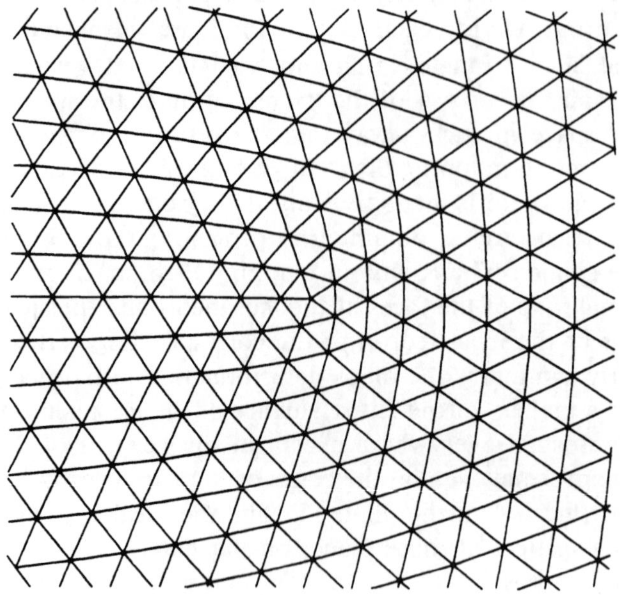

FIGURE 10. The way $P(k)$ (near a vertex of the icosahedron) gets mapped to the plane by h in the proof of Lemma 8. This is also the way that the boundary of a blemish of type (c) gets embedded in the plane in the proof of Theorem 1.

the maximum volume of a tetrahedron), satisfies

$$V_0 - \text{vol}(C(\Delta)) = K_1(|s(\Delta) - \omega|^2) + O(|s(\Delta) - \omega|^3),$$

for a certain constant K_1. Our goal is to show that the cumulative volume deficit is $O(\log(n))$.

We now digress to estimate the deficit of a triangle obtained by applying an arbitrary complex analytic function f to the vertices of an equilateral triangle. Subsequently we will apply this result using the map h.

Consider the equilateral triangle Δ_t with vertices 0, t, and ωt in the complex plane and suppose that f is a holomorphic (complex analytic, or conformal) embedding of the disk of radius $R > |t|$ about 0 into the complex plane. How can we estimate the volume deficit of the tetrahedron spanned by ∞ together with the images of the vertices of Δ_t under f?

In general, we may change f by postcomposing with a translation so that f fixes the origin. Let us expand f as a power series

$$f(z) = az + bz^2 + cz^3 + \cdots$$

and solve for the coefficients of the power series expansion of the shape parameter $s(t) = f(\omega t)/f(t)$ of the image triangle. We can write

$$\begin{aligned} s(t) &= \omega + At + Bt^2 + \cdots \\ &= \frac{a\omega t + b\omega^2 t^2 + c\omega^3 t^3 + \cdots}{at + bt^2 + ct^3 + \cdots} \end{aligned}$$

and solve for the coefficients. We obtain

$$\omega b + Aa = b\omega^2$$

and

$$\omega c + Ab + Ba = c\omega^3.$$

The term of real interest is

$$A = -b/a = -f''(0)/2f'(0)$$

since $\omega^2 - \omega = -1$. The next term is

$$B = \frac{1}{a}\left(c(\omega^3 - \omega) - \frac{-b}{a}b\right) = -(\omega + 1)\frac{c}{a} + \frac{b^2}{a^2}.$$

Thus, the shape of the image triangle is

$$s(t) = \omega - tf''(0)/2f'(0) + O(t^2).$$

How doees the error term depend on f? First, we claim that the error is uniformly bounded by $O(t^2)$ independent of f defined on a fixed disk of radius R. In fact, the set of all holomorphic embeddings of the disk of radius R into $\mathbf{C}$ is compact in the appropriate topology; that is, any sequence of embeddings has a subsequence that converges to an embedding. The errors could not get worse and worse, or else the limit function would not have an estimate of the form $O(t^2)$.

The dependence of the error term on R can now be easily deduced. A disk of radius R can be mapped to a disk of radius S by a complex affine map — that is, a complex linear map followed by a translation. The parameter t is multiplied by the ratio of the radii of the disks under such an affine map. Consequently, the error term above is $O((t/R)^2)$.

Now we return to the estimate of the volume deficit. For any point p on the icosahedron define $inj(p)$, the *injectivity radius* at p, to be the maximum radius of a disk in the Euclidean plane that can be isometrically embedded on the icosahedron with its center at p. Suppose we have a triangle in some subdivision of the icosahedron that has no vertices in common with the icosahedron and no vertex mapped to infinity. Let ε be the size of a side of the triangle (in the metric of the icosahedron). Then its volume deficit is not more than

$$K_2\left(\varepsilon^2\left|\frac{h''}{2h'}\right|^2\right) + O\left(\left(\frac{\varepsilon}{inj(p)}\right)^3\right). \tag{1}$$

Since ε^2 is proportional to the area of the triangle, the total volume deficit can be approximated by an integral.

$$\int_{\text{good triangles}} K_3 \left| \frac{h''}{2h'} \right|^2 dA + O(1). \tag{2}$$

By good triangles we mean those that have no bad vertex. The contribution of the $O((\varepsilon/inj(p))^3)$ term in (1) is at most a constant because on the good triangles $inj(p)$ is at least ε. This has been included in the $O(1)$ term in (2). The contribution of the bad triangles (those with a bad vertex) is also only a constant. The contribution of the part of the icosahedron that is farther than a fixed distance ε_0 away from a bad vertex is also bounded by a constant. This is because the integrand is continuous and bounded except near the bad vertices. The only contribution left to evaluate is that of the annular regions of inner radius ε and outer radius ε_0 centered on the bad vertices (where ε is the triangle mesh size).

Near the bad vertices h behaves like z^β where $\beta = 6/5$ if the bad vertex is an icosahedron vertex and $\beta = -1$ if the bad vertex is the one mapped to infinity. (The local coordinates are chosen so that the bad point is at the origin.) The entire deficit is estimated to within an additive constant by the sum (over bad vertices) of the integrals of $K_3(1/4)(\beta-1)^2(1/|z|^2)$ over annular regions centered at the bad vertices with fixed outer radius ε_0 and inner radius approximately equal to the mesh size ε. The value of each of these integrals is

$$K_3 \tfrac{1}{4}(\beta - 1)^2 2\pi(\log \varepsilon_0 - \log \varepsilon).$$

Since $-\log\varepsilon = \log(1/\varepsilon) = O(\log(n))$, we have bounded the deficit by $O(\log(n))$. This completes the proof that $\text{vol}(P(k)) = nV_0 - O(\log(n))$. □

3.5. The structure of minimal triangulations. We shall now apply the fact that $P(k)$ has a volume deficit that is small compared to the number of tetrahedra to completely determine the minimal extensions of $P(k)$ to the ball, provided that k is sufficiently large.

Theorem 1. *For sufficiently large k, any exposed triangulation of the ball extending the boundary of $P(k)$ having the minimum number of tetrahedra is a cone-type triangulation.*

Proof. Suppose that T is any minimal exposed triangulation of the ball extending the boundary of $P(k)$. For any fixed ε, at most $O(\log(n))$ of the simplicies of T can have volume less than $V_0 - \varepsilon$. In particular, only $O(\log(n))$ of its simplices can touch the boundary on two faces, since such simplices have volume roughly $(2/3)V_0$.

Since T is minimal it can have no more tetrahedra than any cone-type triangulation. By coning to a vertex of degree six we obtain a triangulation with $F-6$ tetrahedra, where F is the number of faces. If there are m tetrahedra in

T that have no face on the boundary then there are at least $m+6$ tetrahedra of T that have two faces on the boundary. Since this quantity is $O(\log(n))$, we conclude that $m = O(\log(n))$. Thus, most of the tetrahedra of T have exactly one face on the boundary.

The three-dimensional triangulation near a vertex v has several possibilities. To analyze the possibilities, let B_v be the union of tetrahedra that have v as a vertex. B_v is a ball and its boundary is composed of two parts: N_v, the union of the triangles of $P(k)$ containing v as a vertex, and Q_v, the union of the remaining triangles. N_v and Q_v are joined along a polygon p, which is a hexagon unless v is one of the twelve vertices of order 5, in which case it is a pentagon. It is convenient to color the faces of Q_v red or white, according to whether they are faces of the boundary $P(k)$ of T. (This happens naturally if T is an apple with a red skin and a white interior.) All the faces of N_v are red.

The vertex v is an *ordinary* vertex if the triangulation of Q_v is isomorphic to that of N_v, with exactly one interior vertex and all triangles having one corner in the interior and an edge along p. It is a *cone* vertex if $B_v = T$, so that the entire triangulation is a cone-type triangulation to v. If v is neither an ordinary vertex nor a cone vertex, it is an *extraordinary* vertex.

We shall show that if v is extraordinary, then either (1) at least one pair of triangles of N_v belong to a common tetrahedron, or (2) there is at least one white triangle of Q_v that does not have an edge on the boundary of Q_v. Suppose that v is extraordinary and that (2) is false; that is, all white triangles of Q_v have an edge on the boundary of Q_v. There is at least one white triangle of Q_v (otherwise v would be a cone vertex). If one of these white triangles has two edges on the boundary of Q_v then two of the other sides of its tetrahedron are triangles of N_v. In this case statement (1) holds.

It remains to consider the case in which all of the white triangles of Q_v have exactly one of their edges on the boundary of Q_v. Two white triangles are said to be adjacent if they share an edge. Consider a maximal set of adjacent white triangles of Q_v. This set is either a cycle of white triangles or it is a sequence of white triangles bounded at each end by a red triangle. If the former case occurs it shows that v must have been an ordinary vertex. In the latter case, let e be the edge on one end of the sequence of white triangles separating it from a red triangle and let f be the edge on the other end of the sequence separating it from a red triangle. A cycle of length four on the boundary of polyhedron $P(k)$ is formed by e and f along with two edges incident on v. In $P(k)$ the only cycles of length four are the boundary of the union of two adjacent triangles. Therefore, the only possibility is that the length of the sequence of white triangles is either 1 or *order* $(v) - 1$. In the first case, the single white triangle would actually have been red (all of its edges are on the boundary of $P(k)$), contrary to assumption. In the latter case, Q_v would have had one red triangle and *order* $(v) - 1$ white triangles, making it an ordinary vertex after all.

It follows that there are at most $O(\log(n))$ extraordinary vertices, since at each extraordinary vertex there is at least one corner of a simplex either touching ∂T on two faces or on no faces, and there are only $O(\log(n))$ such simplices.

Let m be the map from the faces of the polyhedron to the vertices of the polyhedron that associates to a face f the fourth vertex of the tetrahedron of T having f as a face. If f and g are any two faces in N_v for any ordinary vertex v, then $m(f) = m(g)$.

Consider the partition of the faces of the polyhedron according to $m(f)$. Let us assume that there is no cone vertex. Any two distinct partition elements can be separated from each other by a cycle of edges passing only through extraordinary vertices. A simple curve of length m on the icosahedron separates the surface into two regions, at least one of which has area A less than the area of a circle of circumference m in the plane; that is, $A \leq m^2/4\pi$. It follows that one partition element must have $n - O((\log(n))^2)$ faces, and the other $O(\log(n))$ partition elements are of size $O((\log(n))^2)$.

The boundary of $P(k)$ minus the large component is a collection of simply connected regions. Call each of these components a *blemish*.

For each triangle of the big component there is a tetrahedron of T from that triangle to a common vertex v^*. Thus T is almost a cone-type triangulation, with just a few possible bad spots. The places where T may disagree with the cone-type triangulation $C(v^*)$ are in the volumes bounded by the triangulation of the big component and the blemishes. We know the triangulations of the boundaries of these volumes, but the interior triangulations are unknown. To complete the analysis, we shall show that the triangulations of the volumes of these blemishes in fact must also agree with $C(v^*)$, given that they are minimal triangulations.

The blemishes can be sorted into four types, depending on the type of boundary they have. Here are the four types.

(a) The blemish contains no vertex of degree five, and does not contain v^*. In this case the boundary is a portion of the tessellation of the plane by equilateral triangles, with the boundary of these triangles coned to v^*.

(b) The blemish contains no vertex of degree five, and v^* is among the vertices of the blemish. The boundary of the blemish is not a sphere; it may be chosen to be homeomorphic to a sphere with the north pole and south pole identified.

(c) The blemish contains a vertex of degree five, and v^* is not among the vertices of the blemish. The boundary triangulation is obtained from a regular pentagon by first subdividing into five triangles, then subdividing these into congruent subtriangles, repeating this process, taking a subset of this, and then coning its boundary to one extra vertex v^*.

(d) The blemish contains a vertex of degree five, and v^* is among the vertices of the blemish. This case is like (b) and (c) combined.

We shall show that none of these four blemishes can actually exist. That is, the unique minimum way to triangulate each of the blemishes is to cone to vertex v^*. We shall use four proofs, one for each of the blemish types. The impossibility of (b) and (d) follows from that of (a) and (c). The proofs that (a) and (c) are impossible are hyperbolic volume arguments similar to those used to prove the bounds of §§3.3 and 3.4. Each proof describes how the blemish is to be embedded in hyperbolic three-space, evaluates the volume of the embedding, and finally shows that the volume is so large that any other triangulation besides $C(v^*)$ uses more simplices.

Case (a) is the easiest to resolve. We can embed the boundary of the blemish so that all its vertices except v^* agree with the vertices of the equilateral triangulation of the plane and v^* is mapped to the point at infinity. (The boundary of the blemish is the same as the boundary of one of the partial tessellations of space with maximal simplices described in §3.3.) The triangulation of the blemish obtained by coning to v^* contains only simplices of maximal volume, and these are nonoverlapping. Therefore, any triangulation of this blemish that uses a simplex of less than maximal volume would use at least one more simplex. The only way to avoid using a simplex of less than maximal volume is by coning to v^*. It is easy to see this by considering a simplex s with one of its vertices at v^* and having a face on the boundary of the blemish. If s is of maximal volume, its fourth vertex must be at the lattice point on the complex plane nearest its other two vertices. Removing s and iterating this process shows that the only triangulation exclusively using maximal simplices is a cone triangulation to v^*.

Case (b) can be handled in a similar fashion. Again embed the boundary of the blemish so that all its vertices except v^* agree with the vertices of the equilateral triangulation of the plane and v^* is mapped to the point at infinity. The resulting object resembles an annular cylinder in which the hole is hexagonal. Now each simplex of the cone triangulation $C(v^*)$ has maximal volume, and these simplices are nonoverlapping. Again, this is the only possible minimal triangulation extending the boundary of the blemish.

There is another approach to eliminating type (b) blemishes. Let B be a type (b) blemish. By gluing six implices to the boundary of B around v^*, we obtain a blemish B' of type (a). Suppose there were an alternative method of triangulating B that used no more simplices than coning to v^*. Then this alternative triangulation could be made into a noncone-type triangulation of B' using no more simplices than the cone-type triangulation of B'. We have already shown that such alternative triangulations do not exist.

It remains for us to deal with the twelve possible blemishes of types (c) and (d). The same argument used above implies that if we can show that a blemish of type (c) must be triangulated with a cone-type triangulation then one of type (d) must also. Thus it only remains to deal with the type (c) blemishes.

We can embed the boundary of any such blemish in hyperbolic space as follows. First we construct a tiling of the plane with nearly equilateral triangles and one vertex of degree five. This is done by modifying the tessellation of the plane by equilateral triangles. First remove a wedge (with its apex at the origin) containing one sixth of the triangles. Now raise every point to the power $6/5$, which closes the gap left by removing the wedge. (Figure 10 shows a portion of this tiling.) Now select a connected subset of this tiling of the plane that contains the vertex of degree five, such that this subset is isomorphic to the portion of the boundary of the blemish coming from the boundary of $P(k)$. Place the vertex v^* at infinity to complete the embedding of the boundary of the blemish.

In this embedding, the simplices of the cone triangulation $C(v^*)$ are not regular,but they are nearly regular. To evaluate the deficit of this structure one can do a numerical calculation. To simplify matters we embed the blemish in a larger structure, one in which the portion of the tiling of the complex plane is shaped like a regular pentagon. Let j be the number of edges in the subdivision of each edge of the pentagon, so that the pentagon has $5j^2$ triangles. We shall show that the only way to minimally triangulate this extended blemish is to cone to v^*.

When we apply the volume estimate of §3.4 to the above mapping and evaluate the constants, we obtain the following formula for the deficit.

$$Def(j) = \tfrac{\pi}{60}\ln(j) + O(1).$$

Numerical calculation can be used to determine the behavior of the $O(1)$ term. The results are shown in Table 1. The deficit increases by about .0363 each time j is doubled, and the $O(1)$ term is approaching $.08538\cdots$. (Our formula for estimating the deficit did not take into account the deficit from the five tetrahedra around the core of the blemish; note that the error term is always less than the deficit accounted for by these central tetrahedra.)

TABLE 1

j	Triangles	Deficit	Estimated	Error
1	5	.087935	.000000	.087935
2	20	.122353	.036293	.086060
4	80	.158141	.072586	.085554
8	320	.194304	.108879	.085424
16	1280	.230564	.145172	.085391
32	5120	.266849	.181466	.085383
64	20480	.303140	.217759	.085381

If there were any extraordinary vertices in a minimal triangulation of the blemish, then there would have to be at least one simplex with two faces on the base of the blemish, since the total number of simplices is no more than the number of faces of the base. The deficit of a simplex touching two faces along an edge tends to $V_0/3 = 0.3383$. We calculated values of the deficit for simplices having two faces sharing an edge near the vertex of order 5; the smallest deficit of any such tetrahedron is 0.3333.

Consider a triangulation T of the blemish that has exactly one simplex with two faces on the base of the blemish, exactly one simplex with no faces on the base, and all other simplices coned from the base of the blemish to v^*. In order for this triangulation to use no more tetrahedra than the canonical one, the blemish must be large enough so that the deficit of the entire structure is at least 0.3333. Table 1 shows that the radius must therefore be at least 64. Because T has only two deviant tetrahedra (tetrahedra that are not coned to v^*), it has at most eight extraordinary vertices. It therefore must be possible to embed this blemish in a pentagon of radius at most eight. Since eight is less than 64, such a triangulation T is impossible.

Before there is enough volume deficit for a second simplex touching two faces, the radius of the blemish would have to double at least eight more times, to more than 4000. Such a triangulation could be embedded in a pentagon of radius at most 16. The reasoning used above shows that this too is impossible. In general, no matter how many of these deviant simplices there are, the radius required to achieve the necessary deficit is much bigger than required to absorb that many simplices. This completes the proof of Theorem 1. □

If the sizes of the deficits were not quite so small, this argument would not work for small j, and we could only deduce that the minimal number of simplices was within an additive constant of the number for $C(v^*)$. For example, consider the sequence of subdivisions of the tetrahedron or of the octahedron instead of the icosahedron. For a tetrahedron, the numbers do not work out: in fact, the cone-type triangulations of subdivisions of the tetrahedron can be improved by first cutting off the corners. This fact manifests itself in Table 2 in which the deficit is more than 1 even for small values of j.

TABLE 2

j	Triangles	Deficit	Estimated	Error
1	3	1.014942	0.000000	1.014942
2	12	1.520982	0.544397	0.976585
4	48	2.054948	1.088793	0.966155
8	192	2.596627	1.633190	0.963437
16	768	3.140336	2.177586	0.962750
32	3072	3.684560	2.721983	0.962578
64	12288	4.228914	3.266379	0.962535

For an octahedron, the numbers work easily to prove that the cone-type triangulation is minimal: a blemish with deficit V_0 must have radius more than 11, at which size it is easy to see that one can replace it with smaller blemishes. See Table 3. However, in this case, the cone-type triangulation of the blemish is not unique. Already the blemish of radius 1 has deficit 0.395904. This blemish is the octahedron, which admits three different minimal triangulations (related by the symmetry of the octahedron). This octahedron is contained within the cone-type triangulations of all the larger blemishes, so none of them are unique.

TABLE 3

j	Triangles	Deficit	Estimated	Error
1	4	0.395904	0.000000	0.395904
2	16	0.566991	0.181466	0.385525
4	64	0.745656	0.362931	0.382725
8	256	0.926398	0.544397	0.382002
11	484	1.009655	0.627768	0.381887
16	1024	1.107681	0.725862	0.381819
32	4096	1.289101	0.907328	0.381774
64	16384	1.470555	1.088793	0.381762

Theorem 1 applies to a far wider variety of triangulations than $P(k)$. Our proofs of both Lemma 8 and Theorem 1 apply with almost no changes to any triangulation satisfying the following conditions: (1) All the vertices are of order five or six. (2) The triangulation can be drawn with equilateral triangles on the surface of a convex polytope whose shape is distorted from a sphere by only a "constant amount." (3) Within a radius of $O(\log n)$ of any vertex of order five, the triangulation is isomorphic to Figure 10, and elsewhere it is isomorphic to the standard tessellation of the plane by equilateral triangles.

To prove a result analogous to Lemma 8 for this class of triangulations we first need to map the triangulation to a sphere, with a map that is conformal everywhere except at the vertices of order five. In the proof of Lemma 8 we used the Riemann mapping theorem. Here we must use a more general theorem called the uniformization theorem. It tells us that there is a map from the surface of any polytope to a sphere that is conformal everywhere except at the vertices of the polytope. Furthermore the fact that the polytope is distorted from a sphere by only a constant amount allows this map to have its derivative bounded below and its second derivative bounded above away from the vertices. These are the bounds required for the proof.

Given condition (3) on the triangulation and the $O(\log n)$ bound on the deficit of the embedding, the same reasoning employed in the proof of Theorem 1 shows that the cone-type triangulation is the minimal for these triangulations. To prove that $d(n) = t(n) = 2n - 10$ for all sufficiently large values of n, it only remains to show how to construct an appropriate triangulation for every sufficiently large n. We shall now give this construction.

First consider the triangulation of the plane by equilateral triangles, arranged so that one vertex is at the origin. Pick any other two vertices v_1 and v_2 so that the three are not collinear and mark off the sublattice they generate. There is a group of symmetries of the triangulation generated by 180° rotations about 0, v_1, and v_2. Each of these rotations preserves the lattice we have marked off. This group preserves not only the tessellation of the plane by equilateral triangles, but also a tessellation by parallelograms congruent to the one spanned by v_1 and v_2. If the parallelograms are colored black and white in a checkerboard pattern, the coloring is also preserved by the group. If we fold up the plane by the group action, we obtain as quotient space a polyhedron that has four vertices of order three (coming from 0, v_1, v_2, and $v_1 + v_2$), while all the other vertices are of order six. It can be realized in space as a tetrahedron, with a finer triangulation by equilateral triangles drawn on its surface. Its area is the sum of the area of the black and the white parallelograms. The number of vertices is the area of the tetrahedron, divided by the area of the smallest lattice parallelogram, plus two. Using this information, it is not hard to construct an example of this type with any even number of vertices bigger than two and where the vertices of order three are fairly far apart.

A slight modification of this construction gives rise to triangulated polyhedra with n vertices for any $n \geq 8$. In the new construction superimpose on the original equilateral triangulation a triangulation by equilateral triangles of half the edge length, arranged so that the origin is in the middle of an edge of a larger triangle. Call the vertices of the larger triangles *coarse vertices*, and all the remaining vertices *fine vertices*. Let the equilateral triangles formed by three neighboring coarse vertices be called *coarse triangles*, and define *fine triangles* analogously. Choose v_1 and v_2 so that v_1, v_2, and $v_1 + v_2$ are all fine vertices. Mark out the lattice generated by v_1 and v_2. It is again the case that a rotation by 180° about any lattice point preserves both the coarse and the fine vertices. Now cut out from the plane all coarse triangles that touch a lattice point. The picture is now the plane minus a collection of parallelograms (each formed by a pair of missing coarse triangles). Form the quotient space by the group generated by 180° rotations about lattice points, as before. The four edges of each of the missing parallelograms fold up to form a bi-gon. Glue together the two edges of the bi-gon to form a single edge. The desired triangulation is that obtained by considering the coarse vertices and the coarse triangles. Each vertex is of order six except those near the missing parallelograms. Each of the four missing parallelograms creates a vertex of order five and a vertex of order four. (The process described here breaks down if n is less than eight, because for these small numbers the deleted parallelograms are not disjoint.) Figure 11(a) shows the first stage of the constructure of a polyhedron of nine vertices. The fine triangles are not shown. Figure 11(b) shows the triangulation that results from this choice of lattice vertices.

The number of vertices in a triangulation formed by this process is just twice the area of a lattice parallelogram as measured in units of coarse parallelograms (each of which is two coarse triangles). To prove that we can obtain triangulations with any number of vertices, we need to show that there is a basis parallelogram for the lattice that has any desired half-integral area. The following paragraph describes how to do this in such a way that the lengths of the basis vectors are $\Theta(n^{1/2})$.

Choose a fine vertex as the origin. Then choose nonorthogonal x and y axes such that (1) they go through the origin, (2) they avoid all coarse vertices, (3) they are parallel to one of the sides of each coarse triangle, and (4) the x

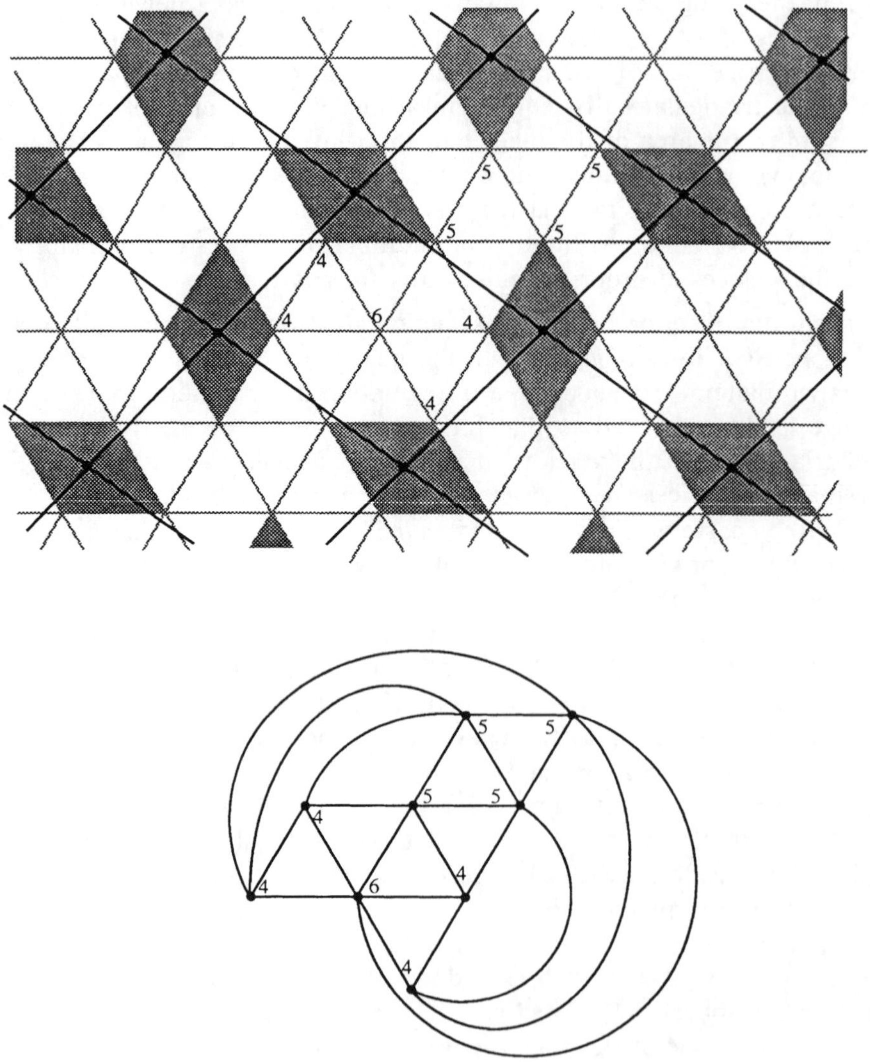

FIGURE 11. Constructing a triangulation of nine vertices, one of order six, four of order five, and four of order four. (a) shows the coarse triangles, the lattice points, and the lines between the lattice points. The shaded parallelograms are the ones that are removed. (b) shows the triangulation that results.

axis is 60° clockwise of the y axis. In the x direction let the unit distance be the length of a side of a coarse triangle. In the y direction let the unit distance be the length of a side of a fine triangle. Each point with integral coordinates is a fine vertex (although not all fine vertices have integral cordinates in this framework). We are seeking two vectors $v_1 = (a, b)$ and $v_2 = (c, d)$ such that the area of the resulting parallelogram (in the metric of this coordinate system) is the desired value n and the parallelogram is nearly square. Let

$$a = 1, \qquad b = \lfloor n^{1/2} \rfloor, \qquad c = \lceil \tfrac{n}{b} - \tfrac{1}{2} \rceil, \qquad d = cb - n.$$

The area of the parallelogram is $bc - ad = bc - (cb - n) = n$. It is also easy to see that all the sides are $\Theta(n^{1/2})$ in length.

A further modification gives rise to triangulations with twelve order five vertices and all others of order six, as follows. First, consider a triangle T with coarse vertices and with sides a, b, and c in counterclockwise order. Let p be the vertex between sides a and b. Around the vertex p, one can arrange six triangles, alternating between equilateral triangles and congruent copies of T arranged so that one of the other corners is at p. The union of the six triangles covers a hexagon H with sides of type a, b, c, a, b, and c in counterclockwise order, with opposite sides parallel. The hexagon admits a 180° symmetry. The point of symmetry is always on a coarse or fine vertex; which type depends on the original triangle T. For present purposes, it is desirable that T be nearly equilateral and that the point of symmetry be a fine vertex. We can construct such a T by starting with a large equilateral triangle on the coarse vertices and moving one of the vertices of this triangle to one of the six nearby coarse vertices. Figure 12 shows how the hexagon is formed from three copies of a triangle a, b, c, and three equilateral triangles.

Use H and T to construct the desired triangulation as follows. As in the previous construction, let the origin be a fine vertex, and choose vectors v_1 and v_2 on fine vertices so that $v_1 + v_2$ is also a fine vertex. Form the lattice from these two vectors. Now, instead of removing the two coarse triangles near each lattice point, remove a copy of H centered there. Form the quotient by the same group of symmetries, generated by order two rotations about lattice points. These symmetries carry the copies of H to themselves so that the quotient is a surface with boundary, where each boundary component has three edges of types a, b, and c. Glue a copy of T to each one. (In the previous construction, each missing parallelogram became a bi-gon, which we then closed up. Here the missing hexagon becomes a missing triangle, and we patch the hole with a copy of T.) The polyhedron we obtain has twelve vertices of degree five (one at each corner of each copy of T that is inserted), and all the other vertices have degree six.

It is easy to obtain such a polyhedron for any sufficiently large n. First construct a T that contains about αn coarse vertices, where α is some small fraction, say 1/20. Let $|T|$ be the number of coarse vertices in T, not counting boundary vertices. Construct H, and define $|H|$ similarly. Let $n' = n - 2|H| + 4|T|$. Now choose v_1 and v_2 such that they would give a construction with n' vertices without replacing the hexagons by triangles. When this replacement is done, the number of vertices left is n.

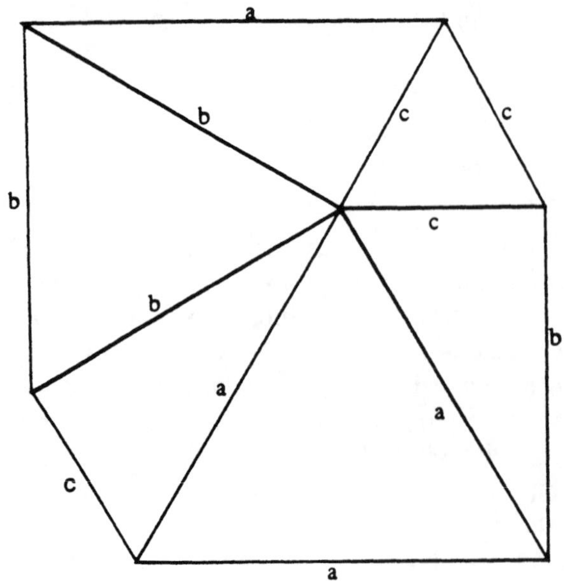

FIGURE 12.The way three copies of triangle (a, b, c) are glued together, along with three equilateral triangles, to form a hexagon. Note that the hexagon has 180° symmetry.

It is easy to see that the resulting triangulation has all the properties required for our proof of Theorem 1 to apply, and it has any sufficiently large number of vertices. Thus we have proved the following theorem.

Theorem 2. $t(n) = d(n) = 2n - 10$ *for all sufficiently large values of* n.

4. Remarks and questions

Our results say nothing about small values of n. We conjecture that $t(n) = 2n - 10$ for all $n > 12$, but we have been unable to extend our proof to show this. Probably some more concrete calculations of volumes and triangulations, for polyhedra with low values of n, would show this.

Empirically, it seems that for very many triangulations of the sphere, cone-type triangulations give the best possible extensions to a ball. It would be nice to have a good criterion of when this is the case. Even when cone-type triangulations are not the best possible, they provide a seemingly efficient method of finding a minimum triangulation: begin with a cone-type triangulation to a vertex v^* of maximal order. Now look for subpolyhedra where the cone vertex is not the vertex of minimal order for the boundary triangulation; replace the triangulation in the subpolyhedron by a different cone-triangulation. For instance, if one begins with a triangulation of the sphere having vertices v_1 and v_2, of orders say fifteen and twelve, separated by a simple cycle of edges p of length eleven or less, then one can begin by constructing the cone triangulation to v_1. The union of tetrahedra that touch v_2's side of p form a subpolyhedron Q, which has a cone-type triangulation to v_1 — but on the boundary of Q, v_1 has order the length of p, which is not more than eleven. Replacing the triangulation of Q by the cone to v_2 reduces the number of tetrahedra. This idea can be iterated. There are sometimes situations where several moves of this type that do not reduce the number of tetrahedra, but maintain it at a constant, are necessary before a move of this type can reduce the number. It is possible to get from any triangulation to any other if we allow moves of this type that may increase the number of tetrahedra. We know of no instance, however, where in order to decrease the number of tetrahedra, it is first necessary to increase the number.

A more complete development of our method would allow other types of blemishes that are not nearly round and that touch the boundary triangulation in a surface that is not necessarily a disk. The estimates would be much stronger in this case. We conjecture that such a method would show that if there are no vertices of order less than four, and if the injectivity radius is always bigger than some fixed constant, that the cone-type triangulation is a minimal triangulation.

The role that geometry plays in this problem may seem mysterious. There should be an analysis that is entirely combinatorial. It may help clarify the relation to combinatorics if we point out the relation of this question to network flow problems, and to the "max flow min cut" principle.

To develop this idea, consider the $(n-1)$-simplex Δ^{n-1}, which has n vertices $v_0, \dots, v_{n-1}$. This circular order defines a Hamiltonian circuit in the 1-skeleton of Δ^{n-1}, which we can think of as identified with the boundary of our original polygon. The sum of the oriented edges of this circuit is a 1-cycle α. A triangulation of the disk defines a 2-chain σ such that $\partial\sigma = \alpha$, since each triple of vertices of α spans a unique triangle in the 2-skeleton of Δ^{n-1}.

Consider two different triangulations of the disk, giving two different 2-chains σ_1 and σ_2. Then $z = \sigma_1 - \sigma_2$ is a 2-cycle. Define the quantity $g(z)$ to be the minimum L^1 norm of a 3-chain T such that $\partial T = z$. (The L^1 norm of a chain is the sum of the absolute values of its coefficients.) Both the minimum number of elementary moves to get from σ_1 to σ_2 and the minimum number of 3-simplices it takes to extend the triangulation they define to the ball give upper bounds for $g(z)$. Geometrically, if z comes from a triangulation of the boundary of a ball, then any 3-chain T such that $\partial T = z$ can be mapped into the ball, giving a possibly multiple-valued triangulation of the ball, where tetrahedra with fractional weights are allowed, but the weights of overlapping tetrahedra must add up to 1 at each point.

Unlike the problem of finding minimal triangulations, the problem of finding a chain T such that $\partial T = z$ with minimum L^1 norm is a linear problem: it is a question of minimizing a convex function subject to a linear constraint. Therefore there is a dual problem. The dual space of the space C_k of k-chains of a simplicial complex is denoted C^k, the space of k-cochains. A k-cochain is a linear function on the k-chains, and it is determined by its values on the oriented simplices, since they span the space (and a choice of an orientation on each simplex gives a basis). The coboundary map δ maps C^k to C^{k+1}; when β is an oriented $k+1$-simplex and c is a k-cochain, $\delta(\beta)$ is c applied to the sum $\partial\beta$ of the oriented faces of β. A cochain whose coboundary is 0 is called a cocyle.

The dual problem is this: given z, what is the minimum L^∞ norm $h(z)$ of a 3-cocycle V such that $\langle V, T\rangle = 1$, where T is any chain such that $\partial T = z$? Here, $\langle\ ,\ \rangle$ is the dual pairing of cochains with chains. This formulation depends on the fact that the simplex Δ^{n-1} is *acyclic*; that is, in any dimension except 0, every cycle is a boundary. Therefore, if T' is any other chain such that $\partial T' = z$, the difference $T - T'$ is a 3-cycle, so there exists a 4-chain a with $\partial a = T - T'$. Therefore

$$\langle V, T\rangle - \langle V, T'\rangle = \langle V, \partial a\rangle = \langle \delta V, a\rangle = \langle 0, a\rangle = 0.$$

The solutions to the dual problems satisfy $g(z)\cdot h(z) = 1$. A good example of a chain T gives an upper bound for $g(z)$; a good example of a cocycle V gives a lower bound. The quantity $g(z)$ is like a minimum cut; the quantity $1/h(z)$ is like a maximum flow.

A mapping f of Δ^{n-1} into hyperbolic space defines a 3-cocycle V_f, which assigns to any 3-chain the algebraic volume of its image. (We should divide V_f by the total volume to make it literally fit the formulation above.) If we allowed ourselves to push the triangles around so that they were curved, we could obtain every possible 3-cocycle in this way. In most cases, no single map that takes triangles to geodesic triangles gives the best cocycle for our problem. Presumably, some convex combination of cocycles V_f of the form we used would give a cocycle of norm even closer to the minimum. Our actual argument made use of the fact that we only allowed integral cycles T, with no fractional weighting.

The quantity $g(z)$ was introduced and studied in a more general topological context by Gromov (see [12]). He proved that for a closed oriented hyperbolic manifold M^n, the minimum L^1 norm of an n-cycle representing the fundamental class of M is exactly the volume of M divided by the maximal volume of a hyperbolic n-simplex. To approach the best bound in this case of a closed manifold, it is definitely necessary to use simplices of fractional weights. Perhaps this work will be useful in analyzing best possible actual triangulations of 3-manifolds.

References

1. H. M. S. Coxeter, *Non-Euclidean geometry*, Univ. Toronto Press, Toronto, 1942.
2. K. Culik and D. Wood, *A note on some tree similarity measures*, Inform. Process. Lett. **15** (1982), 39–42.
3. A. K. Dewdney, *Wagner's theorem for torus graphs*, Discrete Math. **4** (1973), 139–149.

4. M. Gardner, *Time travel and other mathematical bewilderments*, Freeman, New York, NY, 1987.
5. D. E. Knuth, *The art of computer programming, Vol.* 3: *Sorting and searching*, Addison-Wesley, Reading, MA, 1973.
6. J. Milnor, *Hyperbolic geometry: the first* 150 *years*, Bull. Amer. Math. Soc. **6** (1982), 9–24.
7. J. Pallo, *On rotation distance in the lattice of binary trees*, Inform. Process. Lett. **25** (1987), 369–373.
8. D. D. Sleator and R. E. Tarjan, *Self-adjusting binary search trees*, J. Assoc. Comput. Mach. **32** (1985), 652–686.
9. D. D. Sleator, R. E. Tarjan, and W. P. Thurston, *Rotation distance, triangulations, and hyperbolic geometry*, Proc. 18th Annual ACM Symposium on Theory of Computing, May 28–30, 1986, Berkeley, CA, Association for Computing Machinery, NY, pp. 122–135.
10. R. E. Tarjan, *Data structures and network algorithms*, Soc. Indust. Appl. Math., Philadelphia, PA, 1983.
11. W. P. Thurston and J. R. Weeks, *The mathematics of three-dimensional manifolds*, Scientific American **251** (1984), 108–120.
12. W. P. Thurston, *The geometry and topology of three-dimensional manifolds*, Princeton Univ., Dept. of Math., 1979.
13. W. T. Tutte, *A theorem on planar graphs*, Trans. Amer. Math. Soc. **82** (1956), 99–116.
14. K. Wagner, *Bemerkungen zum Vierfarbenproblem*, J. Deutschen Math.-Verein. **46** (1936), 26–32.
15. H. Whitney, *A theorem on graphs*, Ann. of Math. **32** (1931), 378–390.

ABSTRACT. A *rotation* in a binary tree is a local restructuring that changes the tree into another tree. Rotations are useful in the design of tree-based data structures. The *rotation distance* between a pair of trees is the minimum number of rotations needed to convert one tree into the other. In this paper we establish a tight bound of $2n - 6$ on the maximum rotation distance between two n-node trees for all large n. The hard and novel part of the proof is the lower bound, which makes use of volumetric arguments in hyperbolic 3-space. Our proof also gives a tight bound on the minimum number of tetrahedra needed to dissect a polyhedron in the worst case and reveals connections among binary trees, triangulations, polyhedra, and hyperbolic geometry.

COMPUTER SCIENCE DEPARTMENT, CARNEGIE MELLON UNIVERSITY, PITTSBURGH, PENNSYLVANIA 15213

COMPUTER SCIENCE DEPARTMENT, PRINCETON UNIVERSITY, PRINCETON, NEW JERSEY 08544

AT&T BELL LABORATORIES, MURRAY HILL, NEW JERSEY 07974

MATHEMATICS DEPARTMENT, PRINCETON UNIVERSITY, PRINCETON, NEW JERSEY 08544

SIAM J. DISC. MATH.
Vol. 5, No. 3, pp. 428–450, August 1992

012

SHORT ENCODINGS OF EVOLVING STRUCTURES*

DANIEL D. SLEATOR†¶, ROBERT E. TARJAN‡¶, AND WILLIAM P. THURSTON§

Abstract. A derivation in a transformational system such as a graph grammar may be redundant in the sense that the exact order of the transformations may not affect the final outcome; all that matters is that each transformation, when applied, is applied to the correct substructure. By taking advantage of this redundancy, we can develop an efficient encoding scheme for such derivations. This encoding scheme has a number of diverse applications. It can be used in efficient enumeration of combinatorial objects or for compact representation of program and data structure transformations. It can also be used to derive lower bounds on lengths of derivations. It is shown, for example, that $\Omega(n \log n)$ applications of the associative and commutative laws are required in the worst case to transform an n-variable expression over a binary associative, commutative operation into some other equivalent expression. Similarly, it is shown that $\Omega(n \log n)$ "diagonal flips" are required in the worst case to transform one n-vertex numbered triangulated planar graph into some other one. Both of these lower bounds have matching upper bounds. An $O(n \log n)$ upper bound for associative, commutative operations was known previously, whereas here an $O(n \log n)$ upper bound for diagonal flips is obtained.

Key words. graph grammar, graph transformations, associativity, commutativity, diagonal flips, triangulations

AMS(MOS) subject classifications. 05, 68

1. Introduction.

The object of this paper is to study succinct representations of derivations in transformational systems. To model transformational systems, we use *graph grammars* [2]. Roughly speaking, a graph grammar consists of a finite set of productions $\{L_i \to R_i\}$. (Section 2 gives a precise definition of the form of graph grammar that we use.) Each production $L_i \to R_i$ consists of a connected graph L_i, called the *left side* of the production, and a graph R_i, called the *right side* of the production. A production $L_i \to R_i$ is *applicable* to a graph G if G contains a subgraph isomorphic to L_i. The production is applied to G by replacing an occurrence of L_i in G by a copy of R_i. (There may be more than one way of applying a production to G, since G may contain more than one copy of the left side.) A *derivation* is a sequence of graphs $G = G_0, G_1, G_2, \cdots, G_m = G'$ such that each G_i is obtained from G_{i-1} by applying one production once. The derivation *transforms* graph G into graph G'. A particular application of a production during a derivation is called an *action*.

Let Γ be a fixed graph grammar, and let G be a fixed starting graph of size n. Consider the collection $R(G, \Gamma, m)$ of all graphs obtainable from G by derivations

* Received by the editors June 12, 1990; accepted for publication (in revised form) August 7, 1991.

† School of Computer Science, Carnegie Mellon University, Pittsburgh, Pennsylvania 15213. This research was supported in part by National Science Foundation grant CCR-8658139 and by the Defense Advanced Research Projects Agency, monitored by the Space and Naval Warfare Systems Command under contract N00039-87-C-0251.

‡ Computer Science Department, Princeton University, Princeton, New Jersey 08544 and NEC Research Institute, Princeton, New Jersey 08540. This author's research at Princeton University was partially supported by National Science Foundation grant CCR-8920505 and Office of Naval Research contract N00014-91-J-1463.

§ Department of Mathematics, University of California at Berkeley, Berkeley, California 94720. This author's research was done while at the Department of Mathematics, Princeton University, Princeton, New Jersey 08544 and was partially supported by National Science Foundation grant DMS-8806067.01.

¶ This author's research was partially supported by DIMACS (Center for Discrete Mathematics and Theoretical Computer Science), a National Science Foundation Science and Technology Center, grant NSF-STC88-09648.

of length m or less. Our main result is an efficient method of encoding any graph in $R(G,\Gamma,m)$. To encode any such graph, use $O(m+\log\binom{n}{m})$ bits. (In most cases of interest, $m>n$, and the second term in this bound is zero.) This bound is an improvement by a logarithmic factor over the obvious bound of $O(m\log s)$, where $s\geq n$ is the size of the largest graph occurring in the derivation [9]. This logarithmic improvement is crucial in obtaining the tight lower bounds discussed below.

Our encoding represents an equivalence class of derivations obtained by permuting commutative applications of the productions. The efficiency of such an encoding arises from the fact that there may be many derivations equivalent to any given one, a fact that follows from the localized nature of applications of the production rules. For simplicity, we formulate our result in the setting of labeled undirected graphs; it holds for more general combinatorial structures such as hypergraphs and simplicial complexes, however.

Our result has a number of general and specific applications, both theoretical and practical. Our main theoretical application is in demonstrating the existence of pairs of graphs that are far apart, in the sense that any derivation of one graph of the pair from the other must take many actions. If $N(G)$ is a lower bound on the number of graphs derivable from a graph G of size n, then there is a graph G' such that any derivation of G' from G has length $\Omega(\log N(G)-n)$. This is because our encoding scheme implies that the number of graphs derivable from G' by derivations of length m or less is at most c^{n+m}, for some constant c that depends only on the grammar and not on m and n.

Our first application involves transformations of arithmetic expressions. Consider the collection of fully parenthesized expressions of n variables over an associative, commutative binary operation. A *move* consists of applying either the commutative law (exchanging two subexpressions that are combined by the operation) or the associative law (erasing a pair of matching parentheses to put three expressions at the same level, and adding a new pair of parentheses to alternatively regroup this triple). We show that, given any n-variable expression E, there is an equivalent expression whose distance from E in this metric is $\Omega(n\log n)$. This solves an open problem of Culik and Wood [1], who obtained a matching upper bound. Thus the worst-case distance between two equivalent expressions is $\Theta(n\log n)$. This contrasts with the corresponding bound of $2n-O(1)$ if commutativity is not allowed [6].

As a second application of our lower bound, we consider the collection of numbered triangulations of the plane, transformed by the "flip" operation. This operation removes an edge, thereby creating a quadrilateral face, and replaces it with the other diagonal of the face. A flip is only allowed if it does not create a multiple edge. Our encoding method proves that there exist pairs of n-vertex triangulations that are $\Omega(n\log n)$ flips apart. We show, furthermore, that this bound is tight by giving a method for converting any n-vertex triangulation into any other in $O(n\log n)$ flips. This improves the previous $O(n^2)$ upper bound of Wagner [8].

We envision several other applications of our technique. First, it can be used to efficiently encode graphs or other combinatorial structures that are close to a given one (in the sense of being obtainable by a small number of transformations). Such encodings may be useful in situations that require the representation of multiple versions of a structure, as in program transformation systems and other applications of *persistent* data structures [3]. Second, it provides a way to enumerate graphs of various kinds that are generated by graph grammars or other such transformational systems. By enumerating our encodings rather than enumerating sequences of productions, all of the desired graphs are generated, but with far fewer redundant copies of isomorphic graphs.

The remainder of this paper consists of five sections. In §2 we give a precise formulation of graph grammars and graph grammar derivations, describe our encoding scheme for derivations, and use this to prove upper bounds on the number of graphs obtainable by short derivations. Section 3 gives several refinements and improvements of our method. Sections 4 and 5 show how the encoding scheme applies to prove our lower bound results for expressions and plane triangulations. Section 6 contains our upper bound on the distance between plane triangulations; it is independent of the rest of the paper.

2. Encoding graph derivations. We are concerned with graphs that are undirected and of degree at most b (a fixed constant independent of n). Each end of each edge is labeled with an integer called an *edge-end label*. The edge-end labels incident on a vertex are distinct and between 1 and b inclusive. It is useful to be able to refer to *half* of an edge. Each such *half-edge* has one end vertex from the original edge, and one edge-end label. We allow the graph to have multiple edges between the same pair of vertices, and even to have self-loops. (It is easy to modify our construction to disallow such structures, although doing so would only weaken our lower bounds.)

A *graph grammar* (usually denoted by Γ) is a finite set of productions $\{L_i \to_i R_i | i = 1, 2, \cdots\}$. The ith production is comprised of three parts: L_i, the *left side* of the production; R_i, *the right side* of the production; and $\to_i$, the *correspondence* of the production. The three parts of a production have the following characteristics:

L_i: This is a connected, undirected, edge-end labeled graph, with degree bounded by b. Strictly speaking, L_i is not a graph, because it has a set of half-edges that have only one end vertex. The one end vertex of each half-edge that is attached to a vertex of L_i has an edge-end label.

R_i: This is also a graph with edge-end labels, and half-edges, of which it has the same number as L_i.

$\to_i$: This is a one-to-one map between the half-edges of L_i and those of R_i.

The production $L_i \to_i R_i$ applies to a graph G if G contains a set of vertices S such that $G(S)$ (the subgraph induced by S in G) is isomorphic (including edge-end labels) to L_i. The induced subgraph $G(S)$ is most simply defined by retaining half of every edge incident to a vertex in S. The half-edges of $G(S)$ come from the edges of G with one endpoint in S and one not in S. The production is applied by replacing this occurrence of L_i in G by R_i, where each half-edge of R_i is attached to a half-edge of $G - G(S)$ just as the corresponding half-edge of L_i was attached. Sections 3 and 4 give examples of specific graph grammars.

Each vertex occurring in an L_i has a unique *position number* from the set $\{1, 2, \cdots, c\}$, where c is the total number of vertices in all left sides. The position numbers are used to uniquely specify a vertex in a production. The vertices of each R_i are also numbered $1, 2, \cdots$ within each production. These numbers are the *right position numbers*.

A *derivation* is a sequence of graphs $G = G_0, G_1, \cdots, G_m = G'$ such that each G_i is obtained from G_{i-1} by applying one production once. An *action* is a particular application of a production during the derivation. The derivation *transforms* G into G'. The *length* of such a derivation is m, the number of actions in it.

We construct a pair of functions $\text{ENCODE}_{G,\Gamma}$ and $\text{DECODE}_{G,\Gamma}$. The function $\text{ENCODE}_{G,\Gamma}$ takes a derivation D that transforms G into some other graph G' and returns a string of symbols from the alphabet $\Sigma = \{0, 1, 2, \cdots, c\}$ of length $n + r \cdot m$. Here n is the number of vertices of G, m is the length of the derivation D, c is the number of vertices in left sides of Γ, and r is the number of vertices in the largest right side of Γ. This sequence is called the *encoding* of the derivation. The function $\mathrm{DECODE}_{G,\Gamma}$ takes as input such an encoding and returns the graph G'. That is, $\mathrm{DECODE}_{G,\Gamma}(\mathrm{ENCODE}_{G,\Gamma}(D)) = G'$.

For our purposes, it is useful to think of the process of applying a production as *destroying* vertices (the ones that are matched to the vertices of L_i) and *creating* new and different ones (the ones introduced by R_i). The actions of a derivation D of length m are numbered $1, 2, \cdots, m$ in the order in which they occur. Each vertex that is created during the derivation can be identified uniquely by specifying the number of the action that created it and the position number of the vertex in R_i that created it. This is the *name* of the vertex. The *required vertices* of an action are the vertices that are destroyed by it. An action is said to be *ready* at some time during a derivation if all of its required vertices exist at that time. Readiness implies that the entire copy of L_i that is to be replaced (including all of its edges and half-edges) is present as well.

LEMMA 2.1. *Consider a derivation D that transforms G into G'. If the actions of D are reordered in any way so that each production is ready when it is applied, then the new derivation also transforms G into G'.*

Proof. By induction, it is sufficient to prove that, if a_t and a_{t+1} (two consecutive actions of D) are such that none of the required vertices of action a_{t+1} is created by a_t, then, if these actions are swapped, the resulting derivation also transforms G into G'. Since either order is allowed, we know that those vertices created by a_t are not used by a_{t+1} and those created by a_{t+1} are not used by a_t. It follows that the actions commute, since they do not involve any of the same vertices. □

We can now give an explicit algorithm for computing the encoding of a derivation D. First, the actions of D are numbered, the vertices of the derivation are named, and the required vertices of each action are computed.

Our encoding algorithm assigns to each vertex of the derivation a unique *number*. First, the vertices of G are numbered $\{1, 2, \cdots, n\}$ in an arbitrary order. (The same ordering must be used by the decoding procedure described below.) The remaining vertices are numbered in conjunction with the construction of a *canonical* derivation D', which is a reordering of the actions of D.

The actions of the canonical derivation D' are computed one at a time. At any time, it is easy to determine which actions are ready; these are the ones whose required vertices exist. Let q be the ready action that destroys the lowest-numbered vertex among all ready actions. This action is the one chosen to be the next action of D'. This action is applied to the graph, and the vertices created by it are now numbered consecutively starting after the largest vertex number used thus far. (If several vertices are created by q, they are numbered in the order of the right position numbers in the right side of the production that created them.)

After computing the canonical derivation D', the algorithm proceeds to compute a *label* for every vertex that occurs in the derivation. The label of a vertex v is zero if v is not destroyed by any action in the derivation. Otherwise, it is the position number of the role played by v in the production that destroys it. The desired encoding is a list of, at most, $n+r\cdot m$ labels of all the vertices in increasing order by vertex number.

We can now describe the decoding procedure. This algorithm takes the graph G (with vertex numbers that agree with those of the encoding procedure), the grammar Γ, and the encoding (the list of labels), and determines G'. The procedure works by constructing the canonical derivation D', from which it is easy to get G'. As in most data compression/decompression methods, the decoding algorithm mimics the behavior of the encoding algorithm step by step.

The crucial fact concerning the labels of the vertices existing at any time during the canonical derivation is that from these it is possible to determine exactly which actions were ready at the corresponding stage of the encoding process. This is accomplished by the following *matching* procedure.

If a vertex v has a nonzero label, the label determines i, the production that eventually destroys v, and also the role v plays in this production. For each such vertex, check its neighborhood to see if it is isomorphic to L_i (including edge-end labels). This check is easy, since we know which vertex of L_i must match v, L_i is connected, and that there are edge-end labels to follow. (Recall that the edge-end labels incident to a vertex are disjoint.) If such a subgraph is found, then the labels of these vertices are checked to see if they match the position numbers of the roles that they are supposed to play in the proposed action. If all of these tests are passed, then the action is ready.

LEMMA 2.2. *The matching procedure determines the ready productions that existed at the corresponding stage of the encoding process.*

Proof. If an action is ready, then the matching procedure certainly finds it because the vertices corresponding to the left side of the ready action will exist and are labeled in a way consistent with all the conditions checked above.

On the other hand, suppose that the above check is satisfied starting from some vertex v. Let i be the production indicated by the label of v, and let S be the set of vertices that form the subgraph isomorphic to L_i. We claim that, in any continuation of this derivation, all vertices of S must be destroyed simultaneously by a single action. Since all these vertices are destroyed by one action, this action must now be ready.

It remains to show that all vertices of S must be destroyed simultaneously. Consider the first action a in some continuation of the derivation that destroys some vertex w of S. Since a is the first action involving the vertices of S, at the moment action a is applied, all of the vertices of S exist (and have the same labels). From the vertex w, the matching algorithm described above constructs the set S. There is no other possible matching pattern involving w. Therefore the action a destroys all the vertices of S simultaneously. $\square$

Now, given that we know the ready productions and the numbering of the vertices of the current graph, it is easy to find q (the next production of D') because it is the ready action that destroys the lowest-numbered vertex. This action is applied to the graph. The vertices created by it are numbered sequentially (as in the encoding procedure) and are labeled as specified by the encoding. This step is repeated to determine all of the productions of D'. The process terminates when there are no ready productions.

The following theorem, which bounds the number of graphs obtainable from a given one as a function of the length of a derivation, is a consequence of our encoding scheme.

THEOREM 2.3. *Let G be a graph of n vertices, Γ be a graph grammar, c be the number of vertices in left sides of Γ, and r be the maximum number of vertices in any right side of a production of Γ. Let $R(G,\Gamma,m)$ be the set of graphs obtainable from G by derivations in Γ of length at most m. Then $|R(G,\Gamma,m)| \le (c+1)^{n+r\cdot m}$.*

Proof. Encode the derivation using the scheme described above. The length of the encoding is, at most, $n + r \cdot m$ symbols. This encoding can be padded with zeros so that its length is exactly $n + r \cdot m$. (This does not interfere with the decoding process, since it is self-terminating.) The alphabet is of size $c + 1$, so the number of such encodings is $(c+1)^{n+r\cdot m}$. Each graph reachable by m or fewer actions is the outcome of applying the decoding procedure to one of these encodings. Therefore the number of such graphs is, at most, the number of such encodings. □

3. Generalizations and improvements. This section describes various extensions and improvements to our encoding scheme, most of which are used later in this paper.

3.1. Encoding short derivations. Our encoding scheme can be modified to make it more efficient when the length of the derivation is short compared to the size of the starting graph. In this case, most of the labels of the vertices of the initial graph are zero. The more efficient encoding specifies which vertex labels are nonzero and only includes labels for these in the vertex label list. Let k be the number of vertices that have nonzero label in the initial labeling of G, and let m, n, r, and c be defined as above. Then the size of this encoding (in bits) is

$$\lceil \log_2 n \rceil + \lceil \log_2 \binom{n}{k} \rceil + (k + mr)\lceil \log_2(c+1) \rceil.$$

The first term is for bits to encode k, and the second term encodes the subset of vertices with nonzero labels.

THEOREM 3.1. *It holds that*

$$\log |R(G,\Gamma,m)| = O(\log \binom{n}{m} + m),$$

where $R(G,\Gamma,m)$ is the number of graphs obtainable by derivations of length at most m in grammar Γ starting from a graph G of n vertices. (If $m > n$, then $\log \binom{n}{m}$ is interpreted as zero.)

Proof. Theorem 2.3 shows that $\log |R(G,\Gamma,m)| = O(n+m)$. If $r \cdot m > \frac{1}{2}n$, then $O(n+m) = O(m) = O(\log \binom{n}{m} + m)$. If $r \cdot m \leq \frac{1}{2}n$, then we use the above encoding scheme. Since each action causes, at most, r vertices of G to have nonzero labels, we know that

$$k \leq r \cdot m \leq \tfrac{1}{2}n.$$

It follows that

$$(k + mr)\lceil \log_2(c+1) \rceil = O(m)$$

and that

$$\log_2 \binom{n}{k} \leq \log_2 \binom{n}{rm} \leq \log_2 \binom{n}{m}^r = r \log_2 \binom{n}{m}.$$

Finally, we know that $\log_2 n \leq \log_2 \binom{n}{m}$. The theorem follows from these inequalities and the bound on the number of bits used by the efficient encoding scheme. □

3.2. Leaders and followers. The labels on the set of vertices destroyed by an action contain redundant information. For example, each label of this set has sufficient information to determine which production is the one that destroys all of them. There is a way to eliminate this redundancy and thereby reduce the size of the encoding in most cases.

The new encoding algorithm begins by computing the standard encoding described above. It then applies a map f to each symbol of the encoded string, giving the new encoding. It remains to define the map f.

Let one vertex of each L_i be chosen to be the *leader*, and let all the other vertices be *followers*. For each L_i, choose a spanning tree. (This can be done, since each left side is connected.) For each follower vertex v, let $\mathrm{DIR}(v)$ be the value of the edge-end label of the v end of the first edge on the path (in the spanning tree) from v to the leader of L_i. (In other words, starting from any vertex in L_i, following the $\mathrm{DIR}(\cdot)$ direction repeatedly leads to the leader.)

The map f is defined as follows ($|\Gamma|$ is the number of productions of Γ, and $v(x)$ is the vertex of a left side with position number x):

$$f(x) = \begin{cases} 0 & \text{if } x = 0, \\ \mathrm{i} & \text{if } v(x) \text{ is the leader of } L_i, \\ |\Gamma| + \mathrm{DIR}(v(x)) & \text{if } v(x) \text{ is a follower.} \end{cases}$$

The decoding algorithm must be modified to accommodate this new encoding. The only difference is in the matching step, which is revised as follows. For each vertex v that is a leader, check its neighborhood to see if it is isomorphic to L_i. If such an isomorphic subgraph is found, then the labels of these vertices are checked to see if they are all followers, and, if a directed edge is drawn from each follower w in the direction of $\mathrm{DIR}(w)$ (which is the label of w minus $|\Gamma|$), then the result is a directed spanning tree rooted at v. If all of these tests are passed, then the action is ready.

We now must verify that Lemma 2.2 still holds, that is, that the sets of vertices satisfying the new matching procedure above exactly correspond to the ready actions. The first part of the proof remains easy; any ready action of the original derivation results in a match in the above procedure. On the other hand, a match also indicates that the corresponding action is ready. Let S be the set of matched vertices. Starting from any follower vertex $w \in S$, the entire set S can be constructed uniquely. Similarly, from the leader vertex v of S, the set S can be uniquely constructed. This is a sufficient condition to guarantee that all vertices of S are destroyed simultaneously, which (as shown above) is the condition that we need to prove that the action is ready.

It may be possible to further reduce the alphabet size by making use of the flexibility that exists in choosing which spanning tree to use on each left side. The number of labels can be reduced from $|\Gamma|+b+1$ to $|\Gamma|+d+1$, where d is the number of different directions used in the directed spanning trees of the left sides.

The leader-follower technique applies in any situation in which there is a production with more than one vertex on the left side. It may decrease the size of the label alphabet, but it can never increase it. If the technique applies, then it can be used in conjunction with the next technique to further reduce the alphabet size.

3.3. Eliminating the zero label. Suppose that, for any graph occurring in a derivation using Γ, there exists a way of labeling it with nonzero labels, so that no production is ready. Then the zero label can be eliminated. The encoding procedure must be modified slightly to eliminate the zero labels, while the decoding procedure remains the same.

We now describe how the encoding procedure is revised. First, compute the labeling of all the vertices as before. The vertices with zero labels are exactly those that are eventually in G', the final graph of the derivation. These are called the *terminal* vertices. Compute the labeling of G' with nonzero labels, so that no production is ready. For each terminal vertex, replace its label with that terminal label just computed in G'.

It is easy to see that this works by reviewing the proof of Lemma 2.2. The proof only differs at the point where it is shown that if the labels match the pattern of some left side L_i, then the production i applied to that set of vertices S is ready. The crucial statement is that, if this situation occurs, then all the vertices of S must be destroyed simultaneously. This is still true. All of the vertices cannot be terminal ones, since their labels admit the application of a production. The set cannot be comprised of both nonterminals and terminals because then the nonterminals would never be allowed to change. Therefore all the vertices of S must be nonterminals, and the previous argument shows that the production is ready.

Note that in any situation in which the leader-follower technique applies, we can eliminate the zero label. This is done by labeling all the terminal vertices as followers.

3.4. Tags. It is sometimes useful to carry extra information along during a derivation. (Sections 4 and 5 give examples of this.) To accommodate this, we allow each vertex to have a *tag* associated with it. Each production also supplies an arbitrary function that is used to define the values of the tags of the vertices created in terms of the tags of the vertices destroyed. Because the tags are computed locally (as a function only of tags of the vertices on the left side of the production), the commutativity that we have exploited in constructing our encoding is still present. Therefore our encoding method and theorems apply to tagged graphs without any changes.

4. Expressions over an associative, commutative operation. Let $X = \{x_1, x_2, \cdots, x_n\}$ be a fixed set of variables, let $\oplus$ be a binary operation, and let E_n be the set of fully parenthesized expressions over $\oplus$ in which each variable x_i occurs exactly once. We consider the problem of estimating how many applications of the associative and commutative laws are required to transform any expression in E_n into any other.

To make this problem somewhat more concrete, we restate it as a problem on binary trees. Our binary tree terminology is that of Knuth [5]. Let T_n be the set of full binary trees with n external nodes, numbered $1, 2, \cdots, n$. Any permutation of $1, 2, \cdots, n$ is allowed; thus $|T_n| = n!\binom{2n-2}{n-1}\frac{1}{n} = (n-1)!\binom{2n-2}{n-1}$ [4]. We permit two transformations of a tree $T \in T_n$: a *twist*, in which the left and right subtrees of a specified internal node are exchanged, and a *rotation*, in which an internal node changes places with one of its children while symmetric order in the tree is preserved. (See Fig. 4.1.) The problem is to estimate the minimum number of twists and rotations needed to transform any tree in T_n into any other. We denote this number by R_n.

This problem is equivalent to the expression transformation problem. The isomorphism (also shown in Fig. 4.1) between expressions and trees is the standard one—an external node labeled i corresponds to the expression "x_i"; an internal node corresponds to the expression $(E_\ell \oplus E_r)$, where E_ℓ and E_r are the expressions corresponding to the left and right children of the node. A twist corresponds to the application of the commutative law; a rotation, to an application of the associative law.

Culik and Wood [1] derived an $O(n \log n)$ bound on R_n. We derive a matching $\Omega(n \log n)$ bound. (Culik and Wood actually worked with a slightly different transformational system, but their result applies to our system, and vice versa.)

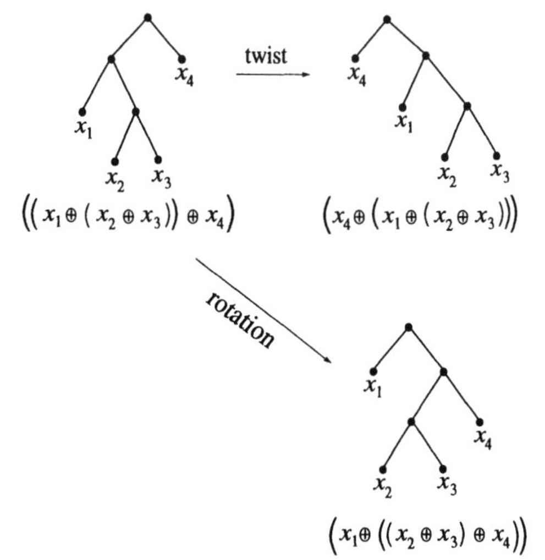

FIG. 4.1. *Illustrating a twist and a rotation.*

These transformations can be represented as productions in a graph grammar. The graphs that we consider differ slightly from the above binary trees. To transform a tree into the corresponding graph, add an extra node of degree one, called the *superroot*, and connect it to the root of the tree. The edge-end labels of the three edges incident to an internal node are 1, 2, and 3, for the edges connecting the node to its left child, right child, and parent, respectively. (The superroot is the parent of the root.) The $n + 1$ edge-ends that are incident on vertices of degree one are irrelevant, since these are never involved in any production. The vertices of degree one are tagged with a name that is carried along during the derivation. Figure 4.2 shows an expression tree and the corresponding graph.

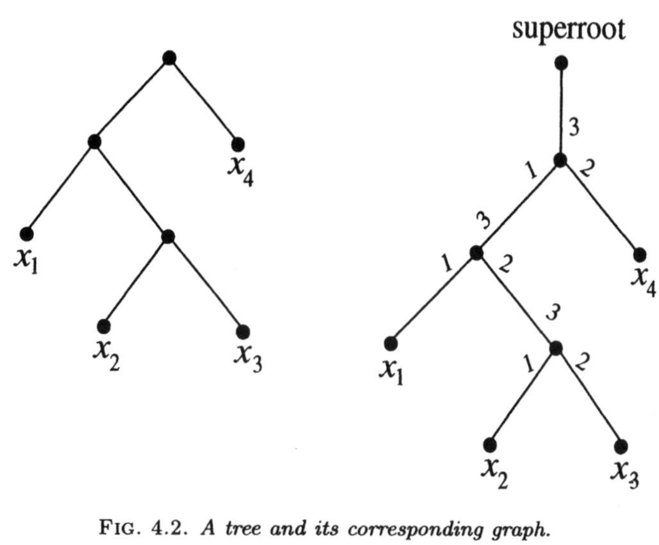

FIG. 4.2. *A tree and its corresponding graph.*

The grammar to represent this process has three productions: one for a twist, one for a left rotation, and one for a right rotation. These productions are shown in Fig. 4.3.

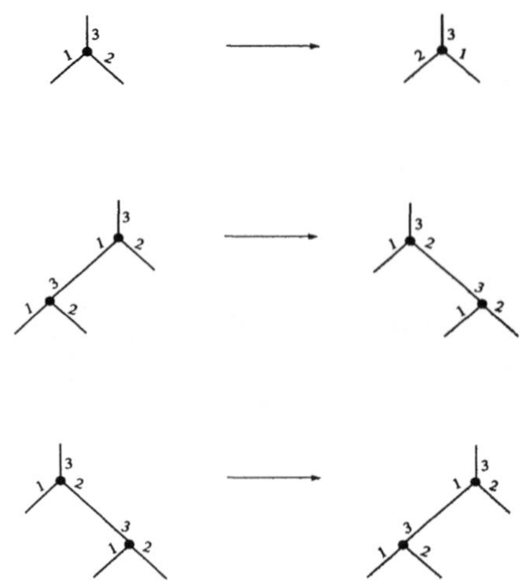

FIG. 4.3. *The productions for a twist and rotations. The correspondence between the half-edges is obtained by pairing the topmost edges and walking clockwise simultaneously around the left and right diagrams.*

From Theorem 2.3, it immediately follows that, starting from a tree with n external nodes, the number of trees reachable in m or fewer twists and rotations is, at most, $6^{2n+2m+1}$. The leader-follower technique can be used to prove a tighter bound. By choosing the upper vertex of the left side of each rotation to be the leader, and the other to be the follower, the label alphabet size is reduced to 5. The zero elimination technique now applies. This reduces the alphabet size to 4, and the bound to $4^{2n+2m+1}$. This can be further improved by specializing the encoding and decoding procedures for this application. We do not need to encode the labels for the n leaves or the superroot because these are not involved in any actions. This improves the bound to 4^{n+2m-1}. The total number of bits needed to encode any tree derivable in m or fewer moves is, at most, $2n + 4m - 2$.

We summarize this result in the following theorem.

THEOREM 4.1. *For any expression E of n variables:*

1. *The number of different arithmetic expressions obtainable by m applications of the commutative and associative laws starting from E is, at most, $2^{2n+4m-2}$;*
2. *There exists an expression E' such that the number of operations required to transform E into E' is $\Omega(n \log n)$.*

Proof. Part 1 follows from the prior discussion. Part 2 follows from the fact that there are $(n-1)!\binom{2n-2}{n-1}$ expressions obtainable starting from E. To obtain all of them in m moves, we must have that

$$2^{2n+4m-2} \geq (n-1)!\binom{2n-2}{n-1} = \Omega(n!),$$

$$2n + 4m - 2 = \Omega(n \log n),$$

$$m = \Omega(n \log n) . \qquad \square$$

5. Numbered plane triangulations: A lower bound. A *numbered plane triangulation* (henceforth, just called a triangulation) is an undirected graph embedded in the plane, all of whose faces are triangles and whose vertices are numbered sequentially from 1. We denote by P_n the set of all n-vertex triangulations. A *flip* of an edge in a triangulation is the operation of removing an edge, thereby forming a quadrilateral face, and adding the other diagonal of the face. (See Fig. 5.1.) A flip is allowed only if it does not introduce a multiple edge.

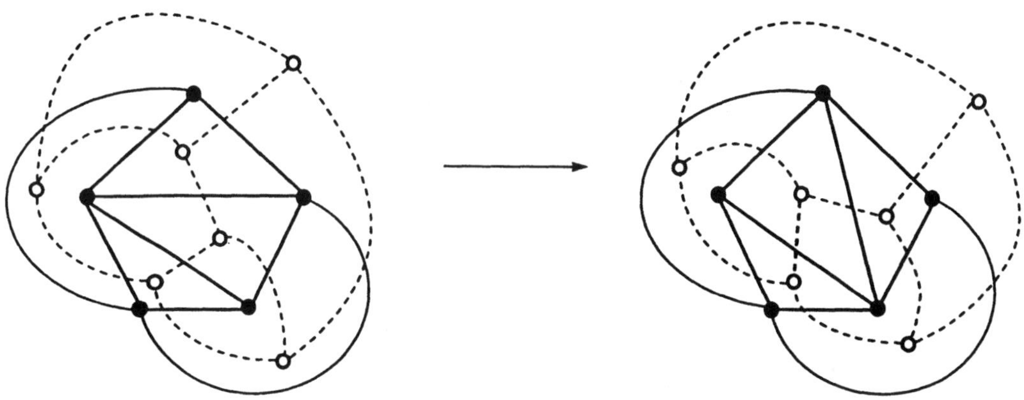

FIG. 5.1. *A flip in a triangulated graph and the corresponding operation in the dual graph.*

Let F_n be the minimum number of flips needed to convert any n-vertex triangulation into any other. We wish to estimate F_n. It is easy to establish that F_n is $O(n^2)$; Wagner [8] gave a construction. We show in §6 that F_n is $O(n \log n)$; in this section, we use our succinct encoding approach to prove that F_n is $\Omega(n \log n)$.

There is no upper bound on the degree of a vertex in a plane triangulation. Therefore, to apply our technique, we work in the space of planar graphs that are dual to plane triangulations. In such a graph, every vertex has degree 3. (Each vertex of the dual graph (a face in the original graph) maintains as a tag the set of vertex numbers of the vertices in the original graph to which it is incident. These tags along with the dual graph are sufficient to reconstruct the original numbered plane triangulation. This observation is required to get a reliable upper bound on the number of reachable numbered plane triangulations.) The edge-end labels of the initial graph are chosen arbitrarily, subject to the constraint that walking one step clockwise around a vertex increases the label by 1 (modulo 3). This ordering of the edge-end labels encodes the embedding of the plane triangulation.

There are several different ways to represent sequences of diagonal flips as derivations in a graph grammar. One way is shown in Fig. 5.2.

This method uses two productions: one for doing the flip, and the other for preparing the edge-end labels to allow the flip. Each flip in the original derivation may correspond to as many as five actions: two to cycle the edge-end labels on one end, two for the other end, and one for the actual flip. A sequence of m flips becomes

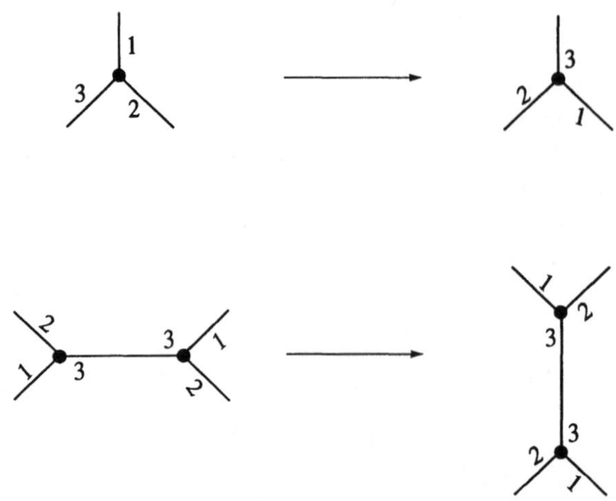

FIG. 5.2. *Two productions for representing flip sequences as graph grammar derivations.*

a sequence of as many as $5m$ actions. A plane triangulation of n vertices has $2n-4$ faces. Therefore the dual graphs in which the derivations take place have $2n-4$ vertices. The number of vertices in left sides of productions (c) is 3, and the number of vertices in the largest right side (r) is 2.

We can now apply Theorem 2.3 to bound the number of n node numbered plane triangulations reachable in m flips by $4^{2n+10m-4}$. This implies that, for any triangulation P, at most $4^{2n+10m-4}$ distinct triangulations can be obtained by doing m or fewer flips. Since P_n contains at least $(n-3)!$ triangulations (there are this many different sorted wheels; see §6), there must be at least two triangulations, and indeed many pairs of triangulations, that are $\Omega(n \log n)$ flips apart; that is, $F_n = \Omega(n \log n)$.

The bound on the number of reachable configurations can be tightened significantly by the use of a different graph grammar. This grammar is shown in Fig. 5.3.

Because this grammar includes each of the six ways that the ends of the edge to be flipped can be labeled, there is a one-to-one correspondence between diagonal flips in the plane triangulation and applications of one of the productions to the dual graph. Using the leader-follower trick and eliminating the zero label reduces the number of different labels to 9. Each production creates two new labels, so our improved encoding scheme proves that the number of graphs reachable in m moves is at most $9^{2n+2m-4}$.

Leader vertices can be avoided entirely. An encoding without leaders can be made to work by using the convention that a production involving a pair of adjacent vertices is ready if and only if their labels mutually point at each other. (That is, following the DIR(v) edge from v leads to w, and following the DIR(w) edge from w leads to v.) To verify that the zero label (indicating a terminal vertex) is not necessary, we must show that there exists a labeling of any planar graph of degree 3 with follower labels such that no pair of adjacent vertices point to each other. This can be done as follows. If the graph is a tree, choose a place in the middle of some edge, and make all the vertices point away from this. If the graph has a cycle, choose the labels of the vertices on the cycle to point consistently around it. Now choose a subset of the remaining edges, so that these edges plus the cycle form a subgraph with all of the vertices and exactly one cycle. (This is a spanning tree with one extra edge.) The

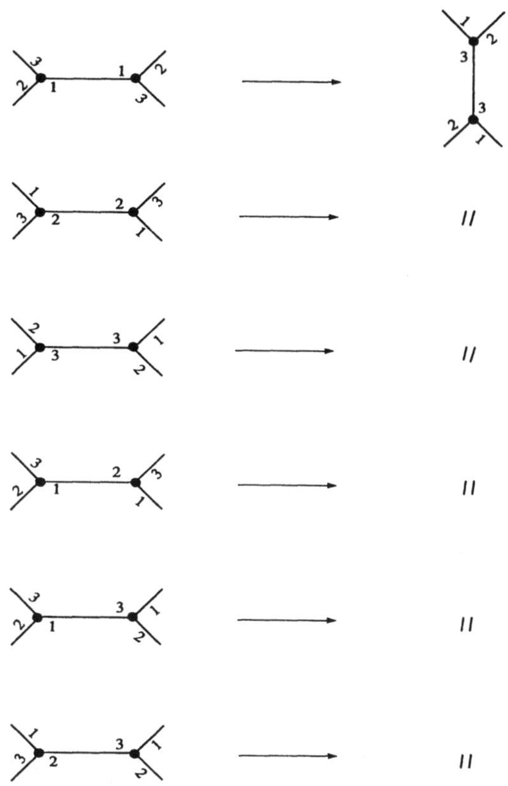

FIG. 5.3. *Six productions give a tighter bound on flip distance.*

follower label on a noncycle vertex points toward the cycle along the path in the tree. This gives the required match-free labeling. This argument bounds the number of reachable configurations by $3^{2n+2m-4}$.

The set of configurations reachable in m or fewer flips is not changed if we do not allow a sequence to make a flip, then immediately make another flip that cancels it out. This observation means that of the nine possible labelings of the pair of vertices resulting after a move, we can restrict our attention to eight of them. This improves the bound to $3^{2n-4}8^m$.

We summarize the results of this section in the following theorem.

THEOREM 5.1. *For any plane triangulation T of n vertices, the following hold:*

1. *The number of different plane triangulations obtainable by m or fewer flips starting from T is, at most, $3^{2n-4}8^m$;*
2. *There exists a plane triangulation T' such that the number of flips required to transform T into T' is $\Omega(n \log n)$.*

6. Numbered plane triangulations: An upper bound.

THEOREM 6.1. *Let G_1 and G_2 be two n-vertex numbered plane triangulations (with no multiple edges). If $n \geq 5$, then there is a sequence of $O(n \log n)$ diagonal flips that transforms G_1 into G_2 in such a way that there are no multiple edges in any* *intermediate state.*

Proof. We show that any such triangulation G can be transformed into a particular canonical form called a *sorted wheel* in $O(n \log n)$ diagonal flips. Using this transformation, we can transform G_1 into the sorted wheel, then transform the sorted wheel into G_2 (using the transformation in reverse).

A *wheel* of $n \geq 5$ vertices is a planar graph that has two special vertices called *hubs* and $n-2$ other vertices called *rim vertices*. There is an edge from each hub to each rim vertex (these are the *spokes*). There are $n-2$ other edges in the graph, and these form a simple cycle through all of the rim vertices. There is a unique way of embedding the wheel in a sphere.

A *sorted wheel* of n vertices is a wheel with labeled vertices embedded in the sphere. The hubs are labeled 1 and n, and the vertices of the rim are labeled $2, 3, \cdots, n-1$ in clockwise order when viewed from hub 1.

We first consider the special case of $n = 5$. Any graph G of five vertices satisfying the hypotheses of the theorem is a wheel. We show this by first applying Euler's formula, which implies that G must have six triangular faces and nine edges, and that the sum of the degrees of the vertices is 18. No vertex can have degree two, because then its two neighbors would be connected by two different edges, which violates the assumption that G has no multiple edges. Furthermore, no vertex can have degree greater than four. It follows that the multiset of the degrees of the vertices is $\{3, 3, 4, 4, 4\}$. The three vertices of degree four must be attached to all the other vertices in the graph. This accounts for all of the edges incident on the vertices of degree three, which therefore must not be neighbors. It follows that the graph is a wheel in which the vertices of degree three are the hubs, and the vertices of degree four are the rim.

We finish the proof for $n = 5$ in two stages. First, we show that we can make vertices 1 and 5 the hubs of the wheel. Second, we show that if the resulting structure is not the sorted wheel (it must be its mirror image), then it can be transformed into the sorted wheel.

If vertices 1 and 5 are on the rim, then a diagonal flip of the edge between them makes them the two hubs. If 1 is a hub and 5 is on the rim, then we flip the edge between the other two rim vertices creating a configuration where both 1 and 5 are on the rim, which we handle as above. A similar technique suffices if 5 is a hub and 1 is on the rim.

Figure 6.1 shows how the mirror image of the sorted wheel of five vertices is transformed into the sorted wheel by the application of five diagonal flips.

We are now ready to consider the case where $n \geq 6$. The transformation of the graph G into a sorted wheel is broken up into three phases: constructing a Hamiltonian circuit, transforming the Hamiltonian circuit into a wheel with hubs 1 and n, and sorting the rim of the wheel. These three steps are described in the following three sections.

6.1. Constructing a Hamiltonian circuit. By Tutte's theorem on planar graphs [7] (and by a theorem of Whitney [10]), any 4-connected planar graph has a Hamiltonian circuit. The graph G under consideration is 3-connected, since it is planar, triangulated, and has no multiple edges. Unfortunately, it may not be 4-connected. If it is not 4-connected, then it must have a *separating triangle*; that is, a triangle whose removal separates the graph. We show how to transform the given graph G into one that has no separating triangles by making $O(n)$ diagonal flips. This completes our construction of the Hamiltonian circuit.

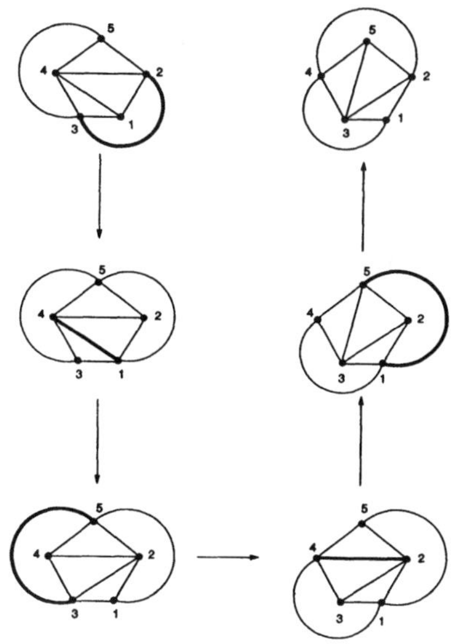

FIG. 6.1

The graph G is given to us embedded on a sphere. We choose a face arbitrarily and map the embedding on the sphere to an embedding on the plane such that the chosen face is infinite. Each separating triangle of G partitions the faces and remaining vertices of G into two components. The *interior* component is the one not containing the infinite face. Let I_1 be the set of faces interior to a separating triangle T_1, and let I_2 be the set of faces interior to a separating triangle T_2. Either I_1 and I_2 are disjoint, or satisfy $I_1 \subset I_2$ or $I_2 \subset I_1$. It follows from these relations that there must always be a set of innermost separating triangles, i.e., those that do not contain another separating triangle in their interior.

Our algorithm for eliminating separating triangles works from innermost separating triangles outward. A diagonal flip operation is applied to an edge of one of the innermost separating triangles. The chosen edge is any one that does not introduce a new separating triangle. We prove below that there always exists such an edge. It follows immediately that this algorithm eliminates all of the separating triangles in $O(n)$ diagonal flips because each flip reduces by at least one the number of edges that are in separating triangles.

It remains to show that there is always an edge of an innermost separating triangle such that if that edge is flipped then no new separating triangle is created. The following case analysis shows this. Consider an innermost separating triangle with vertices a, b, and c. Let d be the vertex inside the triangle such that triangle (a, b, d) is empty. (There must be such a vertex since triangle (a, b, c) is a separating triangle, and there must be something inside of it.) Similarly, there must be a vertex e outside of triangle (a, b, c) such that (a, b, e) is an empty triangle. Figure 6.2 shows the situation.

We assume that flipping edge (a, b) creates a new separating triangle and show that flipping one of the other edges does not create one. We know that the separating triangle that was created by flipping (a, b) must be (d, c, e), and that (d, c) and (c, e) are edges of the original graph. Triangles (a, d, c) and (b, d, c) must be empty;

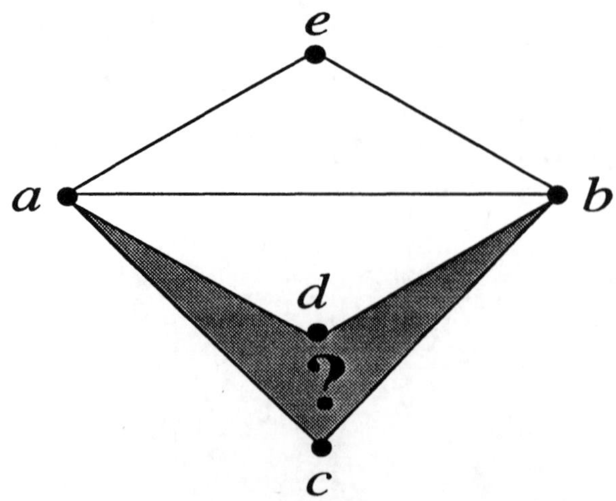

FIG. 6.2

otherwise, (a, b, c) would not be an innermost separating triangle. We now know that the structure of the graph near triangle (a, b, c) is as shown in Fig. 6.3.

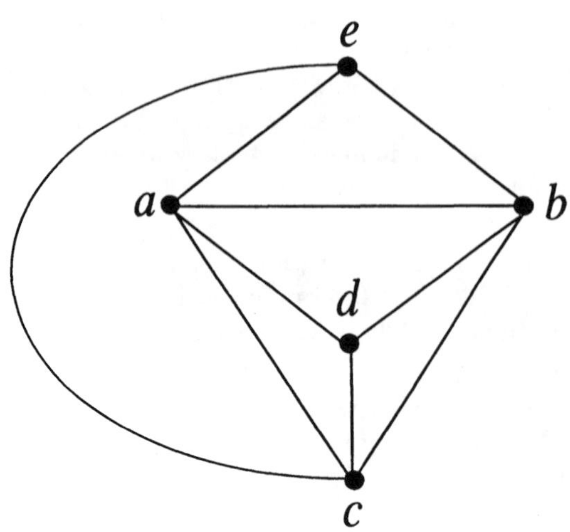

FIG. 6.3

Since the graph has at least six vertices, we know that there must be another vertex f outside of triangle (e, b, c) such that (b, f, c) is an empty triangle. Now it is clear that flipping edge (b, c) cannot create a separating triangle. This completes our construction of a Hamiltonian circuit.

6.2. Transforming the Hamiltonian circuit into a wheel with hubs 1 and n. Given that there is a Hamiltonian circuit, we can regard the graph as consisting of a cycle and two triangulations of an n-gon, one on each side of the cycle. By definition, a triangulation of a polygon has no interior vertices. A *coning* triangulation of a polygon is one in which all of the interior edges of the polygon are incident to the same vertex. We use several facts about diagonal flips in triangulations of a polygon [6].

FACT 1. *Any triangulation of an n-gon can be transformed into the coning triangulation with all edges incident on a vertex v by making at most $n-2$ diagonal flips, each of which increases the degree of v by one.*

FACT 2. *Any triangulation of an n-gon can be transformed into any other in at most $2n-4$ diagonal flips.*

FACT 3. *In any triangulation of an n-gon, there is a vertex v such that v is incident to only two edges, and those are the boundary edges that connect v to its two neighbors around the polygon.*

Call the two triangulations of the n-gon that comprise the current version of G the *top* triangulation and the *bottom* triangulation. Let v be a vertex such that Fact 3 holds for v in the top triangulation. Now we can apply Fact 1 to vertex v in the bottom triangulation to transform that triangulation into a coning triangulation to vertex v. This process will never introduce multiple edge because all the new edges added to the bottom side of the triangulation are incident to vertex v, which has no edges on the top side. The situation is depicted in Fig. 6.4.

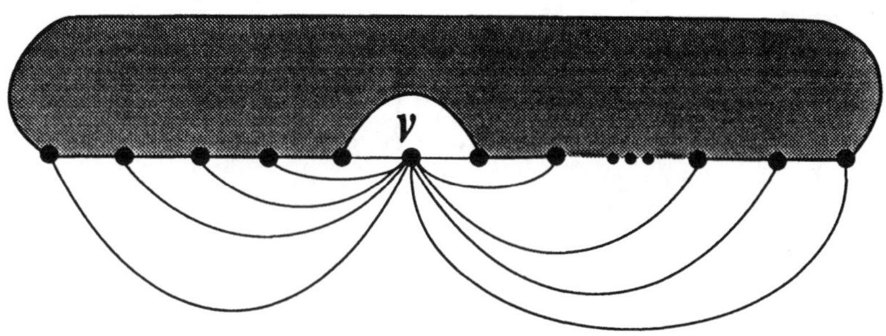

FIG. 6.4

We now change our definition of the top and bottom sides. We view vertex v as belonging to the interior of the bottom side, which is a hub with $n-1$ spokes connecting v to all other vertices. The top side becomes a triangulation of an $(n-1)$-gon. At least one of vertices 1 or n is on this $(n-1)$-gon. Without loss of generality, assume that 1 is on this $(n-1)$-gon. (If 1 is not on this polygon, then the following construction can be fixed by swapping the roles of n and 1.) Now we transform the triangulation of the $(n-1)$-gon (the top side) into a coning triangulation to vertex 1. The result is shown in Fig. 6.5.

We now flip edge $(v,1)$ and move 1 into the top side. The result is a wheel with hubs 1 and v, as shown in Fig. 6.6.

It remains to transform this wheel into one with vertices 1 and n as the hubs. If $v=n$, then we are done; otherwise, it only remains to replace v by n. We begin by flipping any edge around the rim of the wheel, resulting in the situation shown in Fig. 6.7.

Now we retriangulate the bottom $(n-1)$-gon, so that it is a coning to n. Then we flip edge $(n,1)$ and move n into the bottom half to give the triangulation shown

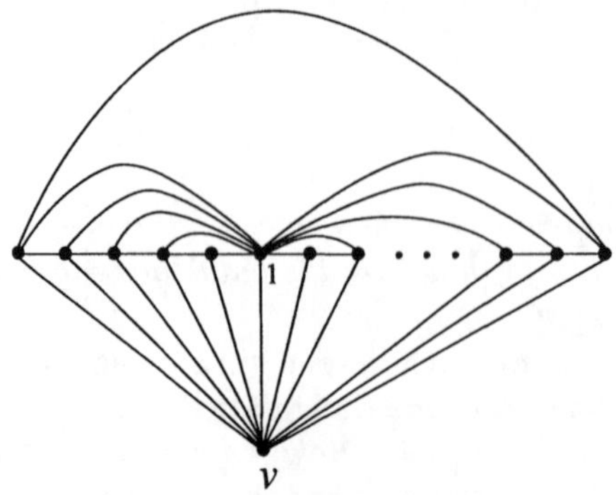

FIG. 6.5

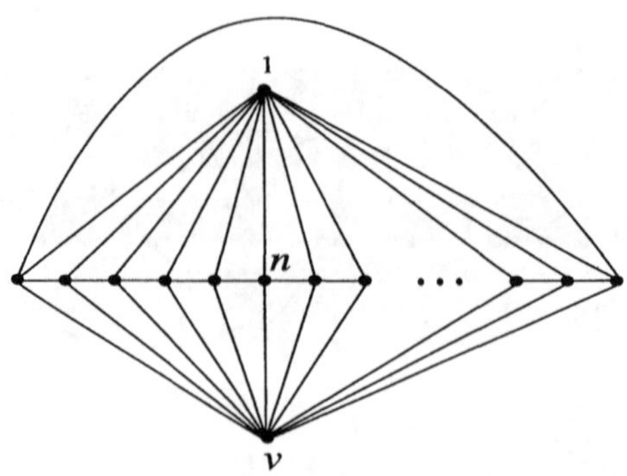

FIG. 6.6

in Fig. 6.8.

This construction works without creating multiple edges, as long as the rim of the wheel is always at least of size four. This is certainly the case, since $n \geq 6$.

6.3. Sorting the rim of the wheel. We first give a sequence of four flips that can exchange any pair of adjacent vertices around the rim of the wheel. Figure 6.9 shows this sequence. This works as long as the number of vertices on the rim is at least 4.

If $6 \leq n \leq 15$, we can use repeated transpositions to sort the wheel. We henceforth assume that $n \geq 16$. A *double wheel* is a wheel with two rims, as shown in Fig. 6.10.

The number of vertices in the *top rim* differs from the number in the *bottom rim* by at most one. Furthermore, all the edges in the region bounded by the two rims cross from one rim to the other.

We now show how to use $O(n)$ diagonal flips to transform a double wheel into a single wheel. We call this transformation a *merge step*. The merge allows us to form

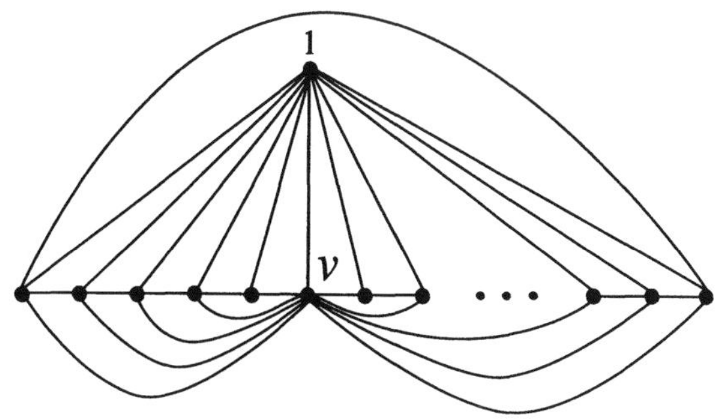

FIG. 6.7

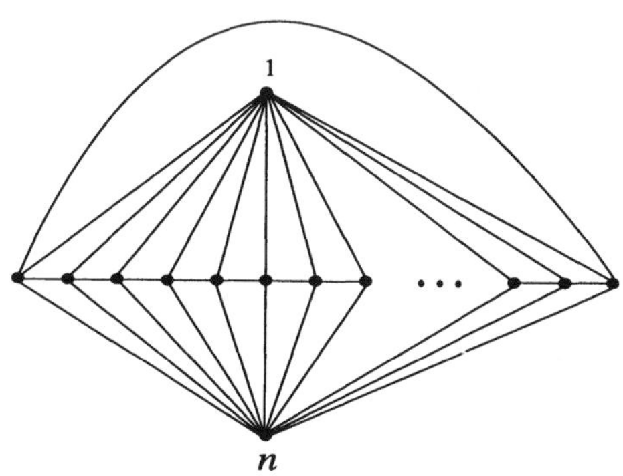

FIG. 6.8

any ordering of the vertices around the rim of the wheel subject to the constraint that the order is consistent with that defined by the orderings on the two rims of the double wheel. That is, if we traverse the rim of the wheel in clockwise order (from the point of view of, say, vertex 1), then the traversal encounters all the vertices that came from the bottom rim (top rim) of the double wheel in the same cyclic order in which they occurred in the bottom rim (top rim) of the double wheel. A more intuitive way to think of this process is to imagine two decks of cards (the double wheel) that are shuffled into one (the rim of the wheel). This is also analogous to the way a merge-sorting algorithm combines two sorted subfiles into a sorted file.

We can also apply the merge step in reverse (an *unmerge*) to split a wheel into a double wheel. A wheel can be sorted by applying a sequence of $\lceil \log_2(n-2) \rceil$ unmerge-merge pairs. (Observe that a merge sort can be implemented using these primitives. Each pass of the sorting algorithm through all of the data corresponds to one of the unmerge-merge pairs.)

It remains to show how to implement the merge step (never introducing multiple edges) in $O(n)$ diagonal flips. An edge (i, j) is called an *amicable edge* if (1) i is on

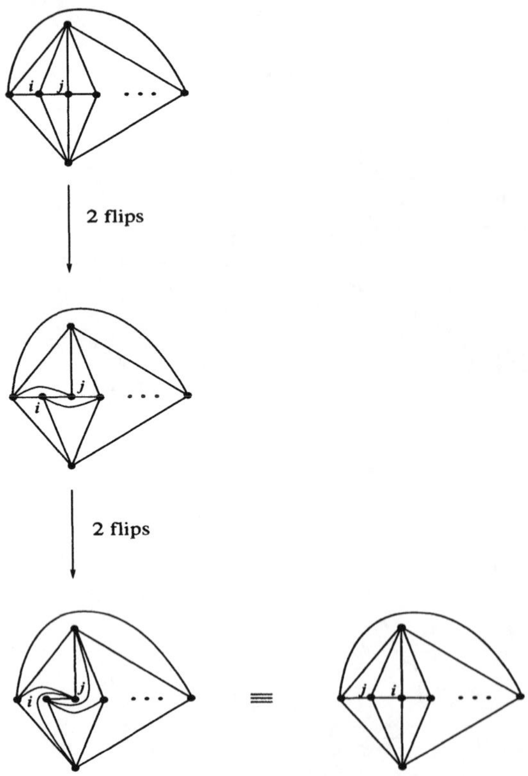

FIG. 6.9

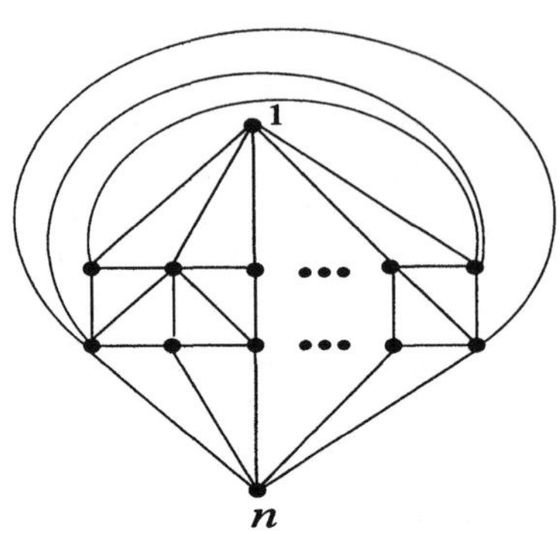

FIG. 6.10

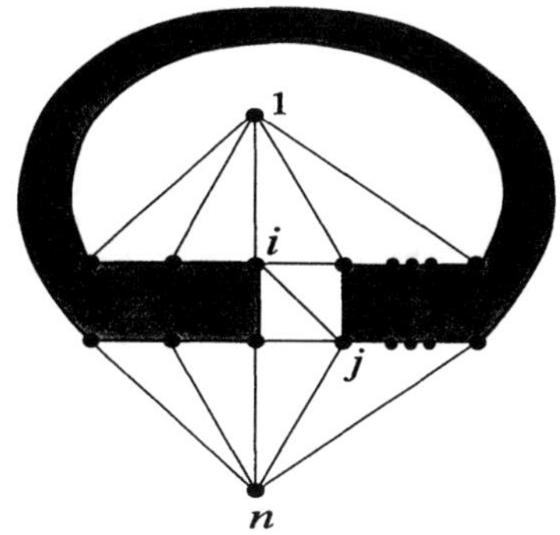

FIG. 6.11

one side of the rim of a double wheel and j is on the other side, (2) the quadrilateral obtained by removing edge (i, j) has one edge on each rim of the wheel and two edges crossing from one side of the rim to the other, and (3) the other vertex of the quadrilateral on the same side of the rim as i is counterclockwise from i (with respect to 1). Figure 6.11 shows an amicable edge (i, j).

In any double wheel there must be an amicable edge. By flipping three edges in the vicinity of an amicable edge, we can create an $(n-2)$-gon such that the edges on the outside of the polygon do not connect any pair of vertices of the polygon. When this operation is applied to the above diagram, the result is shown in Fig. 6.12.

It is now the case that we can apply any algorithm for retriangulating the $(n-2)$-gon between the two rims without fear of creating multiple edges.

We can apply this technique three times to transform a double wheel with one triangulation between the rims into a double wheel with any other triangulation between the two rims. Let (i, j) be an amicable edge of the initial double wheel. Let (k, l) be an amicable edge of the desired final triangulation. (These pairs of vertices may or may not be disjoint.) Let x and x_c be two neighbors on the top rim of the wheel such that x_c is a counterclockwise neighbor of x, and neither x nor x_c is i, or k, or a counterclockwise neighbor of either i or j. Since the length of the rim is at least 7, there must be such a pair. Define y and y_c similarly on the bottom rim.

To transform the triangulation between the rims from any one to any other, we first cut the double rim at amicable edge (i, j), as shown in Fig. 6.12. We then retriangulate the region between the two rims such that (x, y) is an amicable edge. Then we close up the cut of amicable pair (i, j) and open up the one for amicable edge (x, y). We then retriangulate the polygon between the rims, so that the pair (k, l) becomes an amicable edge. We then close up the cut at amicable edge (x, y) and open up the cut at pair (k, l). Now we triangulate the $(n-2)$-gon as specified by the desired final triangulation between the rims. Closing up the cut at amicable pair (k, l) completes the construction of the desired triangulation between the rims. This

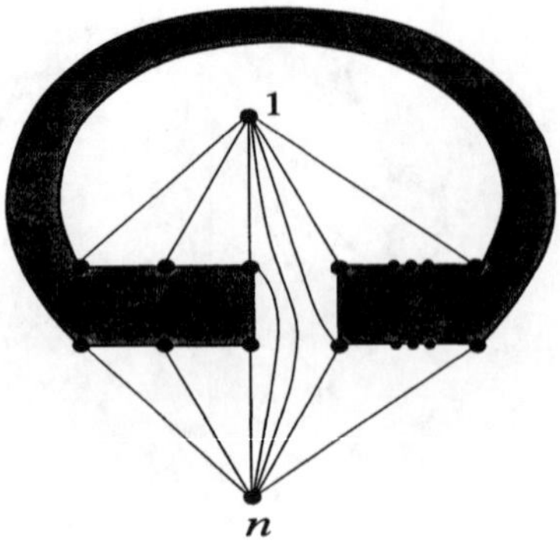

FIG. 6.12

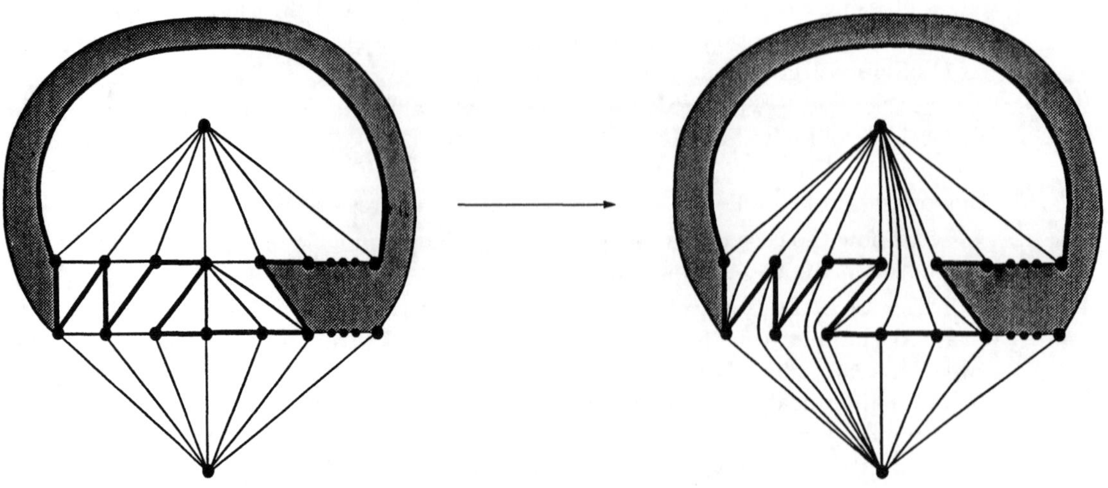

FIG. 6.13

process can never introduce any multiple edges, and it uses $O(n)$ diagonal flips.

As the desired triangulation between the rims, we choose any one such that there is an edge joining each pair of vertices that are adjacent on the rim of the desired wheel. It is easy to see that there must be such a triangulation between the rims.

The last step of the process is to convert such a double wheel into a single wheel. Figure 6.13 illustrates how this is done. The highlighted edges are those of the rim of the wheel.

This step does at most one flip for each vertex of the rim of the wheel and completes the merging process. This also completes the proof of the theorem. □

COROLLARY 6.2. *Let G_1 and G_2 be two n-vertex numbered plane triangulations (possibly with multiple edges). There exists a sequence of $O(n \log n)$ diagonal flips that transforms G_1 into G_2 in such a way that no flip ever creates a self-loop.*

Proof. An easy case analysis proves the result for $n = 4$. We show that multiple edges can be eliminated by flipping them one at a time. This takes only a linear number of flips, since each edge is flipped at most once. The corollary result follows by applying Theorem 6.1 to the graphs obtained by eliminating the multiple edges from G_1 and G_2.

We now prove our claim that if any multiple edge in a plane triangulation is flipped, the number of multiple edges is reduced. Let e_1 and e_2 be a pair of edges between vertices v and w in a plane triangulation. The cycle (e_1, e_2) divides the vertices (except v and w) into two disjoint sets: those on one side of the cycle and those on the other side. Neither of these two sets can be empty, since every face of the graph is a triangle. If either edge e_1 or e_2 is flipped, it is replaced by an edge that connects a vertex on one side of the cycle to one on the other side. Since before the flip there were no edges between vertices in these two sets (they are separated by a cycle), the edge created by the flip cannot be a multiple edge. □

REFERENCES

[1] K. CULIK AND D. WOOD, *A note on some tree similarity measures*, Inform. Process. Lett., 15 (1982), pp. 39–42.

[2] H. EHRIG, M. NAGL, G. ROZENBERG, AND A. ROSENFELD, EDS., *Graph-Grammars and Their Application to Computer Science*, Lecture Notes in Computer Science, 291, Springer-Verlag, Berlin, New York, 1987.

[3] J. R. DRISCOLL, N. SARNAK, D. D. SLEATOR, AND R. E. TARJAN, *Making data structures persistent*, J. Comput. System Sci., 38 (1989), pp. 86–124.

[4] D. E. KNUTH, *The Art of Computer Programming, Vol 1: Fundamental Algorithms*, Addison–Wesley, Reading, MA, 1968.

[5] ———, *The Art of Computer Programming, Vol 3: Sorting and Searching*, Addison–Wesley, Reading, MA, 1973.

[6] D. D. SLEATOR, R. E. TARJAN, AND W. P. THURSTON, *Rotation distance, triangulations, and hyperbolic geometry*, J. Amer. Math. Soc., 1 (1988), pp. 647–681.

[7] W. T. TUTTE, *A theorem on planar graphs*, Trans. Amer. Math. Soc., 82 (1956), pp. 99–116.

[8] K. WAGNER, *Bemerkungen zum Vierfarbenproblem*, J. Deutschen Math.–Verein., 46 (1936), pp. 26–32.

[9] E. WELZL, *Encoding graph derivations and implications for the theory of graph grammars*, Lecture Notes in Computer Science, 172, Springer-Verlag, Berlin, New York, 1984, pp. 503–513.

[10] H. WHITNEY, *A theorem on graphs*, Ann. Math., 32 (1931), pp. 378–390.

AUTOMATIC MESH PARTITIONING

GARY L. MILLER *, SHANG-HUA TENG †, WILLIAM THURSTON ‡,
AND STEPHEN A. VAVASIS §

Abstract This paper describes an efficient approach to partitioning unstructured meshes that occur naturally in the finite element and finite difference methods. This approach makes use of the underlying geometric structure of a given mesh and finds a provably good partition in random $O(n)$ time. It applies to meshes in both two and three dimensions. The new method has applications in efficient sequential and parallel algorithms for large-scale problems in scientific computing. This is an overview paper written with emphasis on the algorithmic aspects of the approach. Many detailed proofs can be found in companion papers.

Keywords: Center points, domain decomposition, finite element and finite difference meshes, geometric sampling, mesh partitioning, nested dissection, radon points, overlap graphs, separators, stereographic projections.

1. Introduction. Many large-scale problems in scientific computing are based on unstructured meshes in two or three dimensions. Examples of such meshes are the underlying graphs of finite volume methods in computational fluid dynamics or graphs of the finite element and finite difference methods in structural analysis. These meshes may have millions of nodes. Quite often the mesh sizes used are determined by the memory available on the machine rather than the physics of the problem to be solved. Thus, the larger the memory the larger the mesh used and, hopefully, the better the simulation produced.

The main goal of this paper is to describe our work on how and under what conditions unstructured meshes will have partitions into two roughly equal sized pieces with a small boundary (called small separators to be defined later). When these partitions exist they have several important applications to the finite element and finite difference methods. We list some of them here.

One approach to achieving the large memory and computation power requirements for large-scale computational problems is to use massively parallel distributed-memory machines. In such an approach, the underlying computational mesh is divided into submeshes, inducing a subproblem to be stored on each processor in the parallel system and boundary information to communicated [67]. To fully utilize a massively parallel machine, we need a subdivision in which subproblems have approximately equal size and the amount of communication between subproblems is relatively small. This approach will decrease the time spent per iteration. There are also methods

* School of Computer Science, Carnegie Mellon University, Pittsburgh, PA 15213. Supported in part by National Science Foundation grant CCR-9016641.

† Xerox Corporation, Palo Alto Research Center, Palo Alto, CA 94304. Part of the work was done while the author was at Carnegie Mellon University. Current address: Department of Mathematics, Massachusetts Institute of Technology, Cambridge, MA 02139.

‡ Department of Mathematics, University of California, Berkeley CA 94720.

§ Department of Computer Science, Cornell University, Ithaca, NY 14853. Supported by an NSF Presidential Young Investigator award. Revision work on this paper was supported by the Applied Mathematical Sciences program of the U.S. Department of Energy under contract DE-AC04-76DP00789 while the author was visiting Sandia National Laboratories.

which use good partitioning to either decrease the number of iterations used or the time used by direct methods.

Several numerical techniques have been developed using the partitioning method to solve problems on a parallel system. Examples include *domain decomposition* and *nested dissection.* Domain decomposition divides the nodes among processors of a parallel computer. An iterative method is formulated that allows each processor to operate independently. See Bramble, Pasciak and Schatz [11], Chan and Resasco [13], and Bjørstad and Widlund [9]. *Nested dissection* is a divide-and-conquer node ordering for sparse Gaussian elimination, proposed by George [34] and generalized by George and Liu [36] and Lipton, Rose and Tarjan [49]. Nested dissection was originally a sequential algorithm, pivoting on a single element at a time, but it is an attractive parallel ordering as well because it produces blocks of pivots that can be eliminated independently in parallel. Parallel nested dissection was suggested by Birkhoff and George [8] and has been implemented in several settings [12, 21, 35, 84]; its complexity was analyzed by Liu [52] (for the regular square grid) and Pan and Reif [63] (in the general case).

Vaidya has produced results which indicate that the quality of good preconditioners may also be linked to the existence of good partitions [78].

Therefore, one of the key problems in solving large-scale computational problems on a parallel machine is the question of how to partition the underlying meshes in order to reduce the total communication cost and to achieve load balance.

If a mesh has a sufficiently regular structure, then it is easy to decide in advance how to distribute it among the processors of a parallel machine. However, meshes of many applications are irregular and unstructured, making the partition problem much more difficult. In general, there are meshes in three dimensions which have no small partition [59]. These examples are not the type that would naturally arise in the finite element methods, but they are meshes. One important goal is to understand which meshes do and which do not have small partitions.

Various heuristics have been developed and implemented [65, 68, 82]. However, none of the prior mesh partitioning algorithms is both efficient in practice and provably good, especially for meshes from three dimensional problems. Leighton and Rao [46] have designed a partitioning algorithm based on multicommodity flow problems, which finds a separator that is optimal within logarithmic factors. But their algorithm runs in superlinear time and it remains to be seen if it could be used in practice for large-scale problems.

1.1. A new method. In a series of papers, the authors (Vavasis [81]; Miller and Thurston [59]; Miller and Vavasis [60]; Miller and Teng [55]; Miller, Teng, and Vavasis [56]) have developed an efficient and provably good mesh partitioning method. This overview paper describes this new approach. It is written with emphasis on the algorithmic aspects of the approach. Many detailed proofs can be found in companion papers [57, 58].

This method applies to meshes in both two and three dimensions. It is based on the following important observation: graphs from large-scale problems in scientific computing are often defined geometrically. They are meshes of elements in a fixed dimension (typically two and three dimensions), that are *well shaped* in some sense, such as having elements of bounded aspect ratio or having elements with angles that are not too small. In other words, they are graphs embedded in two or three dimensions that come with natural geometric coordinates and with structures.

Our approach makes use of the underlying geometric structure of a given mesh and finds a provably good partition efficiently. The main ingredient of this approach is a novel geometrical characterization of graphs embedded in a fixed dimension that have a small *separator*, which is a relatively small subset of vertices whose removal divides the rest of the graph into two pieces of approximately equal size. By taking advantage of the underlying geometric structure, we also develop an efficient algorithm for finding such a small separator.

In contrast, all previous separator results (see Section 1.2) are combinatorial in nature. They not only characterize the small separator property combinatorially, but also find a small separator based only on the combinatorial structure of the given graph. When applied to unstructured geometric meshes, they simply discard the valuable geometric information. The result has been that they are either too costly to use or they do not find a separator as good as it should be. Worst of all, none of the earlier separator results is useful for graphs in three dimensions.

1.2. Separators and earlier work. DEFINITION 1.1 (SEPARATORS). *A subset of vertices C of a graph G with n vertices is an $f(n)$-**separator** that δ-**splits** if $|C| \leq f(n)$ and the vertices of $G - C$ can be partitioned into two sets A and B such that there are no edges from A to B, $|A|$, $|B| \leq \delta n$, where f is a function and* $0 < \delta < 1$.

Two of the most well-known families of graphs that have small separators are trees and planar graphs. Every tree has a single vertex separator that 2/3-splits [44]. Lipton and Tarjan [50] proved that every planar graph has a $\sqrt{8n}$-separator that 2/3-splits. Their result improved an earlier one by Ungar [77]. Some extensions of their work have been made [19, 31, 32, 54], and separator theorems have also been obtained for graphs with bounded genus [38, 43] and graphs with bounded excluded minor [2]. In particular, Gilbert, Hutchinson, and Tarjan showed that all graphs with genus bounded by g have an $O(\sqrt{gn})$-separator, and Alon, Seymour, and Thomas proved that all graphs with an excluded minor isomorphic to the h-clique have an $O(h^{3/2}\sqrt{n})$-separator.

Interestingly, all the characterizations above are combinatorial, not geometric, as are their proofs!

Separator results for families of graphs closed under the subgraph operation immediately lead to divide-and-conquer recursive algorithms for many applications. In general, the efficiency of such algorithms depends on δ being bounded away from 1 and $f(n)$ being a slowly-growing function.

Perhaps the most classical application of small separator results is nested dissection, a widely used technique for solving a large class of sparse linear systems. This approach was pioneered by George [34], who designed the first $O(n^{1.5})$-time nested dissection algorithm for linear systems on regular grids using the fact that the $\sqrt{n} \times \sqrt{n}$ grid has a $\sqrt{n}$-separator. His result was extended to planar linear systems by Lipton, Rose, and Tarjan [49]. Gilbert and Tarjan [40] examined several variants of the nested dissection algorithms. It has been demonstrated, in theory and in practice, that nested dissection can be implemented efficiently in parallel.

In the analysis of sparse matrix algorithms, *a priori* upper bounds on operation counts are rare in the literature (aside from the trivial dense-matrix upper bounds). The major exception is nested dissection. The *a priori* bounds attained by nested dissection, which in many cases are asymptotically the best possible, always depend on the associated bounds of the underlying graph-separator algorithm. This means that a careful analysis of separator sizes is an important aspect of nested dissection.

Small separator results have found fruitful applications in VLSI design (Leiserson [47]; Leighton [45]; Valiant [79]) and efficient message routing (Fredrickson and Janardan [28]). They have also been used in proving several complexity-theoretic results (Paterson [62]; Lipton and Tarjan [51]), and have been used to design efficient graph algorithms such as parallel construction of breadth-first-search trees (Pan and Reif [63]), testing graph isomorphism (Gazit [33]), and approximating NP-complete problems (Lipton and Tarjan [51]).

1.3. Outline of the paper. Section 2 defines a new class of geometric graphs, the overlap graphs, and describes our main separator theorem. This class has a simple definition and contains many important classes of graphs as special cases. Section 3 studies meshes from the finite element and finite difference methods. We show that overlap graphs include "well shaped" meshes. We also show that planar graphs are a special case of overlap graphs in two dimensions. Section 4 presents a partitioning algorithm for overlap graphs. The algorithm first uses the geometric information of the input graph to find a "continuous" separator, then uses the combinatorial structure to compute a "discrete" separator from its continuous counterpart. The central step of the algorithm is to find a *center point* of a point set in a fixed dimensions, where a center point is a point such that every hyperplane passing through it about evenly divides the point set. We show that center points always exist and can be computed in polynomial time using linear programming. We also show that the step of computing a "discrete" separator from a continuous one can be performed in linear time. Section 5 introduces *geometric sampling*, a technique that reduces the problem size and simultaneously guarantees a provably good approximation of the larger problem. Using geometric sampling, we can compute an approximate center point in random constant time and find a "good" separator of an overlap graph in random linear time. We further give a practical heuristic for approximating a center point. We then extend the partitioning algorithm for unstructured meshes. Section 6 gives the proof outline of the main separator theorem. We demonstrate how to use geometric arguments to prove separator properties for graphs embedded in fixed dimension. Section 7 summarizes the paper and gives open questions.

2. Neighborhood systems and overlap graphs. Our geometric characterization of graphs that have small separators is based on the following elementary concept.

2.1. Neighborhood systems. DEFINITION 2.1. *Let* $P = \{p_1, \ldots, p_n\}$ *be points in* $\mathbb{R}^d$. *A* **k-ply neighborhood system** *for* P *is a set,* $\{B_1, \ldots, B_n\}$*, of closed balls* *such that (1) B_i is centered at p_i and (2) no point $p \in \mathbb{R}^d$ is strictly interior to more than k balls from B.*

A 3-ply neighborhood system in two dimensions is illustrated in Figure 1.

FIG. 1. *A 3-ply neighborhood system*

The following notation will be used throughout this paper. For each positive real α, if B is a ball of radius r in $\mathbb{R}^d$, then $\alpha \cdot B$ denotes the ball with the same center as B but radius αr.

We now state an important property of neighborhood systems [58].

LEMMA 2.2 (BALL INTERSECTION). *Suppose $\{B_1, \ldots, B_n\}$ is a k-ply neighborhood system in $\mathbb{R}^d$. For each d-dimensional ball B with radius r, for all constant $\beta : 0 < \beta \leq 1$,*

$$|\{i : B_i \cap B \neq \emptyset \text{ and } r_i \geq \beta r\}| \leq \beta^{-d} 3^d k,$$

where r_i is the radius of B_i.

2.2. Overlap graphs. DEFINITION 2.3. *Let $\alpha \geq 1$ and let $\{B_1, \ldots, B_n\}$ be a k-ply neighborhood system for $P = \{p_1, \ldots, p_n\}$. The (α, k)-**overlap graph** for the k-ply neighborhood system $\{B_1, \ldots, B_n\}$ is the undirected graph with vertices $V = \{1, \ldots, n\}$ and edges*

$$E = \{(i,j) : (B_i \cap (\alpha \cdot B_j) \neq \emptyset) \textbf{ and } ((\alpha \cdot B_i) \cap B_j \neq \emptyset)\}.$$

For simplicity, we call a $(1, k)$-overlap graph a *k-intersection graph.* In the case that $\alpha = 1$ and $k = 1$, and no two balls in the neighborhood system have a common point in their interior, we have the family of graphs known as *sphere-packings*; this interesting class of graphs will be discussed in the next section.

2.3. Main separator theorem. THEOREM 2.4 (MAIN). *Let G be an (α, k)-overlap graph for some fixed d. Then G has an*

$$O\left(\alpha \cdot k^{\frac{1}{d}} \cdot n^{\frac{(d-1)}{d}} + q(\alpha, k, d)\right)\text{-separator}$$

that $(d+1)/(d+2)$-splits. Furthermore, such a separator that $(d+1+\epsilon)/(d+2)$-splits can be computed in random linear time sequentially and in random constant time, using n processors, for any $1/n^{1/2d} < \epsilon < 1$.

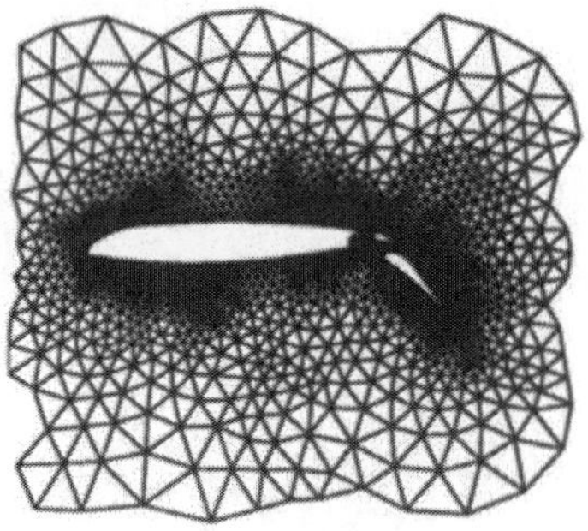

FIG. 2. *Airplane wing (Barth and Jespersen)*

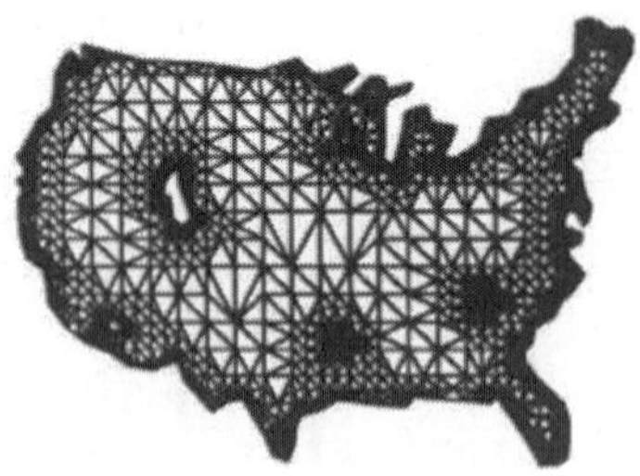

FIG. 3. *US map*

The function $q(\alpha, k, d)$ depends exponentially on d but is independent of n. Since the interesting cases are when $d = 2$ or $d = 3$ and when n is large, this term should be considered low order.

It has been shown in the companion paper [58] that the bound of Theorem 2.4 is tight up to a constant factor. Section 6 will outline the proof of this theorem.

3. Finite element and finite difference meshes. One important aspect that distinguishes a finite element or finite difference mesh from a regular graph is that it has two structures: the combinatorial structure and the geometric structure. In general, it can be represented by a pair (G, xyz) where G describes the combinatorial structure of the mesh and xyz gives the geometric information.

3.1. Meshes from the finite element method. The finite element method is a collection of numerical techniques for approximating a continuous problem by a finite structure [69]. To approximate a continuous function, the finite element method subdivides the domain (a subset of $\mathbb{R}^d$) into a *mesh* of polyhedral *elements* (Figures 2 and 3), and then approximates the continuous function by a piecewise polynomial on the elements.

A common choice for an element in the finite element method is a d-dimensional *simplex*, which is the convex hull of $(d+1)$ affinely independent points in $\mathbb{R}^d$, e.g., a triangle in two dimensions and a tetrahedron in three dimensions. A d-dimensional *simplicial complex* is defined to be a collection of d-dimensional simplices that meet only at shared faces [6, 7, 59]. So a 2-dimensional simplicial complex is a collection of triangles that intersect only at shared edges and vertices.

For most applications, a mesh is given as a list of its elements, where each element is given by the information describing the hierarchical structure of the elements, its lower dimensional structures such as its faces, edges, and vertices. Moreover, each vertex has geometric coordinates in two or three dimensions.

Associated with each simplicial complex is a natural graph, its 1-skeleton. For example, the 1-skeleton of a 2-dimensional simplicial complex is a planar graph. Conversely, every planar graph can be embedded in the plane such that each edge is mapped to a straight line segment (Fáry [25]; Tutte [74, 75]; Thomassen [71]; Fraysseix, Pach, and Pollack [27]).

In the finite element method, a linear system is defined over a mesh, with variables representing physical quantities at the nodes. Let *finite element graph* refer to the nonzero structure of the coefficient matrix of such a linear system. In the case of linear finite elements based on a triangulation, such as in Figures 2 and 3, the nodes of the finite element graph are exactly the nodes of the mesh, and hence the finite element graph is the same as the 1-skeleton of the simplicial complex. In the case of higher-order elements, the finite element graph usually contains the 1-skeleton as a proper subset. It can be obtained from the finite element mesh as follows: Identify certain points (vertices, points on edges, points in faces, and points in elements) as "nodes." Add edges between every pair of nodes that share an element.

To properly approximate a continuous function, in addition to the conditions that a mesh must conform to the boundaries of the region and be fine enough, each individual element of the mesh must be *well shaped.* A common shape criterion for elements is the condition that the angles of each element are not too small, or the aspect ratio of each element is bounded [6, 29].

Several definitions of the *aspect ratio* have been used in literature. We list some of them.

1. The ratio of the longest dimension to the shortest dimension of the simplex S, denoted by $A_1(S)$. For a triangle in $\mathbb{R}^2$, it is the ratio of the longest side divided by the altitude from the longest side.
2. The ratio of the radius of the smallest containing sphere to the radius of the inscribed sphere of S, denoted by $A_2(S)$.
3. The ratio of the radius of the circumscribing sphere to the radius of the

FIG. 4. *Meshes derived from quadtrees (S. Mitchell)*

inscribed sphere of S, denoted by $A_3(S)$.

4. The ratio of the diameter to the dth root of the volume of the simplex S, denoted by $A_4(S)$, where the *diameter* of a d-simplex S is the maximum distance between any pair of points in S.

Examples of a simplicial complex with bounded aspect ratio are illustrated in Figure 4 as well as in Figures 2 and 3. The above definitions of the aspect ratio are polynomially related to each other. They are also closely related to the *smallest angle* of the simplex, which is the smallest angle among all the angles between pairs of supporting hyperplanes of the simplex. Using elementary geometric arguments, one can prove the following set of inequalities: There are constants $c_1 < c_2$, c_3, $c_4 < c_5$, depending only on d, such that if the smallest angle of S is θ, then

$$\begin{aligned}
\frac{1}{|\sin\theta|} &\leq A_1(S) \leq \frac{2}{|\sin\theta|} \\
c_1 A_1(S) &\leq A_2(S) \leq c_2 A_1(S) \\
A_1(S) &\leq A_3(S) \leq c_3(A_1(S))^2 \\
c_4 A_1(S) &\leq (A_4(S))^d \leq c_5(A_1(S))^{d-1}.
\end{aligned}$$

Therefore, if one of the above parameters is bounded by a constant, then all of them are bounded.

3.2. Graphs from the finite difference method. The finite difference method is another useful technique for solving computational problems in scientific computing. It also uses a finite and discrete structure, *a finite difference mesh*, to approximate a continuous problem.

Finite difference meshes are often produced by inserting a uniform grid of $\mathbb{R}^2$ or $\mathbb{R}^3$ into the domain via a boundary-matching conformal mapping. In general, the derivative of the conformal transformation must be slowly varying with respect to the mesh size in order to produce good results. See, for example [72]. This means that the mesh will probably satisfy a *density condition* [5, 60].

Let G be an undirected graph and let π be an embedding of its nodes in $\mathbb{R}^d$. We say π is an embedding of *density* α if the following inequality holds for all vertices v in G. Let u be the closest node to v. Let w be the farthest node from v that is

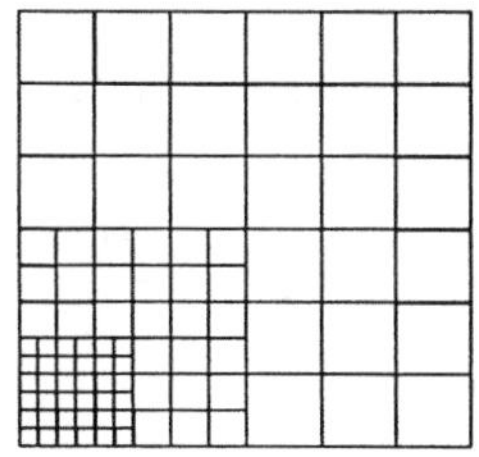

FIG. 5. *Berger and Bokhari's example of a density graph.*

connected to v by an edge. Then

$$\frac{||\pi(w) - \pi(v)||}{||\pi(u) - \pi(v)||} \leq \alpha.$$

In general, G is an *α-density graph* in $\mathbb{R}^d$ if there exist an embedding of G in $\mathbb{R}^d$ with density α. It can be easily shown that there is a $\Delta(\alpha, d)$ depending only on α and d such that the maximum degree of an α-density graph is bounded by $\Delta(\alpha, d)$.

Furthermore, a finite difference mesh may not be a collection of simplices or elements as a finite element mesh, so we can not analyze it as a triangulation. Finite difference meshes are often locally refined by further subdividing some mesh cells. See Figure 5. This means that nodes occur on the sides of some mesh cells and that interpolation must be used in the finite difference approximation. For numerical accuracy of the interpolation, the usual practice is that mesh cells are refined to a level no more than a constant factor smaller than their neighboring cells [5]. In the presence of such refinement, the finite difference mesh will still satisfy a constant density condition.

3.3. Overlap graphs and well shaped meshes. One of the most valuable aspects of the class of overlap graphs is that it enables us to give a unified geometric characterization of graphs with the small separator property. The set of overlap graphs in $\mathbb{R}^d$ contains all finite subgraphs of infinite grids, planar graphs and sphere packing graphs. Moreover, overlap graphs include graphs associated with finite element and finite difference methods, as special cases. The parameter α, in a strong sense, measures the degree to which the mesh is well-shaped.

We now show that for each well-shaped mesh, there is an overlap graph with a pair α and k that contains the graph defined by (G, xyz) as a subgraph. We say a graph G_1 is a *spanning subgraph* of another graph G_2 if G_1 can be obtained from G_2 by deleting edges. A graph G is *(α, k)-embeddable* in $\mathbb{R}^d$ if it is a spanning subgraph of an (α, k)-overlap graph in $\mathbb{R}^d$. Notice that the small separator property is preserved under spanning subgraphs.

LEMMA 3.1. *If G is an α-density graph in $\mathbb{R}^d$, then G is $(2\alpha, 1)$-embeddable.*

Proof: Let π be an embedding of G with density α in $\mathbb{R}^d$. Without loss of generality, assume that G has vertex set $V = \{1, 2, ..., n\}$. Let $P = \{\pi(1), \pi(2), ..., \pi(n)\}$. For each $p \in P$, let $c(p)$ denote the point of $P - \{p\}$ closest to p. Let $\Gamma = \{B_1, \ldots, B_n\}$, where for each $i : 1 \leq i \leq n$, B_i is a ball centered at $\pi(i)$, whose radius r_i is $0.5||c(p) - p||$. Clearly, balls from Γ do not intersect each other. We claim that G is a subgraph of the 2α-overlap graph G' of Γ.

For each edge (u, v) of P, we need to show that $(\pi(u), \pi(v))$ is an edge of G'. Without loss of generality, assume $r_u \leq r_v$. Because π is an α-embedding of G, we have

$$\frac{||\pi(u) - \pi(v)||}{2r_u} \leq \alpha.$$

So $\pi(v) \in (2\alpha) \cdot B_u$, and therefore $(\pi(u), \pi(v))$ is an edge of G', completing the proof. □

LEMMA 3.2. *Suppose G is the 1-skeleton of a simplicial complex K in $\mathbb{R}^d$. Let $\tilde{G}$ be the subgraph of G obtained by removing all vertices that appear on the external boundary of K. Then if the aspect ratio of K is bounded by a constant α, then there is a constant c depending only on d and α such that $\tilde{G}$ is $(c, 1)$-embeddable.*

Proof: Because the aspect ratio of the complex K is bounded by α, there is a Θ depending only on α and d such that the smallest angle of each simplex of K is at least Θ. Therefore, there is a constant Δ depending only on d and α, such that the degree of G is bounded by Δ. Furthermore, the bounded aspect ratio implies that the ratio of the longest edge over the shortest edge of any simplex in K is also bounded. Thus, there is a constant c_1 depending only on d and α, such that for each vertex v of K, the ratio of the longest edge to the shortest edge connecting v is bounded by c_1. Moreover, for each vertex v of K that is not on the external boundary, the closest vertex of v in K is also connected with v by an edge, because no vertex can appear in the interior of an simplex. Therefore, if we remove all vertices of K on the external boundary, we obtain a c_1-density graph $\tilde{G}$. Let $c = 2c_1$. It follows from Lemma 3.1 that $\tilde{G}$ is $(c, 1)$-embeddable. □

Therefore, the following theorem follows from Theorem 2.4.

THEOREM 3.3.

- *If G is the 1-skeleton of a simplicial complex K with bounded aspect ratio, letting $\tilde{n}$ be the number of exterior vertices of K, then G has an*

$$O\left(n^{\frac{(d-1)}{d}} + \tilde{n}\right)\textit{-separator.}$$

- *If G is a graph with bounded density α, then G has an*

$$O\left(\alpha \cdot n^{\frac{(d-1)}{d}}\right)\textit{-separator.}$$

Moreover, such separators that $(d + 1 + \epsilon)/(d + 2)$-split can be computed in random $O(n)$ time and in random $O(1)$ time using n processors, for any $1/n^{1/2d} < \epsilon < 1$.

An algorithm for partitioning finite element and difference meshes is given in Section 5.4.

3.4. Overlap graphs include all planar graphs. The proof that planar graphs are a special case of overlap graphs relies on the following theorem of Andreev and Thurston [3, 4, 73] characterizing all planar graphs in a novel geometric fashion.

THEOREM 3.4 (ANDREEV AND THURSTON). *Each triangulated planar graph G is isomorphic to a 2-dimensional sphere packing graph.*

Simply from the definition, each sphere packing graph is a $(1,1)$-overlap graph. Therefore, planar graphs are a special case of overlap graphs.

4. A randomized partitioning algorithm. In this section, we describe a randomized algorithm for computing a small separator of a given overlap graph. In the next section, we shall show how to make it efficient by using geometric sampling. In Section 6, we will outline a correctness proof of the algorithm to derive a constructive proof of Theorem 2.4.

4.1. The algorithm. Given a k-ply neighborhood system $\Gamma = \{B_1, \ldots, B_n\}$ and the α-overlap graph G of Γ, let $P = \{p_1, \ldots, p_n\}$ be the centers of Γ. The following algorithm first finds a $(d-1)$-sphere S with some desired properties to be stated later and then computes a vertex separator of G from S. In the following algorithm, let U_d be the unit d-sphere in $\mathbb{R}^{d+1}$. We define $ST : \mathbb{R}^d \to U_d$ to be the standard stereographic projection mapping. This mapping can be described as follows. Assume $\mathbb{R}^d$ is embedded in $\mathbb{R}^{d+1}$ as the $x_{d+1} = 0$ coordinate plane, and assume U_d is also embedded in $\mathbb{R}^{d+1}$ centered at the origin. Given a point p in $\mathbb{R}^d$, construct the line L in $\mathbb{R}^{d+1}$ passing through p and through the north pole of U_d (that is, the point $(0, .., 0, 1)$). Line L must pass through one other point q of U_d; we define $ST(p)$ to be q. For a set $P = \{p_1, \ldots, p_n\}$ in $\mathbb{R}^d$, we denote $\{ST(p_1), ST(p_2), ..., ST(p_n)\}$ by $ST(P)$. Recall that the center point of a point set is the one such that every hyperplane passing through it about evenly divides the point set. We will define center point formally in Section 4.2.

Algorithm 1 (Generic Geometric Partitioning)
Input: (a neighborhood system Γ and the geometric coordinates of its centers P).
1. Compute $Q = ST(P)$;
2. Find a δ-center point c of Q;
3. Compute the rotation $\pi_1 : U_d \to U_d$ that maps c to c_1, a point on the diameter between the south and the north poles, say $c_1 = (0, ..., 0, r)$.
4. Let π_2 be the dilation of $\mathbb{R}^d$ by a factor of $\sqrt{(1-r)/(1+r)}$.
5. Choose a random great circle GC of U_d;
6. Transform GC back $\mathbb{R}^d$ using the inverse of the above transformations to obtain a separating sphere S, i.e., $S = [ST \circ \pi_2 \circ ST^{-1} \circ \pi_1 \circ ST]^{-1}(GC)$;
7. Compute a vertex separator of G from S.

Algorithm 1 defines some point sets that are not explicitly computed. We introduce them below only for the purpose of explaining the algorithm.

- Let $Q_1 = \pi_1(Q)$ in Step 3 above;
- Let $P_1 = ST^{-1}(Q_1)$, the pre-image of Q_1 in $\mathbb{R}^d \cup \{\infty\}$. The pre-image of the north pole is defined to be a point at infinity;
- Let $P_2 = \pi_2(P_1)$ in Step 4;

- Let $Q_2 = ST(P_2)$. Note that the origin $(0, 0, ..., 0)$ is a center point of Q_2. See further comments below.

4.2. Center points. Suppose P is a finite set of points in $\mathbb{R}^d$. A hyperplane H in $\mathbb{R}^d$ divides P into three subsets: $P^+ = H^+ \cap P$, $P^- = H^- \cap P$, and $P \cap H$. The *splitting ratio* of H over P, denoted by $\phi_H(P)$, is defined as

$$\phi_H(P) = \max\left(\frac{|P^+|}{|P|}, \frac{|P^-|}{|P|}\right)$$

For each $0 < \delta < 1$, a point $c \in \mathbb{R}^d$ is a δ-center point of P if every hyperplane containing c δ-splits P. Each $d/(d+1)$-center point is called a *center point* of P, and the set of all center points is denoted by Center(P). The balanced separation property of a center point makes it very useful for designing efficient divide and conquer algorithms [16, 30, 55, 83].

Given a set of points $P \subset \mathbb{R}^d$, the question of whether P has a center point is always affirmative. This follows from Helly's Theorem [18].

THEOREM 4.1 (HELLY). *Suppose $\mathcal{K}$ is a family of at least $d+1$ convex sets in $\mathbb{R}^d$, and $\mathcal{K}$ is finite or each member of $\mathcal{K}$ is compact. Then if each $d+1$ members of $\mathcal{K}$ have a common point, there is a point common to all members of $\mathcal{K}$.*

LEMMA 4.2 (CENTER POINTS). *For each set $P \subseteq \mathbb{R}^d$,* Center(P) $\neq \emptyset$.

Proof: [1] We prove the lemma by induction on d. When $d = 1$, the lemma is clearly true. We now assume that the lemma holds for all $d' < d$. If all points of P lie in a $(d-1)$-dimensional affine space, then we can reduce the dimension by one and apply the induction hypotheses to prove that a better center point exists.

So without loss of generality, assume that P does not lie in a $(d-1)$-dimensional affine space. Notice that P induces an equivalence relation on the set of closed halfspaces in $\mathbb{R}^d$: those halfspaces which contains the same subset of points from P are equivalent. Each equivalence class can be identified with a halfspace whose supporting hyperplane passes through d affinely independent points from P.

Let $\mathcal{H}$ be the set of all closed half-spaces with supporting hyperplane passing through d affinely independent points of P that contain more than $\lfloor d|P|/(d+1)\rfloor$ points of P. We want to show that

$$\text{Center}(P) = \bigcap_{H\in\mathcal{H}} H \neq \emptyset.$$

We first show that $\bigcap_{H\in\mathcal{H}} H \neq \emptyset$. Clearly, each element from $\mathcal{H}$ is convex and $\mathcal{H}$ is finite. By Helly's theorem, it is sufficient to show that for each $H_1, \ldots, H_{d+1} \in \mathcal{H}$, $\bigcap_{i=1}^{d+1} H_i \neq \emptyset$.

Note that

$$\bigcap_{i=1}^{d+1} H_i = \mathbb{R}^d - \bigcup_{i=1}^{d+1}(\mathbb{R}^d - H_i) \supseteq P - \bigcup_{i=1}^{d+1}(\mathbb{R}^d - H_i).$$

[1] We present this proof to indicate that there is an $O(n^d)$ time algorithm for computing a center point. Similar proofs can be found in many previous works, e.g., [18].

Note also

$$|\bigcup_{i=1}^{d+1}((\mathbb{R}^d - H_i)\cap P)| \le \sum_{i=1}^{d+1}|((\mathbb{R}^d - H_i)\cap P)| < (d+1)\lfloor\frac{1}{d+1}|P|\rfloor < |P|.$$

Hence, $P - \bigcup_{i=1}^{d+1}(\mathbb{R}^d - H_i) \neq \emptyset$.

We now show that each point c in $\bigcap_{H\in\mathcal{H}} H$ is a center point of P. Suppose c is not a center point of P. Then there is a hyperplane h passing through c defining a halfspace H such that the interior of H contains at least $\lceil d|P|/(d+1)\rceil$ points of P. Thus, there is a closed halfspace H' contained in the interior of H that has at least $\lceil d|P|/(d+1)\rceil$ points of P, contradicting the assumption that $c \in H'$. Therefore, every point in $\bigcap_{H\in\mathcal{H}} H$ is a center point of P. Similarly, we can show that every center point of P is in $\bigcap_{H\in\mathcal{H}} H$. □

Immediately following from the above proof is an $O(n^d)$ time algorithm for computing a center point of a set P. This algorithm uses linear programming. It forms a collection of $O(n^d)$ linear inequalities by considering the set of hyperplanes passing through d affinely independent points of P, and finding the common intersection of the halfspaces that contain at least $dn/(d+1)$ points from P. The intersection of the $O(n^d)$ halfspaces can be found in $O(n^d)$ time using Megiddo's linear programming algorithm [22, 53]. We will refer this algorithm as the *LP algorithm*. Of course, this algorithm is too slow for applications in practice. An efficient algorithm will be presented in Section 6.

If c in Algorithm 1 is a δ-center point of Q, then the origin o is also a δ-center point of Q_2 [58]. First of all, the point c_1 is a δ-center point of Q_1. Now intuitively, a dilation of $\mathbb{R}^d$ moves a center point on the diameter between the south and the north poles along this diameter either up or down depending on the dilation factor. We will prove in our companion paper [58] that the dilation of by factor $\sqrt{(1-r)/(1+r)}$ indeed makes o a δ-center point of Q_2. So, any hyperplane passing through o δ-splits Q_2, and hence GC δ-splits Q_2. Because all transformations used in the above partitioning algorithm preserve the splitting ratio of spheres, S also δ-splits P.

4.3. Separating spheres. We now explain how to choose a random great circle in Algorithm 1.

A *great circle* of U_d is the intersection of U_d with a hyperplane passing through the center of U_d.

Let $randn(m)$ be a function that generates m normally distributed random numbers with mean 0.0 and variance 1.0. A random point p from U_d can be chosen as $p = q/||q||_2$, where $q = randn(d+1)$. A random great circle of U_d is then the great circle normal to the vector p.

Each $(d-1)$-sphere S separates int(S) from ext(S): any segment connecting a point in int(S) and one in ext(S) must intersect S. In analogy to vertex separators in graph theory, we say that S is called a *separating sphere* in d-space.

More specifically, for a set of points $P = \{p_1, \ldots, p_n\}$ in $\mathbb{R}^d$ and a constant $0 < \delta < 1$, we say that S *δ-splits* P if both $|\text{int}(S)\cap P| \le \delta n$ and $|\text{ext}(S)\cap P| \le \delta n$.

4.4. Computing a vertex separator from a separating sphere. We now show how to compute a vertex separator of an overlap graph G from a separating sphere S.

One approach is to remove one of the two endpoints of each edge cut by S. We say S *cuts* an edge (B_i, B_j) of the overlap graph if the line segment between the centers of B_i and B_j has a common point with S. Let E_S be the set of edges cut by S. A ball B_i is a *boundary* ball with respect to S if there is an edge cut by S incident to it. Let U be the set of all boundary balls and let $G_S = (U, E_S)$ be the subgraph of G *induced* by S. Clearly, G_S is a bipartite graph, with boundary balls from the interior of S on one side and boundary balls from the exterior of S on the other side.

The discussion in this section assumes that no point of P lies exactly on S. Because we choose S at random in Algorithm 1, the occurrence of a point of P exactly on S is a zero-probability event. Even if this event were to occur, a slight generalization of the results in this section would cover that case. In fact, we can first put all points of P that appear exactly on S in the vertex separator.

Recall that a *vertex cover* of a graph G is a subset C of vertices such that each edge G has an endpoint in C. In other words, deleting C from G removes all edges of G. Simply from the definition of vertex cover, we have

LEMMA 4.3. *Suppose $\Gamma = \{B_1, \ldots, B_n\}$ is a k-ply neighborhood system in $\mathbb{R}^d$ and G is the α-overlap graph of Γ. If S is a $(d-1)$-sphere that δ-splits Γ, then each vertex cover C of G_S δ-splits G.*

Therefore, the best way to compute a vertex separator from S is perhaps to take a minimum vertex cover of G_S. In fact, a minimum vertex cover of a bipartite graph can be computed in polynomial time, using Dulmage-Mendelsohn decomposition [20] (see [64] for related applications of Dulmage-Mendelsohn decomposition in sparse matrix computations).

On the hand, a faster way to compute a vertex separator from S is to put all the boundary balls from either the interior of S or the exterior of S, whichever has smaller cardinality, into the separator (also see [84]).

However, in the above construction, one has to check the structure of an overlap graph to find a vertex separator from a sphere separator. In some applications, only the neighborhood system is given and it is relatively expensive to compute the overlap graph. We now show how to find a small vertex separator directly from the neighborhood system and the separating sphere.

For each edge (B_i, B_j) cut by S, let $q_{i,j}$ be the common point of S and line segment between the centers of B_i and B_j. Let r_i be the radius of B_i. Without loss of generality, assume $r_i \leq r_j$. Notice that $q_{i,j}$ is either in B_j or in $\alpha \cdot B_i$. If $q_{i,j}$ is in B_j, then we put B_j in D. If $q_{i,j}$ is not in B_j (in which case it must be in $\alpha \cdot B_i$), we put B_i in D. Clearly, since at least one endpoint of every cut edge is in D, D is a vertex cover of G_S.

We now introduce the notion of overlap neighbor. Let S be a $(d-1)$-sphere in $\mathbb{R}^d$, whose radius is r. A ball B_i is an *overlap neighbor* of S if one of the following conditions is true.

1. $B_i \cap S \neq \emptyset$;
2. $\alpha \cdot B_i \cap S \neq \emptyset$ and $r_i \leq r$.

The number of overlap neighbors of S is called the *overlap number* of S. The set of overlap neighbors of a sphere can be computed in $O(n)$ time directly from the neighborhood system.

LEMMA 4.4. *The set of all overlap neighbors of S is a vertex cover of G_S.*

Proof: From the discussion above, D is a vertex cover of G_S. We want to establish that each ball from D is an overlap neighbor of S to prove the lemma.

We partition the set D into two subsets D_1 and D_2, with

$$\begin{aligned} D_1 &= \{B_i \in D : B_i \cap S \neq \emptyset\} \\ D_2 &= D - D_1. \end{aligned}$$

Clearly, each ball from D_1 is an overlap neighbor. Now we need to show that for all $B_i \in D_2$, $r_i \leq r$.

If $B_i \in D_2$, then $B_i \cap S = \emptyset$. There are two possible cases:

- **Case 1:** If $p_i \in \text{int}(S)$, then it simply follows from $B_i \cap S = \emptyset$, that $r_i \leq r$;
- **Case 2:** If $p_i \in \text{ext}(S)$, then from $B_i \in D_2$, it follows that $\alpha \cdot B_i \cap S \neq \emptyset$, and there is a ball B_j in the neighborhood system such that (1) $p_j \in \text{int}(S)$; (2) $B_j \cap S = \emptyset$; (3) $r_i \leq r_j$; and (4) $\alpha \cdot B_i \cap B_j \neq \emptyset$. Because $q_{i,j}$ is not in B_j (otherwise B_i would not be in D) there is no intersection between B_j and S, and hence condition (2) holds. Hence $r_j \leq r$, and $r_i \leq r_j \leq r$.

Thus, in each case, we have $r_i \leq r$, i.e., B_i is an overlap neighbor, completing the proof of the lemma. □

So, our second method is to remove the set of all overlap neighbors. The method is efficient when the overlap graph G is not given. Section 6 shows that the expected number of overlap neighbors generated by the above algorithm is small.

Notice that the definition of overlap neighbors implicitly removes the assumption that no point of P lies exactly on S. If the center of B_i appears exactly on S, then $B_i \cap S \neq \emptyset$. Hence B_i is an overlap neighbor and is placed in the vertex separator.

5. Making the method practical. The run time of the algorithm above crucially depends on the time needed to compute a center point in $(d+1)$-space. All other steps of the algorithm can be performed in $O(n)$ time, and in constant parallel time using $O(n)$ processors.

Unfortunately, no linear-time algorithm is known for computing center points. As shown in Section 4.2, there is a method that requires solving a set of $\Theta(n^d)$ linear inequalities. The only improved result, due to Cole, Sharir, and Yap [16], is that a center point in two dimensions can be computed in $O(n \log^5 n)$ time, and in three dimensions in $O(n^2 \log^7 n)$ time. No subquadratic algorithm is known that always returns even an approximate center point.

In this section, we show that an approximate center point can be found efficiently using geometric sampling [15, 42, 66], which is an important algorithmic technique for designing efficient geometric algorithms.

5.1. Geometric sampling for efficiency. To illustrate the idea, we first show how to use random sampling to compute an approximate center point in one dimension. In this case, the input is a set of $2n$ integers $P = \{p_1, \ldots, p_{2n}\}$. If $p_i < p_j$ for all $i < j$ then center$(P) = [p_n, p_{n+1}]$.

Now suppose we randomly select an element from P, say p. The probability that $p \in \{p_n, p_{n+1}\}$ is $1/n$, while the probability that $p_{\lceil n/2 \rceil} \leq p \leq p_{\lceil 3n/2 \rceil}$ is 0.5. So, with probability 0.5, a randomly selected element from P is an $\epsilon = 3/4$ center point.

We can improve ϵ using larger samples! Suppose l random elements $S = \{r_1, \ldots, r_l\}$ are selected and their median r is the output. Letting $I(r)$ be the rank of r in P, it follows from a simple analysis that $E[I(r)] = n$, and $V[I(r)] = (2n+1)(2n-1-2l)/(8k+6)$. By Chebyshev's inequality,

$$P(|I(r) - n|) \geq t) \leq \frac{n^2}{2lt^2}$$

Thus, with probability at least 0.5, $|I(r) - n| \leq n/\sqrt{l}$, i.e., r is a $1/2\sqrt{l}$ center point of $|P|$. A $1/2 + 1/2\sqrt{l}$ center point of $|P|$ can be computed in $O(l)$ time. A similar sampling idea was used by Floyd and Rivest [26] in their fast selection algorithm.

The algorithm can be generalized to higher dimensions. In d dimensions, the randomized δ-center point algorithm has the following form.

Algorithm 2: (The Sampling Algorithm for Center Points)
Input:(a point set $P \subset \mathbb{R}^d$)

1. Select a subset S of P with size l uniformly at random;
2. Compute a center point c_S of S, using the LP algorithm for center points, given in Section 4.2;
3. Output c_S.

The feasibility of Algorithm 2 above is specified in the following question: What is the probability that c_S computed above is an ϵ-center point?

We now introduce a notation which will be very useful in quantifying the quality of the c_S computed by Algorithm 2. Recall that $\phi_h(P)$ is the ratio in which hyperplane h splits a point set P.

DEFINITION 5.1 (ϵ-GOOD SAMPLE). *Suppose P is a set of points in $\mathbb{R}^d$. $S \subseteq P$ is an ϵ-good sample if for all hyperplanes h, $|\phi_h(S) - \phi_h(P)| \leq \epsilon$.*

The following lemma shows the importance of the ϵ-good sample in approximating center points. Its proof is straightforward.

LEMMA 5.2. *For each $P \subset \mathbb{R}^d$, if $S \subseteq P$ is an ϵ-good sample, then each δ-center point of S is a $(\delta + \epsilon)$-center point of P.*

Now the question becomes: how often does a set of l randomly chosen points form an ϵ-good sample? This is not a trivial question, but was in fact answered by Vapnik and Chervonenkis [80] (see [70] for a detailed proof).

THEOREM 5.3 (VAPNIK AND CHERVONENKIS). *There is a constant c_d depending only on d such that for each $0 < \epsilon \leq 1$ and $l \geq 2/\epsilon^2$, if S is a set of l randomly chosen points from P, then*

$$Pr[S \text{ is an } \epsilon\text{-good sample}] \geq 1 - c_d l^{d+1} e^{\frac{-\epsilon^2 l}{8}}$$

The Vapnik-Chervonenkis bound implies that we only need to sample about $O(d \log d)$ points to compute an approximate center point with high probability.

THEOREM 5.4. *For all $P \in \mathbb{R}^d$, Algorithm 2 computes a $(\lambda_d + \epsilon)$-center point of P in*

$$O\left(\left[\frac{d}{\epsilon^2}\log\frac{d}{\epsilon} + \log\frac{1}{\eta}\right]^d\right)$$

time, with probability at least $1 - \eta$, where $\lambda_d = d/(d+1)$.

Notice that the computation above can be efficiently implemented in parallel.

5.2. Separators using sampling. We now incorporate random sampling into the partitioning algorithm for overlap graphs.

Algorithm 3 (Fast Geometric Partitioning)
Input: (a neighborhood system Γ and the geometric coordinates of its centers P).

1. Choose a random sample P' of size given by Theorem 5.4;
2. Let $Q = ST(P')$;
3. Compute a δ-center point c of Q using the LP algorithm;
4. Compute the rotation π_1 and the dilation π_2 that conformally map c to the origin;
5. Choose a random great circle GC of U_d;
6. Let $S = [ST \circ \pi_2 \circ ST^{-1} \circ \pi_1 \circ ST]^{-1}(GC)$;
7. Induce a vertex separator of G from S.

According to our experiments, about 800 points work very well for meshes in two dimensions, and 1100 points work very well for meshes in three dimensions.

THEOREM 5.5. *Algorithm 3 computes S in random constant time, and a vertex separator of an overlap graph in random $O(n)$ time. Using p processors, the time can be reduced to n/p.*

Algorithm 3 demonstrates the usefulness of geometry, sampling, and randomization in mesh partitioning. The random sampling in the above algorithm reduces the problem size and simultaneously guarantees a provably good approximation of the larger problem. It is the underlying geometric structure that ensures the quality of the partition.

5.3. A fast heuristic for center points. Although the sampling algorithm (Algorithm 2) for center point (in fixed dimensions) is efficient from theoretical viewpoint, it uses linear programming to solve the center point problem on a smaller sample point set. The use of linear programming becomes a serious concern in practical implementation. For example, the experimental results show that the sampling algorithm need to choose a sample of about five hundred to eight hundred points in two dimensions. The sampling algorithm thus needs to solve $\binom{500}{3} \approx 20$ million linear inequalities! Worse, the state-of-art linear programming algorithms (for fixed dimensions) have a large constant. The sample size would be larger for higher dimensions. The seemingly efficient sampling algorithm is too expensive for practical applications.

To overcome this difficulty, we have developed a heuristic for approximating center points [55]. The heuristic uses randomization and runs in linear time in the number of sample points. Most importantly, it does not use linear programming. Our algorithm is based on the notion of a radon point. Let P be a set of points in $\mathbb{R}^d$. A point $q \in \mathbb{R}^d$ is a *radon point* [18] if P can be partitioned into 2 disjoint subsets P_1 and P_2 such that q is a common point of the convex hull of P_1 and the convex hull of P_2. Such a partition is called a *radon partition.*

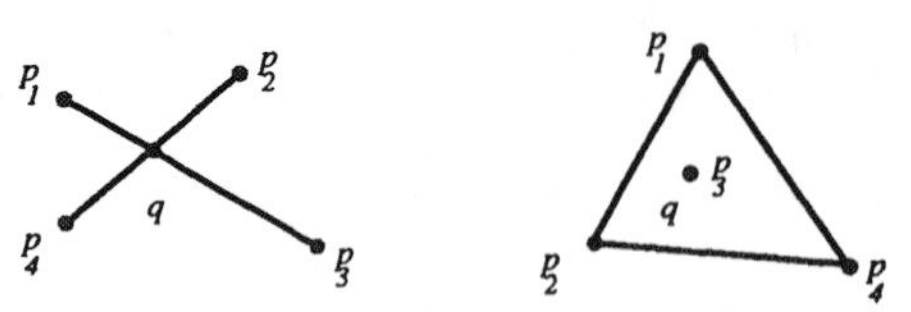

FIG. 6. *The radon point of four points in $\mathbb{R}^2$. When no point is in the convex hull of the other three (the left figure), then the radon point is the unique cross of two linear segments. Otherwise (the right figure), the point that is in the convex hull of the other three is a radon point.*

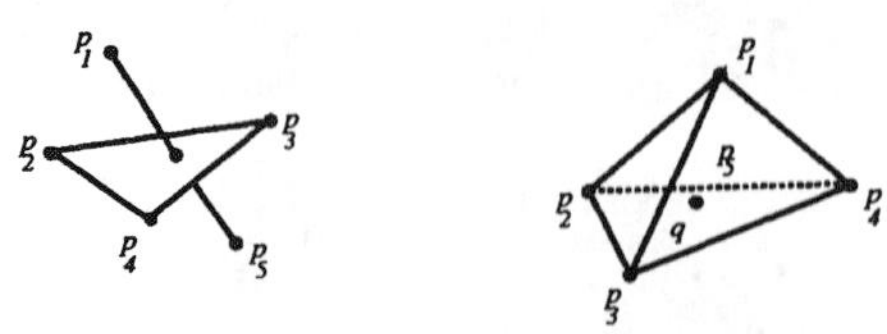

FIG. 7. *The radon point of five points in $\mathbb{R}^3$. Two cases are similar to those in two dimensions.*

The following theorem shows that if $|P| \geq d+2$, then a radon point always exists. Moreover, it can be computed efficiently.

THEOREM 5.6 (RADON [18]). *Let P be a set of points in $\mathbb{R}^d$. If $|P| \geq d+2$, then there is a partition (P_1, P_2) of P such that the convex hull of P_1 has a point in common with the convex hull of P_2.*

Proof: Suppose $P = \{p_1, \ldots, p_n\}$ with $n \geq d+2$. Consider the system of $d+1$ homogeneous linear equations

$$\sum_{i=1}^{n} \alpha_i = 0 = \sum_{i=1}^{n} \alpha_i p_i^j \qquad (1 \leq j \leq d),$$

where $p_i = (p_i^1, ..., p_i^d)$ are the usual coordinates of in $\mathbb{R}^d$. Since $n \geq d+2$, the system has a nontrivial solution $(\alpha_1, \ldots, \alpha_n)$. Let U be the set of all i for which $\alpha_i \geq 0$, and V the set for which $\alpha_i \leq 0$, and $c = \sum_{i\in U} \alpha_i > 0$. Then (U, V) is a partition of P, and $\sum_{i\in V} \alpha_i = -c$ and $\sum_{i\in U}(\alpha_i/c)p_i = \sum_{i\in V}(\alpha_i/c)p_i$. Let $q = \sum_{i\in U}(\alpha_i/c)p_i = \sum_{i\in V}(\alpha_i/c)p_i$. The point q is simultaneously written as a convex combination of points in U and a convex combination of points in V. Hence, q is in the convex hull of U and the convex hull of V, completing the proof. □

To compute a radon point of P, we need only to compute a radon point for the first $d+2$ points. It follows from the proof above that a radon point can be computed in $O(d^3)$ time.

We now describe our heuristic for approximating center points.

Algorithm 4: (Fast Center Points)
Input:(a point set $P \subset \mathbb{R}^d$)

1. Construct a complete balanced $(d+2)$-way tree T of L leaves (for an integer L);
2. For each leaf of T, choose a point from P uniformly at random, independent of other leaves;
3. Evaluate tree T in a bottom-up fashion to assign a point in $\mathbb{R}^d$ to each internal node of T such that the point of each internal node is a radon point of the points with its $(d+2)$ children;
4. Output the point associated with the root of T.

A complete $(d+2)$-way tree of L leaves has at most $L/(d+2)$ internal nodes. Algorithm 4 takes $O(d^2L)$ time, with a small constant. Our experimental results suggest that, independent of the size of the mesh, $L = 900$ is sufficient for meshes from two dimensions and $L = 1200$ for meshes from three dimensions. Moreover, about 10 to 30 tries give a small cost separator that approximately 0.52-splits a mesh.

On the theoretical side, recently, Eppstein, Miller, Sturtivant, and Teng [23] gave a proof that Algorithm 4 finds a $(1 - 1/d^2)$-center point with high probability.

5.4. A practical algorithm for partitioning unstructured meshes. A finite element mesh is not given by a neighborhood system or an overlap graph. Fortunately, our partitioning algorithms do not require such a neighborhood system representation; only the proof does. To cope with the new setting, we show how to adapt our partitioning algorithm.

Algorithm 5 (Partitioning Finite Element Meshes)
Input: (the combinatorial structure of a mesh G, and the geometric coordinates of the mesh xyz).

1. Choose a random sample P of size given by Theorem 5.4 from xyz;
2. Apply Algorithm 3 (Fast Geometric Partitioning) to compute a separating sphere S for P; In practice, we replace the LP algorithm of Step 3 in Algorithm 3 by Algorithm 4, the Radon partition based fast center point algorithm.
3. Partition G into two subgraphs G_1 and G_2 by removing the set of edges $\tilde{E}$ that cut this sphere. This set of edges can be found using the structure of G; each vertex is placed into G_1 or G_2, depending on whether it is mapped to the interior or the exterior of S;
4. Compute the density α of the embedding given by (G, xyz). Then, for each v that is an endpoint of an edge in $\tilde{E}$, find the ball centered at v whose radius is given in the proof of Lemma 3.1. Use the rule of overlap neighbors of Section 4.4 to compute a vertex separator of (G, xyz).

However, in practice, we do not calculate the density of the embedding to compute a vertex separator. We will use more direct and practical subroutines for finding a vertex separator of (G, xyz) in the last step of the algorithm above. For example, a faster way to compute a vertex separator is to put every vertex v in the interior or exterior of S that is an endpoint of an edge in $\tilde{E}$, whichever has smaller cardinality, into the separator. Notice that the degree of a density graph or a finite element graph is bounded. So, this heuristic also finds a vertex separator that satisfies the separator bound in Theorem 3.3. To compute the smallest vertex separator induced by S, we can use the minimum bipartite vertex cover procedure (see Section 4.4). Although the worst case time complexity of such a procedure is quite expensive, its expected (average) time complexity is much lower, especially since in our case the bipartite graph induced by $\tilde{E}$ is much smaller than the original graph G.

Figure 8 shows the mesh of Figures 2 and 3 partitioned using our experimental implementation [39] of Algorithm 5 recursively.

Our mesh partitioning algorithms first use the geometric information to compute a continuous separator, then use the combinatorial structure to find a vertex separator. One major reason making the above partitioning algorithm suitable for efficient practical implementation is the use of *geometric sampling*, a technique that reduces the problem size and simultaneously guarantees a provably good approximation of the larger problem.

6. Proof outline for the main separator theorem. We now outline the proof of the Main Theorem 2.4, to show how to use geometric arguments to prove separator properties for graphs embedded in fixed dimension. The detailed proofs are presented in the companion paper [58]. In Section 6.1, we present a continuous separator theorem, based on which we give a geometric method for proving a small separator theorem in Section 6.2. We then apply this method to prove Theorem 2.4 in the remainder of this section.

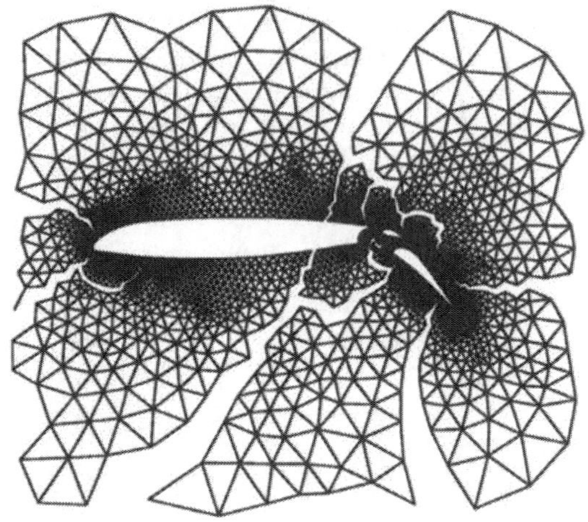

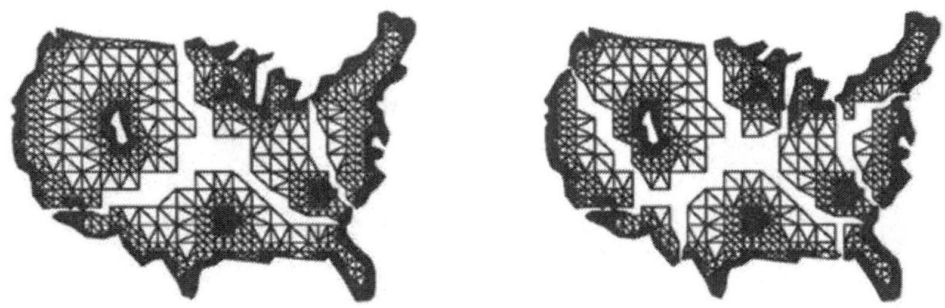

FIG. 8. *Recursive partitioning using Algorithm 5.*

6.1. A continuous separator theorem. In the partitioning algorithms presented above, a "continuous" separator, in the form of a sphere, is computed first, and then a "discrete" separator is deduced from its continuous counterpart. The quality of the continuous separating sphere, in a strong sense, determines the quality of the vertex separator.

Suppose $f(x)$ is a real-valued nonnegative function defined on $\mathbb{R}^d$ such that f^k is integrable for all $k = 1, 2, 3, \ldots$. Such an f is called a *cost function*. The *total volume* of the function f is defined as

$$\text{Total-Volume}(f) = \int_{v \in \mathbb{R}^d} (f(v))^d (dv)^d$$

Suppose S is a $(d-1)$-sphere in $\mathbb{R}^d$. The *surface area* of S is then

$$\text{Area}(f, S) = \int_{v \in S} (f(v))^{d-1} (dv)^{d-1}$$

Let π denote a map from $\mathbb{R}^d$ to U_d, which is formed by a stereographic projection, followed by a rotation, followed by an inverse stereographic projection, followed by a dilation, followed by a stereographic projection. Such a map is called an *H-map*. Our partitioning algorithms compute an H-map, choose a random great circle, and use the inverse of the H-map to transform the great circle to a sphere in $\mathbb{R}^d$.

For each great circle GC of U_d, let S_{GC} be the sphere defined by $\pi^{-1}(GC)$. Let $\text{Cost}(GC) = \text{Area}(f, S_{GC})$. Let $\text{Avg}(f)$ be the average cost of all great circles of U_d. We will use the following lemma, whose proof can be found in the companion paper [58].

LEMMA 6.1. *Suppose f is a cost function in $\mathbb{R}^d$. Then*

$$\text{Avg}(f) = O\left((\text{Total-Volume}(f))^{\frac{d-1}{d}}\right).$$

Consequently, we have the following continuous separator theorem.

THEOREM 6.2 (CONTINUOUS SEPARATOR). *Suppose f is a cost function on $\mathbb{R}^d$ and P is a set of n distinct points in $\mathbb{R}^d$. Let S be a sphere chosen by the random process described in Algorithm 1. Then S $(d+1)/(d+2)$-splits P, and with high probability,*

$$\text{Area}(f, S) = O\left((\text{Total-Volume}(f))^{\frac{d-1}{d}}\right).$$

The splitting ratio of the separator in the theorem above is $(d+1)/(d+2)$ instead of $d/(d+1)$. This is because points are mapped from $\mathbb{R}^d$ to the unit sphere in $\mathbb{R}^{d+1}$, and the center of the unit sphere is a $(d+1)$-dimensional center point of the image rather than a d-dimensional center point.

6.2. A new approach to proving small separator theorems. The continuous separator theorem of the last section provides the following generic approach to prove that a graph G embedded in $\mathbb{R}^d$ has an $O(c^{(d-1)/d})$-separator that $(d+1)/(d+2)$-splits.

A Geometric Approach to Proving Small Separator Theorems
1. Define a real-valued function f based on the structure of G so that Total-Volume(f) is bounded by a function c;
2. Find a $(d-1)$-dimensional separating sphere S that $(d+1)/(d+2)$-splits the vertices of G and has Area$(f, S) = O\left(c^{(d-1)/d}\right)$, by Theorem 6.2.
3. Deduce a vertex separator of G from the separating sphere S.

In order to deduce a vertex separator from its continuous counterpart, the function f must be *faithful* in the sense that the cost of a continuous separator models the size of a vertex separator of the underlying graph. In other words, the continuous function f faithfully encodes some combinatorial properties related to separators of the underlying graph.

We will follow the basic steps above to prove Theorem 2.4. To this end, for each α-overlap graph of a k-ply neighborhood system $\Gamma = \{B_1, \ldots, B_n\}$ in $\mathbb{R}^d$, we construct a real-valued function f based on Γ and prove that

$$\text{Total-Volume}(f) \leq O\left(\alpha^{\frac{d}{d-1}} k^{\frac{1}{d-1}} n\right);$$

then we show that from each separating sphere S we can deduce a vertex separator of size linearly bounded by Area(f, S).

6.3. Local cost functions. Just as each overlap graph is defined from its neighborhood system, $B_1, \ldots, B_n$, the cost function f itself is defined from the *local cost functions*, $f_1, \ldots, f_n$, with f_i based on B_i.

Let $P = \{p_1, \ldots, p_n\}$ be the set of centers of $\{B_1, \ldots, B_n\}$, and suppose that the radius of B_i is r_i. We define f_i as

$$f_i(x) = \begin{cases} 1/(2\alpha r_i) & \text{if } x \in (2\alpha)\cdot B_i, \text{ i.e., } \|x - p_i\| \leq 2\alpha r_i \\ 0 & \text{otherwise} \end{cases}$$

Intuitively, f_i sets up a cost on each $(d-1)$-sphere S such that the closer S is to B_i, the more B_i contributes to the surface area of S. The function f_i measures the cost of a sphere passing through B_i and its vicinity.

Notice that the function f_i defined above has Total-Volume(f_i) equal to V_d, the volume of a unit ball in d dimensions.

6.4. Putting local cost functions together. Now we need to put the local cost functions together into a global cost function. Perhaps the simplest way is to take the sum, $f = \sum_i f_i$. But this is not the best choice. To see this, just check the extreme case where the neighborhood is a collection of n identical balls and $\alpha = 1$. In this case $k = n$. The total volume of the sum is $n^d V_d$, while we need a cost function of total volume $O\left(n^{d/(d-1)}\right)$ to establish Theorem 2.4.

To achieve a tight bound, we make use of the "slight" difference between the various p-norms when applied to high-dimensional vectors. This is a technique that appears to be new and is interesting in its own right. Recently, Mitchell and Vavasis

[61] used a cost function similar to ours to analyze their three dimensional mesh generation algorithm.

Suppose $a_1, \ldots, a_n$ are reals. For each positive integer p, the L_p *norm* of $a_1, \ldots, a_n$, denoted $L_p(a_1, \ldots, a_n)$, is

$$L_p(a_1, \ldots, a_n) = \left(\sum_{i=1}^{n} |a|^p\right)^{\frac{1}{p}}.$$

The following lemma states the relationship between different norms.

LEMMA 6.3. *Let $a_1, \ldots, a_n$ be real. If $p \leq q$, then $L_p(a_1, \ldots, a_n) \geq L_q(a_1, \ldots, a_n)$.*

Proof: See Hardy, Littlewood and Pólya [41] (pages 26 and 144). □

We define the *global cost function* of the overlap graph to be the L_{d-1} norm of $f_1, \ldots, f_n$, i.e.,

$$f(x) = L_{d-1}(f_1, \ldots, f_n) = \left(\sum_{i=1}^{n} (f_i(x))^{d-1}\right)^{\frac{1}{d-1}}.$$

Notice that the L_d norm of f_i is not a good choice, because its total volume is nV_d for all neighborhood systems. The following lemma, proved in [58], bounds the total volume of the function f.

LEMMA 6.4. *Let $\Gamma = \{B_1, \ldots, B_n\}$ be a k-ply neighborhood system in $\mathbb{R}^d$. If $f_1, \ldots, f_n$ are the local cost functions of Γ and f is its global cost function, then*

$$\text{Total-Volume}(f) = O(\alpha^{\frac{d}{d-1}} k^{\frac{1}{d-1}} n).$$

Consequently, by Theorem 6.2, we have the following lemma.

LEMMA 6.5. *Suppose $\{B_1, \ldots, B_n\}$ is a k-ply neighborhood system in $\mathbb{R}^d$. Then there exists a $((d+1)/(d+2))$-splitting sphere S of $\{B_1, \ldots, B_n\}$ with* $\text{Area}(f, S) = O\left(\xi \alpha k^{1/d} n^{(d-1)/d}\right)$, *where $\xi = 2^{d-1} V_d$.*

6.5. A vertex separator from a continuous one. Recall that a ball B_i is an *overlap neighbor* of a sphere S if one of the following conditions is true.

1. $B_i \cap S \neq \emptyset$;
2. $\alpha \cdot B_i \cap S \neq \emptyset$ and $r_i \leq r$.

The number of overlap neighbors of S is called the *overlap number* of S, denoted $\vartheta_\Gamma(S)$. The following lemma bounds the overlap number $\vartheta_\Gamma(S)$ of a k-ply neighborhood system in $\mathbb{R}^d$.

LEMMA 6.6. *Suppose $\Gamma = \{B_1, \ldots, B_n\}$ is a k-ply neighborhood system in $\mathbb{R}^d$. Let $f_1, \ldots, f_n$ be local cost functions defined for the α-overlap graph of Γ and let $f = L_{d-1}(f_1, \ldots, f_n)$. For each $(d-1)$-sphere S,*

$$\vartheta_\Gamma(S) = O(\alpha^d k) + O(\text{Area}(f, S)).$$

The constant in the big-O notation depends only on d.

Theorem 2.4 follows from Lemma 6.4, Theorem 6.2, and Lemma 6.6.

7. Final remarks. We have demonstrated that geometric structure is useful for mesh partitioning. We have shown that geometric sampling can be used to reduce the problem size, making our approach feasible for practical applications in large-scale computation. We have implemented the random linear-time separator algorithm of Section 6 [39] and have experimented with various examples. The numerical results are encouraging. With the help of some heuristics to speed up the geometric transformation and the local optimization, the program is very fast. In practice, our program generates partitions much better than what the theoretical results predict; the partitions are competitive with such previous methods as those based on an expensive eigenvector computation [65].

Recently, Eppstein, Miller and Teng [24] have showed that a small separator for the intersection graph of a k-ply neighborhood system can be in fact found in deterministic linear time.

We conclude the paper with the following two open questions.

1. What is the computational complexity of deciding whether a graph G is k-embeddable or (α, k)-embeddable?
2. Is there a polynomial time algorithm for computing the disk packing of a planar graph?

Acknowledgments. We would like to thank David Applegate, Marshall Bern, David Eppstein, John Gilbert, Bruce Hendrickson, Ravi Kannan, Tom Leighton, Mike Luby, Oded Schramm, Doug Tygar, and Kim Wagner for invaluable help and discussions. We would like to thank John Gilbert especially for his editorial contribution which greatly improved this paper.

REFERENCES

[1] A. Agrawal and P. Klein. Cutting down on fill using nested dissection: Provably good elimination orderings. In *Sparse Matrix Computations: Graph Theory Issues and Algorithms, IMA Volumes in Mathematics and its Applications*, (this book), A. George, J. Gilbert and J. Liu, Springer-Verlag, New York. 1992.

[2] N. Alon, P. Seymour, and R. Thomas. A separator theorem for non-planar graphs. In *Proceedings of the 22th Annual ACM Symposium on Theory of Computing*, Maryland, May 1990. ACM.

[3] E. M. Andreev. On convex polyhedra in Lobacevskii space. *Math. USR Sbornik*, 10(3):413–440, 1970.

[4] E. M. Andreev. On convex polyhedra of finite volume in Lobacevskii space. *Math. USR Sbornik*, 12(2):270–259, 1970.

[5] M. J. Berger and S. Bokhari. A partitioning strategy for nonuniform problems on multiprocessors. *IEEE Trans. Comp.*, C-36:570-580, 1987.

[6] M. Bern, D. Eppstein and J. R. Gilbert. Provably good mesh generation. In *31st Annual Symposium on Foundations of Computer Science, IEEE*, 231–241, 1990, (to appear JCSS).

[7] M. Bern and D. Eppstein. Mesh generation and optimal triangulation. In *Computing in Euclidean Geometry*, F. K. Hwang and D.-Z. Du editors, World Scientific, 1992.

[8] G. Birkhoff and A. George. Elimination by nested dissection. *Complexity of Sequential and Parallel Numerical Algorithms*, J. F. Traub, Academic Press, 1973.

[9] P. E. Bjørstad and O. B. Widlund. Iterative methods for the solution of elliptic problems on regions partitioned into substructures. *SIAM J. Numer. Anal.*, 23:1097-1120, 1986.

[10] G. E. Blelloch. *Vector Models for Data-Parallel Computing.* MIT-Press, Cambridge MA, 1990.

[11] J. H. Bramble, J. E. Pasciak, and A. H. Schatz. An iterative method for elliptic problems on regions partitioned into substructures, *Math. Comp.* 46:361-9, 1986.

[12] D. Calahan. Parallel solution of sparse simultaneous linear equations. in *Proceedings of the 11th Annual Allerton Conference on Circuits and Systems Theory*, 729-735, 1973.

[13] T. F. Chan and D. C. Resasco. A framework for the analysis and construction of domain decomposition preconditioners. UCLA-CAM-87-09, 1987.

[14] L. P. Chew. Guaranteed quality triangular meshes, Department of Computer Science, Cornell University TR 89-893, 1989.

[15] K. Clarkson. Fast algorithm for the all-nearest-neighbors problem. In *the 24th Annual Symposium on Foundations of Computer Science*, 226–232, 1983.

[16] R. Cole, M. Sharir and C. K. Yap. On k-hulls and related problems. *SIAM J. Computing*, 61, 1987.

[17] J. H. Conway, and N. J. A. Sloane. *Sphere Packings, Lattices and Groups.* Springer-Verlag, 1988.

[18] L. Danzer, J. Fonlupt and V. Klee. Helly's theorem and its relatives. *Proceedings of Symposia in Pure Mathematics*, American Mathematical Society, 7: 101-180, 1963.

[19] H. N. Djidjev. On the problem of partitioning planar graphs. *SIAM J. Alg. Disc. Math.*, 3(2):229–240, June 1982.

[20] A. L. Dulmage and N. S. Mendelsohn. Coverings of bipartite graphs. *Canadian J. Math.* 10, pp 517-534, 1958.

[21] I. S. Duff. Parallel implementation of multifrontal schemes. *Parallel Computing*, 3, 193–204, 1986.

[22] M. E. Dyer. On a multidimensional search procedure and its application to the Euclidean one-centre problem. *SIAM Journal on Computing* 13, pp 31–45, 1984.

[23] D. Eppstein, G. L. Miller, C. Sturtivant and S.-H. Teng. Approximating center points with and without linear programming. Manuscript, Massachusetts Institute of Technology, 1992.

[24] D. Eppstein, G. L. Miller and S.-H. Teng. A deterministic linear time algorithm for geometric separators and its applications. Manuscript, Xerox Palo Alto Research Center, 1991.

[25] I. Fáry. On straight line representing of planar graphs. *Acta. Sci. Math.* 24: 229–233, 1948.

[26] R. W. Floyd and R. L. Rivest. Expected time bounds for selection. *CACM* 18(3): 165–173, March, 1975.

[27] H. de Fraysseix, J. Pach, and R. Pollack. Small sets supporting Fáry embeddings of planar graphs. In *Proceedings of the 20th Annual ACM Symposium on Theory of Computing*, 426–433, 1988.

[28] G. N. Fredrickson and R. Janardan. Separator-based strategies for efficient message routing. In *27st Annual Symposium on Foundation of Computation Science, IEEE*, 428–237, 1986.

[29] I. Fried. Condition of finite element matrices generated from nonuniform meshes. *AIAA J.* 10, pp 219–221, 1972.

[30] A. M. Frieze, G. L. Miller and S.-H. Teng. Separator based divide and conquer in computational geometry. *Proceedings of the 1992 ACM Symposium on Parallel Algorithms and Architectures*, 1992.

[31] H. Gazit. An improved algorithm for separating a planar graph. Manuscript, Department of Computer Science, University of Southern California, 1986.

[32] H. Gazit and G. L. Miller. A parallel algorithm for finding a separator in planar graphs. In *28st Annual Symposium on Foundation of Computation Science, IEEE*, 238–248, Los Angeles, October 1987.

[33] H. Gazit. A deterministic parallel algorithm for planar graph isomorphism. In *32nd Annual Symposium on Foundations of Computer Science, IEEE*, to appear, 1991.

[34] J. A. George. Nested dissection of a regular finite element mesh. *SIAM J. Numerical Analysis*, 10: 345–363, 1973.

[35] A. George, M. T. Heath, J. Liu, E. Ng. Sparse Cholesky factorization on a local-memory multiprocessor. *SIAM J. on Scientific and Statistical Computing*, 9, 327–340, 1988.

[36] J. A. George and J. W. H. Liu. An automatic nested dissection algorithm for irregular finite element problems. *SIAM J. on Numerical Analysis*, 15, 1053–1069, 1978.

[37] J. A. George and J. W. H. Liu. *Computer Solution of Large Sparse Positive Definite Systems.* Prentice-Hall, 1981.

[38] J. R. Gilbert, J. P. Hutchinson, and R. E. Tarjan. A separator theorem for graphs of bounded genus. *J. Algorithms*, 5 pp391–407, 1984.

[39] J.R. Gilbert, G.L. Miller, and S.-H. Teng. Geometric mesh partitioning: Implementation and experiments. Technical Report, Xerox Palo Alto Research Center, to appear, 1992.

[40] J. R. Gilbert and R. E. Tarjan. The analysis of a nested dissection algorithm. *Numerische Mathematik*, 50(4):377–404, 1987.

[41] G. Hardy, J. E. Littlewood and G. Pólya. *Inequalities.* Second edition, Cambridge University Press, 1952.

[42] D. Haussler and E. Welzl. ϵ-net and simplex range queries. *Discrete & Computational Geometry*, 2: 127–151, 1987.

[43] J. P. Hutchinson and G. L. Miller. On deleting vertices to make a graph of positive genus planar. In *Discrete Algorithms and Complexity Theory - Proceedings of the Japan-US Joint Seminar, Kyoto, Japan*, pages 81–98, Boston, 1986. Academic Press.

[44] C. Jordan. Sur les assemblages de lignes. *Journal Reine Angew. Math*, 70:185–190, 1869.

[45] F. T. Leighton. *Complexity Issues in VLSI.* Foundations of Computing. MIT Press, Cambridge, MA, 1983.

[46] F. T. Leighton and S. Rao. An approximate max-flow min-cut theorem for uniform multicommodity flow problems with applications to approximation algorithms. In *29th Annual Symposium on Foundations of Computer Science*, pp 422-431, 1988.

[47] C. E. Leiserson. *Area Efficient VLSI Computation.* Foundations of Computing. MIT Press, Cambridge, MA, 1983.

[48] C. E. Leiserson and J. G. Lewis. Orderings for parallel sparse symmetric factorization. in *3rd SIAM Conference on Parallel Processing for Scientific Computing*, 1987.

[49] R. J. Lipton, D. J. Rose, and R. E. Tarjan. Generalized nested dissection. *SIAM J. on Numerical Analysis*, 16:346–358, 1979.

[50] R. J. Lipton and R. E. Tarjan. A separator theorem for planar graphs. *SIAM J. of Appl. Math.*, 36:177–189, April 1979.

[51] R. J. Lipton and R. E. Tarjan. Applications of planar separator theorem. *SIAM J. Comput*, 9(3): 615–627, August 1981.

[52] J. W. H. Liu. The solution of mesh equations on a parallel computer. in 2nd Langley Conference on Scientific Computing, 1974.

[53] N. Megiddo. Linear programming in linear time when the dimension is fixed. *SIAM Journal on Computing* 12, pp 759–776, 1983.

[54] G. L. Miller. Finding small simple cycle separators for 2-connected planar graphs. *Journal of Computer and System Sciences*, 32(3):265–279, June 1986.

[55] G. L. Miller and S.-H. Teng. Centerpoints and point divisions. Manuscript, School of Computer Science, Carnegie Mellon University, 1990.

[56] G. L. Miller, S.-H. Teng, and S. A. Vavasis. A unified geometric approach to graph separators. In *32nd Annual Symposium on Foundations of Computer Science, IEEE*, pp538–547, 1991.

[57] G. L. Miller, S.-H. Teng, W. Thurston and S. A. Vavasis. Separators for sphere-packings and nearest neighborhood graphs. in progress 1992.

[58] G. L. Miller, S.-H. Teng, W. Thurston and S. A. Vavasis. Finite element meshes and geometric separators. in progress 1992.

[59] G. L Miller and W. Thurston. Separators in two and three dimensions. In *Proceedings of the 22th Annual ACM Symposium on Theory of Computing*, pages 300–309, Maryland, May 1990. ACM.

[60] G. L. Miller and S. A. Vavasis. Density graphs and separators. In *Second Annual ACM-SIAM Symposium on Discrete Algorithms*, pages 331–336, San Francisco, January 1991. ACM-SIAM.

[61] S. A. Mitchell and S. A. Vavasis. Quality mesh generation in three dimensions. Proc. ACM Symposium on Computational Geometry, pp 212-221, 1992.

[62] M. S. Paterson. Tape bounds for time-bounded Turing machines. *J. Comp. Syst. Sci.*, 6:116–124, 1972.

[63] V. Pan and J. Reif. Efficient parallel solution of linear systems. In *Proceedings of the 17th Annual ACM Symposium on Theory of Computing*, pages 143–152, Providence, RI, May 1985. ACM.

[64] A. Pothen and C.-J. Fan. Computing the block triangular form of a sparse matrix. *ACM Transactions on Mathematical Software* 16 (4), pp 303-324, 1990.

[65] A. Pothen, H. D. Simon, K.-P. Liou. Partitioning sparse matrices with eigenvectors of graphs. *SIAM J. Matrix Anal. Appl.* 11 (3), pp 430–452, July, 1990.

[66] J. H. Reif and S. Sen. Polling: A new randomized sampling technique for computational geometry. In *Proceedings of the 21st annual ACM Symposium on Theory of Computing.* 394–404, 1989.

[67] E. Schwabe, G. Blelloch, A. Feldmann, O. Ghattas, J. Gilbert, G. Miller, D. O'Hallaron, J. Schewchuk and S.-H. Teng. A separator-based framework for automated partitioning and mapping of parallel algorithms in scientific computing. In *First Annual Dartmouth Summer Institute on Issues and Obstacles in the Practical Implementation of Parallel Algorithms and the use of Parallel Machines*, 1992.

[68] H. D. Simon. Partitioning of unstructured problems for parallel processing. *Computing Systems in Engineering* 2:(2/3), pp135-148.

[69] G. Strang and G. J. Fix. *An Analysis of the Finite Element Method*, Prentice-Hall, 1973.

[70] S.-H. Teng. *Points, Spheres, and Separators: A Unified Geometric Approach to Graph Partitioning.* PhD thesis, School of Computer Science, Carnegie Mellon University, Pittsburgh, 1991. CMU-CS-91-184.

[71] C. Thomassen. Planarity and duality of finite and infinite graphs. *Journal of Combinatorial Theory*, Series B, 29: 244–271, 1980.

[72] J. F. Thompson, Z. U. A. Warsi and C. W. Mastin. *Numerical Grid Generation: Foundations and Applications.* New York, North Holland, 1985.

[73] W. P. Thurston. *The geometry and topology of 3-manifolds.* Princeton University Notes, 1988.

[74] W. T. Tutte. Convex representations of graphs. *Proc. London Math. Soc.* 10(3): 304–320, 1960.

[75] W. T. Tutte. How to draw a graph. *Proc. London Math. Soc.* 13(3): 743–768, 1963.

[76] J. D. Ullman. *Computational Aspects of VLSI.* Computer Science Press, Rockville MD, 1984.

[77] P. Ungar. A theorem on planar graphs. *Journal London Math Soc.* 26: 256–262, 1951.

[78] P. M. Vaidya. Constructing provably good cheap preconditioners for certain symmetric positive definite matrices. *IMA Workshop on Sparse Matrix Computation: Graph Theory Issues and Algorithms*, Minneapolis, Minnesota, October 1991.

[79] L. G. Valiant. Universality consideration in VLSI circuits. *IEEE Transaction on Computers*, 30(2): 135–140, February, 1981.

[80] V. N. Vapnik and A. Ya. Chervonenkis. On the uniform convergence of relative frequencies of events to their probabilities. *Theory Probab. Appl.*, 16: 264-280, 1971.

[81] S. A. Vavasis. Automatic domain partitioning in three dimensions. *SIAM J. Sci. Stat. Comp.*, 12 (1991) 950-970.

[82] R. D. Williams. Performance of dynamic load balancing algorithms for unstructured mesh calculations. Technical Report, California Institute of Technology, 1990.

[83] F.-F. Yao. A 3-space partition and its application. In *Proceedings of the 15th Annual ACM Symposium on Theory of Computing, ACM*, 258–263, 1983.

[84] E. E. Zmijewski. *Sparse Cholesky Factorization on a Multiprocessor.* PhD thesis, Department of Computer Science, Cornell University, 1987.

Separators for Sphere-Packings and Nearest Neighbor Graphs

GARY L. MILLER

Carnegie Mellon University, Pittsburgh, Pennsylvania

SHANG-HUA TENG

University of Minnesota, Minneapolis, Minnesota

WILLIAM THURSTON

University of California, Berkeley, California

AND

STEPHEN A. VAVASIS

Cornell University, Ithaca, New York

Abstract. A collection of n balls in d dimensions forms a k-ply system if no point in the space is covered by more than k balls. We show that for every k-ply system Γ, there is a sphere S that intersects at most $O(k^{1/d}n^{1-1/d})$ balls of Γ and divides the remainder of Γ into two parts: those in the interior and those in the exterior of the sphere S, respectively, so that the larger part contains at most $(1 - 1/(d + 2))n$ balls. This bound of $O(k^{1/d}n^{1-1/d})$ is the best possible in both n and k. We also present a simple randomized algorithm to find such a sphere in $O(n)$ time. Our result implies that every k-nearest neighbor graphs of n points in d dimensions has a separator of size $O(k^{1/d}n^{1-1/d})$. In conjunction with a result of Koebe that every triangulated planar graph is isomorphic to the intersection graph of a disk-packing, our result not only gives a new geometric proof of the planar separator theorem of Lipton and Tarjan, but also generalizes it to higher dimensions. The separator algorithm can be used for point location and geometric divide and conquer in a fixed dimensional space.

G. L. Miller was supported in part by National Science Foundation grant CCR 90-16641.

S.-H. Teng was supported by an NSF CAREER award (CCR 95-02540), an Alfred P. Sloan Research Fellowship, and an Intel research grant. Part of the work was done while S.-H. Teng was at MIT, Xerox Corporation (PARC), and Carnegie Mellon University.

S. A. Vavasis was supported by an NSF Presidential Young Investigator award.

Authors' addresses: G. L. Miller, School of Computer Science, Carnegie Mellon University, Pittsburgh, PA 15213; S.-H. Teng, Department of Computer Science, University of Minnesota, Minneapolis, MN 55455; W. Thurston, Department of Mathematics, University of California, Berkeley, CA 94720; S. A. Vavasis, Department of Computer Science, Cornell University, Ithaca, NY 14853.

Journal of the ACM, Vol. 44, No. 1, January 1997, pp. 1–29.

Categories and Subject Descriptors: [**Data Structures**]: *graphs*, trees; F.2.1 [**Analysis of Algorithms and Problem Complexity**]: Numerical Algorithms and Problems—*computation of transforms (e.g., Fast Fourier transform); computations on matrices*; F.2.2 [**Analysis of Algorithms and Problem Complexity**]: Nonnumerical Algorithms and Problems—*computations on discrete structures; geometrical problems and computations; sorting and searching*; G.2.1 [**Discrete Mathematics**]: Combinatorics; G.2.2 [**Discrete Mathematics**]: Graph Theory—*graph algorithms; trees*; G.3 [**Probability and Statistics**]: *probabilistic algorithms (including Monte Carlo); random number generation*; G.4 [**Mathematical Software**]: *algorithm analysis; efficiency*

General Terms: Algorithms, Theory

Additional Key Words and Phrases: Centerpoint, computational geometry, graph algorithms, graph separators, partitioning, probabilistic method, point location, randomized algorithm, sphere-preserving mapping

1. *Introduction*

Motivations of this work are the planar separator theorem of Lipton and Tarjan [1979], the geometric characterization of planar graphs of Koebe [1936] (see also Andreev [1970a; 1970b] and Thurston [1988]), and geometric divide and conquer.

In 1979, Lipton and Tarjan [1979] gave a linear time algorithm that divides any n-vertex planar graph into two disconnected subgraphs each has size no more than $(2/3)n$ by removing at most $\sqrt{8n}$ vertices. Their result improved a theorem of Ungar [1951] who showed it is sufficient to remove $O(\sqrt{n}\log n)$ vertices to partition a planar graph. A subset of vertices whose removal divides a graph into two subgraphs of roughly equal size, as given in the results above, is called a *separator* of the graph (see Section 3 for the definition).

Separators are most useful for designing efficient divide and conquer graph algorithms. The planar separator theorem of Lipton and Tarjan has been used in the solution of planar linear systems [Lipton et al., 1979], in the design of efficient graph algorithms [Lipton and Tarjan 1979] and in VLSI layout [Leighton 1983; Leiserson 1983; Valiant 1981].

Building on Lipton and Tarjan's planar separator theorem, Gilbert et al. [1984] showed that every graph with genus bounded by g has an $O(\sqrt{gn})$-separator. Another generalization was obtained by Alon et al. [1990] who showed that graphs that exclude minor isomorphic to the h-clique have an $O(h^{3/2}\sqrt{n})$-separator. Plotkin et al. [1994] reduced the dependency on h from $h^{3/2}$ to h, but in the process, they picked up a factor of $\sqrt{\log n}$. Perhaps the oldest separator result is that every tree has a 1-separator [Jordan 1869].

However, these results are apparently not applicable to geometric graphs such as nearest neighbor graphs [Preparata and Shamos 1985] when the dimension d is higher than 2.

In this paper, we give a geometrical condition of graphs embedded in d dimensions that have a small *separator*. We also present efficient linear time sequential algorithms and optimal $O(\log n)$ time parallel algorithms for finding such a small separator. The method and condition that we propose, unlike previous works, assume that the graph G comes with an embedding of its nodes in $\mathbb{R}^d$. This is a very natural assumption for nearest neighbor graphs. Our algorithm is randomized. It always splits the graph into pieces of roughly equal size, and we show that with high probability, the separator size satisfies an upper

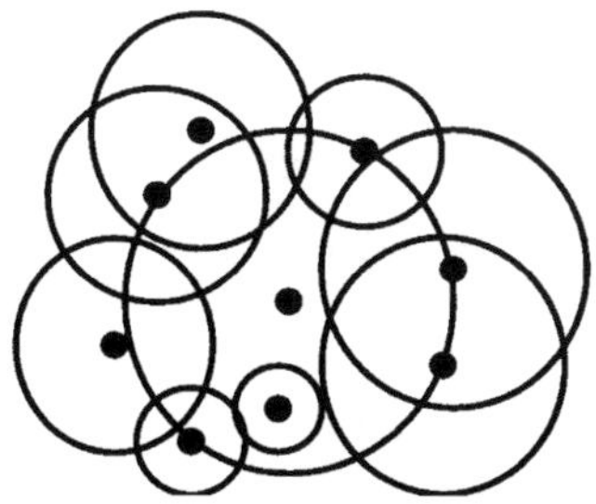

FIG. 1. A 3-ply system.

bound that is the best possible bound for this class of graphs that we consider. In conjunction with a result of Koebe [1936] that every triangulated planar graph is isomorphic to the intersection graph of a disk-packing, our result not only gives a new geometric proof of the planar separator theorem of Lipton and Tarjan [1979], but also generalizes it to higher dimensions. The separator algorithm can be used for point location and geometric divide and conquer in a fixed dimensional space.

We now review the relationship between this paper and other papers by the same authors. This paper and its companion paper [Miller et al. 1997] either extend or explain several short conference papers [Miller and Thurston 1990; Miller et al. 1991; 1993], a thesis [Teng 1991] and one journal paper [Vavasis 1991]. The focus of this paper is on graphs arising in computational geometry and the application of our separator theorem to geometric divide and conquer; the companion paper [Miller et al. 1997] focuses on finite element meshes. The authors have also jointly written a survey paper [Miller et al. 1993] that surveys the results from this paper, the companion, and several additional results by various authors on efficient centerpoint computation.

The remainder of this paper is organized as follows: In Section 2, we introduce neighborhood systems and prove our main separator theorem. We also give our randomized separator algorithm. In Section 3, we define the intersection graph of a neighborhood system and apply our main results to planar graphs and nearest neighbor graphs. In Section 4, we develop a separator-based divide-and-conquer paradigm and apply it to solve several problems in computational geometry. In Section 5, we give several open questions motivated by this research.

2. *Sphere Separators of Neighborhood Systems*

Throughout the paper we regard the dimension d as a small constant.

The class of geometric graphs that we consider is defined by the intersection of a collection of balls in d-dimensional Euclidean space. We will refer such a collection of balls as a *neighborhood system*. In this section, we will prove a geometric separator theorem for neighborhood systems. Its applications to planar graphs, k-nearest neighbor graphs, and geometric divide-and-conquer will be given in the subsequent sections.

2.1. NEIGHBORHOOD SYSTEMS

Definition 2.1.1 (*k-ply Neighborhood Systems*). A *k-ply neighborhood system in d dimensions* is a set $\{B_1, \ldots, B_n\}$ of closed balls in $\mathbb{R}^d$ such that no point in $\mathbb{R}^d$ is strictly interior to more than k of the balls.

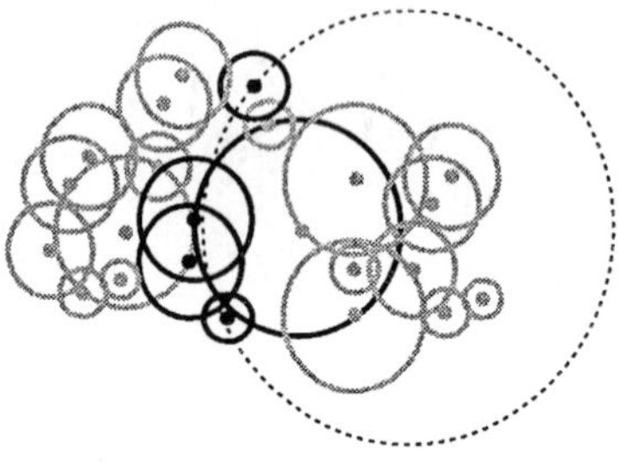

FIG. 2. The dotted circle is S. The circles with dark color are the circles in $\Gamma_O(S)$ and other circles are either in $\Gamma_E(S)$ or in $\Gamma_I(S)$.

For example, the neighborhood system given in Figure 1 is a 3-ply system.

In this definition, we used n for the number of points and d for the dimension of the embedding. We continue to use this notation throughout the paper. We also use the following notation: if $\alpha > 0$ and B is a ball of radius r, we define $\alpha \cdot B$ to be a ball with the same center as B but radius αr. In this paper, a $(d-1)$-sphere is the boundary of a d-dimensional ball.

2.2. A GEOMETRIC-SEPARATOR THEOREM. We now state our main separator theorem with respect to neighborhood systems. Each $(d-1)$-sphere S divides a neighborhood system $\Gamma = \{B_1, \ldots, B_n\}$ in $\mathbb{R}^d$ into three subsets: $\Gamma_E(S)$, the set of all balls of Γ in the exterior of S; $\Gamma_I(S)$, the set of all balls of Γ in the interior of S; and $\Gamma_O(S)$, the set of all balls of Γ that intersect S (see Figure 2).

THEOREM 2.2.1 (SPHERE-SEPARATOR THEOREM). *Suppose* $\Gamma = \{B_1, \ldots, B_n\}$ *is a k-ply system in* $\mathbb{R}^d$. *Then there is a sphere S such that*

$$|\Gamma_O(S)| = O(k^{1/d}n^{1-1/d}), \text{ and}$$

$$|\Gamma_I(S)|, |\Gamma_E(S)| \leq \frac{(d+1)n}{d+2}.$$

Furthermore, for any constant ϵ *in the range* $0 < \epsilon < 1/(d+2)$ *we can compute sphere S such that* $|\Gamma_I(S)|, |\Gamma_E(S)| \leq ((d+1)/(d+2) + \epsilon)n$, *and* $|\Gamma_O(S)| = O(k^{1/d}n^{1-1/d})$ *with probability at least* $1/2$.

The running time of this algorithm is bounded by $c(\epsilon, d) + O(nd)$, *where* $c(\epsilon, d)$ *is a constant depending only on* ϵ *and d.*

In this theorem and for the rest of the paper, for each finite set A, $|A|$ denotes the cardinality of A.

Remark 2.2.2. The randomization in the algorithm is over random numbers chosen by the algorithm irrespective of the input. Therefore, by rerunning the algorithm a constant number of times, we can increase the probability of success from 1/2 to $1 - \delta$ for an arbitrary $\delta > 0$.

Remark 2.2.3. The running time of the algorithm is linear, but almost all the work is done on a constant-sized subset (a subset whose size depends only on ϵ and d). The $O(nd)$ term in the running time arises from a final pass over the input to determine which of the B_i's goes into each set.

The remainder of Section 2 is devoted to proving Theorem 2.2.1.

2.3. Centerpoints and Conformal Maps. To prove Theorem 2.2.1 we need two geometric concepts: centerpoints and sphere-preserving (conformal) maps.

2.3.1. *Centerpoints*. A *centerpoint* of a given set P of points in d dimensions is a point $\mathbf{c} \in \mathbb{R}^d$ (not necessarily one of the given points) such that every hyperplane through $\mathbf{c}$ divides the given points approximately evenly (in the ratio $d : 1$ or better) [Edelsbrunner, Section 4]. It follows from Helly's theorem that every finite point set in $\mathbb{R}^d$ has a centerpoint. Various proofs can be found in Danzer et al. [1963]; Edelsbrunner [1987]; Miller et al. [1993]. It follows directly from these proofs that such a centerpoint can be found by Linear Programming on $O(n^d)$ linear inequalities of d variables. Unfortunately, no linear-time algorithm is known for computing centerpoints in higher dimensions. The following sampling algorithm can efficiently compute an approximate centerpoint:[1]

> Algorithm: (*Sampling for Approximate Centerpoints*)
>
> Input: (a point set $P \subset \mathbb{R}^d$)
>
> 1. Select a subset S of P with size l uniformly at random;
> 2. Compute a centerpoint $\mathbf{c}_S$ of S, using the Linear Programming algorithm for centerpoints;
> 3. Output $\mathbf{c}_S$.

It can be shown [Haussler and Welzl 1987; Teng 1991; Vapnik and Chervonenkis 1971] that for any constant $\epsilon < 1$, the above algorithm will compute a $(d + \epsilon) : 1$ centerpoint with high probability provided that $l > q(\epsilon, d)$, where q is a function that does not depend on n. Therefore, we can approximate a centerpoint in random constant time. In practice, we use an even simpler centerpoint approximation algorithm (see Gilbert et al. [1997], Miller et al. [1993], and Clarkson et al. [1993]).

2.3.2. *Conformal Mappings*. In our separator algorithm and our proof of Theorem 2.2.1, we map a neighborhood system from $\mathbb{R}^d$ to the unit d-sphere S^d in $\mathbb{R}^{d+1}$. An example of such a map is *stereographic projection.*

We will use notations that are consistent with our companion paper [Miller et al. 1997]. Let $\Pi(\mathbf{x})$ be the stereographic projection mapping from $\mathbb{R}^d$ to S^d. Geometrically, this map may be defined as follows: Given $\mathbf{x} \in \mathbb{R}^d$, append '0' as the final coordinate yielding $\mathbf{x}' \in \mathbb{R}^{d+1}$. Then compute the intersection of S^d with the line in $\mathbb{R}^{d+1}$ connecting $\mathbf{x}'$ to the north pole of S^d, $(0, 0, \ldots, 0, 1)^T$. This intersection point is $\Pi(\mathbf{x})$. Note that superscript T indicates transpose; thus the inner product of two vectors $\mathbf{x}, \mathbf{y}$ is denoted $\mathbf{x}^T\mathbf{y}$.

Algebraically, the mapping is defined as

$$\Pi(\mathbf{x}) = \begin{pmatrix} 2\mathbf{x}/\chi \\ 1 - 2/\chi \end{pmatrix} \tag{1}$$

[1] See, for example, Clarkson [1983], Haussler and Welzl [1987], Vapnik and Chervonenkis [1971], and Teng [1991].

where $\chi = \mathbf{x}^T\mathbf{x} + 1$. It is also simple to write down a formula for the inverse of Π. Let $\mathbf{u}$ be a point on S^d. Then

$$\Pi^{-1}(\mathbf{u}) = \frac{\overline{\mathbf{u}}}{1 - \mathbf{u}_{d+1}},$$

where $\bar{\mathbf{u}}$ denotes the first d entries of $\mathbf{u}$ and u_{d+1} is the last entry.

We will prove Theorem 2.2.1 using the fact that stereographic projection is *sphere-preserving*, that is, it maps spheres and hyperplanes (degenerate spheres) of $\mathbb{R}^d$ to spheres on S^d. A direct proof of this fact is given in Miller et al. [1997]. We will now give a somewhat indirect proof that stereographic projection preserves spheres. The purpose of this indirect proof is to explain some interesting properties of conformal maps in high dimensions as well as their connections back to two-dimensional conformal maps.

Define the inverse map from $\mathbb{R}^d$ to $\mathbb{R}^d$ to be

$$R(\mathbf{v}) = \frac{\mathbf{v}}{\mathbf{v}^T\mathbf{v}},$$

for all $\mathbf{v} \in \mathbb{R}^d$. We adopt the convention that $R(\vec{0}) = \infty$ and $R(\infty) = \vec{0}$, so that R is defined on $\mathbb{R}^d \cup \infty$.

Notice that each point $\mathbf{v}$ on the unit sphere S^{d-1} in $\mathbb{R}^d$ is mapped to itself. We now show that the inverse map R preserves spheres. In this statement, we will regard a hyperplane as a sphere as well.

Every sphere in $\mathbb{R}^d$ can be expressed by a quadratic equation of the following form:

$$a\mathbf{x}^T\mathbf{x} + b\mathbf{x}^T\mathbf{v}_0 + \mathbf{c} = 0,$$

where $\mathbf{v}_0$ is a constant d-vector. When $a = 0$, it is a hyperplane and when $\mathbf{c} = 0$ it is a sphere containing the origin, $\vec{0} \in \mathbb{R}^d$.

Notice that for each $\mathbf{v} \in \mathbb{R}^d$,

$$R(R(\mathbf{v})) = \frac{(\mathbf{v}/\mathbf{v}^T\mathbf{v})}{(\mathbf{v}/\mathbf{v}^T\mathbf{v})^T(\mathbf{v}/\mathbf{v}^T\mathbf{v})} = \frac{(\mathbf{v}/\mathbf{v}^T\mathbf{v})}{(1/\mathbf{v}^T\mathbf{v})} = \mathbf{v}.$$

Therefore, the inverse map is an involution.

PROPOSITION 2.3.2.1. *The inverse map preserves spheres.*

PROOF. Let $a\mathbf{y}^T\mathbf{y} + b\mathbf{y}^T\mathbf{v}_0 + c = 0$ be a sphere and let

$$\mathbf{y} = R(\mathbf{x}) = \frac{\mathbf{x}}{\mathbf{x}^T\mathbf{x}}.$$

After the substitution, we have

$$a\left(\frac{\mathbf{x}}{\mathbf{x}^T\mathbf{x}}\right)^T\left(\frac{\mathbf{x}}{\mathbf{x}^T\mathbf{x}}\right) + b\left(\frac{\mathbf{x}}{\mathbf{x}^T\mathbf{x}}\right)\mathbf{v}_0 + \mathbf{c} = 0,$$

yielding

$$a + b\mathbf{x}^T\mathbf{v}_0 + c\mathbf{x}^T\mathbf{x} = 0. \qquad \square$$

We now express stereographic projection (from $\mathbb{R}^d$ to S^d) and its inverse (from S^d to $\mathbb{R}^d$) in term as the inverse map from $\mathbb{R}^{d+1}$ to $\mathbb{R}^{d+1}$. Because S^d is embedded in $\mathbb{R}^{d+1}$, we need to first embed $\mathbb{R}^d$ in $\mathbb{R}^{d+1}$. Let L be the "natural" map from $\mathbb{R}^d$ to the hyperplane $x_{d+1} = 0$ in $\mathbb{R}^{d+1}$, that is, L sends each point $\mathbf{x} \in \mathbb{R}^d$ to a point in $\mathbb{R}^{d+1}$ by appending '0' as the final coordinate. Let $\mathbf{e} = (0, 0, \ldots, 0, 1) \in \mathbb{R}^{d+1}$. Let G be a map from $\mathbb{R}^{d+1}$ to $\mathbb{R}^{d+1}$ such that for each $\mathbf{u} \in \mathbb{R}^{d+1}$,

$$G(\mathbf{u}) = \frac{2(\mathbf{u} - \mathbf{e})}{(\mathbf{u} - \mathbf{e})^T(\mathbf{u} - \mathbf{e})} + \mathbf{e} = 2R(\mathbf{u} - \mathbf{e}) + \mathbf{e}. \tag{2}$$

Notice that G is an involution, because $G(\mathbf{u}) = 2R(\mathbf{u} - \mathbf{e}) + \mathbf{e}$, $R(G(\mathbf{u}) - \mathbf{e}) = R(2R(\mathbf{u} - \mathbf{e}))$. From the fact that $R(2\mathbf{v}) = R(\mathbf{v})/2$ for all $\mathbf{v} \in \mathbb{R}^{d+1}$, we have

$$R(G(\mathbf{u}) - \mathbf{e}) = R(2R(\mathbf{u} - \mathbf{e})) = \frac{\mathbf{R}(\mathbf{R}(\mathbf{u} - \mathbf{e}))}{2} = \frac{(\mathbf{u} - \mathbf{e})}{2},$$

implying

$$\mathbf{u} = 2R(G(\mathbf{u} - \mathbf{e})) + \mathbf{e} = G(G(\mathbf{u})).$$

Notice that for each $\mathbf{x} \in \mathbb{R}^d$, $(L(\mathbf{x}) - \mathbf{e})^T(L(\mathbf{x}) - \mathbf{e}) = \mathbf{x}^T\mathbf{x} + 1$, and thus we have

$$\Pi(\mathbf{x}) = G(L(\mathbf{x})).$$

Similarly, for each $\mathbf{y} \in S^d$,

$$\Pi^{-1}(\mathbf{y}) = L^{-1}(G(\mathbf{y})).$$

Because the inverse map and translations preserve spheres, G preserves spheres as well, implying stereographic projection Π preserves spheres. Thus, Π maps a ball (or a halfspace) of $\mathbb{R}^d$ to a cap on S^d, where a cap of S^d is the intersection of a closed halfspace in $\mathbb{R}^{d+1}$ with S^d.

2.3.3. *Conformally Mapping Centerpoints.* The translation of $\mathbb{R}^{d+1}$ by a vector $\mathbf{v}_0$ is a map that sends $\mathbf{v}$ to $\mathbf{v} - \mathbf{v}_0$. The dilation of $\mathbb{R}^{d+1}$ by a factor α is a map that sends $\mathbf{v}$ to $\alpha\mathbf{v}$. Clearly, they are all sphere-preserving. Other basic sphere preserving maps in $\mathbb{R}^{d+1}$ include the rigid rotations of $\mathbb{R}^{d+1}$, reflections, the inverse map, stereographic projection and its inverse.

LEMMA 2.3.3.1 (CENTERPOINT). *Let $P = \{\mathbf{p}_1, \ldots, \mathbf{p}_n\}$ be a point set in $\mathbb{R}^d$. There is a sphere-preserving map Φ from $\mathbb{R}^d$ to S^d such that the origin is a centerpoint of $\Phi(P) = \{\Phi(\mathbf{p}_1), \ldots, \Phi(\mathbf{p}_n)\}$.*

PROOF. Let $\mathbf{c} \in \mathbb{R}^{d+1}$ be a centerpoint of $\Pi(P) = \{\Pi(\mathbf{p}_1), \ldots, \Pi(\mathbf{p}_n)\}$, where Π is stereographic projection. Let $U_\mathbf{c}$ be a rotation or a (Householder) reflection such that $U_\mathbf{c}(\mathbf{c}) = (0, \ldots, 0, \|\mathbf{c}\|)$, where $\|\mathbf{c}\| = \sqrt{\mathbf{c}^T\mathbf{c}}$ denotes the standard l_2 norm of $\mathbf{c}$. Clearly, $U_\mathbf{c}(\mathbf{c})$ is a centerpoint of $U_\mathbf{c} \circ \Pi(P)$.

For any positive α, let $D_\alpha = \Pi \circ (\alpha I) \circ \Pi^{-1}$, where I is the $d \times d$ identity matrix, that is, αI is a dilation map. Let $Q = U_\mathbf{c} \circ \Pi(P)$.

We now show that if $\alpha = \sqrt{(1 - \|\mathbf{c}\|)/(1 + \|\mathbf{c}\|)}$, then the center of S^d, $\vec{0}$, is a centerpoint of $D_\alpha(Q)$. Thus, $\Phi = D_\alpha \circ U_\mathbf{c} \circ \Pi$ satisfies the lemma.

We first consider the case when $d \geq 2$.

Let $\mathbf{c}' = U_\mathbf{c}(\mathbf{c}) = (0, \ldots, 0, \|\mathbf{c}\|)$. A *circle* of S^d is given by the intersection of a hyperplane in $\mathbb{R}^{d+1}$ with S^d. Let $C_{\mathbf{c}'}$ be all circles on S^d whose hyperplanes contain $\mathbf{c}'$. For each circle $H \in C_{\mathbf{c}'}$, let $D_\alpha(H)$ be the image of H under D_α. Because D_α preserves circles, $D_\alpha(H)$ is also a circle in S^d.

We first consider the circle $H_0 \in C_{\mathbf{c}'}$ whose hyperplane is normal to axis x_{d+1} (which connects the north pole with the south pole). Notice that $\Pi^{-1}(H_0)$ is a sphere in $\mathbb{R}^d$ centered at the origin. The radius of $\Pi^{-1}(H_0)$ is $\sqrt{(1 + \|\mathbf{c}\|)/(1 - \|\mathbf{c}\|)}$. By our choice of α, $D_\alpha(H_0)$ is the equator of S^d and hence is normal to axis x_{d+1} and contains the origin.

Suppose $\mathbf{q}_1$ and $\mathbf{q}_2$ are two points in $H_0 \cap S^d$. If the line segment between them is a diameter of H_0, then this line segment contains $\mathbf{c}'$. Moreover, the line segment between $D_\alpha(\mathbf{q}_1)$ and $D_\alpha(\mathbf{q}_2)$ is a diameter of $D_\alpha(H_0)$ and hence it contains the origin.

We now show for each $H \in C_{\mathbf{c}'}$ that the hyperplane of $D_\alpha(H)$ contains the origin as well. By doing this, we can conclude from definition of a centerpoint that the origin is a centerpoint of $D_\alpha \circ U_\mathbf{c} \circ \Pi(P)$.

The intersection of the hyperplanes of H and H_0 is an affine set of dimension $d - 1$. Because, we assume $d \geq 2$, this set has dimension at least 1 and contains $\mathbf{c}'$. Thus, it must contain a diameter of H_0. In other words, there exist two points $\mathbf{q}_1$ and $\mathbf{q}_2$ in $H_0 \cap H$ such that the line segment between them is a diameter of H_0. $D_\alpha(\mathbf{q}_1)$ and $D_\alpha(\mathbf{q}_2)$ are both in $D_\alpha(H_0) \cap D_\alpha(H)$, and the line segment between them contains the origin. Therefore, the hyperplane of $D_\alpha(H)$ also contains the origin.

When $d = 1$, we can embed $\mathbb{R}^1$ in $\mathbb{R}^2$. The proof as given above for $\mathbb{R}^2$ also shows that the lemma is true for $d = 1$. ☐

2.4. A RANDOMIZED ALGORITHM. We now present our separator algorithm. The algorithm uses randomization, and it chooses the separating sphere at random from a distribution that is carefully constructed so that the separator will satisfy the conclusions of Theorem 2.2.1 with high probability. The distribution is described in terms of sphere-preserving maps in $\mathbb{R}^{d+1}$.

Algorithm *Sphere Separator*

Input: a k-ply system $\{B_1, \ldots, B_n\}$ in $\mathbb{R}^d$ with centers $P = \{\mathbf{p}_1, \ldots, \mathbf{p}_n\}$.

- **Project Up.** Compute $\Pi(P) = \{\Pi(\mathbf{p}_1), \ldots, \Pi(\mathbf{p}_n)\}$.
- **Find Centerpoint.** Compute a *centerpoint* $\mathbf{z}$ of $\Pi(P)$ in $\mathbb{R}^{d+1}$.
- **Conformal Map I: Rotate.** Compute an orthogonal $(d + 1) \times (d + 1)$ matrix $U_\mathbf{z}$ such that $U_\mathbf{z}(\mathbf{z}) = \mathbf{z}'$ where

$$\mathbf{z}' = (0, \ldots, 0, \|\mathbf{z}\|).$$

  Note that $\mathbf{z}'$ is a centerpoint of $U_\mathbf{z} \circ \Pi(P)$.
- **Conformal Map II: Dilate.** Let $D_\alpha = \Pi \circ (\alpha I) \circ \Pi^{-1}$, where $\alpha = \sqrt{(1 - \|z\|)/(1 + \|z\|)}$. As shown in Lemma 2.3.3.1, the origin $\vec{0}$ is a centerpoint of $D_\alpha \circ U_z \circ \Pi(P)$.
- **Find Great Circle.** Choose a random great circle C on S^d.
- **Unmap and Project Down.** Transform the great circle C to a sphere S in $\mathbb{R}^d$ by undoing the dilation, rotation, and stereographic projection:

$$S = \Pi^{-1} \circ U_\mathbf{z}^{-1} \circ D_\alpha^{-1}(C).$$

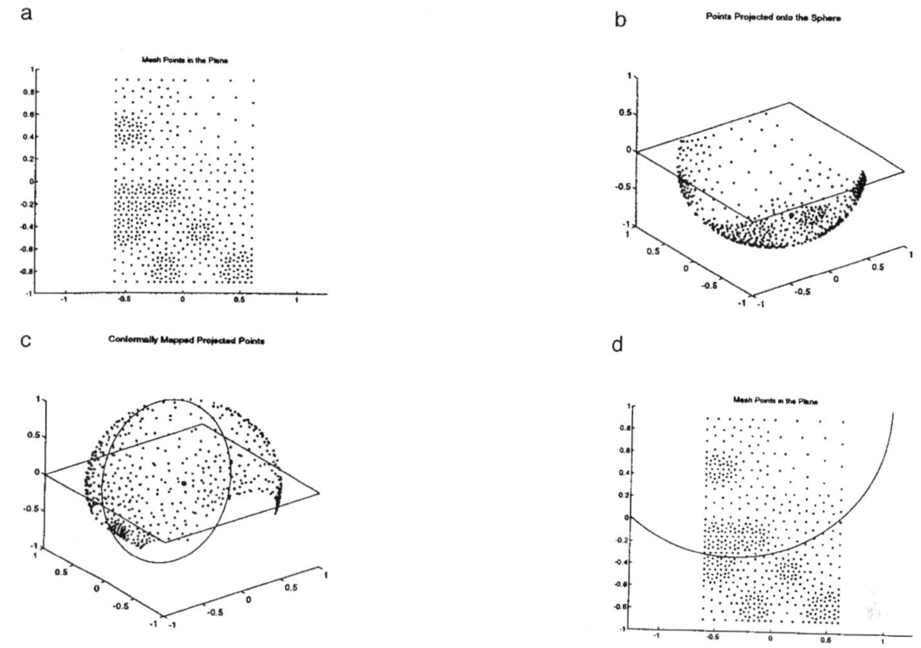

FIG. 3. (a) The point set of a neighborhood system. (b) Project Up and Find Centerpoint; the largest dot in the figure is a centerpoint. (c) Conformal Map II and Find Great Circle. (d) Unmap and Project Down. (Generated by the Matlab Geometric Separator Tool-box developed by Gilbert and Teng.

- Return S and $\Gamma_O(S)$, $\Gamma_I(S)$, and $\Gamma_E(S)$.

Figure 3 depicts the basic steps of our separator algorithm. It is generated by the Matlab Geometric Separator Tool-box of Gilbert and Teng [Gilbert et al. 1997]. The example geometric graph is generated by Eppstein. The neighborhood system of this point set is not explicitly shown in the figure. Each point defines a ball which is the largest ball centered at the point whose interior contains no other points. The ply of this neighborhood system is 6.

Notice that we can use an approximate centerpoint in the algorithm above.

In the next section, we will prove Theorem 2.2.1 by showing that $|\Gamma_O(S)| \leq O(k^{1/d}n^{1-1/d})$ with high probability.

2.5. A HIGH-LEVEL DISCUSSION. We now give a high level description of our approach to prove Theorem 2.2.1.

Let $\Gamma = \{B_1, \ldots, B_n\}$ be a k-ply neighborhood system in $\mathbb{R}^d$. Let $\mathbf{p}_i$ be the center of B_i, for $1 \leq i \leq n$. Let S^d be the unit sphere in $\mathbb{R}^{d+1}$ whose center is the origin. In the previous subsection, we have shown that there is a *sphere-preserving* map Φ from $\mathbb{R}^d$ to S^d such that the origin is a centerpoint of $\{\Phi(\mathbf{p}_1), \ldots, \Phi(\mathbf{p}_n)\}$. Let $\Phi(B_i)$ be the image of B_i on S^d; $\Phi(B_i)$ is a cap on S^d.

A *great circle* of S^d is the intersection of S^d with a hyperplane that passes through the center of S^d. Clearly, every great circle divides S^d into two open half-spheres (hemispheres). Each great circle C of S^d divides $\{\Phi(B_1), \ldots, \Phi(B_n)\}$ into three sets, C_1, C_{-1} and C_0 where C_1 and C_{-1} respectively contains those caps that are completely in each of the open hemisphere, and C_0 contains those caps that intersect C. Because the center of S^d is a centerpoint of

$\{\Phi(\mathbf{p}_1), \ldots, \Phi(\mathbf{p}_n)\}$, we have $|C_{-1}|, |C_1| \le (d + 1)/(d + 2)n$. In order to prove Theorem 2.2.1, it is then sufficient to show that there exists a great circle C of S^d such that $|C_0| = O(k^{1/d}n^{1-1/d})$. We will prove this by arguing that the expected size of C_0 is $O(k^{1/d}n^{1-1/d})$ when C is chosen uniformly among all great circles of S^d. Since the result of this random choice is always a nonnegative number, we conclude that the probability of exceeding the expected value by more than a factor of 2 is at most 0.5.

2.6. A GEOMETRIC TECHNIQUE FOR PROVING SEPARATOR THEOREMS. A much simplified proof has been obtained, independently, by Agarwal and Pach [1995] and Spielman and Teng [1996a] since the conference publication of our result.[2] We will present our original proof here because its technique might be useful for other problems. Readers interested in the simpler proof should skip the remainder of Section 2 and refer to Pach and Agarwal [1995] and Spielman and Teng [1996a]. We will also give a high-level explanation of the simpler proof in Section 5.

Our approach is to first design a continuous function and apply a continuous version of the separator theorem (given in the next subsection) to show that some "weighted surface area" of S is "small," from which we then show the number of balls that intersect S is "small."

2.6.1. *A Continuous Separator Theorem.* Suppose $f(x)$ is a real-valued non-negative function defined on $\mathbb{R}^d$ such that f^k is integrable for all $k = 1, 2, 3, \ldots$. Such an f is called a *cost function*. The *total volume* of the function f is defined as

$$\text{Total-Volume}(f) = \int_{\mathbf{v} \in \mathbb{R}^d} (f(\mathbf{v}))^d (\mathbf{dv})^d$$

Suppose S is a $(d - 1)$-sphere in $\mathbb{R}^d$. The *surface area* of S is then

$$\text{Area}(f, S) = \int_{\mathbf{v} \in S} (f(\mathbf{v}))^{d-1} (\mathbf{dv})^{d-1}$$

Let $P = \{\mathbf{p}_1, \ldots, \mathbf{p}_n\}$ be a point set in $\mathbb{R}^d$. Let Φ be a sphere-preserving map from $\mathbb{R}^d$ to S^d so that the center of S^d is a centerpoint of $\Phi(P)$ (see Lemma 2.3.3.1). Recall that our sphere separator algorithm computes such a map and then uses its inverse to map a random great circle back to a sphere in $\mathbb{R}^d$.

Let S be a sphere in $\mathbb{R}^d$. The weighted surface area of S equal to $\text{Area}(f, S)$. Because Φ carries S in $\mathbb{R}^d$ to a circle $\Phi(S)$ in S^d, we define $\text{Cost}(\Phi(S)) = \text{Area}(f, S)$. Let $\text{Avg}_\Phi(f)$ be the average cost of all great circles of S^d.

The following theorem has been proven in our companion paper [Miller et al. 1997].

THEOREM 2.6.1.1 (CONTINUOUS SEPARATOR). *Suppose f is a cost function on $\mathbb{R}^d$. Let Φ be the map from $\mathbb{R}^d$ to S^d constructed in Lemma 2.3.3.1. Then,*

$$Avg_\Phi(f) = O((\textit{Total-Volume}(f))^{1-1/d}).$$

[2] See, for example, Miller et al. [1991], Miller and Thurston [1990], and Miller and Vavasis [1991].

Remark 2.6.1.2. The conformality of the mapping from $\mathbb{R}^d$ to S^d is necessary in Theorem 2.6.1.1 in order to deal with the volume elements in the high-dimensional integrations. We refer the reader to our companion paper [Miller et al. 1997] for a discussion.

2.6.2. *Construction of a Cost Function.* To prove Theorem 2.2.1, it is sufficient to construct, for each k-ply system $\Gamma = \{B_1, \ldots, B_n\}$, a continuous function whose total volume is $O(k^{1/(d-1)}n)$ such that each sphere S in $\mathbb{R}^d$ intersects at most $O(\text{Area}(f, S))$ balls of Γ.

Let r_i be the radius of B_i and let $\gamma_i = 2r_i$, and define

$$f_i(\mathbf{x}) = \begin{cases} 1/\gamma_i & \text{if } \|\mathbf{x} - \mathbf{p}_i\| \leq \gamma_i \\ 0 & \text{otherwise.} \end{cases}$$

Intuitively, f_i sets up a cost on each $(d - 1)$-sphere S such that the closer S is to B_i, the larger B_i contributes to the surface area of S. Using a simple geometric argument (see the companion paper [Miller et al. 1997]), we can show that for any sphere S in $\mathbb{R}^d$,

$$|\Gamma_O(S)| = O(\text{Area}_f(S)). \tag{3}$$

The function f_i is called the *local* function of B_i. We define our continuous function f as

$$f(\mathbf{x}) = l_{d-1}(f_1(\mathbf{x}), \ldots, f_n(\mathbf{x})) = \left(\sum_{i=1}^{n} (f_i(\mathbf{x}))^{d-1} \right)^{1/(d-1)},$$

where for each positive integer p, l_p denotes the standard pth norm in Euclidean space, that is, for each $a_1, \ldots, a_n$,

$$l_p(a_1, \ldots, a_n) = \left(\sum_{i=1}^{n} \left| a_i \right|^p \right)^{1/p}.$$

2.6.3. *Bounding the Total Volume of the Cost Function.* Now we give an upper bound on the total volume of f. Let V_d be the volume of a unit ball in $\mathbb{R}^d$. Clearly, $\int_{\mathbf{x}\in \mathbb{R}^d} (f_i(\mathbf{x}))^d (d\mathbf{x})^d = V_d$.

Consequently, letting

$$g(\mathbf{x}) = L_d(f_1, \ldots, f_n) = \left(\sum_{i=1}^{n} (f_i(\mathbf{x}))^d \right)^{1/d},$$

we have

$$\text{Total-Cost}(g) = \int_{\mathbf{x} \in \mathbb{R}^d} (g(\mathbf{x}))^d (d\mathbf{x})^d = V_d n$$

LEMMA 2.6.3.1. *Suppose $\Gamma = \{B_1, \ldots, B_n\}$ is a k-ply system in $\mathbb{R}^d$. Let $f_1, \ldots, f_n$ and f be functions defined as above. Then*

$$\textit{Total-Cost}(f) = O(k^{1/(d-1)}n).$$

PROOF. Because Total-Cost$(g) = V_d n$, it is sufficient to show that for all $\mathbf{x} \in \mathbb{R}^d$,

$$(f(\mathbf{x}))^d \leq c_d k^{1/(d-1)} \cdot (g(\mathbf{x}))^d.$$

We focus on a particular point $\mathbf{p} \in \mathbb{R}^d$. Notice that if $g(\mathbf{p}) = 0$, then, $f(\mathbf{p}) = 0$ as well. The inequality follows. Now, assume $g(\mathbf{p}) > 0$ and define

$$M_l = \{i \in \{1, \ldots, n\} : 2^{-l} \leq f_i(\mathbf{p}) < 2^{-l+1}\},$$

for all l such that $-\infty < l < \infty$.

Because $\cup_{-\infty \leq l \leq \infty} M_l = \{i : f_i(\mathbf{p}) \neq 0\}$ and M_l's are pairwise disjoint, each index i such that $f_i(\mathbf{p}) \neq 0$ occurs in exactly one of M_l's. Let $m_l = |M_l|$. We claim $m_l \leq 6^d k$.

We now prove the claim. For each $i \in M_l$, by the definition of M_l and f_i, $2^{l-1} \leq \gamma_i \leq 2^l$, where $\gamma_i = 2r_i$. Let B be a ball centered at $\mathbf{p}$ with radius $2^l + 2^{l-1}$. Since $\|\mathbf{p} - \mathbf{p}_i\| \leq \gamma_i$, it follows $B_i \subset B$. Because the neighborhood system has ply k, we have

$$k \cdot \mathrm{vol}(B) \geq \sum_{j \in M_l} \mathrm{vol}(B_j).$$

Let $V_d(r)$ be the volume of a ball in $\mathbb{R}_d$ of radius r. Because for all $j \in M_l$, $\mathrm{vol}(B_j) \geq V_d(2^{l-2})$,

$$k \cdot V_d(2^l + 2^{l-1}) \geq |M_l| V_d(2^{l-2}),$$

which implies $|M_l| \leq 6^d k$, completing the proof of the claim. Now, we have

$$\begin{aligned}(f(\mathbf{p}))^d &= \left(\sum_{l=-\infty}^{\infty} \sum_{i \in M_l} f_i(\mathbf{p})^{(d-1)}\right)^{d/(d-1)} \\ &\leq \left(\sum_{l=-\infty}^{\infty} m_l (2^{-l+1})^{d-1}\right)^{d/(d-1)} \\ &\leq 2^d \left(\sum_{l=-\infty}^{\infty} m_l (2^{-l})^{d-1}\right)^{d/(d-1)},\end{aligned}$$

where $m_l \leq 6^d k$.

We now use Inequality (4) below, established in our companion paper [Miller et al. 1997] to bound the right hand side of the equation above.

Let $\ldots, m_{-1}, m_0, m_1, m_2, \ldots$ be a doubly infinite sequence of nonnegative numbers such that each m_i is bounded above by θ and such that at most a finite number of m_i's are nonzero. Let $d \geq 2$ be an integer. Then

$$\left(\sum_{k=-\infty}^{\infty} m_k 2^{-k(d-1)}\right)^{d/(d-1)} \leq c_d \theta^{1/(d-1)} \sum_{k=-\infty}^{\infty} m_k 2^{-kd}. \tag{4}$$

where c_d is a positive number depending on d.

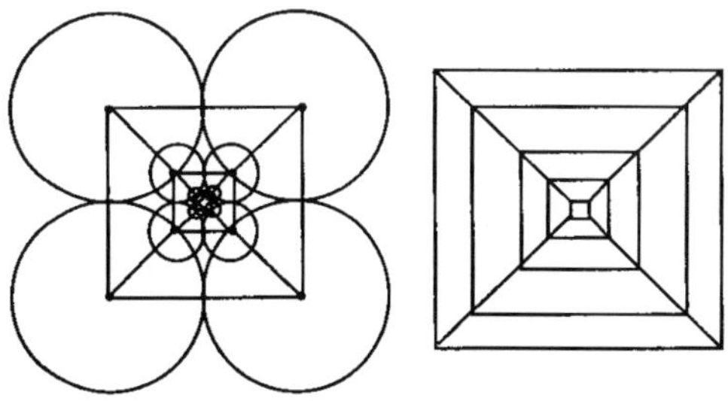

FIG. 4. Why not hyperplane separators.

Setting $\theta = 6^d k$ and applying Inequality (4), we obtain

$$f(\mathbf{p})^d \leq c_d 2^d (6^d k)^{1/(d-1)} \sum_{l=-\infty}^{\infty} m_l 2^{-ld}.$$

This summation is a lower bound on $g(\mathbf{p})^d$ because for each $i \in M_l$, $f_i(\mathbf{p})^d \geq 2^{-ld}$. This concludes the proof of the lemma. □

Consequently, by Theorem 2.6.1.1, there exists a $(d+1)/(d+2)$-splitting sphere S of Γ with

$$\text{Area}_f(S) = O(k^{1/d} n^{1-1/d}).$$

From the definition of centerpoint, we have

$$|\Gamma_I(S)|,\ |\Gamma_E(S)| \leq \frac{(d+1)n}{d+2}.$$

By Inequality (3), we have

$$|\Gamma_O(S)| = O(\text{Area}_f(S)) = O(k^{1/d} n^{1-1/d}).$$

We thus proved Theorem 2.2.1.

2.7. SPHERES VS HYPERPLANES. The simplest way to split a set of points in d-space is to use a $(d-1)$-dimensional hyperplane. Notice that a $(d-1)$-dimensional hyperplane is just a degenerate $(d-1)$-sphere. Like a $(d-1)$-dimensional sphere, a $(d-1)$-dimensional hyperplane $\mathbf{h}$ partitions $\mathbb{R}^d$ into three subsets, $\mathbf{h}^+$, those that are above $\mathbf{h}$, $\mathbf{h}^-$, those below $\mathbf{h}$, and $\mathbf{h}$ itself, respectively. We now show, for some k-ply neighborhood systems, that it is necessary to use sphere to achieve the bound given in Theorem 2.2.1.

One such an example is given in Figure 4. It is a 2-ply system. Notice that any hyperplane that divides the neighborhood system into two of a constant ratio must intersect $\Omega(n)$ balls of the 2-ply system.

In contrast, the above 2-ply system does have a sphere that intersects $O(n^{1-1/d})$ balls and divides the rest of the balls into two sets of ratio no worst than $1 : (d+1)$ as illustrated in Figure 5. The ratio of the two partitioned sets is $1 : 1$ in the example above.

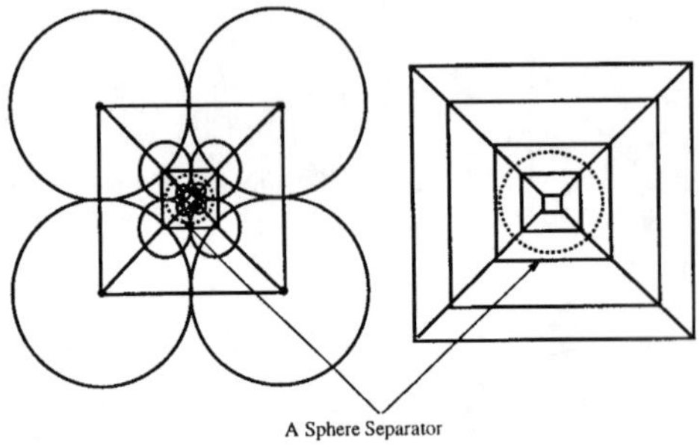

FIG. 5. Why sphere separators.

3. *Intersection Graphs*

In Section 2, we analyzed separators of an abstract geometric arrangement called a k-ply neighborhood system. The purpose of this section is to apply this abstraction to concrete classes of graphs. The most straightforward such application is to the intersection graph of k-ply systems. We also show in this section that the theory applies to planar graphs, sphere packings, and nearest neighbor graphs. Other classes of graphs not detailed here, such as finite subgraphs of the regular d-dimension grid-graph, are also covered by the theory developed in this section.

We will use the following definition of graph separator.

Definition 3.1 (*Separators*). A subset of vertices C of a graph G with n vertices is an $f(n)$*-separator* that δ*-splits* if $|C| \leq f(n)$ and the vertices of $G - C$ can be partitioned into two sets A and B such that there are no edges from A to B, and $|A|$, $|B| \leq \delta n$, where f is a function and $0 < \delta < 1$.

Given a neighborhood system, it is possible to define the intersection graph associated with the system (see Figure 6).

Definition 3.2 (*Intersection Graphs*). Let $\Gamma = \{B_1, \ldots, B_n\}$ be a neighborhood system. The *intersection graph* of Γ is the undirected graph with vertices $V = \Gamma$ and edges

$$E = \{(B_i, B_j) : B_i \cap B_j \neq \emptyset.\}$$

It follows directly from Theorem 2.2.1, that the intersection graph of a k-ply neighborhood system has a small separator.

THEOREM 3.3. *Suppose* $\Gamma = \{B_1, \ldots, B_n\}$ *is a k-ply neighborhood system in* $\mathbb{R}^d$. *Then the intersection graph of* Γ *has an* $O(k^{1/d}n^{1-1/d})$ *separator that* $(d + 1)/(d + 2)$*-splits.*

The separator bound of Theorem 3.3 is the best possible in both k and n up to a constant factor. An $\Omega(n^{1-1/d})$ bound on the intersection graph of a 1-ply neighborhood system appeared in Vavasis [1991]. Let P be the set of all points of the $m \times m \times \cdots \times m$ regular grid in $\mathbb{R}^d$, where $n = m^d$. It has been shown in Teng [1991], for a sufficiently large n, that the k-nearest neighbor graph of P has no separator of size $o(k^{1/d}n^{1-1/d})$.

3.1. SPHERE-PACKINGS AND PLANAR GRAPHS. A graph $G = (V, E)$ is *planar* if we can "draw" it in the plane in such a way that each vertex is represented by

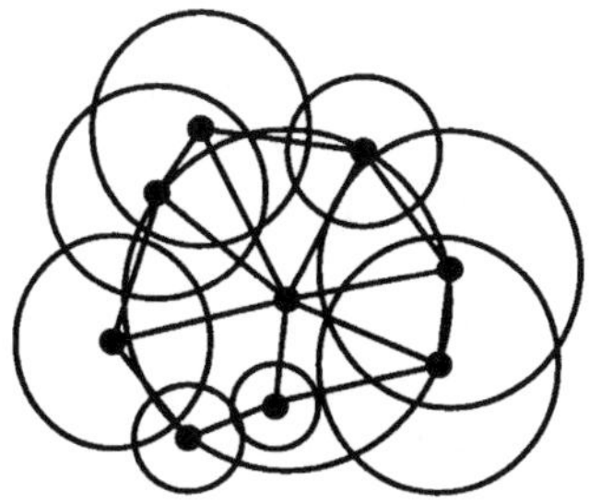

FIG. 6. The intersection graph of a 3-ply system.

a point; each edge is represented by a continuous curve connecting the two points which represent its end vertices, and no two curves share any points, except at their ends. We will only consider simple planar graphs which are graphs that do not have self-loops nor multiple edges between any pair of vertices.

We now show that Theorem 2.2.1 in conjunction with the beautiful Koebe-embedding result of planar graphs [Koebe 1936; Andreev 1970a; 1970b; Thurston 1988] gives a geometric proof of the Lipton and Tarjan planar separator theorem.

Let a *disk-packing* be a set of disks $D_1, \ldots, D_n$ that have disjoint interiors. Notice that every disk-packing is a 1-ply neighborhood system. We call the intersection graph of a disk-packing a *disk-packing graph.* It is not hard to see that every disk-packing graph is a planar graph. Koebe [1936] showed that in fact every planar graph can be represented as the intersection graph of a disk-packing. We call such a realization a *Koebe*-embedding of the planar graph. For a history of this result, including a comparison of Koebe's original result versus the Andreev–Thurston's proof, see Ziegler [1988].

THEOREM 3.1.1 (KOEBE). *Every triangulated planar graph G is isomorphic to a disk-packing graph.*

Because disk-packings are 1-ply systems in two dimensions, it follows from the existence of Koebe-embedding of planar graphs and our sphere separator Theorem 2.2.1 that every planar graph has an $O(\sqrt{n})$-separator that 3/4-splits.

We can extend the disk-packing to high dimensions: A *sphere-packing* is a neighborhood system $\Gamma = \{B_1, \ldots, B_n\}$ in $\mathbb{R}^d$ whose balls have disjoint interiors. Clearly, each sphere packing is a 1-ply neighborhood system. Therefore, the intersection graph of a sphere packing has an $O(n^{1-1/d})$ separator that $(d + 1)/(d + 2)$-splits.

Remark 3.1.2. Koebe's result strengthens Fáry's [1948] and Tutte's [1960; 1963] theorem that every planar graph can be embedded in the plane such that each edge is mapped to a straight line segment (see Thomassen [1980] and De Fraysseix et al. [1988].

Remark 3.1.3. Spielman and Teng [1996a] have recently demonstrated that the application of Theorem 2.2.1 on the Koebe-embedding of a planar graph finds a $1.84\sqrt{n}$-separator that 3/4-splits. The two constants 1.84 and 3/4 occurring in Spielman and Teng [1996a] lead to the best known bound for the constants in planar nested dissection, improving on Lipton et al. [1979] and other subsequent improved constants.

Given a Koebe-embedding, our geometric algorithm runs in random linear time with a constant smaller than that of Lipton–Tarjan's linear-algorithm [Lipton and Tarjan 1979]. It is still quite expensive to compute a Koebe embedding. Mohar [1993] has recently developed a polynomial time algorithm.

3.2. NEAREST NEIGHBOR GRAPHS. The *nearest neighbor graph* is an important class of graphs in computational geometry [Preparata and Shamos 1985]. The nearest neighbor graph arises naturally in practical applications such as image reconstruction and pattern recognition.

Let $P = \{\mathbf{p}_1, \ldots, \mathbf{p}_n\}$ be a point set in $\mathbb{R}^d$. For each $\mathbf{p}_i \in P$, let $N_k(\mathbf{p}_i)$ be the set of k points closest to $\mathbf{p}_i$ in P (where ties are broken arbitrarily). A *k-nearest neighbor graph* [Preparata and Shamos 1985] of P is a graph with vertex set $\{\mathbf{p}_1, \ldots, \mathbf{p}_n\}$ and edge set

$$E = \{(\mathbf{p}_i, \mathbf{p}_j) : \mathbf{p}_i \in N_k(\mathbf{p}_j) \text{ or } \mathbf{p}_j \in N_k(\mathbf{p}_i)\}.$$

In this section, we show that every k-nearest neighbor graph in $\mathbb{R}^d$ is a subgraph of the intersection graph of a $O(k)$-ply neighborhood system.

For each point $\mathbf{p}_i$, let $B_i^{(k)}$ be the largest ball centered at $\mathbf{p}_i$ that contains at most k points from P, counting $\mathbf{p}_i$ itself, in the interior of $B_i^{(k)}$. Clearly the radius of $B_i^{(k)}$ is equal to the distance from $\mathbf{p}_i$ to its kth nearest neighbor in P. We call $N_k(P) = \{B_1^{(k)}, \ldots, B_n^{(k)}\}$ the k-nearest neighborhood system for P.

Notice that the k-nearest neighbor graph of P is a subgraph of the intersection graph of $\{B_1^{(k)}, \ldots, B_n^{(k)}\}$ because if $\mathbf{p}_j$ is one of $\mathbf{p}_i$'s k nearest neighbors, then $\mathbf{p}_j$ is contained in $B_i^{(k)}$ and hence $B_i^{(k)}$ must intersect with $B_j^{(k)}$.

We now show that the ply of $N_k(P)$ is at most $\tau_d k$, where τ_d is the *kissing number* in d dimensions, which is the maximum number of nonoverlapping unit balls in $\mathbb{R}^d$ that can be arranged so that they all touch a central unit ball [Conway and Sloane 1988]. It is known that $\tau_1 = 2$, $\tau_2 = 6$, $\tau_3 = 12$, $\tau_8 = 240$, and $\tau_{24} = 196560$. Although there is no explicit formula known for the kissing number τ_d for a general choice of d, it can be bounded from above and below by the following inequalities.

$$2^{0.2075..d(1+o(1))} \le \tau_d \le 2^{0.401d(1+o(1))}.$$

The first inequality was given by Kabatiansky and Levenshtein [1978] and the second one by Wyner [1965].

LEMMA 3.2.1 (PLY LEMMA). *Let $P = \{\mathbf{p}_1, \ldots, \mathbf{p}_n\}$ be a point set in $\mathbb{R}^d$. Then the ply of $N_k(P)$ is bounded by $\tau_d k$.*

PROOF. Denote the balls in $N_k(P)$ by $\{B_1, \ldots, B_n\}$. We first prove the lemma when $k = 1$. In this case no ball contains the center of other balls in its interior.

Let $\mathbf{p}$ be the point in $\mathbb{R}^d$ with the largest ply. Without loss of generality, let $\{B_1, \ldots, B_t\}$ be the set of all balls that contain $\mathbf{p}$. Let C_i be the ball centered at $\mathbf{p}_i$ with radius $\|\mathbf{p}_i - \mathbf{p}\|$, and hence $\mathbf{p}$ is on the boundary of C_i for each i in the range $1 \le i \le t$ (see Figure 7). Clearly, C_i is contained in B_i and C_i does not contains the center of any other balls.

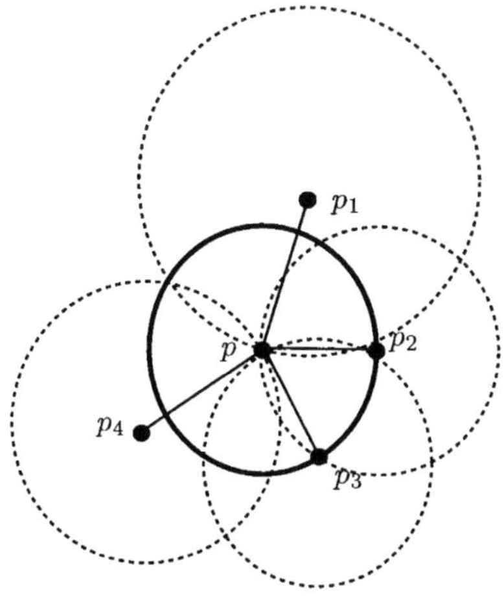

FIG. 7. The set of balls that touch a point.

Let $\delta = \min\{\|\mathbf{p} - \mathbf{p}_i\| : 0 \le i \le t\}$. Let $S_\mathbf{p}$ be the sphere centered at $\mathbf{p}$ with radius δ. Let $\mathbf{q}_i$ be the intersection of the ray $\mathbf{pp}_i$ with the sphere $S_\mathbf{p}$. We claim that for each pair i, j, $i \ne j$, in the range $1 \le i, j \le t$, $\|\mathbf{q}_i - \mathbf{q}_j\| \ge \delta$.

Without loss of generality, assume $\|\mathbf{p} - \mathbf{p}_i\| \ge \|\mathbf{p} - \mathbf{p}_j\|$. Let s be a point on the ray $\mathbf{pp}_i$ such that $\|\mathbf{p} - \mathbf{s}\| = \|\mathbf{p} - \mathbf{p}_j\|$ (see Figure 8). It follows $\|\mathbf{p} - \mathbf{p}_i\| = \|\mathbf{p} - \mathbf{s}\| + \|\mathbf{s} - \mathbf{p}_i\|$.

By the triangle inequality, we have $\|\mathbf{s} - \mathbf{p}_i\| + \|s - \mathbf{p}_j\| \ge \|\mathbf{p}_i - \mathbf{p}_j\|$.

Because $\mathbf{p}_j \notin C_i$, and the radius of C_i is $\|\mathbf{p} - \mathbf{p}_i\|$, we have $\|\mathbf{p} - \mathbf{p}_i\| \le \|\mathbf{p}_i - \mathbf{p}_j\|$. Thus $\|\mathbf{p} - \mathbf{s}\| \le \|\mathbf{s} - \mathbf{p}_j\|$.

By the similarity of triangles $\Delta\mathbf{pq}_i\mathbf{q}_j$ and $\Delta\mathbf{psp}_j$, we have $\|\mathbf{q}_i - \mathbf{q}_j\| \ge \|\mathbf{p} - \mathbf{q}_i\| = \delta$. Notice that the kissing number τ_d is equal to the maximum number of points that can be arranged on a unit $(d - 1)$-sphere (the boundary of a unit d-ball), such that the distance between each pair of points is at least 1. Therefore, $t \le \tau_d$, completing the proof of the lemma when $k = 1$.

We now prove the lemma for any $k > 1$. Without loss of generality, assume $B_1, \ldots, B_t$ contain $\mathbf{p}$. Define a subset Q of $\{\mathbf{p}_1, \ldots, \mathbf{p}_t\}$ by the following procedure. Initially, let $P = \{\mathbf{p}_1, \ldots, \mathbf{p}_t\}$ and $Q = \emptyset$.

while $P \ne \emptyset$

(1) Let $\mathbf{q}$ be the point in P with the largest $\|\mathbf{q} - \mathbf{p}\|$, let $Q = Q \cup \{\mathbf{q}\}$;

(2) Let $P = P - \mathrm{int}(B_\mathbf{q})$, (where $B_\mathbf{q}$ stands for the closed ball centered at $\mathbf{q}$).

Because no ball contains more than k points from $\{\mathbf{p}_1, \ldots, \mathbf{p}_t\}$ in its interior, we have $m \ge \lceil t/k \rceil$, where m denotes $|Q|$.

We now show that for all $\mathbf{q} \in Q$, $\mathrm{int}(B_q) \cap Q = \{\mathbf{q}\}$. Suppose $Q = \{\mathbf{q}_1, \ldots, \mathbf{q}_m\}$ such that for all $i < j$, $\mathbf{q}_i$ is put Q in the above procedure before $\mathbf{q}_j$. Notice that for all $j > i$, $\mathbf{q}_j \notin \mathrm{int}(B_{\mathbf{q}_i})$. Also because $\|\mathbf{q}_i - \mathbf{q}_j\| \ge \|\mathbf{q}_i - \mathbf{p}\| \ge \|\mathbf{q}_j - \mathbf{p}\|$, we have for all $i < j$, $\mathbf{q}_i \notin \mathrm{int}(B_{\mathbf{q}_j})$. So $\mathrm{int}(B_\mathbf{q}) \cap Q = \{\mathbf{q}\}$. Thus, $m \le \tau_d$ which implies $t \le \tau_d k$. $\square$

Consequently, by Theorem 2.2.1:

THEOREM 3.2.2. *Every k-nearest neighbor graph in d dimensions has an* $O(k^{1/d}n^{1-1/d})$ *separator that* $(d + 1)/(d + 2)$*-splits.*

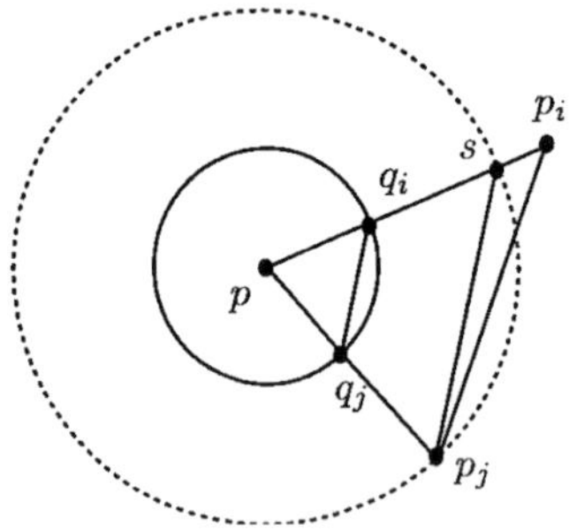

FIG. 8. The distance between q_i and q_j.

The following is an interesting consequence of the Ply Lemma 3.2.1:

COROLLARY 3.2.3. *The degree of all k-nearest neighbor graphs in d dimensions is bounded above by* $(\tau_d + 1)k$.

PROOF. For each point $\mathbf{p}_i$ in a given set P, let B_i be the largest ball centered at $\mathbf{p}_i$ such that the interior of B_i contains no more than k points from P. Notice that if $(\mathbf{p}_i, \mathbf{p}_j)$ is an edge in a k-nearest neighbor graph of P, then either $\mathbf{p}_i \in B_j$ or $\mathbf{p}_j \in B_i$. By Lemma 3.2.1, we have the degree of k-nearest neighbor graphs is bounded above by $(\tau_d + 1)k$. □

Remark 3.2.4. Toussaint [1988] called the intersection graph of $N_k(P)$ the *kth sphere-of-influence graph* of P and showed that this class of graphs can be used in image processing. It follows from Lemma 3.2.1 and Theorem 3.3.1 that the kth sphere-of-influence graph of any set of n points in $\mathbb{R}^d$ has an $O(k^{1/d} n^{1-1/d})$ separator that $(d + 1)/(d + 2)$-splits.

3.3. INDUCTIVITY OF INTERSECTION GRAPHS. In this subsection, we show that the intersection graph of a k-ply neighborhood system has a linear number of edges. Because the star graph can be realized as the intersection graph of some 1-ply neighborhood system, the maximum degree of these intersection graphs can be unbounded.

For any integer δ, a graph is *δ-inductive* if its vertices can be numbered such that each vertex has at most δ edges to higher numbered vertices. Clearly, a δ-inductive graph with n vertices has at most $(\delta \cdot n)$ edges. For example, every tree is 1-inductive and each simple planar graph is 5-inductive. The latter can be shown by observing that each planar graph has at least one vertex of degree less than 6 (by Euler's formula). So a 5-inductive numbering can be obtained by assigning the smallest number to such a vertex and inductively numbering other vertices.

THEOREM 3.3.1. *The intersection graph of a k-ply neighborhood system in* $\mathbb{R}^d$ *is* $3^d k$*-inductive.*

To prove the theorem, it is sufficient to show that each ball in a k-ply neighborhood system in $\mathbb{R}^d$ intersects at most $3^d k$ other balls of larger or equal radius. Then, we can number the balls by sorting the radius of the balls in the increasing order. Therefore, Theorem 3.3.1 follows directly from the following lemma.

LEMMA 3.3.2 (BALL INTERSECTION). *Suppose* $\Gamma = \{B_1, \ldots, B_n\}$ *is a k-ply system in* $\mathbb{R}^d$. *Then for each d-dimensional ball B with radius r,* $|\{i : B_i \cap B \neq \emptyset$ *and* $r_i \geq r\}| \leq 3^d k$.

PROOF. Without loss of generality, let $B_1, \ldots, B_t$ be the set of all balls in Γ of radius at least r that intersect B. For each i in the range $1 \leq i \leq t$, if $\mathbf{p}_i$, the center of B_i, is in $2 \cdot B$, let B_i' be the ball of radius r centered at $\mathbf{p}_i$; if $\mathbf{p}_i$ is not in $2 \cdot B$, let $\mathbf{p}_i'$ be the point common of the ray $\mathbf{pp}_i$ and the boundary of $2 \cdot B$, and let B_i' be the ball centered at $\mathbf{p}_i'$ and of radius r. In either case, $B_i' \subseteq B_i$ and B_i' intersects B, and if B_i is replaced by B_i', the ply of the resulting neighborhood system does not increase and thus is bounded above by k. Notice that each ball $B_i'(1 \leq i \leq t)$, is contained in the ball $3 \cdot B$. We have

$$\sum_{i=1}^{t} \text{Volume}(B_i') \leq k\text{Volume}(3 \cdot B),$$

which implies $t \leq 3^d k$. □

Remark 3.3.3. Lemma 3.3.2 was given in the conference publication of this work [Miller et al. 1991] and a proof appeared in Teng's dissertation [1991]. A few years later, a variant of it was independently proved by Eppstein and Erickson [1994].

Notice that any k-inductive graph G is $(k + 1)$-colorable by the following greedy algorithm. Suppose the vertex set of G labeled by a k-inductive labeling $\{1, \ldots, n\}$. Color the vertices $n - k, \ldots, n$ by colors $1, \ldots, k + 1$, respectively. We color the remainder of the vertex set in the order of $n - k - 1, \ldots, 1$. Because each vertex is connected to at most k vertices of higher labels, we can always assign it a color that is not used by its neighbors with higher labels. So this greedy algorithm is guaranteed to use $k + 1$ colors.

COROLLARY 3.3.4. *The intersection graph of a k-ply neighborhood system in* $\mathbb{R}^d$ *is* $(3^d k + 1)$*-colorable.*

4. *Geometric Divide and Conquer*

In this section, we present a divide-and-conquer paradigm that uses Theorem 2.2.1. We will demonstrate the usefulness of this paradigm in computational geometry. The new paradigm is compared with a commonly used paradigm for solving geometry problems, the multi-dimensional divide and conquer of Bentley [1980]. We will show that this paradigm outperforms multi-dimensional divide and conquer on various geometry problems. The new paradigm also provides a good support for designing efficient parallel algorithms for geometry problems in fixed dimensions (see Frieze et al. [1992]).

4.1. POINT LOCATION. The point location problem for a neighborhood system can be defined as: given a neighborhood system $\Gamma = \{B_1, \ldots, B_n\}$ in d-space, preprocess the input to organize it into a search structure so that queries of the form "output all neighborhoods that contain a given point $\mathbf{p}$" can be answered efficiently.

Like other geometry query problems, there are three costs associated with this point location problem: the *preprocessing* time $T(n, d)$ required to build the search structure, the query time $Q(n, d)$ required to answer a query, and the *space* $S(n, d)$ required to represent the search structure in memory.

If Γ is an arbitrary neighborhood system, then there may exist some point $\mathbf{p}$ that is covered by $\Omega(n)$ balls. In this case, just to print the output would require $\Omega(n)$ work. However, if Γ is restricted to be a k-ply system, then the number of balls in the output is bounded by k. Using separator based divide and conquer, we are able to construct a search structure with the following properties:

$$T(n, d) = \text{random } O(n \log n),$$

$$Q(n, d) = O(k + \log n),$$

$$S(n, d) = O(n).$$

By saying an algorithm runs in random $t(n)$ time, we mean that the algorithm never gives a wrong output but may not terminate in the claimed time bound. The probability of success, namely, that it produces a correct output in $t(n)$ steps, is at least $1 - \lambda$ for any $\lambda > 0$. To simplify the discussion, in the following sections, we assume that the ply k is a constant.

The main idea is to use a sphere separator which intersects an $O(k^{1-\beta}n^{\beta})$ number of balls for any constant $(\beta < 1)$ to partition the neighborhoods into two subsets of roughly equal size, and then recursively build search structures for each subsets.

Given a neighborhood system Γ with ply k, we will build a binary tree of height $O(\log n)$ to guide the search in answering a query. Associated with each leaf of the tree is a subset of neighborhoods in Γ, and the search structure has the property that for all $\mathbf{p} \in \mathbb{R}^d$, the set of neighborhoods that covers $\mathbf{p}$ can be found in one of the leaves.

In the following construction, we will use sphere separators that have the following useful properties

- It can be represented with $O(1)$ space.
- It takes $O(1)$ time to test whether a point is in the interior or the exterior of the sphere.

The algorithm is very simple. It first finds a sphere S that intersects $c \cdot k^{1-\beta}n^{\beta}$ balls that δ-split Γ. In the remainder of this section we assume that β, δ and c are constants with the property that $0 < \beta < 1$, $0 < \delta < 1$ and c is a positive real that only depends on the dimension d, β and δ. To apply Theorem 2.2.1, we can use $\beta = 1 - 1/d$, $\delta = (d + 1)/(d + 2) + \epsilon$ for any constant ϵ in the range $0 < \epsilon < 1/(d + 2)$, and c is the constant term of the separator size given in Theorem 2.2.1. Because testing whether a sphere intersects more than $c \cdot k^{1-\beta}n^{\beta}$ balls can be done in linear time, our randomized separator algorithm can guarantee the quality of its separators.

Let Γ_0 be the subset of balls which intersect either S or the interior of S, and Γ_1 the subset of balls which intersect either S or the exterior of S. Clearly $|\Gamma_0|$, $|\Gamma_1| \leq \delta n + c \cdot k^{1-\beta}n^{\beta}$, and $|\Gamma_1| + |\Gamma_2| \leq n + c \cdot k^{1-\beta}n^{\beta}$. We store the information about S, namely its center and radius, in the root of the search tree, and recursively build binary search trees for Γ_0 and Γ_1, respectively. The roots of the tree for Γ_0 and Γ_1 are respectively the left and right children of the node associated with S. The recursive construction stops when the subset has cardinality smaller than $m_0 = \alpha k$ for a constant α that depends on β, δ, c but not n and k. The precise requirements for α (and hence m_0) will be determined below.

To answer a query when given a point $\mathbf{p} \in \mathbb{R}^d$, we first check $\mathbf{p}$ against S, the sphere separator associated with the root of the search tree. There are three cases:

Case 1. If $\mathbf{p}$ is in the interior of S, then recursively search on the left subtree of S;

Case 2. If $\mathbf{p}$ is in the exterior of S, then recursively search the right subtree of S;

Case 3. If $\mathbf{p}$ is on S, then recursively search on the left subtree of S.

When reaching a leaf, we then check $\mathbf{p}$ against all balls associated with the leaf and print all those that cover $\mathbf{p}$.

The correctness of the search structure and the above searching procedure is obvious and can be proved by induction: if $\mathbf{p}$ is in the interior (exterior) of S, then all balls that cover $\mathbf{p}$ must intersect either S or the interior (exterior) of S, and hence are in the left (right) subtree of S. The time complexity to answer a query is bounded by $O(h_k(n) + m_0)$, where $h_k(n)$ is the worse-case height of the search tree for n k-ply balls. We can bound $h_k(n)$ from above by the following recurrence.

$$h_k(m) \le \begin{cases} 1 & \text{if } \; m \le m_0 \\ h_k(\delta m + ck^{1-\beta}m^{\beta}) + 1 & \text{if } \; m \ge m_0. \end{cases} \tag{5}$$

The following lemma gives an upper bound on $h_k(n)$. In proof, we will give the first condition on $m_0 = \alpha k$.

LEMMA 4.1.1. *Let h_k be a function defined above. Then $h_k(n) = O(\log n)$ for a sufficiently large constant α that depends only on d, δ, β, and c.*

PROOF. We will choose $m_0 = \alpha k$ such that for all $m \ge m_0$, $ck^{1-\beta}m^{\beta} \le ((1 - \delta)/2)m$. This condition is true if

$$\alpha \ge \left(\frac{2c}{1-\delta}\right)^{1/(1-\beta)}.$$

Because $h_k(m)$ is a nondecreasing function in m, we have

$$h_k(m) \le \begin{cases} 1 & \text{if } \; m \le m_0 \\ h_k\left(\left(\frac{1+\delta}{2}\right)m\right) + 1 & \text{if } \; m > m_0. \end{cases}$$

Since $\delta < 1$ and hence $2/(1 + \delta) > 1$, we can infer $h_k(n) = \lceil \log_{2/(1+\delta)} n \rceil = O(\log n)$. □

Consequently,

$$Q(n, d) = O(\log n + m_0) = O(\log n + k).$$

We now analyze the space requirement of the search structure. First, observe that each internal node requires a constant amount of space and each leaf requires $O(m_0)$ space. To bound the total space, it is sufficient to bound the total number of leaves in the tree.

Let $s_k(m)$ denote the maximum number of leaves in the search tree for m balls. The sphere separator decomposes the data structure for m balls into two substructures, one for those balls intersecting the interior of the sphere and one for those balls intersecting the exterior of the sphere. Our separator results guarantee that number of balls in each two substructures is no more than $\delta m + ck^{1-\beta}m^{\beta}$ and the sum of number of balls from both sides is no more than $m + ck^{1-\beta}m^{\beta}/2$. Hence, there is a δ_1 such that (1) $1 - \delta_1 \leq \delta_1 \leq \delta$; and (2) the smaller side (either interior or exterior) has no more than $(1 - \delta_1)m$ balls. The number of balls in the larger side is at most $\delta_1 m + ck^{1-\beta}m^{\beta}$. Notice that we implicitly charge the additional term for the separator (which is bounded by $ck^{1-\beta}m^{\beta}$ in $(1 - \delta_1)m$). Thus, $s_k(m)$ is given by the following recurrence.

$$s_k(m) \leq \begin{cases} 1 & \text{if } m \leq m_0 \\ s_k(\delta_1 m + ck^{1-\beta}m^{\beta}) + s_k((1-\delta_1)m) & \text{if } m > m_0. \end{cases} \tag{6}$$

The following lemma gives an upper bound on $s_k(n)$. In proof, we will give the second condition on $m_0 = \alpha k$.

LEMMA 4.1.2. *Let s_k be the function defined above. Then $s_k(n) = O(n/k)$ for a sufficiently large constant α that depends only on d, δ, β, and c.*

PROOF. For any constant γ such that $\beta < \gamma < 1$, we use induction to establish $s_k(n) \leq C(n/k - (n/k)^{\gamma})$ for a sufficiently large constant α that depends only on d, δ, β, and c and an appropriate choice of the constant $C > 1$.

Because $s_k(m) = 1$ for $m \leq m_0 = \alpha k$, we need to choose C, and m_0 such that $C(m_0/k - (m_0/k)^{\gamma}) \geq 1$. Because $\gamma < 1$, this condition holds for a sufficiently large α, establishing the base for the induction. Now assuming the lemma is true for all $m < n$, using the substitution method of Cormen et al. [1990], we have

$$\begin{aligned} s_k(n) &\leq s_k(\delta_1 n + ck^{1-\beta}n^{\beta}) + s_k((1-\delta_1)n) \\ &\leq \frac{C(\delta_1 n + ck^{1-\beta}n^{\beta})}{k} + C\frac{((1-\delta_1)n)}{k} - C\left(\frac{(\delta_1 n + ck^{1-\beta}n^{\beta})}{k}\right)^{\gamma} - C\left(\frac{(1-\delta_1)n}{k}\right)^{\gamma} \\ &= \frac{Cn}{k} - C\left(\frac{n}{k}\right)^{\gamma} + C\left(\frac{n}{k}\right)^{\gamma} + Cc\left(\frac{n}{k}\right)^{\beta} - C\left(\delta_1\frac{n}{k} + c\left(\frac{n}{k}\right)^{\beta}\right)^{\gamma} - C\left(\frac{(1-\delta_1)n}{k}\right)^{\gamma} \\ &\leq \frac{Cn}{k} - C\left(\frac{n}{k}\right)^{\gamma} + C\left(\frac{n}{k}\right)^{\gamma} + Cc\left(\frac{n}{k}\right)^{\beta} - C\left(\frac{\delta_1 n}{k}\right)^{\gamma} - C\left((1-\delta_1)\frac{n}{k}\right)^{\gamma} \\ &\leq C\left(\frac{n}{k} - \left(\frac{n}{k}\right)^{\gamma}\right), \end{aligned}$$

as long as we choose C and $m_0 = \alpha k$ such that for all $n \geq m_0$,

$$C\left(\frac{\delta_1 n}{k}\right)^{\gamma} + C\left((1-\delta_1)\frac{n}{k}\right)^{\gamma} - C\left(\frac{n}{k}\right)^{\gamma} - Cc\left(\frac{n}{k}\right)^{\beta} \geq 0. \tag{7}$$

By a Taylor expansion of $(1-x)^{\gamma}$ around the point 0, we have

$$(\delta_1)^{\gamma} + (1-\delta_1)^{\gamma} \geq (\delta_1)^{\gamma} + 1 - \gamma\delta_1.$$

Because $0 < \gamma < 1$ and $1 < \delta_1 < \delta < 1$, the above inequality (7) holds if

$$C[(\delta^{\gamma} - \gamma\delta)(n/k)^{\gamma} - c(n/k)^{\beta}] \geq 0. \tag{8}$$

Because $0 < \beta < \gamma < 1$, inequality (8) holds for all

$$n \geq m_0 \geq \left(\frac{C}{\delta^{\gamma} - \gamma\delta}\right)^{1/(\gamma-\beta)} k.$$

Therefore, the lemma is true for sufficiently large constants α and C that only depend on d, δ, β, and c. □

Because the number of nodes in a proper binary tree is no more than twice the number of leaves, we have for each sufficiently large constant α satisfying the conditions given in the proof of Lemma 4.1.2,

$$s(n) = O\left(\frac{n}{k}\right).$$

Therefore, the total space requirement of the above search structure is bounded by

$$S(n, d) = O(ks_k(n)) = O(n).$$

Now let us look at the time required in building such a search structure.

From Theorem 2.2.1, each k-ply system of m balls in $\mathbb{R}^d$ has a sphere separator that intersects $O(k^{1/d}n^{1-1/d})$ balls and $(d+1)/(d+2)$-splits the system. If we could compute such a sphere separator in deterministic $O(n)$ time, then the worst-case time required in computing such a search structure, $T_k(m)$, would be given by the following recurrence.

$$T(m) \leq \begin{cases} 1 & \text{if} \quad m \leq m_0 \\ T_k(\delta_1 m + ck^{1-\beta}m^{\beta}) + T_k((1-\delta_1)m) + O(m) & \text{if} \quad m > m_0, \end{cases} \tag{9}$$

where $\delta_1 \leq \delta$.

The following lemma gives an upper bound on $T_k(n)$. In proof, we will give the third condition on $m_0 = \alpha k$:

LEMMA 4.1.3. *Let T_k be the function defined above. Then $T_k(n) = O(n \log n)$ for a sufficiently large constant α that depends only on d, δ, β, and c.*

PROOF. We use induction to establish $T_k(n) \leq Cn \log n$ for an appropriate choice of the constant $C > 1$.

Clearly, $T_k(m) = 1 \leq C$ for $m \leq m_0$. This is the base of the induction. Now, assuming the lemma is true for all $m < n$, we have

$$\begin{aligned} T_k(n) &\leq T_k(\delta_1 n + ck^{1-\beta}n^{\beta}) + T_k((1-\delta_1)n) + c_2 n \\ &\leq C(\delta_1 n + ck^{1-\beta}n^{\beta})\log(\delta_1 n + ck^{1-\beta}n^{\beta}) + C((1-\delta_1)n)\log((1-\delta_1)n) + c_2 n \\ &\leq Cn\log(\delta n + ck^{1-\beta}n^{\beta}) + Cck^{1-\beta}n^{\beta}\log(\delta + ck^{1-\beta}n^{\beta}) + c_2 n \\ &\leq Cn\log\left(\frac{1+\delta}{2}n\right) + Cck^{1-\beta}n^{\beta}\log(\delta n + ck^{1-\beta}n^{\beta}) + c_2 n \\ &= Cn\log n - Cn\log\left(\frac{2}{1+\delta}\right) + Cck^{1-\beta}n^{\beta}\log(\delta + ck^{1-\beta}n^{\beta}) + c_2 n \\ &\leq Cn\log n, \end{aligned}$$

as long as we choose C and $m_0 = \alpha\, k$ such that for all $n \geq m_0$,

$$Cn\log\left(\frac{2}{1+\delta}\right) - Cck^{1-\beta}n^{\beta}\log(\delta + ck^{1-\beta}n^{\beta}) - c_2 n \geq 0. \tag{10}$$

Because $\beta < 1$ and $\delta < 1$, inequality (10) holds for sufficiently large α and $C > c_2$ which only depend on d, δ, β, and c. □

Consequently,

$$T_k(n, d) = O(n\log n).$$

Notice that, however, our algorithm is randomized. As shown in our main theorem, if $\beta = 1 - 1/d + \epsilon$ for some constant ϵ such that $0 < \epsilon < 1/d$, then the probability such a randomized algorithm outputs a sphere separator that intersects $O(k^{1-\beta}n^{\beta})$ balls is at least $1 - (1/n^{\epsilon})$. Moreover, in linear time, we can check whether the number of balls a sphere separator intersects is $O(k^{1-\beta}n^{\beta})$. Frieze et al. [1992] in a parallel extension of this algorithm, shown that for a sufficiently large constant m_0 the search structure can be constructed in random $O(n\log n)$ time with a probability of success $1 - 1/n$.

4.2. CONSTRUCTING INTERSECTION GRAPHS. The problem of this section is to construct the intersection graph of a given neighborhood system. There is a simple solution for this problem: test each pair of balls to decide whether they intersect. Since there are $O(n^2)$ pairs and the testing of each pair can be performed in constant time, the whole construction can be performed in $O(n^2)$ time. If we require the algorithm to report all edges of the intersection graph, the above algorithm is optimal if we are working with general neighborhood systems. This is because that there could be as much as $\Omega(n^2)$ number of edges in some intersection graphs. However, every k-intersection graph has at most $O(kn)$ edges. We present a randomized $O(kn + n\log n)$ time construction algorithm.

To illustrate the idea, let us view the graph construction problem as a search problem: a problem of exploring the structure of an unknown graph with the help of some oracles. First, assume that n, the number of vertices, is known in advance and we have an oracle – the *edge oracle* – which answers the question of the form "is there an edge between vertex u and v?" in constant time. It is not hard to see that even though the number of edges is known in advance, $\Omega(n^2)$ queries have to be asked in the worst case.

Now suppose that there is more information available: it is known in advance that the graph has a $ck^{1-\beta}n^{\beta}$-separator that δ-splits and moreover each subgraph of $m > m_0$ vertices also has a $ck^{1-\beta}m^{\beta}$-separator that δ-splits, for some constant c, α, and $0 < \beta < 1$, where $m_0 = \alpha k$. Can the number of queries be reduced? It is remain to be seen whether this is true.

Now, suppose in addition, we have an oracle – a *separator oracle* – which, when presented with a subset of m vertices, delivers three sets, A, B, and C, where C is an $ck^{1-\beta}m^{\beta}$-separator that δ-splits the subgraph induced by those m vertices into A and B. Then it is sufficient to consult with the oracle $O(n)$ times to compute the structure of the unknown graph G, if k is much smaller than n.

The strategy is divide and conquer. We first present the separator oracle with the whole set of vertices and get back from the oracle three sets A, B, C, where C is a $ck^{1-\beta}n^{\beta}$-separator that δ-splits G into A and B. We then recursively search the structure of subgraphs induced by $A \cup C$ and $B \cup C$ until the size of subproblems is below m_0. Finally, we use the edge oracle to complete the graph.

The total number of query $q_k(n)$ to the separator oracle is clearly given by the following recurrence.

$$q_k(n) \leq \begin{cases} 0 & \text{if } \; n \leq m_0 \\ q_k(\delta_1 n + ck^{1-\beta}n^{\beta}) + q((1-\delta_1)n) + 1 & \text{if } \; n > m_0, \end{cases}$$

where $\delta_1 \leq \delta$.

By a similar argument as Lemma 4.1.2, it can be shown $q_k(n) = O(n/k)$.

Now suppose each query to the separator oracle costs $O(m)$ time, where m is the size of query. It is not hard to see that the total time $T_k(n)$ needed to search the structure of the graph is given by the following recurrence.

$$T_k(n) \leq \begin{cases} O(1) & \text{if } \; n \leq m_0 \\ T_k(\delta_1 n + ck^{1-\beta}n^{\beta}) + T_k((1-\delta_1)n) + O(n) & \text{if } \; n > m_0. \end{cases}$$

By Lemma 4.1.3, we have $T_k(n) = O(n \log n)$.

The divide-and-conquer algorithm for constructing intersection graphs is based on the following interesting observation: We do not need to have the intersection graph in order to compute a small separator efficiently – all we need is the neighborhood system!

To see this, let us recall how we compute a small separator of an intersection graph. First, we find a sphere separator S of low cost. This step involves computing an approximate centerpoint and a conformal map. We then compute a vertex separator from the sphere separator S. The rule of choosing vertices is very simple: If ball B_i has a common point with S, then the vertex corresponding to B_i is placed in the separator. The time complexity of the above step is $O(n)$.

THEOREM 4.2.1. *The intersection graph of a k-ply system in* $\mathbb{R}^d$ *can be computed in random* $O(kn + n \log n)$ *time. Moreover, the algorithm uses* $O(n)$*-space.*

In contrast, one can only derive an $O(kn \log^{d-1} n)$ time, $O(n \log^{d-1} n)$-space algorithm for computing an intersection graph using the multidimensional divide-and-conquer paradigm.

Remark 4.2.2. Guibas et al. [1994] gave an

$$O(n^{2-\frac{2}{(1+\lfloor(d+2)/2\rfloor)+o(1)}} + kn \log^2 n)$$

time deterministic algorithm for constructing the kth sphere of influence graph of a point set in d dimensions. Our construction, using randomization, reduces the time complexity to expected $O(kn + n \log n)$.

Remark 4.2.3. Eppstein et al. [1993] gave a deterministic linear time algorithm for finding a small cost sphere separator of a k-ply system. Therefore, in theory, all results presented in this section can be made deterministic. However, the randomized construction presented in this section is much faster in practice. More practical deterministic algorithms are desirable.

Eppstein et al. also showed how to apply this divide-and-conquer method to approximate the ply k. Hence, we can apply separator based divide and conquer to neighborhood systems without knowing its ply k a-priori.

5. *Final Remarks and Open Questions*

Recently, using the duality on S^d suggested in Teng [1991] and Eppstein et al. [1993], Agarwal and Pach [1995] and Spielman and Teng [1996a], independently, gave a much simpler geometric proof that bounds the expected number of balls of a k-ply systems on S^d that a random great circle intersects (a la Sections 2.6.2 and 2.6.3): Let $\Gamma = (B_1, \ldots, B_n)$ be a k-ply neighborhood system in R^d, and let Φ be the sphere-preserving map used in Lemma 2.3.3.1. let r_i be the radius of $\Phi(B_i)$. As shown in Teng [1991], Eppstein et al. [1993], Pach and Agarwal [1995], and Spielman and Teng [1996a], it follows from the duality between points and great circles of S^d that the expected number of caps of $\Phi(\Gamma)$ that a random great circle of S^d intersect is equal to

$$c_1 \sum_{i=1}^{n} r_i,$$

for a constant c_2 depends only on d. Because $\Phi(\Gamma)$ is k-ply, the total volume of $\{\Phi(B_1), \ldots, \Phi(B_n)\}$ is at most kA_d, where A_d is the surface area of a unit d-sphere in $\mathbb{R}^{d+1}$. Therefore, there is a constant c_2 depending only on d such that

$$\sum_{i=1}^{n} r_i^d \leq c_2 k.$$

By a convexity argument or Lagrange's method, one can show that

$$c_1 \sum_{i=1}^{n} r_i = O(k^{1/d} n^{1-1/d}).$$

An important problem is, given a graph without an embedding, can its nodes be embedded in $\mathbb{R}^d$ to make it a subgraph of an intersection graph of a k-ply system?

Recently, Linial et al. [1995] studied the problem of embedding graphs in the Euclidean space so that (1) the dimension is kept as small as possible, and (2) the distances among vertices of the graph are closely matched with the distances between their geometric images. They showed that such an embedding of a graph in $\mathbb{R}^d$ in conjunction with the partitioning technique of this paper implies that the graph has a separator of size $O(n^{1-1/d})$.

In contrast to our approach, the results of Linial et al. [1995] can be applied to any graph, not only graphs arising from neighborhood systems. On the other hand, their algorithm in general requires the embedding dimension to be as large as $\Omega(\log n)$, so the bounds they attain ours are weaker than ours. It would be very interesting if there were an embedding algorithm for general graphs that would be able to find very low-dimensional embedding in the special case that the graph admits such an embedding.

It is interesting to point out that Vavasis [1991] has defined a class of geometric graphs called local graphs and showed that any local graph of n vertices has a hyperplane based separator of size $n^{1-1/d}$. The class of local graphs is properly contained in the class of overlap graphs defined in our companion paper [Miller et al. 1996] and is much weaker than k-intersection graphs. Previously, hyperplanes have been used in the recursive coordinate bisection heuristic.

Recently, Plotkin et al. [1994] improved the bounded forbidden minor separator theorem of Alon et al. [1990]. Their work, in conjunction with a structure lemma of Teng [1994], gives a combinatorial proof of a much weaker version of the separator theorem for the class of intersection graphs presented in this paper.

It is also interesting to determine how our algorithm performs in practice compared to other current algorithms such as the spectral method. This is the subject of current work by Gilbert et al. [1997]. Very recently, using Koebe-embedding and sphere-preserving mapping in a similar way to the work of this paper, Spielman and Teng [1996b] showed that the spectral method can be used to find edge bisectors of size $O(\sqrt{n})$ for k-nearest neighbor graphs and bounded degree planar graphs.

ACKNOWLEDGMENTS. We would like to thank David Applegate, Marshall Bern, David Eppstein, Alan Frieze, John Gilbert, Bruce Hendrickson, Ravi Kannan, Tom Leighton, Mike Luby, Oded Schramm, Dan Spielman, Dafna Talmor, Doug Tygar, and Kim Wagner for invaluable help and discussions. We would like to thank both referees for their very helpful comments and suggestions.

REFERENCES

ALON, N., SEYMOUR, P., AND THOMAS, R. 1990. A separator theorem for graphs with an excluded minor and its applications. In *Proceedings of the 22th Annual ACM Symposium on Theory of Computing* (Baltimore, Md., May 14–16). ACM, New York, pp. 293–299.

ANDREEV, E. M. 1970a. On convex polyhedra in Lobacevskii space. *Math. USSR Sbornik 10*, 3, 413–440.

ANDREEV, E. M. 1970b. On convex polyhedra of finite volume in Lobacevskii space. *Math. USSR Sbornik, 12*, 2, 270–259.

BENTLEY, J. L. 1980. Multidimensional divide-and-conquer. *Commun. ACM 23*, 4 (Apr.), 214–229.

CLARKSON, K. 1983. Fast algorithm for the all-nearest-neighbors problem. In *Proceedings of the 24th Annual Symposium on Foundations of Computer Science*. IEEE, New York, pp. 226–232.

CLARKSON, K., EPPSTEIN, D., MILLER, G., STURTIVANT, C., AND TENG, S.-H. 1993. Approximating center points with iterated radon points. In *Proceedings of 9th Annual ACM Symposium on Computational Geometry* (San Diego, Calif., May 19–21). ACM, New York, pp. 91–98.

CONWAY, J. H., AND SLOANE, N. J. A. 1988. *Sphere Packings, Lattices and Groups*. Springer-Verlag, New York.

CORMEN, T. H., LEISERSON, C. E., AND RIVEST, R. L. 1990. *Introduction to Algorithms*. MIT Press, Cambridge, Mass.

DANZER, L., FONLUPT, J., AND KLEE, V. 1963. Helly's theorem and its relatives. In *Proceedings of Symposia in Pure Mathematics*, vol. 7. American Mathematical Society, 7:101–180.

DE FRAYSSEIX, H., PACH, J., AND POLLACK, R. 1988. Small sets supporting Fáry embeddings of planar graphs. In *Proceedings of the 20th Annual ACM Symposium on Theory of Computing* (Chicago, Ill., May 2–4). ACM, New York, pp. 426–433.

EDELSBRUNNER, H. 1987. *Algorithms in Combinatorial Geometry*. Springer-Verlag, New York.

EPPSTEIN, D., AND ERICKSON, J. 1994. Iterated nearest neighbors and finding minimal polytopes. *Disc. Comput. Geometry, 11,* (3), 321–350.

EPPSTEIN, D., MILLER, G. L., AND TENG, S.-H. 1993. A deterministic linear time algorithm for geometric separators and its applications. In *Proceedings of 9th Annual ACM Symposium on Computational Geometry* (San Diego, Calif., May 19–21). ACM, New York, pp. 99–108.

FÁRY, I. 1948. On straight line representing of planar graphs. *Acta. Sci. Math. 24*, 229–233.

FRIEZE, A., MILLER, G., AND TENG, S.-H. 1992. Separator-based parallel divide and conquer in computational geometry. In *Proceedings of the 4th Annual ACM Symposium on Parallel Algorithms and Architectures* (San Diego, Calif., June 29–July 1). ACM, New York, pp. 422–430.

GILBERT, J. R., MILLER, G. L., AND TENG, S.-H. 1997. Geometric mesh partitioning: Implementation and experiments. *SIAM J. Sci. Comput.*, to appear.

GILBERT, J. R., HUTCHINSON, J. P., AND TARJAN, R. E. 1984. A separation theorem for graphs of bounded genus. *J. Algorithms 5*, 391–407.

GRÖTSCHEL, M., LOVÁSZ, L., AND SCHRIJVER, A. 1988. *Geometric Algorithms and Combinatorial Optimization*. Springer-Verlag, New York.

GUIBAS, L., PACH, J., AND SHARIR, M. 1994. Sphere-of-influence graphs in higher dimensions. In *Intuitive Geometry*, K. Boroczky and G. Fejes Toth, eds. Colloquia Mathematica Societatis J. Bolyai, vol. 63. North-Holland, Amsterdam, The Netherlands, pp. 131–137.

HAUSSLER, D., AND WELZL, E. 1987. ϵ-net and simplex range queries. *Disc. Comput. Geometry 2*, 127–151.

JORDAN, C. 1869. Sur les assemblages de lignes. *J. Reine Angew. Math. 70*, 185–190.

KABATIANSKY, G. A., AND LEVENSHTEIN, V. I. 1978. Bounds for packings on a sphere and in space. *Prob. Per. Info. 14*, 1, 3–25.

KOEBE, P. 1936. Kontaktprobleme der konformen Abbildung. *Ber. Verh. Sächs. Akademie der Wissenschaften Leipzig, Math.-Phys. Klasse 88*, 141–164.

LEIGHTON, F. T. 1983. *Complexity Issues in VLSI*. Foundations of Computing. MIT Press, Cambridge, Mass.

LEISERSON, C. E. 1983. *Area-Efficient VLSI Computation*. Foundations of Computing. MIT Press, Cambridge, Mass.

LINIAL, N., LONDON, E., AND RABINOVICH, Y. 1995. The geometry of graphs and some of its algorithmic applications. *Combinatorica 15*, 2, 215–245.

LIPTON, R. J., AND TARJAN, R. E. 1979. A separator theorem for planar graphs. *SIAM J. Appl. Math. 36*, (Apr.), 177–189.

LIPTON, R. J., ROSE, D. J., AND TARJAN, R. E. 1979. Generalized nested dissection. *SIAM J. Numer. Anal.* 16, 346–358.

MILLER, G. L., TENG, S.-H., AND VAVASIS, S. A. 1991. An unified geometric approach to graph separators. In *Proceedings of the 32nd Annual Symposium on Foundations of Computer Science*. IEEE, New York, pp. 538–547.

MILLER, G. L., TENG, S.-H., THURSTON, W., AND VAVASIS, S. A. 1993. Automatic mesh partitioning. In *Sparse Matrix Computations: Graph Theory Issues and Algorithms*, A. George, J. Gilbert, and J. Liu, eds. *IMA Volumes in Mathematics and Its Applications*. Springer-Verlag, New York.

MILLER, G. L., TENG, S.-H., THURSTON, W., AND VAVASIS, S. A. 1997. Geometric separators for finite element meshes. *SIAM J. Sci. Comput.*, to appear.

MILLER, G. L., AND THURSTON, W. 1990. Separators in two and three dimensions. In *Proceedings of the 22th Annual ACM Symposium on Theory of Computing*, (Baltimore, Md., May 14–16). ACM, New York, pp. 300–309.

MILLER, G. L., AND VAVASIS, S. A. 1991. Density graphs and separators. In *Proceedings of the 2nd Annual ACM-SIAM Symposium on Discrete Algorithms* (San Francisco, Calif., Jan. 28–30). ACM, New York, pp. 331–336.

MOHAR, B. 1993. A polynomial time circle packing algorithm. *Disc. Comput. Geom.* 241–250.

PACH, J., AND AGARWAL, P. 1995. *Combinatorial Geometry*. Wiley-Interscience, New York.

PLOTKIN, S., RAO, S., AND SMITH, W. D. 1994. Shallow excluded minors and improved graph decomposition. In *Proceedings at the 5th Symposium Discrete Algorithms*. SIAM, New York, pp. 462–470.

PREPARATA, F. P., AND SHAMOS, M. I. 1985. *Computational Geometry: An Introduction*. Texts and Monographs in Computer Science. Springer-Verlag, New York.

SOMMERVILLE, D. M. Y. 1958. *An Introduction to the Geometry of n Dimensions*. Dover, New York.

SPIELMAN, D. A., AND TENG, S.-H. 1996a. Disk packings and planar separators. In *Proceedings of 12th Annual ACM Symposium on Computational Geometry* (Philadelphia, Pa., May 24–26). ACM, New York, pp. 349–359.

SPIELMAN, D. A., AND TENG, S.-H. 1996b. Spectral partitioning works: planar graphs and finite element meshes. UCB//CSD-96-898, U.C. Berkeley, Berkeley, California.

TENG, S.-H. 1991. Points, Spheres, and Separators: A unified geometric approach to graph partitioning. Ph.D. dissertation. CMU-CS-91-184. School of Computer Science, Carnegie-Mellon Univ., Pittsburgh, Pa.

TENG, S.-H. 1994. Combinatorial aspects of geometric graphs. MIT Tech. Rep. TR-612. May. In *Lecture Notes in Computer Science*. Springer-Verlag, New York.

THOMASSEN, C. 1980. Planarity and duality of finite and infinite graphs. *J. Combin. Theory, Series B 29*, 244–271.

TOUSSAINT, G. 1988. A graph-theoretical primal sketch. In *Computational Morphology*, G. Toussaint, ed., Elsevier-North Holland, Amsterdam, The Netherlands, pp. 229–260.

THURSTON, W. P. 1988. The geometry and topology of 3-manifolds. Princeton University Notes, Princeton, N.J.

TUTTE, W. T. 1960. Convex representations of graphs. *Proc. London Math. Soc. 10*, 3, 304–320.

TUTTE, W. T. 1963. How to draw a graph. *Proc. London Math. Soc. 13*, 3, 743–768.

UNGAR, P. 1951. A theorem on planar graphs. *J. London Math Soc. 26*, 256–262.

VALIANT, L. G. 1981. Universality consideration in VLSI circuits. *IEEE Trans. Comput. 30*, 2 (Feb.). 135–140.

VAPNIK V, N. AND CHERVONENKIS, A. YA. 1971. On the uniform convergence of relative frequencies of events to their probabilities. *Theory Prob. Appl. 16*, 264–280.

VAVASIS, S. A. 1991. Automatic domain partitioning in three dimensions. *SIAM J. Sci. Stat. Comput. 12*, 950–970.

WYNER, A. D. 1965. Capabilities of bounded discrepancy decoding. *Bell System Tech. J. 44*, 1061–1122.

ZIEGLER, G. M. 1988. Lectures on polytopes. In *Graduate Texts in Mathematics*, vol. 152. Springer-Verlag, New York.

RECEIVED APRIL 1995; REVISED NOVEMBER 1996; ACCEPTED SEPTEMBER 1996

SIAM J. SCI. COMPUT.
Vol. 19, No. 2, pp. 364–386, March 1998

003

GEOMETRIC SEPARATORS FOR FINITE-ELEMENT MESHES*

GARY L. MILLER†, SHANG-HUA TENG‡, WILLIAM THURSTON§, AND STEPHEN A. VAVASIS¶

Abstract. We propose a class of graphs that would occur naturally in finite-element and finite-difference problems and we prove a bound on separators for this class of graphs. Graphs in this class are embedded in d-dimensional space in a certain manner. For d-dimensional graphs our separator bound is $O(n^{(d-1)/d})$, which is the best possible bound. We also propose a simple randomized algorithm to find this separator in $O(n)$ time. This separator algorithm can be used to partition the mesh among processors of a parallel computer and can also be used for the nested dissection sparse elimination algorithm.

Key words. graph separators, finite elements, mesh partitioning, domain decomposition, computational geometry, sparse matrix computations, conformal mapping, center points

AMS subject classifications. 65F50, 68Q20

PII. S1064827594262613

1. Domain partitioning.

One motivation for this work is numerical solution of boundary value problems. Let Ω be an open connected region of $\mathbb{R}^d$. Suppose one is given a real-valued map f on Ω and is interested in finding a map $u : \Omega \to \mathbb{R}$ such that

$$\triangle u = f \text{ on } \Omega \quad \text{and} \quad u = 0 \text{ on } \partial\Omega.$$

This problem, Poisson's equation, arises in many physical applications. Two common techniques for this problem are finite differences and finite elements. These techniques grow out of different analyses, but the end result is the same. In particular, a discrete set of nodes is inserted into Ω and a sparse system of linear equations is solved in which there is one node point and one equation for each node interior to Ω. Moreover, the sparsity pattern of the system reflects interconnections of the nodes. Let the nodes and their interconnections be represented as an undirected graph G.

Two numerical techniques for solving this system are *domain decomposition* and *nested dissection*. Domain decomposition divides the nodes among processors of a parallel computer. An iterative method is formulated that allows each processor to

*Received by the editors January 31, 1994; accepted for publication (in revised form) April 1, 1996.

http://www.siam.org/journals/sisc/19-2/26261.html

†School of Computer Science, Carnegie Mellon University, Pittsburgh, PA 15213 (glmiller@theory.cs.cmu.edu). The research of this author was supported in part by NSF grant CCR-9016641.

‡Department of Computer Science, University of Minnesota, Minneapolis, MN 55455 (steng@cs.umn.edu). The research of this author was supported in part by NSF CAREER award CCR-9502540. Part of this work was done while the author was at the Department of Mathematics and the Laboratory for Computer Science, MIT, Cambridge, MA, where the research was supported in part by Air Force Office of Scientific Research grant F49620-92-J-0125 and Advanced Research Projects Agency grant N00014-92-J-1799.

§Department of Mathematics, University of California, Berkeley, CA 94720 (wpt@math.berkeley.edu). The research of this author was supported in part by the Geometry Center in Minneapolis, the NSF Science and Technology Center for Computation and Visualization of Geometric Structures, NSF grant DMS-8920161, and by the Mathematical Sciences Research Institute, NSF grant DMS-9022140.

¶Department of Computer Science, Upson Hall, Cornell University, Ithaca, NY 14853 (vavasis@cs.cornell.edu). The research of this author was supported in part by an NSF Presidential Young Investigator award, with matching funds received from Xerox Corporation and AT&T. Part of this work was done while the author was visiting Xerox Palo Alto Research Center, Palo Alto, CA.

operate independently; see Bramble, Pasciak, and Schatz [5]. *Nested dissection*, due to George [13], George and Liu [14] and Lipton, Rose, and Tarjan [25], is a node ordering for sparse Gaussian elimination. Although originally a sequential algorithm, nested dissection also parallelizes well. For instance, Pan and Reif [37] parallelized it by writing it as sequence of matrix factors. Parallel multifrontal methods (see, e.g., [27]) are also often based on nested dissection.

For either technique it is necessary to first partition the region into subdomains. For the purpose of efficiency in both domain decomposition and nested dissection, it is important that the number of nodes in each subdomain be roughly equal, and it is also important that the size of the separator be as small as possible. For a general graph, such a decomposition may not be possible. Accordingly, it is necessary to restrict attention to classes of graphs that occur in practice in finite-difference and finite-element computations. This class is defined in the next section.

The finite-element method can be applied to many boundary value problems other than Poisson's equation; see, e.g., Johnson [22]. In most of these settings, the nested dissection and domain decomposition ideas carry over. The partitioning technique described in this paper applies to any boundary value problem posed on a spatial mesh provided the mesh satisfies quality bounds described below, and provided the pattern of nonzero entries in the discretized operator is in correspondence with the mesh topology.

First, we review the relationship between this paper and other papers by the same authors. This paper and its companion paper [31] either extend or explain several short conference papers [33, 34, 35] and one journal paper [45]. The focus of this paper is finite-element meshes; the companion paper focuses on problems arising in computational geometry.

The authors have also jointly written a survey paper [30] that surveys the results from this paper, the companion, and several additional results by various authors on efficient center point computation.

The main contributions of this paper (beyond our previous work) are

- an analysis showing that well-behaved finite-element meshes in any dimension are "overlap" graphs (defined below), and
- a complete proof of our main separator theorem showing that overlap graphs have a small separator that can be efficiently computed.

Other authors have recently looked at the mesh-partitioning problem. For instance, Pothen, Simon, and Liou [39] have a "spectral-partitioning" method based on eigenvalues of the "Laplacian" matrix of the graph. This method seems to work well in practice but does not come with any known guarantees for finite-element meshes. Hendrickson and Leland [20] have improved the spectral method with a multilevel heuristic; the improved version is much faster in practice. The Chaco package from Sandia implements this spectral method and a newer multilevel Kernighan–Lin algorithm. The original Kernighan–Lin partitioning algorithm [23] improves the partition by moving individual nodes from one subdomain to the other, and the multilevel Kernighan–Lin is able to move entire connected subgraphs [21].

Another mesh-partitioning algorithm used in practice is a graph-search heuristic due to George and Liu [14]. Leighton and Rao [24] have a partitioning method guaranteed to return a split whose separator size (see below) is within logarithmic factors of optimal, but the technique, based on flow algorithms, currently appears to be too expensive for application to large-scale meshes.

The method that we propose, unlike these previous works, assumes that the graph G comes with an embedding of its nodes in $\mathbb{R}^d$. This is a very reasonable assumption

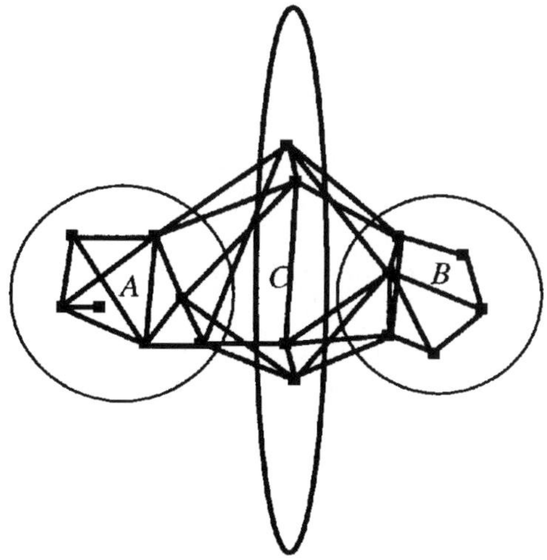

FIG. 1. *Example of* A, B, C *in Definition* 2.1.

for finite-difference and finite-element meshes. This geometric embedding is used extensively by our algorithm. Our algorithm is randomized. It always splits the graph into pieces of roughly equal size, and we show that with high probability the separator size satisfies an upper bound that is the best possible bound for the class of graphs we consider.

Throughout the paper we regard the dimension d as a small constant. The interesting cases for applications are commonly $d = 2$ or $d = 3$.

The remainder of the paper is organized as follows. In section 2 we introduce the concept of graph separators and define a class of graphs called overlap graphs. We also state the main theorem of the paper in section 2: there is an efficient algorithm for computing good separators of overlap graphs. In section 3 we prove that finite-element meshes belong to the class of overlap graphs. In section 4 we state our main algorithm for finding separators. The proof that this algorithm finds small separators is the focus of sections 5 through 7. In sections 8 and 9 we consider some practical issues associated with mesh partitioning.

2. Separators and overlap graphs. We now formally define the concept of separator.

DEFINITION 2.1. *A subset* C *of vertices of an* n*-vertex graph* G *is an* $f(n)$-separator *that* δ-splits G *if* $|C| \leq f(n)$ *and the vertices of* $G - C$ *can be partitioned into two sets* A *and* B *such that there are no edges from* A *to* B *and* $|A|, |B| \leq \delta n$. *Here,* f *is a function and* $0 < \delta < 1$.

In this definition and for the rest of the paper, $|A|$ denotes the cardinality of a finite set A. The type of separator defined here is sometimes called a "vertex separator," that is, a subset C of vertices of G whose removal disconnects the graph into two or more graphs of smaller size; see Fig. 1. A related concept is an "edge separator," that is, a set of edges whose removal disconnects the graph. Edge separators are useful for the problem of partitioning the computational tasks of a traditional iterative algorithm (such as conjugate gradient) for a finite-element problem among the processors of a parallel computer.

Our algorithm can compute an edge separator as effectively and efficiently as it computes vertex separators. We have decided to state our theoretical bounds in terms of vertex separators, however, to be consistent with previous literature.

One of the most well known separator results is Lipton and Tarjan's result [26] that any planar graph has a $\sqrt{8n}$-separator that 2/3-splits, which improved on Ungar's [43] result. Building on this result, Gilbert, Hutchinson, and Tarjan [15] showed that all graphs with genus bounded by g have an $O(\sqrt{gn})$-separator, and Alon, Seymour, and Thomas [1] proved that all graphs with an excluded minor isomorphic to the h-clique have an $O(h^{3/2}\sqrt{n})$-separator. These results are apparently not applicable to graphs arising as finite-element meshes when the dimension d is higher than two.

The class of graphs we consider is defined by a *neighborhood* system.

DEFINITION 2.2. *Let $P = \{\mathbf{p}_1, \ldots, \mathbf{p}_n\}$ be points in $\mathbb{R}^d$. A* k-ply neighborhood system *for P is a set $\{B_1, \ldots, B_n\}$ of closed balls such that* (1) *B_i centered at $\mathbf{p}_i$ and* (2) *no point $\mathbf{p} \in \mathbb{R}^d$ is interior to more than k of $B_1, \ldots, B_n$.*

In this paper we will focus exclusively on the case that $k = 1$, i.e., the interiors of the balls are disjoint. The case when $k > 1$ is interesting for a number of geometric problems and is considered in our other paper [31].

In this definition we used n for the number of points and d for the dimension of the embedding. We continue to use this notation throughout the paper. We also use the following notation: if $\alpha > 0$ and B is a ball of radius r, we define $\alpha \cdot B$ to be a ball with the same center as B but radius αr.

Given a neighborhood system, it is possible to define the *overlap graph* associated with the system.

DEFINITION 2.3. *Let $\alpha \geq 1$ be given, and let $\{B_1, \ldots, B_n\}$ be a* 1*-ply neighborhood system. The* α-overlap graph *for this neighborhood system is the undirected graph with vertices $V = \{1, \ldots, n\}$ and edges*

$$E = \{(i,j) : B_i \cap (\alpha \cdot B_j) \neq \emptyset \text{ and } (\alpha \cdot B_i) \cap B_j \neq \emptyset\}.$$

The main result that we establish in this paper is as follows.

THEOREM 2.4. *Let G be an α-overlap graph, and assume d is fixed. Then G has an*

$$O(\alpha \cdot n^{(d-1)/d} + q(\alpha, d))$$

separator that $(d+1)/(d+2)$ splits. A separator of the same size that $(d+1+\epsilon)/(d+2)$-splits can be computed with high probability by a randomized linear-time algorithm or randomized constant-time parallel algorithm provided that $\epsilon > 1/n^{1/2d}$.

3. Finite-element and finite-difference meshes. The main result of this section is that finite-element meshes that satisfy a "shape" criterion are overlap graphs. The *finite-element method* [41] is a collection of techniques for solving boundary value problems on irregularly shaped domains. The finite-element method subdivides the domain (a subset of $\mathbb{R}^d$) into a mesh of polyhedral elements. A common choice for an element is a d-dimensional simplex. These simplices are arranged in a *simplicial complex*; that is, they meet only at shared subfaces. Based on the finite-element mesh, a coefficient matrix called the "assembled stiffness matrix" is defined, with variables representing unknown quantities in the mesh. Let the *finite-element graph* refer to the nonzero structure of this matrix.

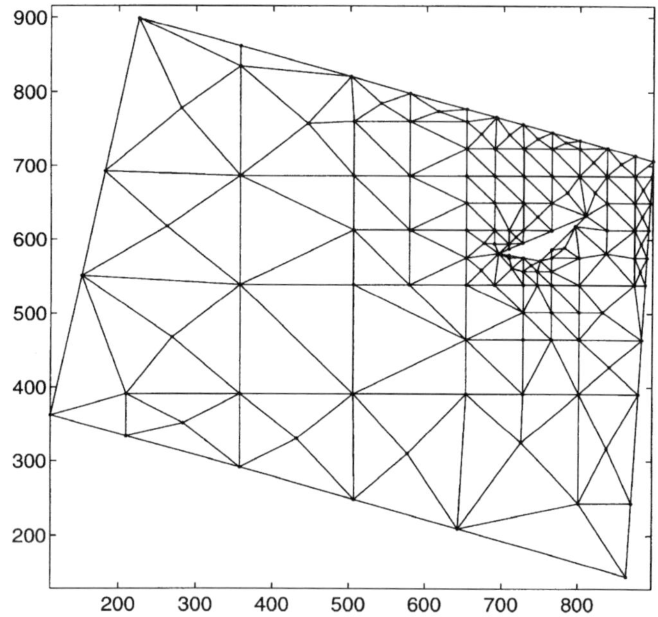

FIG. 2. *A mesh produced by a mesh generator that guarantees bounded aspect ratio.*

Associated with such a simplicial complex is its 1-*skeleton*, that is, the set of nodes accompanied by one-dimensional edges joining them. It is well known that, in the case of a piecewise-linear finite-element approximation for Poisson's equation, the nodes and edges in the finite-element graph defined in the last paragraph are in one-to-one correspondence with the nonboundary nodes and edges of the 1-skeleton of the complex. Notice that this 1-skeleton carries geometric information about the positions of the nodes and edges. In the case of higher-order elements, the finite-element graph is obtained from the 1-skeleton by introducing additional nodes interior to the faces of the simplices, and edges are introduced between every pair of nodes that share an element.

It is usually a requirement for numerical accuracy that the simplices are *well shaped* [4, 11, 46]. A common shape criterion used in mesh generation [4, 6, 36, 40] is an upper bound on the *aspect ratio* of the simplices. This term has many definitions that are all roughly equivalent [30]. One definition of aspect ratio of a simplex T is the radius of the smallest sphere containing T divided by the radius of the largest sphere that can be inscribed in T. Denote these two radii by $R(T)$ and $r(T)$ so that the aspect ratio is $R(T)/r(T)$. Figure 2 shows an example of a finite-element mesh generated by Mitchell's mesh generator and based on the algorithm in [36]. The algorithm guarantees a fixed upper bound on the aspect ratio of all triangles it produces, provided that the input polygon has no sharp angles. A three-dimensional version of that algorithm has recently been implemented by Vavasis [47].

We can show that the interior nodes in a 1-skeleton form an overlap graph. This result is easily generalized to higher-order elements. First, we prove a preliminary geometric lemma. This lemma is also used later on. For the lemma and rest of the paper, let v_d be the volume of the d-dimensional unit ball embedded in $\mathbb{R}^d$, and let s_{d-1} be the surface area of $d-1$-dimensional unit sphere embedded in $\mathbb{R}^d$. These are well known to be

$$v_d = \frac{\pi^{d/2}}{(d/2)!}$$

for d even,

$$v_d = \frac{2^{(d+1)/2}\pi^{(d-1)/2}}{1 \cdot 3 \cdot 5 \cdots d}$$

for d odd, and

$$s_{d-1} = dv_d.$$

LEMMA 3.1. *Let B be a ball and S a sphere both embedded in $\mathbb{R}^d$. Let the radius of B be γ and the radius of S be r such that $r \geq \gamma$. Assume that S and $0.5 \cdot B$ have a common point. Then the surface area of $S \cap B$ is at least*

$$\left(\frac{\sqrt{7}}{4}\gamma\right)^{d-1} v_{d-1}.$$

Proof. It suffices to prove the lemma in the special case that $\gamma = 1$ (and hence $r \geq 1$) because we can initially scale each coordinate of $\mathbb{R}^d$ by $1/\gamma$. Furthermore, without loss of generality, let B be centered at the origin, and let S be centered at the point $(p, 0, 0, \ldots, 0)$ with $p \geq 0$. The assumption that S and $0.5 \cdot B$ have a common point means that $r - 0.5 \leq p \leq r + 0.5$. Let us consider the $(d-2)$-dimensional sphere S' that is the intersection of ∂B and S, that is, the solution to the equations

$$x_1^2 + \cdots + x_d^2 = 1,$$
$$(x_1 - p)^2 + x_2^2 + \cdots + x_d^2 = r^2.$$

Clearly, these equations have a solution if and only if

$$x_1^2 - 1 = (x_1 - p)^2 - r^2,$$

which has as its unique solution

$$x_1^* = \frac{p^2 - r^2 + 1}{2p} = \frac{p}{2} + \frac{1 - r^2}{2p}.$$

Now, let us consider the minimum and maximum possible values of x_1^* over all choices of r, p satisfying these constraints. First, we consider the maximum possible value. Note that both terms in the formula for x_1^* are increasing as p increases (because $1 - r^2 \leq 0$). Therefore, to maximize x_1^* we would pick p maximally to be $r + 1/2$. Substituting this in the formula for x_1^* yields

$$x_{1,\max}^* = \frac{r}{2} + \frac{1}{4} + \frac{1 - r^2}{2r + 1} = \frac{1}{2} + \frac{\frac{3}{4}}{2r + 1}.$$

This is maximized for r as small as possible, i.e., $r = 1$. In this case we have $x_{1,\max}^* = 3/4$.

Now, let us consider the minimum possible value of x_1^*. In this case, we want to pick $p = r - 1/2$, yielding

$$x_{1,\min}^* = \frac{r}{2} - \frac{1}{4} + \frac{1 - r^2}{2r - 1} = -\frac{1}{2} + \frac{\frac{3}{4}}{2r - 1}.$$

This is minimized by taking r as large as possible, yielding $x^*_{1,\min} = -1/2$. Combining the upper and lower bound, we conclude that $|x_1^*| \leq 3/4$.

Now, notice that $S \cap B$, a spherical cap, is the solution to the system

$$(x_1 - p)^2 + \cdots + x_d^2 = r^2,$$
$$x_1 \leq x_1^*.$$

Consider projecting this cap orthogonally onto the plane $x_1 = x_1^*$. The projection B' of $S \cap B$ contains $(x_1^*, x_2, \ldots, x_d)$ if and only if

$$x_2^2 + \cdots + x_d^2 \leq r^2 - (x_1^* - p)^2$$
$$= 1 - (x_1^*)^2.$$

Since orthogonal projection reduces area, the area of $S \cap B$ is at least the area of B'. We have already proved that $|x_1^*| \leq 3/4$, so the radius of B' is at least $(1 - 9/16)^{1/2}$, i.e., at least $\sqrt{7}/4$. Thus, the area of $S \cap B$ is at least the area of B', which is at least $(\sqrt{7}/4)^{d-1} v_{d-1}$. □

We now present the main theorem of this section.

THEOREM 3.2. *Let H be a simplicial complex embedded in $\mathbb{R}^d$. Assume that every simplex has aspect ratio bounded by c_1. Let G be the finite-element graph of H, that is, the 1-skeleton of interior nodes of H. Then G is a subgraph of an α-overlap graph for an α bounded in terms of c_1 and d.*

Proof. Let the interior nodes of the complex be $\{\boldsymbol{p}_1, \ldots, \boldsymbol{p}_n\}$. Fix a particular i and consider $\boldsymbol{p}_i$. Let $T_1, \ldots, T_q$ be the simplices adjacent to $\boldsymbol{p}_i$. Define $r_i = \min(r(T_1), \ldots, r(T_q))$. Surround node $\boldsymbol{p}_i$ with a ball B_i of radius r_i. Carry out this construction of r_i and B_i for each node $\boldsymbol{p}_i$.

Note that r_i is at most half the distance to the facet of T opposite $\boldsymbol{p}_i$ for any simplex T containing $\boldsymbol{p}_i$. Since $\boldsymbol{p}_i$ is interior to H, the shortest altitude to the facets of $T_1, \ldots, T_q$ opposite $\boldsymbol{p}_i$ is shorter than the distance from $\boldsymbol{p}_i$ to any other node. This shows that B_i does not intersect any of the other balls B_j for j not equal to i. Thus, $B_1, \ldots, B_n$ form a 1-ply system.

Now, we prove that the edges of the 1-skeleton adjacent to $\boldsymbol{p}_i$ are covered by $\alpha \cdot B_i$. First, we argue that the number of simplices adjacent to any particular node $\boldsymbol{p}_i$ is bounded above in terms of c_1. The argument for this bound is as follows. Because of the aspect ratio bound, there is a lower bound on the solid angle of each simplex adjacent to $\boldsymbol{p}_i$ and therefore an upper bound q^* on the number of such simplices.

Stating this argument in more detail, let $C_1, \ldots, C_q$ be the balls of radii $r(T_1), \ldots, r(T_q)$ inscribed in $T_1, \ldots, T_q$. For a particular j, $1 \leq j \leq q$, let S_j be the sphere centered at $\boldsymbol{p}_i$ and passing through the center of C_j. By the lemma, the surface area of $S_j \cap C_j$ is at least $(\sqrt{7} r(T_j)/4)^{d-1} v_{d-1}$.

Now, consider the sphere S of radius 1 centered at $\boldsymbol{p}_i$. For each j, S_j is also a sphere centered at $\boldsymbol{p}_i$; hence we can expand or contract its radius to make it coincide with S. Let ρ_j be this radius. This dilation also carries $S_j \cap C_j$ to a subregion U_j of S. These subregions are disjoint (or perhaps have common boundary points only) for $j = 1, \ldots, q$ because $S_j \cap C_j$ and also its dilation U_j lie inside the convex cone centered at $\boldsymbol{p}_i$ defined by T_j. The surface area of U_j is at least $(\sqrt{7} r(T_j)/(4\rho_j))^{d-1} v_{d-1}$.

Note that $\rho_j \leq 2R(T_j)$ because all of T_j is contained inside the sphere of radius $2R(T_j)$ centered at $\boldsymbol{p}_i$. Therefore, S_j must have a smaller radius than this sphere. Thus, the surface area of U_j is at least $(\sqrt{7} r(T_j)/(8R(T_j)))^{d-1} v_{d-1}$, i.e., at least $(\sqrt{7}/(8c_1))^{d-1} v_{d-1}$.

Thus, we have q disjoint subsets $U_1, \ldots, U_j$ of S each with surface area $(\sqrt{7}/(8c_1))^{d-1} v_{d-1}$. Since the surface area of S is s_{d-1}, this shows that

$$q \leq \frac{s_{d-1}(8c_1)^{d-1}}{7^{(d-1)/2} v_{d-1}}.$$

Let us call this upper bound q^*.

Let us say that two simplices are neighbors if they share a $(d-1)$-facet. By the assumption that $\boldsymbol{p}_i$ is an interior node, we know that all of the simplices adjacent to $\boldsymbol{p}_i$ are "connected" under the transitive closure of the "neighbor" relation. Now, we claim that if two simplices $T_j, T_{j'}$ are neighbors, then $R(T_j) \geq r(T_{j'})$. This follows immediately because $r(T_{j'})$ is shorter than half the length of the shortest edge of the common face, where $R(T_j)$ is greater than half the length of the longest edge. The inequality $R(T_j) \geq r(T_{j'})$ implies that $c_1 R(T_j) \geq R(T_{j'})$.

Recall that r_i is the minimum $r(T_j)$ for $j = 1, \ldots, q$; say the minimum is achieved at $j = 1$. Then the arguments in the last two paragraphs show that $R(T_j)$ for any j is bounded by $c_1^{q^*} R(T_1)$. Thus, the ball of radius $2c_1^{q^*} R(T_1)$, i.e., radius $2c_1^{q^*+1} r_i$, contains $T_1, \ldots, T_q$. This ball is $(2c_1^{q^*+1}) \cdot B_i$.

This shows that G is indeed a subgraph of an α-overlap graph because all the nodes connected to $\boldsymbol{p}_i$ are vertices of $T_1, \ldots, T_q$. □

Other shape criteria weaker than an aspect ratio bound have appeared in the literature; for instance, Babuška and Aziz [2] have shown that the two-dimensional finite-element approximation converges to the true solution in the case that the largest angle of the mesh is bounded away from π (this is a weaker condition than bounded aspect ratio). Miller, Talmor, Teng, and Walkington [29] have shown a similar result about three-dimensional Delaunay triangulations satisfying a radius-edge ratio bound. In such a triangulation, the radius of the circumscribing circle of each simplex divided by its shortest edge is bounded above by a constant, and the triangulation is a Delaunay triangulation.

In the case of a two-dimensional triangulation with an upper bound on the largest angle, the 1-skeleton of the triangulation is not necessarily an overlap graph with bounded α. However, the bounded radius-edge Delaunay triangulation is an α-overlap graph as argued by [29].

Another type of discretization used in solving PDEs is a finite-difference mesh with adaptive refinement; see, e.g., Fig. 3 based on a paper by Berger and Bokhari [3]. In such a mesh, it is a common rule to require that no node has neighbors more than twice as far away as its closest neighbor (otherwise the extra interpolations lead to numerical inaccuracy). It is very easy to see that such a graph is an α-overlap graph with $\alpha = 2$.

4. The main algorithm and stereographic projection. We now describe our separator algorithm. Then we describe some of the details of the implementation and explain its complexity. The correctness proof of the algorithm is the subject of sections 5 through 7.

We start with two preliminary concepts. We let Π denote the *stereographic projection* mapping from $\mathbb{R}^d$ to S^d, where S^d is the unit d-sphere embedded in $\mathbb{R}^{d+1}$. Geometrically, this map may be defined as follows. Given $\boldsymbol{x} \in \mathbb{R}^d$, append "0" as the final coordinate yielding $\boldsymbol{x}' \in \mathbb{R}^{d+1}$. Then compute the intersection of S^d with the line in $\mathbb{R}^{d+1}$ passing through $\boldsymbol{x}'$ and $(0, 0, \ldots, 0, 1)^T$. This intersection point is $\Pi(\boldsymbol{x})$.

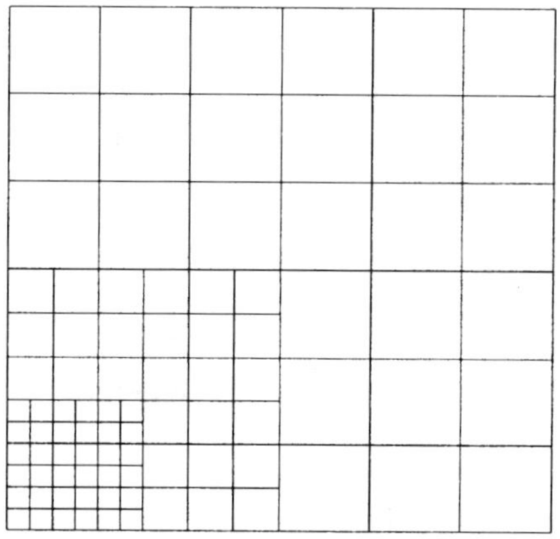

FIG. 3. *An example of finite-difference mesh with refinement.*

Algebraically, the mapping is defined as

$$\Pi(\boldsymbol{x}) = \begin{pmatrix} 2\boldsymbol{x}/\chi \\ 1-2/\chi \end{pmatrix},$$

where $\chi = \boldsymbol{x}^T\boldsymbol{x}+1$. It is also simple to write down a formula for the inverse of Π. Let $\boldsymbol{u}$ be a point on S^d. Then

$$\Pi^{-1}(\boldsymbol{u}) = \frac{\bar{\boldsymbol{u}}}{1-u_{d+1}},$$

where $\bar{\boldsymbol{u}}$ denotes the first d entries of $\boldsymbol{u}$ and u_{d+1} is the last entry. The stereographic mapping, besides being easy to compute, has a number of important properties proved below.

A second crucial concept for our algorithm is the notion of a *center point.* Given a finite subset $P \subset \mathbb{R}^d$ such that $|P| = n$, a *center point* of P is defined to be a point $\boldsymbol{x} \in \mathbb{R}^d$ such that if H is any open half-space whose boundary contains $\boldsymbol{x}$, then

$$|P \cap H| \leq dn/(d+1). \tag{1}$$

It can be shown from Helly's theorem [8] that a center point always exists. Note that center points are quite different from centroids. A center point is largely insensitive to "outliers" in P. On the other hand, a single distant outlier can cause the centroid of P to be displaced by an arbitrarily large distance. In the $d = 1$ case, a center point is the same as a median for n odd and is any point between the two medians for n even.

Main Separator Algorithm.

Let $P = \{\boldsymbol{p}_1, \ldots, \boldsymbol{p}_n\}$ be the input points in $\mathbb{R}^d$ that define the overlap graph.

1. Given $\boldsymbol{p}_1, \ldots, \boldsymbol{p}_n$, compute $P' = \{\Pi(\boldsymbol{p}_1), \ldots, \Pi(\boldsymbol{p}_n)\}$ so that $P' \subset S^d$.
2. Compute a center point $\boldsymbol{z}$ of P'.
3. Compute an orthogonal $(d+1) \times (d+1)$ matrix Q such that $Q\boldsymbol{z} = \boldsymbol{z}'$, where

$$\boldsymbol{z}' = \begin{pmatrix} 0 \\ \vdots \\ 0 \\ \theta \end{pmatrix}$$

such that θ is a scalar.

4. Define $P'' = QP'$ (i.e., apply Q to each point in P'). Note that $P'' \subset S^d$, and the center point of P'' is $\boldsymbol{z}'$.
5. Let D be the matrix $[(1-\theta)/(1+\theta)]^{1/2}I$, where I is the $d \times d$ identity matrix. Let $P''' = \Pi(D\Pi^{-1}(P''))$. Below we show that the origin is a center point of P'''.
6. Choose a random great circle S_0 on S^d.
7. Transform S_0 back to a sphere $S \subset \mathbb{R}^d$ by reversing all the transformations above, i.e., $S = \Pi^{-1}(Q^{-1}\Pi(D^{-1}\Pi^{-1}(S_0)))$.
8. From S compute a set of vertices of G that split the graph as in Theorem 2.4. In particular, define C to be vertices embedded "near" S, define A to be vertices of $G - C$ embedded outside S, and define B to be vertices of $G - C$ embedded inside S. (This step is described in section 7.)

We can immediately make the following observation: because the origin is a center point of P''', and the points are split by choosing a plane through the origin, then we know that $|A| \leq (d+1)n/(d+2)$ and $|B| \leq (d+1)n/(d+2)$ regardless of the details of how C is chosen. (Notice that the constant factor is $(d+1)/(d+2)$ rather than $d/(d+1)$ because the point set P' lies in $\mathbb{R}^{d+1}$ rather than $\mathbb{R}^d$.) Thus, one of the claims made in Theorem 2.4 will follow as soon as we have shown that the origin is indeed a center point of P''' at the end of this section.

We now provide additional details about the steps of the algorithm and also its complexity analysis. We have already defined stereographic projection used in step 1. Step 1 requires $O(nd)$ operations.

Computing a true center point in step 2 appears to a very expensive operation (involving a linear programming problem with n^d constraints), but by using random (geometric) sampling, an approximate center point can be found in random constant time (independent of n but exponential in d) [44, 19]. An *approximate* center point satisfies (1) except with $(d+1+\epsilon)n/(d+2)$ on the right-hand side, where $\epsilon > 0$ may be arbitrarily small. Alternatively, a deterministic linear-time sampling algorithm can be used in place of random sampling [28, 42], but one must again compute a center of the sample using linear programming in time exponential in d. See [7] and [30] for more discussion on center points and efficient algorithms for approximately computing them; see [16] for practical behavior of these randomized algorithms.

In step 3, the necessary orthogonal matrix may be represented as a single Householder reflection; see [17] for an explanation of how to pick an orthogonal matrix to zero out of all but one entry in a vector. The number of floating point operations involved is $O(d)$ independent of n.

In step 4 we do not actually need to compute P''; the set P'' is defined only for the purpose of analysis. Thus, step 4 does not involve computation. Note that $\boldsymbol{z}'$ is the center point of P'' after this transformation because when a set of points is transformed by any orthogonal transformation, a center point moves according to the same transformation (more generally, center points are similarly moved under any affine transformation). This is proved below.

In step 6 we choose a random great circle, which requires time $O(d)$. This is equivalent to choosing a plane through the origin with a randomly selected orientation. (This step of the algorithm can be made deterministic; see [10].) Step 7 is also seen to require time $O(d)$.

Finally, there are two possible alternatives for carrying out step 8, which are both described in section 7 in more detail. One alternative is that we are provided with the neighborhood system of the points (i.e., a list of n balls in $\mathbb{R}^d$) as part of the input. In this case step 8 requires $O(nd)$ operations, and the test to determine which points belong in A, B, or C is a simple geometric test involving S. Another possibility is that we are provided with the nodes of the graph and a list of edges. In this case we determine which nodes belong in A, B, or C based on scanning the adjacency list of each node, which requires time linear in the size of the graph.

We conclude this section by proving lemmas about stereographic projection and center points. These lemmas establish the claims made within the statement of the algorithm; in particular, they establish that the origin in step 5 is indeed a centerpoint of P'''.

LEMMA 4.1. *The mapping* Π *is conformal.*

Proof. Recall that a differentiable mapping $F : \mathbb{R}^d \to \mathbb{R}^{d'}$ is said to be *conformal* if $F'(\boldsymbol{x})^T F'(\boldsymbol{x}) = \beta(\boldsymbol{x})^2 I$ for all $\boldsymbol{x}$, where $\beta(\boldsymbol{x})^2$ is a positive scalar and I is the $d \times d$ identity matrix, i.e., the columns of F' form an orthonormal basis multiplied by a scalar. The proof of this lemma is a straightforward computation; observe that

$$\Pi'(\boldsymbol{x}) = (2/\chi^2) \begin{bmatrix} -2\boldsymbol{x}\boldsymbol{x}^T + \chi I \\ 2\boldsymbol{x}^T \end{bmatrix},$$

and therefore

$$\begin{aligned} \Pi'(\boldsymbol{x})^T \Pi'(\boldsymbol{x}) &= (4/\chi^4)(4\boldsymbol{x}\boldsymbol{x}^T\boldsymbol{x}\boldsymbol{x}^T - 4\chi\boldsymbol{x}\boldsymbol{x}^T + \chi^2 I + 4\boldsymbol{x}\boldsymbol{x}^T) \\ &= (4/\chi^4)(4\boldsymbol{x}\boldsymbol{x}^T(\boldsymbol{x}^T\boldsymbol{x} - \chi + 1) + \chi^2 I) \\ &= (4/\chi^4)(\chi^2 I). \end{aligned}$$

To obtain the last line we used the equation $\boldsymbol{x}^T\boldsymbol{x} - \chi + 1 = 0$ by definition of χ. □

The following lemma concerns planes in $\mathbb{R}^{d+1}$; a *plane* is defined to be the set of points $\boldsymbol{x}$ satisfying one linear equation $\boldsymbol{a}^T\boldsymbol{x} = b$, where $\boldsymbol{a}$ is a nonzero vector.

LEMMA 4.2. *Let* $V = \{\boldsymbol{y} \in \mathbb{R}^{d+1} : \boldsymbol{a}^T\boldsymbol{y} = b\}$ *be a plane in* $\mathbb{R}^{d+1}$ *that intersects* S^d. *The function* Π^{-1} *maps* $S^d \cap V$ *to a sphere in* $\mathbb{R}^d$.

Proof. (Degenerate cases of "spheres" in $\mathbb{R}^d$ include points and planes.) Let $\boldsymbol{u}$ be a point in $S^d \cap V$. Partition $\boldsymbol{u}$ as $(\bar{\boldsymbol{u}}, u_{d+1})$, where $\bar{\boldsymbol{u}} \in \mathbb{R}^d$. Partition $\boldsymbol{a}$ in the same way. Then $\bar{\boldsymbol{a}}^T\bar{\boldsymbol{u}} = b - a_{d+1}u_{d+1}$. Assume that $\boldsymbol{u} = \Pi(\boldsymbol{x})$ for some $\boldsymbol{x} \in \mathbb{R}^d$; then we have $2\bar{\boldsymbol{a}}^T\boldsymbol{x}/\chi = b - a_{d+1}(1 - 2/\chi)$, i.e., $2\bar{\boldsymbol{a}}^T\boldsymbol{x} = (b - a_{d+1})\chi + 2a_{d+1}$, where $\chi = \boldsymbol{x}^T\boldsymbol{x} + 1$. If $b - a_{d+1} = 0$, then this set defines a plane in $\mathbb{R}^d$. Else assume $b - a_{d+1}$ is nonzero. Then the above equation may be written as

$$\boldsymbol{x}^T\boldsymbol{x} - \frac{2\bar{\boldsymbol{a}}^T\boldsymbol{x}}{b - a_{d+1}} + 1 + \frac{2a_{d+1}}{b - a_{d+1}} = 0.$$

This is the equation of a sphere in $\mathbb{R}^d$. □

LEMMA 4.3. *Let* ρ *be the scalar in step* 5, *i.e.,* $\rho = [(1-\theta)/(1+\theta)]^{1/2}$, *and let* $D = \rho I$. *(Note that* ρ *is well defined because* $-1 < \theta < 1$. *The center point of at least* $d+3$ *distinct points on the sphere must be interior to the sphere itself.) Planes passing through* $\boldsymbol{z}'$ *are mapped by the transformation* $\Pi \circ D \circ \Pi^{-1}$ *to planes passing through the origin, and similarly for half-spaces.*

Remark. This lemma shows that the origin is the center point of P''. Furthermore, this lemma also shows that if an approximate center point is used in step 2 instead of an exact center point, then the transformation of step 5 also preserves approximate centering.

Proof. Let V be a plane passing through $\boldsymbol{z}'$; such a plane has the form $V = \{\boldsymbol{u} : \bar{\boldsymbol{a}}^T\bar{\boldsymbol{u}} = a_{d+1}(\theta - u_{d+1})\}$, following the notation of the preceding lemma. (We will prove the lemma just for the case of planes; to prove the case of half-spaces, we would instead start with V defined to be $\{\boldsymbol{u} : \bar{\boldsymbol{a}}^T\bar{\boldsymbol{u}} > a_{d+1}(\theta - u_{d+1})\}$ and then carry out the same analysis.)

If we apply Π^{-1} to such a plane, as in the last lemma, the image is

$$V' = \{\boldsymbol{x} : 2\bar{\boldsymbol{a}}^T\boldsymbol{x} = a_{d+1}(1 + \theta - \boldsymbol{x}^T\boldsymbol{x}(1-\theta))\}.$$

Now, we apply D; the image DV' is

$$V'' = \{\boldsymbol{x} : 2\bar{\boldsymbol{a}}^T\boldsymbol{x}/\rho = a_{d+1}(1 + \theta - \boldsymbol{x}^T\boldsymbol{x}(1-\theta)/\rho^2)\}.$$

Finally, we apply Π; if the image point in S^d is $(\bar{\boldsymbol{z}}, z_{d+1})$, then we know from the stereographic formulas that the preimage point in $\mathbb{R}^d$ is $\boldsymbol{x} = \bar{\boldsymbol{z}}/(1 - z_{d+1})$ and that $\boldsymbol{x}^T\boldsymbol{x} = 2/(1-z_{d+1}) - 1$. Thus, the points $\boldsymbol{z}$ in the image $V''' = \Pi \circ D \circ \Pi^{-1}(V)$ satisfy the equation

$$\frac{2\bar{\boldsymbol{a}}^T\bar{\boldsymbol{z}}}{\rho(1 - z_{d+1})} = a_{d+1}\left[1 + \theta - \frac{(2/(1 - z_{d+1}) - 1)(1-\theta)}{\rho^2}\right].$$

Multiplying through by $1 - z_{d+1}$ yields

$$\frac{2\bar{\boldsymbol{a}}^T\bar{z}}{\rho} = a_{d+1}\left[(1+\theta)(1 - z_{d+1}) - \frac{(2 - (1 - z_{d+1}))(1-\theta)}{\rho^2}\right].$$

This equation is linear in $\boldsymbol{z}$, showing that V''' is a plane in $\mathbb{R}^{d+1}$. To verify that it is a plane passing through the origin, we substitute $\bar{\boldsymbol{z}} = 0$ and $z_{d+1} = 0$ to see if we get an equation:

$$0 = a_{d+1}\left[(1+\theta) - \frac{(1-\theta)}{\rho^2}\right].$$

It is now seen that the choice $\rho = [(1-\theta)/(1+\theta)]^{1/2}$ used in step 5 does indeed make this equation hold. □

There are a few things to note about this separator algorithm.

1. For all steps except the last, the only information used about the nodes is their geometric embedding. We do not need to know their balls B_i's defining the 1-ply neighborhood, we do not need to know α, and we do not need to know explicit edges in the graph. This means that our algorithm can be applied to graphs that are suspected to be α-overlap graphs without actually computing α. In some circumstances (such as k-nearest neighbor graphs) we can construct the k-ply neighborhood system in linear time starting from the coordinates of the points [31].
2. For all steps except the last, a random sample of P can be used in place of P. The size of the random sample depends on d; for example, for $d = 3$ we have used sample sizes of about 1200. This means that the running time of most of the algorithm is independent of n; see [30] and [16].

5. Construction of a cost function. In this section we begin the proof of the main result in Theorem 2.4, which is that $|C|$ is bounded by $O(\alpha n^{(d-1)/d} + \text{const})$ with high probability. Before starting into the details of the proof, let us provide a proof sketch.

- In this section we construct a *cost* function $g(\boldsymbol{x})$ that maps $\mathbb{R}^d$ to the nonnegative real numbers. This construction is based on the neighborhood system. The value of $g(\boldsymbol{x})$ is large if many nodes of the neighborhood system P are "near" $\boldsymbol{x}$. The support of g is finite.
- Most of this section is devoted to establishing the result that the integral of g^d over all of $\mathbb{R}^d$ is bounded by $O(\alpha^{d/(d-1)}n)$. We call this integral the "total cost" of g.
- In section 6 we define the cost of a random sphere S in step 6 of the algorithm by the integral of g^{d-1} over the image of S back in $\mathbb{R}^d$. We show that the expected value of the cost is $O(\alpha n^{(d-1)/d})$. To prove this requires the Hölder inequality, and one of the factors in the Hölder estimate is the total cost that will have been analyzed in section 5.
- Finally, in section 7 we explain which nodes should be placed in C based on the neighborhood system. We then show that every node placed in C (with the exception of a few nodes whose number is bounded above independently of n) can be "charged" against the integral defining the cost of S. There is a constant lower bound on the amount that each node in C charges against the cost of S. Therefore, the total number of nodes in C (besides the exceptional set) is bounded above by the cost of S. But we will have already shown in section 6 that the cost of S is expected to be $O(\alpha n^{(d-1)/d})$, so this yields the upper bound on the expected value of $|C|$.

The intuitive reason that we can "charge" the cost of a node in C against the integral of g^{d-1} over S is that g^{d-1} gets larger wherever S passes near a dense cluster of nodes. But such a cluster is precisely the place where more nodes will have to be put in C to separate the graph.

As mentioned above, for the proof of Theorem 2.4 we construct a nonnegative real-valued cost function g on $\mathbb{R}^d$ based on the neighborhood system. As above, let the nodes be $\boldsymbol{p}_1, \ldots, \boldsymbol{p}_n$ and let the α-overlap graph be defined by 1-ply neighborhood system $B_1, \ldots, B_n$. Let the radii of $B_1, \ldots, B_n$ be $r_1, \ldots, r_n$, and define $\gamma_i = 2\alpha r_i$ for $i = 1, \ldots, n$.

For each $\boldsymbol{p}_i$ we define f_i as follows:

$$f_i(\boldsymbol{x}) = \begin{cases} 1/\gamma_i & \text{if } \boldsymbol{x} \in (2\alpha) \cdot B_i, \text{ i.e., } \|\boldsymbol{x} - \boldsymbol{p}_i\| \le \gamma_i, \\ 0 & \text{otherwise.} \end{cases}$$

Notice that

$$\int_{\mathbb{R}^d} f_i^d \, dV = v_d.$$

(Recall from section 3 that v_d denotes the volume of the d-dimensional unit sphere.) Here, and for the rest of the paper, integrations over volumes in $\mathbb{R}^d$ or d-dimensional surfaces in $\mathbb{R}^{d+1}$ are denoted by dV, and integrations over $(d-1)$-dimensional surfaces are denoted by dS. Next, define f and g, nonnegative functions on $\mathbb{R}^d$, as follows:

$$f(\boldsymbol{x}) = \left(\sum_{i=1}^{n} f_i(\boldsymbol{x})^d \right)^{1/d}$$

and

$$g(\boldsymbol{x}) = \left(\sum_{i=1}^{n} f_i(\boldsymbol{x})^{d-1} \right)^{1/(d-1)}.$$

We notice immediately that

$$\int_{\mathbb{R}^d} f^d \, dV = v_d n \tag{2}$$

because this integral is equal to the sum of the integrals of the f_i^d.

The rest of this section is devoted to establishing an $O(n)$ upper bound on the integral of g^d. For this we need a series of lemmas. We use these lemmas to establish that f^d and g^d are always within a constant factor of each other at every point. Once this fact is established, we can then easily estimate the integral of g^d since (2) is an exact formula for the integral of f^d.

Because the distinction between f and g has to do with a slight shift in the exponents, we need lemmas that establish some basic properties concerning powers of sums and sums of powers.

This first lemma is an auxiliary lemma used to prove Lemma 5.2.

LEMMA 5.1. *Let $a_1, \ldots, a_n$ be nonnegative numbers, and suppose $p \geq 1$. Then*

$$\left(\sum_{i=1}^{n} a_i\right)^p \leq p \sum_{i=1}^{n} \left[a_i \left(\sum_{j=i}^{n} a_j\right)^{p-1} \right].$$

Proof. Define the function

$$\phi(x_1, \ldots, x_n) = \left(\sum_{i=1}^{n} x_i\right)^p.$$

We notice that

$$\frac{\partial \phi}{\partial x_j} = p \left(\sum_{i=1}^{n} x_i\right)^{p-1}$$

for any j. Let $\mathbf{a}^{(i)}$ be the vector in $\mathbb{R}^n$ given by

$$\mathbf{a}^{(i)} = (0, \ldots, 0, a_i, a_{i+1}, \ldots, a_n).$$

Then

$$\begin{aligned}
\phi(a_1, \ldots, a_n) &= \phi(\mathbf{a}^{(1)}) - \phi(\mathbf{a}^{(n+1)}) \\
&= \sum_{i=1}^{n} \left[\phi(\mathbf{a}^{(i)}) - \phi(\mathbf{a}^{(i+1)}) \right] \\
&= \sum_{i=1}^{n} \int_0^{a_i} \frac{\partial \phi}{\partial x_i}(0, \ldots, 0, t, a_{i+1}, \ldots, a_n) \, dt \\
&= p \sum_{i=1}^{n} \int_0^{a_i} \left(t + \sum_{j=i+1}^{n} a_j \right)^{p-1} dt \\
&\leq p \sum_{i=1}^{n} a_i \cdot \left(a_i + \sum_{j=i+1}^{n} a_j \right)^{p-1} \\
&= p \sum_{i=1}^{n} a_i \cdot \left(\sum_{j=i}^{n} a_j \right)^{p-1}. \qquad \square
\end{aligned}$$

The next lemma relates two different sums of powers involving a sequence of numbers $\ldots, m_{-1}, m_0, m_1, m_2, \ldots$. Below we will express f^d and g^d in terms of sums of this kind.

LEMMA 5.2. *Let* $\ldots, m_{-1}, m_0, m_1, m_2, \ldots$ *be a doubly infinite sequence of nonnegative numbers such that each* m_i *is bounded above by* θ *and such that at most a finite number of* m_i*'s are nonzero. Let* $d \geq 2$ *be an integer. Then*

$$\left(\sum_{k=-\infty}^{\infty} m_k 2^{-k(d-1)}\right)^{d/(d-1)} \leq c_d \theta^{1/(d-1)} \sum_{k=-\infty}^{\infty} m_k 2^{-kd},$$

where c_d *is a positive number depending on* d.

Proof. Since at most a finite number of the m_k are nonzero, then we can apply the preceding lemma because the above sums are actually finite. Applying the lemma, we see that

$$\begin{aligned}
\left(\sum_{k=-\infty}^{\infty} m_k 2^{-k(d-1)}\right)^{d/(d-1)} &\leq \frac{d}{d-1} \sum_{k=-\infty}^{\infty} m_k 2^{-k(d-1)} \cdot \left(\sum_{j=k}^{\infty} m_j 2^{-j(d-1)}\right)^{1/(d-1)} \\
&\leq \frac{d}{d-1} \sum_{k=-\infty}^{\infty} m_k 2^{-k(d-1)} \cdot \left(\sum_{j=k}^{\infty} \theta \cdot 2^{-j(d-1)}\right)^{1/(d-1)} \\
&= \frac{d}{d-1} \sum_{k=-\infty}^{\infty} m_k 2^{-k(d-1)} \cdot \left(\frac{\theta \cdot 2^{-k(d-1)}}{1-2^{-(d-1)}}\right)^{1/(d-1)} \\
&= c_d \theta^{1/(d-1)} \sum_{k=-\infty}^{\infty} m_k 2^{-k(d-1)} \cdot 2^{-k} \\
&= c_d \theta^{1/(d-1)} \sum_{k=-\infty}^{\infty} m_k 2^{-kd}. \qquad \square
\end{aligned}$$

The next lemma is used to establish one direction on the relation between f^d and g^d.

LEMMA 5.3. *Let* $a_1, \ldots, a_n$ *be nonnegative numbers, and* $d \geq 2$. *Then*

$$\left(\sum_{i=1}^{n} a_i^d\right)^{1/d} \leq \left(\sum_{i=1}^{n} a_i^{d-1}\right)^{1/(d-1)}.$$

Proof. See [12]. $\square$

We now come to the main result for this section, which uses the preceding lemmas.

THEOREM 5.4. *For all* $\boldsymbol{x} \in \mathbb{R}^d$, *the following inequalities hold:*

$$f(\boldsymbol{x})^d \leq g(\boldsymbol{x})^d \leq c'_d \alpha^{d/(d-1)} f(\boldsymbol{x})^d,$$

where c'_d *is a constant depending on* d.

Proof. The first inequality follows immediately from the definitions of f and g and Lemma 5.3.

For the second inequality we focus on a particular point $\boldsymbol{x} \in \mathbb{R}^d$. If $f(\boldsymbol{x}) = 0$, then $g(\boldsymbol{x}) = 0$ as well, so the inequality follows. Otherwise, define for all integers k

$$M_k = \{i \in \{1, \ldots, n\} : 2^{-k} \leq f_i(\boldsymbol{x}) < 2^{-k+1}\}.$$

Notice that the M_k's are pairwise disjoint, and their union is the set of indices i such that $f_i(\boldsymbol{x}) \neq 0$.

Let m_k denote the cardinality of M_k. We claim that $m_k \leq c''_d \alpha^d$, where c''_d is a constant.

To prove this, observe that if $i \in M_k$, then $2^{-k} \leq 1/\gamma_i \leq 2^{-k+1}$, i.e., $2^{k-1} \leq \gamma_i \leq 2^k$. This means that $r_i \geq 2^{k-1}/(2\alpha)$, where r_i is the radius of B_i. Also, since $f_i(\boldsymbol{x}) > 0$, $\|\boldsymbol{x} - \boldsymbol{p}_i\| \leq \gamma_i$, which implies $\|\boldsymbol{x} - \boldsymbol{p}_i\| \leq 2^k$. Let B be the ball of radius $(1 + 1/(2\alpha))2^k$ centered at x. Since the ball of radius 2^k contains all the $\boldsymbol{p}_i$'s such that $i \in M_k$, we see that B contains all the B_i's for $i \in M_k$.

On the other hand, these balls have disjoint interiors because they define a 1-ply system. Accordingly, there are m_k balls of radius at least $2^{k-2}/\alpha$ lying in a sphere of radius $(1 + 1/(2\alpha))2^k$, so a straightforward volume-counting argument shows

$$\begin{aligned} m_k &\leq \frac{((1+1/(2\alpha))2^k)^d}{(2^{k-2}/\alpha)^d} \\ &\leq \frac{(2 \cdot 2^k)^d}{(2^{k-2}/\alpha)^d} \\ &= c''_d \alpha^d. \end{aligned}$$

Now, we observe that

$$\begin{aligned} g(\boldsymbol{x})^d &= \left(\sum_{k=-\infty}^{\infty} \sum_{i \in M_k} f_i(\boldsymbol{x})^{d-1} \right)^{d/(d-1)} \\ &\leq \left(\sum_{k=-\infty}^{\infty} m_k (2^{-k+1})^{d-1} \right)^{d/(d-1)} \\ &= 2^d \left(\sum_{k=-\infty}^{\infty} m_k (2^{-k})^{d-1} \right)^{d/(d-1)} \end{aligned}$$

with m_k bounded by $c''_d \alpha^d$. Now, we can apply Lemma 5.2 with the choice $\theta = c''_d \alpha^d$ to deduce that

$$g(\boldsymbol{x})^d \leq c_d 2^d (c''_d \alpha^d)^{1/(d-1)} \sum_{k=-\infty}^{\infty} m_k 2^{-kd}.$$

This summation is a lower bound on $f(\boldsymbol{x})^d$ because for each $i \in M_k$, $f_i(\boldsymbol{x})^d \geq 2^{-kd}$. This concludes the proof of the theorem. □

Therefore, g^d is no more than a constant multiple of f^d, where the constant is $c'_d \alpha^{d/(d-1)}$. By (2) we have a bound of the form

$$\int_{\mathbb{R}^d} g^d \, dV \leq c_d \alpha^{d/(d-1)} n, \tag{3}$$

where c_d is a different constant depending on d.

6. Analysis of a random great circle. Let S be a sphere in $\mathbb{R}^d$. We define

$$\operatorname{cost}(S) = \int_S g^{d-1} \, dS.$$

The rationale for this definition will be provided in section 7, where we prove that cost(S) is proportional to the cost of separating the graph with sphere S, i.e., proportional to the number of graph vertices that must be removed to break all connections in G from the interior of S to the exterior of S in step 8 of our algorithm. In this section we obtain an upper bound on the expected value of cost(S) if S is chosen (at random) by step 6 in our algorithm.

Recall that our separator algorithm computes a conformal mapping $F : \mathbb{R}^d \to S^d$ given by

$$F = \Pi \circ D \circ \Pi^{-1} \circ Q \circ \Pi.$$

Recall that the columns of $F'(\boldsymbol{x})$ for any $\boldsymbol{x}$ form an orthonormal basis multiplied by a nonzero scalar which we will denote $\beta(\boldsymbol{x})$. By continuity, $\beta(\boldsymbol{x})$ has the same sign for all $\boldsymbol{x} \in \mathbb{R}^d$, so without loss of generality, let us say that $\beta(\boldsymbol{x}) > 0$. Then it can be checked that for any real-valued integrable function r defined on $\mathbb{R}^d$

$$\int_{\mathbb{R}^d} r(\boldsymbol{x})\, dV = \int_{S^d} r(F^{-1}(\boldsymbol{u})) \cdot \beta(F^{-1}(\boldsymbol{u}))^{-d}\, dV$$

because $\beta(\boldsymbol{x})^d$ is the determinant of $F'(\boldsymbol{x})$ when interpreted as a basis for the tangent space of S^d at $F(\boldsymbol{x})$. Thus, in particular,

$$\int_{\mathbb{R}^d} g(\boldsymbol{x})^d\, dV = \int_{S^d} g(F^{-1}(\boldsymbol{u}))^d \beta(F^{-1}(\boldsymbol{u}))^{-d}\, dV, \tag{4}$$

where g is the cost function from the last section. Let $h : S^d \to \mathbb{R}$ be defined as

$$h(\boldsymbol{u}) = g(F^{-1}(\boldsymbol{u}))/\beta(F^{-1}(\boldsymbol{u})).$$

Then we can conclude from (3) and (4) that

$$\int_{S^d} h(\boldsymbol{u})^d\, dV \leq c'_d \alpha^{d/(d-1)} n.$$

Next, let us consider a randomly chosen great circle S_0 in S^d. The procedure for defining such a great circle is as follows. First, pick a unit-length vector $\boldsymbol{a}$ uniformly at random. Let the plane through the origin whose normal vector is $\boldsymbol{a}$ be denoted as $\boldsymbol{a}^\perp$, i.e., $\boldsymbol{a}^\perp = \{\boldsymbol{u} : \boldsymbol{a}^T\boldsymbol{u} = 0\}$. Finally, the great circle S_0 is $\boldsymbol{a}^\perp \cap S^d$. Thus, the set of all great circles of S^d is itself parameterized by S^d because $\boldsymbol{a} \in S^d$ is chosen uniformly at random.

Next, let us note that if $F(S) = S_0$ as in step 7 of our algorithm, then

$$\int_S g(\boldsymbol{u})^{d-1}\, dS = \int_{S_0} g(F^{-1}(\boldsymbol{u}))^{d-1} \beta(F^{-1}(\boldsymbol{u}))^{-(d-1)}\, dS$$

because F restricted to S is still a conformal mapping of one lower dimension. This shows that

$$\text{cost}(S) = \int_{\boldsymbol{u} \in S_0} h(\boldsymbol{u})^{d-1}\, dS.$$

Accordingly, we now analyze the expected value of the integral on the right-hand side of the preceding equation. This expected value is equal to

$$E[\text{cost}(S)] = \frac{1}{s_d} \int_{\boldsymbol{a} \in S^d} \int_{\boldsymbol{u} \in \boldsymbol{a}^\perp \cap S^d} h(\boldsymbol{u})^{d-1}\, dS\, dV.$$

We now interchange the order of integration; note that $\boldsymbol{u} \in \boldsymbol{a}^{\perp}$ iff $\boldsymbol{a} \in \boldsymbol{u}^{\perp}$. (To fully justify the interchange of integrals also requires an argument from differential geometry, which is in [32], concerning the volume elements in the two integrations.) We obtain

$$E[\text{cost}(S)] = \frac{1}{s_d} \int_{\boldsymbol{u} \in S^d} \int_{\boldsymbol{a} \in \boldsymbol{u}^{\perp} \cap S^d} h(\boldsymbol{u})^{d-1} \, dS \, dV$$

$$= \frac{s_{d-1}}{s_d} \int_{\boldsymbol{u} \in S^d} h(\boldsymbol{u})^{d-1} \, dV. \tag{5}$$

The second line was obtained by noting that the integrand in the first line is independent of $\boldsymbol{a}$ and hence integration over $\boldsymbol{a}$ reduces to a constant factor.

Now, we apply the Hölder inequality [12]. The Hölder inequality says that for nonnegative functions ϕ and ψ suitably integrable on a measurable set V and for positive real numbers p, q such that $1/p + 1/q = 1$, the following relation holds:

$$\int_V \phi\psi \leq \left(\int_V \phi^p\right)^{1/p} \cdot \left(\int_V \psi^q\right)^{1/q}.$$

We apply this inequality to our problem with $V = S^d$, $p = d/(d-1)$, $q = d$, $\phi = h^{d-1}$, and $\psi = 1$ (constant function) to obtain

$$\int_{S^d} h^{d-1} \, dV \leq \left[\int_{S^d} h^d\right]^{(d-1)/d} \cdot \left[\int_{S^d} 1^d\right]^{1/d}$$

$$= \left[\int_{S^d} h^d\right]^{(d-1)/d} \cdot s_d^{1/d}$$

$$\leq [c'_d \alpha^{d/(d-1)} n]^{(d-1)/d} \cdot s_d^{1/d}$$

$$= c_d'^{(d-1)/d} \alpha n^{(d-1)/d} \cdot s_d^{1/d}. \tag{6}$$

Combining (5) with (6) yields the following result.

THEOREM 6.1. *Let S correspond to a randomly chosen great circle in the separator algorithm. Then*

$$E[\text{cost}(S)] \leq c''_d \alpha n^{(d-1)/d}.$$

Note that this is a bound on the expected value of $\text{cost}(S)$. Since $\text{cost}(S)$ is a nonnegative random variable, we know that with probability 0.5 a random trial will yield a choice of S_1 satisfying $\text{cost}(S_1) \leq 2E[\text{cost}(S)]$. Therefore, if we conduct, for instance, 10 random trials, and keep the best choice for S, then with probability exceeding 0.999 the cost will be bounded by $2E[\text{cost}(S)]$.

7. Constructing a vertex separator from S. In this section we explain how to construct a vertex separator of the overlap graph G given the sphere S. In other words, we will partition the nodes of G into A, B, C to prove Theorem 2.4. For this section, we assume S is a true sphere, and the degenerate case that S is a plane is not analyzed. This degenerate case can be easily handled with a variant of the arguments in this section.

Recall that by the center point property (1) and Lemma 4.2 the number of vertices of G strictly inside S is bounded by $(d+1)n/(d+2)$, as is the number of vertices strictly outside. (The weaker bound $(d+1+\epsilon)n/(d+2)$ holds if an approximate center point was computed in step 2 above.) For our construction, A will be a subset of the vertices lying outside S, B will be a subset of the vertices lying inside S, and C will be vertices lying "close" to S. Because of this choice, we immediately establish the bounds on $|A|, |B|$ stated in Theorem 2.4.

In this section we show how to construct C so that the number of vertices in C, other than a constant-sized exceptional set, is proportional to $\text{cost}(S)$. Since we already have established an upper bound on $\text{cost}(S)$, the argument in this section suffices to establish an upper bound on $|C|$.

Let us assume that we are given the neighborhood system $B_1, \ldots, B_n$ and the value of α. Another possibility is that we are given edges of the graph G instead; we comment on this other possibility later on.

Recall that the radii of $B_1, \ldots, B_n$ are denoted $r_1, \ldots, r_n$. Let r denote the radius of S.

We define $C = C_1 \cup C_2$, where

$$C_1 = \{i : (\alpha \cdot B_i) \cap S \neq \emptyset \text{ and } \boldsymbol{p}_i \in \text{int}(S)\}$$

and

$$C_2 = \{i : B_i \cap S \neq \emptyset \text{ and } \boldsymbol{p}_i \in S \cup \text{ext}(S)\}.$$

LEMMA 7.1. *Let A be the set of nodes of $G - C$ outside S, and let B be the nodes of $G - C$ inside. Then G has no edge between A and B. (Note that nodes exactly on S are in C_2 and hence in C.)*

Proof. Let i, j be two nodes of $G - C$ such that $\boldsymbol{p}_i$ is inside S and $\boldsymbol{p}_j$ is outside S. Then $(\alpha \cdot B_i) \cap S = \emptyset$ (because $i \notin C_1$); hence $\alpha \cdot B_i$ is entirely interior to S. Similarly, B_j is entirely exterior to S. By definition of the overlap graph, there is no (i, j) edge in G. □

We now come to the main theorem of this section, which also establishes Theorem 2.4.

THEOREM 7.2. *With this choice of C, $|C| \leq (4\alpha)^d + \text{cost}(S)/c_d'''$.*

Proof. Partition $C_1 = C_1' \cup C_1''$ and $C_2 = C_2' \cup C_2''$, where for $p = 1, 2$ we define

$$C_p' = \{i \in C_p : 2\alpha r_i > r\}$$

and

$$C_p'' = \{i \in C_p : 2\alpha r_i \leq r\}.$$

We bound $|C_1' \cup C_2'|$ and $|C_1'' \cup C_2''|$ separately.

First, we analyze $C_1' \cup C_2'$, which we write as C'. Note that for every $i \in C'$, $r_i > r/(2\alpha)$. We replace each ball B_i for $i \in C'$ with a smaller ball B_i' of radius exactly $r/(2\alpha)$ such that $B_i' \subset B_i$. For $i \in C_1'$ we simply define $B_i' = r/(2\alpha r_i) \cdot B_i$ to obtain this result. For $i \in C_2'$ we shrink the radius of B_i by factor $r/(2\alpha r_i)$, and also we displace the center so as to maintain the property that $B_i' \cap S \neq \emptyset$.

Let $n_1 = |C'|$. Observe that we have constructed n_1 balls of radius exactly $r/(2\alpha)$, all of whose centers are within distance $r + r/(2\alpha)$ of the center of S. This means that all of these balls are contained in a ball of radius $2r$ centered at the center of S. Note also that these balls B_i' for $i \in C'$ are pairwise disjoint because the original B_i's have disjoint interiors. Therefore, a volume argument shows that

$$n_1 \leq \frac{(2r)^d}{(r/(2\alpha))^d} = (4\alpha)^d.$$

Next, we examine $C_1'' \cup C_2''$, which we write as C''. Fix a particular $i \in C''$. Notice that B_i has radius at most $r/(2\alpha)$. Let $B_i' = (2\alpha) \cdot B_i$, so that B_i' has radius γ_i (recall that γ_i was defined in section 5). Observe that, by definition of C_1 and C_2 above, we are guaranteed that $(0.5 \cdot B_i') \cap S$ (which is the same as $(\alpha \cdot B_i) \cap S$) is nonempty. Furthermore, by construction of C'' we know that $\gamma_i \le r$. Therefore, we can apply Lemma 3.1 to conclude that the area of $B_i' \cap S$ is at least $(\sqrt{7}\gamma_i/4)^{d-1} v_{d-1}$. Note that the value of f_i on B_i' is precisely $1/\gamma_i$. Therefore,

$$\begin{aligned}\int_S f_i^{d-1} &= \int_{B_i' \cap S} f_i^{d-1} \\ &= \frac{1}{\gamma_i^{d-1}} \cdot \operatorname{area}(B_i' \cap S) \\ &\ge \frac{1}{\gamma_i^{d-1}} \cdot \left(\frac{\sqrt{7}\gamma_i}{4}\right)^{d-1} v_{d-1} \\ &= c_d'''.\end{aligned}$$

Here, c_d''' is a positive constant depending only on d.

Therefore,

$$\begin{aligned}\operatorname{cost}(S) &= \int_S g^{d-1}\, dS \\ &= \int_S \sum_{i=1}^n f_i^{d-1}\, dS \\ &= \sum_{i=1}^n \int_S f_i^{d-1}\, dS \\ &\ge \sum_{i \in C''} \int_S f_i^{d-1}\, dS \\ &\ge |C''| \cdot c_d'''.\end{aligned}$$

Combining the upper bounds on $|C'|$ and $|C''|$ proves the theorem. □

We have shown how to carry out step 8 of the main algorithm, namely, deducing a vertex separator C from the sphere S. We have also established the bound on C. The construction so far seems to require explicit knowledge of $B_1, \ldots, B_n$ and of α. If we are not given these items as part of the input, but instead we have edges of G represented explicitly, then clearly we can disconnect G into two pieces by removing all edges connecting the interior of S to the exterior. In this paper we are focusing on vertex separators, so we must find a set of vertices that disconnects G. If we let E_1 be the set of edges passing through S, then we can find a vertex separator C by arbitrarily taking one endpoint of every edge in E_1. For the special case of overlap graphs arising from finite-element methods, which have bounded degree, it can be shown that this simple heuristic does indeed produce a vertex set C with the bound stated in Theorem 2.4. Alternatively, we can find in polynomial time the minimum set of vertices C of G that "cover" E_1, where "cover" means that at least one endpoint of every edge in E_1 is in C; see [9] or [38] for this algorithm.

It can be shown that $\alpha n^{(d-1)/d}$ is the best possible bound on a separator set for an n-vertex α-overlap graph. Define graph G, whose nodes are an $m \times m \times \cdots \times m$ array of nodes arranged in a d-dimensional unit-spaced lattice (so that $n = m^d$) and whose edges connect all neighbors within distance α. This is clearly seen to be an overlap graph, and Vavasis [45] shows that any partitioning of this graph into constant-sized pieces must involve a separator set of at least const $\cdot\, \alpha n^{(d-1)/d}$ nodes.

8. Practical issues. Let us call the algorithm defined in section 4 Weak-Split; it produces a partition in which the ratio of the size of the larger of G_1, G_2 to the smaller is at most $d+1+o(1)$. In practice, one often wants a split in which G_1 and G_2 have no more than half the nodes, i.e., a ratio of $1+o(1)$. (Experimental results [16] suggest that for practical examples, our algorithm produces approximately a 45%–55% split after 10 trials for the $d=3$ case, much better than the worst case 20%–80% split claimed in Theorem 2.4.) There is a standard technique originally due to [26] that derives an algorithm Strong-Split using Weak-Split as a subroutine. Strong-Split yields an even split at the expense of a greater running time (by a constant factor) and a larger constant factor in the bound on the size of the separator.

Another practical issue is a splitting into more than one subdomain. If the number of domains desired p is a power of 2, this is accomplished by applying Strong-Split recursively to get domains of the desired size. The total separator size in this case is $O(p^{1/d}\alpha n^{(d-1)/d})$. This approach can be generalized for a number of subdomains p not a power of 2.

9. Conclusions and open questions. An important question is, given a graph without an embedding, can its nodes be embedded in $\mathbb{R}^d$ to make it a subgraph of an overlap graph?

Also, fast deterministic algorithms for computing approximate center points would be very useful. Deterministic linear-time approximate algorithms are known but are not efficient enough for practical use.

It is also interesting to determine how our algorithm performs in practice compared with other current algorithms, such as the spectral method. This is the subject of recent work by Gilbert, Miller, and Teng [16], who also propose some additional heuristics not described here. The results of [16] can be summarized as follows. Typically, 30 random choices for the sphere separator sufficed. The spectral method used by [16] came from [20], where heuristic local improvement is used as well as spectral partitioning. Our geometric separation algorithm usually ran faster than spectral partitioning (but in both cases there are many possible heuristics that could speed up either one). The quality of the separators (in terms of the balance and the size of the cut) from the geometric method were about the same as from the spectral method for most test cases. In some cases spectral did better; in others, geometric did better. Though the spectral method compares favorably with other heuristics, Guattery and Miller [18] recently found a class of graphs on which the spectral method performs poorly.

Acknowledgments. We would like to thank David Applegate, Marshall Bern, David Eppstein, John Gilbert, Bruce Hendrickson, Ravi Kannan, Michael Klugerman, Tom Leighton, Mike Luby, Oded Schramm, Doug Tygar, and Kim Wagner for invaluable help and discussions.

REFERENCES

[1] N. Alon, P. Seymour, and R. Thomas, *A separator theorem for graphs with an excluded minor and its applications*, in Proc. of the 22th Annual ACM Symposium on Theory of Computing, ACM, New York, 1990, pp. 293–299.

[2] I. Babuška and A. K. Aziz, *On the angle condition in the finite element method*, SIAM J. Numer. Anal., 13 (1976), pp. 214–226.

[3] M. J. Berger and S. Bokhari, *A partitioning strategy for nonuniform problems on multiprocessors*, IEEE Trans. Comput., C-36 (1987), pp. 570–580.

[4] M. Bern, D. Eppstein, and J. R. Gilbert, *Provably good mesh generation*, in Proc. 31st Annual Symposium on Foundations of Computer Science, IEEE, Piscataway, NJ, 1990, pp. 231–241.

[5] J. H. Bramble, J. E. Pasciak, and A. H. Schatz, *An iterative method for elliptic problems on regions partitioned into substructures*, Math. Comp., 46 (1986), pp. 361–9.

[6] L. P. Chew, *Guaranteed Quality Triangular Meshes*, Tech. report 89–893, Department of Computer Science, Cornell University, Ithaca, NY, 1989.

[7] K. Clarkson, D. Eppstein, G. L. Miller, C. Sturtivant, and S.-H. Teng, *Approximating center points with iterated radon points*, Internat. J. Comput. Geom. Appl., 6 (1996), pp. 357–377.

[8] L. Danzer, J. Fonlupt, and V. Klee, *Helly's theorem and its relatives*, in Proc. of Symposia in Pure Mathematics, American Mathematical Society, Providence, RI, 7 (1963), pp. 101–180.

[9] A. L. Dulmage and N. S. Mendelsohn, *Coverings of bipartite graphs*, Canadian J. Math., 10 (1958), pp. 517–534.

[10] D. Eppstein, G. L. Miller, and S.-H. Teng, *A deterministic linear time algorithm for geometric separators and its applications*, Fund. Inform., 22 (1995), pp. 309–330.

[11] I. Fried, *Condition of finite element matrices generated from nonuniform meshes*, AIAA J., 10 (1972), pp. 219–221.

[12] J. E. L. G. Hardy and G. Pólya, *Inequalities*, 2nd ed., Cambridge University Press, Cambridge, 1952.

[13] J. A. George, *Nested dissection of a regular finite element mesh*, SIAM J. Numer. Anal., 10 (1973), pp. 345–363.

[14] J. A. George and J. W. H. Liu, *An automatic nested dissection algorithm for irregular finite element problems*, SIAM J. Numer. Anal., 15 (1978), pp. 1053–1069.

[15] J. Gilbert, J. Hutchinson, and R. Tarjan, *A separation theorem for graphs of bounded genus*, J. Algorithms, 5 (1984), pp. 391–407.

[16] J. Gilbert, G. Miller, and S.-H. Teng, *Geometric mesh partitioning: implementation and experiments*, in Proc. of the 9th International Parallel Processing Symposium, IEEE, Piscataway, NJ, 1995, pp. 418–427; also Tech. report CSL-94-13, Xerox Palo Alto Research Center, Palo Alto, CA; SIAM J. Sci. Comput., to appear.

[17] G. H. Golub and C. F. V. Loan, *Matrix Computations*, 2nd ed., The Johns Hopkins University Press, Baltimore, MD, 1989.

[18] S. Guattery and G. L. Miller, *On spectral partitioning methods*, in Proc. of the 6th ACM–SIAM Symposium on Discrete Algorithms, SIAM, Philadelphia, PA, 1995, pp. 233–242.

[19] D. Haussler and E. Welzl, *ϵ-net and simplex range queries*, Discrete Comput. Geom., 2 (1987), pp. 127–151.

[20] B. Hendrickson and R. Leland, *An Improved Spectral Graph Partitioning Algorithm for Mapping Parallel Computations*, SIAM J. Sci. Comput., 16 (1995), pp. 452–469.

[21] B. Hendrickson and R. Leland, *A multilevel algorithm for partitioning graphs*, in Proc. Supercomputing '95, ACM, New York, 1995.

[22] C. Johnson, *Numerical Solution of Partial Differential Equations by the Finite Element Method*, Cambridge University Press, Cambridge, 1987.

[23] B. Kernighan and S. Lin, *An efficient heuristic procedure for partitioning graphs*, Bell System Technical Journal, 29 (1970), pp. 291–307.

[24] F. T. Leighton and S. Rao, *An approximate max-flow min-cut theorem for uniform multicommodity flow problems with applications to approximation algorithms*, in Proc. 29th Annual Symposium on Foundations of Computer Science, IEEE, Piscataway, NJ, 1988, pp. 422–431.

[25] R. J. Lipton, D. J. Rose, and R. E. Tarjan, *Generalized nested dissection*, SIAM J. Numer. Anal., 16 (1979), pp. 346–358.

[26] R. J. Lipton and R. E. Tarjan, *A separator theorem for planar graphs*, SIAM J. Appl. Math., 36 (1979), pp. 177–189.

[27] J. W. H. Liu, *The multifrontal method for sparse matrix solution: theory and practice*, SIAM Rev., 34 (1992), pp. 82–109.

[28] J. Matoušek, *Approximations and optimal geometric divide-and-conquer*, in Proc. 23rd ACM Symposium Theory of Computing, ACM, New York, 1991, pp. 512–522.

[29] G. L. MILLER, D. TALMOR, S.-H. TENG, AND N. WALKINGTON, *A Delaunay based numerical method for three dimensions: generation, formulation, and partition*, in Proc. of the 27th Annual ACM Symposium on the Theory of Computing, ACM, New York, 1995, pp. 683–692.

[30] G. L. MILLER, S.-H. TENG, W. THURSTON, AND S. A. VAVASIS, *Automatic mesh partitioning*, in Graph Theory and Sparse Matrix Computation, A. George, J. Gilbert, and J. Liu, eds., Vol. 56 of IMA Vol. Math. Appl., Springer, New York, 1993, pp. 57–84.

[31] G. L. MILLER, S.-H. TENG, W. THURSTON, AND S. A. VAVASIS, *Separators for sphere-packings and nearest neighborhood graphs*, J. ACM, 44 (1997), pp. 1–29.

[32] G. L. MILLER, S.-H. TENG, W. THURSTON, AND S. A. VAVASIS, *Geometric Separators for Finite Element Meshes: Appendix Concerning Volume Elements*, 1994, unpublished; available via ftp://ftp.cs.cornell.edu/pub/vavasis/papers/diffgeo.ps.

[33] G. L. MILLER, S.-H. TENG, AND S. A. VAVASIS, *A unified geometric approach to graph separators*, in Proc. 31nd Annual Symposium on Foundations of Computer Science, IEEE, Piscataway, NJ, 1991, pp. 538–547.

[34] G. L. MILLER AND W. THURSTON, *Separators in two and three dimensions*, in Proc. of the 22th Annual ACM Symposium on Theory of Computing, Baltimore, May 1990, ACM, New York, pp. 300–309.

[35] G. L. MILLER AND S. A. VAVASIS, *Density graphs and separators*, in Proc. of the ACM–SIAM Symposium on Discrete Algorithms, SIAM, Philadelphia, PA, 1991, pp. 331–336.

[36] S. A. MITCHELL AND S. A. VAVASIS, *Quality mesh generation in three dimensions*, in Proc. of the ACM Computational Geometry Conference, 1992, ACM, New York, pp. 212–221; also Tech. report C.S. TR 92-1267, Cornell University, Ithaca, NY.

[37] V. PAN AND J. REIF, *Efficient parallel solution of linear systems*, in Proc. of the 17th Annual ACM Symposium on Theory of Computing, ACM, New York, 1985, pp. 143–152.

[38] C. H. PAPADIMITRIOU AND K. STEIGLITZ, *Combinatorial Optimization: Algorithms and Complexity*, Prentice–Hall, Englewood Cliffs, NJ, 1982.

[39] A. POTHEN, H. D. SIMON, AND K.-P. LIOU, *Partitioning sparse matrices with eigenvectors of graphs*, SIAM J. Matrix Anal. Appl., 11 (1990), pp. 430–452.

[40] J. RUPPERT, *A new and simple algorithm for quality* 2*-dimensional mesh generation*, in Proc. 4th Symp. Discrete Algorithms, SIAM, Philadelphia, PA, 1993, pp. 83–92.

[41] G. STRANG AND G. J. FIX, *An Analysis of the Finite Element Method*, Prentice–Hall, Englewood Cliffs, NJ, 1973.

[42] S.-H. TENG, *Points, Spheres, and Separators: A Unified Geometric Approach to Graph Partitioning*, Ph.D. thesis, CMU-CS-91-184, School of Computer Science, Carnegie Mellon University, Pittsburgh, PA, 1991.

[43] P. UNGAR, *A theorem on planar graphs*, J. London Math. Soc., 26 (1951), pp. 256–262.

[44] V. N. VAPNIK AND A. Y. CHERVONENKIS, *On the uniform convergence of relative frequencies of events to their probabilities*, Theory Probab. Appl., 16 (1971), pp. 264–280.

[45] S. A. VAVASIS, *Automatic domain partitioning in three dimensions*, SIAM J. Sci. Statist. Comput., 12 (1991), pp. 950–970.

[46] S. A. VAVASIS, *Stable finite elements for problems with wild coefficients*, SIAM J. Numer. Anal., 33 (1996), pp. 890–916.

[47] S. A. VAVASIS, *QMG Version* 1.0, available via ftp://ftp.cs.cornell.edu/pub/vavasis/qmg1.0.tar.gz and http://www.cs.cornell.edu/home/vavasis/qmg-home.html (1995).

Papers for General Audiences

GENERAL AUDIENCE

Throughout his life Thurston was deeply involved with the teaching, education and exposition of mathematics. For example, while a New College undergrad he tutored local disadvantaged students, he reorganized teaching assistant training while a grad student, guest taught at his children's elementary school and created (with John Conway and Jane Gilman) the innovative Geometry and Imagination course at Princeton and the Minnesota Geometry Center [GK]. In the papers [T1] and [BT] Thurston offers manifold and perhaps surprising ways that people think, process and communicate mathematics. He addresses basic questions such as; what is mathematics and what is proof. Thurston also gives wide ranging views on teaching, training mathematicians, computers and the mathematical community. Regarding his own contributions he states, "What mathematicians most wanted and needed from me was to learn my ways of thinking, and not in fact to learn my proof of the geometrization conjecture for Haken manifolds". The editors hope that through the papers in these volumes the reader catches some of Thurston's way of thinking. For example, see paper [T3] presented to an audience composed largely of cosmologists.

Paper [T2] offers eight mathematical views of the Klein Quartic, a genus-3 surface with a symmetry group of 336 elements, including reflections. This paper was written on the occasion of the November 13, 1993 installation of the sculpture of that surface called "The Eightfold Way" by Helaman Ferguson at the Mathematical Sciences Research Institute at Berkeley, where Thurston was the Director. Ferguson was in part inspired by the beautiful stitched model of the Klein Quartic by Thurston's mother Margaret Thurston, made from 24 heptagons [FF].

References

[BT] J. P. Bourguignon and W. Thurston, Interview de William Thurston, *Gaz. Math.* No. 65 (1995), 11–18.

[FF] H. Ferguson and C. Ferguson, Eightfold Way: The Sculpture, *MSRI Publ.* **35** (1998), 133–173.

[GK] D. Gabai and S. Kerckhoff (coordinating editors), "Remembering Bill Thurston", Notices of the AMS, 62 (2015), 1318–1332.

[T1] W. Thurston, On Proof and Progress in Mathematics, *Bulletin. of Amer. Math. Soc.* **30** (1994), 161–173.

[T2] W. Thurston, "The Eightfold Way: A Mathematical Sculpture by Helaman Ferguson", MSRI Publications, **35** (1998), 1–7.

[T3] W. Thurston, "How to see 3-manifolds", Topology of the Universe Conference (Cleveland OH, 1997), *Classical Quantum Gravity***15** (1998), 2545–2571.

BULLETIN (New Series) OF THE
AMERICAN MATHEMATICAL SOCIETY
Volume 30, Number 2, April 1994

ON PROOF AND PROGRESS IN MATHEMATICS

WILLIAM P. THURSTON

This essay on the nature of proof and progress in mathematics was stimulated by the article of Jaffe and Quinn, "Theoretical Mathematics: Toward a cultural synthesis of mathematics and theoretical physics". Their article raises interesting issues that mathematicians should pay more attention to, but it also perpetuates some widely held beliefs and attitudes that need to be questioned and examined.

The article had one paragraph portraying some of my work in a way that diverges from my experience, and it also diverges from the observations of people in the field whom I've discussed it with as a reality check.

After some reflection, it seemed to me that what Jaffe and Quinn wrote was an example of the phenomenon that people see what they are tuned to see. Their portrayal of my work resulted from projecting the sociology of mathematics onto a one-dimensional scale (speculation versus rigor) that ignores many basic phenomena.

Responses to the Jaffe-Quinn article have been invited from a number of mathematicians, and I expect it to receive plenty of specific analysis and criticism from others. Therefore, I will concentrate in this essay on the positive rather than on the contranegative. I will describe my view of the process of mathematics, referring only occasionally to Jaffe and Quinn by way of comparison.

In attempting to peel back layers of assumptions, it is important to try to begin with the right questions:

1. What is it that mathematicians accomplish?

There are many issues buried in this question, which I have tried to phrase in a way that does not presuppose the nature of the answer.

It would not be good to start, for example, with the question

How do mathematicians prove theorems?

This question introduces an interesting topic, but to start with it would be to project two hidden assumptions:

(1) that there is uniform, objective and firmly established theory and practice of mathematical proof, and
(2) that progress made by mathematicians consists of proving theorems.

It is worthwhile to examine these hypotheses, rather than to accept them as obvious and proceed from there.

Received by the editors October 26, 1993.
1991 *Mathematics Subject Classification*. Primary 01A80.

The question is not even

> How do mathematicians make progress in mathematics?

Rather, as a more explicit (and leading) form of the question, I prefer

> How do mathematicians advance human understanding of mathematics?

This question brings to the fore something that is fundamental and pervasive: that what we are doing is finding ways for *people* to understand and think about mathematics.

The rapid advance of computers has helped dramatize this point, because computers and people are very different. For instance, when Appel and Haken completed a proof of the 4-color map theorem using a massive automatic computation, it evoked much controversy. I interpret the controversy as having little to do with doubt people had as to the veracity of the theorem or the correctness of the proof. Rather, it reflected a continuing desire for *human understanding* of a proof, in addition to knowledge that the theorem is true.

On a more everyday level, it is common for people first starting to grapple with computers to make large-scale computations of things they might have done on a smaller scale by hand. They might print out a table of the first 10,000 primes, only to find that their printout isn't something they really wanted after all. They discover by this kind of experience that what they really want is usually not some collection of "answers"—what they want is *understanding*.

It may sound almost circular to say that what mathematicians are accomplishing is to advance human understanding of mathematics. I will not try to resolve this by discussing what mathematics is, because it would take us far afield. Mathematicians generally feel that they know what mathematics is, but find it difficult to give a good direct definition. It is interesting to try. For me, "the theory of formal patterns" has come the closest, but to discuss this would be a whole essay in itself.

Could the difficulty in giving a good direct definition of mathematics be an essential one, indicating that mathematics has an essential recursive quality? Along these lines we might say that mathematics is the smallest subject satisfying the following:

- Mathematics includes the natural numbers and plane and solid geometry.
- Mathematics is that which mathematicians study.
- Mathematicians are those humans who advance human understanding of mathematics.

In other words, as mathematics advances, we incorporate it into our thinking. As our thinking becomes more sophisticated, we generate new mathematical concepts and new mathematical structures: the subject matter of mathematics changes to reflect how we think.

If what we are doing is constructing better ways of thinking, then psychological and social dimensions are essential to a good model for mathematical progress. These dimensions are absent from the popular model. In caricature, the popular model holds that

D. mathematicians start from a few basic mathematical structures and a collection of axioms "given" about these structures, that

T. there are various important questions to be answered about these structures that can be stated as formal mathematical propositions, and

P. the task of the mathematician is to seek a deductive pathway from the axioms to the propositions or to their denials.

We might call this the definition-theorem-proof (DTP) model of mathematics.

A clear difficulty with the DTP model is that it doesn't explain the source of the questions. Jaffe and Quinn discuss speculation (which they inappropriately label "theoretical mathematics") as an important additional ingredient. Speculation consists of making conjectures, raising questions, and making intelligent guesses and heuristic arguments about what is probably true.

Jaffe and Quinn's DSTP model still fails to address some basic issues. We are not trying to meet some abstract production quota of definitions, theorems and proofs. The measure of our success is whether what we do enables *people* to understand and think more clearly and effectively about mathematics.

Therefore, we need to ask ourselves:

2. How do people understand mathematics?

This is a very hard question. Understanding is an individual and internal matter that is hard to be fully aware of, hard to understand and often hard to communicate. We can only touch on it lightly here.

People have very different ways of understanding particular pieces of mathematics. To illustrate this, it is best to take an example that practicing mathematicians understand in multiple ways, but that we see our students struggling with. The derivative of a function fits well. The derivative can be thought of as:

(1) Infinitesimal: the ratio of the infinitesimal change in the value of a function to the infinitesimal change in a function.

(2) Symbolic: the derivative of x^n is nx^{n-1}, the derivative of $\sin(x)$ is $\cos(x)$, the derivative of $f \circ g$ is $f' \circ g * g'$, *etc.*

(3) Logical: $f'(x) = d$ if and only if for every ϵ there is a δ such that when $0 < |\Delta x| < \delta$,

$$\left|\frac{f(x+\Delta x) - f(x)}{\Delta x} - d\right| < \delta.$$

(4) Geometric: the derivative is the slope of a line tangent to the graph of the function, if the graph has a tangent.

(5) Rate: the instantaneous speed of $f(t)$, when t is time.

(6) Approximation: The derivative of a function is the best linear approximation to the function near a point.

(7) Microscopic: The derivative of a function is the limit of what you get by looking at it under a microscope of higher and higher power.

This is a list of different ways of *thinking about* or *conceiving of* the derivative, rather than a list of different *logical definitions*. Unless great efforts are made to maintain the tone and flavor of the original human insights, the differences start to evaporate as soon as the mental concepts are translated into precise, formal and explicit definitions.

I can remember absorbing each of these concepts as something new and interesting, and spending a good deal of mental time and effort digesting and practicing with each, reconciling it with the others. I also remember coming back to revisit these different concepts later with added meaning and understanding.

The list continues; there is no reason for it ever to stop. A sample entry further down the list may help illustrate this. We may think we know all there is to say about a certain subject, but new insights are around the corner. Furthermore, one person's clear mental image is another person's intimidation:

37. The derivative of a real-valued function f in a domain D is the Lagrangian section of the cotangent bundle $T^*(D)$ that gives the connection form for the unique flat connection on the trivial $\mathbf{R}$-bundle $D \times \mathbf{R}$ for which the graph of f is parallel.

These differences are not just a curiosity. Human thinking and understanding do not work on a single track, like a computer with a single central processing unit. Our brains and minds seem to be organized into a variety of separate, powerful facilities. These facilities work together loosely, "talking" to each other at high levels rather than at low levels of organization.

Here are some major divisions that are important for mathematical thinking:

(1) Human language. We have powerful special-purpose facilities for speaking and understanding human language, which also tie in to reading and writing. Our linguistic facility is an important tool for thinking, not just for communication. A crude example is the quadratic formula which people may remember as a little chant, "ex equals minus bee plus or minus the square root of bee squared minus four ay see all over two ay." The mathematical language of symbols is closely tied to our human language facility. The fragment of mathematical symbolese available to most calculus students has only one verb, "=". That's why students use it when they're in need of a verb. Almost anyone who has taught calculus in the U.S. has seen students instinctively write "$x^3 = 3x^2$" and the like.

(2) Vision, spatial sense, kinesthetic (motion) sense. People have very powerful facilities for taking in information visually or kinesthetically, and thinking with their spatial sense. On the other hand, they do not have a very good built-in facility for inverse vision, that is, turning an internal spatial understanding back into a two-dimensional image. Consequently, mathematicians usually have fewer and poorer figures in their papers and books than in their heads.

An interesting phenomenon in spatial thinking is that scale makes a big difference. We can think about little objects in our hands, or we can think of bigger human-sized structures that we scan, or we can think of spatial structures that encompass us and that we move around in. We tend to think more effectively with spatial imagery on a larger scale: it's as if our brains take larger things more seriously and can devote more resources to them.

(3) Logic and deduction. We have some built-in ways of reasoning and

putting things together associated with how we make logical deductions: cause and effect (related to implication), contradiction or negation, *etc.*

Mathematicians apparently don't generally rely on the formal rules of deduction as they are thinking. Rather, they hold a fair bit of logical structure of a proof in their heads, breaking proofs into intermediate results so that they don't have to hold too much logic at once. In fact, it is common for excellent mathematicians not even to know the standard formal usage of quantifiers (for all and there exists), yet all mathematicians certainly perform the reasoning that they encode.

It's interesting that although "or", "and" and "implies" have identical formal usage, we think of "or" and "and" as conjunctions and "implies" as a verb.

(4) Intuition, association, metaphor. People have amazing facilities for sensing something without knowing where it comes from (intuition); for sensing that some phenomenon or situation or object is like something else (association); and for building and testing connections and comparisons, holding two things in mind at the same time (metaphor). These facilities are quite important for mathematics. Personally, I put a lot of effort into "listening" to my intuitions and associations, and building them into metaphors and connections. This involves a kind of simultaneous quieting and focusing of my mind. Words, logic, and detailed pictures rattling around can inhibit intuitions and associations.

(5) Stimulus-response. This is often emphasized in schools; for instance, if you see 3927×253, you write one number above the other and draw a line underneath, *etc.* This is also important for research mathematics: seeing a diagram of a knot, I might write down a presentation for the fundamental group of its complement by a procedure that is similar in feel to the multiplication algorithm.

(6) Process and time. We have a facility for thinking about processes or sequences of actions that can often be used to good effect in mathematical reasoning. One way to think of a function is as an action, a process, that takes the domain to the range. This is particularly valuable when composing functions. Another use of this facility is in remembering proofs: people often remember a proof as a process consisting of several steps. In topology, the notion of a homotopy is most often thought of as a process taking time. Mathematically, time is no different from one more spatial dimension, but since humans interact with it in a quite different way, it is psychologically very different.

3. How is mathematical understanding communicated?

The transfer of understanding from one person to another is not automatic. It is hard and tricky. Therefore, to analyze human understanding of mathematics, it is important to consider **who** understands **what**, and **when**.

Mathematicians have developed habits of communication that are often dysfunctional. Organizers of colloquium talks everywhere exhort speakers to explain things in elementary terms. Nonetheless, most of the audience at an average colloquium talk gets little of value from it. Perhaps they are lost within the first 5 minutes, yet sit silently through the remaining 55 minutes. Or perhaps they quickly lose interest because the speaker plunges into technical details without presenting any reason to investigate them. At the end of the talk, the few mathematicians who are close to the field of the speaker ask a question or two to avoid embarrassment.

This pattern is similar to what often holds in classrooms, where we go through the motions of saying for the record what we think the students "ought" to learn, while the students are trying to grapple with the more fundamental issues of learning our language and guessing at our mental models. Books compensate by giving samples of how to solve every type of homework problem. Professors compensate by giving homework and tests that are much easier than the material "covered" in the course, and then grading the homework and tests on a scale that requires little understanding. We assume that the problem is with the students rather than with communication: that the students either just don't have what it takes, or else just don't care.

Outsiders are amazed at this phenomenon, but within the mathematical community, we dismiss it with shrugs.

Much of the difficulty has to do with the language and culture of mathematics, which is divided into subfields. Basic concepts used every day within one subfield are often foreign to another subfield. Mathematicians give up on trying to understand the basic concepts even from neighboring subfields, unless they were clued in as graduate students.

In contrast, communication works very well within the subfields of mathematics. Within a subfield, people develop a body of common knowledge and known techniques. By informal contact, people learn to understand and copy each other's ways of thinking, so that ideas can be explained clearly and easily.

Mathematical knowledge can be transmitted amazingly fast within a subfield. When a significant theorem is proved, it often (but not always) happens that the solution can be communicated in a matter of minutes from one person to another within the subfield. The same proof would be communicated and generally understood in an hour talk to members of the subfield. It would be the subject of a 15- or 20-page paper, which could be read and understood in a few hours or perhaps days by members of the subfield.

Why is there such a big expansion from the informal discussion to the talk to the paper? One-on-one, people use wide channels of communication that go far beyond formal mathematical language. They use gestures, they draw pictures and diagrams, they make sound effects and use body language. Communication is more likely to be two-way, so that people can concentrate on what needs the most attention. With these channels of communication, they are in a much better position to convey what's going on, not just in their logical and linguistic facilities, but in their other mental facilities as well.

In talks, people are more inhibited and more formal. Mathematical audiences are often not very good at asking the questions that are on most people's minds, and speakers often have an unrealistic preset outline that inhibits them from addressing questions even when they are asked.

In papers, people are still more formal. Writers translate their ideas into symbols and logic, and readers try to translate back.

Why is there such a discrepancy between communication within a subfield and communication outside of subfields, not to mention communication outside mathematics?

Mathematics in some sense has a common language: a language of symbols, technical definitions, computations, and logic. This language efficiently conveys some, but not all, modes of mathematical thinking. Mathematicians learn to translate certain things almost unconsciously from one mental mode to the other, so that some statements quickly become clear. Different mathematicians study papers in different ways, but when I read a mathematical paper in a field in which I'm conversant, I concentrate on the thoughts that are between the lines. I might look over several paragraphs or strings of equations and think to myself "Oh yeah, they're putting in enough rigamarole to carry such-and-such idea." When the idea is clear, the formal setup is usually unnecessary and redundant—I often feel that I could write it out myself more easily than figuring out what the authors actually wrote. It's like a new toaster that comes with a 16-page manual. If you already understand toasters and if the toaster looks like previous toasters you've encountered, you might just plug it in and see if it works, rather than first reading all the details in the manual.

People familiar with ways of doing things in a subfield recognize various patterns of statements or formulas as idioms or circumlocution for certain concepts or mental images. But to people not already familiar with what's going on the same patterns are not very illuminating; they are often even misleading. The language is not alive except to those who use it.

I'd like to make an important remark here: there are some mathematicians who are conversant with the ways of thinking in more than one subfield, sometimes in quite a number of subfields. Some mathematicians learn the jargon of several subfields as graduate students, some people are just quick at picking up foreign mathematical language and culture, and some people are in mathematical centers where they are exposed to many subfields. People who are comfortable in more than one subfield can often have a very positive influence, serving as bridges, and helping different groups of mathematicians learn from each other. But people knowledgeable in multiple fields can also have a negative effect, by intimidating others, and by helping to validate and maintain the whole system of generally poor communication. For example, one effect often takes place during colloquium talks, where one or two widely knowledgeable people sitting in the front row may serve as the speaker's mental guide to the audience.

There is another effect caused by the big differences between how we think about mathematics and how we write it. A group of mathematicians interacting with each other can keep a collection of mathematical ideas alive for a period of years, even though the recorded version of their mathematical work differs from their actual thinking, having much greater emphasis on language, symbols, logic and formalism. But as new batches of mathematicians learn about the subject they tend to interpret what they read and hear more literally, so that the more easily recorded and communicated formalism and machinery tend to gradually take over from other modes of thinking.

There are two counters to this trend, so that mathematics does not become entirely mired down in formalism. First, younger generations of mathematicians are continually discovering and rediscovering insights on their own, thus reinjecting diverse modes of human thought into mathematics.

Second, mathematicians sometimes invent names and hit on unifying definitions that replace technical circumlocutions and give good handles for insights. Names like "group" to replace "a system of substitutions satisfying ... ", and "manifold" to replace

> We can't give coordinates to parametrize all the solutions to our equations simultaneously, but in the neighborhood of any particular solution we can introduce coordinates
>
> $(f_1(u_1, u_2, u_3), f_2(u_1, u_2, u_3), f_3(u_1, u_2, u_3), f_4(u_1, u_2, u_3),$
> $f_5(u_1, u_2, u_3))$
>
> where at least one of the ten determinants
>
> ...[ten 3×3 determinants of matrices of partial derivatives]...
>
> is not zero

may or may not have represented advances in insight among experts, but they greatly facilitate the *communication* of insights.

We mathematicians need to put far greater effort into communicating mathematical *ideas*. To accomplish this, we need to pay much more attention to communicating not just our definitions, theorems, and proofs, but also our ways of thinking. We need to appreciate the value of different ways of thinking about the same mathematical structure.

We need to focus far more energy on understanding and explaining the basic mental infrastructure of mathematics—with consequently less energy on the most recent results. This entails developing mathematical language that is effective for the radical purpose of conveying ideas to people who don't already know them.

Part of this communication is through proofs.

4. What is a proof?

When I started as a graduate student at Berkeley, I had trouble imagining how I could "prove" a new and interesting mathematical theorem. I didn't really understand what a "proof" was.

By going to seminars, reading papers, and talking to other graduate students, I gradually began to catch on. Within any field, there are certain theorems and certain techniques that are generally known and generally accepted. When you write a paper, you refer to these without proof. You look at other papers in the field, and you see what facts they quote without proof, and what they cite in their bibliography. You learn from other people some idea of the proofs. Then you're free to quote the same theorem and cite the same citations. You don't necessarily have to read the full papers or books that are in your bibliography. Many of the things that are generally known are things for which there may be no known written source. As long as people in the field are comfortable that the idea works, it doesn't need to have a formal written source.

At first I was highly suspicious of this process. I would doubt whether a certain idea was really established. But I found that I could ask people, and they could produce explanations and proofs, or else refer me to other people or to written sources that would give explanations and proofs. There were published theorems that were generally known to be false, or where the proofs were generally known to be incomplete. Mathematical knowledge and understanding were embedded in the minds and in the social fabric of the community of people thinking about a particular topic. This knowledge was supported by written documents, but the written documents were not really primary.

I think this pattern varies quite a bit from field to field. I was interested in geometric areas of mathematics, where it is often pretty hard to have a document that reflects well the way people actually think. In more algebraic or symbolic fields, this is not necessarily so, and I have the impression that in some areas documents are much closer to carrying the life of the field. But in any field, there is a strong social standard of validity and truth. Andrew Wiles's proof of Fermat's Last Theorem is a good illustration of this, in a field which is very algebraic. The experts quickly came to believe that his proof was basically correct on the basis of high-level ideas, long before details could be checked. This proof will receive a great deal of scrutiny and checking compared to most mathematical proofs; but no matter how the process of verification plays out, it helps illustrate how mathematics evolves by rather organic psychological and social processes.

When people are doing mathematics, the flow of ideas and the social standard of validity is much more reliable than formal documents. People are usually not very good in checking *formal correctness* of proofs, but they are quite good at detecting potential weaknesses or flaws in proofs.

To avoid misinterpretation, I'd like to emphasize two things I am *not* saying. First, I am *not* advocating any weakening of our community standard of proof; I am trying to describe how the process really works. Careful proofs that will stand up to scrutiny are very important. I think the process of proof on the whole works pretty well in the mathematical community. The kind of change I would advocate is that mathematicians take more care with their proofs, making them really clear and as simple as possible so that if any weakness is present it will be easy to detect. Second, I am *not* criticizing the mathematical study of formal proofs, nor am I criticizing people who put energy into making mathematical arguments more explicit and more formal. These are both useful activities that shed new insights on mathematics.

I have spent a fair amount of effort during periods of my career exploring mathematical questions by computer. In view of that experience, I was astonished to see the statement of Jaffe and Quinn that mathematics is extremely slow and arduous, and that it is arguably the most disciplined of all human activities. The standard of correctness and completeness necessary to get a computer program to work at all is a couple of orders of magnitude higher than the mathematical community's standard of valid proofs. Nonetheless, large computer programs, even when they have been very carefully written and very carefully tested, always seem to have bugs.

I think that mathematics is one of the most intellectually gratifying of human activities. Because we have a high standard for clear and convincing thinking and because we place a high value on listening to and trying to understand each other, we don't engage in interminable arguments and endless redoing of our mathematics. We are prepared to be convinced by others. Intellectually, mathematics moves very quickly. Entire mathematical landscapes change and change again in amazing ways during a single career.

When one considers how hard it is to write a computer program even approaching the intellectual scope of a good mathematical paper, and how much greater time and effort have to be put into it to make it "almost" formally correct, it is preposterous to claim that mathematics as we practice it is anywhere near formally correct.

Mathematics as we practice it is much more formally complete and precise than other sciences, but it is much less formally complete and precise for its content than computer programs. The difference has to do not just with the amount of effort: the kind of effort is qualitatively different. In large computer programs, a tremendous proportion of effort must be spent on myriad compatibility issues: making sure that all definitions are consistent, developing "good" data structures that have useful but not cumbersome generality, deciding on the "right" generality for functions, *etc.* The proportion of energy spent on the working part of a large program, as distinguished from the bookkeeping part, is surprisingly small. Because of compatibility issues that almost inevitably escalate out of hand because the "right" definitions change as generality and functionality are added, computer programs usually need to be rewritten frequently, often from scratch.

A very similar kind of effort would have to go into mathematics to make it formally correct and complete. It is not that formal correctness is prohibitively difficult on a small scale—it's that there are many possible choices of formalization on small scales that translate to huge numbers of interdependent choices in the large. It is quite hard to make these choices compatible; to do so would certainly entail going back and rewriting from scratch all old mathematical papers whose results we depend on. It is also quite hard to come up with good technical choices for formal definitions that will be valid in the variety of ways that mathematicians want to use them and that will anticipate future extensions of mathematics. If we were to continue to cooperate, much of our time would be spent with international standards commissions to establish uniform definitions and resolve huge controversies.

Mathematicians can and do fill in gaps, correct errors, and supply more detail and more careful scholarship when they are called on or motivated to do so. Our system is quite good at producing reliable theorems that can be solidly backed up. It's just that the reliability does not primarily come from mathematicians formally checking formal arguments; it comes from mathematicians thinking carefully and critically about mathematical ideas.

On the most fundamental level, the foundations of mathematics are much shakier than the mathematics that we do. Most mathematicians adhere to foundational principles that are known to be polite fictions. For example, it is a theorem that there does not exist any way to ever actually construct or even define a well-ordering of the real numbers. There is considerable evidence (but no proof) that we can get away with these polite fictions without being caught out, but that doesn't make them right. Set theorists construct many alternate and mutually contradictory "mathematical universes" such that if one is consistent, the others are too. This leaves very little confidence that one or the other is the right choice or the natural choice. Gödel's incompleteness theorem implies that there can be no formal system that is consistent, yet powerful enough to serve as a basis for all of the mathematics that we do.

In contrast to humans, computers are good at performing formal processes. There are people working hard on the project of actually formalizing parts of mathematics by computer, with actual formally correct formal deductions. I think this is a very big but very worthwhile project, and I am confident that we will learn a lot from it. The process will help simplify and clarify mathematics. In not too many years, I expect that we will have interactive computer programs that can help people compile significant chunks of formally complete and correct mathematics (based on a few perhaps shaky but at least explicit assumptions), and that they will become part of the standard mathematician's working environment.

However, we should recognize that the humanly understandable and humanly checkable proofs that we actually do are what is most important to us, and that they are quite different from formal proofs. For the present, formal proofs are out of reach and mostly irrelevant: we have good human processes for checking mathematical validity.

5. What motivates people to do mathematics?

There is a real joy in doing mathematics, in learning ways of thinking that explain and organize and simplify. One can feel this joy discovering new mathematics, rediscovering old mathematics, learning a way of thinking from a person or text, or finding a new way to explain or to view an old mathematical structure.

This inner motivation might lead us to think that we do mathematics solely for its own sake. That's not true: the social setting is extremely important. We are inspired by other people, we seek appreciation by other people, and we like to help other people solve their mathematical problems. What we enjoy changes in response to other people. Social interaction occurs through face-to-face meetings. It also occurs through written and electronic correspondence, preprints, and journal articles. One effect of this highly social system of mathematics is the tendency of mathematicians to follow fads. For the purpose of producing new mathematical theorems this is probably not very efficient: we'd seem to be better off having mathematicians cover the intellectual field much more evenly. But most mathematicians don't like to be lonely, and they have trouble staying excited about a subject, even if they are personally making progress, unless they have colleagues who share their excitement.

In addition to our inner motivation and our informal social motivation for doing mathematics, we are driven by considerations of economics and status. Mathematicians, like other academics, do a lot of judging and being judged. Starting with grades, and continuing through letters of recommendation, hiring decisions, promotion decisions, referees reports, invitations to speak, prizes, ... we are involved in many ratings, in a fiercely competitive system.

Jaffe and Quinn analyze the motivation to do mathematics in terms of a common currency that many mathematicians believe in: credit for theorems.

I think that our strong communal emphasis on theorem-credits has a negative effect on mathematical progress. If what we are accomplishing is advancing human understanding of mathematics, then we would be much better off recognizing and valuing a far broader range of activity. The people who see the way to proving theorems are doing it in the context of a mathematical community; they are not doing it on their own. They depend on understanding of mathematics that they glean from other mathematicians. Once a theorem has been proven, the mathematical community depends on the social network to distribute the ideas to people who might use them further—the print medium is far too obscure and cumbersome.

Even if one takes the narrow view that what we are producing is theorems, the team is important. Soccer can serve as a metaphor. There might only be one or two goals during a soccer game, made by one or two persons. That does not mean that the efforts of all the others are wasted. We do not judge players on a soccer team only by whether they personally make a goal; we judge the team by its function as a team.

In mathematics, it often happens that a group of mathematicians advances with a certain collection of ideas. There are theorems in the path of these advances that will almost inevitably be proven by one person or another. Sometimes the group of mathematicians can even anticipate what these theorems are likely to be. It is much harder to predict who will actually prove the theorem, although there are usually a few "point people" who are more likely to score. However, they are in a position to prove those theorems because of the collective efforts of the team. The team has a further function, in absorbing and making use of the theorems once they are proven. Even if one person could prove all the theorems in the path single-handedly, they are wasted if nobody else learns them.

There is an interesting phenomenon concerning the "point" people. It regularly happens that someone who was in the middle of a pack proves a theorem that receives wide recognition as being significant. Their status in the community—their pecking order—rises immediately and dramatically. When this happens, they usually become much more productive as a center of ideas and a source of theorems. Why? First, there is a large increase in self-esteem, and an accompanying increase in productivity. Second, when their status increases, people are more in the center of the network of ideas—others take them more seriously. Finally and perhaps most importantly, a mathematical breakthrough usually represents a new way of thinking, and effective ways of thinking can usually be applied in more than one situation.

This phenomenon convinces me that the entire mathematical community would become much more productive if we open our eyes to the real values in what we are doing. Jaffe and Quinn propose a system of recognized roles divided into "speculation" and "proving". Such a division only perpetuates the myth that our progress is measured in units of standard theorems deduced. This is a bit like the fallacy of the person who makes a printout of the first 10,000 primes. What we are producing is human understanding. We have many different ways to understand and many different processes that contribute to our understanding. We will be more satisfied, more productive and happier if we recognize and focus on this.

6. Some personal experiences

Since this essay grew out of reflection on the misfit between my experiences and the description of Jaffe and Quinn's, I will discuss two personal experiences, including the one they alluded to.

I feel some awkwardness in this, because I do have regrets about aspects of my career: if I were to do things over again with the benefit of my present insights about myself and about the process of mathematics, there is a lot that I would hope to do differently. I hope that by describing these experiences rather openly as I remember and understand them, I can help others understand the process better and learn in advance.

First I will discuss briefly the theory of foliations, which was my first subject, starting when I was a graduate student. (It doesn't matter here whether you know what foliations are.)

At that time, foliations had become a big center of attention among geometric topologists, dynamical systems people, and differential geometers. I fairly rapidly proved some dramatic theorems. I proved a classification theorem for foliations, giving a necessary and sufficient condition for a manifold to admit a foliation. I proved a number of other significant theorems. I wrote respectable papers and published at least the most important theorems. It was hard to find the time to write to keep up with what I could prove, and I built up a backlog.

An interesting phenomenon occurred. Within a couple of years, a dramatic evacuation of the field started to take place. I heard from a number of mathematicians that they were giving or receiving advice not to go into foliations—they were saying that Thurston was cleaning it out. People told me (not as a complaint, but as a compliment) that I was killing the field. Graduate students stopped studying foliations, and fairly soon, I turned to other interests as well.

I do not think that the evacuation occurred because the territory was intellectually exhausted—there were (and still are) many interesting questions that remain and that are probably approachable. Since those years, there have been interesting developments carried out by the few people who stayed in the field or who entered the field, and there have also been important developments in neighboring areas that I think would have been much accelerated had mathematicians continued to pursue foliation theory vigorously.

Today, I think there are few mathematicians who understand anything approaching the state of the art of foliations as it lived at that time, although there are some parts of the theory of foliations, including developments since that time, that are still thriving.

I believe that two ecological effects were much more important in putting a damper on the subject than any exhaustion of intellectual resources that occurred.

First, the results I proved (as well as some important results of other people) were documented in a conventional, formidable mathematician's style. They depended heavily on readers who shared certain background and certain insights. The theory of foliations was a young, opportunistic subfield, and the background was not standardized. I did not hesitate to draw on any of the mathematics I had learned from others. The papers I wrote did not (and could not) spend much time explaining the background culture. They documented top-level reasoning and conclusions that I often had achieved after much reflection and effort. I also threw out prize cryptic tidbits of insight, such as "the Godbillon-Vey invariant measures the helical wobble of a foliation", that remained mysterious to most mathematicans who read them. This created a high entry barrier: I think many graduate students and mathematicians were discouraged that it was hard to learn and understand the proofs of key theorems.

Second is the issue of what is in it for other people in the subfield. When I started working on foliations, I had the conception that what people wanted was to know the answers. I thought that what they sought was a collection of powerful proven theorems that might be applied to answer further mathematical questions. But that's only one part of the story. More than the knowledge, people want *personal understanding*. And in our credit-driven system, they also want and need *theorem-credits*.

I'll skip ahead a few years, to the subject that Jaffe and Quinn alluded to, when I began studying 3-dimensional manifolds and their relationship to hyperbolic geometry. (Again, it matters little if you know what this is about.) I gradually built up over a number of years a certain intuition for hyperbolic three-manifolds, with a repertoire of constructions, examples and proofs. (This process actually started when I was an undergraduate, and was strongly bolstered by applications to foliations.) After a while, I conjectured or speculated that all three-manifolds have a certain geometric structure; this conjecture eventually became known as the geometrization conjecture. About two or three years later, I proved the geometrization theorem for Haken manifolds. It was a hard theorem, and I spent a tremendous amount of effort thinking about it. When I completed the proof, I spent a lot more effort checking the proof, searching for difficulties and testing it against independent information.

I'd like to spell out more what I mean when I say I proved this theorem. It meant that I had a clear and complete flow of ideas, including details, that withstood a great deal of scrutiny by myself and by others. Mathematicians have many different styles of thought. My style is not one of making broad sweeping but careless generalities, which are merely hints or inspirations: I make clear mental models, and I think things through. My proofs have turned out to be quite reliable. I have not had trouble backing up claims or producing details for things I have proven. I am good in detecting flaws in my own reasoning as well as in the reasoning of others.

However, there is sometimes a huge expansion factor in translating from the encoding in my own thinking to something that can be conveyed to someone else. My mathematical education was rather independent and idiosyncratic, where for a number of years I learned things on my own, developing personal mental models for how to think about mathematics. This has often been a big advantage for me in thinking about mathematics, because it's easy to pick up later the standard mental models shared by groups of mathematicians. This means that some concepts that I use freely and naturally in my personal thinking are foreign to most mathematicians I talk to. My personal mental models and structures are similar in character to the kinds of models groups of mathematicians share—but they are often different models. At the time of the formulation of the geometrization conjecture, my understanding of hyperbolic geometry was a good example. A random continuing example is an understanding of finite topological spaces, an oddball topic that can lend good insight to a variety of questions but that is generally not worth developing in any one case because there are standard circumlocutions that avoid it.

Neither the geometrization conjecture nor its proof for Haken manifolds was in the path of any group of mathematicians at the time—it went against the trends in topology for the preceding 30 years, and it took people by surprise. To most topologists at the time, hyperbolic geometry was an arcane side branch of mathematics, although there were other groups of mathematicians such as differential geometers who did understand it from certain points of view. It took topologists a while just to understand what the geometrization conjecture meant, what it was good for, and why it was relevant.

At the same time, I started writing notes on the geometry and topology of 3-manifolds, in conjunction with the graduate course I was teaching. I distributed them to a few people, and before long many others from around the world were writing for copies. The mailing list grew to about 1200 people to whom I was sending notes every couple of months. I tried to communicate my real thoughts in these notes. People ran many seminars based on my notes, and I got lots of feedback. Overwhelmingly, the feedback ran something like "Your notes are really inspiring and beautiful, but I have to tell you that we spent 3 weeks in our seminar working out the details of $\S n.n$. More explanation would sure help."

I also gave many presentations to groups of mathematicians about the ideas of studying 3-manifolds from the point of view of geometry, and about the proof of the geometrization conjecture for Haken manifolds. At the beginning, this subject was foreign to almost everyone. It was hard to communicate—the infrastructure was in my head, not in the mathematical community. There were several mathematical theories that fed into the cluster of ideas: three-manifold topology, Kleinian groups, dynamical systems, geometric topology, discrete subgroups of Lie groups, foliations, Teichmüller spaces, pseudo-Anosov diffeomorphisms, geometric group theory, as well as hyperbolic geometry.

We held an AMS summer workshop at Bowdoin in 1980, where many mathematicans in the subfields of low-dimensional topology, dynamical systems and Kleinian groups came.

It was an interesting experience exchanging cultures. It became dramatically clear how much proofs depend on the audience. We prove things in a social context and address them to a certain audience. Parts of this proof I could communicate in two minutes to the topologists, but the analysts would need an hour lecture before they would begin to understand it. Similarly, there were some things that could be said in two minutes to the analysts that would take an hour before the topologists would begin to get it. And there were many other parts of the proof which should take two minutes in the abstract, but that none of the audience at the time had the mental infrastructure to get in less than an hour.

At that time, there was practically no infrastructure and practically no context for this theorem, so the expansion from how an idea was keyed in my head to what I had to say to get it across, not to mention how much energy the audience had to devote to understand it, was very dramatic.

In reaction to my experience with foliations and in response to social pressures, I concentrated most of my attention on developing and presenting the infrastructure in what I wrote and in what I talked to people about. I explained the details to the few people who were "up" for it. I wrote some papers giving the substantive parts of the proof of the geometrization theorem for Haken manifolds—for these papers, I got almost no feedback. Similarly, few people actually worked through the harder and deeper sections of my notes until much later.

The result has been that now quite a number of mathematicians have what was dramatically lacking in the beginning: a working understanding of the concepts and the infrastructure that are natural for this subject. There has been and there continues to be a great deal of thriving mathematical activity. By concentrating on building the infrastructure and explaining and publishing definitions and ways of thinking but being slow in stating or in publishing proofs of all the "theorems" I knew how to prove, I left room for many other people to pick up credit. There has been room for people to discover and publish other proofs of the geometrization theorem. These proofs helped develop mathematical concepts which are quite interesting in themselves, and lead to further mathematics.

What mathematicians most wanted and needed from me was to learn my ways of thinking, and not in fact to learn my proof of the geometrization conjecture for Haken manifolds. It is unlikely that the proof of the general geometrization conjecture will consist of pushing the same proof further.

A further issue is that people sometimes need or want an accepted and validated result not in order to learn it, but so that they can quote it and rely on it.

Mathematicians were actually very quick to accept my proof, and to start quoting it and using it based on what documentation there was, based on their experience and belief in me, and based on acceptance by opinions of experts with whom I spent a lot of time communicating the proof. The theorem now is documented, through published sources authored by me and by others, so most people feel secure in quoting it; people in the field certainly have not challenged me about its validity, or expressed to me a need for details that are not available.

Not all proofs have an identical role in the logical scaffolding we are building for mathematics. This particular proof probably has only temporary logical value, although it has a high motivational value in helping support a certain vision for the structure of 3-manifolds. The full geometrization conjecture is still a conjecture. It has been proven for many cases, and is supported by a great deal of computer evidence as well, but it has not been proven in generality. I am convinced that the general proof will be discovered; I hope before too many more years. At that point, proofs of special cases are likely to become obsolete.

Meanwhile, people who want to use the geometric technology are better off to start off with the assumption "Let M^3 be a manifold that admits a geometric decomposition," since this is more general than "Let M^3 be a Haken manifold." People who don't want to use the technology or who are suspicious of it can avoid it. Even when a theorem about Haken manifolds can be proven using geometric techniques, there is a high value in finding purely topological techniques to prove it.

In this episode (which still continues) I think I have managed to avoid the two worst possible outcomes: either for me not to let on that I discovered what I discovered and proved what I proved, keeping it to myself (perhaps with the hope of proving the Poincaré conjecture), or for me to present an unassailable and hard-to-learn theory with no practitioners to keep it alive and to make it grow.

I can easily name regrets about my career. I have not published as much as I should. There are a number of mathematical projects in addition to the geometrization theorem for Haken manifolds that I have not delivered well or at all to the mathematical public. When I concentrated more on developing the infrastructure rather than the top-level theorems in the geometric theory of 3-manifolds, I became somewhat disengaged as the subject continued to evolve; and I have not actively or effectively promoted the field or the careers of the excellent people in it. (But some degree of disengagement seems to me an almost inevitable by-product of the mentoring of graduate students and others: in order to really turn genuine research directions over to others, it's necessary to really let go and stop oneself from thinking about them very hard.)

On the other hand, I have been busy and productive, in many different activities. Our system does not create extra time for people like me to spend on writing and research; instead, it inundates us with many requests and opportunities for extra work, and my gut reaction has been to say 'yes' to many of these requests and opportunities. I have put a lot of effort into non-credit-producing activities that I value just as I value proving theorems: mathematical politics, revision of my notes into a book with a high standard of communication, exploration of computing in mathematics, mathematical education, development of new forms for communication of mathematics through the Geometry Center (such as our first experiment, the "Not Knot" video), directing MSRI, *etc.*

I think that what I have done has not maximized my "credits". I have been in a position not to feel a strong need to compete for more credits. Indeed, I began to feel strong challenges from other things besides proving new theorems.

I do think that my actions have done well in stimulating mathematics.

INTERVIEW DE WILLIAM THURSTON

La Gazette recueille traditionnellement les échos des débats agitatnt le monde des mathématiques : débat sur leur enseignement, leur avenir, leurs modes de communication. Il nous a semblé intéressant d'interroger sur ces thèmes William Thurston. Cette interview a été réalisé à Berkeley par Jean-Pierre Bourguignon que nous remercions ici.

M. William Thurston
Directeur du MSRI - Berkeley
et
M. Jean-Pierre Bourguignon
Directeur de Recherches - IHES

BOURGUIGNON– *Quite often mathematics is presented as having three main approaches : algebraic, analytic and geometric. At the ICMI Congress in Berkeley in 1980, Jean Dieudonné gave a lecture entitled "La domination universelle de la géométrie". This was part of a panel discussion during which Bob Osserman was supposed to respond to him. Bob was somewhat "terrorized" by the title. In fact the lecture was not so imperialistic. You often present yourself as a Geometer. How do you view this possible division of mathematics, and how do you position your own work with respect to it ?*

THURSTON.– Sometimes I think that this traditional division of mathematics has more to do with how the human mind works than with any kind of logical division of the subject. In fact any part of mathematics could be reduced to nearly any other part of mathematics. However, it would be misguided to conclude from this that one branch of mathematics dominates another branch. For example, all of mathematics ultimately reduces to logic, but it does not follow that therefore everything should be put in a completely logical form. Similarly, all of mathematics can be reduced to geometry, but it is misguided to conclude that geometry is the most powerful form of thinking and therefore we should reduce everything to geometry. These reductions are possible, and it is even productive and interesting to study them, but they do not represent how mathematics *ought* to be, and they do not represent a hierarchy of types of mathematics. To me, the main point is that the human mind has many very powerful faculties. One of these faculties is the ability to think geometrically. Another powerful faculty is our linguistic faculty. Different people have different preferred modes of thinking, but really, the most powerful thinking comes from combining all of our resources and trying to think with all of our faculties together.

But I'm not quite sure how algebra and analysis map to faculties of the mind; there's no single simple correspondence.

JPB.– *The analytic approach has often to do with inequalities or estimations. This is quite unusual in Algebra.*

WT.– That's true. I think that algebraic thinking is very linguistically oriented, with sequences of symbols connected to language even though most algebraists, like anybody else, think in many different ways.

JPB.– *Just one precise question concerning geometry : how do you explain why hyperbolic geometry has been playing such a role in bringing together people from so many different fields? This was something which was envisioned early in the development of the theory, but not to the extent to which it's happened in recent years.*

WT.– I learned about hyperbolic geometry almost accidentally. I was an undergraduate at New College, a place where there weren't very many mathematicians; I learned a lot of mathematics, but I wasn't at all in touch with mathematical fashions, so what I learned was somewhat accidental. I think there are just a huge number of opportunities in mathematics where one group of people might know (or have known) about ideas that would be extremely interesting if there were a connection to another group of people but because of barriers in culture and communication, the connections often tend not to be made. Hyperbolic geometry is an area where I had a role connecting several groups of mathematicians. But I think there's a lot more, I think there are many parts of mathematics that could make important connections; it's just a question of people making them.

JPB.– *So did you have any inkling as to what other parts of mathematics you expect to play such a role?*

WT.– It's very hard to say. I've learned as a mathematician that it's important to be humble. It is very hard to make predictions, except for things that you already know how to do. My experience is that almost anything that I've learned connects to almost anything else I've learned, in ways that I did not predict before I saw the connections. I can only take this as a sign that there are many possible connections.

JPB.– *You are very keen on using computers for your research or other purposes. How would you summarize the impact of computers in presenting mathematics?*

WT.– I think computers have begun to have a major impact and often in ways that aren't always formally recognized. There are many kinds of computer experiments that people do that produce results that are in themselves, or give people insights that lead to proving theorems. I believe that some of the most important effects of computers may be in many small uses rather than a few large, major uses. People can do many simple things that allow them to think more clearly, or think about things that they wouldn't otherwise have thought about. For example, many mathematicians use symbolic mathematics programs such as Mathematica and Maple pretty often; they use them in ways that aren't essentially different from what they might have done by hand, but it enables them to be much more efficient. A good use of computers is to concentrate human thinking on things that humans are good for. Computers are certainly useful to me, because I've always had a mind set of wanting to understand mathematics as concretely as I can, as constructively as I can. On many occasions when I have thought of some mathematical possibility, I have done computer experiments, not necessarily very profound experiments, but experiments that have increased my intuition enough so I would know in very real ways what the mathematical phenomenon was. I believe this happens quite a lot with a lot of mathematicians, particularly the younger generations of mathematicians. It's not always something that they can talk about or take credit for very easily.

JPB.– *What future endeavors do you foresee in this direction ? Clearly computers are getting more and more powerful.*

WT.– I think they're going to radically transform mathematics, if mathematics continues to exist.

JPB.– *Do you have doubts about the fact that mathematics continues to exist ?*

WT.– I don't have serious doubts, except that I think there are dangers of the mathematical community not adapting well enough to changing needs of society and changing conditions. These are dangers, but mathematicians are pretty adaptable, so it makes it unlikely. After all, I have doubts about the future of humanity, too.

Things are changing. I think computers are radically changing everything. They're certainly going to change mathematics radically, but more explicitly there are some aspects of mathematics that involve detailed and accurate, long calculations. I mean, I don't know whether to call them calculations or not, but there's a certain mode that mathematicians have of going through formal processes that involves peripheral and complex record-keeping. Computers can greatly enhance this type of mathematical activity, acting as mental tools, as extensions of the human mind.

One of the most important needs in mathematics is to change and improve the ways we communciate. Computers have begun to have a big impact in this, and they have the potential for a far more profound impact. Perhaps that would be the most important thing, because I believe that limitations of communications have a profound effect on mathematics. I think there have been a lot of great mathematical thinkers who have come to certain views and understanding of some kind of mathematical phenomenon, but where their understanding does not come across very directly to others; in the process of explaining, it is changed. I believe that one very systematic modification through communication is the diminution of geometric modes of thought and the emphasis of symbolic modes of thinking, because our primary means of communication has been language. Even though it is possible to talk about geometric things, in formal discussions, language tends to be a weak reflection of the real ideas. When two mathematicians talk informally they are often able to communicate with many kinds of signals, and if they develop a rapport, they communicate other modes of thinking reasonably effectively. But when a mathematician gives a lecture, or even more, writes a paper, there are strong filters. It has been a lot easier to write an equation for some geometric object or some process than to draw a picture of suitable quality for publication. Computers are quickly expanding our ability to produce images that have a high enough professional quality to include in papers.

But much more, the notion of the paper in the process of change. Twenty-five years from now it will be thought of as an archaic medium; maybe much sooner than that, communication will be preferably electronic. Electronic communication offers many more possibilities than paper, for one thing. Paper is essentially linearly-ordered; it's like you have this complicated mathematical concept, essentially a paper is often like a Hamiltonian circuit : in the one-skeleton of a mathematical structure that has rich connectivity. With computers, there are evolving techniques for trying to show the shape of the structures, so that you show many connections, not just connections one at a time. Actually trees are pretty standard representations in a linear paper, but it still doesn't capture the connectivity of the ideas. It's the ability to put pictures, it's the ability to put animations, it's the ability to put programs, with a hypertext content of being able to do easy crossreferencing. It so often happens when I write something, if anybody writes something, it's a big dilemma, because you don't have a single potential audience. There are multiple audiences : for some audiences you want to explain one thing, and for other audiences, explanation really detracts from the thought and you want to suppress it. So, being able to do both, at least for something that's important to communicate, to be able to have different levels of explanation, and different parts of explanations suitable for different people is a real opportunity. With computers, a communication can become much more like an interactive conversation, which can be much quicker and more efficient.

JPB.– *Moving to a slightly different topic. In the US at this moment a number of mathematical institutions make a point of including in their titles "mathematical sciences". MSRI is one. How do you, as director of MSRI and also as mathematician, view these efforts to reach out to communities who use, practice and need mathematics ?*

WT.– What I think is extremely important is for people to form connections.

I think mathematicians are very isolated from each other as well as from other scientists who have quite interesting potential interactions with mathematicians. So I think the Internet, the tools like Mosaic are going to have a major impact on the productivity of the academic landscape in that people can pursue things that they're interested in independent of whether they are personally in touch with mathematicians. Anyway, I believe that the basic fabric of all the sciences, at least all the natural sciences, is really mathematics, and that we have tremendous opportunities, a tremendous amount to contribute and also have a tremendous amount to learn from better connections.

JPB.– Do you feel that having the proper institutions will be enough? Don't you think that this will also involve training students differently?

WT.– Is it a question of training students or training professors? It may be that the younger generations are more equipped to train the older generations.

JPB.– *That could be.*

WT.– This is a problem. How do we go about training students? I mean we could encourage students to learn things that we ourselves don't have a grip on. It's hard to teach them. We may have to rely on nature taking its course.

One of the ways would be through the Internet, because the Internet is a new means of communication that's open to everyone. As the experience has been at a place like the Geometry Center, with which I've been associated.

JPB.– *In Minneapolis?*

WT.– In Minneapolis, when they set up the Mosaic and Web server, immediately they started getting hundreds of inquiries per day, and many people would load it down to see demonstrations of hyperbolic geometry or of how to get 2-dimensional quasi-periodic tilings from four dimensions, just by finding the things over the Internet. So many people who wouldn't ordinarily be in the audience of mathematicians start to see it. There's a danger of information overload. However, computers allow for a potentially very effective searching and filtering function. I really think this is one of the most important things for the mathematical community to concentrate on right now, moving into the future of communication.

JPB.– *Do you see the multiplication of institutes such as MSRI around the world as a coincidence or as a fashion?*

WT.– I think that's a good question, and I don't understand the phenomenon. It seems that there's a very high demand for institutes, certainly MSRI continues to have a very high demand for people who like to come here despite the growth of other institutes around the world. Maybe it's just a natural evolution, that people see something that works well, that's a good idea.

JPB.– *So, as director of MSRI, you see a very high demand. You have no doubt that it should go on ?*

WT.– I have no doubt that it's been a good activity, because people who come here say that it is, because there's a great demand for people to come here to participate, a much greater demand than we can accommodate ourselves. I haven't seen any sign yet there's an oversupply of institutes, but I could be wrong.

JPB.– *During my stay here I heard one of the deputy directors of MSRI saying that MSRI is mainly a center for postgraduate studies. Can you comment on that ?*

WT.– I would not put it quite so strongly. A very central purpose is as a center for postgraduate studies. At MSRI we have a group of usually about 25 postdoctoral fellows of one sort or another. They're our largest group of long-term, fully-funded visitors and my impression is that it makes a big difference in their careers. It makes a big difference because they make connections, they become connected with the mathematical community in a strong sense, particularly when we have their area of concentration and they're young postdocs in that field that may meet a large portion of the world experts in that field and get a real sense of what's going on and why and what's important and what's interesting to the community. They also develop a great deal of sophistication, they make a big improvement in their sophistication through that process. Somehow they broaden their perspective and they make connections, that's the biggest function.

It's also an important experience for older mathematicians, but it's not as dramatic because they already know many of the other people with whom they work when they're here, and they know how to get in touch with other people, other mathematicians. They already go to conferences, other places that renew the mathematical contacts. So I think that portion of our activities is more portable, more replaceable, they could go somewhere else instead of here; it's very different from the experience of a postdoc. Over the course of a year we have about a thousand mathematicians who come through MSRI for formal visits. That's a very large flow of people, it's a large number of people compared to the postdocs, but I still think that one of the main benefits is to the postdocs.

JPB.– *You've been interested for a while in establishing contact between secondary schools and research mathematicians. MSRI is pursuing a program in this direction. How do you see the possibility of such programs, which have been for the moment more or less confined to initiatives of isolated individuals, having a sizable impact on mathematical education ?*

WT.– I think in the US it's a very complicated problem to have a sizable impact on mathematical education because the decisions about what happens in schools are local. School boards have a lot of autonomy, and there are thousands of different local school boards all over the country. There are also many other interest groups such as the publishers of text books and schools of education, so it's a very hard problem to change the system. Nevertheless there has been a very vigorous mathematical education reform movement in the US, that is having an impact. It hasn't been primarily run by mathematical researchers, but mathematical researchers have been involved and I think it's been a healthy mix.

I think there's a major problem in the US about a tiered, multi-layered system that's got people who do mathematical research who are not in very close contact even with people who teach in small colleges and neither of these groups of people are in very close contact with high school teachers of mathematics, let alone teachers of younger ages. In a sense these different groups of people involved in mathematics could easily go in different directions. A student goes to high school before college, and to college before graduate school, but the people in charge of these different parts of a student's education often know very little about what happens in other places. One function that I think is valuable is just to have the right set-up for some degree of contact and discussion between these different groups of people.

JPB.– *Do you think the Internet can play a role in that for example ?*

WT.– I think that it's already begun to play a role. I think that's one good instance. There's something called the geometry forum which is an Internet project, a discussion group. I think it's been spear-headed by Gene Klotz at Swarthmore University. Teachers and students, mathematics researchers, have been involved in it, and certainly high school students are now looking at things like the world wide web server and the geometry center's film and I think it helps. There's also a discussion group called K-12 Math Education.

JPB.– *Kindergarten through twelfth grade ?*

WT.– Yes, kindergarten through twelfth grade. So that has a certain group of teachers. I think it's not yet widespread, but it probably will become rapidly widespread since the telephone companies have decided to help the schools with Internet connections. I guess it's hard to find time to look very much at mathematics news groups. We need better filter mechanisms. I happened to look at this particular one two or three months ago, and noticed that somebody there was posting portions of an article I had written a few years ago, paragraph by paragraph, and having long discussions about it— healthy discussions, I thought.

JPB.– *Do you see a time coming where for example departments would have a definite commitment, almost like a contract, to really participate in activities like the geometry forum you mentioned?*

WT.– There's a question of what accomplishments of mathematicians are recognized and appreciated. It's pretty clear that in the United States there are I think about 20,000 members of the American Math Society, more than that many members of the Mathematical Association of America. The main value of the professional involvement for most of these mathematicians is not through contributing to mathematical research, but mathematical research is about the only kind of accomplishment that is visible anywhere outside of the classroom.

JPB.– *Another thing which would be visible would be proofs that mathematics is at work in some objects or even some phenomena of the everyday life. Revealing this presence of mathematics would also be something in which mathematicians could be involved, and at the moment there is not much information on that.*

WT.– I think that there are other kinds of creative contributions that mathematicians could make to society other than proving another theorem. One of these kinds of things is finding connections, not just within mathematics, but with mathematics to everyday life or mathematics to other sciences and so forth. These aren't things that would be appropriate to publish in the Annals of Mathematics. If it was published there it wouldn't reach the audience that it would interest, that would be interested or would find it useful. But on the Internet I think there's many more– In essence I think publication will become much more democratic, much more widespread. We'll have to have means of filtering it, collating information, but I think that will happen, since physically or electronically there are no obstacles, it's just a question of culture, of how people do things. So I think what this will do is bring out many more kinds of creative contributions, a wider appreciation.

JPB.– *So that, if somebody is in a small school in a small village, he could also take a very active part in the wider network.*

WT.– Yes. People will be able to find out what's of interest to them, in some way. They'll be able to pursue the interest even if it's not in their immediate circle.

JPB.– *And even intervening in the system.*

WT.– Right. I think good ideas about teaching will spread, are spreading already but they will spread more via these means of communication, and good ideas about lots of things. It's going to radically transform society.

JPB.– *Thank you for spending some time with us.*

Class. Quantum Grav. **15** (1998) 2545–2571. Printed in the UK PII: S0264-9381(98)95879-8

How to see 3-manifolds

William P Thurston†

Mathematics Department, University of California at Davis, Davis, CA 95616, USA

Received 10 July 1998

Abstract. There have been great strides made over the past 20 years in the understanding of three-dimensional topology, by translating topology into geometry. Even though a lot remains to be done, we already have an excellent working understanding of 3-manifolds. Our spatial imagination, aided by computers, is a critical tool, for the human mind is surprisingly well equipped with a bit of training and suggestion, to 'see' the kinds of geometry that are needed for 3-manifold topology.

This paper is not about the theory but instead about the phenomenology of 3-manifolds, addressing the question 'What are 3-manifolds like?' rather than 'What facts can currently be proven about 3-manifolds?'

The best currently available experimental tool for exploring 3-manifolds is Jeff Weeks' program SnapPea. Experiments with SnapPea suggest that there may be an overall structure for the totality of 3-manifolds whose backbone is made of lattices contained in $PSL(2, \mathbb{Q})$.

PACS number: 0240

1. Introduction

Training the imagination. Our mental facilities for geometry and vision are remarkable. From an impersonal perspective, the act of walking through a crowd to meet up with someone on the other side is truly astounding. It is a far greater achievement than any merely intellectual achievement such as writing a PhD thesis in mathematics. Writing a PhD thesis is a great *intellectual* challenge, but the powerful intelligence needed to walk through a crowd is something that pooled human *intellect* has not come close to matching, despite years of effort and massive investment in related technologies.

Of course, we are no more able to program computers to 'do mathematics' than we are able to program them to walk down the street. Computers are powerful tools in mathematics, particularly for the symbolic aspects of mathematics (including numerical computations). It is ironic that the human use of symbols has often been touted as the unique human characteristic that makes us special. But our minds are complex organs, composed of many different cooperating modules. It is not the equations, symbols or logic that are hardest for computers, but the seemingly 'low-level' foundations of perception that prove hardest to match. 'Imagination', 'intuition' and 'instinct' are some of the words that are often used to allude to some of these perceptual foundations.

Geometric imagination is a powerful tool for three-dimensional geometry and topology—provided we teach it the foreign imagery it wants and needs to work with our intellect instead of rebelling. Our spatial–geometric instincts are rather strong-headed, and if we do not bring them along, they are bound to rebel.

† E-mail address: wpt@math.ucdavis.edu

0264-9381/98/092545+27$19.50

Hyperbolic manifolds. Hyperbolic geometry, also called Lobachevskian geometry or non-Euclidean geometry, is the geometry of a complete Riemannian metric on $\mathbb{R}^3$ with constant curvature -1. (A metric is *complete* when you cannot get to the edge of space in a finite distance, that is, a metric ball of any finite radius is compact.) Hyperbolic space is symbolized as $\mathbb{H}^3$. This geometry is the crucial tool for seeing three-dimensional topology. We will visit $\mathbb{H}^3$ a little later and look around.

Most 3-manifolds are hyperbolic

A manifold is *hyperbolic* when it has been given a metric that locally is identical to $\mathbb{H}^3$. In other words, a small neighbourhood of any point matches hyperbolic space on the nose. This is the same as a metric of constant sectional curvature -1. But a more revealing description is that a hyperbolic 3-manifold comes from $\mathbb{H}^3$ modulo a discrete group of isometries, where all points equivalent under the group are identified to a single point of the manifold. One can think of the group 'rolling up' space into a compact bundle, as if it is a big floppy rug but much neater.

The geometry of our minds. Whether by design or accident, we have geometric modules in our minds that are remarkably well suited for use inside hyperbolic space and inside hyperbolic 3-manifolds. Three modules in particular—our sense of perspective, our sense of scaling and our sense of symmetry—connect directly to what we see when we visit a hyperbolic 3-manifold. Our minds must adjust their interpretations of these perceptual modules, but our brains are plastic and can readily make the needed adjustments.

Geometry gives us an understanding of 3-manifolds that we topologists of 25 and more years ago never imagined possible. This paper aims to present some of the basic geometric vision that lets us see, know and understand 3-manifolds. Predominantly, this means to see, know and understand hyperbolic 3-manifolds.

Disclaimer 1: other geometries. Not *all* 3-manifolds are hyperbolic. There are actually eight different flavours of three-dimensional geometry, describing eight different classes of 3-manifolds. Maybe the reason that for a long time nobody suspected that hyperbolic 3-manifolds are plentiful is that the simplest 3-manifolds are not hyperbolic, but typically have one of the seven non-hyperbolic flavours of geometry. It is easy to deduce that a manifold that has one of the seven non-hyperbolic flavours is topologically quite special. It is natural to assume by analogy that hyperbolic manifolds are special. They are indeed special, but they are plentiful.

We are lucky that most flavours are rare, because the geometry modules adapt much better to Euclidean space and hyperbolic space than to any of the other geometries†.

A 3-manifold is *geometric* if it has a Riemannian metric for which any two points have neighbourhoods that are identical. A metric satisfying this condition is a *locally homogeneous* metric. Any compact manifold with a locally homogeneous metric has other locally homogeneous metrics: you can always scale the metric by a constant factor. The total volume changes, so the new metric cannot be isometric with the old one. Some locally homogeneous metrics can be modulated in additional ways, while remaining locally

† Spherical geometry (S^3) is the runner-up, but S^3 when naively rendered is quite disorienting to visit because very distant objects look as big as very close objects. Our perception continues to rebel against this phenomenon long after our intellect has accepted it. Real-time special effects with fog and focus would undoubtedly remedy much of this difficulty.

homogeneous. A locally homogeneous metric is *isotropic* if there is a local isometry taking any given direction to any other direction, and *anisotropic* if there is some proper subset of directions distinguishable by the local geometry. In dimension 3, the tangent space to an anisotropic manifold has a line field constructible from the local geometry, and an orthogonal field of 2-planes. These scale factors in the line and the plane can be modulated independently. These variations are why we talk about 'flavours' of geometry rather than about the exact shapes of metrics.

Only three of the eight flavours are isotropic: spherical geometry, Euclidean geometry and hyperbolic geometry. It is a curious fact that all spherical 3-manifold (manifolds with metrics of constant positive curvature $+1$) also possess anisotropic locally homogeneous metrics. Euclidean 3-manifolds do not have anisotropic locally homogeneous metrics, but they all have a line field that is globally parallel to itself. In a certain sense, hyperbolic manifolds are the only ones that are genuinely anisotropic.

The anisotropic flavours of geometry, and 3-manifolds having those flavours, have the form either of a product or a fibre bundle, combining one of the two-dimensional geometries with one-dimensional geometry perhaps with one or another type of twisting.

Disclaimer 2: decomposable manifolds. Not *all* 3-manifolds are even geometric. There are two topological processes for joining geometric 3-manifolds, to form new, 'compound', 3-manifolds that do not have geometric structures.

The first compounding process is the *connected sum*, which means to remove balls from two 3-manifolds, and join their bounding spheres together—in other words, connecting them by a tunnel or wormhole whose cross section is the 2-sphere S^2.

The second process involves joining 3-manifolds with a boundary, to obtain a new manifold that may or may not have a boundary. A surface inside or on the boundary of a 3-manifold is *incompressible* if a curve on the surface cannot be the boundary of a disc in the 3-manifold unless it is already the boundary of a disc on the surface. For example, the surface of a doughnut in space is an example of a *compressible* torus, because for example a disc that cuts across the doughnut hole has a boundary that is a non-trivial curve on the torus. A *torus sum* of 3-manifolds is something that results from gluing together two incompressible component tori that are components of the boundary. This creates a second kind of tunnel or wormhole, whose cross section is a torus, T^2. Tunnels with cross sections that are multi-holed tori with two or more holes have no special status, because these surfaces do not interfere with geometric structures.

There is a complete topological theory of how to analyse 3-manifolds that have been combined by these two processes: topological criteria to recognize and undo these two ways of joining and to recover the original pieces. Undoing connected sums is called the *prime decomposition* of a 3-manifold. The theory of the prime decomposition was analysed by Kneser in the 1930s. A *prime* 3-manifold is a manifold which cannot be expressed as a connected sum except in a trivial way. Every compact 3-manifold is the connected sum of finitely many prime pieces, and the prime pieces are determined up to homomorphism by the 3-manifold.

The second decomposition process, undoing tunnels whose cross section is a torus, is called the *torus decomposition*, and was analysed in the 1970s, by Jaco, Shalen and Johannson. The theory is analogous to the prime decomposition, but we will skip giving the exact specifications, which are a little more involved.

Disclaimer 3: unproven. It has not been proved that *all* 3-manifolds are composed of geometric pieces. A number of years ago I proposed the *geometrization conjecture*, that the primitive parts produced by the prime decomposition and torus decomposition of any 3-manifold are always geometric.

To prove the geometrization conjecture in full generality is a great challenge in topology. One special case is the Poincaré conjecture, which has long been notorious as a tempting lure surrounded by hidden traps.

If you are a consumer rather than developer of 3-manifold theory, the status of the geometrization conjecture is probably fairly academic, because for most purposes you may as well assume it is true of the manifolds you need to use. The conjecture is overwhelmingly supported by the evidence we have seen to date. There is theoretical evidence, which has established the existence of the geometric decomposition for several broad classes of 3-manifolds. In addition, large numbers of particular examples have been tested. Mathematics is full of surprises, so we can never be certain of the geometrization conjecture until and unless it is proven. Nonetheless, the geometrization conjecture is a safe working hypothesis, and if you need to actually know for a particular case, you should probably just compute its structure anyway.

The words and the reality. The full description of the geometrization conjecture sounds complicated, since it involves two different decomposition processes and the eight geometries. Do not be fooled by this. The complexity of the verbal description is mismatched with the actual complexity of 3-manifolds. The prime decomposition and the torus decomposition may sound complex, but they are actually very orderly and straightforward processes. Similarly, the geometric manifolds having any of the seven non-hyperbolic flavours have been completely classified in an orderly and understandable way.

The true complexity of the structure of 3-manifolds—at least if the geometrization conjecture holds—is the structure of hyperbolic 3-manifolds.

The geometry of a hyperbolic 3-manifold is a topological invariant, according to the Mostow rigidity theorem (extended to the non-compact lattices by Prasad). This fact allows one to extract a great deal of topological information; it makes it quite easy to tell whether or not two hyperbolic 3-manifolds are homeomorphic. The volume, in particular, gives an interesting measure that describes in a certain way the three-dimensional complexity of the manifold. Volumes seem to all be irrational. There are some rational relationships among volumes of different manifolds, but most pairs of manifolds have volumes that do not seem to be in any rational ratio. The set of all possible volumes is a countable, closed subset of $\mathbb{R}$. Each volume that occurs does so for only a finite set of examples. The accumulation points of volume are manifolds which have 'cusps'—they are non-compact, with exponentially shrinking tubes going out to infinity whose cross sections are tori. The phenomenon of this convergence is part of the theory of continuous surgery on hyperbolic manifolds, which we will be experiencing.

Appearance, reality and imagination. It is a wonderful dream to see the topology of the universe some day. However, this paper is not about the topology of the physical universe, but about topology in our minds. We will imagine bending space, but this has the effect of bending the imagination. The purpose is not for science fiction diversions, but to develop true vision meeting the precise standards demanded by three-dimensional geometry and topology. Our instincts about appearance and reality are strong, so our minds need time and exposure to adjust to new possibilities.

The scale in our imagination can make a big difference in our thinking. An effective strategy is to think about 3-manifolds on the scales we might inhabit: perhaps the size of a house, the size of a stadium, or the size of a town. It is harder to attend as seriously to objects the size you might hold in your hand. It is interesting to sit back and imagine your surroundings—the streets and the land in the neighbourhood where you live—and then think of the same degree of imagination about a teacup. At the opposite pole, very huge objects—the size of the universe, or even of the earth—are so far removed from everyday experience that our imagination on those scales tends to be abstract and distant.

Other resources. The rather unique video 'Not Knot' [Geo81] journeys through territory that overlaps some of where we will go; it is worth seeking it out (A K Peters is the current publisher). Clips and pointers to this and other relevant resources, including some good downloadable software, can be found on the Geometry Center website at www.geom.umn.edu. My book *Three-dimensional Geometry and Topology* [Thu97] develops the broad sweep of three-dimensional topology from the geometric perspective. A 1982 review article [Thu82] summarized the geometric theory of 3-manifolds at that time. Theory has advanced since then, the broad picture has not changed, but the basic picture is close enough that it is still recommended as a summary of the geometric theory of 3-manifolds. See also the online version [Thu79] of lecture notes from seminars I gave in Princeton 1978–81. There is a large literature on the geometry of 3-manifolds that I will not even attempt to review; MathSciNet (http://www.ams.org/mathscinet/) is a much better window than anything I could write. I will only mention three of my own primary contributions [Thu86, Thu98a, Thu98b].

2. Geometry from the inside

Imagine walking in a barren desert when you see the space in front of you begin to shift. You are startled, and stop. You see a vertical, straight fracture where the left side does not quite match the right: the images overlap ever so slightly. At first you think your vision has gone bad, maybe you have become cross-eyed. However, when you turn your head and move from side to side, the fracture does not turn or move with your head and eyes. When you circle around at a wide distance, you see that the fracture is not fixed on the ground or on the distance scenery, but is localized on a line going straight up into the sky.

Paper models. When you get back to camp, you can make a model to help explain what you have seen. Cut a 350° sector of paper (i.e. a disc with a 10° angle removed) with edges joined to form a blunt cone. This is the cone with *cone angle* 350° or *curvature* 10°. You can trace geodesics on this cone either by stretching pieces of string, by flattening portions of it and drawing straight lines with a ruler or by using a folded strip of paper as a ruler that is flexible enough to fit the surface. If you draw geodesics that emanate from a point on the cone at an angle less than 10° and aimed to the left and right of the apex, they will cross again behind the apex. This behaviour is identical with the behaviour of geodesics in a certain distorted metric in the plane, as shown in figure 1.

In the same way, we can think of a distorted metric in space that is equivalent to the metric we would get from a sector, of say 359° around our singular line. This affects what you see, since light 'bends' to follow the geodesics in the metric; it is as if light near the axis is slowed down.

Return to the axis. Overcome by curiosity, you return to gaze at the singular axis. The singularity has progressed dramatically, and as you watch, the scenes along the two sides

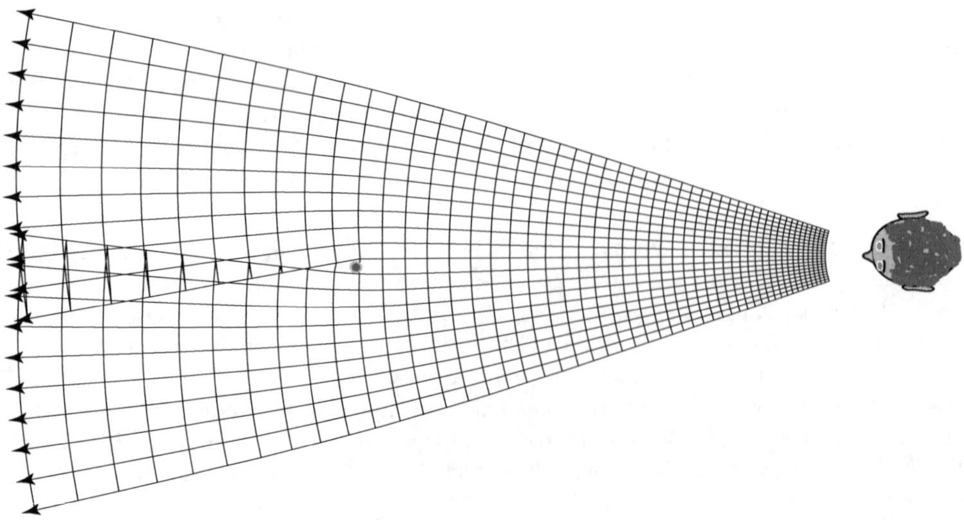

Figure 1. This figure shows paths of light near the apex of a cone. The metric of the cone is represented as a distortion of the metric in the plane. The long, nearly straight curves are geodesics, and the orthogonal trajectories can be thought of as wavefronts.

of the singular line scroll outward, squeezing into the surrounding panorama at your sides and behind you. You are startled to see arms and shoulders growing outward from the axis, then feet, a torso—and suddenly, there is your own head staring back! The image snaps into perfect alignment, and the motion stops. In a daze, you slowly turn around to take stock. Every single thing is repeated twice. For each boulder, there is an opposite boulder. Each mountain has an opposite twin mountain. There are even twin suns in the sky, shining from directions 180° apart, lighting up every shadow.

You realize that the singular line is a 180° cone axis (figure 2). Light 'bends' so much that you see another image of yourself directly behind the axis—lines of sight that go just to the right of the singular axis make a complete U-turn, heading almost straight back to you, so that you see the left side of your face. Similarly, your line of sight just to the left of the singular line, you see the right side of your face. With cone angle π, the two halves fit together seamlessly.

This is an example of an *orbifold*, which is a space locally modelled on $\mathbb{E}^n/\Gamma$, where Γ is a finite group of symmetries. Here $\mathbb{E}^n$ symbolizes Euclidean space equipped with its standard metric, and has a connotation somewhat different than $\mathbb{R}^n$. Orbifolds are generalizations of manifolds having the advantage that many phenomena are illustrated with much simpler examples of orbifolds than of manifolds.

Now the cone angle starts decreasing even further. The images to the left and right of the cone axis go out of sync once again (figure 3), but when the cone angle reaches $2\pi/3$, they are again coordinated and match each other seamlessly. You are in the orbifold $\mathbb{E}^3/C_3$, where C_3 is the cyclic group with three elements (that is, the group of integers modulo 3). As the cone angle continues to decrease, the images match when the cone angle is $2\pi/4, 2\pi/5, \dots, 2\pi/n$. Figure 4 shows the case of cone angle $2\pi/7$.

The effect is reminiscent of two mirrors that meet at an edge, when the angle between them is varied. A single mirror can be thought of as a model for the orbifold $\mathbb{E}^3/C_2$, where

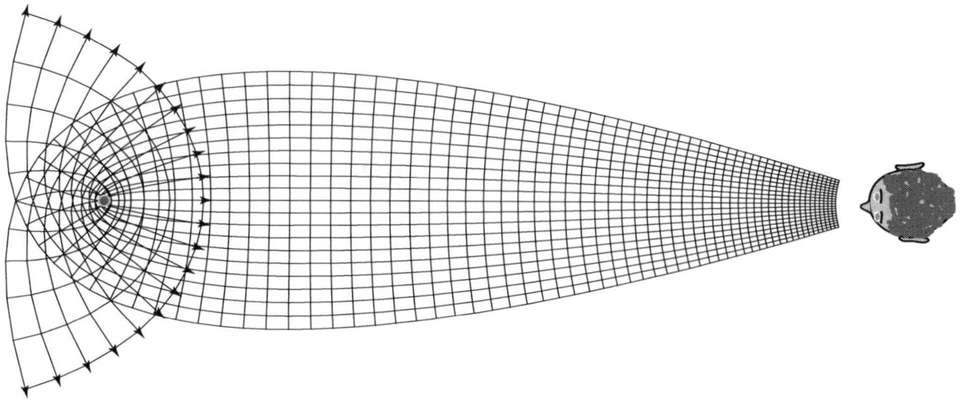

Figure 2. The behaviour of geodesics near the apex of a cone with cone angle π. The geodesics just to the left and right of the apex align perfectly. What you see is a perfect image of yourself. Unlike your reflection in a mirror, this image has the same orientation that you do.

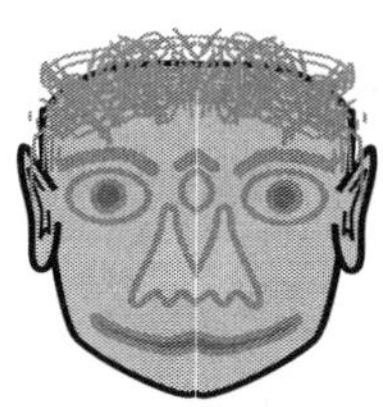

Figure 3. When there is a cone axis with angle $\pi - \epsilon$, you see an image of yourself on either side of the cone axis that is doubly covered in a narrow zone in the middle.

C_2 acts by reflection through the plane of the mirror. Two mirrors at an angle π/n give a model for the orbifold $\mathbb{E}^3/D_n$, where D_n is the dihedral group of $2n$ elements, generated by reflection in the two mirrors. Since $C_n \subset D_n$ is a subgroup of index 2, the optics of a cone angle of $2\pi/n$ can be visualized by ignoring every other image as seen in two mirrors that meet at an angle π/n (that is, ignore all images that have the opposite orientation from you).

Reality and appearance. Despite *Through the Looking Glass*, there is a significant distinction between the effects of a cone axis and the effects of mirrors. When you go with a companion to visit a cone axis with cone angle $\pi/2$, you see four images of your companion and three of yourself, plus your identity image. *The four images of your companion are all equally real.* You can walk in a straight line toward any of the images of your companion and shake hands. You can turn around and head toward any other image and shake hands again.

It is not quite the same if you try to shake hands with one of your own images, because whenever you move toward your image, your image moves away. If you stubbornly keep following anyway, you and your three images chase each other in an inward spiral until you eventually are close enough to the cone axis that you can reach out with your left hand

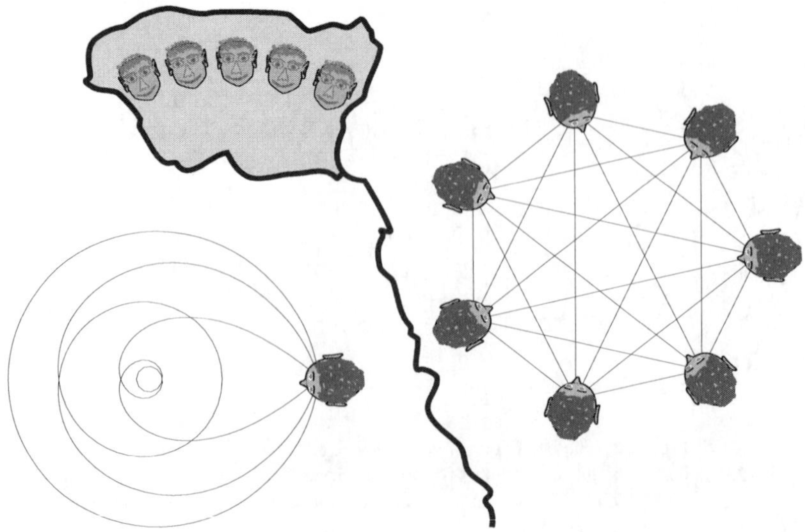

Figure 4. When a cone axis has cone angle $2\pi/7$, you see six images of yourself coming from geodesics that go ± 1, ± 2 or ± 3 turns around the axis before reaching you. The appearance is the same as if you are part of a ring of septuplets. At the top is a snapshot of you near an n-fold axis, taken by a friend standing on a boulder behind you. Since the images are in perfect alignment, the cone axis is not visible.

and shake your right, forming a ring around the axis. You know that really there is only one of you, but it will take some time for your brain to readjust its model of how the world relates to seeing and touching.

The limit. With the right choice of normalization, the cone angle can keep shrinking all the way to 0 without crushing or impossibly distorting the space around you. The trick is that as the cone angle decreases, the cone axis recedes away from you further and further into the distance. To normalize, watch the cylinder through you around the cylinder axis. Initially, if the line of singularities is a distance r away from you, the cylinder has circumference $2\pi r$. For cone angle α, if the cone axis moves to a distance $2\pi r/\alpha$, the circumference of the cylinder in the cone manifold is still $2\pi r$; the intrinsic geometry of this cylinder stays the same, but the cylinder becomes less curved as $\alpha \to 0$. In the limit, the cone axis moves infinitely far away and vanishes and the two half-planes converge to parallel planes, and the cylinder you are on flattens out and become planar. The resulting space is now a manifold, $\mathbb{E}^3/\mathbb{Z}$, which can be thought of as a different metric for $\mathbb{R}^3 - \mathbb{R}$. What you see is that every object has an infinite repeating sequences of images lined up in a horizontal straight line, It is reminiscent of a barber shop with two opposite and parallel mirrors, although different because there is no obstacle to reaching an image of anything other than yourself.

Curvature can buffer from crushing. This choice of normalization protects you, but rather unfortunately, most of $\mathbb{E}^3$ is cataclysmically distorted. If you want to protect the rest of space from experiencing unbounded distortion, the only way is to give up local Euclidean geometry and allow space to become curved. With curvature, it is easy. To distort $\mathbb{E}^2$ in this way, just rest a cone on the plane with its point pointing up and round it off like a conical mountain. As the cone angle deforms to 0, the surface develops a long tube, growing like an asparagus sprout until it is asymptotically a cylinder.

These surfaces have a net negative curvature that exactly balances the curvature at the cone point; in the limit, the total curvature is -2π. In three dimensions, the product of the two-dimensional metrics with a line accomplishes a similar result. Whenever a cone angle along a singular axis decreases, the axis develops an increasing concentration of positive curvature, and it 'wants' to build a balancing cloud of negative curvature in planes transverse to the axis.

The process will work with great fluidity once we turn to the hyperbolic space, whose fabric is negatively curved.

A new cone axis appears. The space around us has transformed into $\mathbb{E}^3/\mathbb{Z}$, obtained by periodically identifying all points in $\mathbb{E}^3$ by a translation along a horizontal axis. The translation axis wraps around to form a closed geodesic loop C. Make sure you are standing some distance away from C. Now watch as space starts to develop a new cone singularity along C whose cone angle slowly decreases from 2π. The geometry of the changing metric can be represented concretely by slicing along a 'half-plane' radiating from C—that is, the surface has the local geometry of a half-plane, but it actually wraps around to form a half-infinite cylinder. After the singular metric is sliced along one of these planar cylinders, it matches the metric of an almost 360° wedge in the non-singular metric of $\mathbb{E}^3/\mathbb{Z}$ bounded by two planar cylinders. The singular metric is reconstructed by gluing the two cylinders back together.

The cone angle keeps on decreasing, while the geometry of space away from the singular axis maintains its locally Euclidean nature. When the cone angle reaches π, the cone manifold becomes an orbifold. At this instant, all visual images of any object are coordinated. You see two rows of images of yourself, rotated by 180° about the axis. Since the axis is horizontal, the 'other' row is upside down.

Carrying on, whenever the cone angle is $2\pi/n$, the images are again perfect: you see n rows of images that are symmetric by order n rotations about C, and repeating by translations along C. We make sure that the cone axis moves further and further away as $\alpha \to 0$, so that there is a limit for cone angle 0, normalized in a way that keeps the neighbouring rows of images so they are spaced a constant distance apart. The cylinder of points a constant distance from the axis is actually wrapped around to form a torus. In the limit, this torus flattens out and becomes planar. What one actually sees is a doubly periodic set of images of any object, repeating by an action of $\mathbb{Z}^2$ acting as a discrete group of translations of $\mathbb{E}^2$.

The Euclidean geometry of $\mathbb{E}^3$ is like a harbour on the edge of a small island in the middle of a vast hyperbolic ocean of hyperbolic manifolds. So far, we have taken a short stroll along the edge of the harbour. There are many fascinating tourist attractions here on the island—for instance, it is only a short trip from our hotel ($T^2 \times \mathbb{E}^1$) to the top of the mountain, S^3 that looms over us—but time is limited and we have far to go, so we will be wiser to work on preparations for going to sea.

3. Moving in two directions: continuous surgery

There are so many 3-manifolds that it is very easy to lose your bearings. It will help if we upgrade our navigational capabilities to be able to modulate the geometry at singular axes in the 'sideways' directions, in addition to increasing and decreasing cone angles. Without the new capability we would still be able to go wherever we like, just as even someone who is stubborn about choice of airline could still fly the 90 miles from San Francisco to Sacramento—by detouring thousands of miles through Chicago or Atlanta.

The new process is similar to what we have already encountered. Take a wedge of space between two half-planes, and glue one to the other, but shift one side up slightly with respect to the other. Optically, this creates a vertical displacement between images seen on the two sides of the axis. If the original cone angle was 2π, for example, and you shift the side to your right upward before identifying it with the left, then you see an image of yourself through the axis, with the portion of the image to your right shifted down (see figure 5).

Figure 5. This is the image seen when there an axis with cone angle π slightly shifted parallel to the vertical axis, with the side to your right joined to a slightly higher level on the right. Your lines of sight to the right of the axis whip around and come back at a higher level, so the topmost portion of your image is lower.

If we shift along an axis that is closed in a loop, as in $\mathbb{E}^3/2\mathbb{Z}$, the same effect occurs, but the detailed behaviour near the singular axis has a somewhat bizarre description. To effect the shift, cut along a half-infinite cylinder whose boundary is the axis and reglue with a vertical shift of a. Every point on the axis itself is identified with the point on the axis a distance a higher, so also with $2a$ higher, etc. If a is incommensurable with the length L of the axis, then these identifications are dense along the axis, so the axis collapses to a single point. But if we shift by a rational multiple $a = (p/q)L$ of the length of the axis, the axis now wraps around itself q times. Each point on the resulting axis has a neighbourhood formed from q wedges of the original axis, so the cone angle is multiplied by q. If the axis was non-singular before the shift, this results in a cone axis with a large cone angle $2\pi q$, bigger than 2π. On the other hand, if we start with a $2\pi/q$ (orbifold) axis and shift by $a = (p/q)L$, the new cone angle is 2π, and the axis becomes non-singular!

We can watch this process happening, starting with say with a sevenfold cone axis $\mathbb{E}^3/(\mathbb{Z} \times C_7)$. Let us orient the axis vertically. You are part of a ring of septuplets circled around the axis. Up above, you see another ring of images of yourself, and above them, yet another. Now the geometry begins to shear along the axis. Your ring now spirals slightly upward to your right, so the third septuplet to your right is slightly higher than the next one around, the third to your left. Since these images come from light paths that actually wrap several times around the axis before coming back to you, they shifted every time they hit the vertical half-cylinder which has been sliced and reglued. the vertical misalignment of images you see is a sevenfold amplification of the shift of the regluing. The shearing increases; when it reaches $L/7$, your images align again in an infinite spiral, very similar to a steep spiral staircase. The shearing continues. When the vertical shift reaches $2L/7$, the

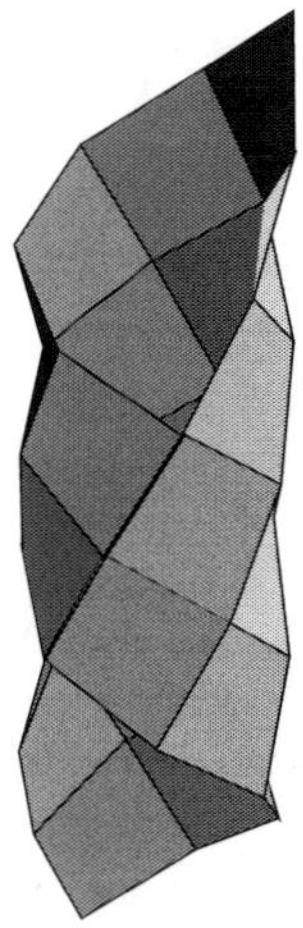

Figure 6. Left: $\mathbb{E}^3/\mathbb{Z}$ has undergone a compound shifting along a circular axis, first developing a $2\pi/5$ cone angle, then shearing vertically by $\frac{3}{5}$ the length of the axis. The central axis is not singular, but it is only $\frac{1}{5}$ as long as it used to be. You see multiple images of a single square whose sides are joined to form a torus. Before space shifted, this square extended full height and full circle around the original non-singular circular axis. Right: you have gone to investigate another compound axis close up. It is reassuring to belong to a whole flock of like-minded souls.

images go full circle after a vertical distance of $2L$, and you see two intertwined spirals, each twice as steep as the staircase before. The spirals are diametrically opposite. No two people are on the same height, but they alternate between the two spirals. When $a = 3L/7$, the images once again are all in alignment. You can make out three intertwined right-handed spirals, but this visual organization is not as automatic; you can see it instead as a parallelogram-grid of images wrapped around a cylinder. Perhaps you even see four intertwined left-handed spirals.

Instead of cutting and regluing, we can think of this operation as deforming a metric on a fixed manifold, while always retaining its locally Euclidean character. Focus on the set of distance r from the cone axis; topologically this is a torus. Consider how we can modify the three-dimensional metric (singular at the core), while maintaining a constant two-dimensional intrinsic geometry on our chosen torus. The deformation in three dimensions will be determined by how the torus is bent; the bending of the torus always matches a cylinder. The bending is determined by specifying the two principal directions of curvature and the two principal curvatures. One of these principal directions has curvature 0 and points along the generating lines for the cylinder, so the geometry of bending is parametrized by a single tangent vector to the torus, up to a sign. (Note that the bending at one point determines the bending elsewhere along the torus: the principal directions are parallel in the intrinsic geometry of the torus, and the principal curvature is constant.)

If we go inward toward the singular core in the family of parallel tori on the inside of our given one, the metric on these tori changes by shrinking lengths in the curved direction at a steady rate while maintaining length in the perpendicular direction. If these lines of curvature close up, then ultimately the torus shrinks to a circle, which is a cone axis with cone angle equal to the integral of principal curvature around the closed line of curvature.

Otherwise, the torus shrinks to a point†.

Surgery coefficients. The convention is to describe surgery in terms of a choice of generators μ and λ (a *meridian* and *longitude*) for the fundamental group of our chosen torus. If we identify the torus with $\mathbb{R}^2/\mathbb{Z}^2$, we can identify μ and λ with lattice vectors in the plane. The bending is determined by a pair $\pm(m, l)$ of real numbers, determined up to a sign, such that the vector $m\mu + l\lambda$ points has total curvature 2π, i.e. if you go along that path you will have bent around full circle. The parameters (m, l) are the *continuous surgery coefficients* or *real surgery coefficients* for the singular geometry.

In the important special case that (m, l) is a pair of relatively prime integers, the tori shrink to a circle with cone angle 2π, so the geometry is in fact non-singular, and we have a manifold. This case is called (m, l) *Dehn surgery*, or sometimes m/l *rational surgery*. If (m, l) is a pair of integers with greatest common divisor n, the tori shrink to a cone axis with cone angle n, so we have an orbifold. In either of these two cases, what one actually sees is a regular array of images of any object, wrapped around in a cylindrical array: when the continuous surgery coefficients are integral, the paths of geodesics match up continuously on the two sides of the singular axis, so images align perfectly. In any other case, what one would actually see would be a linear dislocation or discontinuity along the axis—this is the same phenomenon as a branch cut in an analytic function. The connection is not superficial: continuous surgery is closely related to families of analytic functions of one complex variable that have singularities like z^a, where z is a complex variable and a is a complex power.

4. Hyperbolic space

Our vision has made quick progress in learning to see repetition and symmetry as topology. With only a little more training to recognize some common hyperbolic objects, and a little adaptation to seeing scaling as distance, our eyes will be ready to lead us into hyperbolic manifolds.

Hyperbolic perspective. Hyperbolic space has a succinct description as the projective geometry of the interior of the unit ball in $\mathbb{E}^3$. Hyperbolic lines map to Euclidean straight lines in the ball, and hyperbolic planes map to Euclidean planes; it is just that hyperbolic distances are longer than Euclidean distances, very much larger near the edge of the ball, which is actually at an infinite distance. The unit ball makes a good, compact 'cheat sheet' that is easy to carry in our heads for quick reference when we need it.

Isometries of the hyperbolic geometry are the same as projective automorphisms of the ball, i.e. homeomorphisms of the ball that take planes to planes. A hyperbolic plane is determined by its circle of intersection with the unit sphere. The angle between two planes can be seen as the angle between these two circles.

In a dimension one lower, this means that constructions using only a straight edge use only a straight edge in a plane where a single circle has been drawn translate into hyperbolic geometry†.

† It is interesting that this family of metrics constitutes a geodesic in the hyperbolic metric on the Teichmüller space for a torus. The 3-manifold metric has a cone axis if and only if the geodesic tends to a cusp in the modular orbifold $\mathbb{H}^2/PSL_2(\mathbb{Z})$.

† The tangent to the circle at a given point is an important ingredient for many hyperbolic constructions. It is a nice exercise to construct this tangent from a point on a circle using a straight edge alone. Of course if we could

In other words, the projective geometry of hyperbolic space is a fragment of the projective geometry of Euclidean space. Imagine yourself in the centre of a ball perhaps 100 m in radius. Objects have the same visual geometry whether interpreted as Euclidean or hyperbolic space: hyperbolic straight lines appear straight. Even the stereoscopic effect of parallax is correct: hyperbolic parallax makes everything look a bounded distance away to Euclidean-adjusted eyes.

A hyperbolic plane appears to you as a visually round disc where a Euclidean plane slices the ball. In hyperbolic perspective, figures on this plane are identical with the projective model of the hyperbolic plane.

With computer geometry tools such as Geometer's Sketch Pad, one can readily perform these constructions and imagine them to be hyperbolic perspective drawings. As we move control points, the projectively constructed figures move with them, and keys into our sense of perspective, so we can imagine looking at a hyperbolic plane while flying over it in hyperbolic space (figure 7).

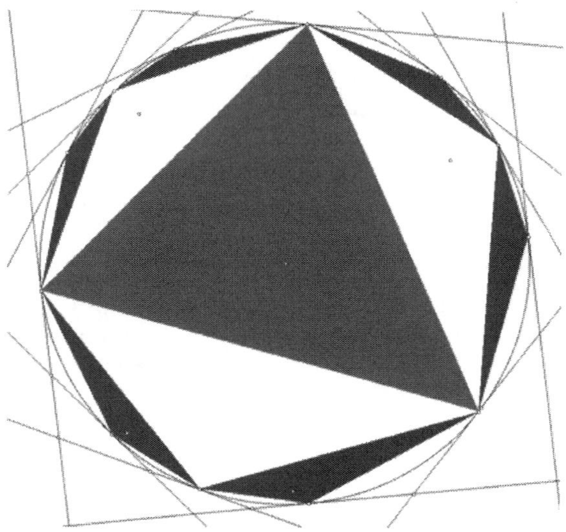

Figure 7. This is a screen shot from Geometer's Sketch Pad, showing the hyperbolic plane as seen in perspective from hyperbolic 3-space, showing the first stages of the construction of a tiling by congruent triangles. The construction starts with one circle and subsequently only requires a ruler, although it is more convenient to have automatic construction of tangents to the circle. The visual horizon of any plane in hyperbolic space is a circle like this. In hyperbolic perspective distant parts of the plane near the horizon circle are greatly foreshortened, since you have only a glancing view. In the program, moving corners of the centre triangle with a mouse helps trigger your sense of perspective and motion; the effect is stronger when the circle is large (or you put your eye closer to it) so that it more closely matches Euclidean perspective.

Hyperbolic space has lots of room. The big qualitative difference between $\mathbb{E}^3$ and $\mathbb{H}^3$ is that in $\mathbb{H}^3$ the area of a sphere of radius r grows incredibly faster: instead of scaling quadratically as in $\mathbb{E}^3$, the area grows exponentially fast. One consequence is that figures you see are closer than they appear, very much closer as their apparent size decreases.

not do it, we would dispense with the handicap and add it as a primitive operation.

This phenomenology is well captured in another model for $\mathbb{H}^3$, the upper half-space model, consisting of points in the half-space $z > 0$ in $\mathbb{R}^3$, where the hyperbolic arc length ds is given by the formula

$$ds^2 = (1/z^2)(dx^2 + dy^2 + dz^2).$$

What this says is that hyperbolic lengths scale in proportion to height above the xy-plane. Any similarlity of $\mathbb{E}^3$ that preserves the xy-plane also preserves this metric. In upper half-space, a geometric series of figures shrinking by factors of 2 toward the origin is a sequence of equally spaced hyperbolic figures.

All vertical lines are geodesics in upper half-space; the other geodesics are semicircles perpendicular to the xy-plane. The visual image of a figure in upper half-space, as seen by an observer high above the xy-plane, is very close to the perpendicular projection of the figure to the xy-plane. As an object moves downward, away from this observer, its visual image shrinks by similarities.

Thus for nearby objects in hyperbolic space, our Euclidean sense of perspective is a good guide to the geometry, while for distant objects, our sense of similarities can be mentally re-interpreted to give a good sense of the geometry. You can get a good feeling for this by using a hyperbolic-space viewer such as geomview (available free from http://www.geom.umn.edu), which enables you to put objects here and there, rotate them, translate them and fly around among them, all with the correct intrinsic hyperbolic geometry. To Euclidean-adjusted eyes, the distant scenery appears flattened. When we move toward distant objects, the appearance is very much like zooming a two-dimensional image. Large objects resemble Euclidean views that are in exaggerated perspective, taken with wide-angle lenses that are very close to their subject. However, this exaggerated perspective does not diminish when the object moves away; it simply records the exponential growth of the sphere of increasing radius, which corresponds to rapid shrinking of visual images with distance.

Complex coordinates. It is convenient to think of the xy-plane as the complex plane $\mathbb{C}^1$, because any orientation-preserving isometry of hyperbolic space extends to a meromorphic map, a fractional linear transformation $w \longmapsto (aw + b)/(cw + d)$ where $w = x + iy$. Algebraically, the composition of fractional linear transformations is the same as matrix multiplication of their arrays of coefficients $\left(\begin{smallmatrix} a & b \\ c & d \end{smallmatrix}\right)$, thus giving an isomorphism with the group $PSL_2(\mathbb{C})$. Because of this, everything in three-dimensional hyperbolic geometry tends to have natural complex parameters. For example, a typical isometry of $\mathbb{H}^3$ to itself is a screw motion, translating a distance d along some axis and then rotating by an angle θ. The complex number $d + i\theta$ is the *complex* translation length. Angles and length merge into a unified concept that yield holomorphic formulae for the geometry of any figure that varies freely in $\mathbb{H}^3$.

Banana cylinders. There are some specific surfaces that are important for understanding continuous surgery in the hyperbolic realm. Any line in $\mathbb{H}^3$ is the axis of a family of concentric cylinders of radius $\mathbb{R}$. In the upper half-space model, the cylinders about a vertical line appear as vertical cones. For the other (semi-circle) geodesics, these surfaces look something like bananas (figure 8), of varying thinness or fatness.

We can use the upper half-space to see how a banana-cylinder actually appears to a hyperbolic observer; simply think of the observer as stationed high above the xy-plane.

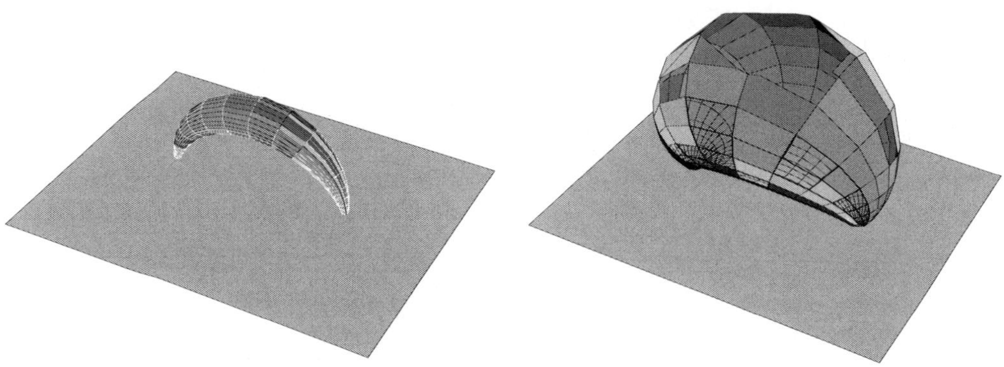

Figure 8. These bananas are actually hyperbolic cylinders concentric to a hyperbolic straight line, as seen in the upper half-space model. The banana is a union of arcs of the family of Euclidean circles that intersect the xy-plane at a fixed angle in a fixed pair of points.

A Euclidean cylinder of radius r curves in one direction only: its principal curvatures are 0 and $1/r$. A hyperbolic cylinder of radius r is shaped rather differently. In a cylindrical coordinate system (θ, t), where t measures arc length along the core axis and θ is the angle around it, the element of arc length $\mathrm{d}s$ on the hyperbolic cylinder satisfies

$$\mathrm{d}s^2 = (\sinh(r)\,\mathrm{d}\theta)^2 + (\cosh(r)\,\mathrm{d}t)^2.$$

As r grows large, the ratio of the scale factors $\sinh(r)$ and $\cosh(r)$ quickly converges to 1. The principal curvatures are the logarithmic derivatives of these scale factors: $\cosh(r)/\sinh(r)$ in the θ direction, and $\sinh(r)/\cosh(r)$ in the t direction. Before r is very large, both principal curvatures nearly equal 1, and the surface looks quite round even though it has the local intrinsic geometry of $\mathbb{E}^2$.

The limiting shape for a hyperbolic cylinder of radius $r \to \infty$, is a *horosphere*, with all principal curvatures equal to 1. The horosphere is also the limiting shape for spheres of radius tending to ∞, as well as the limiting shape for a surface at constant distance r from a plane. The reason these limiting shapes are all the same is that in $\mathbb{H}^3$, distant lines and distant planes all look like tiny dots, so the surfaces of constant radius about them are almost identical.

The intrinsic geometry of a horosphere is Euclidean, just like the intrinsic local geometry of a cylinder. In the upper half-space rendition of $\mathbb{H}^3$, most horospheres look like spheres tangent to $\mathbb{C}$, but there is a special case of horospheres tangent to ∞ that look like horizontal planes. In terms of hyperbolic geometry, these horizontal Euclidean planes are bent upward (that is, geodesics tangent to one of these horospheres tend downward from it).

No matter how far you are away from a horosphere in $\mathbb{H}^3$, most of it is hidden: the portion that is visible is a disc on the horosphere whose limiting radius as you go far away is 1, so the amount of area you can see from the 'outside' of a horosphere is always less than π. A similar effect is seen in the hyperbolic views of cylinders, as evident from the cut-away views in figure 9.

In $\mathbb{E}^3$, just to create one singular cone axis while maintaining locally Euclidean geometry requires a globally distributed distortion of space. In $\mathbb{H}^3$, cone axes can be arranged so that the distortion they force damps out exponentially with increasing distance from the axis. To construct a cone angle α along an axis, start with the map that adjusts angles and nothing

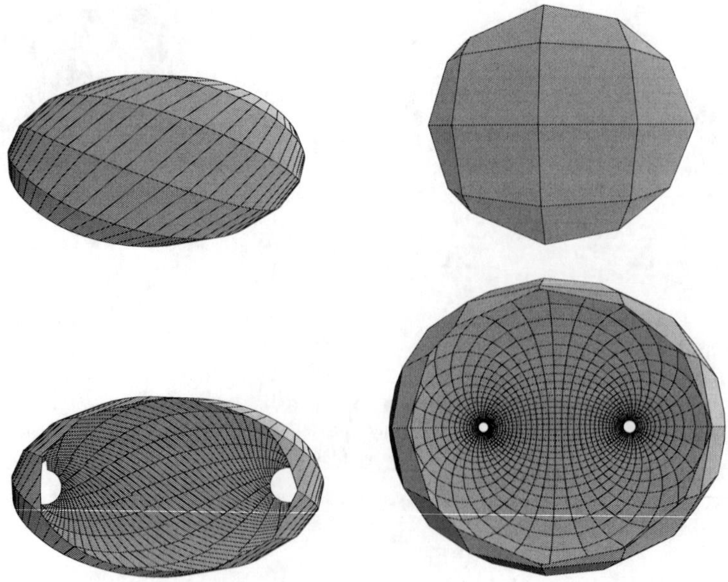

Figure 9. Two hyperbolic cylinders seen from within hyperbolic space. Each cylinder has been sliced into two pieces arranged one above the other so you can see inside. Left: a cylinder of radius 0.65, tiled with a skew grid of congruent parallelograms. Visualize the hyperbolic line that runs through the central axis of this banana. This line, like all hyperbolic geodesics, recedes from you quite quickly at its two ends. The turning away is forced by the explosive growth of the area of spheres of increasing radii about you. Staring at these pictures can help you sense the huge amount of 'stuff' at increasing distances in hyperbolic space. Right: a cylinder of radius 1.44, tiled with rectangles. In contrast to the analogous situation for Euclidean space, wider hyperbolic cylinders do not have the same visual shape as smaller cylinders, but appear rounder; this effect is exponentially rapid as a function of radius. When you are reasonably far away (as in these pictures) the diameter of a hyperbolic cylinder can be estimated as the log of ratio of the visual scaling factors from the front to the back, measured at the centre of the figure. Compare with figure 6, an analogous figure in $\mathbb{E}^3$.

else, given in cylindrical coordinates by the formula

$$(\theta, t, r) \longmapsto ((2\pi/\alpha)\theta, t, r).$$

Now adjust the value of r to get a new map that preserves the area element $\cosh(r)\sinh(r)\,\mathrm{d}\theta\,\mathrm{d}t$ on cylinders. As r increases, this converges to a motion of each cylinder by a constant distance. Since the ratio of scales in the two directions is very nearly 1, this is very nearly an isometry. Transporting the new metric by the map creates a cone axis by a distortion that decays exponentially away from the axis.

5. Embarking from T^3

The 3-torus T^3 has the simplest intrinsic description of any 3-manifold. It can be constructed by identifying each face of a cube directly to the opposite face. When you are in this geometric manifold, what you see is an array of multiple images, repeating in a regimented pattern in every direction.

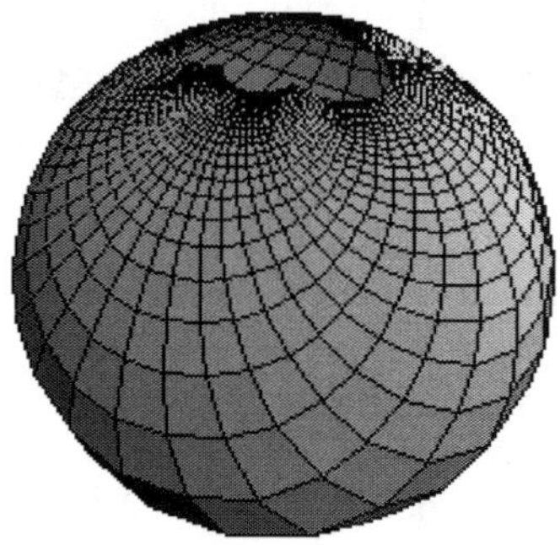

Figure 10. This is a horosphere in hyperbolic space, shown upside down (that is, in the lower half-space model). The tiles are congruent squares in the intrinsic geometry of the horosphere. Horospheres have the limiting shape of a sphere of large radius, and also the limiting shape of a cylinder of large radius.

This appearance is very special, and is not indicative of the qualitative appearance of a typical 3-manifold. Another way to understand the regimented quality of the 3-torus is to look at what happens when you move along a straight line within the 3-torus. It is unlikely that your journey is periodic, returning to exactly how it started, but your journey is almost periodic. With reasonably accurate initial data, you can predict where you will be for a long time to come. You can think of this like a ripple spreading in a pond. The set of possible places you could be after going a distance r has the local geometry of a sphere of radius r in $\mathbb{E}^3$, although in the torus wraps around and through itself. The sphere of radius r has area $4\pi r^2$: its area grows only quadratically as a function of r. Doubling r only quadruples the number of trajectories that separate within that time by at least some given small amount ϵ.

We will now watch as the metric of the torus deforms. Imagine yourself inside a cube (perhaps 20 m on a side) whose faces are identified to form a torus. A vertical line bisects the face of the cube in front of you; in the torus, it closes up to form a closed geodesic loop A, which also appears on the face of the cube behind you. A horizontal line bisects the faces to your left and right; in the torus, they are identified to make a closed geodesic circle. The floor and ceiling of the cube are bisected by a line running left to right, which identify to a closed geodesic loop C.

Now the metric of space starts to distort near loops A, B and C creating singularities along them with cone angles slowly decreasing from 2π. Geodesics in the new metric appear bent in the original Euclidean metric, bending toward the axes when they come near so that they cross just behind. What you see is a discontinuity of visual images along axis, with a double image for a narrow band (as in figures 1 and 3).

When we watched this happen earlier, we were in manifolds where the cone axes had single images. That is no longer the case: in the 3-torus, multiple images of axes A, B and C appeared, like three families of parallel lines in $\mathbb{E}^3$, translated in a pattern to interlace rather than intersect. As the metric deforms, the optical effect of image doubling occurs near *every* image of the axis. The effects are cumulative. We can understand the situation by thinking about how ripples (wavefronts) spread in the new metric, as a function of time t. Whenever the spreading ripple hits a singular axis, a thin wedge the space behind the axis is traversed doubly, because geodesics can arrive there after going to either side of the axis. This effect is compounded recursively. Every small area of the spreading ripple encounters singular axes and has portions that are doubled, at a small but regular rate. The compounding of growth of the area of the spreading ripple means that in the new metric, the area grows as an exponential function of time. In other words, the geodesic flow for the new metric now has positive topological entropy.

Another way to understand what is happening is to think about what happens if you trace a geodesic in the new metric but visualize it in the original flat metric that looks $\mathbb{E}^3$. Every time the trajectory comes near an image of one of the singular axes, it is deflects a little, in a way that a small change in the distance from an axis changes the angle, which causes a large change in position after some time. This means that successive deflections are poorly correlated with each other. After a large number of close encounters with axes, the trajectory is completely unpredictable: on a large scale in the image $\mathbb{E}^3$ image, the trajectory looks like a random walk, or a trajectory of Brownian motion.

So far this discussion of exponential growth and unpredictability applies equally well to many different geometries for the torus with singular cone axes as described. However, the geometry of our 3-torus happens to be following a very special path that is far superior to any other. As we look off in the distance, images of round objects appear round, except where they are sliced by images of the singular axes. This feature of our geometry is very special and rare; a generic, variably curved geometry would show local astigmatic effects, oriented this way and that, and growing more pronounced the further we look. The shapes of distant objects would be unrecognizable.

Our torus accomplishes this by using hyperbolic metrics, of constant negative curvature. When the cone angles at the three axes decrease, they are 'trying' to create surrounding clouds of negative curvature in their transverse directions. This negative curvature has been homogenized and spread evenly around everywhere! Metrics of this quality can be constructed from polyhedra in hyperbolic space that are combinatorially described as cubes with six faces each divided into half, making 12 faces in all. Each of these seemingly rectangular faces is actually a pentagon, when you take into account the extra vertex in one of its sides. The pattern of subdivision is the same as for a regular dodecahedron with regular pentagonal faces. It is not hard to construct a hyperbolic dodecahedron where the six edges that identify to the three loops A, B and C take a value $0 < \alpha/2 < \pi$, and all other angles are $\pi/2$. The shapes of these hyperbolic polyhedra are determined by their angles; when the faces are glued together, it gives a hyperbolic cone-metric for the torus. When a is very small, the polyhedron is very small, and its shape is nearly cubical. When rescaled to constant diameter (longest edge length 20 m), these tori converge to our cubical torus. Figure 11 show some examples of dodecahedra of this form.

The cone angles along axes A, B and C keep decreasing, until they eventually attain the value π. At this moment, the metric can be constructed from a regular, right-angled hyperbolic dodecahedron with all angles equal to $\pi/2$. Suddenly, all the double images align with each other so that every image is perfect. What we see is a pattern repeating in hyperbolic space. Our right-angled dodecahedron is a fundamental domain, or basic tile; we can fill up $\mathbb{H}^3$ with non-overlapping dodecahedra, matching up on their faces. We see one image of ourself in each of these dodecahedral cells. The sizes of the images decreases exponentially fast with distance, since there are exponentially many at a given distance. Since the long-term behaviour of geodesics is random, distant images face every which way, no matter in what direction we look.

The cone angles start to decrease again. As they decrease, the three axes A, B and C also decrease in length. When the cone angles reach $\pi/3$, all images are in alignment; then they go out again. They come back in alignment for an infinite sequence of values π/n, the

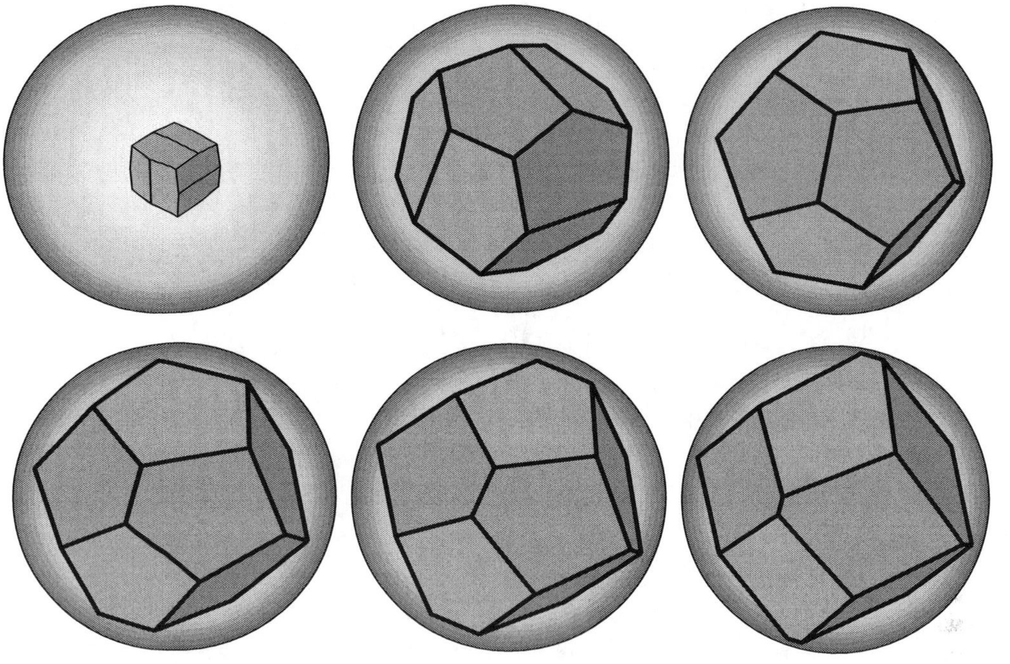

Figure 11. Six hyperbolic dodecahedra with varying angles, seen in the projective model of $\mathbb{H}^3$. Each dodecahedron has a set of six edges which decrease, in successive frames, through values $\pi - \epsilon$, $2\pi/3$, $\pi/4$, $\pi/5$, $\pi/6$ and $\pi/12$, while the other 24 edges remain at $\pi/2$. When opposite pairs of like-coloured faces are glued, one obtains T^3 with evolving cone angles at three edges. Notice that for the last five dodecahedra an observer in $\mathbb{H}^3$ would never be able to see more than three faces at a time without going inside the dodecahedron. A hyperbolic perspective drawing facing the centre of the ball is the same as the Euclidean perspective drawing from the same point, but for the correct Euclidean interpretation of the drawing you need to stand close to it, while for the hyperbolic interpretation you stand far away. The first figure is just a sketch, while the others are traced from output of SnapPea's Dirichlet domain module.

axes growing shorter and shorter every time, and receding off into the distance from where we are standing. There is a limit, when the cone angles go to 0. The limiting manifold is homeomorphic to $T^3 \setminus A \cup B \cup C$. The limiting metric is *complete*, meaning that the holes where A, B and C have vanished are infinitely deep, and it is not possible to reach the edge of the universe in a finite distance. These holes are surrounded by a family of concentric tori, bent into the shape of horospheres. The intrinsic metric of each torus is flat. If you proceed inward down the hole, the concentric tori all have the same shape, but they shrink exponentially fast. It follows that the total volume for this metric is finite. If the metric is scaled so that its curvature is -1 (that is, the hyperbolic radian is used as a unit of distance), the metric is uniquely determined by the topology; in particular, its volume is a topological invariant. Our current manifold has volume $7.327\,724\,753\ldots$.

The program SnapPea, when given the topological description of a manifold such as this, can almost instantaneously compute its complete hyperbolic metric (which is unique), computes numerical invariants such as its volume and displays various pictures. The most convenient way to describe the topology is usually by specifying a Dehn surgery to produce it from a link in S^3. In this case, $T^3 \setminus A \cup B \cup C$ is homeomorphic with $S^3 \setminus B$, where B

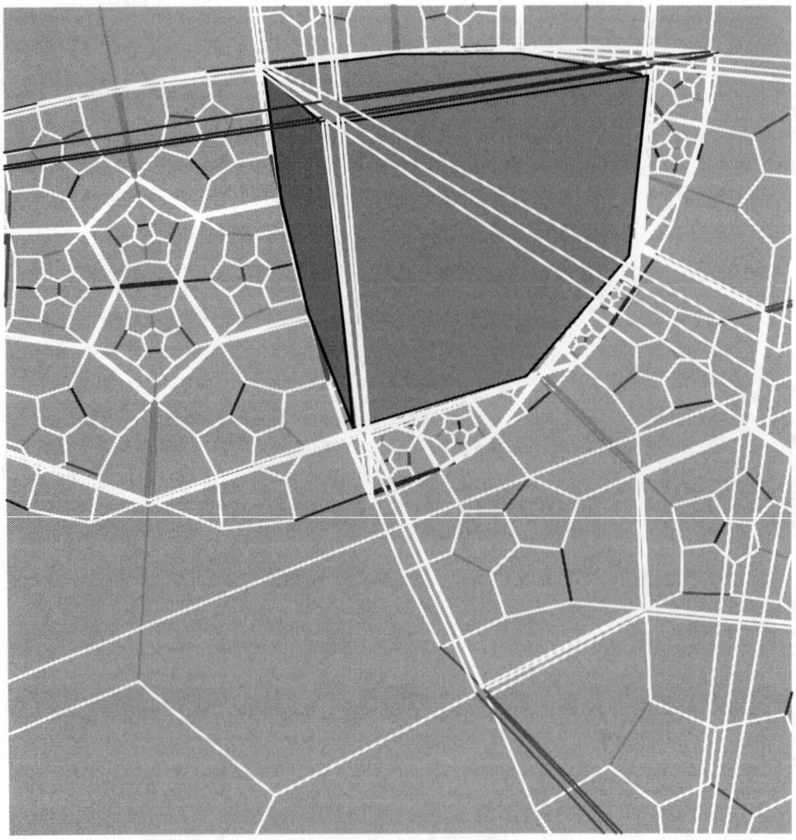

Figure 12. Hyperbolic space tiled by right-angled dodecahedra. This is a true perspective drawing inside T^3 when it has a hyperbolic metric with three singular axes of cone angle π. Compare with figure 11, frame 3. The fundamental domain (the dodecahedron) has many images, which appear in distorted perspective to Euclidean-adjusted eyes. The prominent circle is the horizon circle for one of the planar sheets of faces. Can you pick out any 'bananas', or rough hyperbolic cylinders, made from strings of dodecahedra? This figure is a postscript snapshot from geomview, a program available for common unix platforms from http://www.geom.umn.edu. Click on the Not Knot Flythrough module (written by Charlie Gunn) and you will be conducted on a flying tour through this scene (minus the solid dodecahedron). From the File menu open data/geom/dodec; with patience you can translate and rotate dodec in one of the bays. Fancier versions of scenes like this occur in the video NotKnot.

is a famous link of three components called the Borromean rings. These three interlocked circles have the property that when any one is removed, the other two are unlinked. T^3 is obtained from S^3 by doing (0, 1) surgery on the Borromean rings.

Perhaps the most informative picture in SnapPea is the cusp view. A *cusp* is the jargon for a deep 'hole' in a hyperbolic manifold that is enclosed by a family of concentric horospherical-shaped tori. SnapPea has sliders to choose one horospherical torus concentric to each cusp, identifying which torus by the volume of the portion of the cusp it encloses. For each cusp, a display can be produced of the view from deep within that cusp of the images of all other horospherical tori, except for the one you are inside (otherwise it would block the view of everything). The cusp views can have quite distinctive qualitative features. By studying and comparing them, you can 'triangulate', and see how the geometry of the entire manifold fits together.

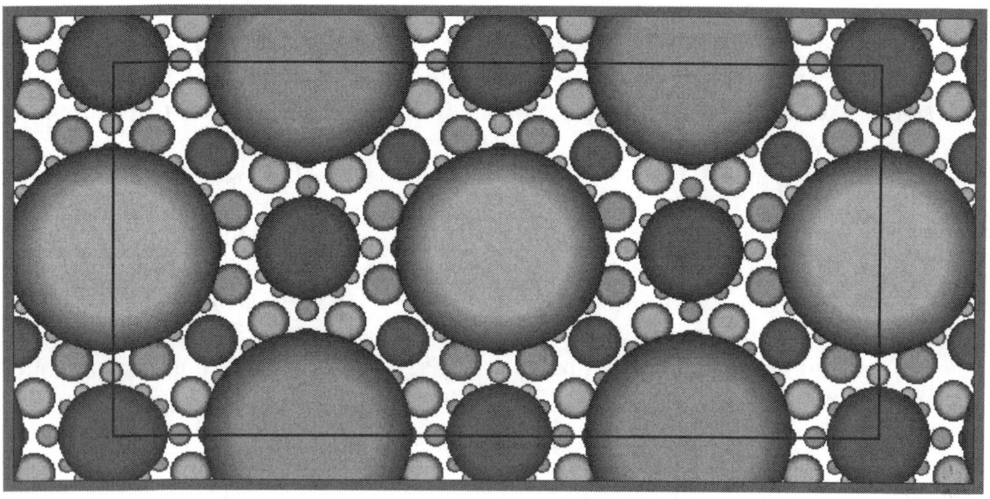

Figure 13. Axes A, B and C in T^3 have gone off to infinity as their cone angles went to 0. This is the view from deep inside cusp A, looking out, showing multiple images of three horospherical tori that enclose axes A, B and C. We are inside the region enclosed by the A torus, whose fundamental domain is indicated by the rectangle. Luckily, it is made of one-way glass so we see out to the rest of space. This view is similar to the panorama wrapping around the vertical pillar A in the front of the cube from which the original Euclidean geometry of T^3 was formed. Large images of horizontal cylinders B on the left and right of the room appear at mid-height. Large images of cylinders C on the floor and ceiling appear at the top and bottom of the rectangle. In between these large images we can make out images of our own cusp A, at the front and back of the room. The view from each cusp is the same. Imagine how it looks to an observer stationed in another cusp, and match the visual geometry as she would see it to the geometry you see. The sky above each horosphere has a closest layer of horospheres, arranged in a diamond pattern just like what you see.

Continuous hyperbolic surgery works in quite a nice way. Hyperbolic manifolds with singular axes are controlled by holomorphic parameters, one complex parameter for each cusp that enables its singular behaviour to change in two real dimensions, in the cone angle direction and in the shear direction. SnapPea implements continuous surgery in a way that enables one to navigate quite readily among 3-manifolds, and to identify where you are. For a hyperbolic manifold, a view such as in figure 13 is determined by the geometry, plus the user-controlled parameters of how big the cusp neighbourhoods are. The topology determines the geometry, by the rigidity theorem of Mostow and Prasad. Therefore, it is mathematically a straightforward computation to decide whether or not two hyperbolic manifolds are homeomorphic; SnapPea's Isometry module answers a particular instance of this question in short order. A human looking at the cusp-view pictures can also usually tell at a glance whether or not two 3-manifolds are the same or different, although there are some tricky special cases when manifolds are hard to compare accurately by eye, involving 3-manifolds are built out of identical polyhedra that are glued together in different ways.

To make sense of the view inside a hyperbolic 3-manifold, it helps to keep the imagery of hyperbolic cylinders (bananas) in the front of the mind.

Starting with a complete hyperbolic manifold of finite volume, all but a bounded (compact) set of surgery coefficients yield a singular hyperbolic manifold, and in particular, all but a finite number of Dehn fillings for each cusp give hyperbolic manifolds. Often there are no exceptions; at most there are a handful. The volume of a manifold obtained by Dehn filling is always less than the volume of the original. The set of all hyperbolic 3-manifolds with volume less than a constant can all be obtained by Dehn filling from a finite set of manifolds with cusps.

The formula for the volume of a hyperbolic polyhedron involves the dilogarithm function, which is a non-elementary but readily computed function. Much simpler than the volume is the formula for the derivative of volume, as the polyhedron varies while maintaining a combinatorial type. The formula, discovered by Schlafly, says that the derivative of volume of a polyhedron is the sum over all its edges of the edge length times the derivative of its exterior angle, i.e. as the edges get sharper, the volume gets bigger. The same formula applies when polyhedra are glued together to form a hyperbolic cone-manifold: the derivative of its hyperbolic volume is the sum, over all cone axes, of the length of the derivative of the curvature concentrated at the axis (curvature is 2π minus its cone angle). It is as if there is a secret tunnel to another world along each cone axis; volume flows in or out of a segment of a cone axis in direct proportion to the change in its cone angle.

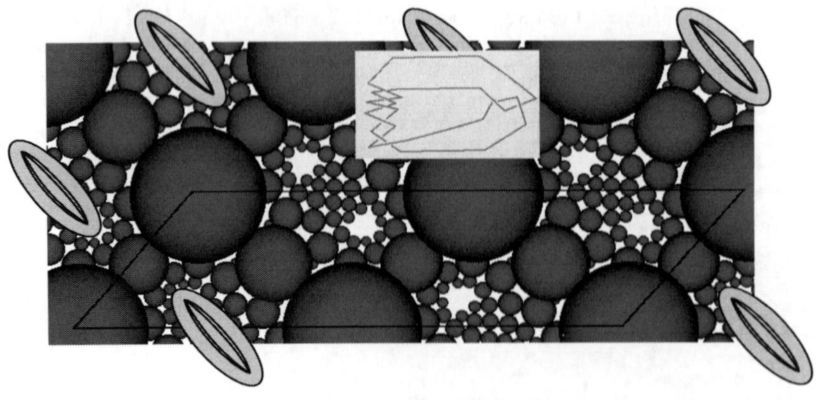

Figure 14. The cusp view of a knot (inset) from SnapPea, with bananas. The cut-away form of hyperbolic cylinders is plainly visible in the hollows where bananas were omitted: cf figure 9. Each banana has many images of horospheres spiralling around it. All horospheres wrap up to a single torus in the manifold, so the picture strongly suggests that the axis of the bananas wrap into a fairly short geodesic. The knot diagram suggests a short geodesic looping around the twisty area. spiralling fairly tightly (see figure 15).

The knot in the inset of figure 14 has two strands that twist three times around each other. The twisting is reflected by the geometry of the complete hyperbolic structure for its complement. If you have memorized the appearance of a cylinder in $\mathbb{H}^3$, you can plainly make out the shape of a fairly fat cylinder in the cusp view of the knot. This strongly suggests that there is a short geodesic. In fact, a loop encircling the two twisted strands is homotopic to a closed geodesic. This can be calculated in SnapPea by drawing the

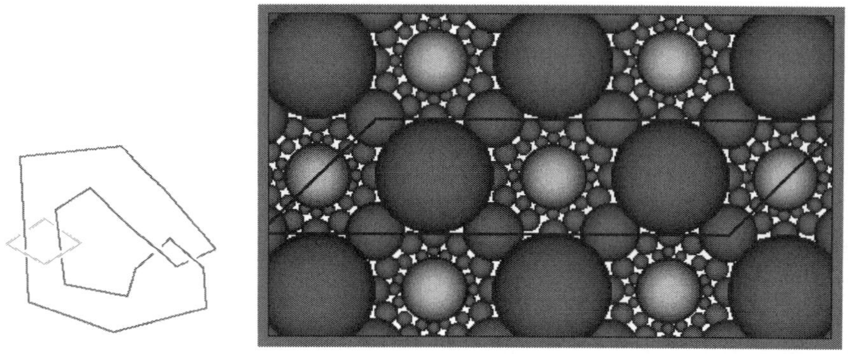

Figure 15. Left: the Whitehead link consists of two circles, which separately are unknotted. The link in figure 14 is obtained by $(1, 3)$ Dehn filling of the shorter loop. This is the same as slicing along a horizontal disc spanning this loop, twisting three times and gluing back. Right: the cusp view of the Whitehead link, as seen from the cusp for the 'long' loop. The bananas have turned into horospheres. SnapPea's sliders for cusp volumes have been adjusted to maximize the 'big' cusp and keep the other one small, for better comparison with figure 14. (It is not obvious from these pictures, but the two strands are actually symmetric with each other.)

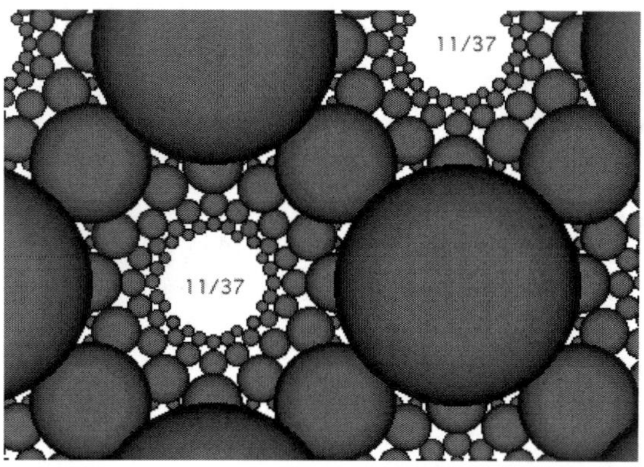

Figure 16. This is the cusp view of $(11, 37)$ surgery on the small component of the Whitehead link. There is a closed geodesic that is quite short (length is $0.000\,989\,647\ldots$) and far away, so that the cylindrical neighbourhood is not easy to distinguish from a horosphere. This view bears a striking resemblance to comparable cusp views for the complement of the Borromean rings. A picture in figure 13 would be obtained by adjusting the objects that are drawn: shrinking the single horosphere in the present picture, and drawing a hyperbolic cylinder of appropriate diameter (a very bloated banana) around the short geodesic that is the core of the $(11, 37)$ filling. The nearly congruent geometry reflects the fact that the complement of the Borromean rings is a twofold cover of the complement of the Whitehead link.

loop in question, forming the complement of the resulting link, filling the new component using $(1, 0)$ surgery (which means filling that component back the way it was). There is a core geodesic view in SnapPea which now will tell you the length of the filled-in core: it indeed reports a short hyperbolic length $0.1535\ldots$ with a twist of $2.2071\ldots$ rad. In comparison, SnapPea's length spectrum module reports a second shortest length of

1.138 084 551 958 0791 for a closed geodesic—still a reasonably short hyperbolic length.

A famous theorem of Gordon and Luecke says that knots are equivalent if and only if there is a homomorphism between their complements. However, the situation is quite different for links with more than one component. Any time there is an unknotted component to a link, you can find a disc that has that unknot as its boundary, intersecting other strands in some collection of points. If you slice along the disc, twist one side of the cut n full times, then glue the two cuts back together, you obtain a new link, usually different from the old one since you have added some extra twists. But the complement of the new link is homeomorphic to the old one, since apart from discontinuous behaviour at the boundary, the new gluing map is the same as the old. Thus, when a new unknotted component is added to our link, we can untwist it to make a simpler picture, known as the Whitehead link (figure 15). The $(11, 37)$ Dehn surgery on the second component of the Whitehead link is shown in figure 16. The length of the core geodesic in that case is quite short, only 0.000 989 647, and the geometry is almost indistinguishable from that of the Whitehead link except near the central part of the core.

6. The ocean of hyperbolic manifolds

The complement of the Borromean rings is a twofold cover of the complement of the Whitehead link. This kind of relationship is not an isolated phenomenon. Further exploration of 3-manifolds via continuous surgery reveals quite a lot of convergence, if one aims toward bright beacons. it is as if there are just a few major shipping channels marked by powerful beacons, but with large numbers of interesting destinations off to the sides.

In SnapPea there is a module for *drilling* out geodesics loops that are identified geometrically. Given a hyperbolic manifold, the program will present a list of all loops up to a specified combinatorial complexity (as determined by an internal representation of the manifold), sorted in order of hyperbolic length. SnapPea can remove any of these geodesics from the manifold, in essence deforming its cone angle to 0 so as to obtain a complete hyperbolic structure for its complement.

Hyperbolic volume increases, by an amount that is typically in rough proportion to the length of the geodesic you drill.

Here is one phenomenon that seems to always occur. This is a striking effect that others (among them Jeff Weeks) have also noticed, but I do not know if it has found its way into print:

Conjecture 6.1. When the shortest simple closed geodesics are repeatedly removed from any complete hyperbolic 3-manifold of finite volume, eventually one obtains a manifold whose shortest closed geodesic has length $2.122\,55\ldots$ with angle of twist $\pm 1.809\,11\ldots$, and has self-replicating behaviour when removed.

This length is one that occurs in the complement of the Borromean rings; it is the length of a loop that links around two of the rings where they cross, and comes from an element of $SL_2(\mathbb{C})$ whose trace is $2 + 2i$. This length also occurs in the complement of the Whitehead link. If you cut the Whitehead link open along a twice-punctured horizontal disc with boundary the 'small' loop in figure 15, you obtain a fragment of a hyperbolic manifold bounded by two twice-punctured discs. This particular fragment has identical geometry in any manifold where it occurs. This module is a *clasp*, and is fairly common in 3-manifold topology. If a short curve encircling the two arcs of the clasp is removed, the result has the

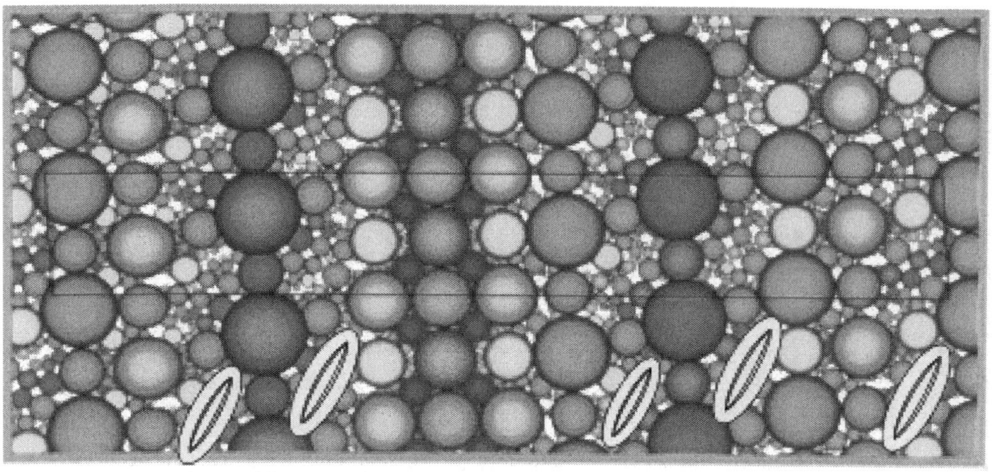

Figure 17. One of the cusp views of a nine-cusped hyperbolic manifold of volume 44.0227. The horosphere that walls off this cusp encloses a volume of 7.5; the next largest cusp volume is 4.2. Features that look like hyperbolic cylinders are scattered around the image, as indicated by the bananas sketched in at the bottom. Notice how the bananas together with the horoballs seem to fit in a semi-regular pattern, as for the cusp view for the Borromean rings but with dislocations where the grid turns 45°. Patterns like this are common.

same topology as two clasps strung together, joined on twice-punctured discs. Like Hydra, it only multiplies when it is zapped.

There are other patterns that occur under drilling. The Borromean rings is an instance of a phenomenon called a *universal link*: every possible 3-manifold can be obtained by Dehn filling of some covering space of finite degree of the Borromean rings. To a surprising degree, one can learn about covering spaces and certain relatives of covering spaces by drilling geodesics based only on the patterns of their lengths. For the simplest manifolds repeated drilling of the shortest geodesic soon arrives at the complement of an n-link chain, in a traditional circular chain arrangement. The topology of such a chain depends on the number of links and the amount of twist. The least-twisted chain of five links is particularly important for the simpler hyperbolic manifolds. Its complement has volume 10.149.... Its cusp view looks like figure 18, except that its volume is $\frac{10}{28}$ the size and the colouring repeats on a shorter scale. This manifold has an amazing symmetry group of order 120 that performs all possible permutations of the cusps. Very many of the simpler hyperbolic manifolds arrive at this topology/geometry after recursively drilling out shortest geodesics until there are five cusps (see the discussion of the census of the simplest 3-manifolds whose data are available within SnapPea [HW89]). When one more shortest geodesic is drilled out, the manifold recrystallizes into a 6-cups manifold with the same cusp pattern as the Borromean rings.

More regularities occur for larger examples. Repeatedly drilling out of relatively short geodesics (but *not* always the very shortest, since the shortest geodesics eventually start to be confined to certain regions) always seems to start 'crystallizing' a 3-manifold, so that its cusp neighbourhoods can be adjusted in size to pack together in regular crystalline patterns, most commonly in a square-looking pattern as for the Borromean rings. There are several other patterns that occur frequently, but less frequently, notably the equilateral pattern of

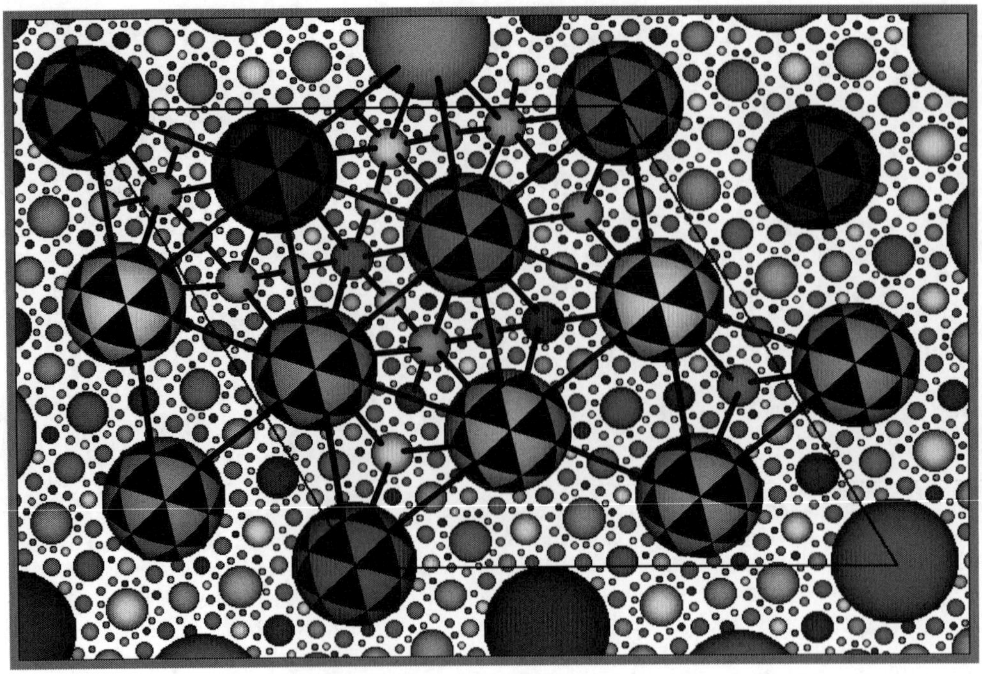

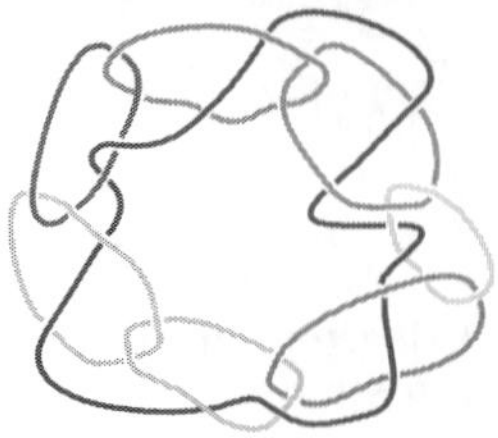

Figure 18. Eight linked full tori (right) when inflated fill S^3 so that each meets each other along a single (curved) hexagon. The complement of the link is a hyperbolic manifold (above) of volume 28.418..., having extraordinary 336-fold symmetry. The seven nearest cusps describe a seven-colour map of the torus surrounding our own cusp. As seen from a neighbouring cusp, we form part of a triangular grid along with a hexagon of common neighbours, as indicated by the painted triangles. Beams join pairs of nearest neighbours. The closest images of ourselves are in the fourth tier, quite distant for a manifold of this volume.

figure 18. There are often dislocations, where two different patterns meet or where the crystalline grids are at different angles.

The patterns, constructed and viewed with SnapPea, are striking. The phenomena are similar even for large manifolds of volume 100 or 200. It is often hard to formally capture and describe patterns that we see plainly with our eyes, but these observations suggest that there might be a comprehensible systematic organization for hyperbolic crystallography. Possibly every compact hyperbolic 3-manifold can be thought of as a 'crystallized' arrangement of solid tori surrounding relatively short geodesics, fitting with each other in a small repertoire of local arrangements similar to figures in this paper. Any universal pattern of this nature would need to match certain manifolds in more than one way, but with luck the ambiguity would be manageable.

Acknowledgment

Partially supported by NSF grant DMS-9704135.

References

[Geo81] The Geometry Center (University of Minnesota) 1981 *Not Knot* (Boston: Jones and Bartlett) video tape (16 min)

[HW89] Hildebrand M and Weeks J 1989 A computer generated census of cusped hyperbolic 3-manifolds *Computers and Mathematics (Cambridge, MA)* (New York: Springer) pp 53–9

[Thu79] Thurston W P 1979 *Geometry and Topology of Three-manifolds (Princeton Lecture Notes)* http://www.msri.org/publications/books/gt3m

[Thu82] Thurston W P 1982 Three-dimensional manifolds, Kleinian groups and hyperbolic geometry *Bull. Am. Math. Soc.* 6

[Thu86] Thurston W P 1986 Hyperbolic structures on 3-manifolds, I: Deformation of acylindrical manifolds *Ann. Math.* **124** 203–46

[Thu97] Thurston W P 1997 *Three-dimensional Geometry and Topology* vol 1 *(Princeton Mathematical Series 35)* ed S Levy (Princeton, NJ: Princeton University Press)

[Thu98a] Thurston W P 1998 Hyperbolic structures on 3-manifolds, III: deformations of 3-manifolds with incompressible boundary *Preprint*

[Thu98b] Thurston W P 1998 Hyperbolic structures on 3-manifolds, II: surface groups and 3-manifolds which fiber over the circle *Preprint* (orig. *Preprint* 1981, revised 1986)

The Eightfold Way
MSRI Publications
Volume **35**, 1998

The Eightfold Way: A Mathematical Sculpture by Helaman Ferguson

WILLIAM P. THURSTON

This introduction to *The Eightfold Way* and the Klein quartic was written for the sculpture's inauguration. On that occasion it was distributed, together with the illustration on Plate 2, to a public that included not only mathematicians but many friends of MSRI and other people with an interest in mathematics. Thurston was the Director of MSRI from 1992 to 1997.

Mathematics is full of amazing beauty, yet the beauty of mathematics is far removed from most people's everyday experience. The Mathematical Sciences Research Institute is committed to the search for ways to convey the beauty and spirit of mathematics beyond the circles of professional mathematicians.

As a step in this effort, MSRI (pronounced "Emissary") has installed a first mathematical sculpture, *The Eightfold Way*, by Helaman Ferguson. The sculpture represents a beautiful mathematical construction that has been studied by mathematicians for more than a century, from many points of view: geometry, symmetry, group theory, algebraic geometry, topology, number theory, complex analysis. The surface depicted by the sculpture was discovered, along with many of its amazing properties, by the German mathematician Felix Klein in 1879, and is often referred to as the Klein quartic or the Klein curve in his honor.

The abstract surface is impossible to render exactly in three-dimensional space, so the sculpture should be thought of as a kind of topological sketch. Ridges and valleys carved into the white marble surface divide it into 24 regions. Each region has 7 sides, and represents the ideal of a regular heptagon (7-gon). The 24 heptagons fit together in triples at 56 vertices. It is the pattern of the division of the surface into heptagons that carries the essence of the mathematics. The Klein quartic thus is an extension of the concept of a regular polyhedron, of which the dodecahedron, the cube and the tetrahedron are examples:

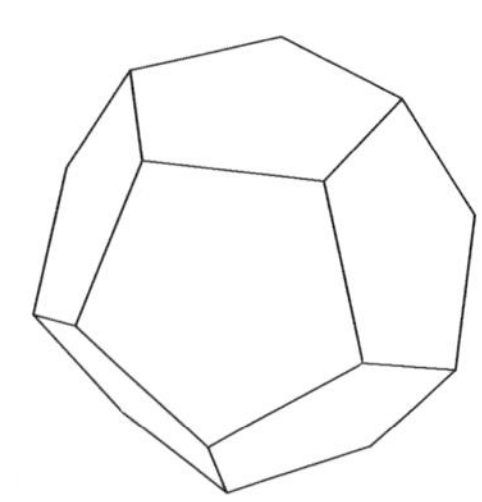

Dodecahedron

1

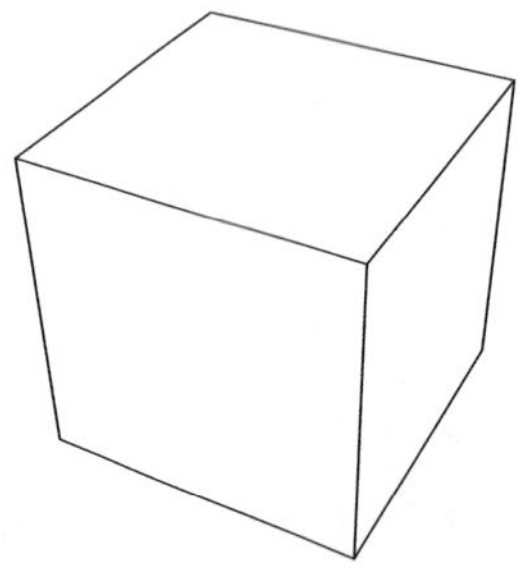

Cube

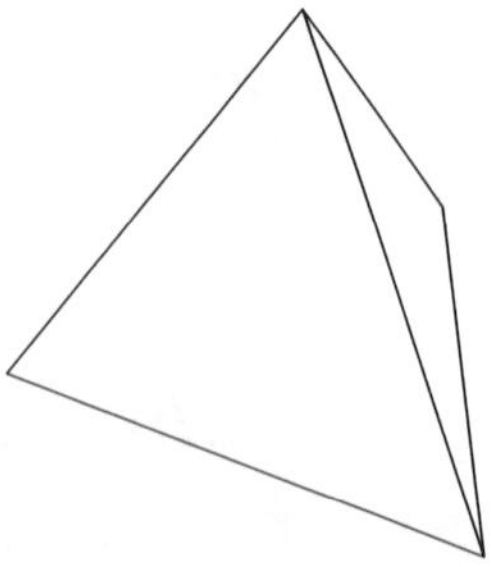

Tetrahedron

Even though the heptagons on the physical surface are not regular, the *pattern* of heptagons on the surface is completely symmetric — in fact, the pattern is just as symmetric as the pattern of pentagons on a dodecahedron. One way to get a sense of the symmetry is to place a finger on any edge. Trace out along the edge to the next intersection, and turn left. Now proceed to the next intersection and turn right. Continue in this way, making a total of 8 turns, LRLRLRLR. If you do this carefully, with concentration and contortion, you arrive back where you started. It doesn't matter where you start or in which direction you go: in 8 alternating turns, you always arrive back at the beginning. (Question: what happens when you do this on a tetrahedron, cube, or dodecahedron?)

In the pattern of heptagons on the surface, and of the 24 heptagons is equivalent to any other heptagon. Furthermore, if any heptagon is rotated by $\frac{1}{7}$th of a revolution, it still fits into the pattern in an identical way. This makes $24 \times 7 = 168$ ways that the pattern of heptagons on the surface can be mapped to itself. Mathematically, the pattern has order 168. When a heptagon is reflected along any of its altitudes, it still fits into the pattern in an identical way, making a total of 336-fold symmetry when the mirror-image transformations are allowed.

The circular base area of the sculpture is also tiled by heptagonal tiles, in a regular geometric pattern that resembles a honeycomb. The sides of the heptagons are arcs of circles; when these arcs are continued, they meet the boundary at a 90° angle. This circular base area is a map of the hyperbolic or non-Euclidean plane. In hyperbolic geometry, it is possible to construct regular heptagons whose angles are exactly 120°; these heptagons fit together to tile the hyperbolic plane. The physical map of the hyperbolic plane is distorted, but in hyperbolic geometry itself all the heptagons have an identical size and shape.

The heptagonal tiling of the base and the heptagonal tiling of the surface are closely related. The 7-sided column that supports the sculpture starts off this relationship: it sweeps up from the central heptagon in the hyperbolic plane to one of the 24 heptagons on the surface. Imagine continuing this relationship. The 7 heptagons adjacent to the foot of the column sweep up and stretch to cover the 7 heptagons that border the top of the column, and so forth. The base area stretches out and wraps around the surface to completely encompass it; it continues stretching and wrapping around and around, infinitely often.

Plate 2 following page 150 shows the heptagonal hyperbolic honeycomb with a pattern superimposed to indicate what happens when it wraps around the surface. The infinite hyperbolic honeycomb is divided into 3 kinds of groups of 8 cells each, where each group is composed of a heptagon together with its 7 neighbors. There are red rings surrounding one person, green groups surrounding another person, and white groups with letters.

When the honeycomb is wrapped around the surface, equivalent groups wrap up to the same place on the surface. In other words, the pattern superimposed on the surface would have only one green group, one red group, and one white group, making 24 heptagons in all. You can check this out by testing the LRLRLRLR rule on the hyperbolic honeycomb. For instance, if you start on an edge that points in toward the central white area and has a red group on its left and a green group on its right, and proceed LRLRLRLR, you will arrive at another edge with red on its right and green on its left. If the initial edge pointed toward and "a" (say), the final edge also points toward an "a".

It is interesting to watch what happens when you rotate the pattern by a $\frac{1}{7}$ revolution about the central tile: red groups go to red groups, green groups go to green groups and white groups go to white groups. The person in the center of a green group rotates by $\frac{2}{7}$ revolution, and the person in the center of a red group rotates by $\frac{4}{7}$ revolution. The interpretation on the surface is that the 24 cells are grouped into 8 affinity groups of 3 each. The symmetries of the surface always take affinity groups to affinity groups. This is analogous to the dodecahedron, whose twelve pentagonal faces are divided into 6 affinity groups of 2 each, consisting of pairs of opposite faces.

The name "Klein quartic" or "Klein curve" refers to an algebraic description of the ideal surface that the sculpture represents, determined by the equation $x^3y + y^3 + x = 0$. (This equation is called a quartic or 4th-degree equation because the highest term x^3y has 3 x's and 1 y, making degree 4 in all.) The solutions to this equation in the (x, y)-plane form the curve shown at the top of the next page. [A more symmetric view is presented in Figure 10 on page 326. –Ed.]

But when x and y are allowed to be complex numbers, there are many more solutions; in fact, the set of solutions forms a 2-dimensional surface in 4-dimensional space. The symmetry of the surface is reflected algebraically by the phenomenon that there are many possible substitutions that keep the equation the same. For instance, if you replace x by Y/X and y by $1/X$, the equation becomes $Y^3/X^4 + 1/X^3 + Y/X = 0$; if you multiply both sides by X^4 to clear denominators, you get the original equation. There are 168 essentially different algebraic "substitutions" that preserve the equation, one for each of the orientation-preserving symmetries of the surface. (Coordinates can be chosen so that the

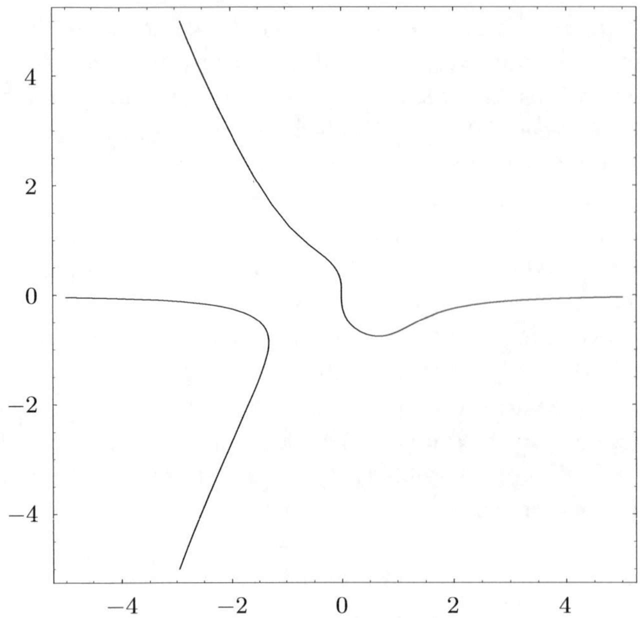

$x^3y + y^3 + x = 0$

center of the central tile of the hyperbolic honeycomb maps to the solution $(0, 0)$ to the equation, and a vertical line through that point maps to the curve graphed above. The substitution just described takes the integral sign to the person in a red ring, the person in a red ring to the person in a green ring, and the person in the green ring back to the integral sign.) Most of the substitutions are more complicated, involving complex algebraic numbers, and we won't describe them.

Topologically, the Klein curve is called a surface of genus 3 or a 3-holed torus. Why 3? Here is a standard picture of a 3-holed torus:

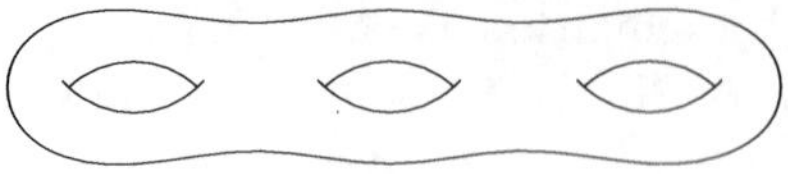

In topology, two figures are equivalent if one can be deformed into the other. So we can rearrange the holes (stretching but not tearing or gluing) however we like without changing the topology, for instance into

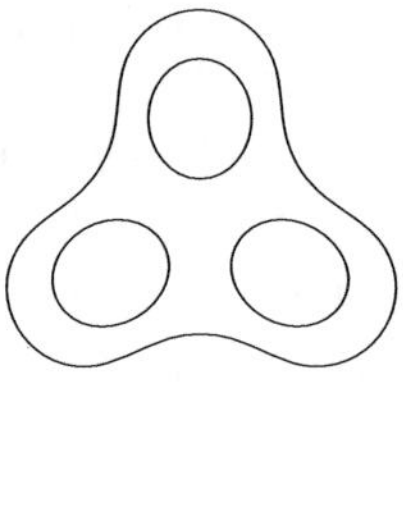

(to reveal a kind of 3-fold symmetry that was not evident before), and further into

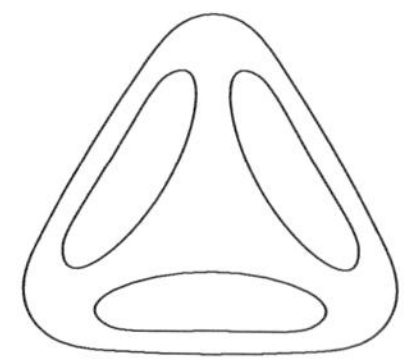

which looks like the frame of a tetrahedron as seen from above. This is the approximate form of the sculpture, and it displays the maximal amount of the symmetry of the ideal surface that can be made directly visible in space.

The heptagonal hyperbolic honeycomb has an interesting relationship to the Fibonacci sequence

$$1, 1, 2, 3, 5, 8, 13, 21, 34, 55, \ldots,$$

in which each number, starting with the third, is the sum of the preceding two. Imagine growing the hyperbolic honeycomb like a crystal, starting with the central white group of 8 heptagons as a seed. At each unit of time, adjoin a heptagon wherever there is a concave angle—that is, adjoin all the heptagons that touch at least two of the heptagons already present. At the first step, you will add 7 heptagons. Second, you will add the 14 green heptagons that fill out the next complete ring. On the third step, you add 21 heptagons, consisting of 14 red and 7 white heptagons (the centers of green groups). The sequence, if we include the ring of 7 white heptagons in the initial seed, goes

$$7, 7, 14, 21, 35, 56, 91, 147, 238, 385, 623, \ldots.$$

Each term is the sum of the preceding two: this sequence is 7 times the Fibonacci sequence!

The number of tiles grows very rapidly as you add additional layers. That is why the tiles around the edges must get quite small in the map of the hyperbolic plane: there are so many of them that otherwise they wouldn't fit. The base of the sculpture includes tiles corresponding to the first 7 terms of the sequence, making 231 tiles in all (or 232 if you include the spot where the column fits). The cover diagram shows the tiles for two additional terms plus a few scattered heptagons, making over 617.

Instructions for how to glue the 24 heptagons of the surface together can be constructed as follows. Label the 24 heptagons with different labels, say the letters `a` through `x`. Arrange these letters, together with a sharp sign (`#`), in a grid pattern in the plane that repeats very 7 units across, every 7 units up, and is symmetric about each `#`. All the information is contained in a 7×7 table. Notice that the 7×7 table is filled out by the `#` together with 2 occurrences of each of the 24 letters:

```
# a b c c b a # a b c c b a #
d e f g h i j d e f g h i j d
k l m n o p q k l m n o p q k
r s t u v w x r s t u v w x r
r x w v u t s r x w v u t s r
k q p o n m l k q p o n m l k
d j i h g f e d j i h g f e d
# a b c c b a # a b c c b a #
d e f g h i j d e f g h i j d
k l m n o p q k l m n o p q k
r s t u v w x r s t u v w x r
r x w v u t s r x w v u t s r
k q p o n m l k q p o n m l k
d j i h g f e d j i h g f e d
# a b c c b a # a b c c b a #
d e f g h i j d e f g h i j d
k l m n o p q k l m n o p q k
r s t u v w x r s t u v w x r
r x w v u t s r x w v u t s r
k q p o n m l k q p o n m l k
d j i h g f e d j i h g f e d
# a b c c b a # a b c c b a #
```

To determine what heptagons to glue to a given heptagon (call it `z`), find the letter of the heptagon in the table. It's always possible to construct a line segment that connects some `#` to `z` without going through any intermediate letters. Draw a line parallel to `#z` that is as close as possible to the right while still going through letters in the table. The letters along this line are the heptagons adjacent to `z`, in counterclockwise order. For example, the `a` heptagon is glued to the 7 heptagons in the second row of the table: `defghij`. The `e` heptagon is glued to `adltvnf`. The neighbors of `t` are slightly harder to determine: `t` is 2 units to the left and 3 units up from a `#`, so starting with an `l`, which is 2 units to the left and 2 units up, one can repeatedly go 2 over and 3 up, reading off `loirbve`.

Try labeling the blank heptagonal honeycomb on the next page, using this rule.

The spirit of mathematics and the essence of its beauty is remarkably fragile, because mathematics is about ideas and about thought. Mathematics takes place in the mind, and no two minds are the same. After many years of study and work, a mathematician may stumble on a vast and beautiful vista that unifies and simplifies many things that once appeared disparate and complicated. Mathematicians can share a beautiful mathematical vista with one another, but there is no camera that can easily capture an image of such a vista to convey it in full to people who have not trudged along many of the same trails.

We have only touched on a small part of the mathematical vista associated with this sculpture, but we hope that you can get form it some glimpse of the unity, the beauty, and the spirit of mathematics.

William P. Thurston
University of California Davis
Department of Mathematics
565 Kerr Hall
Davis, CA 95616
United States
wpt@math.ucdavis.edu

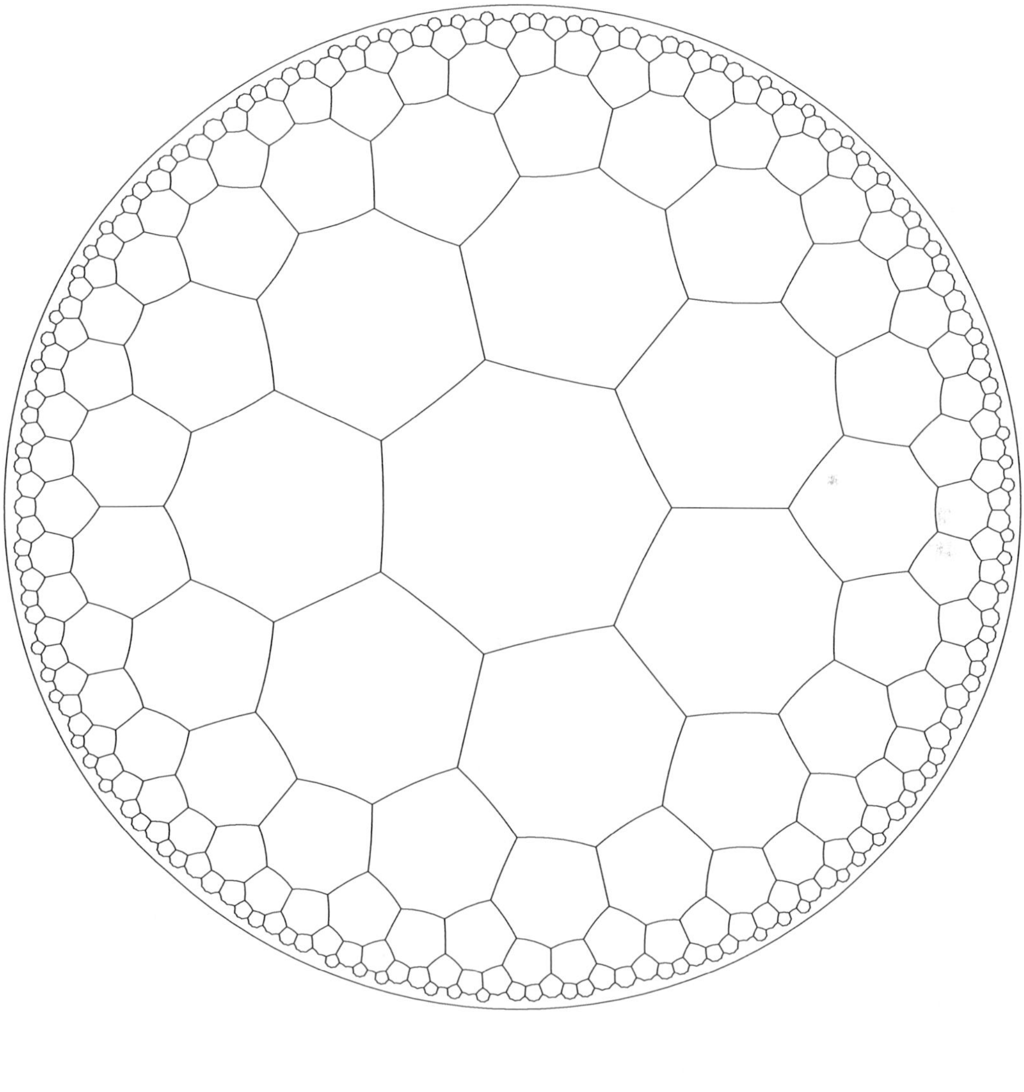

Miscellaneous

MISCELLANEOUS PAPERS

This Chapter contains various papers of Thurston that do not fit neatly into any of the other mathematical topics through which the chapters of these collected works have been delineated. They essentially speak for themselves, but we comment on one in particular: Thurston's 1967 New College undergraduate thesis [Th1] : "A Constructive Foundation for Topology". The topic of the thesis is intuitionistic topology.

From the hindsight of what Thurston became, and the way he changed mathematics, the thesis is notable in several ways. In it we see opinions and themes that would recur throughout Thurston's career. For example, in the following quote we see Thurston's willingness to move forward without having learned what came before; his willingness to start from scratch and to understand a subject in his own way. As he states in [Th1]:

> In this paper I will develop an alternate foundation for topology, based on a non-classical approach to mathematics: roughly, intuitionism, but since the primary sources of intuitionism are unavailable to me (Brouwer and Heyting), the underlying philosophy will partly be based on some of my own ideas.

We see Thurston's emphasis on understanding and imagination:

> The criteria for the reasoning [in intuitionistic topology] can be taken as intuitive clearness and constructibility : together, these might be summarized as imaginability.

We see Thurston's thinking hard about what the practice of doing math should be. This was a topic that he pondered throughout his life, as he tried different approaches to learning and communicating mathematics: from building physical models to computer animation to sewing. This strong opinion about how math should be done, a topic that Thurston would come back to time and time again, for example in print 40 years later in [Th2], is already on view in the following remarkable paragraph from [Th1]:

> ...it should be recognized from the beginning of a development that it will never be possible to formalize completely the ideas involved in the theory, nor will it ever be possible to attain full generality. Thus, each step in the development of a theory must be well-motivated and intuitively clear in its own right. The process is evolutionary: each developmental step is an attempt to better deal with the environment or concrete mathematical ideas and images, and as the theory develops the environment meanwhile adapts to it.

References

[Th1] W.P. Thurston, A Constructive Foundation for Topology, Senior Thesis, New College, June 14, 1967.

[Th2] W.P. Thurston, Thurston, William P. On proof and progress in mathematics *Bull. Amer. Math. Soc. (N.S.)* 30 (1994), no. 2, 161–177.

A Constructive Foundation for Topology

Bill Thurston

June 14, 1967

New College Senior Thesis,

submitted to Roger Renne

Introduction

In modern mathematics--abstract mathematics, that is, for applied mathematics uses as its criterion workability--certain things are generally taken for granted.

Classical logic is taken as the standard for proofs, in particular, the principle of *tertium non datur*: either a proposition is true, or its denial is true; in another form, to prove a proposition it is sufficient to show that its denial would be absurd.

A strong form of set theory is taken as the underlying substance upon which a mathematical structure is imposed. It is assumed that from any set, a new set can be formed which contains all the subsets of that set ; the axiom of subsets. The axiom of choice is also assumed, which together with the axiom of subsets implies the well-ordering principle: the elements in any set can be put in an order analogous to a counting order. That is, in any subset there is one element you come to first in the ordering--elements are adjoined one at a time to the back end of the order.

Finally, a mathematical theory proceeds in its development abstractly and formalistically. In the beginning there was chaos, and then certain axioms were created. The rest of the history of a mathematical theory is the exploration of the consequences of these original decrees.

One outstanding example of the achievements of this approach is modern general topology. In this paper I will develop an alternate foundation for topology, based on a non-classical approach to mathematics: roughly, intuitionism, but since the primary sources of intuitionism are unavailable to me (Brouwer and Heyting), the underlying philosophy will partly be based on some of my own ideas.

A development of a formal system for basic intuitionistic logic, along with classical logic, can be found in Kleene, Introduction to Metamathematics, pp 69-213 and a more philosophical discussion of mathematical reasoning pp 36-68. That will be the basis of logical reasoning used here. It will not be developed formally, but occasional discussions will point out deviations from classical reasoning. The criteria for the reasoning can be taken as intuitive clearness and constructibility: together, these might be summarized as imaginability. Intuitionists claim it to be impossible to lay out ahead of time all valid principals of reasoning; Godel's theorem on the incompleteness of any formal system which includes number theory has provided strong backing for this claim.

I will use four different mathematical concepts to replace classical set theory.

1) Finite set theory. This is much like the finite case of classical set theory. The principal of tertium non datur holds for subsets of a finite set: given a finite set and a

subset of it, a complementary subset can be formed, so that any element in the set is either in the subset or in its complementary subset (but not both). It is supposed that a list could be formed which would be known to contain exactly the members of a finite set. Finite sets are to be distinguished from non-infinite denumerated sets, and even from finitely bounded denumerated sets. An illustrative example will be given soon.

2) Denumerated sets. The basic kind of denumerated set I call an array, to emphasize its dependence on its ordering (and thus distinguishing it from the classical denumerable set). The intuitionist image of a denumerated set is not the classical completed infinite, but instead a potential infinity: a denumerated set is seen in an unterminating process of growing. One fairly liberal formal system for arrays could be developed as follows. The basic pool of symbols is , ( ) 0 1. The basic pool of objects which may belong to an array consists of the finite sequences of these symbols. This pool is cut down by requiring that the subsequences (, ,) ,, do not occur, and that parentheses be closed in the proper manner. Then an array is defined as an algorithm which decides to accept or reject any sequence of symbols from the pool. In order for it to be possible to form a disjoint copy of any array, it is also required that there be a (known) finite upper bound to the number of compositions of parentheses. An algorithm can be found which determines the ordered pairs of members of two given arrays: this is the

Cartesian product of the two given arrays. Functions can be defined either as in classical set theory--a subarray of the Cartesian product of two arrays fulfilling the obvious condition--or as a rule (algorithm) which assigns to any member of one array a member of a second array. One divergence from classical set theory at this point is that the image of an array under a function need not be a subarray of the range. Godel's theorem provides an example of one such function, if we limit our reasoning to a given formal system: this function maps the natural numbers into the finite set consisting of 0 and 1, and it is impossible to prove within the formal system either that 0 is in the image of the function or that it is not. Intuitionistic reasoning shows however that it is not. If we had not stumbled upon this reasoning, the function would be an example of the assertion. There are examples, however, of functions whose images we know not to be arrays (see, for instance, Kleene's Intro. to Metamathematics, section 61), and there are even Diophantine equations whose images are not arrays. There is no general method for finding an algorithm to determine the image of a function, even when the image is bounded (as for the function in Godel's theorem); so intuitionistically, it is denied that for any function, the image of the function can bedenumerated as an array. However, the image of a function is a denumerated set, for it can be denumerated in the order of its domain.

The intractability of such a denumerated set is due to the basic divergence between the order of its denumeration and the order of the array which contains it. It is not in general possible, for instance, to rearrange a denumerated subset of the natural numbers to coincide with the order of the natural numbers.

By the very definition of array, given a subarray it is possible to find a complementary subarray. But it is not in general possible, given a denumerated subset of an array, to find a complement, even in the form of a denumerated subset.

3) Spreads. Finite sets can be thought of as lists of objects, and denumerated sets are the generalizations of such lists obtained by allowing the lists to grow indefinitely (possibly with repetitions): by adjunction of objects. Finite sets sometimes have other natural structures. For instance, the natural way to describe the (all too finite) set of books in the New College Library would be to make use of the cataloguing system. That is, first to divide the books into rough categories, then to divide each of these categories still further, and so on. The set is thus described by division, rather than by adjunction. If more books were to be added, they would go into existing sections, but these sections could be refined into further categories. The generalization of this dividing process obtained by allowing the refinements of categories to continue indefinitely is a spread. This particular kind of spread, for which the branches at any stage form a finite set, is called a fan. In the general spread,

the branches at any stage of the cataloguing form an array, and branches may terminate after a finite number of stages.

4) Species. Sometimes it would be impractical either to compile a list of the members of a finite set, or to classify the m embers; but it is easy to identify one of the members, by certain characteristics. The generalization of this kind of set is the species. The intuitionistic notion of a species is a rule for recognizing certain (pre-formed) mathematical objects. Thus, the algorithm for determining an array would be a species--a very special kind, because an array algorithm also recognizes when an object is not a member of the array. Mathematical manipulation of a species is simply manipulation of the predicate (rule for recognizing the members) which defines it.

Notice that each of these generalizations of the finite set includes the previous generalizations, or rather includes mathematical systems which are equivalent to any example of one of the previous generalizations.

A mathematical theory has a historical development, and another, static, kind of development. The latter is the ordering of a theory by the dependence of concepts within the theory. Neither kind of dev elopment should attempt to begin with full generality and then make specializations and applications, or even begin with special cases and yet proceed teleological, to the final theory. Such a method leads to stagnation. Instead, it should be recognized from the beginning of a development that it will never be possible to formalize completely the ideas involved in the theory, nor will it ever be possible to attain full generality. Thus, each step in the development of a theory must be well-motivated and intuitively clear in its own right. The process is evolutionary: each developmental step is an attempt to better deal with the environment of concrete mathematical ideas and images, and as the theory develops the environment meanwhile adapts to it.

The terminology of classical general topology reflects the teleological approach. Thus, a 'topological space' is the most general, while ordinary cases of topologies must be highly qualified: 'first countable locally compact Hausdorff topological space'. Unnecessary provinciality in definitions should be avoided on the one hand, but too much generality on the other hand can cause glossing over of important local deviations.
We have divided omnipresent set theory into four parts, and by doing this we will find it possible to avoid using the very glib axiom of choice. The same policy--not always to try for maximum generality--will be carried out in the future. The definition of topology to be given has simple generalizations which, however, are hard to handle. Instead of applying qualifying adjectives to the practical forms of the definition, adjectives can be applied to such generalizations.

Topology: Basic Definitions

The following logical notation will be used. The four propositional connectives do not require explanation:

Implies: $\Rightarrow$ And: & Or: $\vee$ Not: $\neg$

Propositional sentence order is like English sentence order if "that" before a clause is replaced by parentheses around the clause, with the exception that not always precedes the clause it modifies.

The two quantifiers $\forall x_{x\in S}$ and $\exists x_{x\in S}$ are to be interpreted as follows. S is supposed to be a species, and x is then a variable which ranges over the members of the species. Then $\forall x_{x\in S}(A(x))$ means we have a method that will provide a demonstration, given an arbitrary object x of the species, of the predicate A(x). This does not coincide with the classical idea of the generality quantifier, which is interpreted more like, 'for every object x in the species S, A(x) holds'. Intuitionistically, it is not claimed that it is possible to examine every member of a (potentially) infinite species, so the analogue of the classical generality quantifier must make the stronger statement, that we can show that A(x) holds for an object x using in our proof only the assumption that x is in the species S: the defining predicate of the species S. $\exists x_{x\in S}(A(x))$ means we have in mind a specific object t such that A(t); or at least, we have a method of finding or constructing such an object in the species. Then $\forall x_{x\in S}(\exists x_{x\in S}(A(x,y)))$ would mean that, given an arbitrary $x\in S$, we can prove that we have a method for finding a y such that A(x,y)--

which means we have a rule which assigns to each x in the species a y in the species such that A(x,y). This rule may be part of the defining predicate of the species, for instance. The formula $\neg \forall x_{x \in S} A(x)$ would mean we cannot find a general method such that, given an arbitrary x in S, we can demonstrate A(x); which is not as much as to say that we can find a specific x such that not A(x): $\exists x_{x\in S}(\neg A(x))$. However, the following provable formulas are not hard to reason through intuitively:

$$\forall x_{x\in S} \neg A(x) \Leftrightarrow \neg \exists x_{x\in S} A(x)$$
$$\exists x_{x\in S} \neg A(x) \Rightarrow \neg \forall x_{x\in S} A(x)$$

Arrays from now on will be understood as explained in the introduction. Notation for arrays will be similar to notation for classical set theory:

$A \cup B$: an object is accepted if it is accepted by A or by B, rejected if rejected by A and by B.

$A \cap B$: an object is accepted if it is accepted by A and by B, rejected if rejected by A or by B.

A*B: ('A complement in B') an object is accepted if accepted by B and rejected by A, rejected if accepted by A or rejected by B.

$x \in B$: the object x is accepted by B. $\emptyset$: x is rejected by $\emptyset$

$x \notin B$: the object x is rejected by B.

$A \subset B$: $(x \in A) \Rightarrow (x \in B)$ for arbitrary x in the basic pool for arrays.

The basic set-like structure upon which a topology is imposed is formed by division: a spread. The definition of spread to be given for the purposes of topology will be slightly de-generalized by the requirement that a branch be non-terminating. This is to correlate with the image that a topology is indefinitely divisible. A line of descent in a spread could even so never split into multiple branches; but the only case when this turns out to be topologically non-trivial is when this line of descent is not topologically "close" to any other line of descent, that is, the line of descent represents an isolated point.

Definition. A spread is a sequence of arrays $\{A_n\}$, and a rule S which assigns to each $x \in A_n$ a nonempty array $S(x) \subset A_{n+1}$ This rule must fulfill the conditions

i) $\forall x_{x \in A_n} \forall y_{y \in A_n} (x \neq y \Rightarrow S(x) \cap S(y) = \emptyset)$;

ii) $\forall x_{x \in A_{n+1}} \exists y_{y \in A_n} S(y) = x$.

The mth iteration of the assignment S applied to $t \in A_n$ is then a subarray of A_{n+m}, because, by ii) given any $x \in A_p$ we can find a (unique by i)) $y \in A_{p-1}$ such that $x \in S(y)$. If we begin with any $w \in A_{n+m}$ and repeat this process iteratively m times, the result is a unique $z \in A_n$ such that $w \in S^m(z)$. Then w is accepted or rejected according as $z = t$ or $z \neq t$. We abbreviate '$\exists n_{n \text{ is a natural number}} y \in S^n(x)$' as $D(x,y)$, and say that x is an ancestor of y or that y is a descendant of x.

A spread is a special case of an inverse system, as defined in Eilenberg and Steenrod, Foundations of Algebraic Topology, pp. 212-213. Indeed, an alternative definition for a spread would have been as a sequence of arrays, and for each pair of natural numbers n and m, with $n \leq m$, a map π^m_n taking A_m onto A_n. However, even though such a definition would be simpler and equivalent, the definition given is more easily modified for the important special case that S(x) is a finite set, or for the introduction of more general rules Se, where $x_1 \neq x_2$ does not necessarily mean that $Se(x_1)$ and $Se(x_2)$ are disjoint.

The rule S is called the successor assignment, and elements of the array S(x) are called successors, or immediate descendants, of x.

A spread for which S assigns a finite set to every x is said to be locally compact. If A_n is a finite set for every n, then the spread is said to be compact. It is easy to see that any subarray of a finite set not only is finite, but also is itself a finite set, so a compact spread is necessarily locally compact. Compact spreads are also called fans.

When we t alk of an object x in some A_n of a spread, we mean to consider x in its role as a member of A_n, and not as a member of some other A_m. If this were to be formalized, it would probably simplest to label the objects of A_n, so that they would be in the form (n,x). If this were to be done, there would be an algorithm to decide when an object x is in some A_n and when it is not, and the objects in the spread could this be formed into an array. When we use the formula $x \in Z$ for a spread Z, we will mean that Z is to be taken in the interpretation as an array.

The Cartesian product $\times_{i\leq n} A_i$ of n arrays $A_1 \ldots A_n$ is defined as the array which accepts exactly the ordered n-tuples $(x_1, \ldots, x_n)$ with x_i accepted by A_i. The Cartesian product $\times_{i\leq n} Z_i$ of n spreads $Z_1 \ldots Z_n$ is defined as the spread consisting of the sequence $\times_{i\leq n} A_{i;m}$ $(m = 0, 1, \ldots)$ and the successor assignment $\times_{i\leq n} S_i$ which assigns to each n-tuple $(x_1, \ldots, x_n) \in \times_{i\leq n} Z_i$ the Cartesian product of the component successor arrays: $\times_{i\leq n} S_i(x_i)$. Here, $A_{i;m}$ is the nth array in the sequence Z_i, and S_i is the successor assignment of Z_i. It is readily verified that this sequence of arrays with this successor assignment is a spread. Note that the Cartesian product of spreads is not the Cartesian product of the arrays which correspond to them.

A subspread Z' of a spread Z is a sequence consisting of a subarray A_i' of each A_i, and a successor function S' generated by S: $S'(x) = S(x) \cap Z'$. The subarrays must be chosen in such a way that each $x \in A'_{n+1}$ is a successor of some y in A'_n, and so that for any $x \in Z'$, $S'(x) \cap Z' \neq \emptyset$.

Definition. A topology is a spread Z, and a sequence consisting of a subspread C^m of $\times_{i\leq m} Z_{(i)}$ for each m, where $Z_{(i)} = Z$, to fulfill the following conditions:

(1) $C^1 = Z$ ('every piece is close to itself'. The objects of a a topology Z are called pieces.)

(2) $(x_1, \ldots, x_m) \in C^m \Rightarrow$ for any permutation p of $\{1, \ldots, m\}$, $(x_{p(1)}, \ldots, x_{p(m)}) \in C^m$. ('If a list of pieces is close, any permutation of the order of the list forms a list which also is close.')

(3) $(x_1, \ldots, x_i, \ldots, x_m) \in C^m \Rightarrow (x_1, \ldots, \hat{x}_i, \ldots, x_m) \in C^m$ where a caret $\hat{}$ over an object means that object is to be omitted. ('any sublist of a close list is close.')

(4) $(x_1, \ldots, x_i, \ldots, x_m) \in C^m \Rightarrow (x_1, \ldots, x_i, \ldots, x_i, \ldots, x_m)$ C^{m+1}, where here an additional x_i has been inserted somewhere in the list. ('repetitions don't alter the closeness of a list').

(5) Local finiteness. There is a finite set $C(x)$ associated with each x which includes exactly the objects y such that $(x, y) \in C^2$.

(6) Transitivity in the limit.

a) If $(x, y) \in C^2$, and $(y, z) \in C^2$, and if for every n there are nth order descendants x^n, y^n, z^n of x, y and z respectively, such that $(x^n, y^n) \in C^2$ and $(y^n, z^n) \subset C^2$, then $(x, z) \in C^2$.

b) If each pair of the n-tuple $(x_1, \ldots, x_n)$ is in C^2, and if for every m there are mth order descendants $x_1^m, \ldots, x_n^m$ of $x_1, \ldots, x_n$ respectively such that every pair among them is in C^2, then $(x_1, \ldots, x_n) \in C^n$.

A set fulfilling the condition that every pair of its members is close, is called pairwise close. Any set that is close must also be pairwise close, as is easily shown by repeated applicat ions of 3). Since C^n must be a spread, every n-tuple which is close must have arbitrarily distant close descendants, therefore arbitrarily distant pairwise close descendants: this is the converse of 6b). Classically, it would follow that all the C^m with $m > 2$ would be redundant and unnecessary. But intuitionistically, if we had not explicitly demanded that the C^m be given, we would not generally be able to assert or deny the hypotheses of 6b). A topological theory is in fact feasible without the assumption of the higher order C^m, without the assumption that branches do not terminate and do not mingle, without the assumption of symmetry (2).), without the assumption of local finiteness. But the present definition is clearer and easier to work with.

Example: the unit interv al I. Let the base spread of the topology be the sequence Z of Arrays A_i as follows:

$A_0 =$ 0

$A_1 =$ 0, 1

$A_2 =$ 00, 01, 10, 11

...

$A_n =$ the finite set of n-digit binary numbers.

Let the successor assignment S assign to each binary number b the set consisting of b followed by 0 and b followed by 1.

C^1 is defined by (1): C^1 is Z. Let C^2 contain all repeated pairs (b, b) and all pairs adjacent in counting: (b, b+1), and (b+1, b). (Provided of course that they have the same number of digits.) Let C^n be generated by repetitions of components of pairs in C^2. It is not hard to verify that all the C^i thus defined form spreads, and in fact, fans. All the conditions (1)-(6) are simply checked. Here is a diagram of this topology:

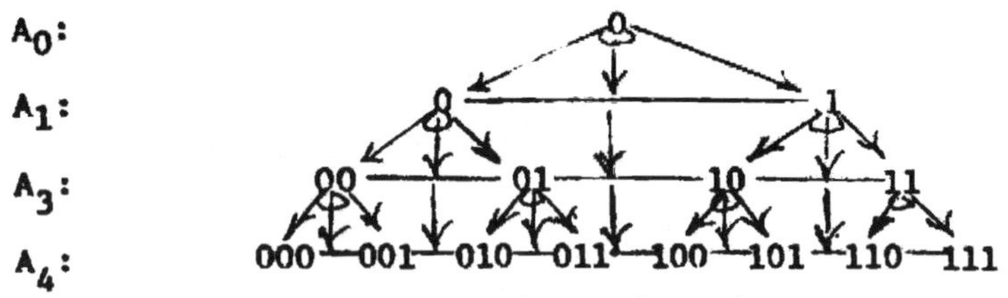

Arrows denote the successor assignment S; horizontal lines denote close pairs.

The geometric interpretation of the topology just defined: Let A_0 correspond to the set containing a line segment of length 1. Let 0 as a member of A_1 correspond to the right half of this line segment; 1 corresponds to the left half. Whenever a correspondence has already been defined between some binary number b in A_n and a line segment, let the binary number b0 correspond to the right half of the line segment for b and b1 correspond to the left half. It is easy to see that the closeness relation between pairs of these intervals corresponds to pairs of intervals of the same length which touch on the ends.

Example: the real line E. Simply place copies of I end-to-end, in a sequence that grows on both ends.

Definition. The Cartesian product $\mathop{X}\limits_{i \leq m} T_i$ of m topologies $T_1, \ldots, T_m$ is the Cartesian product of the corresponding spreads $Z_1, \ldots, Z_m$, and for each n the subspread C^n of $\mathop{X}\limits_{j \leq n} (\mathop{X}\limits_{i \leq m} T_i)_{(j)}$ determined as follows. Form the Cartesian product of the C_i^n corresponding to the T_i; elements of this spread are then m-tuples whose components are n-tuples. Each n-tuple component can be labelled $x_{i;j}$ where $1 \leq i \leq m$, $1 \leq j \leq n$. If the indices are interchanged: $x_{i;j}$ is relabelled $x_{j;i}$, this yields an n-tuple of m-tuples. Let C^n consist exactly of the n-tuples of m-tuples obtained in this way. In other words, an n-tuple is close exactly when the n-tuple obtained from the first components of its components is close, the n-tuple obtained from the second components of its components is close, and so on.

Example. The unit n-cube I^n is $\mathop{X}\limits_{i \leq n} I_{(i)}$, where each $I_{(i)} = I$. In particular, the geometric interpretation of I^2 can be diagrammed as follows:

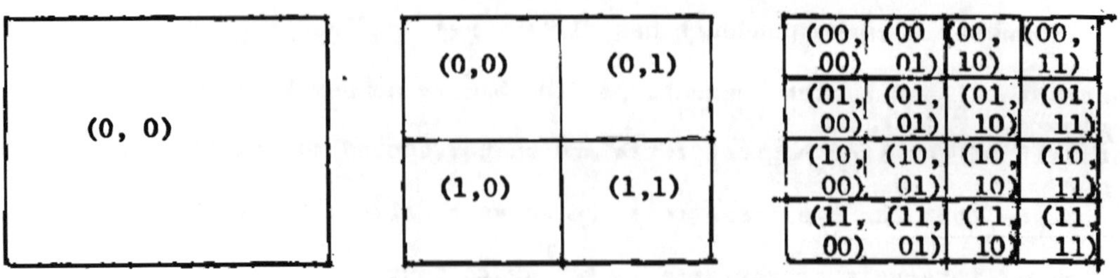

It is elementary to verify that the Cartesian product of topologies is a topology.

The Comparison of Topologies

An interpretation will be presented as motivation for some further developments.

Spreads were presented in the introduction as generalizations of finite cataloguing systems. We interpret a spread, then, as an infinite (growing at the branching tips) cataloguing or describing system. The classical structure which ordinarily takes the place of a given spread is the set of all infinite branches, of the spread. But instead of reifying the ultimate descriptions of our cataloguing system as objects, and then identifying a certain description or category with the set of objects which fir the description, we think of the descriptions and the structure of the descriptive system as the basic information which is given to us. More concretely, we imagine various observers, each able to make observations and give us reports of these observations. We also imagine objects which we may present to the observers to be observed, but these are "objects" in a different sense than above: they function as standards of consistency for the observation--each observation refers to some object, and (we think of the observers as being perfectly reliable) descriptions which may be applied to the same object are consistent with each other. An observer may end up describing two distinct objects in exactly the same way; if we had thought of objects as the ultimate descriptions of a descriptive system, these objects would be the same, but we suppose that we have ways of keeping track of the identity of particular objects without relying on the descriptions of some observer (for instance, through spatio-temporal continuity). On the other hand, we presume ourselves incapable of observing the objects ourselves, apart from keeping track of their identity, and so we must get all our information about them secondhand through the various observers.

Each observer has a definite repertoire of reports, which consists of finite sequences from a finite pool of symbols. (This means he can communicate to us a finite amount of information, in the form of distinct, nonmistakable language.) At request when presented with an object he will give a report of the object, and a definite pool of reports is allotted to such first impressions. Then he can be requested to elaborate, any number of times; for each degree of elaboration there is a definite pool of allotted reports. If his cataloguing system is sharp, he must respond with the same nth elaboration each time he is presented with an object (the categories are mutually exclusive), and thus there is a certain definite pool of reports which can be elaborations of any given report. (This pool is considered known to us and to the observer, because it is part of his conceptual system). The stages of elaboration are considered strict elaborations--if not, we could make them so by replacing a report by the sequence of reports the observer gives up till that one--and so the reports which are permitted to follow distinct reports must be distinct.

The standards of consistency compare various observers, as well as the various degrees of elaboration of the reports of one observer. We considered ourselves to know the conceptual systems of individual observers, and we also consider ourselves to know the relations between conceptual systems of different observers (example--eyes, ears and nose). In fact, the objects are completely unnecessary to the theory of the conceptual systems, for we can obviously not get reports on enough objects to completely define the structure and relationships of the conceptual systems (alias cataloguing systems, descriptive systems); the objects are mentioned only because the cataloguing systems serve as vehicles for handling the objects, and their structure is motivated by this. The comparison of two observers is called a correlation, and is put in the following form. To each category of one stage of elaboration (say the nth) of one system is assigned an array of categories in the same stage of elaboration (the nth) of the other; and vice versa. As usual, these arrays must be given by some rule (since an infinite number of assignments must be made). Then, if on a given level (stage of elaboration) we have a rule which assigns to each category an array of categories of another categorizing (cataloguing, describing, conceptual, spread) system, the inverse rule can be simply derived--a is in b's array iff b is in a's array. A correlation is interpreted as telling us, when observer A gives one report on a given level, exactly which reports it is possible that observer B may give. It may occur that there is only one such possible correlated report, and in that case it is obviously necessary whenever A gives the report that B give the correlated report. From our ideas of consistency, we see that if a report by A is correlated with a report by B, then there must be elaborations of these two reports which are correlated (consistent) with each other; and also, the immediate ancestors of consistent reports must be consistent. Each stage of elaboration can be thought of as a descriptive cross-section of the universe, and a correlation really acts between such cross-sections; so we can think of the successor assignment as a correlation, with the special property that the correlate of any category in the later stage of elaboration is necessary. With this, the last condition is a special case of the condition that necessary correlates of consistent reports be consistent. For the present, we regard descriptive systems all as referring to the same universe. Then, it must hold that the correlates in one cross-section of a category in another cross-section form a non-empty array, and an even stronger condition must hold: if we say that one sequence of reports is correlated to another whenever the representatives on each level of elaboration are correlated, then any sequence of reports by oneobserver must be correlated to at least one sequence of reports by a second observer.

One way to compare various observers is to compare their power of distinguishing between objects. If observer A can distinguish between all pairs of objects which observer B can distinguish, then A is said to be more acute than B; if A and B are each more acute than the other, then they are said to have the same resolving power. Rephrasing in terms of conceptual systems and correlations, A is more acute than B whenever any pair of sequences of reports by A correlated to distinct sequences of reports by B are distinct.

Description can be resolved into two processes: distinguishing and comparing. The categories of a descriptive system are not simply arbitrary groupings, or collections, of distinguished objects--they are also comparisons, and the fact that the categories are formed in a certain way is an assertion in effect that everything within a single category is somehow similar. Observers can be compared in the way they categorize, as well as the way they distinguish. Obviously if observer A classes an object in a certain category, observer B must place the object in a category of which A's category is a possible correlate. But this does not mean that As category is ever a necessary correlate of the sequence of B's reports, even when A and B have the same resolving power. For instance, let the objects to be observed be the natural numbers, and let A's first report specify the natural number, while all A's elaborations are identical with his first report; and let B's first report specify either 1 or 0,2,3,...; Bs second report specify either 1 or 2 or 0, 3, 4, ...; and so on. B and A both distinguish between each pair of natural numbers, but when presented with the object 0, A's first report 0 is never a necessary correlate of any of Bs reports. B always hedges about calling a 0 a 0, while A comes right out and says it. This situation means that B has no way to translate As category 0, while A can translate every category of B. (Arrays must here be allowed as translations; since we think of arrays as being described by a rule, we can think of this rule as the translation, or at least an explication in terms comprehensible to A of B's meaning). If every classification by B is a necessary correlate of classifications by A, then A is said to be more flexible than B; if A and B are each more flexible than the other, then they have the same conceptual power. It is not hard to see that being more flexible implies being more acute.

In many practical cases, a descriptive system is not sharp. One way to obtain blurry descriptive systems is to allow overlapping categories on each level of elaboration. In this type of system, an observer always chooses a definite category at each level of elaboration, but the category is not always uniquely determined by the object, and different reports at one level may have identical elaborations at another: we have a spread with mingling descendants. The description of such a cataloguing system would involve saying which categories at each level are consistent with one another, and for each initial sequence of reports, what following elaborations are consistent. We choose instead to obtain blurry spreads in a different way: by permitting an observer to report indecision between several categories, but requiring that the formal system of permissible elaborations be sharp--with mutually exclusive permissible elaborations. An observer's report need not be uniquely determined by the object itself, so that when an observer makes a report, he is not saying that it is the only possible report (the necessary report), but only that he will be able to continue making elaborations of it indefinitely (like the Devil's Advocate): it is a plausible report. Part of the conceptual system must now involve stating which sets of alternatives are permissible as a report. We call such sets of elementary categories confusible sets, or in topology, close sets. We require that if a set of elementary categories is confusible, then there must be at any stage of elaboration a confusible set of categories including at least one representative of the descendants of each of the original categories--confusibility cannot be resolved. But confusion in the reports themselves can be resolved: an observer may report wavering between several categories at one stage of elaboration, and yet at another stage make a report of which the original categories are not all possible correlates.

This must be allowed, because we regard the process of observation, just like the process of reporting, as never completed. When two categories are not confusible, this means that after a finite amount of observation the observer can definitely place an object in the one or the other if it is in one of the two; but when two categories are confusible, the sequence of reports consisting of an indecision between descendants of the two categories is permissible, and this sequence is never resolved into one or the other of the categories; so when an observer has an indecision between the two categories, he cannot tell whether he will be able to resolve this indecision or not, by a finite amount of observation--so it is pointless for him to try. If he were able to tell exactly when he could resolve indecision between several categories, he might as well use the common borders of groups of categories as categories themselves, for then he would obtain a sharp categorizing system. The same discussion applies to larger sets of categories.

From any category is derived a set of possible correlates of it on the same level: those categories confusible with it. When there are no other categories confusible with it, it is a necessary correlate of itself. Now we look at the first elaboration of this level. The possible correlates of the given category on this lower level form a smaller portion of the universe, in the sense that the only possible correlates of them on the former level were the former possible correlates (from general properties of confusibility, or closeness). And so we can continue: the possible correlates of a given category on all levels together form a subspread of the original spread, and this spread shrinks with respect to the original spread, in the sense that not all possible descendants of members of the spread are generally in the spread. Similarly, on each level there is an array consisting of the categories necessarily in a given category, i.e., the given category is a necessary correlate (on its level) of each member of the array. These do not immediately form a subspread, because a category necessarily in the given category need not have an ancestor necessarily in the given category. However, by rearranging them so that each category which is the first in its line of descent to have the given category as a necessary correlate is placed on the first stage of elaboration, these categories necessarily in the given category do form a spread. And finally, the categories which are permissible or plausible correlates of a given category (i.e., they are in a permissible line of descent with it,) together form a subspread. These three spreads are called the possibility, the necessity, and the plausibility, of the given category.

The concept of a necessary correlate in the case of blurry spreads should be cleared up a little, before we continue. Suppose we superimposed two descriptive cross-sections of the universe; this would be permissible for use in blurry classifying systems. But in such a case, any category would have at least two possible correlates in this derived cross-section. The necessity of any simple category would then be empty; for this and other cases, it is important also to talk of an indecision between several categories as a necessary correlate of another category whenever there are no other possible correlates of this category within the cross-section of the indecision. In the present usage, necessary correlates depend on the particular cross-section in which a category or indecision between categories happens to be embedded, while possible correlates do not.

The categories in the possibility of a given category but not in its necessity are those which are correlated both to the given category and to other categories in the cross-section of the given category. These also must form a subspread, evidently. This subspread is the freedom of the given category. Freedom is locked between possibility and necessity; and as time progresses, necessity grows, possibility shrinks, and freedom is squashed between them.

More conventionally, these last three spreads could be called the interior subspread, the closure subspread, and the boundary subspread of a piece.

In the blurry cataloguing system, different categories do not in general distinguish between objects; two objects, or rather, two sequences of permissible elaborations, are now distinguished whenever they are eventually not confusible with each other--when they are eventually in non-confusible categories. The criterion for comparing the acuity of observers now carries over from the sharp case. Observer A is more acute than observer B whenever any pair of sequences of reports by A correlated to distinguished sequences of reports by B are distinguished.

The comparison of conceptual power also carries over without much change. Previously, every classification by an observer was a necessary classification, but we now need to explicitly require necessary classifications by B in: if every necessary classification by B is a necessary correlate of classifications by A, then A is more flexible than B; if A and B are each more flexible than the other, then A and B have the same conceptual power, or are topologically equivalent. (Under the given correlation)

The identity correlation of a cataloguing system associates all pairs of confusible categories (within constant levels). Under this correlation, a system should be more acute than itself and more flexible than itself. The acuteness condition means in particular that if arbitrary stages of elaboration of a pair of categories are confusible with a common category, then this pair of categories must be confusible (6a). This implies the condition, that if arbitrary stages of elaboration of a pair of categories have categories confusible with confusible categories, then this pair of categories must be confusible under the assumption of local finiteness (5), for then there must be two particular categories confusible respectively with the two given categories and confusible with each other, with arbitrary pairs of descendants confusible respectively with descendants of the two given categories and confusbile with each other--or more simply, there is a chain of length four having adjacent links confusible, and with arbitrary descendant chains of length four having adjacent links confusible. Then (6a) applied to the first three links makes the first and third links confusible, and another application on the first, third and fourth links makes the first and fourth confusible. Brouwer's fan theorem (see Kleene and Vesley) states that if every growing branch of a fan (a spread which is finite on each level) is to eventually have some property, then there must be some finite n such that from level n on, every branch has the property. With the assumption of local finiteness (5), the sequence of all categories confusible with a given sequence of reports forms a fan with terminating branches. For any pair of sequences of elaborations which is eventually distinguished, any pair of branches from the corresponding fans (i.e., any pair of branches always confusible respectively with the two given sequences of reports) must eventually be distinguished as a consequence of (6a). It is easily seen, using a simple argument with the fan theorem, that there must be some finite n after which every pair of branches of the two spreads must be distinguished. Thus, (6a) and (5) imply that sequences of waverings confusible respectively with two distinguished sequences of waverings must eventually be distinguished. (6b) is derived readily from the assumption that the identity correlation preserves acuteness, and with an argument analogous to the above the conditions (6a), (6b) are equivalent to this assumption under the condition (5). Somewhat surprizingly, the fan theorem and local finiteness (5) easily yield that (6a) implies the condition that a categorizing system be more flexible than itself under the identity correlation. This condition is the condition of regularity, in topological terminology, while (6a) is Hausdorffity and (5) is the constructive form of paracompactness.

More general correlations can relate observers who do not necessarily have the same scope of competence: observers may be permitted, when confronted with certain objects, to say, "this is beyond me, I can't make it out." Whenever there is at least one sequence of reports by A correlated to every sequence of reports by B, A has a broader scope of competence than B. (Under the given correlation.) This might more precisely be called a broader scope up to the resolution of B, for there might be a pair of objects which B does not distinguish but one of which A classifies and the other of which is out of his scope of competence; such a situation does not count against A in the comparison of A's scope with B's scope. From this example it is easy to see that having a broader scope is not generally transitive, except when the observers concerned have the same resolving power. Classical continuous functions correspond to the case when the second observer is less flexible but has a broader scope than the first observer.

Roots and Sand
(or, a comparison of the spread and the point set as foundations for topology)

A direct analogue of classical topology, using the point set as a foundation, cannot be defined within intuitionistic mathematics, because the machinery available is not strong enough. But blurry cataloguing systems can be defined and developed within classical mathematics, because classical machinery is usually at least as strong as intuitionistic. To justify our definitions of topology, then, we must make the analogous definitions within classical mathematics and show their equivalence to the ordinary classical definitions; unless, that is, we wished to make a direct metamathematical comparison of the theorems provable within our topological system (using intuitionistic reasoning) with the theorems provable in classical topology, showing in what cases analogous theorems are provable. The latter course is beyond the scope of this paper, since it involves too much metamathematics.

Consider any classical system which fulfills the conditions for a spread-topology. Form the set of all infinite branches of the spread. (This can be done, because this is a subset of the Cartesian product of the arrays in the spread; this set is non-empty in the case of a non-empty spread by the axiom of choice.) Certain pairs of these branches remain always close. This relation between branches is transitive, by (6a), symmetric by (2) and reflexive by (1) and (4). Hence it is an equivalence relation, so we can identify equivalence classes and call them points. We define "piece" in a new sense to mean the set of points with at least one branch in the corresponding "piece" in the original sense. (Or, in the terminology of the interpretation of the preceding section, we are defining "category" in a new sense). We impose a topology on this set of points, by defining basic neighbourhoods: the union of the pieces representing all the branches of a point p at level n is a basic neighbourhood of p. These neighbourhoods for a fixed point p form a (set-theoretically) decreasing sequence, so that finite intersections of basic neighbourhoods of a point contain a basic neighbourhood of that point.

From a correlation under which A is more acute than B, and under which B has a broader scope than A, a function can be derived: take any point of A to the intersection of the unions of the correlates of the categories of any branch of the point. There is exactly one point in this intersection, because A is more acute than B. Even without the two hypotheses on the correlation, a correspondence between the point-sets of A and B is derived from a correlation, by letting p correspond to q when q is always in some correlate of some branch of p.

On the other hand, given any function taking the point set of A to the point set of B, a correlation is derived: correlate a category of A with every category on the same level of B which intersects the closure of its image. It is easily verified that the derived correspondence between categories is a correlation, except for the condition that a category correspond to a an array, and that this corresondence be defined by a rule. However, this condition is unnecessary within classical mathematics, for correspondences can be well-defined without being given constructively; so we drop that condition. How does the correspondence derived from this correlation compare with the original function? Form the Cartesian product A X B. The original function was a subset of the point-set Cartesian product. The derived-derived correspondence is the closure of this subset: any point which has a neighbourhood which does not intersect the original subset is obviously not in the derived-derived correspondence, while any point for which every neighbourhood intersects this original subset must always have representatives of branches--pairs of pieces--which contain points in the graph of the function, and thus are correlated. Then, by definition, this point corresponds to a pair of points in the derived correspondence. Similarly the derived-derived correlation from a correlation is a kind of closure of the original correlation--a pair of level categories are in the derived-derived correlation whenever it always has descendant pairs close to pairs of pieces in the correlation. In summary, these derivations form a Galois connection (see Garrett Birkhoff, Lattice Theory, p. 56) between the lattice of subsets of the Cartesian product of the point-sets and the lattice of subspreads of the Cartesian product of the spreads. In particular, the derived-derived correspondence isthe original correspondence exactly when the original correspondence had a closed graph; so from any continuous function is derived-derived the original function.

Since being more flexible implies being more acute, whenever A is more flexible and narrower in scope than B under a given correlation, a function is derived. Such a function must be continuous--an immediate corollary of the definitions. The converse is also immediate. Taking into account other obvious connections between spreads and point-sets, such as the connection between the two kinds of Cartesian products, we have:

The classical theory of topology on spreads is equivalent to the portion of the classical theory of topology which involves structures and operations definable by spreads--in particular, continuous functions and homeomorphisms (topological equivalences) between topologies definable from spreads, finite (and indeed, denumerable) Cartesian products, and quite a large portion of the other desirable things in point-set topology.

Now we investigate the question of how large a class of point-set topologies are definable from spreads.

First we get some definitions out of the way:

An essential cover of a space is a family of sets such that the interiors of finite unions of closures of sets in the family form a cover. Obviously, an open cover is an essential cover. A family of sets is close, if the intersection of the closures of its members is non-void. From any essential cover an open cover can be derived, by forming the family of interiors of unions of closures of members of finite close families--in other words, by limiting the finite unions in the definition to the case of close families. This forms a cover, for given a point in the interior of some finite union of closures of members of the essential cover, the point is still in the interior of the union when those sets whose closures are disjoint from the point are eliminated; and the remaining sets form a close family, for their closures intersect at least at the point. A close refinement of an essential cover is an essential cover, such that its derived open cover refines the derived open cover of the original essential cover. Obviously, an open refinement of an open cover is also a close refinement. A close: simple refinement of a cover is an essential cover whose derived open cover refines the original cover; etc. Close: simple refinements are similar to star-refinements (a <u>star</u> is the union of a close family, not necessarily finite. A cover V is a star-refinement of U if the family of stars derived from V refines U. See J. L. Kelley, <u>General Topology</u>, p. 170). A <u>clean cover</u> is an essential cover, such that the intersections of interiors of closures of pairs of its members are void. A <u>clean refinement</u> of a clean cover is a close refinement such that each set of the refinement is contained in the closure of one of the original sets.

A family of sets is <u>locally finite</u>, if every point in the space has a neighbourhood which intersects only a finite number of sets in the family. A space is <u>paracompact</u> if every open cover has a locally finite refinement.

Lemma 1: Each paracompact and Hausdorff space is regular, and each paracompact and regular space is normal.

This is a standard theorem, whose proof is analogous to that of the corresponding theorem for compact spaces. See e.g. Hu, <u>Elements of General Topology</u>, propositions 2.19 and 2.20 (p. 68) or Kelley, pp. 156, 159.

Lemma 2: Each regular 2nd countable space is paracompact. See Kelley, p. 156 Theorem 28.

Lemma 3: Each countable locally finite open cover of a normal space has an open close:simple locally finite refinement.

For a sketch of the proof, see my notes to Lecture 2, "A Constructive Theory of Topology,", presented on May 30, 1967; p. 21.

Lemma 4. Every countable open essential cover has a close refinement by a clean cover.

Proof: From a cover $\{A_n\}$ define $B_i = A_n - (\bigcup_{i=1}^{n-1} A_i)^-$. Then $(\bigcup_{i=1}^{n} B_i)^- = (\bigcup_{i=1}^{n} A_i)^-$, and B_n forms a clean cover.

A locally refining sequence of covers is a sequence of covers such that for any point and any neighbourhood of the point there is an n such that for any m greater than n, each member of the mth cover which contains the point is contained in the neighbourhood. A sequence of covers S_1 refines a sequence S_2 if for every n there is an m such that the mth cover of S_1 refines the nth cover of S_2. A clean sequence of clean covers is a sequence of clean covers such that covers later in the sequence cleanly refine covers before them in the sequence.

Theorem. Every regular 2nd countable space has a clean sequence of clean covers, of which the derived sequence of open covers is locally refining.

Proof:

a) Enumerate a base. From this, derive a sequence of essential covers, by letting the nth cover in the sequence consist ofthe 2^n intersections of families of n sets formed by drawing one set from each pair of A_i $(i \leq n)$ and A_i.

The sequence of covers derived this way must be a locally close:simple refining sequence of essential covers; for, given a point and a neighbourhood of the point, there must be some A_n which is contained in this neighbourhood, and by regularity there must be some A_m, a neighbourhood of the point whose closure is contained in A_n; as soon as q exceeds A_n and A_m, the qth cover will have the property that every close set containing the point must be contained in the neighbourhood. Call this sequence S_1.

b) Construct, by Lemma 2, a locally finite simple: close refining sequence S_2 of S_1. Such a sequence must also be locally refining.

c) Find, by Lemma 3, a locally finite close:simple refining sequence S_3 of S_2.

d) Construct a locally finite close-refining sequence S_4 of S_3 consisting of clean covers, by Lemma 4.

e) If S_4 consists of clean covers U_n, define S_5 to be the sequence V_n, where V_n consists of intersections of one set from each of $U_1, \dots, U_n$. S_5 is now a locally finite clean sequence of clean covers, of which the derived sequence of open covers is locally refining.

Thus we can construct a spread from any topology. However, the spread we construct in this way is always compact, so the derived classical topology is compact. If we begin with a compact topology, the derived derived topology is the original,

If we have a locally compact topology, we can find a spread for it using this construction on each set of some decomposition of the space into compact subspaces (possible for any 2nd countable space. It is not possible to represent every paracompact 2nd countable regular space as a spread: a very simple example is the set of rational numbers with the topology as a subspace of the reals. If we attempt to represent them as a spread with this topology, we end up completing them, to form the reals. An example of a paracompact but not locally compact space which it is possible to represent as a spread is the countable Cartesian product of copies of the real line. That is as far as I have investigated the situation; to me, the condition of being representable as a spread seems a very natural way to designate the class of topologies which are representable as spreads.

As a somewhat surprising development, it might be noted that if the local finiteness condition were dropped from the definition of a topology, it appears that classical topologies could be obtained which would not have countable bases, or even countable bases for the neighbourhood systems of individual points.

Bibliography

Garrett Birkhoff, Lattice Theory, American Math. Soc. Colloquium Publications v. XXV, Providence: 1948

Samuel Eilenberg and Norman Steenrod, Foundations of Algebraic Topology, Princeton: 1952

S. T. Hu, Elements of General Topology, Holden-Day, San Francisco: 1964

John L. Kelley, General Topology, D. van Nostrand, Princeton: 1955

Stephen C. Kleene, Introduction to Metamathematics, D. van Nostrand, Princeton: 1952

Stephen C. Kleene and R. S. Vesley, The Foundations of Intuitionistic Mathematics, North-Holland Publishing Co., Amsterdam: 1965

S. Körner, The Philosophy of Mathematics, Hutchinson, London: 1960

Casimir Kuratowski, Topologie, v.1 and v.2 (Quatrième Edition), Panstwowe Wydawnictwo Naukowe, Warszawa (Poland); 1958

E. J. Lemmon, "New Foundations for Lewis Modal Systems", J. Symbolic Logic, 22 (1957) pp. 176-186

J. C. C. McKinsey and A. Tarski, "Some Theorems about the Sentential Calculi of Lewis and Heyting", J. Symbolic Logic 13 (1948) pp. 1-13

A. Mostowski, "Quelques applications de la Topologie à la Logique Mathématique", appendice VIII à Topologie, v.1, C. Kuratowski.

Bill Thurston, "Lecture 1: Two Standards for Mathematics" and "Lecture 2: A Constructive Theory for Topology" (mimeographed notes, New College, May 1967).

Topology Vol. 15, pp. 253–258. Pergamon Press, 1976. Printed in Great Britain

EVERY CONNECTED SPACE HAS THE HOMOLOGY OF A $K(\pi, 1)$†

D. M. Kan and W. P. Thurston

§1. STATEMENT OF RESULTS

In recent years Barratt–Kahn–Priddy[8], Mather[6], Priddy[9], Quillen[11], Thurston[12], Wagoner[13] and others discovered that various *infinite loop spaces* and *spaces of homeomorphisms* have the *homology of a* $K(\pi, 1)$ and those results make one wonder whether this is also the case for other spaces.

The main purpose of this note is to give a positive answer to this question and to show that in fact *every path connected space has the homology of a* $K(\pi, 1)$. More precisely:

1.1 Theorem. *For every path connected space* X *with base point there exists a (Serre) fibration*

$$TX \xrightarrow{\ tX\ } X$$

which is natural with respect to X *and has the following properties.*

(i) *the map* tX *induces an isomorphism on (singular) homology and cohomology with local coefficients*

$$H_*(TX; A) \approx H_*(X; A) \qquad H^*(TX; A) \approx H^*(X; A)$$

for every local coefficient system A *on* X, *and*

(ii) $\pi_i TX$ *is trivial for* $i \neq 1$ *and* $\pi_1 tX$ *is onto.*

Furthermore *the homotopy type of* X *is completely determined by the pair of groups* (G_X, P_X) where

$$G_X = \pi_1 TX \qquad \text{and} \qquad P_X = \ker \pi_1 tX.$$

In fact it is not hard to prove, using Theorem 1.1, the Serre spectral sequence and the commutativity of the diagram

$$\begin{array}{ccccc}
* & \longrightarrow & \pi_1 X & \longrightarrow & \pi_1 X \\
\downarrow & & \downarrow & & \downarrow \\
UX & \longrightarrow & \tilde{T}X & \longrightarrow & \tilde{X} \\
\downarrow & & \downarrow & & \downarrow \\
UX & \longrightarrow & TX & \longrightarrow & X
\end{array}$$

in which $\tilde{X}$ denotes the universal covering of X, $\tilde{T}X$ the induced covering of TX and UX is the fibre of tX, that P_X *is a perfect* (i.e. $H_1 = 0$) *normal subgroup of* G_X and that

1.2. *The space* X *can, up to homotopy, be recovered from the pair of groups* (G_X, P_X) (i) *by applying the* $(\ )^+$ *construction of Quillen* [11, 13] *to* $K(G_X, 1)$ *with respect to the perfect normal subgroup* $P_X \subset G_X$, *or equivalently* (ii) *by applying the fibre-wise* Z*-completion functor of Bousfield–Kan* [1] *to the fibration* $K(G_X, 1) \to K(G_X/P_X, 1)$.

1.3. *The simply connected space* $\tilde{X}$ *can, up to homotopy, be obtained by applying the* Z*-completion functor* [1] *to* $K(P_X, 1)$.

1.4. *The acyclic space* UX *can, up to homotopy, be obtained by applying the acyclic functor of Dror* [3] *to* $K(P_X, 1)$.

†This research was partially supported by the National Science Foundation.

The proof of Theorem 1.1 is rather technical and we therefore will in §2 discuss the ideas behind it and indicate how the actual proof is spread out over the remaining sections.

§2. COMMENTS ON THE PROOF OF THEOREM 1.1

2.1. The ideas behind the proof

We can probably best illustrate these by describing how, for every integer $n \geq 1$, one can obtain a group σ^n such that $K(\sigma^n, 1)$ has the homology of the n-sphere S^n.

Clearly one can take $\sigma^1 = Z$, the integers. Now suppose that we already have σ^{n-1}. Copying the construction of S^n out of S^{n-1} by attaching two cones, we (see 3.3) embed σ^{n-1} in a kind of *homological cone*, i.e. a group $C'\sigma^{n-1}$ such that $K(C'\sigma^{n-1}, 1)$ is acyclic and then glue two copies of $K(C'\sigma^{n-1}, 1)$ together along $K(\sigma^{n-1}, 1)$. The resulting space obviously has the homology of S^n and, by an *asphericity* result of J. H. C. Whitehead (4.1), its higher homotopy groups vanish and its fundamental group is thus the desired group, the free product with amalgamation

$$\sigma^n = C'\sigma^{n-1} \amalg_{\sigma^{n-1}} C'\sigma^{n-1}.$$

2.2. Further comments

It is not hard to see that one can apply essentially the same methods to any CW-complex X, which has only one vertex and for which the attaching map $S^{n-1} \to X$ of any n-cell maps S^{n-1} homeomorphically onto a subcomplex of X. However, if one tries to deal with more general CW-complexes, things become rather complicated as one has to keep track of lots of homotopies.

To get around this we work (*semi-*) *simplicially*. This has the advantage that we now can do everything *functorially* and that, in view of the well known equivalence of simplicial and topological homotopy theories, our results automatically apply to topological spaces as well. But a possible disadvantage is that the groups G_X become very large, because the singular complex of a space is so huge; for instance G_{S^n} is much larger than the group σ^n constructed above.

Although the group G_X we obtain in the proof of Theorem 1.1 is *uncountable* even for a finite CW-complex X, our results imply (see 5.1) that *for every countable CW-complex X there is a countable group G'_X such that $K(G'_X, 1)^+ \simeq X$* where the $(\ )^+$ construction is taken with respect to a suitable perfect normal subgroup. Two other interesting questions along this line of inquiry are:

(i) If X is a *finite CW-complex*, is there a finite CW-complex $T'X$, homologically equivalent to X, where $T'X$ is a $K(\pi, 1)$?

(ii) If X is a *compact n-manifold* $\neq S^2$ or P^2, is there a compact n-manifold $T''X$, homologically equivalent to X, with $T''X$ a $K(\pi, 1)$?

If X is a finite CW-complex and $\dim X \leq 2$, then (i) has a positive answer and (ii) has a positive answer, if X is a manifold of dimension ≤ 3.

2.3. Organization of the proof

§3 deals with *homological cones* for groups as well as for simplicial groups and §4 contains a simplicial group version of J. H. C. Whiteheads *asphericity* result. Both of these are then used in §5 to obtain a simplicial version of Theorem 1.1, from which we finally derive Theorem 1.1 itself.

For details on *simplicial homotopy theory* and its relationship to topological homotopy theory we refer the reader to [7] and [1, Ch. VIII].

§3. HOMOLOGICAL CONES FOR GROUPS AND SIMPLICIAL GROUPS

In this section we show how one can *embed every group and every simplicial group in an acyclic one*, i.e. one which has an acyclic classifying space. We start with constructing

3.1. A homological cone functor for groups

Let Q denote the rationals, let Aut Q be the group of the (set) automorphisms of Q which have compact support, i.e. which are the identity outside some finite interval, and, for every group G, let G^Q denote the group of the functions $Q \to G$ which have compact support, i.e. which go to 1 outside some finite interval. Then Aut Q acts on G^Q by composition and we define CG as the resulting semi-direct product (or split extension), i.e. the elements of CG are the pairs (b, a)

where $b \in G^Q$ and $a \in \operatorname{Aut} Q$ and the multiplication is given by the formula

$$(b, a)(b', a') = (ba(b'), aa') \qquad b, b' \; G^Q, a, a' \in \operatorname{Aut} Q.$$

Clearly this construction is *functorial* and it is easy to obtain a *natural embedding* $G \to CG$ by assigning, for instance, to every element $g \in G$ the element $(b_g, \mathrm{id}) \in CG$, where $b_g 0 = g$ and $b_r r = 1$ for $r \neq 0$.

Finally, applying the argument of Mather[6] to $\operatorname{Aut} Q$, one gets that the group $\operatorname{Aut} Q$ is acyclic, i.e. $H_i(\bar{W}\operatorname{Aut} Q); Z) = 0$ for $i > 0$, where $\bar{W}$ denotes the *simplicial classifying functor* of Eilenberg–MacLane[7, p. 87], and a slight refinement of this argument readily yields:

3.2. PROPOSITION. *The group CG is acyclic, i.e. $H_i(\bar{W}CG; Z) = 0$ for $i > 0$.*

The group CG is always *uncountable*, because $\operatorname{Aut} Q$ is so. However one has

3.3. PROPOSITION. *There is a (not necessarily natural) subgroup $C'G \subset CG$, which contains G, is acyclic and has the same cardinality as G (except if G is finite, in which case $C'G$ may have to be countable).*

Proof. This follows readily, by a direct limit argument, from 3.2 and the fact that for every subgroup $B \subset CG$ and *cycle* z on B there is a *finite* set of elements of CG which together with B generate a subgroup of CG on which z *bounds.*

Applying the functor C *dimension-wise* to a simplicial group Y one gets a simplicial group CY for which the following generalization of 3.2 holds

3.4. PROPOSITION. *The simplicial group CY is acyclic, i.e. $H_i(\bar{W}CY; Z) = 0$ for $i > 0$.*

Furthermore the argument of 3.3 yields

3.5. PROPOSITION. *There is a (not necessarily natural) simplicial subgroup $C'Y \subset CY$, which contains Y, is acyclic and has the same cardinality as Y.*

Proof of 3.4. By applying the classifying functor $\bar{W}$ to each of the groups CY_n one gets a bi-simplicial set. The homology spectral sequence of this bi-simplicial set converges strongly to the homology of the diagonal D [2, 2.9] and 3.2 thus implies that $H_i(D; Z) = 0$ for $i > 0$. It remains to show that D has the same homotopy type as $\bar{W}CY$. For this one applies the functor W of [7, p. 88] to each of the groups CY_n and denotes by E the diagonal of the resulting bi-simplicial group. As each WCY_n is contractible [7, 21.5], so is E [10]. Moreover, it is not hard to verify that the obvious map $E \to D$ is a principal fibration with CY as fibre and thus D has the same homotopy type as $\bar{W}CY$.

A convenient property of the functor C on simplicial groups is that it preserves homotopy in the following strong sense:

3.6. HOMOTOPY LEMMA. *Let $f: Y \to Y'$ be a homomorphism between simplicial groups which induces an isomorphism on the homotopy groups in all dimensions ≥ 0. Then so does the map $Cf: CY \to CY'$.*

A useful consequence for our purposes is

3.7. COROLLARY. *Let Y be an 'aspherical' simplicial group, i.e. $\pi_i Y = 0$ for $i > 0$. Then so is CY and the projection $Y \to \pi_0 Y$ induces an isomorphism $\pi_0 CY \approx C\pi_0 Y$. Furthermore, if $C'Y$ is as in 3.5, then the induced map $\pi_0 Y \to \pi_0 C'Y$ is a monomorphism.*

Proof of 3.6. This lemma follows readily from the observation that, for a group G, the underlying pointed set of CG depends only on the underlying pointed set of G and the fact that (i) a homomorphism between simplicial groups induces an isomorphism on all homotopy groups if and only if it induces a homotopy equivalence between the underlying (pointed) simplicial sets [7, 12.5 and 17.1], and (ii) every dimension-wise functor on (pointed) simplicial sets preserves homotopies [4, 5.3].

3.8. *Remark.* The homological cone functor constructed above is *not* the only one possible. Variants can be obtained by, for instance, considering only continuous automorphisms of Q or replacing Q by a finite dimensional vector space over Q or over the reals. One can also embed a group in its group ring and then take a cone on the latter as in [13]. And, of course, a composition of two cone functors is again a cone functor.

§4. J. H. C. WHITEHEAD'S ASPHERICITY RESULT

In this section we formulate and prove a simplicial group version (4.2) of an asphericity result of J. H. C. Whitehead. We first recall the original version [14, Th. 5].

4.1. PROPOSITION. *Let*

$$\begin{array}{ccc} A & \longrightarrow & A' \\ \downarrow & & \downarrow \\ A'' & \longrightarrow & X \end{array}$$

be a push out diagram of reduced (only one vertex) simplicial sets, such that (i) *A, A' and A'' are aspherical, i.e. their homotopy groups vanish in dimensions >1, and* (ii) *the maps $A \to A'$, $\pi_1 A \to \pi_1 A'$, $A \to A''$, $\pi_1 A \to \pi_1 A''$, are all monomorphisms.*

Then X is also aspherical and the induced diagram

$$\begin{array}{ccc} \pi_1 A & \longrightarrow & \pi_1 A' \\ \downarrow & & \downarrow \\ \pi_1 A'' & \longrightarrow & \pi_1 X \end{array}$$

is a push out diagram of groups, i.e. $\pi_1 X$ is the free product with amalgamation $\pi_1 X = \pi_1 A' \amalg_{\pi_1 A} \pi_1 A''$.

The simplicial group version of this result is

4.2. PROPOSITION. *Let*

$$\begin{array}{ccc} B & \longrightarrow & B' \\ \downarrow & & \downarrow \\ B'' & \longrightarrow & Y \end{array}$$

be a push out diagram of simplicial groups such that (i) *B, B' and B'' are 'aspherical' in the sense that their homotopy groups vanish in dimensions ≥ 1, and* (ii) *the maps $B \to B'$, $\pi_0 B \to \pi_0 B'$, $B \to B''$, $\pi_0 B \to \pi_0 B''$, are all monomorphisms.*

Then Y is also 'aspherical' and the induced diagram

$$\begin{array}{ccc} \pi_0 B & \longrightarrow & \pi_0 B' \\ \downarrow & & \downarrow \\ \pi_0 B'' & \longrightarrow & \pi_0 Y \end{array}$$

is a push out diagram of groups, i.e. $\pi_0 Y$ is the free product with amalgamation

$$\pi_0 Y = \pi_0 B' \amalg_{\pi_0 B} \pi_0 B''.$$

This follows immediately from 4.1 and the following proposition, which seems to be of interest in itself.

4.3. PROPOSITION. *Let*

$$\begin{array}{ccc} B & \longrightarrow & B' \\ \downarrow & & \downarrow \\ B'' & \longrightarrow & Y \end{array}$$

be a push out diagram of simplicial groups such that the maps $B \to B'$ and $B \to B''$ are both monomorphisms. Then the induced map $\bar{W}B' \amalg_{\bar{W}B} \bar{W}B'' \to \bar{W}Y$ is a weak homotopy equivalence.

Here again $\bar{W}$ denotes the simplicial classifying functor of Eilenberg–MacLane [7, p. 87].

Proof. By applying $\bar{W}$ dimension-wise to the simplicial groups B, B', B'' and Y, one gets bi-simplicial sets of which the diagonals D, D', D'' and E have the homotopy type of $\bar{W}B$, $\bar{W}B'$, $\bar{W}B''$ and $\bar{W}Y$ respectively (see the proof of 3.4) and it thus suffices to prove that the induced map $D'\amalg_D D'' \to E$ is a weak homotopy equivalence. But this is not hard to do, using [1, Ch. XII 4.2 and 4.3] and the fact that, in view of 4.2, the induced map $\bar{W}B'_n \amalg_{\bar{W}B_n} \bar{W}B''_n \to \bar{W}Y$ is a weak homotopy equivalence for all $n \geq 0$.

§5. PROOF OF THEOREM 1.1

We prove Theorem 1.1 by first obtaining an analogous result (5.1) for reduced (only one vertex) simplicial sets and then translating this into topological language with the help of the *geometric realization* $\|$ and the *singular functor* Sin.

5.1. THEOREM. *For every reduced simplicial set X there exists a fibration*

$$TX \xrightarrow{tX} X$$

which is natural with respect to X and has the following properties: (i) *the map tX induces an isomorphism on homology and cohomology with local coefficients $H_*(TX; A) \approx H_*(X; A)$, $H^*(TX; A) \approx H^*(X; A)$ for every local coefficient system A on X*, (ii) *$\pi_i TX$ is trivial for $i \neq 1$ and $\pi_1 tX$ is onto, and* (iii) *$\pi_1 TX$ has the same cardinality as (the set of simplicities of) X.*

Proof. We recall from [7, p. 87 and p. 118] that the *loop group functor G* (which assigns to every reduced simplicial set X a free simplicial group which has the homotopy type of the loops on X) is the *left adjoint* of the *simplicial classifying functor* $\bar{W}$ of Eilenberg–MacLane and that the adjunction map $X \to \bar{W}GX$ is a weak homotopy equivalence.

The core of the proof then consists of constructing a (natural) sequence of simplicial groups and homomorphisms

$$1 = G^0X \to G^1X \to \cdots \to G^{n-1}X \to G^nX \to \cdots$$

together with a compatible set of homomorphisms $G^nX \to GX^n$ (where X^n denotes the n-skeleton of X), such that the induced map $\bar{W}G^\infty X = \bar{W} \varinjlim G^nX \to \bar{W}GX = \bar{W} \varinjlim GX^n$ has the properties (i), (ii) and (iii) of Theorem 5.1, as the desired map $tX\colon TX \to X$ then readily can be obtained from the pull back diagram:

$$\begin{array}{ccc} TX & \longrightarrow & \bar{W}G^\infty X \\ \downarrow & & \downarrow \\ X & \longrightarrow & \bar{W}GX \end{array}$$

For every integer $n > 0$ let $D[n]$ denote the reduced simplicial set obtained from the standard n-simplex $\Delta[n]$ by identifying all its vertices. The simplicial groups G^nX then are inductively defined by means of the push out diagrams

$$\text{(I)} \qquad \begin{array}{ccc} \amalg G^{n-1}D[n] & \longrightarrow & G^{n-1}X \pi \amalg C'G^{n-1}D[n] \\ \downarrow & & \downarrow \\ \amalg GD[n] \pi \amalg C'G^{n-1}D[n] & \longrightarrow & G^nX \end{array}$$

(where $C'G^{n-1}D[n]$ is as in 3.5) and the homomorphisms $G^nX \to GX^n$ are induced by the obvious maps from these diagrams to the commutative diagrams

$$\text{(II)} \qquad \begin{array}{ccc} \amalg GD[n]^{n-1} & \longrightarrow & GX^{n-1} \\ \downarrow & & \downarrow \\ \amalg GD[n] & \longrightarrow & GX^n \end{array}$$

In all these diagrams the free products $\amalg$ are taken over all maps $D[n] \to X$. It now follows immediately from 3.5, 3.7 and 4.2 that the induced map $\bar{W}G^\infty X \to \bar{W}GX$ has properties (ii) and (iii) and it thus remains to verify property (i).

To do this we consider the reduced simplicial set $\bar{X}$ which has exactly one *non-degenerate* simplex $\bar{x}$ for every simplex $x \in X$ with, of course, the obvious faces and degeneracies. The map $\bar{X} \to X$ which sends $\bar{x}$ to x then is a weak homotopy equivalence [5] and one clearly has push out diagrams

$$\text{(III)} \qquad \begin{array}{ccc} \amalg GD[n]^{n-1} & \longrightarrow & G\bar{X}^{n-1} \\ \downarrow & & \downarrow \\ \amalg GD[n] & \longrightarrow & G\bar{X}^{n} \end{array}$$

where again the free products $\amalg$ are taken over all maps $D[n] \to X$. Moreover the maps from the diagrams (I) to the diagrams (II) clearly factor through the diagrams (III) and it is not hard to show, using 3.5, 4.3 and a Mayer–Vietoris argument, that the resulting map $G^nX \to G\bar{X}^n$ induces isomorphisms $H_*(\bar{W}G^nX; A) \approx H_*(\bar{W}G\bar{X}^n; A)$, $H^*(\bar{W}G^nX; A) \approx H^*(\bar{W}G\bar{X}^n; A)$ for every local coefficient system A which comes from X. The desired result now follows immediately.

We end with completing the

Proof of theorem 1.1. Let $E_0 \operatorname{Sin} X$ denote the largest subcomplex of $\operatorname{Sin} X$ (the singular complex of X) which has no other vertex than the base point, and consider the composition

$$|TE_0 \operatorname{Sin} X| \xrightarrow{|tE_0 \operatorname{Sin} X|} |E_0 \operatorname{Sin} X| \xrightarrow{\text{incl.}} |\operatorname{Sin} X| \to X$$

where T and t are as in Theorem 5.1 and the map $|\operatorname{Sin} X| \to X$ is the adjunction map. Then it is not hard to see that any natural factorization of this map

$$|TE_0 \operatorname{Sin} X| \to TX \xrightarrow{tX} X$$

for which the first map is a weak homotopy equivalence and the second map is a Serre fibration, yields the desired result.

REFERENCES

1. A. K. Bousfield and D. M. Kan: Homotopy limits, completions and localizations, *Lecture Notes in Math.* **304**, Springer (1972).
2. A. Dold and D. Puppe: Homologie nicht-additive Functoren. Anwendungen, *Ann. Inst. Fourier* **11** (1961), 201–312.
3. E. Dror: Acyclic spaces, *Topology* **11** (1972), 339–348.
4. D. M. Kan: On the homotopy relation for c.s.s. maps, *Bull. Soc. Mat. Mex.* (1957), 75–81.
5. Y. Kodawa: A relation between two realizations of complete semisimplicial complexes, *Proc. Japan Ac.* **33** (1957), 536–540.
6. J. Mather: The vanishing of the homology of certain groups of homeomorphisms, *Topology* **10** (1971), 297–298.
7. J. P. May: Simplicial objects in algebraic topology, Van Nostrand (1967).
8. S. B. Priddy: On $\Omega^\infty S^\infty$ and the infinite symmetric group, *Proc. Symp. Pure Math. AMS* **22** (1971), 217–220.
9. S. B. Priddy: Fredholm permutations and stable homotopy, *Mich. Math. J.* **20** (1973), 187–192.
10. D. G. Quillen: Spectral sequences of a double semi-simplicial group, *Topology* **5** (1966), 155–157.
11. D. G. Quillen: Cohomology of groups, *Proc. Int. Congress of Math.* (1970).
12. W. Thurston: Foliations and groups of diffeomorphisms, *Bull. AMS* **80** (1974), 304–307.
13. J. B. Wagoner: Delooping classifying spaces in algebraic K-theory, *Topology* **11** (1972), 349–370.
14. J. H. C. Whitehead: On the asphericity of regions in a 3-sphere, *Fund. Math.* **32** (1939), 149–166.

Massachusetts Institute of Technology
Princeton University

Topology and its Applications 23 (1986) 145–148
North-Holland

ON K_1-THEORY OF THE EUCLIDEAN SPACE*

W. THURSTON and L. VASERSTEIN

Institute for Advanced Study, Princeton, NJ 08540, USA

Received 30 April 1985

The algebraic functor K_1 of the ring of continuous functions of three variables is computed.

AMS(MOS) Subj. Class.: 18F25

algebraic K-theory of rings of continuous functions
Stone–Čech compactification on the Euclidean space
cohomotopy groups vector fields

Let N, n be natural numbers, $\mathbb{R}$ the field of real numbers, $X = \mathbb{R}^N$ the N-dimensional Euclidean space (with the usual norm $|(x_i)| = (\sum x_i^2)^{1/2}$), $A = \mathbb{R}^X$ the ring of continuous functions $X \to R$, $\mathrm{SL}_n A$ the group of n-by-n matrices over A with determinant 1, $E_n A$ the subgroup generated by elementary matrices $1_n + e_{i,j}(a)$, where $i \neq j$, $a \in A$, and 1_n is the identity matrix.

In topological terms, $\mathrm{SL}_n A$ is the set of all continuous maps $\mathbb{R}^n \to \mathrm{SL}_n \mathbb{R}$, and $\mathrm{SL}_n A / E_n A$ is $\pi_0(\mathrm{SL}_n A)$, where A is endowed with the uniform convergence topology [3].

Matrix groups over rings of functions appear in mathematical physics as gauge groups, and their structure, topology, and representations are of some interest.

The Gramm–Schmidt process gives [3] that $\mathrm{SL}_n A / E_n A = \mathrm{SL}_n A_0 / E_n A_0 = \pi_0 \mathrm{SL}_n A_0 = \pi_0 \mathrm{SO}_n A$, where A_0 is the subring of bounded functions. Since A_0 is also the ring of continuous functions on the Stone–Čech compactification Y_N of $\mathbb{R}^N$, $\mathrm{SL}_n A / E_n A$ can be also identified with the homotopy classes of continuous maps $Y_N \to \mathrm{SL}_n \mathbb{R}$ (or $Y_N \to \mathrm{SO}_n \mathbb{R}$).

It is not hard to show that every finitely generated projective A-module (as well as A_0-module) is free [4] (in geometric terms, every vector bundle over $\mathbb{R}^N$ is trivial). In particular $K_0 Y_N = Z$ for all N. So a similar result for the algebraic functor K_1, namely, that $\mathrm{SL}_n A = E_n A$ would not be surprising. However, it was shown in [3] that $\mathrm{SL}_2 A / E_2 A = A / A_0$ (in particular, Y_N is not contractible), $N \geq 1$.

Recall that for the subring $B \subset A$ of polynomial functions the following is true: $\mathrm{SL}_2 B \neq E_2 B$ when $N \geq 2$ and $\mathrm{SL}_n B = E_n B$ in all other cases. So there is a strange difference between continuous (or smooth) and polynomial functions when $N = 1$. Still one can pose the conjecture that $\mathrm{SL}_n A = E_n A$ when $n \geq 3$ for all N. In fact, this was proved in [3] when $N = 1$. Later Vaserstein proved this also for $N = 2$.

* Work supported in part by National Science Foundation.

In this note we prove the conjecture for $N=3$ (which, of course, implies the cases $N=1$ and 2):

Theorem 1. *If* $X=\mathbb{R}^3$ *and* $A=\mathbb{R}^X$, *then* $\mathrm{SL}_n A = E_n A$ *for all* $n \geq 3$.

In other terms, every real matrix of determinant 1 depending continuously on 3 real parameters can be reduced continuously by additive operations to the identity matrix (or, equivalently, every invertible matrix can be reduced to a upper triangular form). Note that A in Theorem 1 can be replaced by a dense subring [3] such as the subring of smooth functions. As noted above it can also be replaced by the subring of polynomial functions (Suslin). The situation with the subring of analytic functions is not clear.

The main tool in the proof of Theorem 1 below is the following statement about $\pi^N(Y_N)$, the homotopy classes of continuous maps $Y_N \to S^N$ (into the N-sphere):

Theorem 2. $\pi^N(Y_N)=0$ *for all* $N \geq 2$.

Note that for $N=1$ we have $\pi^1(Y_1)=\pi_0(\mathrm{SO}_2 A)=A/A_0 \neq 0$, where $A=\mathbb{R}^X$ with $X=\mathbb{R}^1$.

Proof of Theorem 2. We realize S^N as $\{(x_0,\ldots,x_N)\in\mathbb{R}^{N+1}: x_0^2+\cdots+x_N^2=1\}$. Note that a continuous function $Y_N\to\mathbb{R}$ is the same as a bounded continuous function $\mathbb{R}^N\to\mathbb{R}$. So a continuous map $f: Y_N \to S^N$ is the same as the set $(f_0,\ldots,f_N)$ of $N+1$ continuous functions $\mathbb{R}^N\to\mathbb{R}$ such that $f_0(x)^2+\cdots+f_N(x)^2=1$ for all x in $\mathbb{R}^N$.

We consider on $A_0=\mathbb{R}^{Y_N}$ the topology of uniform convergence so A_0 is a Banach algebra.

We want to prove that every $f: Y_N\to S^N$ is homotopic to a constant map (since S^N is connected, all constant maps are homotopic to the map taking $\mathbb{R}^N$ to the north pole $x_0=1$ of S^N).

A small perturbation of f (in the uniform topology mentioned above) makes it smooth (that is, all f_i are C^∞-functions). Another small smooth perturbation makes the inverse image $f_0^{-1}(1)$ of the north pole discrete in $\mathbb{R}^N$, with the Jacobian of $(f_1,f_2,\ldots,f_N)\neq 0$ at each $x\in V:=f_0^{-1}(1)\cap\mathbb{R}^n$.

Pick $\delta(x)>0$ for each x in V such that f is an immersion of the disc $D_x:=\{y\in\mathbb{R}^N: |y-x|<\delta(x)\}$ for each x in V and the closures of the discs D_x, $x\in V$, are pairwise disjoint.

Now we define $g: R^N\to S^N$ as follows: $g(y)=(-1,0,\ldots,0)$ (the south pole) when y belongs to the complement C in $\mathbb{R}^N$ of all the discs, and $g(y):=(\cos(\pi|z|),$ $(s(x)(y_1-x_1),\ y_2-x_2,\ldots,y_N-x_N)\sin(\pi|z|)/|z|\delta(x))$, when $y\in D_x$, where $s(x)$ is the sign of the Jacobian of $(f_1,\ldots,f_N)$ at x, and $z=(y-x)/\delta(x)$.

It is clear that f is homotopic to g (in particular, when V is empty, f is homotopic to a constant map and we are done).

Now we want to arrange a homotopy of g into a map missing the south pole. To do this, we use a continuous unit vector field F on C with the following boundary conditions: $F(y) := (s(x)(y_1 - x_1),\ y_2 - x_2, \ldots, y_N - x_N)/\delta(x)$ when $|y - x| = \delta(x)$, $x = (x_i) \in V$, $y = (y_i)$. This field would direct each point in C how to leave the south pole $g(C)$.

For example, when V consists of a single point x with $s(x) = 1$, we can just set $F(y) := (y - x)/|y - x|$ for $|y - x| > \delta(x)$. Figure 1 shows a possible picture of integral curves for F when $N = 2$, $V = \{u, v, w\}$ with $s(u) = s(v) = 1 = -s(w)$.

A way to see the existence of such a field F (here the condition $N \geqslant 2$ is essential) is as follows. Pick a locally finite triangulization of C (into curved simpleces) with N-simpleces $S_1, S_2, \ldots$ ordered in such a way that the boundary of S_m is not contained in the union of all S_k with $k < m$ $(m = 2, 3, \ldots)$. Such a triangulization (see Fig. 2) can be based on a triangulization of $\mathbb{R}^N$ using points in V as vertices ordered according to $|x|$.

Then we can construct F inductively on the union of S_k, $k < m$. Namely, assuming we have F on the union, we have given it on a proper part of the boundary of S_m. Then we can extend F continuously on the whole S_m.

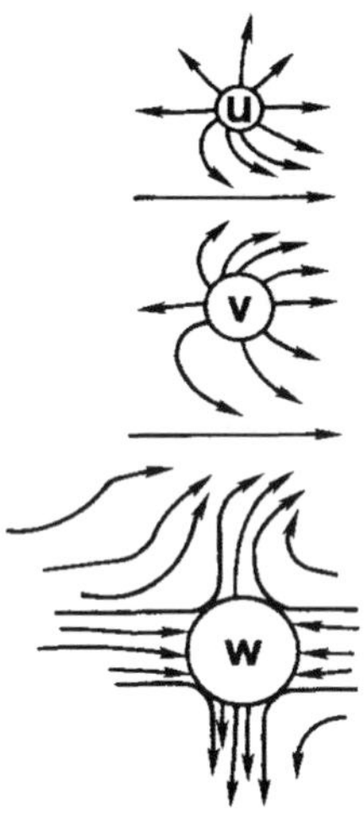

Fig. 1.

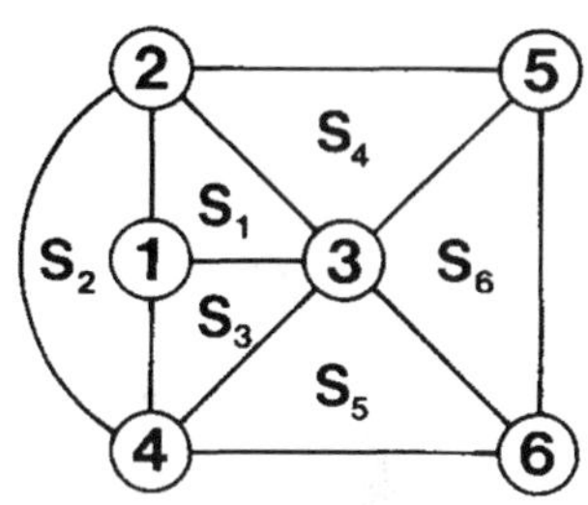

Fig. 2.

Given a continuous unit vector field F on C with the above boundary conditions, we can complete our proof of Theorem 2 as follows.

For any ε such that $0 \leqslant \varepsilon \leqslant 1$ set

$$h_\varepsilon(y) := \begin{cases} (\cos(\pi(1-\varepsilon)),\ F(y)\sin(\pi(1-\varepsilon))) \in S^N & \text{when } y \in C, \\ g(y - (y - x)\varepsilon) & \text{when } y \in D_x. \end{cases}$$

The boundary conditions on F give that $h_\varepsilon(y)$ is continuous function of $y \in \mathbb{R}^N$ for any ε. When $\varepsilon = 0$, $h_0 = g$. It is clear that h depends continuously on ε, $0 \leqslant \varepsilon < 1$. Since $|F(y)| = 1$ and $g(D_x)$ do not contain the south pole when $x \in V$, $h_\varepsilon(\mathbb{R}^N)$ misses the pole for $\varepsilon > 0$. Therefore, all h are homotopic to a constant map when $\varepsilon > 0$. Thus, g and hence f are homotopically trivial. □

Remark 1. Similarly, it can be proved that if X is an N-dimensional CW-complex whose N-cells can be well-ordered in such a way that the boundary of no N-cell is contained in the union of the preceding cells, then $\pi^N(Y) = 0$ for the Stone–Čech compactification of X, provided $N \geq 2$.

Proof of Theorem 1. Homotopically, $SO_3\mathbb{R}$ is the 3-dimensional projective space RP^3. Consider the canonical covering $S^3 \to RP^3$ whose fiber consists of 2 points. Since $\mathbb{R}^3$ is simply connected, any continuous map $f: Y_3 \to RP^3$ can be covered by a continuous map $g: Y_3 \to S^3$. By Theorem 2, this map g is homotopically trivial, so f is trivial too. Therefore $\pi_0 SO_3 A = 0$, hence $SL_3 A_0 = E_3 A_0$ and $SL_3 A = E_3 A$. Since $sr(A) = 3$ (see [2]), the canonical homomorphism $SL_n A / E_n A \to SL_{n+1} A / E_{n+1} A$ is onto for $n \geq 3$ (see [1]). Thus, $SL_n A = E_n A$ for $n \geq 3$. □

Remark 2. If X and Y are as in Remark 1 with $N \leq 3$ and X is simply connected for each base point, then, similarly $SL_n A = E_n A$ for $n \geq 3$, and $SL_2 A / E_2 A = A / A_0$.

Remark 3. For the complex valued functions C^X on $X = \mathbb{R}^3$ we have $SL_2 A / E_2 A = \pi_0 SU_2 A = \pi^3(Y_3) = 0$ by Theorem 2, so $SL_n A = E_n A$ for all $n \geq 1$.

Remark 4. For the ring A of entire functions on C^N it is unknown whether $SL_n A = E_n A$, except for the case $N = 1$ when $sr(A) = 1$.

Remark 5. For any complex commutative Banach algebra B and any n we have $SL_n B / E_n B = SL_n A / E_n A$, where $A = C^Y$, Y the space of maximal ideals of B (B is mapped into A by the Gelfand transform).

References

[1] H. Bass, K-theory and stable algebra, Publ. Math. IHES, 22.

[2] L.N. Vaserstein, The stable range of rings and the dimension of topological spaces, Funnk. An. Pril. 5 (2) (1971) 17–27; in Russian, translated in Funct. An. Appl.

[3] L.N. Vaserstein, On K_1-theory of topological spaces, A.M.S. Cont. Math., to appear.

[4] L.N. Vaserstein, Vector bundles and projective modules, Trans. Amer. Math. Soc. 294 (1986) 749–755.

Published Titles in This Series

27.4 **William P. Thurston:** The Geometry and Topology of Three-Manifolds, 2022

27 Collected Works of William P. Thurston with Commentary, volumes I–III (Benson Farb, David Gabai, and Steven P. Kerckhoff, Editors), 2022

26 **Phillip A. Griffiths:** Selected Works of Phillip A. Griffiths with Commentary, 2017

25 **Richard P. Stanley:** Selected Works of Richard P. Stanley, 2017

24 **John Tate:** Collected Works of John Tate, 2016

23 Collected Works of Hervé Jacquet (Dorian Goldfeld, Editor), 2011

22 The Selected Works of Sigurður Helgason (Gestur Ólafsson and Henrik Schlichtkrull, Editors), 2009

21 Selected Papers of Alberto P. Calderón with Commentary (Alexandra Bellow, Carlos E. Kenig, and Paul Malliavin, Editors), 2008

20 **Yuri I. Manin:** Mathematics as Metaphor, 2007

19 Collected Papers of John Milnor

18 Selected Works of Phillip A. Griffiths with Commentary (Enrico Arbarello, Robert L. Bryant, C. Herbert Clemens, Maurizio Cornalba, Mark L. Green, Joe Harris, David R. Morrison, and Wilfried Schmid, Editors), 2003

17 Selected Works of Eberhard Hopf with Commentaries (Cathleen S. Morawetz, James B. Serrin, and Yakov G. Sinai, Editors), 2002

16 Selected Papers of S. A. Amitsur with Commentary (Avinoam Mann, Amitai Regev, Louis Rowen, David J Saltman, and Lance W Small, Editors), 2001

15 Selected Works of Ilya Piatetski-Shapiro (James Cogdell, Simon Gindikin, and Peter Sarnak, Editors), 2000

14 **E. B. Dynkin:** Selected Papers of E. B. Dynkin with Commentary, 2000

13 Selected Works of Frederick J. Almgren, Jr. (Jean E. Taylor, Editor), 1999

12 Selected Works of Ellis Kolchin with Commentary (Hyman Bass, Alexandru Buium, and Phyllis J. Cassidy, Editors), 1999

11 **V. S. Varadarajan:** The Selected Works of V.S. Varadarajan, 1999

10 Selected Works of Maurice Auslander (Idun Reiten, Sverre O. Smalø, and Øyvind Solberg, Editors), 1999

9 Selected Works of Lipman Bers (Irwin Kra and Bernard Maskit, Editors), 1998

8 **Walter E. Thirring:** Selected Papers of Walter E. Thirring with Commentaries, 1998

7 **Robert Steinberg:** Robert Steinberg Collected Papers, 1997

6 The Collected Works of Julia Robinson (Solomon Feferman, Editor), 1996

5 **Freeman Dyson:** Selected Papers of Freeman Dyson with Commentary, 1996

4 Collected Works of Witold Hurewicz (Krystyna Kuperberg, Editor), 1995

3 A. Adrian Albert Collected Mathematical Papers (Nathan Jacobson, Daniel Zelinsky, Richard E. Block, David J. Saltman, and J. Marshall Osborn, Editors), 1993

2 **Salomon S Bochner and Salomon S Bochner:** Collected Papers of Salomon Bochner, 1992